T0396458

Handbook of Optical Fibers

Gang-Ding Peng

Editor

Handbook of Optical Fibers

Volume 2

With 1544 Figures and 89 Tables

 Springer

Editor
Gang-Ding Peng
Photonics and Optical Communications
School of Electrical Engineering and Telecommunications
University of New South Wales
Sydney, NSW, Australia

ISBN 978-981-10-7085-3   ISBN 978-981-10-7087-7 (eBook)
ISBN 978-981-10-7086-0 (print and electronic bundle)
https://doi.org/10.1007/978-981-10-7087-7

This Springer imprint is published by the registered company Springer Nature Singapore Pte Ltd.
The registered company address is: 152 Beach Road, #21-01/04 Gateway East, Singapore 189721, Singapore

Preface

This research- and application-oriented book covers main topical areas of optical fibers. The selection of the chapters is weighted on technological and application-specific topics, very much a reflection of where research is heading to and what researchers are looking for. Chapters are arranged in a user-friendly format essentially self-contained and with extensive cross-references. They are organized in the following sections:

Optical Fiber Communication
Solitons and Nonlinear Waves in Optical Fibers
Optical Fiber Fabrication
Active Optical Fibers
Special Optical Fibers
Optical Fiber Measurement
Optical Fiber Devices
Optical Fiber Device Measurement
Distributed Optical Fiber Sensing
Optical Fiber Sensors for Industrial Applications
Polymer Optical Fiber Sensing
Photonic Crystal Fiber Sensing
Optical Fiber Microfluidic Sensors

Several sections and chapters of the book show how diverse optical fiber technologies are becoming now. We envisage that many new optical fibers under development will find important future applications in telecommunications, sensing, and so on. Though we have been trying to cover most relevant and important topics in optical fibers, some topics may have not been presented.

All the authors are either pioneers or leading researchers in their respective areas. Their chapters have reflected well the excellent research work, technology deployment, and commercial application of their own and others. Hence this handbook, as a new entry to the Springer Nature's Major Reference Works (MRWs), will be useful for researchers, academics, engineers, and students to access expertly summarized specific topics on optical fibers for research, education, and learning purposes.

There could be technical and grammatical errors in this book. Please feel free to send your correction, advice, and feedback to us. One key feature of the Springer Nature's MRWs is to ensure continuing update and improvements.

I would take this opportunity to express by deepest gratitude to all my colleagues, either as section editors or authors, for their hard work and great contribution to this book. I also would like to thank the Springer Nature editors and staff, especially Dr. Stephen Siu Wai Yeung and Dr. Juby George, for their kind and professional support throughout this book project.

July 2019   Gang-Ding Peng

Contents

About the Editor

Gang-Ding Peng
Photonics and Optical Communications
School of Electrical Engineering
and Telecommunications
University of New South Wales
Sydney, NSW, Australia

Gang-Ding Peng received his B.Sc. degree in physics from Fudan University, Shanghai, China, in 1982, and the M.Sc. degree in applied physics and Ph.D. in electronic engineering from Shanghai Jiao Tong University, Shanghai, China, in 1984 and 1987, respectively. From 1987 through 1988 he was a lecturer at Jiao Tong University. He was a postdoctoral research fellow in the Optical Sciences Centre of the Australian National University, Canberra, from 1988 to 1991. He has been working at the University of NSW in Sydney, Australia, since 1991; was a Queen Elizabeth II Fellow from 1992 to 1996; and is currently a professor in the same university. He is a fellow and life member of both Optical Society of America (OSA) and The International Society for Optics and Photonics (SPIE). His research interests include silica and polymer optical fibers, optical fiber and waveguide devices, optical fiber sensors, and nonlinear optics.

He has worked in research and teaching in photonics and fiber optics for more than 30 years and maintained a high research profile internationally.

Section Editors

Part I: Optical Fiber Communication

Ming-Jun Li   Corning Incorporated, Corning, NY, USA

Chao LU   Department of Electronic and Information Engineering, The Hong Kong Polytechnic University, Hong Kong SAR, China

Part II: Solitons and Nonlinear Waves in Optical Fibers

Boris A. Malomed   Faculty of Engineering, Department of Physical Electronics, School of Electrical Engineering, Tel Aviv University, Tel Aviv, Israel

ITMO University, St. Petersburg, Russia

Part III: Optical Fiber Fabrication

Hairul Azhar Bin Abdul Rashid   Faculty of Engineering, Multimedia University, Cyberjaya, Malaysia

Part IV: Active Optical Fibers

Kyunghwan Oh   Department of Physics, Institute of Physics and Applied Physics, Yonsei University, Seoul, Republic of Korea

Part V: Special Optical Fibers

Perry Shum   Nanyang Technological University, Singapore, Singapore

Zhilin Xu Center for Gravitational Experiments, School of Physics, Huazhong University of Science and Technology, Wuhan, China

Part VI: Optical Fiber Measurement

Jianzhong Zhang Key Lab of In-fiber Integrated Optics, Ministry of Education, Harbin Engineering University, Harbin, China

Part VII: Optical Fiber Devices

John Canning *i*nterdisciplinary Photonics Laboratories (*i*PL), Global Big Data Technologies Centre (GBDTC), Tech Lab, School of Electrical and Data Engineering, University of Technology Sydney, Sydney, NSW, Australia

Tuan Guo Institute of Photonics Technology, Jinan University, Guangzhou, China

Part VIII: Optical Fiber Device Measurement

Yanhua Luo   Photonics and Optical Communications, School of Electrical Engineering and Telecommunications, University of New South Wales, Sydney, NSW, Australia

Key Laboratory of Optoelectronic Devices and Systems of Ministry of Education and Guangdong Province, Shenzhen University, Shenzhen, China

Part IX: Distributed Optical Fiber Sensing

Yosuke Mizuno   Institute of Innovative Research, Tokyo Institute of Technology, Yokohama, Japan

Part X: Optical Fiber Sensors for Industrial Applications

Tong Sun OBE   School of Mathematics, Computer Science and Engineering, City, University of London, London, UK

Part XI: Polymer Optical Fiber Sensing

Ginu Rajan   School of Electrical, Computer and Telecommunications Engineering, University of Wollongong, Wollongong, Australia

School of Electrical Engineering and Telecommunications, UNSW, Sydney, Australia

Part XII: Photonic Crystal Fiber Sensing

D. N. Wang   College of Optical and Electrical Technology, China Jiliang University, Hangzhou, China

Part XIII: Optical Fiber Microfluidic Sensors

Yuan Gong   Key Laboratory of Optical Fiber Sensing and Communications (Ministry of Education of China), University of Electronic Science and Technology of China, Chengdu, Sichuan, China

Contributors

John S. Abbott Corning Incorporated, Corning, NY, USA

Hairul Azhar Bin Abdul Rashid Faculty of Engineering, Multimedia University, Cyberjaya, Malaysia

Kazi S. Abedin OFS Laboratories, Somerset, NJ, USA

Muhammad Rosdi Abu Hassan Centre for Optical Fibre Technology (COFT), School of Electrical, Electronic Engineering, Nanyang Technological University, Singapore, Singapore, Singapore

Claudia Aichele Department of Fiber Optics, Leibniz Institute of Photonic Technology (Leibniz IPHT), Jena, Germany

Shaif-Ul Alam Optoelectronics Research Centre (ORC), University of Southampton, Southampton, UK

Eliathamby Ambikairajah School of Electrical Engineering and Telecommunications, UNSW, Sydney, Australia

Ghafour Amouzad Mahdiraji School of Engineering, Taylor's University, Subang Jaya, Selangor, Malaysia

Flexilicate Sdn. Bhd., University of Malaya, Kuala Lumpur, Malaysia

S. A. Babin Institute of Automation and Electrometry SB RAS, Novosibirsk, Russia

Novosibirsk State University, Novosibirsk, Russia

John Ballato Center for Optical Materials Science and Engineering Technologies (COMSET) and the Department of Materials Science and Engineering, Clemson University, Clemson, SC, USA

A. Barthélémy XLIM, UMR CNRS 7252, Université de Limoges, Limoges, France

Kishore Bhowmik HFC Assurance, Operate and Maintain Network, NBN, Melbourne, VIC, Australia

Scott R. Bickham Corning Incorporated, Corning, NY, USA

Lúcia Bilro Instituto de Telecomunicações, Campus Universitário de Santiago, Aveiro, Portugal

L. Brun Faiveley Brecknell Willis, Somerset, UK

John Canning *i*nterdisciplinary Photonics Laboratories (*i*PL), Global Big Data Technologies Centre (GBDTC), Tech Lab, School of Electrical and Data Engineering, University of Technology Sydney, Sydney, NSW, Australia

J. Carlton City, University of London, London, UK

Christophe Caucheteur Electromagnetism and Telecommunication Department, University of Mons, Mons, Belgium

Quan Chai Key Laboratory of In-Fiber Integrated Optics, Ministry Education of China, Harbin Engineering University, Harbin, China

I. S. Chekhovskoy Novosibirsk State University, Novosibirsk, Russia

Institute of Computational Technologies SB RAS, Novosibirsk, Russia

Jianping Chen State Key Laboratory of Advanced Optical Communication Systems and Networks, Department of Electronic Engineering, Shanghai Jiao Tong University, Shanghai, China

Xin Chen Corning Incorporated, Corning, NY, USA

Y. Chen City, University of London, London, UK

Baokai Cheng Department of Electrical and Computer Engineering, Center for Optical Materials Science and Engineering Technologies (COMSET), Clemson University, Clemson, SC, USA

Yushi Chu Key Laboratory of In-Fiber Integrated Optics, Ministry Education of China, Harbin Engineering University, Harbin, China

Photonics and Optical Communications, School of Electrical Engineering and Telecommunications, UNSW, Sydney, NSW, Australia

*i*nterdisciplinary Photonics Laboratories (*i*PL), Global Big Data Technologies Centre (GBDTC), Tech Lab, School of Electrical and Data Engineering, University of Technology Sydney, Sydney, NSW, Australia

J. Doug Coleman Corning Incorporated, Corning, NY, USA

Matteo Conforti CNRS, UMR 8523 – PhLAM – Physique des Lasers Atomes et Molécules, University of Lille, Lille, France

Kevin Cook *i*nterdisciplinary Photonics Laboratories (*i*PL), Global Big Data Technologies Centre (GBDTC), Tech Lab, School of Electrical and Data Engineering, University of Technology Sydney, Sydney, NSW, Australia

V. Couderc XLIM, UMR CNRS 7252, Université de Limoges, Limoges, France

Katrina D. Dambul Faculty of Engineering, Multimedia University, Cyberjaya, Selangor, Malaysia

S. Das Fiber Optics and Photonics Division, CSIR-Central Glass and Ceramic Research Institute, Kolkata, India

A. Dhar Fiber Optics and Photonics Division, CSIR-Central Glass and Ceramic Research Institute, Kolkata, India

Mingjie Ding Photonics and Optical Communications, School of Electrical Engineering and Telecommunications, University of New South Wales, Sydney, NSW, Australia

Fabrizio Di Pasquale Institute of Communication, Information and Perception Technologies (TECIP), Scuola Superiore Sant'Anna, Pisa, Italy

Yongkang Dong National Key Laboratory of Science and Technology on Tunable Laser, Harbin Institute of Technology, Harbin, China

John D. Downie Corning Incorporated, Corning, NY, USA

Peter Dragic Department of Electrical and Computer Engineering, University of Illinois at Urbana-Champaign, Urbana, IL, USA

D. Dutta Fiber Optics and Photonics Division, CSIR-Central Glass and Ceramic Research Institute, Kolkata, India

M. Fabian City, University of London, London, UK

Ghazal Fallah Tafti Photonics and Optical Communications, School of Electrical Engineering and Telecommunications, UNSW, Sydney, NSW, Australia

Desheng Fan Photonics and Optical Communications, School of Electrical Engineering and Telecommunications, University of New South Wales, Sydney, NSW, Australia

Xinyu Fan State Key Laboratory of Advanced Optical Communication Systems and Networks, Department of Electronic Engineering, Shanghai Jiao Tong University, Shanghai, China

Andrea Fasano DTU Mekanik, Department of Mechanical Engineering, Technical University of Denmark, Lyngby, Denmark

M. P. Fedoruk Novosibirsk State University, Novosibirsk, Russia

Institute of Computational Technologies SB RAS, Novosibirsk, Russia

Dimitrios J. Frantzeskakis Department of Physics, National and Kapodistrian University of Athens, Athens, Greece

Alexander Fuerbach MQ Photonics Research Centre, Department of Physics and Astronomy, Macquarie University, North Ryde, NSW, Australia

C. Gerada The University of Nottingham, Nottingham, UK

Anahita Ghaznavi Photonics and Optical Communications, School of Electrical Engineering and Telecommunications, UNSW, Sydney, NSW, Australia

Chao-Yang Gong Key Laboratory of Optical Fiber Sensing and Communications (Ministry of Education of China), University of Electronic Science and Technology of China, Chengdu, Sichuan, China

Yuan Gong Key Laboratory of Optical Fiber Sensing and Communications (Ministry of Education of China), University of Electronic Science and Technology of China, Chengdu, Sichuan, China

K. T. V. Grattan City, University of London, London, UK

Jian Guo Shandong Key Laboratory of Optical Fiber Sensing Technologies, Qilu Industry University (Laser Institute of Shandong Academy of Sciences), Jinan, China

Tuan Guo Institute of Photonics Technology, Jinan University, Guangzhou, China

Tetsuya Hayashi Optical Communications Laboratory, Sumitomo Electric Industries, Ltd., Yokohama, Kanagawa, Japan

Jun He Key Laboratory of Optoelectronic Devices and Systems of Ministry of Education and Guangdong Province, College of Physics and Optoelectronic Engineering, Shenzhen University, Shenzhen, China

Guangdong and Hong Kong Joint Research Centre for Optical Fibre Sensors, Shenzhen University, Shenzhen, China

Hoi Lut Ho Department of Electrical Engineering, The Hong Kong Polytechnic University, Hong Kong, China

Theodoros P. Horikis Department of Mathematics, University of Ioannina, Ioannina, Greece

Sheng-Lung Huang Graduate Institute of Photonics and Optoelectronics, and Department of Electrical Engineering, National Taiwan University, Taipei, Taiwan

Darren D. Hudson MQ Photonics Research Centre, Department of Physics and Astronomy, Macquarie University, North Ryde, NSW, Australia

Georges Humbert XLIM Research Institute, UMR 7252 CNRS, University of Limoges, Limoges, France

Ezra Ip NEC Laboratories America, Princeton, NJ, USA

Stuart D. Jackson Department of Engineering, MQ Photonics Research Centre, School of Engineering, Macquarie University, North Ryde, NSW, Australia

S. Javdani City, University of London, London, UK

Taofei Jiang National Key Laboratory of Science and Technology on Tunable Laser, Harbin Institute of Technology, Harbin, China

Wei Jin Department of Electrical Engineering, The Hong Kong Polytechnic University, Hong Kong, China

Yongmin Jung Optoelectronics Research Centre (ORC), University of Southampton, Southampton, UK

S. I. Kablukov Institute of Automation and Electrometry SB RAS, Novosibirsk, Russia

Gerd Keiser Boston University, Boston, MA, USA

A. V. Kir'yanov Centro de Investigaciones en Optica, Guanajuato, Mexico

K. Krupa Department of Information Engineering, University of Brescia, Brescia, Italy

A. Kudlinski CNRS, UMR 8523 – PhLAM – Physique des Lasers Atomes et Molécules, University of Lille, Lille, France

Elizabeth Lee Precision Measurements Group, Singapore Institute of Manufacturing Technology, Singapore, Singapore

Ming-Jun Li Corning Incorporated, Corning, NY, USA

Changrui Liao College of Optoelectronic Engineering, Shenzhen University, Shenzhen, China

Sascha Liehr Division 8.6 "Fibre Optic Sensors", Bundesanstalt für Materialforschung und –prüfung (BAM), Berlin, Germany

Chupao Lin College of Optoelectronic Engineering, Shenzhen University, Shenzhen, China

Horng Sheng Lin Universiti Tunku Abdul Rahman, Sungai Long Campus, Kajang, Malaysia

Florian Lindner Department of Fiber Optics, Leibniz Institute of Photonic Technology (Leibniz IPHT), Jena, Germany

Deming Liu School of Optical and Electronic Information, Next Generation Internet Access National Engineering Laboratory (NGIAS), Huazhong University of Science and Technology, Wuhan, Hubei, P. R. China

Tongyu Liu Laser Institute, Qilu University of Technology-Shandong Academy of Science, Jinan, Shandong, China

Xin Long State Key Laboratory of Advanced Optical Communication Systems and Networks, Department of Electronic Engineering, Shanghai Jiao Tong University, Shanghai, China

Jiaqi Luo Precision Measurements Group, Singapore Institute of Manufacturing Technology, Singapore, Singapore

Yanhua Luo Photonics and Optical Communications, School of Electrical Engineering and Telecommunications, University of New South Wales, Sydney, NSW, Australia

Key Laboratory of Optoelectronic Devices and Systems of Ministry of Education and Guangdong Province, Shenzhen University, Shenzhen, China

Faisal Rafiq Mahamd Adikan Flexilicate Sdn. Bhd., University of Malaya, Kuala Lumpur, Malaysia

Integrated Lightwave Research Group, Department of Electrical Engineering, Faculty of Engineering, University of Malaya, Kuala Lumpur, Malaysia

Sergejs Makovejs Corning Incorporated, Ewloe, UK

Boris A. Malomed Faculty of Engineering, Department of Physical Electronics, School of Electrical Engineering, Tel Aviv University, Tel Aviv, Israel

ITMO University, St. Petersburg, Russia

Christos Markos DTU Fotonik, Department of Photonics Engineering, Technical University of Denmark, Lyngby, Denmark

G. Millot ICB, UMR CNRS 6303, Université de Bourgogne, Dijon, France

Fedor Mitschke Institut für Physik, Universität Rostock, Rostock, Germany

S. Z. Muhamad Yassin Photonics Laboratory, Telekom Research and Development, Cyberjaya, Malaysia

A. Mussot CNRS, UMR 8523 – PhLAM – Physique des Lasers Atomes et Molécules, University of Lille, Lille, France

Hossein Najafi Institute for Applied Laser, Photonics and Surface Technologies (ALPS), Bern University of Applied Sciences, Burgdorf, Switzerland

Rogério Nogueira Instituto de Telecomunicações, Campus Universitário de Santiago, Aveiro, Portugal

Ricardo Oliveira Instituto de Telecomunicações, Campus Universitário de Santiago, Aveiro, Portugal

Nasr Y. M. Omar Faculty of Engineering, Multimedia University, Cyberjaya, Malaysia

M. Pal Fiber Optics and Photonics Division, CSIR-Central Glass and Ceramic Research Institute, Kolkata, India

M. C. Paul Fiber Optics and Photonics Division, CSIR-Central Glass and Ceramic Research Institute, Kolkata, India

Gang-Ding Peng Photonics and Optical Communications, School of Electrical Engineering and Telecommunications, University of New South Wales, Sydney, NSW, Australia

Jiankun Peng National Engineering Laboratory for Fiber Optic Sensing Technology (NEL-FOST), Wuhan University of Technology, Wuhan, China

Sönke Pilz Institute for Applied Laser, Photonics and Surface Technologies (ALPS), Bern University of Applied Sciences, Burgdorf, Switzerland

Soo Yong Poh Integrated Lightwave Research Group, Department of Electrical Engineering, Faculty of Engineering, University of Malaya, Kuala Lumpur, Malaysia

Haifeng Qi Shandong Key Laboratory of Optical Fiber Sensing Technologies, Qilu Industry University (Laser Institute of Shandong Academy of Sciences), Jinan, China

Ginu Rajan School of Electrical, Computer and Telecommunications Engineering, University of Wollongong, Wollongong, Australia

School of Electrical Engineering and Telecommunications, UNSW, Sydney, Australia

Yun-Jiang Rao Key Laboratory of Optical Fiber Sensing and Communications (Ministry of Education of China), University of Electronic Science and Technology of China, Chengdu, Sichuan, China

P. H. Reddy Academy of Scientific and Innovative Research (AcSIR), IR-CGCRI Campus, Kolkata, India

A. A. Reduyk Novosibirsk State University, Novosibirsk, Russia

David J. Richardson Optoelectronics Research Centre (ORC), University of Southampton, Southampton, UK

Valerio Romano Institute for Applied Laser, Photonics and Surface Technologies (ALPS), Bern University of Applied Sciences, Burgdorf, Switzerland

Institute of Applied Physics (IAP), University of Bern, Bern, Switzerland

A. M. Rubenchik Lawrence Livermore National Laboratory, Livermore, CA, USA

Kay Schuster Department of Fiber Optics, Leibniz Institute of Photonic Technology (Leibniz IPHT), Jena, Germany

Filipa Sequeira Instituto de Telecomunicações, Campus Universitário de Santiago, Aveiro, Portugal

O. V. Shtyrina Novosibirsk State University, Novosibirsk, Russia

Institute of Computational Technologies SB RAS, Novosibirsk, Russia

O. S. Sidelnikov Novosibirsk State University, Novosibirsk, Russia

D. V. Skryabin Department of Nanophotonics and Metamaterials, ITMO University, St Petersburg, Russia

Department of Physics, University of Bath, Bath, UK

Yang Song Department of Electrical and Computer Engineering, Center for Optical Materials Science and Engineering Technologies (COMSET), Clemson University, Clemson, SC, USA

Zhiqiang Song Shandong Key Laboratory of Optical Fiber Sensing Technologies, Qilu Industry University (Laser Institute of Shandong Academy of Sciences), Jinan, China

Marcelo A. Soto Institute of Electrical Engineering, EPFL Swiss Federal Institute of Technology, Lausanne, Switzerland

Dan Sporea National Institute for Laser, Plasma and Radiation Physics, Center for Advanced Laser Technologies, Măgurele, Romania

Biao Sun Precision Measurements Group, Singapore Institute of Manufacturing Technology, Singapore, Singapore

Qizhen Sun School of Optical and Electronic Information, Next Generation Internet Access National Engineering Laboratory (NGIAS), Huazhong University of Science and Technology, Wuhan, Hubei, P. R. China

Tong Sun OBE School of Mathematics, Computer Science and Engineering, City, University of London, London, UK

Ming Tang Wuhan National Lab for Optoelectronics (WNLO) and National Engineering Laboratory for Next Generation Internet Access System (NGIA), School of Optical and Electronic Information, Huazhong University of Science and Technology (HUST), Wuhan, China

Lei Teng National Key Laboratory of Science and Technology on Tunable Laser, Harbin Institute of Technology, Harbin, China

A. Tonello XLIM, UMR CNRS 7252, Université de Limoges, Limoges, France

Stefano Trillo Department of Engineering, University of Ferrara, Ferrara, Italy

S. K. Turitsyn Novosibirsk State University, Novosibirsk, Russia

Aston Institute of Photonic Technologies, Aston University, Birmingham, UK

Sonja Unger Department of Fiber Optics, Leibniz Institute of Photonic Technology (Leibniz IPHT), Jena, Germany

M. Vidakovic City, University of London, London, UK

S. Wabnitz Novosibirsk State University, Novosibirsk, Russia

Department of Information Engineering, University of Brescia, Brescia, Italy

National Institute of Optics INO-CNR, Brescia, Italy

Chao Wang School of Electrical Engineering, Wuhan University, Wuhan, Hubei, China

D. N. Wang College of Optical and Electrical Technology, China Jiliang University, Hangzhou, China

Min Wang National Engineering Laboratory for Fiber Optic Sensing Technology (NEL-FOST), Wuhan University of Technology, Wuhan, China

School of Electronic and Electrical Engineering, Wuhan Textile University, Wuhan, China

Weijia Wang National Engineering Laboratory for Fiber Optic Sensing Technology (NEL-FOST), Wuhan University of Technology, Wuhan, China

Weitao Wang Shandong Key Laboratory of Optical Fiber Sensing Technologies, Qilu Industry University (Laser Institute of Shandong Academy of Sciences), Jinan, China

Wenyu Wang Photonics and Optical Communications, School of Electrical Engineering and Telecommunications, University of New South Wales, Sydney, NSW, Australia

Yiping Wang Key Laboratory of Optoelectronic Devices and Systems of Ministry of Education and Guangdong Province, College of Physics and Optoelectronic Engineering, Shenzhen University, Shenzhen, China

Guangdong and Hong Kong Joint Research Centre for Optical Fibre Sensors, Shenzhen University, Shenzhen, China

Lei Wei School of Electrical and Electronic Engineering, Nanyang Technological University, Singapore, Singapore

Jianxiang Wen Key Laboratory of Specialty Fiber Optics and Optical Access Networks, Shanghai University, Shanghai, China

Aleksander Wosniok 8.6 Fibre Optic Sensors, Federal Institute for Materials Research and Testing (BAM), Berlin, Germany

Getinet Woyessa DTU Fotonik, Department of Photonics Engineering, Technical University of Denmark, Lyngby, Denmark

Gui Xiao Photonics and Optical Communications, School of Electrical Engineering and Telecommunications, UNSW, Sydney, NSW, Australia

Hai Xiao Department of Electrical and Computer Engineering, Center for Optical Materials Science and Engineering Technologies (COMSET), Clemson University, Clemson, SC, USA

Limin Xiao Advanced Fiber Devices and Systems Group, Key Laboratory of Micro and Nano Photonic Structures (MoE), Department of Optical Science and Engineering Fudan University, Shanghai, China

Key Laboratory for Information Science of Electromagnetic Waves (MoE), Fudan University, Shanghai, China

Shanghai Engineering Research Center of Ultra-Precision Optical Manufacturing, Fudan University, Shanghai, China

Fei Xu National Laboratory of Solid State Microstructures and College of Engineering and Applied Sciences, Nanjing University, Nanjing, Jinagsu, P. R. China

Binbin Yan State Key Laboratory of Information Photonics and Optical Communications, Beijing University of Posts and Telecommunications, Beijing, China

Zhijun Yan School of Optical and Electronic Information, Next Generation Internet Access National Engineering Laboratory (NGIAS), Huazhong University of Science and Technology, Wuhan, Hubei, P. R. China

Fan Yang Department of Electrical Engineering, The Hong Kong Polytechnic University, Hong Kong, China

Jun Yang Key Lab of In-Fiber Integrated Optics, Ministry Education of China, Harbin Engineering University, Harbin, China

College of Science, Harbin Engineering University, Harbin, China

Minghong Yang National Engineering Laboratory for Fiber Optic Sensing Technology (NEL-FOST), Wuhan University of Technology, Wuhan, China

Xia Yu Precision Measurements Group, Singapore Institute of Manufacturing Technology, Singapore, Singapore

Zhangjun Yu Key Lab of In-Fiber Integrated Optics, Ministry Education of China, Harbin Engineering University, Harbin, China

College of Science, Harbin Engineering University, Harbin, China

Lei Yuan Department of Electrical and Computer Engineering, Center for Optical Materials Science and Engineering Technologies (COMSET), Clemson University, Clemson, SC, USA

Libo Yuan Key Lab of In-Fiber Integrated Optics, Ministry Education of China, Harbin Engineering University, Harbin, China

College of Science, Harbin Engineering University, Harbin, China

Zulfadzli Yusoff Multimedia University, Persiaran Multimedia, Cyberjaya, Malaysia

Amirhassan Zareanborji Photonics and Optical Communications, School of Electrical Engineering and Telecommunications, UNSW, Sydney, NSW, Australia

Chen-Lin Zhang Key Laboratory of Optical Fiber Sensing and Communications (Ministry of Education of China), University of Electronic Science and Technology of China, Chengdu, Sichuan, China

Hongying Zhang Institute of Photonics and Optical Fiber Technology, Harbin University of Science and Technology, Harbin, China

Jianzhong Zhang Key Lab of In-fiber Integrated Optics, Ministry of Education, Harbin Engineering University, Harbin, China

Lei Zhang College of Optical Science and Engineering, Zhejiang University, Hangzhou, China

Lin Zhang Aston Institute of Photonic Technologies, Aston University, Birmingham, UK

Chun-Liu Zhao College of Optical and Electrical Technology, China Jiliang University, Hangzhou, China

Qiancheng Zhao Photonics and Optical Communications, School of Electrical Engineering and Telecommunications, UNSW, Sydney, NSW, Australia

Dengwang Zhou National Key Laboratory of Science and Technology on Tunable Laser, Harbin Institute of Technology, Harbin, China

Feng Zhu College of Optoelectronic Engineering, Shenzhen University, Shenzhen, China

E. A. Zlobina Institute of Automation and Electrometry SB RAS, Novosibirsk, Russia

Weiwen Zou State Key Laboratory of Advanced Optical Communication Systems and Networks, Department of Electronic Engineering, Shanghai Jiao Tong University, Shanghai, China

Part V
Special Optical Fibers

Optical Fibers for High-Power Lasers

23

Xia Yu, Biao Sun, Jiaqi Luo, and Elizabeth Lee

Contents

Abstract

Lasers with high output powers are demanded for a wide variety of applications, ranging from material processing, remote sensing, medical surgery, to fundamental science. Across all these application scenarios, there are two main challenges: the scaling of output power and the quality of the laser beam. In the past decade, there have been tremendous research efforts to tackle these two issues in both continuous wave (CW) and pulsed lasers, to improve the power level, wavelength tunability, coherence, line width, etc. Among them, fiber technology has enabled the flexible delivery of high-power laser beams with precision beam quality

X. Yu (✉) · B. Sun · J. Luo · E. Lee
Precision Measurements Group, Singapore Institute of Manufacturing Technology, Singapore, Singapore
e-mail: xyu@SIMTech.a-star.edu.sg; sunb@SIMTech.a-star.edu.sg; JLUO4@e.ntu.edu.sg; EMLEE1@e.ntu.edu.sg

© Springer Nature Singapore Pte Ltd. 2019
G.-D. Peng (ed.), *Handbook of Optical Fibers*,
https://doi.org/10.1007/978-981-10-7087-7_39

877

(Jauregui et al., Nat Photonics 7:861–867, 2013). The technology development could be summarized in two approaches: passive fiber technology and active fiber technology. Passive fibers offer the last step manipulation of high-power laser beams from gas laser, semiconductor lasers, or other solid-state lasers. Active fibers are the gain component in the fiber oscillator or amplifier to generate the optical emission. Compared with traditionally step-index fibers, new fiber structure designs open new horizons in laser technology. In this book chapter, the main content has been arranged according to different fiber structure designs. Typical specialty fibers have been chosen, including double-cladding fibers, large mode area photonic crystal fibers, large pitch fibers, leakage channel fibers, chirally coupled core fibers, pixelated Bragg fibers, and hollow-core fibers. The design principle, manufacturability, and future outlook have been discussed in each subsections.

Introduction

Passive Fibers for High-Power Laser Beam Delivery

Fibers are commonly used as pigtails for different types of lasers, such as semiconductor lasers for telecommunication or high-power pump diodes. These single-mode fused silica fibers not only feature an excellent coupling efficiency but also good performance output beam. This is enabled by an optimum in optical design and precision alignment. An example is shown in Fig. 1. However, diode lasers have finite brightness or irradiance at which light can be coupled into the fiber (Dawson et al. 2008). This intensity has increased steadily with time as technology progresses. Diode lasers with more than 0.02 W/(μm^2-steradian) are commercially available.

In some new wavelength regime, new material processes have been developed for fiber technology to reduce the transmission loss for the flexible delivery of the laser beam to the targeted place, for example, solarization-resistant optical fibers by hydrogen loading for deep UV transmission window and fluoride/tellurite/chalcogenide fibers for mid-IR wavelength regime.

In recent development of high-power ultrafast lasers, hollow fibers infiltrated with pressurized inert gases were used to provide ultrawide band spectrum broadening, before the pulses were compressed to femtosecond temporal width.

Active Fibers for High-Power Laser Beam Generation

In the past two decades, there have been vast advancements in active fiber designs for high-power laser generation, as shown in Fig. 2. The gain media was rare-earth element(s) doped inside the core of an optical fiber. There have been intensive research on conversion efficiency, power scaling, and nonlinear processes. Fiber properties have been investigated for high-power applications, including doping

Fig. 1 Fiber coupling from different laser output

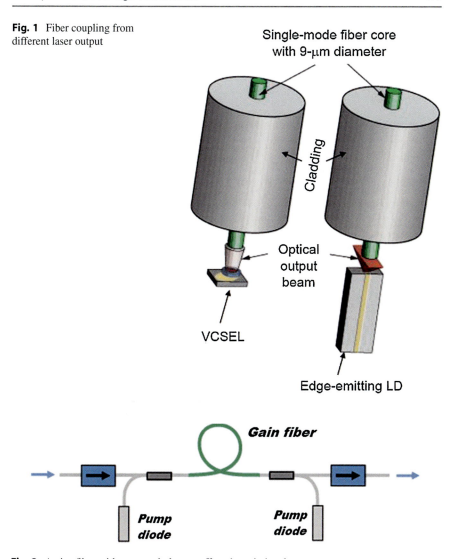

Single-mode fiber core
with 9-μm diameter

Cladding

Optical
output
beam

VCSEL

Edge-emitting LD

Gain fiber

Pump
diode

Pump
diode

Fig. 2 Active fiber with rare-earth dopant offers the gain in a laser system

concentration, photon-darkening effects, Raman scattering, Brillouin scattering, optical damage, thermal lensing effect, etc.

Instead of considering different rare-earth-doped elements for different wavelength emission, this book chapter is organized from specialty fiber design perspective. Detailed summary has been explained in section "Specialty Fibers for High-Power Lasers," including double-cladding fibers, large mode area photonic crystal fibers, large pitch fibers, leakage channel fibers, chirally coupled core fibers, pixelated Bragg fibers, hollow-core fibers, etc.

Specialty Fibers for High-Power Lasers

Double-Cladding Fibers

The typical optical fiber geometry with a large surface-to-volume ratio offers an outstanding heat dissipation efficiency and a long interaction length with the gain medium, which makes fiber-based lasers attractive. The single-mode fiber guarantees a single transverse mode propagation, but it also limits the coupling efficiency of the pump laser to the fiber core. The efficiency degradation is more severe when high-power diodes with poor beam quality is used as the pump laser.

Conventional step-index fibers have a core-cladding structure, with the core having a higher refractive index in order to guide light, as shown in Fig. 3a. An additional low-refractive index layer (may be glass, polymer, or air) is introduced as a secondary cladding or outer cladding in double-cladding fibers (DCFs) (Maurer 1974; Snitzer et al. 1988). The introduction of the secondary cladding allows the original inner cladding to guide light, as there exists a refractive index difference between the inner and outer cladding now.

The outer cladding is designed such that the refractive index contrast between the outer and inner cladding is larger than the contrast between the inner cladding and the fiber core, as shown in Fig. 3b. This, together with the fact that the inner cladding is of a larger diameter than the core, corresponds to a larger numerical

Fig. 3 Typical refractive index profile of single-clad fiber and double-clad fiber

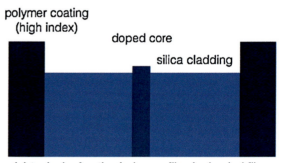

(a) typical refractive index profile single-clad fiber

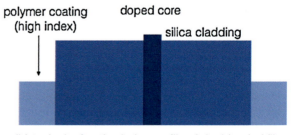

(b) typical refractive index profile of double-clad fiber

aperture (NA) for the inner cladding as compared to the core. Hence, the pump laser coupling efficiency is improved as it is easier to couple the pump light into the inner cladding, mitigating the poor beam quality of pump lasers. The coupled pump light propagating in the inner cladding is then able to exchange energy with the core gain medium when the modes spatially overlap during total internal reflection (TIR).

Typically, the fiber cores are designed for a single fundamental mode (FM) output, where the normalized frequency (V-number) is smaller than 2.405. The V-number is given by:

$$V = \frac{\pi d}{\lambda} \bullet \text{NA} = \pi d / \lambda \bullet \sqrt{n_{\text{core}}^2 - n_{\text{clad}}^2}$$

where d is the diameter, λ is the operating wavelength, NA is the numerical aperture, and n is the refractive index. Using the equation above and the condition of $V = 2.405$ for single-mode operation, the fiber becomes few mode or multimode once the fiber diameter $d > 12.7\lambda$.

Rare-earth-doped glasses have higher refractive indices than the undoped counterparts; hence, rare-earth-doped glass forms the cores of optical waveguides. The core diameter is typically less than 15 μm to maintain single-mode guidance. The inner cladding that allows pump light propagation has varied dimensions in the range of hundreds of micrometers depending on the application. A larger inner cladding size is able to support a higher pump power; however, the pump absorption coefficient decreases with a reducing core-to-cladding ratio ($d_{\text{core}}/d_{\text{cladding}}$), thus requiring longer fiber length for the same amount of pump absorption. Hence, the core-to-cladding ratio is typically kept below 1/20 to use reasonable fiber lengths.

The shape of the inner cladding of DCFs is intentionally designed to be elliptical, D-shaped, or octagonal, away from the original symmetrical circular shape. The cross-sectional asymmetry breaks the cladding mode symmetry and improves energy coupling between the pump light and the rare-earth ions (Javadimanesh et al. 2016). There have been discussions on the broken circular symmetry for efficiency optimization of pump absorption in double-clad fiber amplifiers (Kouznetsov and Moloney 2002).

As mentioned earlier, an extra layer of outer cladding is introduced in DCFs, for light confinement of the pump laser in the inner cladding and also to provide protection and flexibility to the fragile glass structure. The outer cladding of DCFs usually consists of a low-index polymer layer, typically with a refractive index of about 1.37. The refractive index contrast between the polymer outer cladding and the glass inner cladding enables the inner cladding to have a large NA, typically about 0.46 in most commercial DCFs. If the polymer has inherently strong absorption in the pump laser wavelength (in the case of 1.9 μm for holmium-doped fiber pumping), fluorosilicate glass may be used, resulting in a smaller NA due to the smaller refractive index contrast. The double-cladding configuration has enabled some critical milestones in high-power fiber laser (Jeong et al. 2004; Overton 2015).

Deleterious nonlinear effects and lower power damage thresholds plague fibers with small cores, limiting their use at high-average and high-peak powers. To overcome these limitations, especially for constructing high-power fiber lasers and amplifiers, a large mode area (LMA) and high dopant concentration were proposed.

To maintain single-mode operation by preserving the V-number <2.405, enlargement of the core diameter needs to be balanced by the reduction of NA. A lower NA implies a reduction in the refractive index contrast between the core and the cladding. However, a higher dopant concentration of rare-earth elements, such as Nd, Yb, Er, Tm, and Ho, increases the glass refractive index, thus in contrary to the requirement to maintain single-mode guidance for larger core diameters.

To suppress the undesirable increase in refractive index, a "negative" refractive index material, typically silicon tetrafluoride, is added to the rare-earth-doped glass. Alternatively, an additional high-refractive index pedestal is formed by surrounding the core with germanium dioxide-doped glass. In practice, the refractive index contrast cannot be arbitrarily engineered and fabricated; the refractive index contrast is typically in the range of 10^{-3} in commercial fibers. A reliable NA of 0.06 of commercial step-index fibers (SIFs) can be obtained by modified chemical vapor deposition (MCVD) method. The low numerical aperture renders the fiber to be susceptible to bending-induced losses; the losses are especially significant for the higher-order modes (HOMs). Thus, the HOMs can be stripped out by proper coiling of the fiber to a suitable coil diameter, allowing the fundamental mode to propagate with significantly less losses, such that the fiber becomes pseudo-single mode. Otherwise, the fiber can be designed in such a way that the core has a much higher gain for the fundamental mode and a low gain for the HOMs, thus allowing single-mode guidance. Even with these measures, the maximum core diameter is limited at about 30 μm in all-solid LMA fibers. This is not sufficiently large to avoid undesirable nonlinear effects in ultrafast pulsed lasers where the peak power may exceed megawatts. Hence, other fiber structures and designs have been explored to support even higher peak power.

Large Mode Area Photonic Crystal Fibers

In the late 1990s, a new class of optical fibers, the photonic crystal fiber (PCF), was made (Knight et al. 1996). The new fiber had a regular array of air holes along its length. The air-hole structure surrounding the solid-core forms the fiber cladding and causes the cladding effective refractive index to be lower than that of the glass core. These solid-core PCFs which guide light with the modified total internal reflection are known as index-guiding PCFs.

Index-guiding PCFs are widely exploited as LMA gain fibers (Limpert et al. 2004). Instead of using a low-index polymer layer as discussed in the earlier section "Double-Cladding Fibers," an air-hole array is used to create the low-index outer cladding layer as shown in Fig. 4. The silica bridge thickness between air holes can be made as thin as a few hundred nanometers, which make the air filling fraction large enough for the outer cladding to achieve very low effective refractive indices close to that of air. Thus, a large index contrast between the outer cladding and

Fig. 4 Cross section of typical double-clad rare-earth ion-doped LMA PCF

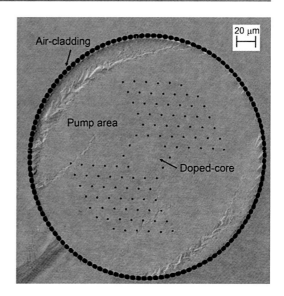

inner cladding exists for LMA PCFs. The NA of the inner cladding can exceed 0.5, in some cases 0.8, due to the large refractive index contrast.

The pump coupling efficiency is improved in the LMA PCF as compared to the conventional all-solid LMA fiber with the same inner cladding diameter but with a lower NA. Correspondingly, similar coupling efficiencies can be maintained for a LMA PCF with a smaller inner cladding diameter than an all-solid LMA fiber. Thus, the core/cladding (signal/pump) overlap ratio is increased, necessitating a shorter gain fiber for an equivalent amount of pump absorption. This is advantageous as it tremendously reduces the accumulated nonlinearity.

As mentioned in the previous section, one of the limitations of all-solid LMA fibers is the precise control of the refractive index contrast, by material selection or doping processes. The air-hole microstructure in LMA PCFs provides an alternative approach in modifying the refractive index by changing the air-hole array geometry. The refractive index contrast can be reliably controlled in the range of 10^{-4} by varying the air-hole-to-hole distance (pitch) and air-hole diameter. The improvement in the control of refractive index contrast enabled the demonstration of a single-mode LMA PCF with a core diameter of $d \approx 50\lambda$.

Ultralow NA LMA fibers suffer from another drawback. Their performance is highly affected by bending, which causes propagation loss and shrinkage of effective mode field area. The bending-induced effective mode field area shrinkage gets more severe when the fiber core is larger – a shrinking factor of 2 will be applied when a LMA PCF with 50 μm core diameter is coiled to a 10 cm radius. Hence, a new class of PCF was devised, the "rod-type" PCF which combines the concept of both rod and fiber (Limpert et al. 2005). These "rod-type" fibers sacrifice the flexibility offered by conventional fibers, by incorporating an extra 1–2 mm layer surrounding the PCF to provide mechanical support. Thus, the "rod-type" fiber is unaffected by macro- and micro-bending, avoiding the bending-induced drawbacks described earlier. Furthermore, the additional layer eliminates the need for a protective polymer coating, increasing the heat loading capacity of these fibers.

Ytterbium-doped "rod-type" PCF with a core diameter as large as 100 μm ($\sim$75 μm MFD) has been fabricated, and single-mode operation in this few-mode fiber has been demonstrated at optimized launch conditions (Limpert et al. 2006). Based on the mode coupling theory, an effective single-mode operation can be realized in few-mode fibers by optimizing the mode excitation at the fiber end, but this becomes more difficult as the core diameter increases. LMA fibers with highly distinguished mode distributions between the FM and HOMs are more feasible solutions for effective single-mode operation by mode-matching techniques.

Large Pitch Fibers

Distinguished mode distributions help to optimize the FM excitation and to minimize coupling to HOMs. This enables effective single-mode operation in very large mode area (VLMA, MFD >50λ) fibers. Based on the original PCF design, a novel PCF with a much large pitch size (hole-to-hole distance, >10λ), namely, large pitch fiber (LPF), has been demonstrated with high delocalization of HOMs from the FM and the doped core (Stutzki et al. 2012, 2014), as shown in Fig. 5. In conventional fibers, nearly all the modes are located in the fiber core, which also means a high overlap ratio between the modes and the doped region. Consequently, all the modes get a relatively even gain factor in the amplification process. The purity of FM follows the initial excitation at the input, even degrades due to mode cross talk. In LPFs, the HOM distributions are strongly deformed and pushed away from the doped core due to the inner cladding structure, while the FM retains its Gaussian shape confinement. Thus, high discrimination between FM and HOMs can be realized by a differential gain instead of the aforementioned differential "loss."

Typically, in a LPF with 50 μm MFD for 1 μm laser, the FM has a two times higher overlap ratio than HOMs (>85% for FM, <40% for HOMs). Furthermore, an earlier study showed that this overlap ratio difference can be extended to close to triple when thermal load is considered. The difference in the overlap ratio may be engineered and scaled up exponentially to suppress the HOMs and purify the FM. As a result, it has been shown that the FM can have a gain that is ten times higher than that of the HOMs. Given a fiber with mode content consisting 90% FM and 10% HOMs excited with a differential gain factor of 50 and 5, respectively, it is possible to have the output with a power extinction ratio of $\sim$20 dB between FM and HOMs.

Among the HOMs, low-order LP_{1x} modes are most likely to be excited in the few-mode LMA/VLMA fibers. These modes theoretically have mode distributions of axial symmetry under a highly symmetrical boundary confinement, such as in a circular or hexagonal-shaped core. Thus, low-symmetry LPF designs have been exploited to increase delocalization of HOMs further. The low-symmetry LPFs

Fig. 5 Cross section of LPF with a core diameter of 135 μm (Limpert et al. 2004)

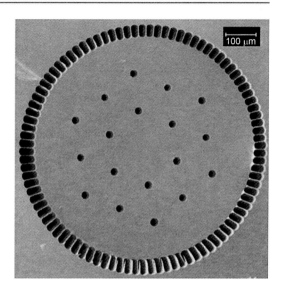

can be designed by having non-hexagonal shapes or by selectively increasing the diameter of the air holes on the hexagonal design without affecting stack-and-draw process. With deforming the HOMs distribution, LPF design also increases the mode instability threshold, which is one of the main limitations in high-average power fiber laser, especially for few-mode fibers.

Leakage Channel Fibers

Microstructure design in PCF is one solution to scale up mode area. However, PCF designs with small diameter-to-pitch ratio lead to weakly guiding. Hence it would be sensitive to the external environment. Leakage channel fiber (LCF) is a kind of emerging LMA fiber which can overcome this problem. It makes use of the fact that robust single-mode propagation can be supported by a multimode fiber with sufficient loss for higher-order modes. Normally, LCFs have solid cores surrounded by a few large holes, which help to improve the bending performance. In the meantime, the large pitch size between holes clearly distinguishes the core and cladding boundary. As a result, the total internal reflection cannot be satisfied everywhere, and the waveguide is leaky. Fortunately, the leaky property is mode dependent. LCFs can be engineered to maximize the confinement loss for unwanted higher-order modes and minimize that for the fundamental mode.

The first LCF was reported in 2005 (Wong et al. 2005). This fiber had a large solid core surrounded by a ring of air holes. The cross section is illustrated in Fig. 6a. The fiber parameters were listed as the diameter $d = 39.6$ μm with pitch size $\Lambda = 51.2$ μm for two smaller holes and $d = 46$ μm with $\Lambda = 51.1$ μm for larger four holes. This passive LCF could provide an effective mode field area (MFA) of

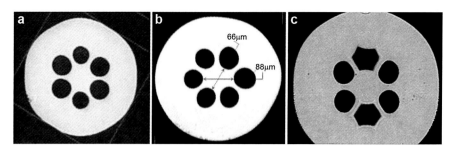

Fig. 6 The cross sections of (**a**) the first LCF (Limpert et al. 2006), (**b**) the first YDLCF (Stutzki et al. 2012), and (**c**) the first PM YDLCF (Stutzki et al. 2014)

1,417 μm^2. It was worth to be noted that this fiber could be coiled to 15 cm with negligible bending loss.

One year later, the first active LCF doped with ytterbium ions was demonstrated (Dong et al. 2006). With 55 μm hole diameter, 67 μm pitch size, and 350 μm fiber outer diameter (Fig. 6b), the ytterbium-doped LCF (YDLCF) could provide effective mode area up to 3,160 μm^2. The measured absorption for 980 nm pump was 3.6 dB/m. A section of 5-m-long fiber loosely coiled with 40 cm diameter achieved 60% slope efficiency. The measure M^2 value of the output beam was 1.3. In the same year, polarization maintaining (PM) LCF was demonstrated as well (Peng and Dong 2007). In order to induce the birefringence to maintain the polarization state, two holes were replaced by boron-doped silica stress rod. Shown in Fig. 6c, the four air holes and two by boron-doped silica formed the ytterbium-doped core with diameter of 50 μm. The MFA was 1,400 μm^2 (mode field diameter (MFD) of 42 μm). Laser operation was demonstrated with slope efficiency of 60%. 15.3 dB polarization extinction ratio (PER) was achieved, and the M^2 value of output beam was 1.2.

Similar with PCF manufacturing, the stack-and-draw method is commonly used to fabricate the LCF with air holes. The appropriate capillaries are arranged in correspondence to the aimed structure. Compared with conventional fibers, however, the manufacturing process is much more complex. It is hard to control the hole dimensions accurately and avoid the hole distortion. In order to eliminate these fabrication drawbacks, Dong et al. reported an all-glass YDLCF (Dong et al. 2009). The fluorine-doped silica was used to replace the air holes. Both PM and non-PM LCFs have been manufactured and tested. Figure 7a shows the cross section for the non-PM YDLCF. The core diameter was 47 μm, feature size to pitch $d/\Lambda = 0.7$. The dark ring which was highly fluorine-doped silica formed the cladding with 0.28 NA. The pump absorption was 12 dB/m at 980 nm. 15 ps pulses with 30 nJ pulse energy at 1 MHz were launched to this fiber to characterize the amplification operation. The 4 m fiber was coiled with 50 cm diameter. The peak power up to 1 MW with corresponding 11.2 μJ pulse energy was achieved. The PM YDLCF shown in Fig. 7b had a core diameter of 80 μm. Pulses with 40 mW average power and 14.2 ps pulse width at 10 MHz repetition rate were amplified to 27.4 W average power in

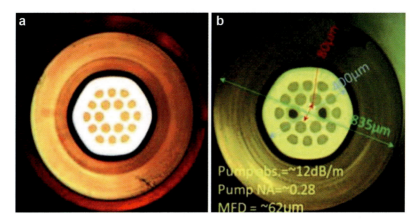

Fig. 7 Cross sections of all-glass YDLCF. (**a**) Non-PM LCF; (**b**) PM LCF (Wong et al. 2005)

the PM YDLCF. The output had 1.35 M^2 value and more than 15 dB polarization extinction ratio (PER) value.

Chirally Coupled Core Fibers

Applications in high-precision material processing using laser require the delivery of high-peak powers while maintaining diffraction-limited beam qualities. Traditionally, LMA fibers were candidates for high-peak power delivery as their large core diameters improve the threshold of the onset of undesirable nonlinear effects. However, it is difficult to achieve diffraction-limited beam qualities using LMA fibers. Hence, chirally coupled core fibers (3CFs) were devised to overcome the limitations of LMA fibers in terms of beam quality, while maintaining large core sizes and solid fiber robustness (Liu et al. 2007).

The 3CF structure consists of a large central core and one or more helical satellite cores surrounding the central core. The 3CF's unique structure serves to efficiently and selectively couple the HOMs into the helical satellite cores, where the HOMs experience high losses. The phase matching condition between the central core modes and satellite core modes is related to the core diameters and the helical pitch size. Proper design of these parameters allows single-mode guidance and even linear polarization qualities for a large core size in 3CFs. These advantages make 3CFs attractive candidates in fiber-based high-power delivery and amplification.

3CFs were first reported by Liu et al. (2007). The schematic and cross section is illustrated in Fig. 8. The center core size was 35 μm with 0.07 NA, and the helical satellite core diameter was 12 μm with 0.09 NA. The helical period was 6.2 mm. At wavelength >1,550 nm, the simulated HOM was effectively suppressed with a loss larger than 130 dB/m. Meanwhile, the fundamental mode loss was around 0.3 dB/m. The measured MFD was 29.5 μm.

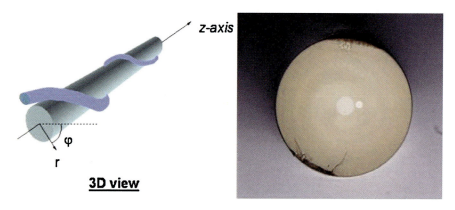

Fig. 8 The geometry of the first 3CF (Dong et al. 2006)

A 3CF with an 85 μm core was used in a fiber-based chirped pulse amplification system, to extract up to 10 mJ energy of stacking pulses with a high beam quality without severe amplified spontaneous emission and nonlinearities (Pei et al. 2017). Such low nonlinearity, provided by the large core area, enabled not only amplification of stacking pulses but also compression of single pulse to femtosecond region. Recently, 2.5 ns pulses were boosted to 1.2 mJ in a two-stage ytterbium-doped 3CF amplifier system (Bai et al. 2017). The system provided 17.8 dB gain with 82% slope efficiency. The output beam had a good beam quality with M^2 value at approximately 1.2. Another interesting characteristic of 3CFs is their ability to have nonlinear polarization switching effects at high-peak intensities (Hu et al. 2015). This leads to either polarization degradation or preservation effects under different input polarization conditions at high-peak powers.

The fabrication of 3CF differs from other fibers, whereby the fiber is spun at high rates during the drawing process. Thus, the satellite cores helically orbit the main central core along the fiber, with a pitch size related to the spin rate. So far, several commercial companies have successfully fabricated 3CFs.

Pixelated Bragg Fibers

Conventional Bragg fiber structure consists of alternating cylindrical layers of material with high and low-refractive indices. The Bragg fiber design originated from the one-dimensional planar Bragg stack, where the confinement mechanism stems from the reflection of the desired optical wavelengths by the alternating dielectric layers while rejecting unwanted wavelengths (Duguay et al. 1986; Mizrahi and Schächter 2004). The continuous layer allows guidance of higher-order linear-polarized (LP) modes within the annular structure, increasing attenuation within the transmission window. To eliminate this drawback while preserving the advantages of conventional Bragg fibers, pixelated Bragg fiber (PiBF) was proposed (Baz et al. 2012).

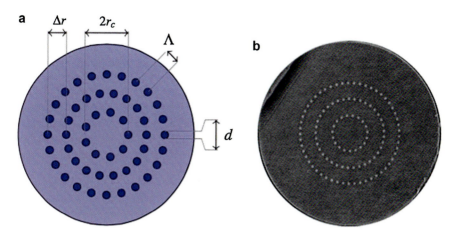

Fig. 9 (**a**) The schematic structure of PiBF, (**b**) SEM picture of the fabricated fiber (Hu et al. 2015)

The continuous high-index rings in the conventional Bragg fiber was altered to discontinuous pixelated rings formed by high-index rods in the PiBF. Illustrated in Fig. 9, the core was surrounded by discontinuous rings made of circular high-index rods. At wavelength of 1,400 nm, the MFD was 26 μm. The transformation improved transmission properties of the PiBF in terms of decreased attenuation and extended transmission window. The elimination of higher-order cladding modes that couple with the core mode made single-mode guidance possible.

Recent work on PiBF focused on realizing very large mode areas to support high-power delivery, as their core sizes can be increased while still maintaining single-mode guidance. Furthermore, hetero-structuring of the cladding can be done to enhance HOM rejection (Yehouessi et al. 2015). The state-of-the-art MFD of 60 μm with fundamental mode loss of less than 0.1 dB/m from a PiBF with single-mode guidance was reported (Yehouessi et al. 2016).

The development of ever higher-power lasers warrants the need for waveguides that are able to support these powers while minimizing nonlinear effects. Conventional step-index fibers require large core diameters to avoid optical damage and nonlinear effects. However, larger core diameters bring about multimode guidance, together with bending-sensitive losses and mode instabilities at high powers. PiBFs circumvent this limitation as their photonic bandgap-based guidance allows large mode field diameters while maintaining single-mode operation. Thus, we expect future research in PiBFs to focus on the improvement of HOM rejection to further push the mode field diameter.

Hollow-Core Fibers

PCFs with hollow cores are termed as hollow-core fibers (HCFs) as shown in Fig. 10. The hollow center is surrounded by layers of periodic microstructure as in Fig. 10a, and light is confined through photonic bandgap effects.

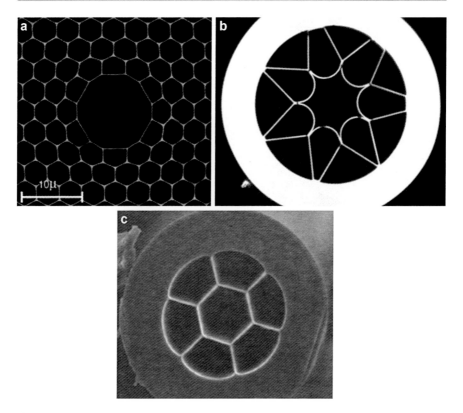

Fig. 10 Cross-sectional view of typical hollow-core fibers (a) Photonic bandgap fiber (b) Negative curvature fiber (c) anti-resonant HCF

The advantage of guiding light within the air holes is the high optical damage threshold of HCFs as compared to conventional solid core-guided PCFs. In high-energy pulse delivery, optical damage is a concern for solid-core fibers, causing damage on the fiber end and within the fiber. The optical damage limit is imposed by the solid material (e.g., fused silica) damage threshold, where most of the energy resides for solid-core fibers. By confining the pulse within the air-filled core instead of a solid core, high-energy pulse delivery is possible. Larger solid cores may be used to reduce the energy density to avoid laser-induced damage; however, larger core supporting multimode operations are more sensitive to bend-induced losses. Thus, HCFs are more attractive for flexible delivery of high-energy pulses as compared to solid-core PCFs (Shephard et al. 2015).

Another advantage of HCFs is the possibility to guide longer-wavelength light where the fused silica glass material reaches its cutoff around 2,200 nm. Moreover, the transmission window and optical properties, such as dispersion, can be engineered through the design of the fiber structure. The hollow air holes of glass capillaries and PCFs provide microchannels for the infiltration of gases or liquids, exploiting light-matter interaction effects with long interaction lengths.

Traditional photonic bandgap HCFs have a limited and narrow light wavelength range due to its inherent photonic bandgap mechanism. The understanding and design of how light propagates within these complex structured photonic bandgap-based HCFs are based on the "photonic tight-binding" model, where there exists three resonators – the interstitial silica apexes, the silica struts, and the hollow air holes – that are associated with the cladding Bloch modes responsible for the photonic bandgap formation (Couny et al. 2007).

The narrow transmission window is useful for applications requiring spectral filtering; however, it also limits the usefulness of these fibers in applications requiring broadband transmission. Additionally, photonic bandgap HCFs exhibit steep dispersion profiles which vary between positive to negative thousands to hundreds of fs^2/cm across its zero-dispersion wavelength (Várallyay and Saitoh 2010). The steep dispersion profile limits the fiber's use in the transmission of ultrafast pulses.

Hence, newer HCF designs have been fabricated to achieve broader transmission range based on the antiresonant effect, as shown in Fig. 10b, c. The guidance mechanism is also called inhibited coupling, as the antiresonance inhibits coupling between the mode that propagates in the central hollow core and the other modes that may exist in the hollow cladding tubes, the cladding glass, and interstitial glass apexes. Due to the small spatial mode overlap and a mismatch in mode effective indices, there is little leakage of the core mode into the cladding (Wei et al. 2017). The power ratio residing in the glass region of antiresonant/inhibited coupling-based HCFs may be less than 0.01%, lowering the nonlinearity and increasing the power damage threshold when compared with their photonic bandgap counterparts (Roberts et al. 2005).

Early designs consist of a large hollow core surrounded by rings of air holes, and these antiresonant HCFs (AR-HCFs) were called Kagome HCFs as their microstructure lattice resembles "Kagome" – the traditional Japanese woven basket pattern. Further efforts to obtain a low-loss AR-HCF resulted in studies to optimize and isolate the key structural features of the antiresonant guidance mechanism. On one hand, simpler designs lacking periodicity were numerically studied and fabricated (Février et al. 2010). On the other hand, studies on the physical structures, such as the core boundary structure, were done.

Kagome HCF with a hypocycloid-shaped core was demonstrated to have lower loss than its counterpart with the conventional circular core. The hypocycloid ring, which formed a negative curvature around the hollow core, increased the antireso-nant mode coupling suppression, allowing the fiber to reach a lower attenuation of 0.4 dB/m as compared to 1.4 dB/m of its circular core counterpart (Wang et al. 2011). Simplified HCF designs incorporating the negative curvature core were developed as well with similar reduced confinement loss effect (Wei et al. 2017).

Another advantage of the negative curvature HCF (NC-HCF) is the relatively flat dispersion profile around the zero-dispersion wavelength. Dispersion relates to the temporal delay between the high and low frequency components in a pulse. Hence, a flatter dispersion profile limits the temporal broadening of the pulse to a manageable level, which makes it suitable for ultrafast pulse transmission. One example of ultrafast pulse delivery where NC-HCF was used is the delivery of femtosecond pulses for multiphoton spectroscopy, introducing little temporal broadening, from 161.3 fs to 162.0 fs after a fiber length of 1.5 m (Popenda et al. 2017).

The first demonstration of pulse compression using gas-filled HCF was done in 1998, by compressing pulses of 140–10 fs duration (Nisoli et al. 1996). The spectral broadening was achieved by self-phase modulation (SPM) effects in the noble gas-filled HCF. Subsequent temporal compression was achieved using dispersive prisms. The dispersion acquired due to propagating through the gas-filled waveguide and through the other optical components must be compensated by additional dispersive components, of equal magnitude but opposite sign to achieve compression. Since then, many experiments have adapted this pulse compression configuration of using a gas-filled waveguide for spectral broadening and a dispersion-compensating component (such as chirped mirrors, spatial light modulators, gratings, and prisms) for temporal compression (Russell et al. 2014).

Another configuration is to use an anomalous dispersive waveguide that allows simultaneous dispersion compensation of the SPM-induced normal dispersion and from the propagation medium (typically gas). This configuration eliminates the need for additional dispersive elements after the gas-filled fiber (Ouzounov et al. 2005).

HCFs are highly engineerable, where the dispersion properties and transmission bands can be designed by controlling the fiber geometry. Moreover, the propagation of light within the hollow core allows high-power transmission, with low attenuation across its transmission bandwidth. The power residing in the glass region of NC-HCFs is typically 0.01% or less, which allows the transmission band to reach wavelengths where material losses are significant, covering the ultraviolet, visible, infrared, and terahertz regions. The hollow air holes also serve as microchannels to facilitate light-matter interaction at long interaction lengths. With these key advantages of HCFs, we expect future research to continue in the field of high-power delivery, ultrafast pulse delivery and compression, nonlinear optics, and sensing and spectroscopy across the ultraviolet, infrared, and terahertz wavelength regions. Industrial applications – such as micromachining, laser welding, and medical applications, such as laser surgery and multiphoton imaging and spectroscopy – will also benefit from the high-power and ultrafast pulse delivery capabilities of HCFs.

Conclusion

Optical fibers are critical elements in a high-power laser system, ranging from a flexible delivery fiber for high energy, an efficient compression cell of ultrafast and high-power laser beam, to a scalable gain media for power amplification. New fiber designs to foster mode selection, bandgap engineering, and polarization/dispersion/nonlinear property management are essential perspectives in high-power laser development. New material study from the hosting glass to the dopants is another important horizon to extend the laser emission to extreme wavelength regimes, including ultraviolet and infrared.

References

J. Bai, J. Zhang, J. Koponen, M. Kanskar, E. Towe, High pulse energy chirally-coupled-core Yb-doped fiber amplifier system, conference on lasers and electro-Optics 2017, OSA technical digest, paper JW2A.88, 2017

A. Baz, G. Bouwmans, L. Bigot, Y. Quiquempois, Pixelated high-index ring Bragg fibers. Opt. Express **20**, 18795–18802 (2012)

F. Couny, F. Benabid, P.J. Roberts, M.T. Burnett, S.A. Maier, Identification of Bloch-modes in hollow-core photonic crystal fiber cladding. Opt. Express **15**, 325–338 (2007)

J.W. Dawson, M.J. Messerly, R.J. Beach, M.Y. Shverdin, E.A. Stappaerts, A.K. Sridharan, P.H. Pax, H.E. Heebner, C.W. Siders, C.P.J. Barty, Analysis of the scalability of diffraction-limited fiber lasers and amplifiers to high average power. Opt. Express **16**, 13240 (2008)

L. Dong, J. Li, X. Peng, Bend resistant fundamental mode operation in ytterbium-doped leakage channel fibers with effective area up to 3160μm2. Opt. Express **14**, 11512–11519 (2006)

L. Dong, H.A. McKay, L. Fu, M. Ohta, A. Marcinkevicius, S. Suzuki, M.E. Fermann, Ytterbium-doped all glass leakage channel fibers with highly fluorine-doped silica pump cladding. Opt. Express **17**, 8962–8969 (2009)

M.A. Duguay, Y. Kukubun, T.L. Koch, L. Pfeiffer, Antiresonant reflecting optical waveguides in SiO2-Si multiplayer structures. Appl. Phys. Lett. **49**, 13–15 (1986)

S. Février, B. Beaudou, P. Viale, Understanding origin of loss in large pitch hollow-core photonic crystal fibers and their design simplification. Opt. Express **18**, 5142–5150 (2010)

I. Hu, C. Zhu, M. Haines, T. McComb, G. Fanning, R. Farrow, A. Galvanauskas, Nonlinear polarization switching and preservation effects in 55 μm core polygonal-CCC fibers, conference on lasers and electro-Optics 2015, OSA technical digest, paper JTh2A.94, 2015

M. Javadimanesh, S. Ghavami Sabouri, A. Khorsandi, The effect of cladding geometry on the absorption efficiency of double-clad fiber lasers. Opt. Appl. **XLVI**, 2 (2016)

Y. Jeong, J.K. Sahu, D.N. Payne, J. Nilsson, Ytterbium-doped large-core fiber laser with 1:36 kW continuous-wave output power. Opt. Express **12**, 6088–6092 (2004)

J.C. Knight, T.A. Birks, P. St, J. Russell, D.M. Atkin, All-silica single-mode optical fiber with photonic crystal cladding. Opt. Lett. **21**, 1547–1549 (1996)

D. Kouznetsov, J.V. Moloney, Efficiency of pump absorption in double-clad fiber amplifiers. II. Broken circular symmetry. J. Opt. Soc. Am. B **19**, 1259–1263 (2002)

J. Limpert, A. Liem, M. Reich, T. Schreiber, S. Nolte, H. Zellmer, A. Tünnermann, J. Broeng, A. Petersson, C. Jakobsen, Low-nonlinearity single-transverse-mode ytterbium-doped photonic crystal fiber amplifier. Opt. Express **12**, 1313–1319 (2004)

J. Limpert, N. Deguil-Robin, I. Manek-Hönninger, F. Salin, F. Röser, A. Liem, T. Schreiber, S. Nolte, H. Zellmer, A. Tünnermann, J. Broeng, A. Petersson, C. Jakobsen, High-power rod-type photonic crystal fiber laser. Opt. Express **13**, 1055–1058 (2005)

J. Limpert, O. Schmidt, J. Rothhardt, F. Röser, T. Schreiber, A. Tünnermann, S. Ermeneux, P. Yvernault, F. Salin, Extended single-mode photonic crystal fiber lasers. Opt. Express **14**, 2715–2720 (2006)

C. Liu, G. Chang, N. Litchinitser, A. Galvanauskas, D. Guertin, N. Jabobson, K. Tankala, Effectively single-mode chirally-coupled core fiber, advanced solid-state photonics 2007, OSA technical digest, paper ME2, 2007

R. Maurer, Optical waveguide light source, U.S. Patent 3,808,549, 1974

A. Mizrahi, L. Schächter, Bragg reflection waveguides with a matching layer. Opt. Express **12**, 3156–3170 (2004)

M. Nisoli, S.D. Silvestri, O. Svelto, Generation of high energy 10 fs pulses by a new pulse compression technique. Appl. Phys. Lett. **68**, 2793–2795 (1996)

D.G. Ouzounov, C.J. Hensley, A.L. Gaeta, N. Venkateraman, M.T. Gallagher, K.W. Koch, Soliton pulse compression in photonic band-gap fibers. Opt. Express **13**, 6153–6159 (2005)

G. Overton, IPG photonics offers world's first 10 kW single-mode production laser, Laser Focus World, 2015.

H. Pei, J. Ruppe, S. Chen, M. Sheikhsofla, J. Nees, Y. Yang, R. Wilcox, W. Leemans, A. Galvanauskas, 10mJ energy extraction from Yb-doped 85μm core CCC fiber using coherent pulse stacking amplification of fs pulses, Laser Congress 2017 (ASSL, LAC), OSA technical digest, paper AW4A.4, 2017

X. Peng, L. Dong, Fundamental-mode operation in polarization-maintaining ytterbium-doped fiber with an effective area of 1400 μm2. Opt. Lett. **32**, 358–360 (2007)

M.A. Popenda, H.I. Stawska, L.M. Mazur, K. Jakubowski, A. Kosolapov, A. Kolyadin, E. Bereś-Pawlik, Application of negative curvature hollow-core fiber in an optical fiber sensor setup for multiphoton spectroscopy. Sensors **17**, 2278 (2017)

P. Roberts, F. Couny, H. Sabert, B. Mangan, D. Williams, L. Farr, M. Mason, A. Tomlinson, T. Birks, J. Knight, Ultimate low loss of hollow-core photonic crystal fibres. Opt. Express **13**, 236–244 (2005)

P.S.J. Russell, P. Holzer, W. Chang, A. Abdolvand, J.C. Travers, Hollow-core photonic crystal fibres for gas-based nonlinear optics. Nat. Photonics **8**, 278–286 (2014)

J.D. Shephard, A. Urich, R.M. Carter, P. Jaworski, R.R. Maier, W. Belardi, F. Yu, W.J. Wadsworth, J.C. Knight, D.P. Hand, Silica hollow core microstructured fibers for beam delivery in industrial and medical applications. Front. Phys. **3**(24) (2015)

E. Snitzer, H. Po, F. Hakimi, R. Tumminelli, B.C. McCollum, Double clad, offset core Nd fibre laser, paper PD5, in Proc. Opt. Fib. Sensors 2, OSA, 1988

F. Stutzki, J. Florian, A. Liem, C. Jauregui, J. Limpert, A. Tünnermann, 26mJ, 130W Q-switched fiber-laser system with near-diffraction-limited beam quality. Opt. Lett. **37**, 1073–1075 (2012)

F. Stutzki, F. Jansen, H.J. Otto, C. Jauregui, J. Limpert, A. Tünnermann, Designing advanced very-large-mode-area fibers for power scaling of fiber-laser systems. Optica **1**, 233–242 (2014)

Z. Várallyay, K. Saitoh, Photonic crystal fibre for dispersion control, in Frontiers in guided wave optics and optoelectronics (InTech), 2010

Y.Y. Wang, N.V. Wheeler, F. Couny, P.J. Roberts, F. Benabid, Low loss broadband transmission in hypocycloid-core Kagome hollow-core photonic crystal fiber. Opt. Lett. **36**, 669–671 (2011)

C. Wei, R. Joseph Weiblen, C.R. Menyuk, J. Hu, Negative curvature fibers. Adv. Opt. Photon. **9**, 504–561 (2017)

W.S. Wong, X. Peng, J.M. McLaughlin, L. Dong, Breaking the limit of maximum effective area for robust single-mode propagation in optical fibers. Opt. Lett. **30**, 2855–2857 (2005)

J.P. Yehouessi, A. Baz, L. Bigot, G. Bouwmans, O. Vanvincq, M. Douay, Y. Quiquempois, Design and realization of flexible very large mode area pixelated Bragg fibers. Opt. Lett. **40**, 363–366 (2015)

J.P. Yehouessi, G. Bouwmans, O. Vanvincq, A. Cassez, R. Habert, Y. Quiquempois, L. Bigot, Ultra large mode area pixelated Bragg fiber, in *Fiber Lasers XIII: Technology, Systems, and Applications*, Proc. SPIE 9728 (2016)

Multicore Fibers

24

Ming Tang

Contents

M. Tang (✉)
Wuhan National Lab for Optoelectronics (WNLO) and National Engineering Laboratory for Next Generation Internet Access System (NGIA), School of Optical and Electronic Information, Huazhong University of Science and Technology (HUST), Wuhan, China
e-mail: tangming@mail.hust.edu.cn

© Springer Nature Singapore Pte Ltd. 2019
G.-D. Peng (ed.), *Handbook of Optical Fibers*,
https://doi.org/10.1007/978-981-10-7087-7_37

Abstract

Optical fibers, especially the silica single mode fibers (SMFs), play essential roles in building the infrastructure of information technology. However, with the great development of Internet services like cloud computing, HD video, and virtual reality, the current optical fiber communication system based on SMF is suffering from severe burden of sharp burst of capacity. The space division multiplexing (SDM), which multiplexes the information in the spatial degree, is able to increase the capacity greatly. Among the SDM techniques, the multicore fiber (MCF)-based SDM transmission system has broken the current system's capacity records again and again. The design, manufacturing, testing, connection, and application of MCFs compatible with the state-of-the-art fiber/telecom industry are indispensable to investigate.

This chapter mainly considers the all-solid silica-based MCF in which all the cores are shared by a single cladding. After the introduction of general description regarding the SDM oriented MCF, detailed information is given about the design, fabrication, and parameter optimization of MCF for efficient data transmission. The advances of highly efficient fan-in/fan-out coupling and splicing techniques for MCF are also reviewed. Recent demonstrations of MCF-enabled fiber transmission experiments are discussed, including the record long-haul large capacity transmission, fiber-radio convergent access system, high speed passive optical network, and real-time data-center interconnections. Finally, the MCF-based SDM fiber sensing technology developed very recently is summarized as a promising solution toward real-world application of fiber sensing.

Introduction

The Necessity of Space Division Multiplexing Technology

From the essence of optical signals, the physical multiplexing dimensions include five aspects, as shown in Fig. 1, which are time, polarization, frequency, amplitude, phase, and space. Today, high-speed signals in optical fiber communication systems are using multiple multiplexing techniques, such as time division multiplexing (TDM), wavelength division multiplexing (WDM), polarization division multiplexing (PDM), and quadrature amplitude modulation (QAM) using coherent detection techniques. Only the spatial dimension in the physical layer of the fiber has not been studied in depth. So space division multiplexing (SDM) becomes an inevitable choice to break through the capacity limitations of nowadays optical fiber communication systems.

Classification of SDM

In 2010, Professor Nakazawa of Tohoku University in Japan presented 3M technology on the EAXT research program on behalf of the Japanese government,

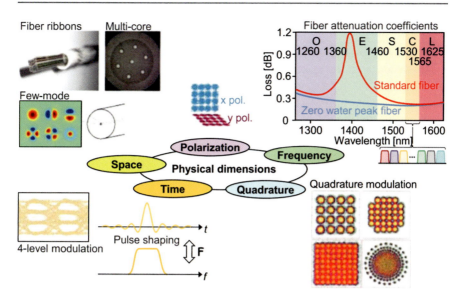

Fig. 1 Optical signal physical multiplexing dimensions (Agrell et al. 2016)

academia, and industry, in the report of the European Conference on Optical Communications (Nakazawa 2010). 3M technology represents multicore, multimode, and multilevel modulation. As a breakthrough technology to improve the capacity of optical fiber communication systems, SDM based on MCF and few-mode fiber (FMF) is regarded as the second fiber transmission technology revolution after WDM.

SDM technology is divided into two technologies based on MCF and FMF. Since its introduction in 2010, it has been studied by scientific research institutions in Europe, America, and Japan. National Institute of Information and Communications Technology (NICT), and Japan Telephone and Telegraph Corporation (NTT), together with Northeastern University, Hokkaido University, Osaka University, Sumitomo Corporation, Fujikura, and other organizations involved in the EXAT program propose to achieve increase the capacity of optical fiber communication systems by thousand times before 2020. In just a few years, a multibatch MCF was designed and fabricated, and a variety of fan-in/fan-out devices were fabricated. The theory of inter-core crosstalk (IC-XT) coupling was studied in depth, and the application transmission experiments were reported in recent Optical Fiber Communication Conference and Exposition (OFC), European Conference on Optical Communication (ECOC), etc. The EU's research on SDM technology has established the MODE-GAP project, which includes several universities and companies, such as Southampton University, Aston University, Eindhoven University of Technology. They focus on the research of SDM technology based on FMF, especially the production and development of fan-in/fan-out devices, and rapidly industrialization. On this basis, the University of Eindhoven has repeatedly broken the record of SDM transmission experiments based on FMF. In the United States, Bell Labs and the University of Central Florida also reported a large number of transmission experiments using the MCF and FMF fabricated by OFS and Corning and realized the real-time transmission experiment of SDM. It marks the SDM transmission moving from the ideal environment of the laboratory to more complex real-time transmission in the field.

Research Progress of SDM Based on MCF

The SDM based on MCF mainly utilizes several parallel channels provided by the MCF itself. Due to the stability of the hexagonal arrangement in the manufacturing process, the early MCFs were arranged in a seven-core hexagonal arrangement with a common step profile cladding. Since there are seven cores in a limited cladding, IC-XT between cores is unavoidable. Even if the core pitch increases, the IC-XT is still as high as −20 dB/2 km (Imamura 2010). In the design of MCF, the trench-assisted structure is added similar to the trench-assisted structure used in G.657 fiber with good bending resistance. Therefore, the IC-XT can be greatly suppressed. In the literature (Takenaga et al. 2011), the IC-XT in MCF is reduced to −35 dB/100 km, which basically meets the communication requirements. On this basis, the researchers proposed the design and drawing methods of homogenous MCF. By using pure silicon as the core material, Sumitomo fabricated ultra-low loss MCF with each core attenuation less than 0.18 dB/km in 2011. By increasing the cutoff wavelength to 1500 nm, the bending loss and additional loss of the outer cores are well managed (Hayashi et al. 2011). In 2012, Furukawa Company produced a MCF with a large effective area of 140 μm^2. At the same time, the IC-XT was well suppressed and the additional losses of the outer cores were reduced to less than 0.001 dB/km (Imamura et al. 2012). After that, in order to increase the number of cores and reduce IC-XT, the researchers have successively designed MCF with various structures, such as 10-core MCF with different core pitches (Matsuo et al. 2011), 12-core MCF with annular arrangement (Matsuo et al. 2012), and a double-ring 12-core MCF (Takenaga 2014). At present, the design and drawing technology of MCF have been developed into square-distributed 24-core MCF and hexagonal multiring 30-core MCF (Amma et al. 2015; Ye et al. 2015).

To apply MCF to SDM transmission, fan-in/fan-out devices are essential. At present, the fan-in/fan-out devices of MCF mainly include: free space lens coupling method, fiber bundle taper method, fiber bundle corrosion fusion method, and polymer waveguide etching method. The amplifier is also one of the key components for long-distance transmission of MCF. In the initial SDM transmission experiment, long-distance transmission based on MCF was de-multiplexed and amplified separately after each span of transmission (Zhu et al. 2011). Such an amplification scheme is not desirable in practice. Especially for long-distance dense SDM transmission systems, the number of amplifiers increases linearly with transmission distance, which is limited by energy consumption and costing.

Therefore, the amplifier of SDM is also one of the key technologies for the practical use of SDM (Abedin et al. 2014a).

The most straightforward technical solution is to fabricate a multicore erbium-doped fiber amplifier (MC-EDFA) and convergence the pump light and signal light through a WDM coupler when amplifying signals in each core by using of fan-in/fan-out devices. Although the scheme avoids seven EDFAs, it is separately amplified for each core. Therefore, it needs several single-mode pump lasers, and additional WDM couplers, isolators, etc. Its cost and complexity are very high. The insertion loss of devices, especially the WDM coupler, affects the noise figure (NF) of the MC-EDFA. In (Abedin et al. 2011), the amplification gain is about 25 dB, and the NF is larger than 6 dB due to the insertion loss of WDM coupler. In order to reduce cost and complexity, as well as to improve amplifier performance, several cladding pump methods have been proposed. It uses a lower cost multimode pump laser to inject into the MC-EDFA center core through a fan-in/fan-out device. Due to the large mode field of the multimode fiber (MMF), the pumping light field is more easily coupled into the outer cores, which needs properly design the MC-EDFA cladding, such as double-cladding structure, the reduce of cladding diameter. Thus, signals in the outer cores are amplified (Abedin et al. 2012). Compared to core pump scheme, this method eliminates the need for passive components, such as WDM couplers and isolators. It reduces the NF due to low insertion loss and also greatly reduces cost by sharing a multimode pump laser. This scheme can achieve a small signal gain of 32 dB and a NF of about 6 dB. In the above amplification schemes, all need to multiplex and de-multiplex the signals in the MC-EDFA. The produced additionally insertion loss will increase the NF of the amplifier. Therefore, in the literature (Abedin et al. 2014b), a side-coupled cladding pump method is proposed, which allows the signals to be directly coupled into the MC-EDFA avoiding the insertion loss of fan-in/fan-out devices. The side-coupled cladding pump method is achieved by using a tapered MMF which is tightly attached to the MC-EDF, and the coupling efficiency of about 67% can be achieved. The gain is increased by 5 dB compared to the edge-coupled cladding pump scheme.

Since the coupling loss of fan-in/fan-out devices is eliminated, the required MC-EDF length is shorter, the pump power can be lower, and the NF is also smaller. In order to further improve the pump efficiency, the latest research has designed a MC-EDF whose cladding is divided into an inner cladding and an outer cladding. And the core is distributed in a ring shape in the outer cladding. Due to the inner cladding has a lower refractive index, the pump light intensity is stronger in the core. And it can further improve the amplification performance (Jin et al. 2015).

Based on the continuous optimizing of structure and drawing process of MCF, solving the problem of fusion splice (Amma et al. 2013), making a multicore amplifier with low insertion loss and compact structure, and improving pump efficiency, the researchers carried out the experiments of SDM transmission. At present, MCF has been used in short distance, medium long distance, unrepeated transmission, and even ultra-long-distance transmission. The transmission capacity is from over 100 Tb/s to more than 1 Pb/s, and the bandwidth-distance product also exceeds Eb/s·km. At the 2012 ECOC, Japanese telecom operator KDDI, together with NEC corporation and Furukawa, conducted the world's first MC-EDFA SDM transmission experiment, which realized 6160 km transmission with seven-core fiber. Each core has 40 wavelengths with 128 Gb/s PDM-QPSK signals, and the bandwidth-distance product reached 177 Pb/s·km (Takahashi et al. 2013). The attenuation of MCF is between 0.188 dB/km~0.2 dB/km, and the length of each span is 55 km. The maximum gain of MC-EDFA is 15 dB and the NF is about 7 dB. At the 2013 ECOC, the team increased this record to 201 wavelengths per core. Using faster-than-Nyquist (FTN) technology, 30 GBaud signals were transmitted 7326 km with seven-core fiber. The bandwidth-distance product reached 1.03 Ebit/s·km (Igarashi et al. 2013). Japan's NICT together with Furukawa reported 19-core fiber transmission in 2013 achieving a capacity of 305 Tb/s for a single fiber (Kanno et al. 2013). Each core transmitted 100 wavelengths with 43 GBaud PDM-QPSK signal. In order to cope with the XT problem caused by the increasing number of cores, NTT Corporation of Japan proposed to use the characteristics of MCF parallel channels for adjacent core bidirectional transmission to overcome the XT problem. And it was reported that a bidirectional transmission of 12-core fiber had a transmission capacity of 344 Tb/s in each direction on the 2013 ECOC. After transmission of 1500 km, the signal was also nonlinearly compensated (Kobayashi et al. 2013). Then the team further realized the 450 km transmission with 12-core fiber and reached capacity of 409 Tb/s + 409 Tb/s (Sano et al. 2013). In 2015, the team reported the world's first unrepeated transmission based on MCF. The average attenuation of MCF is 0.207 dB/km. The splice loss is between 0.01 dB and 0.28 dB. The insertion loss of fan-in device is 0.4~0.6 dB, and the insertion loss of fan-out device is between 0.3~0.9 dB. Through the remote pump amplification method, the MCF has unrepeated transmission of 204 km and the transmission capacity is 120.7 Tb/s, which is more than ten times of the SMF unrepeated transmission capacity record (Takara et al. 2015). SDM transmission based on MCF also provides a unique advantage for nonlocal oscillator coherent receiving transmission systems. The NICT team reported a nonlocal oscillator coherent receiving transmission system based on a 19-core fiber. It used a core of the MCF to transmit continuous light to the receiver as the local oscillator source and the other 18 cores to transmit the signal. At the expense of 1/19 spectrum efficiency, the complexity of the signal processing at the receiving end was reduced (the frequency offset estimation and phase recovery algorithm is omitted), and the local oscillator source at the receiving end was omitted, which reduced the cost (Puttnam et al. 2013). However, a nonlocal oscillator coherent receiving system based on SMF transmits continuous light as a local oscillator in a polarization state orthogonal to the signal, which sacrifices half of the spectral efficiency. At the 2015 ECOC, NICK and Fujikura reported the results based a 22-core fiber, which transmitted 31 km and the capacity reached 2.15 Pb/s. It had 399 wavelengths per core with 24.5 GBaud PDM-64QAM signals per wavelength (Puttnam et al. 2015). It can be seen that the SDM transmission based on MCF is compared to SMF transmission, whose unrepeated amplification technology, coherent processing algorithm, and even nonlinear compensation are all well compatible. And the required fusion splice, amplification, and fan-in/fan-out devices do not significantly affect the transmission results. With the feature of the

MCF parallel channels, the SDM transmission not only easily breaks the capacity limit of SMF transmission but also facilitates the transmission system that is difficult to be realized in SMF communication, such as self-coherent transmission system and bidirectional transmission system.

MCF Design and Fabrication

After the rapid development during the past decades, there are various kinds of MCF for different applications shown as Fig. 2 (Mizuno and Miyamoto 2017). Started with the original 7-core single mode MCF (SM-MCF), MCF has been developed to accommodate up to 32 cores for long haul transmission. In order to avoid differential group delay (DGD) in long distance transmission, each core is designed as single mode. The cladding diameter is limited to a certain value to maintain mechanical reliability (e.g., less than 250 μm) (Mitsunaga et al. 1982; Sakamoto et al. 2016).

When pursuing highest core density, the core pitch between adjacent cores can be reduced to 20 μm or less. Due to the strongly coupling among cores, the electric field would not be limited in one core leading to super mode. Therefore, such fibers could be called coupled core MCF (Xia et al. 2016).

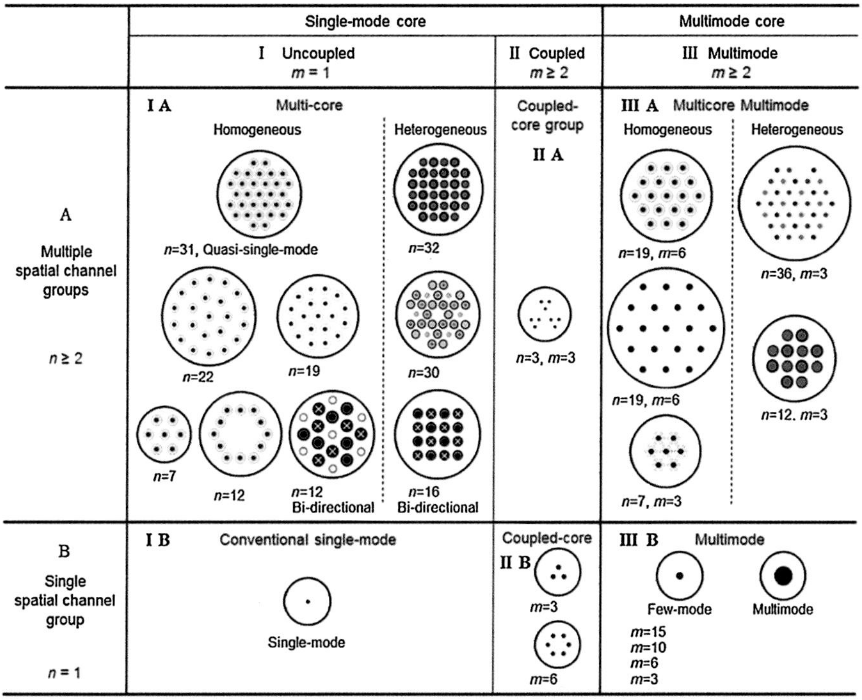

	Single-mode core			II Coupled $m \geq 2$	Multimode core	
	I Uncoupled $m = 1$				III Multimode $m \geq 2$	
A Multiple spatial channel groups $n \geq 2$	I A Multi-core			Coupled-core group II A	III A Multicore Multimode	
	Homogeneous		Heterogeneous		Homogeneous	Heterogeneous
	$n=31$, Quasi-single-mode		$n=32$		$n=19$, $m=6$	$n=36$, $m=3$
	$n=22$	$n=19$	$n=30$	$n=3$, $m=3$	$n=19$, $m=6$	$n=12$. $m=3$
	$n=7$	$n=12$	$n=12$ Bi-directional	$n=16$ Bi-directional	$n=7$, $m=3$	
B Single spatial channel group $n = 1$	I B Conventional single-mode			Coupled-core II B $m=3$ $m=6$	III B Multimode	
	Single-mode				Few-mode $m=15$ $m=10$ $m=6$ $m=3$	Multimode

Fig. 2 Different types of MCF (Mizuno and Miyamoto 2017)

To accommodate more spatial channels in limited cladding diameter, the MCF are combined with few mode fiber (FMF). In few-mode MCF (FM-MCF), each core is designed as few mode, resulting the most spatial channels such as 108 channels (Sakamoto et al. 2017a).

Uncoupled-Core MCF

For uncoupled MCF or weakly coupled MCF, IC-XT is an important evaluation metric. Since all the cores in the MCF share a common cladding, the optical power in one core will be inevitably coupled to adjacent cores. The XT in uncoupled MCF has been widely investigated (Fini et al. 2010, 2012a, b; Hayashi et al. 2010, 2012). The use of low index trench around the core is one approach to reduce XT (Hayashi et al. 2011). The use of different refractive index design between adjacent cores is another approach (Sakaguchi et al. 2016). Alternatively, holey structures such as photonic crystal (Imamura et al. 2009), hole-assisted (Saitoh et al. 2010), and hole-walled MCF (Yao et al. 2012) have also been proposed.

In 2011 (Hayashi et al. 2011), Tetsuya Hayashi et al. fabricated a seven-core fiber with pure silica cores to reduce the attenuation of each core. The cross section of the MCF is shown in Fig. 3. To decrease the XT, the index profile of each core is designed as trench-assisted (TA). The marker is for core identification. The actual core pitch, cladding diameter, and coating diameter were 45 μm, 150 μm, and 256 μm, respectively.

The attenuation spectra of each core are shown in Fig. 4. The attenuation of each core is very low (0.175–0.181 dB/km at optical wavelength 1550 nm, 0.192–0.202 dB/km over the full C + L band), and distinctive degradations for the outer cores were not observed.

The optical properties of each core are shown in Fig. 5. Macrobending losses were very low due to the TA profile and long cutoff wavelength. Polarization mode dispersion (PMD) was also measured for the C + L band.

Along with multiple kinds of MCF development such as 12-core SM-MCF (Matsuo et al. 2012), 19-core SM-MCF (Kanno et al. 2013), the maximum core

Fig. 3 Cross section of seven-core SM-MCF (Hayashi et al. 2011)

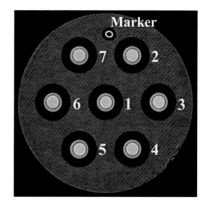

Fig. 4 Attenuation spectra of each core

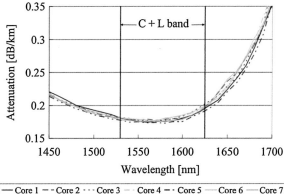

Fig. 5 Optical properties of each core

	Attenuation [dB/km]		MFD [μm]	A_{eff} [μm²]	λ_{cc} [nm]	CD [ps/nm/km]	D. Slope [ps/nm²/km]	Bend loss (R = 5 mm) [dB/turn]		PMD [ps/√km]
λ [nm]	1550	1625	1550	1550	N/A	1550	1550	1550	1625	C + L
Design	N/A	N/A	9.86	79.6	1496	23.3	0.063	0.010	0.021	N/A
Core 1	0.176	0.196	9.83	80.2	1509	22.2	0.062	0.011	0.020	0.132
Core 2	0.179	0.197	9.76	80.2	1550	22.2	0.062	0.010	0.020	0.134
Core 3	0.181	0.202	9.88	81.3	1504	22.2	0.062	0.010	0.020	0.044
Core 4	0.177	0.199	9.85	80.8	1498	22.2	0.062	0.011	0.021	0.093
Core 5	0.175	0.192	9.74	79.0	1485	22.2	0.062	0.011	0.022	0.205
Core 6	0.179	0.200	9.80	79.9	1483	22.1	0.062	0.011	0.022	0.116
Core 7	0.177	0.197	9.72	78.2	1498	22.2	0.062	0.010	0.019	0.106

number reached 32 cores for SM-MCF. In 2017, Takayuki Mizuno et al. designed and fabricated a novel heterogeneous 32-core SM-MCF to realize a low-XT dense space-division-multiplexed (DSDM) long-haul transmission (Mizuno et al. 2017). Figure 6 shows the cross section view of the heterogeneous 32-core SM-MCF. A square lattice arrangement with two types of refractive index designs effectively minimizes IC-XT for heterogeneous MCF. The length of the fabricated MCF spool is 51.4 km, the core pitch is 29.0 μm, and the cladding diameter is 243 μm. The cutoff wavelength at 1 km was less than 1.53 μm. The attenuation and effective area at 1550 nm are 0.24 dB/km and larger than 80.3 μm², respectively. The worst IC-XT was less than −39.4 dB within the same refractive index cores and less than −54.0 dB within different refractive index cores.

Coupled-Core MCF

As the core pitch gradually decreases, the optical power coupling between adjacent cores will be significantly enhanced. The mode field of the coupled MCF's eigenmode will not be limited in single core but distributed in each core according to a certain law (Xia et al. 2016). As shown in Fig. 7, for two coupled cores,

Fig. 6 Cross section of
32-core SM-MCF (Mizuno
et al. 2017)

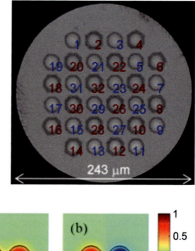

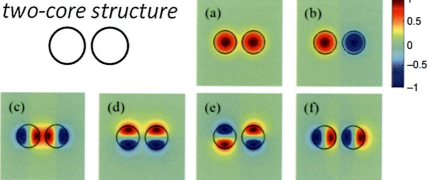

Fig. 7 Modal fields of LP_{01} super modes (**a** and **b**) and LP_{11} super modes (**c–f**) (Xia et al. 2016)

the super mode LP_{01} has two degenerate modes called symmetric super mode and antisymmetric super mode. The super mode LP_{11} has four degenerate modes. Based on coupled mode theory, we believe that the super mode field is the linear combination of eigenmode of each core with certain weights. Each weight could be derived by solving the eigenvalue equation.

It should be noted that when the core pitch is too small, the coupled MCF will gradually exhibit similar properties to the FMF (Sakamoto et al. 2017b). The IC-XT will evolve into differential model delay (DMD) of different super modes. Figure 8 shows the calculated 10 dB down width of the impulse response as a function of the normalized wavelength λ/D, where λ represents optical wavelength and D represents core pitch. The results for the inter-core skew and DMD between super-modes on the assumption that there is no core mode or super-mode coupling are shown as dotted lines for comparison. The figure shows that the impulse response width partially follows the dotted lines in the shorter or longer wavelength region. Therefore, we defined coupled MCF as MCF with a Gaussian-like impulse response, but not individual peaks caused by weak coupling between super modes.

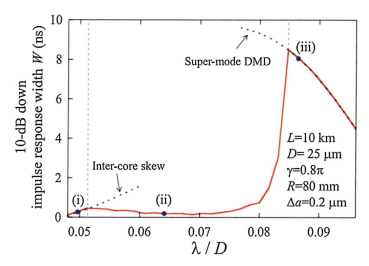

Fig. 8 10 dB down impulse response width as a function of normalized wavelength (Sakamoto et al. 2017b)

Fig. 9 Cross section of three-core coupled MCF (Ryf et al. 2012a)

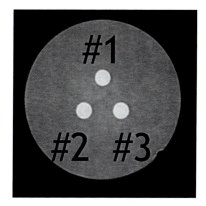

In 2012, R. Ryf et al. designed three-core coupled MCF consisting of three homogeneous cores with diameter 12.4 μm and refractive index difference 0.27% (Ryf et al. 2012a). The attenuation in a 60 km fiber was measured to be 0.181 dB/km at 1550 nm. The cutoff wavelength was designed to be around 1350 nm and chromatic dispersion and dispersion slope are designed to be 21 ps/nm/km and about 0.06 ps/nm²/km, respectively, at 1550 nm. The cladding diameter is 125 μm. The distance between the equidistant cores and the center of the fiber is 17 μm, resulting in a core pitch of 29.4 μm (Fig. 9).

The calculated linearly polarized super-modes of the three-core coupled MCF (upper row) and their far-fields (bottom row) are shown in Fig. 10 (Ryf et al. 2012b). The fundamental mode, which is excited when all core modes have the same phase at coupling, is designated as LP_{01}, in analogy to the features of the LP_{01} mode of

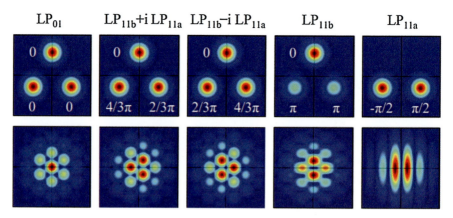

Fig. 10 Linearly polarized super modes (upper row) and corresponding far fields (bottom row) of three-core coupled MCF (Ryf et al. 2012b)

Fig. 11 Cross section of 3*3-core coupled MCF (Ryf et al. 2014b)

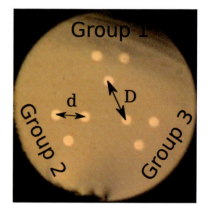

step index FMF. The higher-order mode, designated as LP_{11}, is degenerated, and therefore, the representation is not unique. Note that we assume linearly polarized super modes, and each mode has two orthogonal polarizations. Therefore, a total of six independent SDM transmission channels are available.

In 2014, R. Ryf et al. designed a new fiber which contains nine cores, arranged in three groups. Each group is composed of three coupled cores. The cladding diameter is only 143 μm (Ryf et al. 2014b). Figure 11 shows the cross section of the 35.7 km long 3*3 core coupled MCF. All cores are homogeneous, with a refractive index difference of 0.58% and a diameter of 7.8 μm. The core pitch among cores within a coupled group is 22.3 μm, and the minimum distance between cores from different groups is 34.3 μm.

The effective areas of the nine cores are calculated to be (56 ± 1) μm^2 and the cutoff wavelength is 1410 nm. The core also has a low average attenuation of 0.19 ± 0.01 dB/km, and the dispersion and dispersion slope are found to be 19.9 ps/nm/km and 0.054 ps/nm^2/km, respectively. The impulse responses between

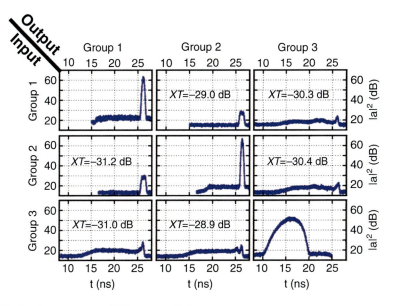

Fig. 12 Impulse response and XT measured between core groups

the core groups after 35.7 km of propagation are measured as shown in Fig. 12. We define the XT as the light power received in the group under measurement divided by the power received in the illuminated group. Group 1 has the smallest XT (about −31 dB), whereas group 2 has the largest XT (about −29 dB). Because of the strong coupling within cores of the same group, almost identical results are obtained when measuring XT from individual cores in different groups.

FM-MCF

In 2014, R.G.H et al. for the first time designed a novel hole-assisted seven-core FM-MCF with linearly polarized mode LP_{01} and LP_{11} modes for each core to realize more spatial channels (van Uden et al. 2014). The coating and cladding diameter of the step index seven-core FM-MCF are 372 μm and 192 μm, respectively. Individual cores with a 13.1 μm diameter, at a core pitch of 40 μm, were arranged on a hexagonal lattice, as shown in Fig. 13a. The air holes have an 8.2 μm diameter and are placed 13.3 μm away, creating an air-hole-to-pitch ratio of 0.62, corresponding to an XT of −80 dB/km. For each core, the LP_{01} mode field diameters are 11.8 μm and have a DMD of 4.6 ps/m. The impulse response measurement is shown in Fig. 13b.

In 2016, Jun Sakaguchi et al. reported the design and characterization of a heterogeneous 3-mode 36-core FM-MCF with three core types, reaching 108 spatial channels (Sakaguchi et al. 2016). All core types are with identical core and trench

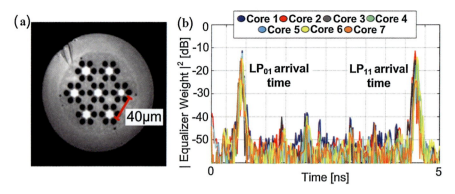

Fig. 13 (**a**) Microscope image of the seven-core FM-MCF end face. (**b**) Measured impulse response from the MIMO equalizer's taps (van Uden et al. 2014)

diameters. All the cores support LP_{01} and LP_{11} modes and the effective index difference between core types was larger than 0.001 for all combinations of LP_{01} and LP_{11} modes. The core pitch is 34 μm, and the cladding diameter is 306 μm as shown in Fig. 14.

Type-A cores have the strongest mode confinement and therefore were arranged on the outermost vertices, closest to the cladding-coating interface. The optical properties of the fabricated 5.5 km FM-MCF are summarized in Fig. 15. It shows similar effective areas for all core types and cutoff wavelengths are suitable for C-band transmission. The inter-core XT for all LP mode combinations was evaluated with coupled power theory based on the measured index profiles of the cores.

To realize a higher spatial density while maintaining a feasible cladding diameter, Taiji Sakamoto et al. designed and fabricated 6-mode 19-core FM-MCF with 114 spatial channels (Sakamoto et al. 2017a). The FM-MCF preform was formed with the stack-and-draw technique. As shown in Fig. 16, the cladding diameter, core pitch, and cladding thickness are 246 μm, 43.4 ± 0.7 μm, and 36.3 ± 0.4 μm, respectively.

The attenuation and mode dependent attenuation are less than 0.24 dB/km and 0.01 dB/km, respectively. The attenuation variation between individual cores is also successfully suppressed below 0.03 dB/km. The maximum DMD for all the cores is well controlled below 0.33 ns/km with a maximum variation of 0.15 ns/km. Figure 17 summarizes the measured optical properties of the FM-MCF.

Fan-In and Fan-Out

The fan-in/fan-out devices (or mux/de-mux devices) are used to connect several SMFs with corresponding cores in one MCF realizing low insertion loss, low XT, and low reflect loss. The MCF-based communication systems will be compatible with current optical fiber communication systems by using the fan-in/fan-out devices.

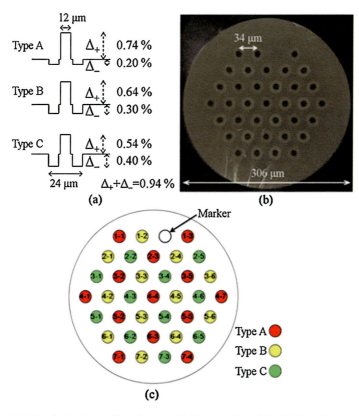

Fig. 14 (**a**) Refractive index profiles of cores. (**b**) Cross-section of the fabricated 3-mode 36-core FM-MCF. (**c**) Schematic design illustrating core arrangement (Sakaguchi et al. 2016)

Measured properties	Mode	Type A			Type B				Type C	
	Core	(4–1)	(4–4)	(5–2)	(4–2)	(5–3)	(6–1)	(4–3)	(5–1)	(6–2)
Transmission loss [dB/km]	$LP_{01} + LP_{11}$	0.242	0.266	0.255	0.283	0.292	0.281	0.262	0.308	0.284
Cable cutoff wavelength [nm]	LP_{21}	1458	1526	1478	1351	1357	1346	1199	1235	1245

Calculation from measured refractive index profiles	Mode	Type A (4–1)	Type B (4–2)	Type C (4–3)
Effective area [μm^2]	LP_{01} / LP_{11}	77 / 105	74 / 102	76 / 110
DMD [ps/m]		7.4	7.1	6.3
		A-B	B-C	C-A
Core-to-core crosstalk [dB] after 5.5 km	LP_{01}-LP_{01} / LP_{01}-LP_{11}	−98 / −61	−86 / −32	−88 / −48
(5 cm correlation length assumed)	LP_{11}-LP_{01} / LP_{11}-LP_{11}	−77 / −54	−60 / −31	−33 / −32

Fig. 15 Optical properties of the fabricated FM-MCF at 1550 nm

The fabrication process of fan-in/fan-out devices is divided into four types: free space coupling type (Klaus et al. 2012; Shimakawa et al. 2014; Tottori et al. 2012, 2013), fused taper type (Uemura et al. 2014), fiber bundle type (Abe et al. 2013; Shimakawa et al. 2013; Watanabe et al. 2014), and integrated waveguide type (Watanabe et al. 2012). Free space coupling-based fan-in/fan-out is bulky and poor

Fig. 16 Cross-section of
19-core FM-MCF

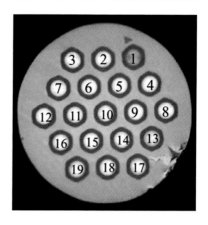

	LP$_{01}$	LP$_{11}$	LP$_{11}$	LP$_{02}$
Loss (dB/km)@1.55 μm	<0.23	<0.24	<0.24	<0.23
A_{eff} (μm^2)@1.55 μm	80.1 ± 5	164$^{(1)}$	203$^{(1)}$	170$^{(1)}$
α_b (dB/turn)@R7.5@1.55 μm		<0.5		
Total corsstalk (dB/100 km)@1.565 μm	< −37.9	< −34.1	< −31.1	< −30.8
22-m cutoff wavelength (μm)		<1.53		
Maximum DMD (ns/km)		<0.33		

Fig. 17 Optical properties of fabricated 6-mode 19-core FM-MCF

in stability but easy to customize. Fused taper-based fan-in/fan-out is low cost, but there is a problem of mode field mismatch. Fiber bundle-based fan-in/fan-out is also low cost, but the fabrication process is more complex. Integrated waveguide-based fan-in/fan-out exists a problem of secondary alignment resulting in complex process.

As shown in Fig. 18a, a single lens is used for coupling all the SMF's outputs to corresponding MCF's cores (Tottori et al. 2012). The end facet of the MCF is placed at the front focal point of the single lens. Each optical beam is spatially separated after passing the back focal point and coupled to the corresponding SMF via a fiber collimator. Compared with direct fiber to fiber coupling, this design is more tolerant to the offset of the core position. Figure 18b shows a picture of the fabricated module. The size was 40 mm (dia.) and 62 mm (length). The insertion loss for each port is less than 0.6 dB. It is noteworthy that the difference of coupling loss was less than 0.4 dB. The XT is less than −50 dB and the measured polarization dependent loss is less than 0.1 dB.

Figure 19a shows a schematic view of fused taper type based fan-in/fan-out device fabrication process (Uemura et al. 2014). 19 SMFs are inserted into a glass capillary which contains 19 holes with hexagonal close-packed structure. It is

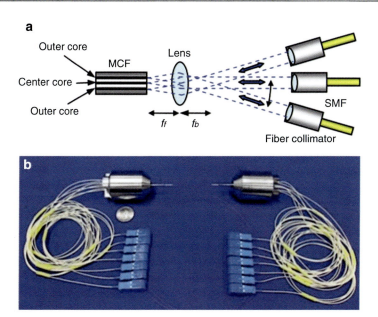

Fig. 18 (**a**) Schematic of the seven-core coupling by lens optics; (**b**) picture of a pair of fabricated optical connection modules (Tottori et al. 2012)

consolidated with the single-core fibers through elongation process. Elongation ratio is controlled to be the same core pitch as MCF. Figure 19b shows a photograph of the fabricated fan-in/fan-out device. The device length and outer diameter are about 35 mm and 0.72 mm, respectively. The average loss is about 0.45 dB with standard deviation of 0.15 dB. Maximum loss of 4.7 dB at C-band and the worst XT of −45 dB at 1550 nm are confirmed in fan-in and fan-out with 20 m long MCF.

Fiber bundle-based fan-in/fan-out device needs chemical etching of each SMFs (Li et al. 2015). Generally, seven bare SSMFs are chemical etched until the cladding diameter matches with the MCF's core pitch. The etched SSMFs are then inserted into a high precision ceramic ferrule. The cross section of fiber bundles fabricated is shown in Fig. 20a. The MCF is also chemical etched to 125 μm to be inserted into another ceramic ferrule with identical size and precision tolerance. After end polishing, both ceramic ferrules are aligned on the six-dimension free-space alignment platform. The photo of fabricated fan-in/fan-out device is shown in Fig. 20b. The average insertion loss is under 1.5 dB, and the XT and reflect is less than −50 dB and −45 dB, respectively.

As shown in Fig. 21a, femtosecond laser is used to inscribe 3D waveguides to couple into and out from the 3*3 MCF (Ryf et al. 2014b). Nine waveguides are routed to map a linear fiber array onto the core arrangement of the 3*3 MCF with an accuracy of ±1 μm. A SMF array with 127 μm spacing between fibers and mounted in a V-groove array is placed on one end, and the 3*3 MCF is on the other end, as

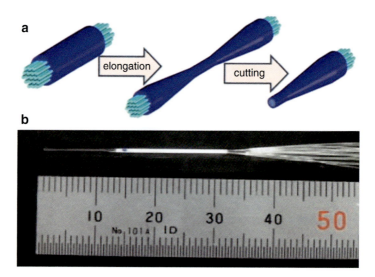

Fig. 19 (**a**) Schematic view of the fabrication process of fan-in/fan-out device. (**b**) Photograph of the fabricated fan-in/fan-out device (Uemura et al. 2014)

Fig. 20 (**a**) Photograph of ferrule end; (**b**) picture of fabricated fan-in device (Li et al. 2015)

shown in Fig. 21b. The fan-in/fan-out has an insertion loss of 2.5 dB and negligible core dependence of the insertion loss is observed.

Splicing Technology for MCF

Fusion splicing is an important technique to connect fibers. High quality fusion splicing is essential for the practical application of MCF. MCF have large cladding diameters and multiple cores, and these features cause several issues that do not exist for fusion splicing of SMF.

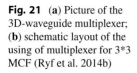

Fig. 21 (**a**) Picture of the 3D-waveguide multiplexer; (**b**) schematic layout of the using of multiplexer for 3*3 MCF (Ryf et al. 2014b)

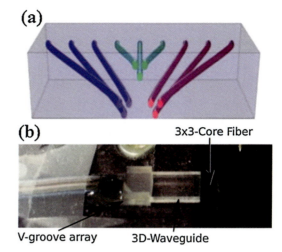

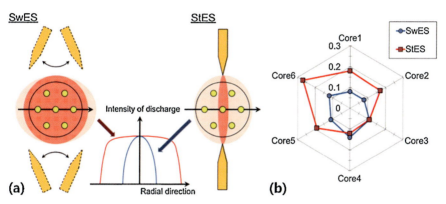

Fig. 22 (**a**) Comparison of discharge by a swing electrode system (SwES) and a static electrode system (StES). (**b**) distribution of measured fusion splice losses of a seven-core MCF

To achieve low and uniform losses for all cores of MCF, a uniform electrode discharge is necessary (Amma et al. 2017). However, this is difficult for a conventional static electrode system (StES) because the discharge intensity concentrates in the center of cladding. Yoshimichi Amma et al. developed a swing electrode system (SwES) (Amma et al. 2013). Figure 22a shows a comparison of the discharged conditions between SwES and StES for a seven-core MCF. The SwES swings the electrodes in the radial direction and realizes an extensively uniform discharge for all cores. In Fig. 22b, it is confirmed that the fusion splice losses by SwES are smaller and uniform than those by StES.

Cleaving angles also have a great influence on the splicing quality of outer cores for MCF (Watanabe et al. 2011). Figure 23a shows an image illustrates the influence of cleaving angle. If the angle is too large, the outer cores become thinner or thicker after fusion splicing. Figure 23b shows the histograms and cumulative probability of

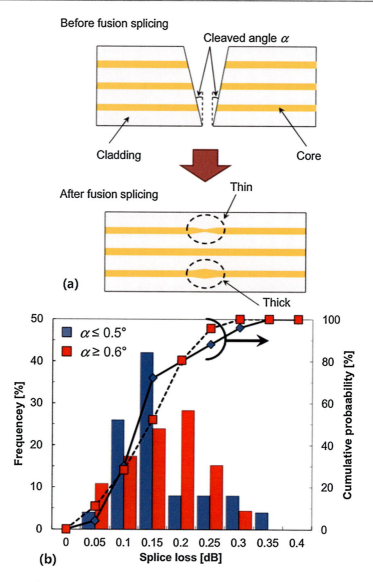

Fig. 23 (**a**) Influence by cleaving angle for fusion splicing of MCFs. (**b**) Histograms and cumulative probability of splice losses for different cleaving angles

the splice losses at 1550 nm for cleaving angles less than 0.5° and larger than 0.6°. The average and standard deviation of splice losses were 0.14 dB and 0.06 dB at both cleaving conditions. However, the probability of splice losses less than 0.15 dB was different. The probability for smaller cleaving angle reached 70%, while for larger cleaving angle was only 50%. This result indicates that it is necessary to reduce the cleaving angle for a stable and low-loss fusion splicing for MCFs.

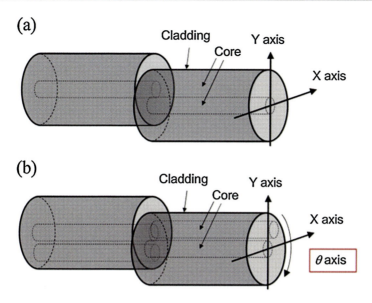

Fig. 24 Direction of core alignment for (**a**) SMF and (**b**) MCF

Figure 24 shows the difference of alignment between SMF and MCF. Aligning SMF only needs to move in the X and Y directions. However, due to complex structures of MCFs, fusion splicing of MCF is much more complicated and additional rotational angle alignment (θ direction) for the outer cores is necessary (Amma et al. 2017).

With the rapid development of splicing techniques, polarization-maintaining fusion splicers equipped with precise fiber rotating function are available, such as Fujikura FSM-100P+ (Yoshida et al. 2012; Zhu et al. 2010a). Applied with image processing algorithms, polarization-maintaining fusion splicers can achieve a convenient passive alignment of MCF. Precise alignment methods using side and end images have been developed based on these splicers.

Kotaro Saito et al. demonstrated a side-view-based alignment method for MCF; Fig. 25 shows the schematic image of the alignment technique (Saito et al. 2016). The side view contrast of MCFs has a rotational angle dependence, so the alignment angle can be acquired by comparing the brightness patterns of the two MCFs. The rotational angle is assumed to be aligned when the correlation coefficient of the two MCFs' brightness patterns has the highest value. The side-view-based alignment method can achieve a splice loss comparable with an active alignment method for a four-core MCF; the maximum splice loss is less than 0.5 dB when the correlation coefficient is higher than 0.975.

With the increasingly complex structure of the fibers, the side-view features will become indistinct, so that it may not be able to achieve reliable angle alignment. End-view-based alignment method can overcome this limitation (Shen et al. 2017). Figure 26 shows a schematic image of end-view-based alignment method. End-view

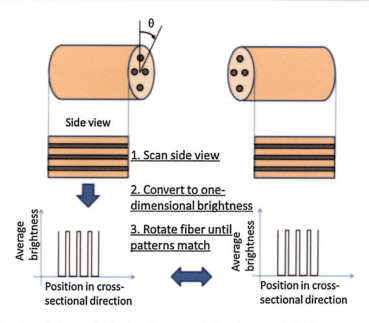

Fig. 25 Schematic image of side-view alignment technique (Saito et al. 2016)

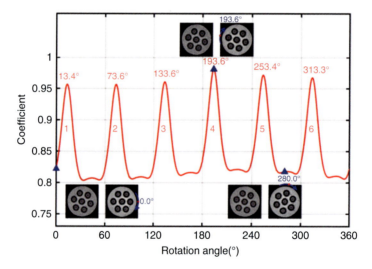

Fig. 26 Schematic image of end-view alignment method (Shen et al. 2017)

alignment technique uses the end-face images for correlation calculation. There are several peaks on the correlation coefficient curve depending on the fiber structures, and high precision angle alignment can be realized when the coefficient reaches maximum. The end-view method can be used for complex end-face structures. For a seven-core MCF, angle alignment accuracy is higher than 0.1°, and the average

Fig. 27 Microscopic image of the cross section of a core pumped MC-EDF

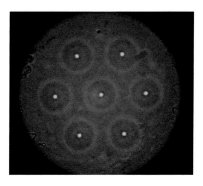

splice losses for the central core and the outer cores are 0.058 dB and 0.295 dB, respectively.

Applying advanced digital image processing method with the fiber rotating module will be promising to achieve high precision angle alignment for the fusion splicing of MCFs. The development of more innovative and simpler alignment methods is expected in the future.

Erbium-Doped Fiber Amplifier for MCF

As a key building block of optical networks, optical amplifiers are inevitable to compensate for signal attenuation. Especially in the MCF-based SDM systems, optical amplifiers compatible with MCF are necessary. During the past several years, multicore erbium-doped fiber amplifier (MC-EDFA) has been explored in the MCF-based long-distance transmission system. MC-EDFA can be classified into two categories according to the pump schemes, the core and cladding pumped MC-EDFA. We will give a detailed review about the MC-EDFA's performance characteristics of these two categories as follows.

Core Pumped MC-EDFA

Figure 27 shows the cross-section of a seven-core MC-EDF using core pumped scheme (Abedin et al. 2011). The cores are arranged in a hexagonal array with a core pitch of 40.9 μm. The core diameter and numerical aperture (NA) are 3.2 μm and 0.23, respectively, and the MFD at 1550 nm is estimated to be about 6 μm. The cladding diameter is 148 μm. The erbium-doped core has an attenuation coefficient of ~2.3 dB/m at 1550 nm, and the Er3+ concentration is estimated to be $6 \times 1024/m^3$.

The architecture of the core-pumped MC-EDFA is shown in Fig. 28 (Abedin et al. 2011). It employs two compact fan-in/fan-out devices to couple signals and pump light into the MC-EDF (Zhu et al. 2010a). SDM signals from MCF are

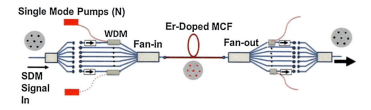

Fig. 28 Architecture of the core-pumped MC-EDFA

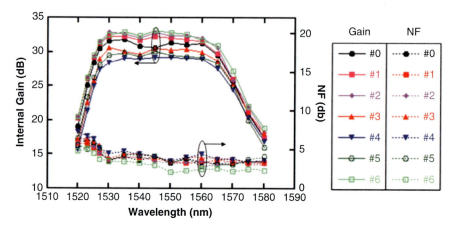

Fig. 29 Internal gain and NF in different cores of the MC-EDFA

de-multiplexed into individual SMFs through a fan-out device first, and then each signal and corresponding single mode 980 nm pump light are combined by WDM coupler and launched into each core of MC-EDF by a fan-in device. Isolators at the input and output end are used to avoid the back-reflection. After the amplification in the MC-EDFA, the signals and pump lights are spatially and wavelength de-multiplexed, and the signals are spatially multiplexed into the MCF again by the fan-in device.

Figure 29 shows the internal gain and NF in the C-band of the seven-core MC-EDFA (Abedin et al. 2011). For an input signal power of −15 dBm, a pump power of 146 mW, and 15 m MC-EDF, it can achieve an average gain of about 30 dB. Considering the losses of couplers and splices, the net gain is about 5 dB lower than the internal gain. Gain variations in different cores are mainly due to couplers' loss difference and it can be further optimized. NF over the whole C-band is less than 4 dB, except around 1520 nm where it increases to 6.4 dB.

Cladding-Pumped MC-EDFA

For the core-pumped MC-EDFA, the number of WDM couplers and single-mode pump diodes is equal to the number of MCF cores. Therefore, the power

Fig. 30 Photograph of the
cross section of the MC-EDF

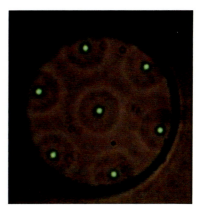

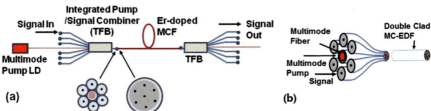

Fig. 31 (**a**) Schematic of edge-coupling cladding pumped MC-EDF. (**b**) Tapered fiber bundle
(TFB) developed for cladding pumped MC-EDFA

consumption and cost are proportional to the count of spatial channels. Compared
with core pumped scheme, cladding pumped scheme uses one cost-effective high-
power multimode laser diode as a common pump source. Because the pump
radiation is shared by all the cores, the cost and complexity of cladding pumped
amplifier are expected to be much lower than core pumped scheme through spatial
integration.

Cladding pumped 12-core fiber amplifier using free-space coupling module to
combine the pump and signal has been demonstrated (Ono et al. 2013). In this
section, we mainly focus on the fiber-based cladding-pumped amplifiers employing
edge- and side-coupling scheme.

Figure 30 shows the cross section of a double cladding MC-EDF in the cladding
pumped scheme (Abedin et al. 2012, 2017). Parameters and profiles of the cores are
similar to those used in the above core pumped scheme. Inner cladding diameter is
reduced to 100 μm to increase the pump intensity. Inner cladding was surrounded by
a lower index polymer layer as outer cladding to form a double cladding structure.
Since the double cladding EDF has a high NA of 0.45, the pump light can be
effectively injected into the inner cladding and distributed across the entire inner
cladding area including cores.

The schematic diagram of edge-coupling cladding pumped MC-EDFA based on
tapered fiber bundle (TFB) is shown in Fig. 31a (Abedin et al. 2012, 2017). A
TFB coupler is used to couple optical signals and the multimode pump light into

Fig. 32 (**a**) The gain measured for the MC-EDFA with input signal power of −20 dBm. (**b**) The NF measured for the MC-EDFA with input signal power of 0 dBm

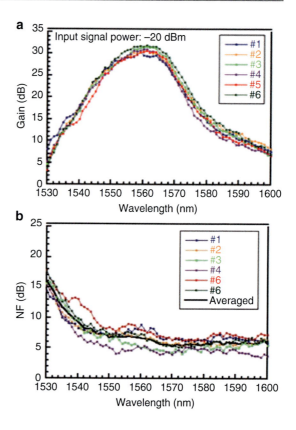

the MC-EDF. The structure of the TFB couplers is shown in Fig. 31b. The TFB coupler is fabricated by tapering a bundle of seven fibers, consisting of a central MMF surrounded by six SMF. The central MMF has a core/cladding diameter of 105/125 μm and is spliced to a 980-nm pump laser diode. The six outer cores of the TFB couplers are connected to the spatially multiplexed signals. During the tapering, as the MMF gets tapered, the pump starts to penetrate into the surrounding fibers. Since the double cladding MC-EDF has a higher NA, the pump light can be effectively guided. The launched pump power is about 10.9 W, of which about 7.6 W is estimated to be coupled into the inner cladding of the MC-EDF (Abedin et al. 2012). Near to the output end of the MC-EDF, the unabsorbed pump is removed through the coating-stripped section of the doped fiber. The amplified signals are de-multiplexed by the other TFB coupler.

Figure 32 shows the net gain and NF versus wavelength for the six outer cores, the coupled pump power 7.6 W. Figure 32a shows the measured gain for small signals (−20 dBm); a maximum gain of 32 dB can be obtained over 50 m long EDF. Considering the passive loss of about 7 dB, the internal gain is estimated to be 39 dB. The internal NF for different cores is shown in Fig. 32b with input signal power of 0 dBm, the NF becomes significantly large for wavelength shorter than

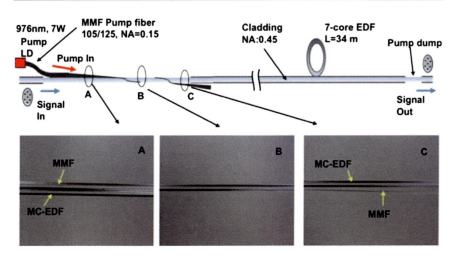

Fig. 33 Schematic diagram of a seven-core EDFA with side pumping

1540 nm, while it tends to decrease at longer wavelengths, average NF is about 6 dB for wavelength longer than 1560 nm.

Compared with core pumped scheme, cladding pumped scheme has a much lower pump intensity due to the pump power distributed across the entire cladding (Takasaka et al. 2013). So it is more difficult to fully invert the gain medium, resulting in a high NF and low gain (Krummrich 2012). Also, when the signal intensity becomes comparable to the pump intensity, the population inversion depletes and the amplifier saturates, which causes the gain to shift towards the L-band. So cladding pumped EDFA has a poor NF in the C-band. There is also another important issue for cladding pumped MC-EDFA. Since the multiple cores share the same pump radiation guided by the cladding, any significant depletion in pump due to the signal in a particular core may affect the gain of the signals in the other cores, introducing unwanted spatial cross-gain modulation (Abedin et al. 2012).

The schematic of the side-coupled pumping is shown in Fig. 33 (Abedin et al. 2014b). The MC-EDF is similar to that used in the edge-coupled scheme. Side coupling of pump radiation is achieved by a tapered multimode fiber (Theeg et al. 2012). A short section of the gain fiber (~8 cm long) near the input end is stripped of its low-index coating. The tapered multimode fiber is wrapped around the uncoated gain fiber to ensure efficient coupling. Along the taper, the light is transferred from the tapered fiber to the inner cladding of the MC-EDF. A coupling efficiency around 67% can be achieved in the demonstration, which could be further improved by optimizing the multimode fiber tapering process. Photographs of different sections of the coupler are shown in Fig. 33.

Figure 34a shows the gain versus wavelength measured in the seven cores for input signal powers of −20 and 0 dBm with the coupled pump power 4.7 W. The maximum small signal (−20 dBm) gain is about 36 dB near 1560 nm, and gain over

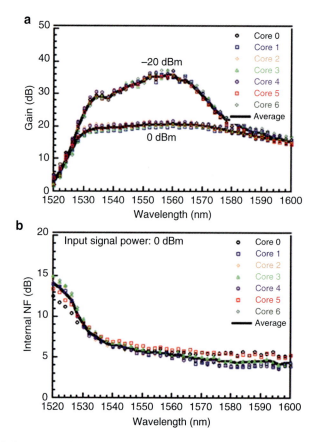

Fig. 34 (**a**) Gain measured for the seven different cores of the MC-EDFA. (**b**) NF measured for the seven different cores of the MC-EDFA

25 dB can be obtained over a bandwidth of ~40 nm. The internal NF for different cores versus wavelength is shown in Fig. 34b with input signal power of 0 dBm. The NF is about 5 dB for the signal wavelength of 1560 nm and it increases to about 8 dB at 1530 nm.

In core pumped and edge-coupled scheme, couplers will cause coupling loss which degrades the net gain and NF. Side coupling can couple multimode pump light into inner cladding efficiently without disturbing the signals in the cores. Due to the absence of TFB couplers, side-coupled amplifier obtains higher external gain and lower NF in comparison with the edge-pumped amplifier, even with a lower pump power (Abedin et al. 2017).

Several investigations have been done to improve the performance of cladding-pumped MC-EDFA. Using an Er-Yb co-doped fiber can obtain a high output power because of high pump-to-signal conversion efficiency. However, Er-Yb co-doping

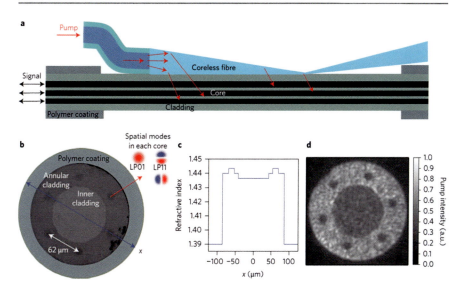

Fig. 35 (**a**) Schematics of integrated cladding-pumped FM-MCF EDFA. (**b**) Image of EDF facet. (**c**) Refractive index profile of EDF. (**d**) Output pump intensity distribution with multimode side pumping at 980 nm

will narrow the gain bandwidth (Ono et al. 2013). Enlarging the core diameter will increase the overlap between signals and pump filed, which results in a higher output power. In addition, reducing the cladding diameter can increase the pump power intensity to achieve a higher level of population inversion (Tsuchida et al. 2016).

Cladding-pumped MC-EDFA consisting of six few mode cores have been reported (Chen et al. 2016). Each core can support three spatial modes, which enables the EDFA to amplify a total of 18 spatial channels (six cores × three modes) simultaneously with only a single pump diode. Integrated cladding-pumped EDFA can greatly increase the transmission capacity per fiber.

The schematic of integrated cladding-pumped FM-MCF EDFA is shown in Fig. 35a. The amplifier employed a side-pumping method, with pump radiation coupled into cladding through a tapered coreless fiber. Adopting annular cladding can enhance the pump intensity around the cores to increase the pump power conversion efficiency. The refractive index profile of annular cladding along the x axis (Fig. 35b) of the six-core EDF is presented in Fig. 35c. The inner cladding is a depressed index region able to prevent pump light from entering the central region due to total internal reflection. This can save more than 25% pump power compared to a uniform cladding. The output pump intensity with multimode side pumping is shown in Fig. 35d. The average pump intensity in the annular cladding is enhanced by a factor of 1.45 compared to the inner cladding.

Each core of the MC-EDF can support the LP_{01} mode and the two degenerated LP_{11} modes in the C-band. To achieve high output power per spatial channel, which

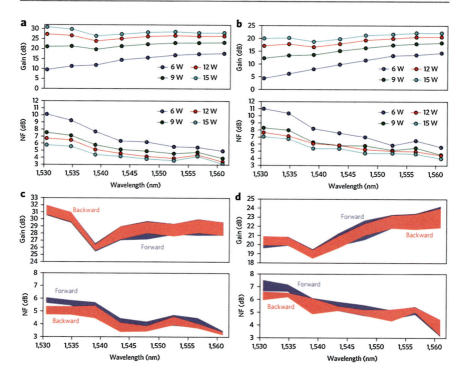

Fig. 36 (**a, b**) Results for LP_{01} mode of one core in a forward-pumping configuration with a total input power of -12 and -2 dBm, respectively, as a function of coupled-pump power. (**c, d**) Range of internal gains and NFs in both forwards and backwards pumping for all six cores with a total input power of -12 and -2 dBm, respectively

can only be accomplished if the EDFA maintains a high ratio of pump intensity to signal intensity throughout the gain medium so that it is not saturated, core and cladding refractive indices providing larger mode areas are designed, thereby reducing the signal intensity without affecting the pump intensity. The calculated mode field areas are 168 μm^2 and 179 μm^2 for the LP_{01} and LP_{11} mode groups, respectively, which are much larger than those of SMFs. The nearly identical mode fields also contribute to minimizing the mode-dependent gain, which is measured to be less than 2 dB.

Figure 36a, b shows the internal gains and NFs for one of the six cores as a function of coupled-pump power in the forward-pumping configuration. Figure 36c, d shows the range of internal gains and NFs for all six cores with a coupled-pump power of 15 W and total input signal power -12 and -2 dBm, respectively. The cladding-pumped six-core EDFA provides an output power >20 dBm per core and <7 dB NF over the C-band for the LP_{01} mode. The gain and NF deviations of 1 dB are mainly attributed to the inaccuracy during the measurements of input and output coupling loss.

MCF-Based Communication Systems

MCF-Based Optical Access Network

In 2010, the first practical trial of MCF based optical access network was demonstrated as shown in Fig. 37 (Zhu et al. 2010a). The seven-core MCF designed and fabricated for that demonstration had a length of 11.3 km, a core diameter of 8 μm, a cladding diameter of 130 μm, a core pitch of 38 μm, and mean XT of −39.0 dB at 1310 nm and −24.8 dB at 1490 nm. Connection to the MCF was realized with help of a tapered MCF connector (TMC) made from seven special SMF by using a taper process. The TMC had an insertion loss of 0.38–1.6 dB and XT of −39.21 to −43.8 dB at the best. Simultaneous transmissions of 1310 nm upstream (US) and 1490 nm downstream (DS) signals at 2.5 Gb/s over 11.3-km of seven-core MCF with a split ratio of 1:64 for passive optical network (PON) were demonstrated and it could serve a total of 448 end-users at the subscriber premises from a single fiber.

In 2015, Z. Feng et al. proposed and experimentally demonstrated a wavelength-space division multiplexing (WSDM) optical access network architecture with centralized optical carrier delivery utilizing MCFs and adaptive modulation based on reflective semiconductor amplifier (RSOA) (Feng et al. 2015). As shown in Fig. 38, five of the outer cores were used for DS transmission only, whereas the remaining outer core was utilized as a dedicated channel to transmit US signals. Optical carriers for US were delivered from the OLT to the ONU via the inner core and then transmitted back to the OLT after amplification and modulation by the RSOA in the colorless ONU side. The mobile backhaul (MB) service was also supported by the inner core. Wavelengths used in US transmission should be different from that of the MB in order to avoid the Rayleigh backscattering effect in bidirectional transmission. With QPSK-OFDM modulation format, the aggregation

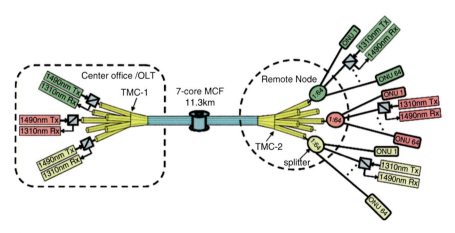

Fig. 37 Schematic diagram of a PON system using a seven-core MCF and two TMCs (Zhu et al. 2010a)

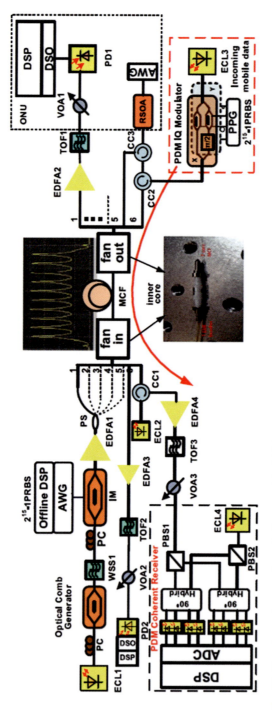

Fig. 38 Experimental setup schematic diagram of the MCF-Enabled WSDM optical access network (Feng et al. 2015)

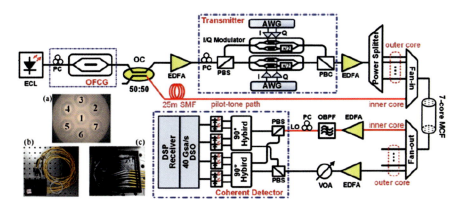

Fig. 39 Experimental setup for the SDM-UDWDM system with SHCD. (**a**) Cross section view of the fabricated seven-core fiber. (**b**) and (**c**) Overall fan-in/fan-out multiplexers with single mode fiber pigtails (Li et al. 2015)

DS capacity reached 250 Gb/s using five outer cores and ten wavelengths, and it could be further scaled to 1 Tb/s using 20 wavelengths modulated with 16 QAM-OFDM. For US transmission, 2.5 Gb/s QPSK-OFDM transmission could be achieved just using a low-bandwidth RSOA, and adaptive modulation was applied to the RSOA to further enhance the US data rate to 3.12 Gb/s. As an emulation of high-speed MB transmission, 48 Gb/s IQ modulated PDM-QPSK signal was transmitted in the inner core of MCF and coherently detected in the OLT side. Both DS and US optical signals exhibited acceptable performance with sufficient power budget.

In 2017, C. Xiong et al. demonstrated an ultra-dense wavelength-division-multiplexing (UDWDM) PON in a spatial-division-multiplexing system utilizing MCF with a self-homodyne coherent detection (SHCD) scheme (Li et al. 2015). As shown in Fig. 39, 12 channels of 40-Gb/s UDWDM-PON signals using PDM-QPSK with a 12-GHz grid were transmitted through the six outer cores of the seven-core MCF with the central core used to transmit the pilot tone. An optical frequency comb generator was employed to supply the multicarriers for cost saving. After 37-km low-XT seven-core MCF long-reach transmission with compact fan-in/fan-out multiplexers, a large transmission capacity of 2.88 Tb/s and a power budget of 26 dB were obtained. At the optical network unit side, a narrow linewidth local oscillator laser was no longer necessary since SHCD was adopted in the system. The complexity of the digital signal processing procedure at the coherent receiver was also greatly reduced as the carrier phase recovery and carrier frequency offset estimation were not needed.

In 2017, J. He et al. demonstrated a bidirectional transmission with wavelength reuse technique enabled coreless CAP-PON system over MCF (He et al. 2017). In the proposed scheme as shown in Fig. 40, the ONU side carrier suppression and OLT side coherent detection for a single wavelength implement was adopted for upstream transmission. The optical carrier was delivered from OLT to ONUs in

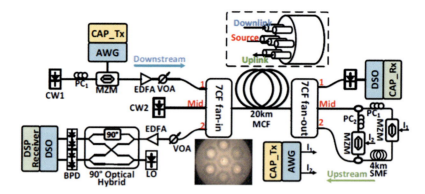

Fig. 40 Experimental setup of the bidirectional MCF-based CAP-PON system with wavelength reuse technique (He et al. 2017)

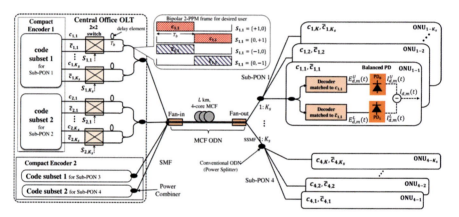

Fig. 41 Schematic architecture of core-multiplexed SAC OCDMA PON adopting CI and bipolar 2-PPM (Farghal and Shalaby 2018)

different cores for upstream transmission to avoid Rayleigh backscattering. In the experiment, 10 Gb/s symmetrical rate of bidirectional CAP signals was successfully transmitted over 20 km MCF.

In 2018, Ahmed E. A. Farghal and Hossam M. H. Shalaby proposed and demonstrated using bipolar M-ary pulse position modulation (PPM) signaling in conjunction with the code-interleaving (CI) technique to address the inter-core XT problem in MCF-based optical code-division multiple-access (OCDMA) PONs with spectral-amplitude coding employed as the coding scheme, as shown in Fig. 41 (Farghal and Shalaby 2018). In addition, a comparison between the performance of core-multiplexed OCDMA PON adopting the newly proposed techniques, namely, unipolar on–off keying (OOK) with CI and bipolar 2-PPM with CI and that adopting unipolar OOK without CI, was presented under both data rate and average photons per bit constraints. The obtained results showed that bipolar 2-PPM and CI were capable of reducing the inter-core XT impact on the performance of MCF-based OCDMA PONs. Furthermore, bipolar 2-PPM with CI outperformed other schemes in terms of the number of supportable users and energy efficiency.

MCF-Based Front-Haul

In 2015, M. Morant et al. firstly proposed and demonstrated the optical wireless fronthaul of a dual M×2×2 multiple-input multiple-output (MIMO) long-term evolution advanced (LTE-A) system employing spatial multiplexing in a single 125 m four-core MCF evaluating different bending radius (Morant et al. 2015). The MCF was configured using the same direction of propagation in the four cores and with two cores in opposite direction corresponding to the fronthaul of a dual LTE-A roof-mounted system. The MCF exhibited an increase of 9.6 dB XT when raising the bending radius from r = 35 cm to r = 67 cm. The 3GPP LTE-A MIMO algorithms compensated successfully the core-to-core XT and backscattering in the MCF (Fig. 42).

In 2016, J. He et al. experimentally demonstrated a bidirectional transmission and massive MIMO enabled radio over seven-core fiber system with centralized optical carrier delivery (He et al. 2016). As shown in Fig. 43, the middle core was used to deliver the seed light to ONU for coreless upstream transmission. By using spectral efficiency OFDM/OQAM modulation technique and hybrid optical and the wireless MIMO channel estimation algorithm, bidirectional transmission of a 4.42 Gb/s 2×2 MIMO-OFDM/OQAM signal over a 20 km seven-core fiber and a 0.4 m wireless link were successfully achieved. These results showed that the MCF could satisfy the transparently transmission of bidirectional RF MIMO signals. Moreover, by using higher core density MCF, this architecture could be easily extended to support massive MIMO ROF transmission in future 5G cellular communication.

In 2017, M. Morant and R. Llorente demonstrated the successful radio-over-multicore fiber transmission of carrier-aggregated M×4×4 MIMO LTE-A signals (Morant and Llorente 2017a). As shown in Fig. 44, carrier aggregation did not affect the signal quality as long as the CCs were not overlapped in spectrum, but the frequency response of the electro-optical devices should be taken into account depending on the 3GPP band used. The robustness of 3GPP 4×4 MIMO processing included in LTE-Advanced standard was evaluated over the different optical paths in a four-core fiber. 3GPP 4×4 MIMO processing provides a 10-dB optical power margin between the four optical paths, given the reference signals were received properly for the correct demodulation of the four spatially multiplexed data streams. The carrier-aggregated 4×4 MIMO LTE-A signal met the 3GPP EVM limit after 150 m MCF for received optical power levels higher than −15 dBm. Compared to the power requirement of −10 dBm for SISO counterpart, MIMO provided +5 dB optical power margin between cores.

Besides, they also proposed and evaluate experimentally radio-over-multicore fiber transmission for flexible and reconfigurable fronthaul links (Morant and Llorente 2017b). As shown in Fig. 45, using MCF they could assign each core

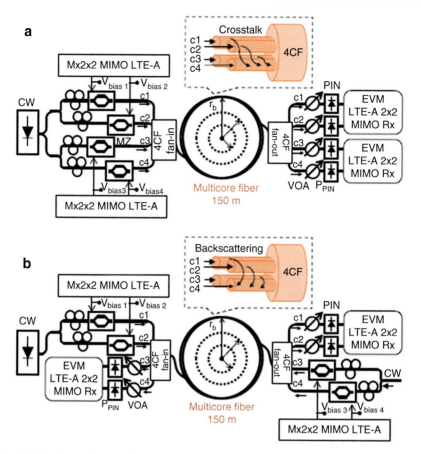

Fig. 42 Experimental setup for dual M×2×2 LTE-A RoF transmission in a four-core fiber considering different bending radius for: (**a**) Core-to-core XT evaluation and (**b**) Rayleigh backscattering evaluation in the MCF (Morant et al. 2015)

to feed different cellular antennas at the RRU providing seamless switching from one 2G, 3G, or 4G service to another using the same frequency band. The experimental evaluation demonstrated that it was feasible to switch GSM, EDGE and EGPRS2-Ausing the same frequency band on a given core, and all the 2G and 3G evaluated signals met the standard requirements after 150 m of MCF for received optical power levels higher than −15 dBm. The signal/core allocation could also feed MIMO signals to multiple antennas in the RRU. For instance, two cores were assigned to 4G spatially multiplexed 2×2 MIMO LTE-A carrier-aggregated signals, doubling the bitrate of SISO transmission when the received optical power was higher than −10 dBm. Moreover, radio-over-multicore fiber could also be used for fronthaul bidirectional links providing DL and UL connectivity in different cores. 3.5G CDMA2000 and 4G LTE-A bi-directional signals were successfully transmitted over 150 m of 4-CF, being the uplink the

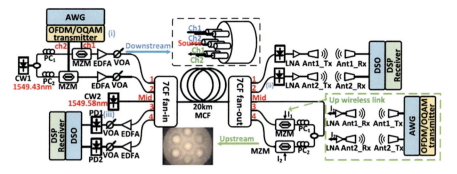

Fig. 43 Experimental setup of the proposed bidirectional N dimensional MIMO radio over 2N + 1 core fiber (He et al. 2016)

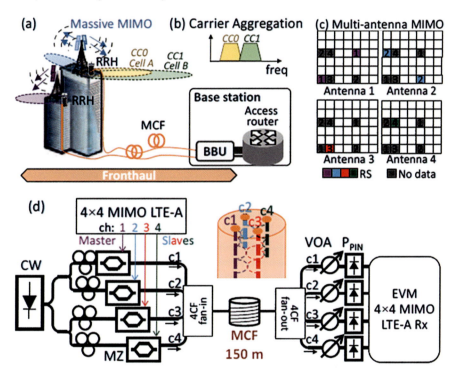

Fig. 44 (**a**) Application scenario of radio-over-multicore fiber fronthaul transmission for (**b**) carrier aggregation and (**c**) multi-antenna MIMO provision using different RS. (**d**) Experimental setup for 4×4 MIMO LTE-A carrier-aggregated RoF transmission over MCF (Morant and Llorente 2017a)

limiting path in both cases. These capabilities confirmed that radio-over-multicore fiber enabled a fully reconfigurable optical fronthaul feeding the different wireless signals to an antenna system capable of seamless supporting 4G, 3G, and 2G services.

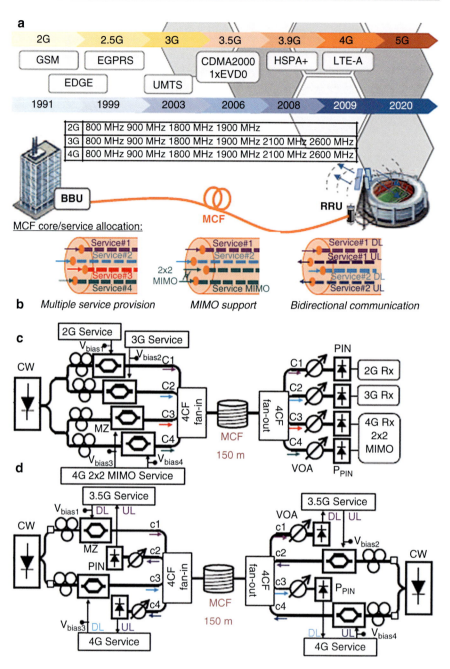

Fig. 45 (**a**) Cellular standards evolution and frequency bands. (**b**) Radio-over-multicore fiber fronthaul application scenario and possible MCF core/services allocation configurations. Experimental setup for: (**c**) multiple cellular services and (**d**) bidirectional services provision over MCF (Morant and Llorente 2017b)

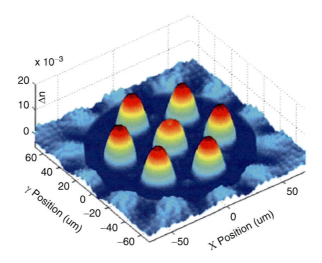

Fig. 46 Refractive-index profile of seven-core fiber by tomographic measurement (Zhu et al. 2010b)

MCF-Based Short-Reach Interconnect

Short-reach applications are suitable for immediate commercialization as IC-XT is the most serious issue in MCFs. Multimode seven-core fiber short-reach transmission was firstly demonstrated as shown in Fig. 46 (Lee 2010; Zhu et al. 2010b). This MCF had a core diameter of 26 μm, a cladding diameter of 125 μm, and a core-to-core distance of 39 μm. The average XT was -42.8 dB over 550 m transmission. In the demonstration, two types of MCF coupling techniques, using a TMC and a VCSEL array, were attempted in the transmission experiment. In the TMC case, 70 Gb/s aggregated capacity transmission through 100 m and 550 m of MCF was demonstrated. In the VCSEL array case, on the other hand, 120 Gb/s aggregated capacity transmission through 100 m of MCF was demonstrated. In the latter case, only six cores were utilized in the seven-core fiber because of layout limitations of the IC.

In 2016, Douglas L. Butler studied two-, four-, and eight-core MCF systems with associated connectors and fan-outs and achieved excellent BER performance (Butler et al. 2017). As shown in Fig. 47, the dual-core system was tested bidirectionally. This transmission system simplified bi-directional traffic and increased the bandwidth density by a factor of two. Overall, MCF systems reduced operational complexity with fewer fibers, cables, and connectors. Bidirectional 25 Gb/s transmission was demonstrated for transmission lengths of 200 m, 2 km, and 10 km with minimal power penalty, offering an upgrade path to higher data rate and longer distances for datacenters and high-performance computers.

In 2017, P. De Heyn et al. co-integrated ultra-dense 16× arrays of 50 GHz GeSi EAMs and GeSi PDs in combination with a 37-channel MCF interface on a single chip to realize an 896 Gb/s GeSi-based SDM transceiver in a total footprint of under

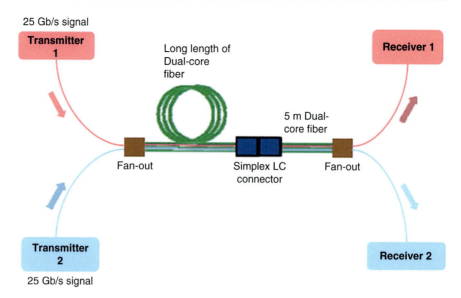

Fig. 47 Experimental setup for measuring BER in dual-core MCF shown here with co-propagating signals (Butler et al. 2017)

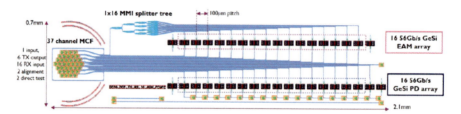

Fig. 48 Layout schematic of the 16-channel SDM transceiver comprising a 37-channel MCF coupling interface, 1-to-16 optical power splitter, and 16-channel 56 Gb/s GeSi waveguide EAM and PD arrays in a footprint of 0.7 mm × 2.1 mm (Srinivasan et al. 2017)

1.5 mm², as shown in Fig. 48 (Srinivasan et al. 2017). Open eye diagrams have been obtained at 56 Gb/s NRZ data rate in a loop-back experiment for all 16 channels, showing the potential of such GeSi EAM-PD arrays for ultra-compact, silicon-based Tb/s-scale optical I/O.

In 2018, Y. Liu et al. designed and fabricated a graded-index multicore fiber (GI-MCF) compatible with both standard multimode and single-mode fiber for high density optical interconnect application in large-scale datacenters as shown in Fig. 49 (Liu et al. 2018). The proposed GI-MCF was capable of reducing the system complexity of the large-scale datacenters. The major parameters of the GI-MCF have been optimized to obtain both a small DMD at 850 nm for multimode transmission, and a small MFD mismatch with SMF of less than 0.5 μm at 1310 nm and 1550 nm for quasi-single mode transmission. In experiment, the multimode

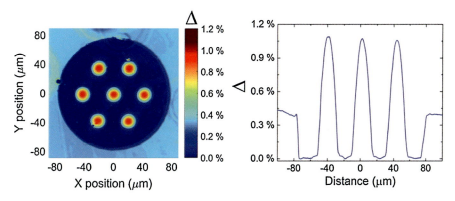

Fig. 49 The measured refractive index profile of the designed GI-MCF (Liu et al. 2018)

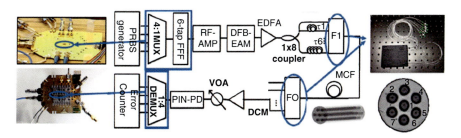

Fig. 50 Experiment setup for real-time 7 × 100 Gbps/λ/core IM/DD transmission over MCFs (Lin et al. 2018)

operation over 1 km-long GI-MCF at 850 nm and the quasi-single mode operation over 12.4 km-long GI-MCF at 1310 nm and 1550 nm were achieved at a data rate of 7 × 10-Gb/s with NRZ signals. The experimental results implied that the proposed GI-MCF satisfies various demands on operating wavelength, accessible distance, and interconnect density and can effectively reduce the fiber numbers and system complexity of large-scale data center.

In 2018, R. Lin et al. experimentally demonstrated real-time 7 × 100 Gbps/λ/core IM/DD transmission over 1 km and 10 km MCF, as shown in Fig. 50 (Lin et al. 2018). Self-developed BiCMOS chipset and monolithic integrated DFB-EAM provide high lane rate and the in-house fabricated MCF and fan-in/fan-out modules were used to scale the lane count in spatial dimension. Over 1 km, EDB communication below KP4 FEC limit was achieved. With optical dispersion compensation, both NRZ and EDB transmission over 10 km MCF achieving 7%-OH-FEC limit were enabled. Simpler Rx configuration could be found for NRZ transmission while better transmission performance of EDB signal could be observed due to the lower bandwidth occupation. The real-time DSP-free SDM system demonstrated its potential to support high-speed DCI. Operation at 1.55 μm band offered high capacity and interoperability throughout the intra-DC and inter-DC networks.

MCF-Based Long-Haul Transmission

In the long-haul transmission scenario, the employed MCFs are mainly based on single-mode MCF and coupled-core MCF (Matsuo et al. 2016; Mizuno and Miyamoto 2017; Mizuno et al. 2016b; Takahashi et al. 2014). During the development of MCF-based long haul transmission, in the early period, the spatial channels need to be demultiplexed first and amplified independently by single-mode EDFA, while this scheme consumes much power and is very complex. The multicore EDFA (MC-EDFA) is the key device to reduce the cost and the power consumption which greatly contributes to the development of the SDM system.

Due to the lack of MC-EDF, early long-haul transmission experiments based on MCF are mainly accomplished by independent single-mode EDFAs. In 2012, S. Chandrasekhar et al. firstly demonstrated 2688-km transmission with ten 50-GHz spaced 128-Gbit/s PDM-QPSK signals, over 76.8-km seven-core fiber per span (Chandrasekhar et al. 2012). To mitigate performance variations among the signals traveling through the multiple cores of a multispan MCF system, a novel concept is proposed and demonstrated, termed core-to-core signal rotation (CCR). The signals launched into each core of a MCF span are routed to a spatially different core in the next MCF span, and it continues along the MCF transmission link. Figure 51 shows the schematic of the experiment setup. The signals after transmission of each span are amplified to compensate for the fiber loss, and WDM channel powers are equalized using a wavelength blocker (WB) array module. After each WB, the DWDM signals from one core are sent to the re-circulating loop input of the next core, in a cyclic fashion. This ensured that the DWDM signals traversed each of the seven cores once every seven round trips.

In 2013, T. Kobayashi et al. demonstrated 12-core fiber bidirectional 1500-km transmission using 11.5-Gbaud Nyquist-pulse-shaped 16QAM signals, with intercore XT management and multicarrier nonlinear compensation. The total capacity is over 300 Tbit/s (Kobayashi et al. 2013). Figure 52 shows the experimental setup. The generated 748 channels 12.5-GHz-spaced WDM 16QAM signals make the line rate 92 Gb/s, resulting in the spectral efficiency of 6.13 b/s/Hz/core, assuming 20% FEC overhead. The transmission line consisted of a 50-km dual-ring-structure MCF with 12-fold re-circulating loops operated synchronously. For propagation-direction interleaving, the assigned propagation direction to each core is shown as Fig. 52b. Demodulation is postprocessed offline as shown in Fig. 52a.

In 2013, H. Takahashi et al. firstly demonstrated 40 × 128-Gbit/s PDM-QPSK signals over 6160-km seven-core MCF transmission with seven-core MC-EDFA (Takahashi et al. 2013). Figure 53 shows the experimental setup. The transmission line in a loop is composed of a 55-km length seven-core MCF span, a seven-core MC-EDFA, gain-flattening filters (GFF), and two single-core EDFAs. In order to obtain high SDM density and low XT, simultaneously, the cladding diameter and the core pitch of the MCF are designed to be less than 200 μm and 56 μm, respectively. Even and odd channels are combined with a 100–50 GHz wavelength interleaver. These signals are fed into a polarization multiplexing emulator (PME). The nominal data-rate after PDM is 128 Gbit/s. This assumes 20% overhead for

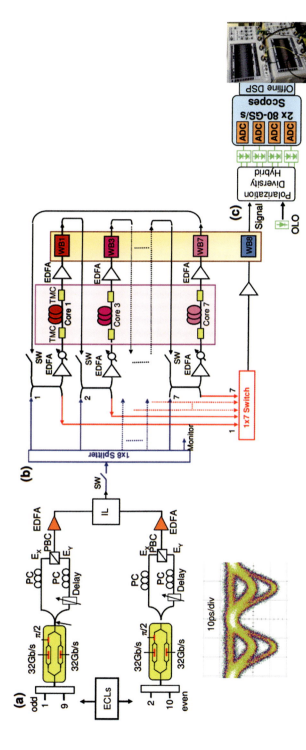

Fig. 51 Schematic of the experimental setup: (**a**) transmitter, (**b**) multiple recirculating loops with core-to-core rotation every round trip, and (**c**) coherent receiver. Inset shows transmitter eye diagram (Chandrasekhar et al. 2012)

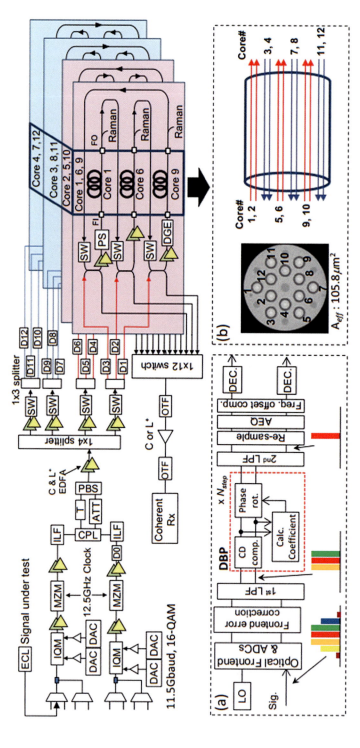

Fig. 52 Experimental setup: (**a**) Rx and offline DSP, (**b**) dual-ring structure 12-core fiber and assignment of propagation direction to each core (Kobayashi et al. 2013)

soft-decision forward-error-correction (SD-FEC) with OTU4 framing over 103.125-Gbit/s Ethernet payload.

Figure 54a–c shows the cross-section, composition, and gain and noise figure of the seven-core MC-EDFA which is prepared to compensate the span loss as shown in Fig. 53. Since the mode field diameter (MFD) of MC-EDF is about 7.3 μm which is much smaller than that of the MCF; a core pitch of 45 μm is sufficient to reduce the XT of the MC-EDF. The MC-EDF has an attenuation coefficient of ∼3.4 dB/m and a small-signal gain of ∼4.3 dB/m at 1550 nm.

At the same time, K. Igarashi et al. demonstrated 140.7-Tbit/s, 7326-km transmission with 7 × 201-channel 25-GHz-spaced Super-Nyquist-WDM 100-Gbit/s optical signals using seven-core fiber and full C-band seven-core EDFAs (Igarashi et al. 2013). For Super-Nyquist WDM transmission, the duobinary-pulse shaping is used in the transmitter to limit the bandwidth of the signal below baud rate. To mitigate the degradation of receiver sensitivity due to the bandwidth limitation, the maximum-likelihood sequence estimation (MLSE) is introduced at the receiver. The impulse response and power spectrum of duobinary-pulse shaping are shown in Fig. 55b, c, respectively. Receiving the duobinary-pulse-shaped signal by decision on the symbol-by-symbol basis gives a 2.1 dB loss in SNR in principle. This penalty can be compensated by MLSE with on memory. Figure 55d shows Trellis for a duobinary-pulse-shaped signal.

Figure 56 shows the experimental setup. 201-channel 25-GHz-spaced Super-Nyquist-WDM 30-GBaud duobinary-pulse-shaped DP-QPSK signals are generated at the transmitter side, resulting in a spectral efficiency of 4 bit/s/Hz assuming a LDPC-based SD-FEC with 20% overhead. For MCF transmission, the 201-channel WDM signals are launched into a specially configured sevenfold recirculating loop consisting of a span of 45.5-km seven-core fiber, a seven-core EDFA, external gain-flattening filters (GFFs), and optical switches (SWs). The output signals from one core are sent to the re-circulating loop input of the next core, in a cyclic fashion.

In 2015, K. Takeshima et al. demonstrated 51.1-Tb/s MCF transmission using cladding pumped seven-core EDFAs and confirmed a reachable distance of 2520 km with 73 × 100-Gbit/s Nyquist-pulse-shaped DP-QPSK signals per core (Takeshima et al. 2015). The cladding pumped MC-EDFA is attractive as it uses a multimode LD that is cheaper than single-mode LD and it can reduce the number of components in the amplifier. The cladding pumped seven-core EDFA is side-coupling pumped by a fiber-based combiner. The insertion loss of the fabricated pump combiner is 0.17 dB and less than 0.5 dB for pump light and signal light, respectively. The schematic configuration of the seven-core MC-EDFA is shown in Fig. 57. A MC-EDF has seven individual cores with core pitch 45 μm. The MC-EDF had glass cladding diameter of 180 μm and polymer cladding (NA 0.36) for double cladding structure. The maximum net gain is about 17 dB and more than 11 dB gain is obtained over a wide bandwidth of 30 nm. The net noise figure is below 7.1 dB in all cores with less than 1 dB deviation.

For MCF transmission, 73-channel WDM signals are launched into a specially configured sevenfold re-circulating loop consisting of a 40-km seven-core fiber span, a CP-7C-EDFA including isolators, external GFF and optical switch (SW).

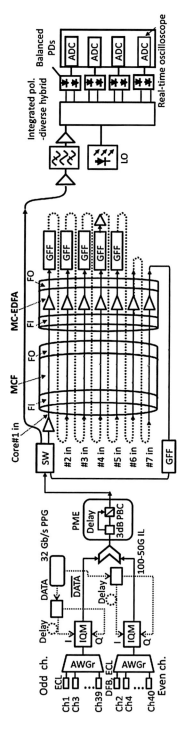

Fig. 53 Experimental setup of MCF loop transmission (Takahashi et al. 2013)

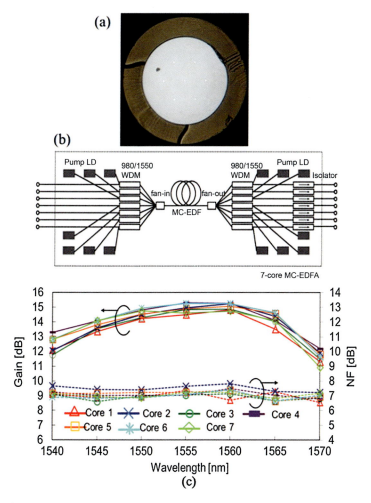

Fig. 54 Seven-core MC-EDFA; (**a**) cross-section of MC-EDF, (**b**) composition, (**c**) gain and noise figure of MC-EDF (Takahashi et al. 2013)

The seven re-circulating loops share a common load switch that launched identical copies of the WDM signals into each of the loops through a power splitter with different delays among cores for signal decorrelation. The output signals from one core are sent to the re-circulating loop input of the next core, in a cyclic fashion. A single-core EDFA and wavelength selective switch are inserted at the output of the seventh core for gain equalization, and the flatness is automatically managed by a wavelength selective switch with an optical spectrum analyzer (Fig. 58).

The latest report by Takayuki Mizuno et al. demonstrates 32-core dense space-division-multiplexed (DSDM) 20 wavelength-division-multiplexed PDM-16QAM unidirectional transmission over 1644.8 km (Mizuno et al. 2016a). The developed

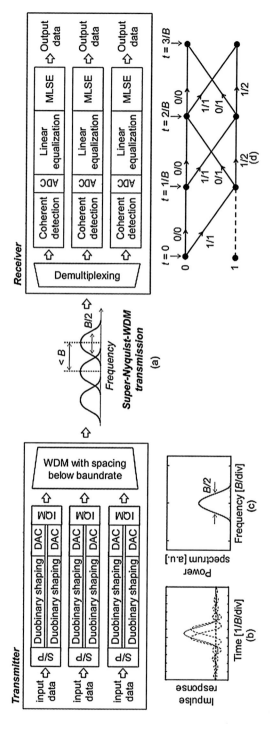

Fig. 55 (**a**) Concept of Super-Nyquist-WDM transmission, (**b**) impulse response, and (**c**) power spectrum of duobinary-pulse shaping. (**d**) Duobinary trellis for two-level amplitude modulation (Igarashi et al. 2013)

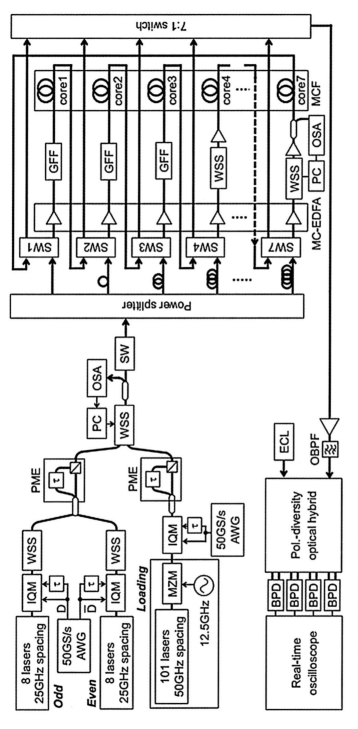

Fig. 56 Experimental setup (Igarashi et al. 2013)

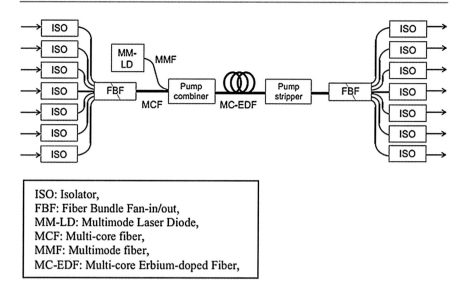

ISO: Isolator,
FBF: Fiber Bundle Fan-in/out,
MM-LD: Multimode Laser Diode,
MCF: Multi-core fiber,
MMF: Multimode fiber,
MC-EDF: Multi-core Erbium-doped Fiber,

Fig. 57 Configuration of cladding pumped seven-core EDFA (CP-7C-EDFA) (Takeshima et al. 2015)

low-XT heterogeneous 32-core MCF is using square lattice arrangement. The span XT of the 51.4-km 32-core transmission line is less than −34.5 dB. A novel partial re-circulating loop system has been introduced as shown in Fig. 59. The signal generated at the transmitter is split into three branches. The first branch is used for the core under measurement in loop #1, and the second branch is further split into four, which are used to provide recirculating XT signals for loops #2 to #5. The third branch is further split into 27, and these signals are used as nonrecirculating signals. The partial recirculating loop system consisted of the first and second 32-channel matrix switches, a 51.4-km spool of the 32-core MCF, 32-channel fan-in/fan-out devices, and five parallel recirculating loops. Signals in the loops are amplified by five channels of a 7C-EDFA.

For coupled-core MCF-based long haul transmission, the first record in 2012 from R. Ryf et al. experimentally demonstrated MIMO transmission of a combined three-space-, and two-polarization-, and five-wavelength-division multiplexed 40-Gbit/s QPSK signals in a three-core coupled-core MCF over 4200 km (Ryf et al. 2012a). The cores are embedded in a standard cladding diameter of 125 μm, and the distance between the equidistant cores and the center of the fiber is 17 μm, resulting in a core pitch of 29.4 μm. A cross-section of the fiber profile is shown in Fig. 60c. The long-distance performance of the three-core coupled-core MCF is measured in a recirculating loop experiment as shown in Fig. 60a. The DFBs are arranged in two groups with odd and even channel numbers, which are combined with a wavelength multiplexer. Three delayed copies are fed to three LiNbO$_3$ switches (LN-SWs) that control the loading of a threefold recirculating loop. The loop consists of a pair of spatial-mode multiplexers (MMUXs) connected by a 60-km 3C-MSF, and 3

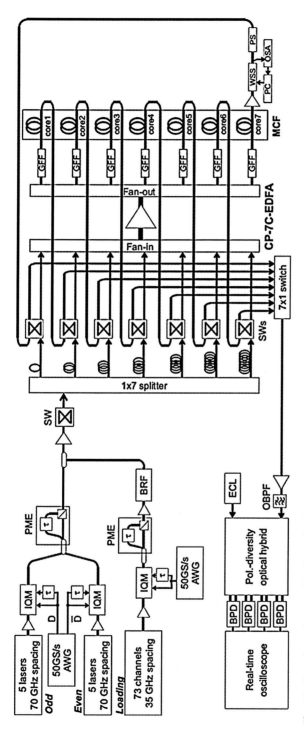

Fig. 58 Experimental setup of re-circulating loops (Takeshima et al. 2015)

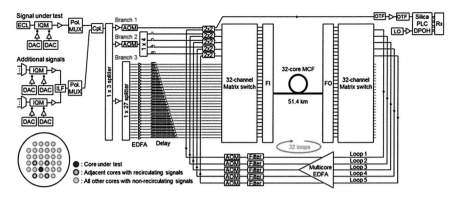

Fig. 59 Experimental setup with a novel partial recirculating loop system (Mizuno et al. 2016a)

two-stage Erbium-doped fiber amplifiers (EDFAs), where a multichannel blocker is inserted between amplification stages in order to spectrally equalize the power in the loop. The transmitted signals were recovered by MIMO DSP, where the number of equalizer taps is increased to 400. The algorithm is based on a 6×6 array of feed-forward equalizers (FFEs), where the FFE coefficients were optimized using the least mean square (LMS) algorithm which was extended with a phase tracking method based on the fourth-power algorithm.

In 2014, R. Ryf et al. for the first time demonstrated transmission combined 2-polarization, 6-SDM and 30-WDM channels with each channel modulated on 30-Gbaud QPSK signals over a distance of 1705 km based on a novel six-core coupled-core MCF (Ryf et al. 2014a). The six-core MCF consists of six homogeneous cores with 11.1 μm diameter and the cladding diameter is 125 μm. The cores are arranged circularly and equally spaced from the fiber center. A cross-section of the MCF is shown in Fig. 61b. The effective area of each core is designed to be 106 μm^2. The transmission performance of the coupled-core MCF was measured using a sixfold recirculating loop experiment as shown in Fig. 61a. 12 de-correlated signals are injected into the sixfold recirculating loop. The loop consists of 6 two-stage EDFA, where wavelength blockers, the electro-optic loop switches, and the 50:50 couplers are used for injecting signal from and to the loop.

At the same time, in 2014 R. Ryf et al. experimentally demonstrated transmission over a novel MCF with nine cores arranged in three groups of three cores, where strong coupling occurs within the groups and weak couplings between groups (Ryf et al. 2014b). The transmission is over 2500 km using a single group with 30-Gbaud PDM-QPSK signals and 715 km when all nine cores are used. Figure 62a shows a cross section of the 35.7 km long 3×3 core MCF used in the experiment. All cores are homogeneous. The XT between groups are reported in Fig. 62d. Group 1 has the smallest XT about −31 dB, whereas group 2 has the largest XT about −29 dB. Because of the strong coupling among cores within the same group, almost identical results are obtained when measuring XT from individual cores from different groups. The femto-second laser inscribed 3D waveguides is used

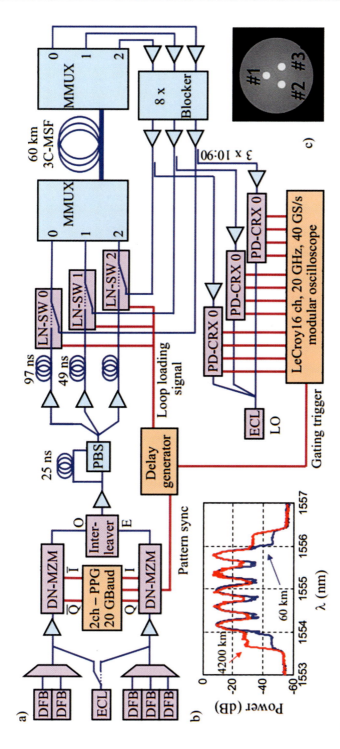

Fig. 60 (**a**) Experimental setup. (**b**) Power spectra after 60 and 4200 km transmission. (**c**) Cross-section of the 3C-MSF (Ryf et al. 2012a)

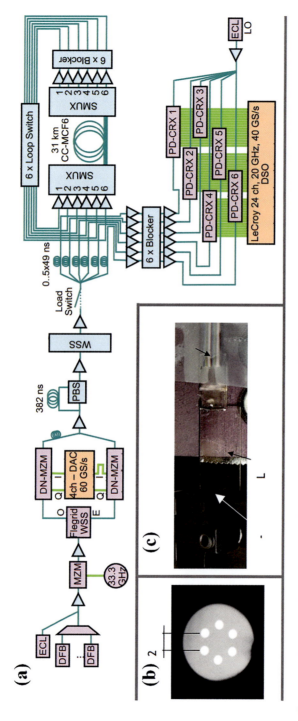

Fig. 61 (**a**) Setup for 12*12 MIMO transmission over six-core coupled-core MCF based on recirculating loop. (**b**) Six-core coupled-core MCF cross section. (**c**) Laser-inscribed waveguide coupler for coupled-core MCF (Ryf et al. 2014a)

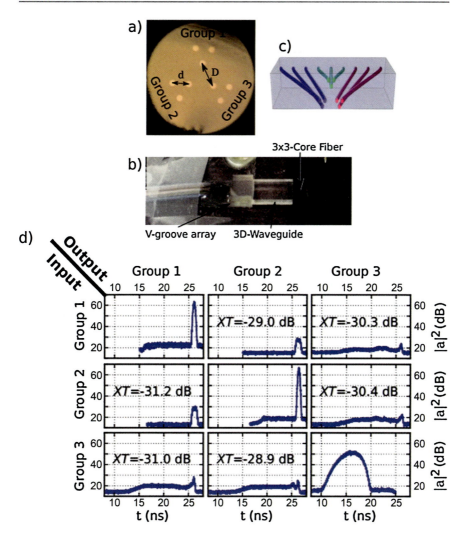

Fig. 62 (**a**) Cross-section of the 3×3 core fiber. (**b**) Picture of the 3D-waveguide multiplexer. (**c**) Schematic layout of the 3D-waveguide. (**d**) Impulse response and XT measured between core groups (Ryf et al. 2014b)

as fan-in/fan-out device, shown in Fig. 62b, and the schematic layout of the 3D waveguide design is shown in Fig. 62c.

The transmission measurement is performed using the setup as shown in Fig. 63. Three delayed copies are fed into a threefold recirculating loop where each copy is further split into three paths, and an additional fiber delay of 147 ns is added to the paths coupled to the "core group under test" in order to provide de-correlated XT signals from the neighbor core groups. The MIMO consists of a network of 6×6 feed-forward equalizers (FFEs) with 900 taps each. The FFE coefficients are

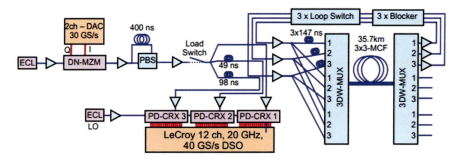

Fig. 63 Experimental setup for coherent MIMO transmission (Ryf et al. 2014b)

determined by a block-based LMS algorithm, and initial convergence was obtained by data aided operation.

MCF-Based Optical Sensing Systems

Over the past three decades, the optical fiber sensing technology has undergone a tremendous development, owing to their outstanding merits, such as lightweight, compact, and chemically inert properties, while offering a total immunity to electromagnetic interference and a good environmental ruggedness. MCFs have also been used for sensing applications, which includes MCF-based discrete fiber sensors and distributed fiber sensors.

Discrete MCF Sensing Technology

So far, MCFs-based discrete fiber sensors mainly consist of two categories. One is MCF-based modal interferometer sensors, and the other one is MCF-based fiber grating sensors.

In MCF-based interferometric structures, bending will lead to phase change of light in the cores. Employing this feature, MCF-based interferometric sensors have been widely used for bending sensing, or quantities that are associated with bending. For example, a four-core fiber-based Mach–Zehnder interferometer (MZI) has been used for two-dimensional bending sensing with its bending angle retrieved from the phase values that are derived from Fourier analysis of the far-field interferogram (Gander et al. 2000b). A two-core fiber-based Michelson interferometer (MI) has also been used for bending sensing, whose two cores act as the arms of the MI (Yuan et al. 2006). By measuring the variation of output power that result from bending-induced phase change of interferometer, bending sensing is achieved. Based on the same principle, the two-core fiber-based MI has then been used for flow velocity sensor (Yuan et al. 2008), accelerometer (Peng et al. 2011), etc.

Thanks to the unique structure of MCF that contains multiple cores within one fiber, MCF offers the flexibility to construct various compact interferometric fiber devices, and they can be used to measure other parameters in addition to bending. For example, a seven-core fiber was used to fabricate a multipath Mach–Zehnder interferometer, and it has been used for high-sensitivity temperature sensing (Zhao et al. 2013). Thanks to the all solid fiber structure, very high temperature sensing is achieved by using a MCF multipath Michelson interferometer (Duan et al. 2016). Temperature and strain cross-sensitivity is one of the major shortcomings that restricts the applications of optical fiber sensors. In order to achieve discriminative measurement between the two parameters, some sensing schemes that utilize MCF have been proposed and demonstrated, e.g., dual-tapered heterogeneous MCF Mach-Zehnder interferometer sensor (Gan et al. 2016), few-mode MCF integrated MZI sensor (Zhan et al. 2018), and helically structured heterogeneous MCF MZI sensor (Zhang et al. 2017b). Moreover, the MCF-based MZI with helical structure has also been used for directional torsion sensing (Zhang et al. 2018). The tapered MCF has also been used for refractive index sensing of liquid (Zhang et al. 2017a) and magnetic field sensing (Tagoudi et al. 2016).

Different from the interferometers in the weakly coupled MCFs, the interferometers constructed by the SC-MCFs are normally based on super-mode interference, and this kind of fiber sensors have been used for very high temperature sensing (Antonio-Lopez et al. 2014), bending/curvature sensing (Salceda-Delgado et al. 2015; Villatoro et al. 2016), simultaneous force and temperature sensing (Newkirk et al. 2015), and vibration sensing (Villatoro et al. 2017), etc.

On the other hand, fiber gratings can be inscribed in MCFs, and it turns out that these functional devices have shown very good feasibility for bending/curvature sensing, due to the reason that bending will generate strain in off-center cores of the MCF, which will lead to the wavelength shift of the gratings. Therefore, by measuring the wavelength shift, bending radius can be retrieved, as shown in Fig. 64 (Zhang et al. 2016). Based on this measurement principle, a variety of MCF grating sensors for measurements of bending/curvature have been developed (Barrera et al. 2015; Flockhart et al. 2003; Gander et al. 2000a, b; Zhang et al. 2016). In addition, fiber Bragg gratings in MCF has also been used for the transverse loading measurement (Silva-Lopez et al. 2004), displacement sensing (Jones et al. 2005), and two-dimensional accelerometer (Fender et al. 2008), etc. Long period gratings written in a heterogeneous seven-core fiber have also been used for simultaneous temperature and strain sensing (Wang et al. 2017). A novel long range distributed fiber sensor based on continuous fiber grating sensor arrays in MCF was proposed and demonstrated recently, which shows great potential for distributed bending and temperature sensing (Westbrook et al. 2017).

Additionally, it must be mentioned that the MCF with fiber Bragg gratings inscribed has been used to achieve three-dimensional (3D) shape sensing (Chan and Parker 2015; Duncan et al. 2007; Moore 2015; Moore and Rogge 2012; Rogge and Moore 2014). The technology relies on the measurement of strains in multiple cores of the MCF along the fiber length through the fiber Bragg gratings, then the differential strains between the cores are used to calculate the local bending angle

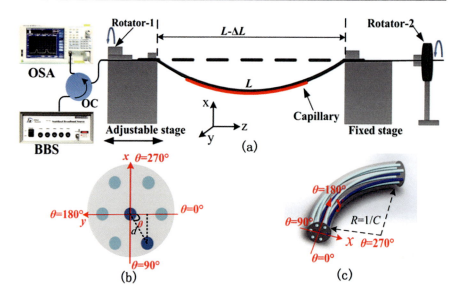

Fig. 64 (a) Experimental setup of the MCF grating sensor; (b) definition of the bending orientation angle θ; (c) schematic diagram of a bending MCF when θ is 270° (Zhang et al. 2016)

and the curvature, and eventually the shape of MCF can be reconstructed with the help of Frenet-Serret equations (Moore and Rogge 2012).

Distributed MCF Sensing Technology

Distributed optical fiber sensor (DOFS) relies on the measurement of backscattered optical signals in optical fiber to acquire the perturbation information (e.g., temperature, strain, and vibration) along the fiber link. The backscattered optical signals include Rayleigh backscattered light, Brilloude backscattered light, and Raman backscattered light. In comparison with the discrete optical fiber sensors, DOFSs are normally able to provide tens of kilometers sensing range with meter scale spatial resolution, so the technology has attracted lots of attention and it has undergone a tremendous development over the last 30 years. So far, most of the DOFSs are using the normal single mode fiber (SMF). Owing to the cross-sensitivity issue, multiparameter discriminative measurement is difficult to be achieved in SMF-based DOFSs. The multiplexing of multiple distributed fiber sensing techniques is difficult to be carried out due to the incompatible pump power levels required for the interrogation of different scattering signals. Thanks to the multiple parallel cores embedded in a single fiber, MCF provides a good platform to implement different sensing techniques simultaneously in a single fiber, which shows great potential to enhance the performance of the traditional DOFS in SMFs (Tang et al. 2016).

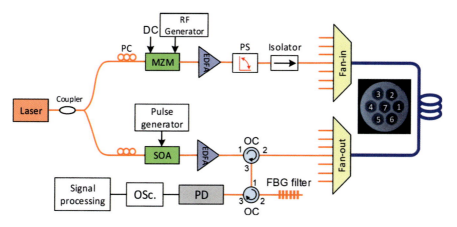

Fig. 65 Experimental setup of MCF based BOTDA. *PC* polarization controller, *MZM* Mach-Zehnder modulator, *SOA* semiconductor optical amplifier, *EDFA* erbium-doped fiber amplifier, *PS* polarization switch, *OC* optical circulator, *FBG* fiber Bragg grating, *PD* photodetector, *OSc.* oscilloscope (Zhao et al. 2016c)

Brillouin scattering in a 1-km long homogeneous MCF was characterized by using a traditional Brillouin optical time-domain analyzer (BOTDA), whose experimental setup is shown in Fig. 65 (Zhao et al. 2016c); 20-cm high spatial resolution was obtained by using the differential pulse-width pairs (DPP) technique.

The experiment result reveals that the Brillouin frequency shift (BFS) in off-center cores is sensitive to bending, as shown in Fig. 66a (Zhao et al. 2016b). This is because bending will result in local tangential strain in off-center positions, and the bending-induced strain ϵ_i in specific core i is given by

$$\epsilon_i = -\frac{d_i}{R} \cos (\theta_b - \theta_i)$$

where d_i is the distance between core i and the fiber center, R is the bending radius, θ_b and θ_i are, respectively, the angle of the bending direction and the angular position of core i, as shown in Fig. 66b. Due to the reason that the central core is located in the neutral axis of the fiber, it is not sensitive to bending.

Employing the feature of bending dependency of BFS in off-center cores, three outer cores of the MCF were used for distributed bending sensing (Zhao et al. 2016c). This is achieved by measuring the Brillouin frequency shifts in the three cores along the entire fiber length, then strains in different cores at any position of the MCF are determined. The strains are then used to calculate the local bending angle and the curvature. Eventually, the experiment demonstrated long range distributed bending sensing with good accuracy, and it also shows great potential to achieve fully 3D shape sensing.

Brillouin distributed fiber sensors are inherently suffering from the cross-sensitivity between strain and temperature. By using the MCF, two solutions have

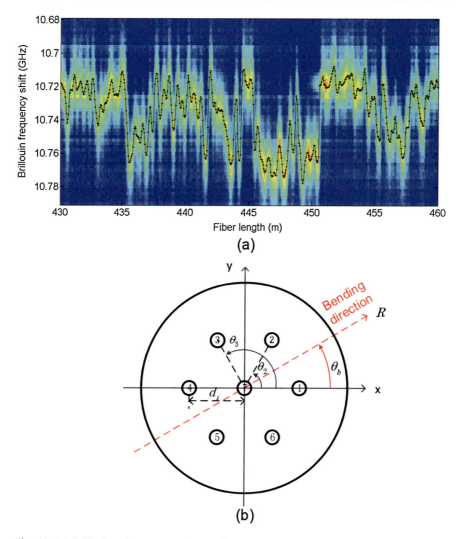

Fig. 66 (**a**) Brillouin gain spectrum in an off-center core of the MCF showing random BFS variations resulting from bending due to coiling. (**b**) Transversal distribution of the cores with the definition of the important geometrical parameters (Zhao et al. 2016b)

been proposed and demonstrated for separating the effects of temperature and strain. One scheme is to implement a MCF-based space-division multiplexed (SDM) Brillouin optical time-domain reflectometry (BOTDR) and Raman optical time-domain reflectometry (ROTDR) hybrid system (Zhao et al. 2016a), as shown in Fig. 67, where a single laser source is used in the setup, and the two reflectometries share the same pulse generation devices, but different spatial cores are used to interrogate the Brillouin scattering signal and the Raman scattering signal. Thus,

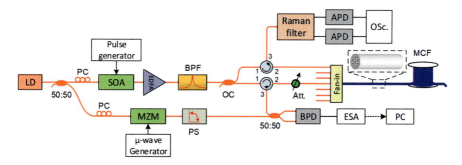

Fig. 67 Experimental setup of the MCF based ROTDR and BOTDR hybrid system; *LD* laser diode, *PC* polarization controller, *SOA* semiconductor optical amplifier, *MZM* Mach-Zehnder modulator, *EDFA* erbium-doped fiber amplifier, *PS* polarization switch, *BPF* band-pass filter, *BPD* balanced photodetector, *Att.* tunable attenuator, *APD* avalanche photodiode, *ESA* electrical spectrum analyzer, *OSc.* oscilloscope (Zhao et al. 2016a)

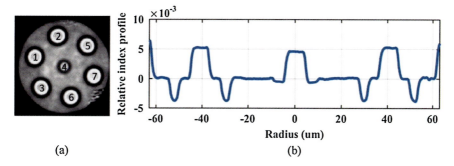

(a)   (b)

Fig. 68 (**a**) Cross section view of the MCF used in this experiment. (**b**) Relative index profile of the MCF (Zhao et al. 2017)

it enables efficient input power management for the two reflectometries, allowing for simultaneous measurement of spontaneous Raman scattering and Brillouin scattering.

In the BOTDR and ROTDR hybrid system, the temperature profile along the sensing fiber is acquired from the measurement of ROTDR, and strain is determined by BOTDR with temperature value extracted by Raman measurement. Different from the hybrid system that is performed in SMF, the implementation of BOTDR and ROTDR are separated into different spatial cores in the proposed SDM hybrid system, so it effectively overcome the incompatible of distinct input pump power level for the hybrid ROTDR and BOTDR in SMFs.

The other approach to achieve discriminative measurement between temperature and strain is to use a heterogeneous MCF (YOFC, China), whose central core is made from G.652 preform and the six outer cores are made from G657.B3 preform, as shown in Fig. 68 (Zhao et al. 2017).

Fig. 69 The measured Brillouin gain spectrum of a heterogeneous MCF, (**a**) the central core with peak at ∼10.81 GHz and (**b**) the outer core with peak at ∼10.74 GHz (Zhao et al. 2017)

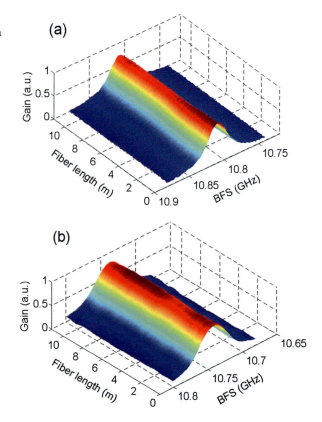

The experiment reveals that the Brillouin frequency shifts are different between the cores that are made from distinct preforms, as shown in Fig. 69. The BFS of the central core is about 10.81 GHz, while that of the outer cores are about 10.74 GHz. The temperature sensitivities and strain sensitivities of the central core and the outer cores were measured. It turns out that the temperature sensitivities of the central core and the outer cores are different, while their strain sensitivities are almost the same.

In order to separate the effects of strain and temperature, a coefficient matrix which involves the measurements of the central core and the outer core can be solved, and in this way the two measurands will be separated. However, due to the reason that the off-center core is also sensitive to bending, so the bending-induced BFS change should be compensated before solving the coefficient matrix. This can be done by averaging the BFSs of two symmetrical outer cores of the MCF. The idea has been verified by experiment, as shown in Fig. 70. The blue and red lines in Fig. 7 are the measured BFSs of the two symmetrical outer cores, respectively, showing random fluctuations due to bending; the green line is the averaged BFS of the two cores, which is very flat. So it demonstrates that the bending effect can be eliminated by this method. The averaged BFS of the two symmetrical outer cores

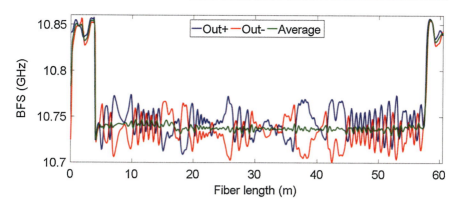

Fig. 70 The measured BFS of two symmetrical outer cores when the MCF was spooled with random orientations (Zhao et al. 2017)

and that of the central core are then used to calculate the coefficient matrix and eventually separate the two measurands successfully (Zhao et al. 2017).

Brillouin distributed fiber sensors have the advantage of large dynamic range of measurement, but they are subjected to the relatively poor temperature/strain measurement resolution. The typical temperature sensitivity and strain sensitivity that are measured by BOTDA in normal silica SMFs are about 1.10 MHz/°C and 0.05 MHz/$\mu\varepsilon$, respectively. For a typical 1 MHz measurement accuracy of BFS, the temperature resolution and strain resolution are estimated to be 1 °C and 20 $\mu\epsilon$, respectively. The poor temperature/strain measurement resolution has hindered its applications in occasions where high temperature/strain resolution is required. On the other hand, phase-sensitive optical time-domain reflectometry (Φ-OTDR) has been demonstrated, which can offer ultrahigh measurement resolutions (Koyamada et al. 2009). Therefore, the combination of BOTDA and Φ-OTDR in one system will enable large dynamic range and high measurement resolution simultaneously.

However, due to the reason that the frequency scanning step and range are not the same for BOTDA and Φ-OTDR, it is difficult to implement the hybrid system in SMF. While MCF turns out to be a perfect solution for the multiplexing of BOTDA and Φ-OTDR. As shown in Fig. 71, the hybrid system uses one laser source; the generated pump pulse is used for both BOTDA and Φ-OTDR, while they are implemented in different cores, and the frequency scanning is carried out separately. In this way, the two sensors are multiplexed spatially. Eventually, the hybrid system achieves about 0.001 °C temperature resolution based on Φ-OTDR, and meanwhile large dynamic range can be ensured by BOTDA (Dang et al. 2017).

Distributed optical fiber sensing has been demonstrated to be the utmost promising solution in pipeline monitoring in oil and gas industry, owning to its outstanding performance with very long sensing range and high spatial resolution. Simultaneous distributed acoustic sensing (DAS) and distributed temperature sensing (DTS) are very attractive in this application field, since it allows for simultaneous vibration and temperature detection. However, Raman optical time-domain reflectometry

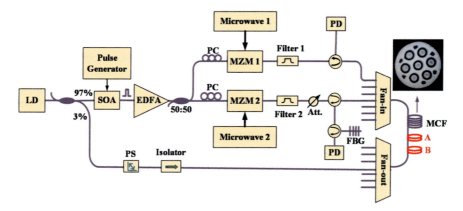

Fig. 71 Experimental setup of the MCF-based SDM hybrid BOTDA and Φ-OTDR system. *LD* laser diode, *PC* polarization controller, *SOA* semiconductor optical amplifier, *MZM* Mach-Zehnder modulator, *EDFA* erbium-doped fiber amplifier, *PS* polarization switch, *Att.* attenuator, *PD* photodetector, *Fan-in* fan-in coupler, *Fan-out* fan-out coupler (Dang et al. 2017)

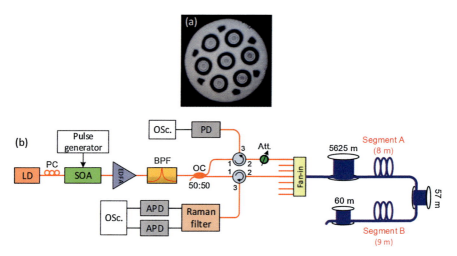

Fig. 72 (**a**) MCF cross sectional view; (**b**) experimental setup; *LD* laser diode, *PC* polarization controller, *SOA* semiconductor optical amplifier, *EDFA* erbium-doped fiber amplifier, *BPF* band-pass filter, *OC* optical coupler, *Att.* tunable attenuator, *APD* avalanche photodiode, *PD* photodetector, *OSc.* oscilloscope (Zhao et al. 2018)

(ROTDR)-based DTS system is difficult to work well with Φ-OTDR-based DAS system in SMF. This is because ROTDR requires very high pump power, but high pump power will cause nonlinear effects (including modulation instability and stimulated Brillouin scattering, etc.) in the fiber, which will distort the Rayleigh backscattering signal and hinder the correct measurement of Φ-OTDR. The problem can be avoided by implementing a space-division multiplexed ROTDR and Φ-OTDR hybrid system by using MCF, as shown in Fig. 72 (Zhao et al. 2018).

Thanks to the SDM configuration, simultaneous detection of Raman backscattering signal and Rayleigh backscattering signal is achieved; thus, it enables simultaneous measurement of DAS and DTS. In the experiment, simultaneous vibration detection and temperature measurement is demonstrated by Φ-OTDR and ROTDR, respectively. The proposed hybrid sensing system shows great potential for long-haul pipelines monitoring in oil and gas industry.

Conclusion

With the significant research efforts world widely, it is evident that the MCF has proved itself as one of the most promising techniques to upgrade the current optical transmission system. However, to move a step forward from laboratory to the real deployment, many challenges are still faced for commercialization. For example, for the long-haul transmission, power-efficient, low-noise, and gain-equalized MCF-EDFA has to be developed. The competition or cooperation with other SDM technologies like few-mode fibers and OAM fibers is another issue. For the short link MCF transmission, the core arrangement, spatial channel number, and connection loss are to be designed carefully according to the networking/switching equipment. The standardization of MCF is still an open area for the research community and industry to be explored.

As a typical artificial channel, the MCF owns the flexibility to engineer its characteristics for various interesting applications, such as imaging, discrete interference-based fiber sensors, and distributed fiber optical sensing systems. It is anticipated to utilize the MCF technology in some niche markets by employing its unique features in the very near future.

References

Y. Abe, K. Shikama, S. Yanagi, T. Takahashi, Low-loss physical-contact-type fan-out device for 12-core multicore fiber, in *European Conference and Exhibition on Optical Communications (ECOC)*, vol. 49, no. 11 (2013)

K.S. Abedin, T.F. Taunay, M. Fishteyn, M.F. Yan, B. Zhu, J.M. Fini, E.M. Monberg, F.V. Dimarcello, P.W. Wisk, Amplification and noise properties of an erbium-doped multicore fiber amplifier. Opt. Express **19**(17), 16715–16721 (2011)

K.S. Abedin, T.F. Taunay, M. Fishteyn, D.J. Digiovanni, V.R. Supradeepa, J.M. Fini, M.F. Yan, B. Zhu, E.M. Monberg, F.V. Dimarcello, Cladding-pumped erbium-doped multicore fiber amplifier. Opt. Express **20**(18), 20191 (2012)

K.S. Abedin, J.M. Fini, T.F. Thierry, V.R. Supradeepa, B. Zhu, M.F. Yan, L. Bansal, E.M. Monberg, D.J. Digiovanni, Multicore erbium doped fiber amplifiers for space division multiplexing systems. J. Lightwave Technol. **32**(16), 2800–2808 (2014a)

K.S. Abedin, J.M. Fini, T.F. Thierry, B. Zhu, M.F. Yan, L. Bansal, F.V. Dimarcello, E.M. Monberg, D.J. Digiovanni, Seven-core erbium-doped double-clad fiber amplifier pumped simultaneously by side-coupled multimode fiber. Opt. Lett. **39**(4), 993–996 (2014b)

K.S. Abedin, M.F. Yan, T.F. Taunay, B. Zhu, E.M. Monberg, D.J. Digiovanni, State-of-the-art multicore fiber amplifiers for space division multiplexing. Opt. Fiber Technol. **35**, 64–71 (2017)

E. Agrell, M. Karlsson, A.R. Chraplyvy, D.J. Richardson, P.M. Krummrich, P. Winzer, Roadmap of optical communications. J. Opt. **18** (2016)

Y. Amma, A. Takahashi, K. Takenaga, S. Matsuo, Low-loss fusion splice technique for multicore fiber with a large cladding diameter, in *Technical Report of IEICE OCS*, vol. 113 (2013), pp. 27–32

Y. Amma, Y. Sasaki, K. Takenaga, S. Matsuo, High-density multicore fiber with heterogeneous core arrangement, in *Optical Fiber Communications Conference and Exhibition* (2015), pp. 1–3

Y. Amma, K. Takenaga, S. Matsuo, K. Aikawa, Fusion splice techniques for multicore fibers. Opt. Fiber Technol. **35**, 72–79 (2017)

J.E. Antonio-Lopez, Z.S. Eznaveh, P. Likamwa, A. Schülzgen, R. Amezcua-Correa, Multicore fiber sensor for high-temperature applications up to 1000°C. Opt. Lett. **39**(15), 4309 (2014)

D. Barrera, I. Gasulla, S. Sales, Multipoint two-dimensional curvature optical fiber sensor based on a nontwisted homogeneous four-core fiber. J. Lightwave Technol. **33**(12), 2445–2450 (2015)

D.L. Butler, M.-J. Li, S. Li, Y. Geng, R.R. Khrapko, R.A. Modavis, V.N. Nazarov, A.V. Koklyushkin, Space division multiplexing in short reach optical interconnects. J. Lightwave Technol. **35**(4), 677–682 (2017)

H.M. Chan, A.R. Parker, Inventors, In-situ three-dimensional shape rendering from strain values obtained through optical fiber sensors. U.S. Patent 8,970,845, (3), (2015)

S. Chandrasekhar, A.H. Gnauck, X. Liu, P.J. Winzer, Y. Pan, E.C. Burrows, T.F. Taunay, B. Zhu, M. Fishteyn, M.F. Yan, et al., WDM/SDM transmission of 10 × 128-Gb/s PDM-QPSK over 2688-km 7-core fiber with a per-fiber net aggregate spectral-efficiency distance product of 40,320 km.b/s/Hz. Opt. Express **20**(2), 706–711 (2012)

H. Chen, C. Jin, B. Huang, N.K. Fontaine, R. Ryf, K. Shang, N. Grégoire, S. Morency, R.J. Essiambre, G. Li, Integrated cladding-pumped multicore few-mode erbium-doped fibre amplifier for space-division-multiplexed communications. Nat. Photonics **10**(8), 529 (2016)

Y. Dang, Z. Zhao, M. Tang, C. Zhao, L. Gan, S. Fu, T. Liu, W. Tong, P.P. Shum, D. Liu, Towards large dynamic range and ultrahigh measurement resolution in distributed fiber sensing based on multicore fiber. Opt. Express **25**(17), 20183–20193 (2017)

L. Duan, P. Zhang, M. Tang, R. Wang, Z. Zhao, S. Fu, L. Gan, B. Zhu, W. Tong, D. Liu, Heterogeneous all-solid multicore fiber based multipath Michelson interferometer for high temperature sensing. Opt. Express **24**(18), 20210 (2016)

R.G. Duncan, M.E. Froggatt, S.T. Kreger, D.K. Gifford, A.K. Sang, High-accuracy fiber-optic shape sensing. Int. Soc. Opt. Eng. **6530**, 65301S-65301S-65311 (2007)

A.E.A. Farghal, H.M.H. Shalaby, Reducing inter-core crosstalk impact via code-interleaving and bipolar 2-PPM for core-multiplexed SAC OCDMA PON. J. Opt. Commun. Netw. **10**(1), 35 (2018)

A. Fender, W.N. Macpherson, R.R.J. Maier, J.S. Barton, D.S. George, R.I. Howden, G.W. Smith, B.J.S. Jones, S. Mcculloch, X. Chen, Two-axis temperature-insensitive accelerometer based on multicore fiber Bragg gratings. IEEE Sensors J. **8**(7), 1292–1298 (2008)

Z. Feng, B. Li, M. Tang, L. Gan, R. Wang, R. Lin, Z. Xu, S. Fu, L. Deng, W. Tong, et al., Multicore-fiber-enabled WSDM optical access network with centralized carrier delivery and RSOA-based adaptive modulation. IEEE Photon. J. **7**(4), 1–9 (2015)

J.M. Fini, B.Y. Zhu, T.F. Taunay, M.F. Yan, Statistics of crosstalk in bent multicore fibers. Opt. Express **18**(14), 15122–15129 (2010)

J.M. Fini, B. Zhu, T.F. Taunay, M.F. Yan, K.S. Abedin, Crosstalk in multicore fibers with randomness: gradual drift vs. short-length variations. Opt. Express **20**(2), 949–959 (2012a)

J.M. Fini, B. Zhu, T.F. Taunay, M.F. Yan, K.S. Abedin, Statistical models of multicore fiber crosstalk including time delays. J. Lightwave Technol. **30**(12), 2003–2010 (2012b)

G.M.H. Flockhart, W.N. Macpherson, J.S. Barton, J.D.C. Jones, L. Zhang, I. Bennion, Two-axis bend measurement with Bragg gratings in multicore optical fiber. Opt. Lett. **28**(6), 387–389 (2003)

L. Gan, R. Wang, D. Liu, L. Duan, S. Liu, S. Fu, B. Li, Z. Feng, H. Wei, W. Tong, Spatial-division multiplexed Mach–Zehnder interferometers in heterogeneous multicore fiber for multiparameter measurement. IEEE Photon. J. **8**(1), 1–8 (2016)

M.J. Gander, W.N. Macpherson, R. Mcbride, J.D.C. Jones, Bend measurement using Bragg gratings in multicore fibre. Electron. Lett. **36**(2), 120–121 (2000a)

M.J. Gander, D. Macrae, E.A.C. Galliot, R. Mcbride, J.D.C. Jones, P.M. Blanchard, J.G. Burnett, A.H. Greenaway, M.N. Inci, Two-axis bend measurement using multicore optical fibre. Opt. Commun. **182**(1), 115–121 (2000b)

T. Hayashi, T. Nagashima, O. Shimakawa, T. Sasaki, E. Sasaoka, Crosstalk variation of multi-core fibre due to fibre bend, in *European Conference and Exhibition on Optical Communications (ECOC)* (2010), p. 34

T. Hayashi, T. Taru, O. Shimakawa, T. Sasaki, E. Sasaoka, Design and fabrication of ultra-low crosstalk and low-loss multi-core fiber. Opt. Express **19**(17), 16576–16592 (2011)

T. Hayashi, T. Taru, O. Shimakawa, T. Sasaki, E. Sasaoka, Characterization of crosstalk in ultra-low-crosstalk multi-core fiber. J. Lightwave Technol. **30**(4), 583–589 (2012)

J. He, B. Li, L. Deng, M. Tang, L. Gan, S. Fu, P.P. Shum, D. Liu, Experimental demonstration of bidirectional OFDM/OQAM-MIMO signal over a multicore fiber system. IEEE Photon. J. **8**(5), 1–8 (2016)

J. He, L. Deng, B. Li, M. Tang, S. Fu, D. Liu, P.P. Shum, Experimental demonstration of MCF enabled bidirectional colorless CAP-PON system with wavelength reuse technique, in *Opto-Electronics and Communications Conference* (2017), pp. 1–3

K. Igarashi, T. Tsuritani, I. Morita, Y. Tsuchida, 1.03-Exabit/skm Super-Nyquist-WDM transmission over 7,326-km seven-core fiber, in *European Conference and Exhibition on Optical Communication* (2013), pp. 1–3

K. Imamura, Design optimization of large Aeff multi-core fiber, in *Opto-Electronics and Communications Conference* (2010)

K. Imamura, K. Mukasa, M. Yu, T. Yagi, Multi-core holey fibers for the long-distance (>100 km) ultra large capacity transmission, in *Optical Fiber Communication Conference and Exposition* (2009), pp. 1–3

K. Imamura, H. Inaba, K. Mukasa, R. Sugizaki, Multi core fiber with large Aeff of 140 mm^2 and low crosstalk, in *European Conference and Exhibition on Optical Communications* (2012), pp. 1–3

C. Jin, B. Ung, Y. Messaddeq, S. Larochelle, Annular-cladding erbium doped multicore fiber for SDM amplification. Opt. Express **23**(23), 29647–29659 (2015)

J.D.C. Jones, D. Zhao, N. Metje, Tunnel monitoring using multicore fiber displacement sensor. Meas. Sci. Technol. **17**(5), 1180 (2005)

A. Kanno, B.J. Puttnam, H. Inaba, J. Sakaguchi, K. Imamura, K. Mukasa, M. Watanabe, N. Wada, R. Sugizaki, T. Kawanishi, 305 Tb/s space division multiplexed transmission using homogeneous 19-core fiber. J. Lightwave Technol. **31**(4), 554–562 (2013)

W. Klaus, J. Sakaguchi, B.J. Puttnam, Y. Awaji, N. Wada, T. Kobayashi, M. Watanabe, Free-space coupling optics for multicore fibers. IEEE Photon. Technol. Lett. **24**(21), 1902–1905 (2012)

T. Kobayashi, H. Takara, A. Sano, T. Mizuno, 2 × 344 Tb/s propagation-direction interleaved transmission over 1500-km MCF enhanced by multicarrier full electric-field digital back-propagation, in *European Conference and Exhibition on Optical Communication* (2013), pp. 1–3

Y. Koyamada, M. Imahama, K. Kubota, K. Hogari, Fiber-optic distributed strain and temperature sensing with very high measurand resolution over long range using coherent OTDR. J. Lightwave Technol. **27**(9), 1142–1146 (2009)

P.M. Krummrich, Efficient optical amplification for spatial division multiplexing. Int. Soc. Opt. Eng. **8284**(1), 13 (2012)

B.G. Lee, 120-Gb/s 100-m transmission in a single multicore multimode fiber containing six cores interfaced with a matching VCSEL array, in *IEEE Summer Topicals* (2010)

B. Li, Z. Feng, M. Tang, Z. Xu, S. Fu, Q. Wu, L. Deng, W. Tong, S. Liu, P.P. Shum, Experimental demonstration of large capacity WSDM optical access network with multicore fibers and advanced modulation formats. Opt. Express **23**(9), 10997–11006 (2015)

R. Lin, J. Van Kerrebrouck, X. Pang, M. Verplaetse, O. Ozolins, A. Udalcovs, L. Zhang, L. Gan, M. Tang, S. Fu, et al., Real-time 100 Gbps/lambda/core NRZ and EDB IM/DD transmission over multicore fiber for intra-datacenter communication networks. Opt. Express **26**(8), 10519–10526 (2018)

Y. Liu, L. Ma, C. Yang, W. Tong, Z. He, Multimode and single-mode fiber compatible graded-index multicore fiber for high density optical interconnect application. Opt. Express **26**(9), 11639–11648 (2018)

S. Matsuo, K. Takenaga, Y. Arakawa, Y. Sasaki, S. Taniagwa, K. Saitoh, M. Koshiba, Large-effective-area ten-core fiber with cladding diameter of about 200 μm. Opt. Lett. **36**(23), 4626–4628 (2011)

S. Matsuo, Y. Sasaki, T. Akamatsu, I. Ishida, K. Takenaga, K. Okuyama, K. Saitoh, M. Kosihba, 12-core fiber with one ring structure for extremely large capacity transmission. Opt. Express **20**(27), 28398–28408 (2012)

S. Matsuo, K. Takenaga, Y. Sasaki, Y. Amma, S. Saito, K. Saitoh, T. Matsui, K. Nakajima, T. Mizuno, H. Takara, et al., High-spatial-multiplicity multicore fibers for future dense space-division-multiplexing systems. J. Lightwave Technol. **34**(6), 1464–1475 (2016)

Y. Mitsunaga, Y. Katsuyama, H. Kobayashi, Y. Ishida, Failure prediction for long length optical fiber based on proof testing. J. Appl. Phys. **53**(7), 4847–4853 (1982)

T. Mizuno, Y. Miyamoto, High-capacity dense space division multiplexing transmission. Opt. Fiber Technol. **35**, 108–117 (2017)

T. Mizuno, K. Shibahara, H. Ono, Y. Abe, Y. Miyamoto, F. Ye, T. Morioka, Y. Sasaki, Y. Amma, K. Takenaga, 32-core Dense SDM unidirectional transmission of PDM-16QAM signals over 1600 km using crosstalk-managed single-mode heterogeneous multicore transmission line, in *Optical Fiber Communications Conference and Exhibition*. Th5C.3 (2016a)

T. Mizuno, H. Takara, K. Shibahara, A. Sano, Y. Miyamoto, Dense space division multiplexed transmission over multicore and multimode fiber for long-haul transport systems. J. Lightwave Technol. **34**(6), 1484–1493 (2016b)

T. Mizuno, K. Shibahara, F. Ye, Y. Sasaki, Y. Amma, K. Takenaga, Y. Jung, K. Pulverer, H. Ono, Y. Abe, et al., Long-haul dense space-division multiplexed transmission over low-crosstalk heterogeneous 32-core transmission line using a partial recirculating loop system. J. Lightwave Technol. **35**(3), 488–498 (2017)

J. Moore, Shape sensing using multi-core fiber, in *Optical Fiber Communications Conference and Exhibition* (2015), pp. 1–3

J.P. Moore, M.D. Rogge, Shape sensing using multi-core fiber optic cable and parametric curve solutions. Opt. Express **20**(3), 2967–2973 (2012)

M. Morant, R. Llorente, Experimental demonstration of LTE-A M×4×4 MIMO radio-over-multicore fiber fronthaul, in *Optical Fiber Communications Conference and Exhibition*. Th4E.4 (2017a)

M. Morant, R. Llorente, Reconfigurable radio-over-multicore optical fronthaul for seamless 2G, UMTS and LTE-A MIMO wireless provision, in *Optical Fiber Communications Conference and Exhibition*. W2A.43 (2017b)

M. Morant, A. Macho, R. Llorente, Optical fronthaul of LTE-advanced MIMO by spatial multiplexing in multicore fiber, in *Optical Fiber Communications Conference and Exhibition* (2015), pp. 1–3

M. Nakazawa, Giant leaps in optical communication technologies towards 2030 and beyond, in *European Conference and Exhibition on Optical Communications (ECOC)* (2010)

A.V. Newkirk, J.E. Antonio-Lopez, G. Salceda-Delgado, M.U. Piracha, R. Amezcua-Correa, A. Schülzgen, Multicore fiber sensors for simultaneous measurement of force and temperature. IEEE Photon. Technol. Lett. **27**(14), 1523–1526 (2015)

H. Ono, K. Takenaga, K. Ichii, S. Matsuo, 12-core double-clad Er/Yb-doped fiber amplifier employing free-space coupling pump/signal combiner module, in *European Conference and Exhibition on Optical Communication* (2013), pp. 1–3

F. Peng, J. Yang, X. Li, Y. Yuan, B. Wu, A. Zhou, L. Yuan, In-fiber integrated accelerometer. Opt. Lett. **36**(11), 2056–2058 (2011)

B.J. Puttnam, J. Sakaguchi, J.M. Mendinueta, W. Klaus, Y. Awaji, N. Wada, A. Kanno, T. Kawanishi, Investigating self-homodyne coherent detection in a 19 channel space-division-multiplexed transmission link. Opt. Express **21**(2), 1561–1566 (2013)

B.J. Puttnam, R.S. Luís, W. Klaus, J. Sakaguchi, J.M.D. Mendinueta, Y. Awaji, N. Wada, Y. Tamura, T. Hayashi, M. Hirano, 2.15 Pb/s transmission using a 22 core homogeneous single-mode multi-core fiber and wideband optical comb, in *European Conference and Exhibition on Optical Communications (ECOC)* (2015), pp. 1–3

M.D. Rogge, J.P. Moore, Inventors; WO, Assignee, Shape sensing using a multi-core optical fiber having an arbitrary initial shape in the presence of extrinsic forces. U.S. Parent 8, 746, 076, (10), (2014)

R. Ryf, R. Essiambre, A.H. Gnauck, S. Randel, Space-division multiplexed transmission over 4200-km 3-core microstructured fiber, in *Optical Fiber Communication Conference and Exposition* (2012a), pp. 1–3

R. Ryf, R.J. Essiambre, S. Randel, M.A. Mestre, C. Schmidt, P.J. Winzer, IEEE, Impulse response analysis of coupled-core 3-core fibers, in *European Conference and Exhibition on Optical Communications (ECOC)* (2012b)

R. Ryf, N.K. Fontaine, B. Guan, R.J. Essiambre, 1705-km transmission over coupled-core fibre supporting 6 spatial modes, in *European Conference and Exhibition on Optical Communications (ECOC)* (2014a), pp. 1–3

R. Ryf, N.K. Fontaine, M. Montoliu, S. Randel, S.H. Chang, H. Chen, S. Chandrasekhar, A.H. Gnauck, R.J. Essiambre, P.J. Winzer, et al., Space-division multiplexed transmission over 3×3 coupled-core multicore fiber, in *Optical Fiber Communication Conference and Exposition* (2014b)

K. Saito, T. Sakamoto, T. Matsui, K. Nakajima, T. Kurashima, Side-view based angle alignment technique for multi-core fiber, in *Optical Fiber Communications Conference and Exhibition*. M3F.3 (2016)

K. Saitoh, T. Matsui, T. Sakamoto, M. Koshiba, Multi-core hole-assisted fibers for high core density space division multiplexing, in *Opto-Electronics and Communications Conference* (2010), pp. 164–165

J. Sakaguchi, W. Klaus, J.M. Delgado Mendinueta, B.J. Puttnam, R.S. Luis, Y. Awaji, N. Wada, T. Hayashi, T. Nakanishi, T. Watanabe, et al., Large spatial channel (36-core × 3 mode) heterogeneous few-mode multicore fiber. J. Lightwave Technol. **34**(1), 93–103 (2016)

T. Sakamoto, T. Matsui, K. Saitoh, S. Saitoh, K. Takenaga, T. Mizuno, Y. Abe, K. Shibahara, Y. Tobita, S. Matsuo, Low-loss and low-DMD few-mode multi-core fiber with highest core multiplicity factor, in *Optical Fiber Communication Conference and Exposition*. Th5A.2 (2016)

T. Sakamoto, T. Matsui, K. Saitoh, S. Saitoh, K. Takenaga, T. Mizuno, Y. Abe, K. Shibahara, Y. Tobita, S. Matsuo, et al., Low-loss and low-DMD 6-mode 19-core fiber with cladding diameter of less than 250 μm. J. Lightwave Technol. **35**(3), 443–449 (2017a)

T. Sakamoto, T. Mori, M. Wada, T. Yamamoto, F. Yamamoto, K. Nakajima, Strongly-coupled multi-core fiber and its optical characteristics for MIMO transmission systems. Opt. Fiber Technol. **35**, 8–18 (2017b)

G. Salceda-Delgado, A. Van Newkirk, J.E. Antonio-Lopez, A. Martinez-Rios, A. Schülzgen, R. Amezcua Correa, Compact fiber-optic curvature sensor based on super-mode interference in a seven-core fiber. Opt. Lett. **40**(7), 1468–1471 (2015)

A. Sano, H. Takara, T. Kobayashi, H. Kawakami, H. Kishikawa, T. Nakagawa, Y. Miyamoto, Y. Abe, H. Ono, K. Shikama, 409-Tb/s + 409-Tb/s crosstalk suppressed bidirectional MCF transmission over 450 km using propagation-direction interleaving. Opt. Express **21**(14), 16777–16783 (2013)

L. Shen, L. Gan, Z. Dong, B. Li, D. Liu, S. Fu, W. Tong, M. Tang, End-view image processing based angle alignment techniques for specialty optical fibers. IEEE Photon. J. **9**(2), 1–8 (2017)

O. Shimakawa, M. Shiozaki, T. Sano, A. Inoue, Pluggable fan-out realizing physical-contact and low coupling loss for multi-core fiber, in *Optical Fiber Communication Conference and Exposition and the National Fiber Optic Engineers Conference* (2013), pp. 1–3

O. Shimakawa, H. Arao, M. Harumoto, T. Sano, Compact multi-core fiber fan-out with grin-lens and micro-lens array, in *Optical Fiber Communications Conference and Exhibition* (2014), pp. 1–3

M. Silva-Lopez, C. Li, W.N. Macpherson, A.J. Moore, J.S. Barton, J.D. Jones, D. Zhao, L. Zhang, I. Bennion, Differential birefringence in Bragg gratings in multicore fiber under transverse stress. Opt. Lett. **29**(19), 2225–2227 (2004)

A. Srinivasan, B. Snyder, N. Dan, G. Lepage, J. Park, J. Singer, J.V. Campenhout, M.S. Wlodawski, M. Pantouvaki, P.D. Heyn, Ultra-dense 16 × 56Gb/s NRZ GeSi EAM-PD arrays coupled to multicore fiber for short-reach 896Gb/s optical links, in *Optical Fiber Communications Conference and Exhibition*. Th1B.7 (2017)

E. Tagoudi, K. Milenko, S. Pissadakis, Intercore coupling effects in multicore optical fiber tapers using magnetic fluid out-claddings. J. Lightwave Technol. **34**(23), 5561–5565 (2016)

H. Takahashi, T. Tsuritani, E.L. de Gabory, T. Ito, W.R. Peng, K. Igarashi, K. Takeshima, Y. Kawaguchi, I. Morita, Y. Tsuchida, First demonstration of MC-EDFA-repeatered SDM transmission of 40 × 128-Gbit/s PDM-QPSK signals per core over 6,160-km 7-core MCF. Opt. Express **21**(1), 789–795 (2013)

H. Takahashi, K. Igarashi, T. Tsuritani, Long-haul transmission using multicore fibers, in *Optical Fiber Communications Conference and Exhibition* (2014), pp. 1–3

H. Takara, T. Mizuno, H. Kawakami, Y. Miyamoto, H. Masuda, K. Kitamura, H. Ono, S. Asakawa, Y. Amma, K. Hirakawa, 120.7-Tb/s MCF-ROPA unrepeatered transmission of PDM-32QAM channels over 204 km. J. Lightwave Technol. **33**(7), 1473–1478 (2015)

S. Takasaka, H. Matsuura, W. Kumagai, M. Tadakuma, Cladding-pumped seven-core EDFA using a multimode pump light coupler, in *European Conference and Exhibition on Optical Communication* (2013), pp. 1–3

K. Takenaga, Multicore fiber with dual-ring structure, in *Opto-Electronics and Communications Conference* (2014), pp. 51–53

K. Takenaga, Y. Arakawa, Y. Sasaki, S. Tanigawa, S. Matsuo, K. Saitoh, M. Koshiba, A large effective area multi-core fiber with an optimized cladding thickness. Opt. Express **19**(26), B543 (2011)

K. Takeshima, T. Tsuritani, Y. Tsuchida, K. Maeda, T. Saito, K. Watanabe, T. Sasa, K. Imamura, R. Sugizaki, K. Igarashi, 51.1-Tbit/s MCF transmission over 2,520 km using cladding pumped 7-core EDFAs, in *Optical Fiber Communications Conference and Exhibition* (2015), pp. 1–3

M. Tang, Z. Zhao, L. Gan, H. Wu, R. Wang, B. Li, S. Fu, S. Liu, D. Liu, H. Wei, W. Tong, Spatial-division multiplexed optical sensing using MCF and FMF, in *Specialty Optical Fibers* (2016)

T. Theeg, H. Sayinc, J. Neumann, L. Overmeyer, D. Kracht, Pump and signal combiner for bi-directional pumping of all-fiber lasers and amplifiers. Opt. Express **20**(27), 28125–28141 (2012)

Y. Tottori, T. Kobayashi, M. Watanabe, Low loss optical connection module for seven-core multicore fiber and seven single-mode fibers. IEEE Photon. Technol. Lett. **24**(21), 1926–1928 (2012)

Y. Tottori, H. Tsuboya, T. Kobayashi, M. Watanabe, Integrated optical connection module for 7-core multi-core fiber and 7 single mode fibers. Paper presented at IEEE Photonics Society Summer Topical Meeting Series, 2013

Y. Tsuchida, K. Maeda, K. Watanabe, K. Takeshima, T. Sasa, T. Saito, S. Takasaka, Y. Kawaguchi, T. Tsuritani, R. Sugizaki, Cladding pumped seven-core EDFA using an absorption-enhanced erbium doped fibre, in *European Conference and Exhibition on Optical Communication* (2016)

H. Uemura, K. Omichi, K. Takenaga, S. Matsuo, K. Saitoh, M. Koshiba, Fused taper type fan-in/fan-out device for 12 core multi-core fiber, in *Opto-Electronics and Communications Conference* (2014), pp. 49–50

R.G.H. van Uden, R.A. Correa, E.A. Lopez, F.M. Huijskens, C. Xia, G. Li, A. Schulzgen, H. de Waardt, A.M.J. Koonen, C.M. Okonkwo, Ultra-high-density spatial division multiplexing with a few-mode multicore fibre. Nat. Photonics **8**(11), 865–870 (2014)

J. Villatoro, A. Van Newkirk, E. Antonio-Lopez, J. Zubia, A. Schülzgen, R. Amezcua-Correa, Ultrasensitive vector bending sensor based on multicore optical fiber. Opt. Lett. **41**(4), 832–835 (2016)

J. Villatoro, E. Antoniolopez, A. Schülzgen, R. Amezcuacorrea, Miniature multicore optical fiber vibration sensor. Opt. Lett. **42**(10), 2022 (2017)

R. Wang, M. Tang, S. Fu, Z. Feng, W. Tong, D. Liu, Spatially arrayed long period gratings in multicore fiber by programmable electrical arc discharge. IEEE Photon. J. **9**(99), 1 (2017)

K. Watanabe, T. Saito, K. Imamura, Y. Nakayama, M. Shiino, Study of fusion splice for single-mode multicore fiber, in *Microopics Conference* (2011)

T. Watanabe, M. Hikita, Y. Kokubun, Laminated polymer waveguide fan-out device for uncoupled multi-core fibers. Opt. Express **20**(24), 26317–26325 (2012)

K. Watanabe, T. Saito, M. Shiino, Development of fiber bundle type fan-out for 19-core multicore fiber, in *Opto-Electronics and Communications Conference* (2014), pp. 44–46

P.S. Westbrook, T. Kremp, K.S. Feder, W. Ko, E.M. Monberg, H. Wu, D.A. Simoff, T.F. Taunay, R.M. Ortiz, Continuous multicore optical fiber grating arrays for distributed sensing applications. J. Lightwave Technol. 1248–1252, 1 (2017)

C. Xia, M.A. Eftekhar, R.A. Correa, J.E. Antonio-Lopez, A. Schulzgen, D. Christodoulides, G. Li, Supermodes in coupled multi-core waveguide structures. IEEE J. Sel. Top. Quantum Electron. **22**(2), 196–207 (2016)

B. Yao, K. Ohsono, N. Shiina, K. Fukuzato, Reduction of crosstalk by hole-walled multi-core fibers, in *Optical Fiber Communication Conference and Exposition* (2012), pp. 1–3

F. Ye, K. Saitoh, H. Takara, R. Asif, T. Morioka, High-count multi-core fibers for space-division multiplexing with propagation-direction interleaving, in *Optical Fiber Communications Conference and Exhibition* (2015), pp. 1–3

K. Yoshida, A. Takahashi, T. Konuma, K. Sasaki, Fusion splicer for specialty optical fiber with advanced functions. Fujikura Tech. R., **41**, 10–13 (2012)

L. Yuan, J. Yang, Z. Liu, J. Sun, In-fiber integrated Michelson interferometer. Opt. Lett. **31**(18), 2692 (2006)

L. Yuan, J. Yang, Z. Liu, A compact fiber-optic flow velocity sensor based on a twin-core fiber Michelson interferometer. IEEE Sensors J. **8**(7), 1114–1117 (2008)

X. Zhan, Y. Liu, M. Tang, L. Ma, R. Wang, L. Duan, L. Gan, C. Yang, W. Tong, S. Fu, Few-mode multicore fiber enabled integrated Mach-Zehnder interferometers for temperature and strain discrimination. Opt. Express **26**(12), 15332–15342 (2018)

H. Zhang, Z. Wu, P.P. Shum, R. Wang, Q.D. Xuan, S. Fu, W. Tong, M. Tang, Fiber Bragg gratings in heterogeneous multicore fiber for directional bending sensing. J. Opt. **18**(8), 085705 (2016)

C. Zhang, T. Ning, J. Li, L. Pei, C. Li, H. Lin, Refractive index sensor based on tapered multicore fiber. Opt. Fiber Technol. **33**, 71–76 (2017a)

H. Zhang, Z. Wu, P.P. Shum, X.Q. Dinh, C.W. Low, Z. Xu, R. Wang, X. Shao, S. Fu, W. Tong, Highly sensitive strain sensor based on helical structure combined with Mach-Zehnder interferometer in multicore fiber. Sci. Rep. **7**, 46633 (2017b)

H. Zhang, Z. Wu, P.P. Shum, X. Shao, R. Wang, X.Q. Dinh, S. Fu, W. Tong, M. Tang, Directional torsion and temperature discrimination based on a multicore fiber with a helical structure. Opt. Express **26**(1), 544 (2018)

Z. Zhao, M. Tang, S. Fu, S. Liu, H. Wei, Y. Cheng, W. Tong, P.P. Shum, D. Liu, All-solid multi-core fiber-based multipath Mach–Zehnder interferometer for temperature sensing. Appl. Phys. B Lasers Opt. **112**(4), 491–497 (2013)

Z. Zhao, Y. Dang, M. Tang, L. Duan, M. Wang, H. Wu, S. Fu, W. Tong, P.P. Shum, D. Liu, Spatial-division multiplexed hybrid Raman and Brillouin optical time-domain reflectometry based on multi-core fiber. Opt. Express **24**(22), 25111–25118 (2016a)

Z. Zhao, M.A. Soto, M. Tang, L. Thévenaz, Curvature and shape distributed sensing using Brillouin scattering in multi-core fibers, in *Optical Sensors* (2016b)

Z. Zhao, M.A. Soto, M. Tang, L. Thévenaz, Distributed shape sensing using Brillouin scattering in multi-core fibers. Opt. Express **24**(22), 25211 (2016c)

Z. Zhao, Y. Dang, M. Tang, B. Li, L. Gan, S. Fu, H. Wei, W. Tong, P. Shum, D. Liu, Spatial-division multiplexed Brillouin distributed sensing based on a heterogeneous multicore fiber. Opt. Lett. **42**(1), 171 (2017)

Z. Zhao, M. Tang, L. Wang, S. Fu, W. Tong, C. Lu, Enabling simultaneous DAS and DTS measurement through multicore fiber based space-division multiplexing, in *Optical Fiber Communication Conference and Exposition*. W2A.7 (2018)

B. Zhu, T.F. Taunay, M.F. Yan, J.M. Fini, M. Fishteyn, E.M. Monberg, F.V. Dimarcello, Seven-core multicore fiber transmissions for passive optical network. Opt. Express **18**(11), 11117–11122 (2010a)

B. Zhu, T.F. Taunay, M.F. Yan, M. Fishteyn, G. Oulundsen, D. Vaidya, 70-Gb/s multicore multimode fiber transmissions for optical data links. IEEE Photon. Technol. Lett. **22**(22), 1647–1649 (2010b)

B. Zhu, T.F. Taunay, M. Fishteyn, X. Liu, S. Chandrasekhar, M.F. Yan, J.M. Fini, E.M. Monberg, F.V. Dimarcello, 112-Tb/s space-division multiplexed DWDM transmission with 14-b/s/Hz aggregate spectral efficiency over a 76.8-km seven-core fiber. Opt. Express **19**(17), 16665–16671 (2011)

Polymer Optical Fibers

25

Kishore Bhowmik and Gang-Ding Peng

Contents

K. Bhowmik
HFC Assurance, Operate and Maintain Network, NBN, Melbourne, VIC, Australia
e-mail: kishorebhowmik@nbnco.com.au

G.-D. Peng (✉)
Photonics and Optical Communications, School of Electrical Engineering and
Telecommunications, University of New South Wales, Sydney, NSW, Australia
e-mail: g.peng@unsw.edu.au

© Springer Nature Singapore Pte Ltd. 2019
G.-D. Peng (ed.), *Handbook of Optical Fibers*,
https://doi.org/10.1007/978-981-10-7087-7_38

Abstract

Polymer Optical Fibers (POF) have been developed as early as silica optical fibers. Because of their significantly larger material attenuation, POFs are limited to lower data rate and shorter distance transmission applications and they have long been overshadowed by the success of silica fibers. Nevertheless, continuing advances and emergence of new POF technologies bring out properties very attractive for many industrial applications. In this chapter, we review the development of POF, POF materials and fabrications, different types of POFs, and some of POF applications.

Keywords

Polymer optical fiber (POF) · Fiber Bragg grating (FBG) · Poly methyl methacrylates (PMMA) · Step-Index polymer optical fiber (SI-POF) · Graded index polymer optical fiber (GI-POF)

Introduction

Optical fibers have been under intensive research and development for over 50 years. The telecommunication infrastructure and capability based on optical fibers have been expanding rapidly around the world. The demand for telecommunication capacity has been increasing with the fast development of the Internet-based modern information society. In addition to telecommunication, optical fibers have found new applications in other industries. Silica optical fibers have so far been the dominating player in almost all of current applications. Polymer or plastic optical fibers (POF) have long been overshadowed by the success of silica optical fibers on one hand and wireless/copper cable on the other (Polishuk 2006). As a matter of fact, the rapid development and deployment of advanced wireless, Wi-Fi and cable technologies in recent years, the long expected booming of POF networks in low data rate, and short distance applications has not appeared, at least not as yet. What will be the important role for the polymer optical fibers to play in future? This remains an important question to ask? To ask this question, we need to review what are real benefits of POFs.

POFs are made of low cost plastic materials such as poly (methyl methacrylate) (PMMA), polystyrene, and polycarbonates with their transmission windows in the visible range (500–800 nm). POFs are much safer and easier to handle and with greater resilience to bending, shock, and vibration, as compared with silica optical fibers which must be handled carefully and safely. Loose fiber tips and pieces must be treated as sharps which is a hazard in handling. Hence, one primary role for POFs to play is in applications in human-involving environments such as home, office, automobile, and production lines. With all the efforts over these years, technological advancements have made POF networks very competitive over a range of important applications for short-distance data communication such as home networks, automobile data links, and industrial controls (Polishuk 2006; Ziemann et al. 2008). Another important role for POFs to play is in fiber sensing areas. Compared with silica fibers, POFs are much more flexible for very large core sizes (typically $0.2 \sim 1$ mm in diameter) and have high breakdown strain. For example, PMMA-based POF can recover well with more than 10% strain while silica fiber tolerates well below 1%. In general, POFs are considered more useful for sensing strain, cracks and bending in composite materials (Peng 2002a; Peters 2010). POFs have also greater compatibility with these functional polymer or biomedical materials, and this means that various synthesis techniques and materials can be used to produce functional POFs for a wide spectrum of application areas including radiation detection, biomedical and chemical sensing and structure health monitoring (Peng 1999; Peng and Chu 2004). Not surprisingly, the development of special POFs for sensing is currently a very active area. In addition, because of low cost, light weight, visible transmission window, and great flexibility, POFs have also been widely used for illumination and lighting purposes (Galatus et al. 2018; Huang et al. 2018). In the following, we first briefly introduce the historic developments, materials and types of POF, then describe different fabrication techniques and applications of POFs.

Development of POF

The first POF was produced by DuPont in early 1960s, earlier than silica optical fibers. Initially DuPont in the United States and shortly after, Mitsubishi Rayon in Japan started the commercial development of POFs. In 1974, Mitsubishi Rayon filed a few patents on the manufacture of POFs (Mitsubish Rayon Co. 1974). Mitsubishi Rayon becomes one of the major supplier of POF after it has acquired DuPont's PMMA-based POF business. Mitsubishi Rayon, after years of technical improvements, reduced the loss of PMMA-POFs to close to its theoretical limit of 100 dB/km at 650 nm in early 1980s. However these POFs are multi-moded with Step Index (SI) profiles that their transmission bandwidth–length product is relatively low, at about 4Mhz-km. Similar to its silica counterpart, the research and development for multi-moded Graded Index (GI) for significantly higher transmission capacity started early on (Ishigure et al. 1992, 1994). While Step Index Polymer Optical Fiber (SI-POF) has been commercially available since early 1980s, PMMA Graded Index Polymer Optical Fiber (GI-POF) were not commercially available until early 2000s.

A remarkable material, perfluorinated (PF) polymer, developed by Asahi Glass, has attracted much attention for its very low loss suitable for POF. PF GI-POF was developed in 1990s and it was not commercially available until 2005 when Chromis Optical Fiber, a spinoff of OFS and Bell Laboratories, licensed the manufacture of PF GI-POF technology from Asahi (Polishuk 2006). Chromis has developed a continuous extrusion process for the manufacture of PF-GI-POF compared to the batch process developed by Asahi. The Chromis fiber process makes it possible to produce large quantity of high quality and broadband PF GI-POF fibers at lower cost. The new generation of PF GI-POF technology based on CYTOP can now achieve very low attenuation (around 10 dB/km) in the near infrared and be used in gigabit transmission systems for a distance of hundreds of meters. The GI-POF can have bandwidths more than 100 times larger than conventional SI-POF.

A quite different and interesting development is on single mode POFs (Kuzuk et al. 1991; Bosc and Toinen 1992; Koike 1992; Tagaya et al. 1993; Peng et al. 1995). Single-mode POFs are application specific, i.e., special POFs developed for special applications. They are made laser-active, photosensitive, photorefractive, or optically nonlinear, either by incorporating functional materials into the fiber cores or by fabricating specially structured fiber cross sections (Kuzuk et al. 1991; Bosc and Toinen 1992; Koike 1992; Tagaya et al. 1993; Peng et al. 1995, 1999; Welker et al. 1998; Xiong et al. 1999a). These single-mode POFs are mainly targeted for fiber-optic devices and fiber sensors. The development of single-mode POFs was started in early 1990s. Earlier work on single-mode POF includes nonlinear optical POFs for optical switching (Kuzuk et al. 1991) and photosensitive POFs for fiber Bragg Gratings (Peng et al. 1999; Xiong et al. 1999a). Both single-mode and multimode polymer optical fibers have been proposed and reported for many fiber sensor applications (Steiger 1998; Peng 2002a; Peters 2010; Kuang et al. 2009; Cennamo et al. 2013; Marques et al. 2015; Zhang et al. 2015; Leal-Junior et al. 2018; Pospori et al. 2017; Liehr 2018; Woyessa et al. 2018).

A brief summary of some milestones in POF development is listed in Table 1.

Key Features of POF

POF has quite different features from those of silica fiber. The comparison of key features of POF, silica fiber, and copper cable is summarized in Table 2. For long distance links where the attenuation and dispersion effects are of prime importance, silica fibers are definitely superior. However, POF is advantageous for short distance networks which are less concerned with attenuation and dispersion and more concerned with flexibility, hazard-free handling, and low-cost installation. In these systems, multimode optical fiber with large core size is preferred over single mode fiber. Conventional multimode silica glass fiber cannot have a very large core size, because large fiber becomes very brittle and inflexible. These properties are important for fiber interfaces within optoelectronic systems where space is usually limited.

Materials for POF

Both material and fabrication techniques have always been the focus of POF research and development. Here we briefly discuss materials and related

Table 1 Historical development of POFs

Year	Event	References
1963	DuPont produced the first POF using PMMA, >1000 dB/km @650 nm	Ziemann et al. (2008), Emslie (1988)
1968	DuPont produced PMMA step-index (SI) POF, >500 dB/km @650 nm	Ziemann et al. (2008), Emslie (1988)
1976	Mitsubishi Rayon produced Eska™, a PMMA SI-POF: >300 dB/km @650 nm	Ziemann et al. (2008), Emslie (1988)
1982	NTT developed deteuterated PMMA with ~55 dB/km @568 nm	Ziemann et al. (2008)
1983	Mitsubishi Rayon produced step-index PMMA-POF: 65 dB/km @570 nm	Emslie (1988)
1990	Keio University reported PMMA GI-POF of bandwidth-length ~ 0.3 GHz-km	Polishuk (2006)
1992	Washington State University reported first single-mode POF	Kuzuk et al. (1991)
1995	UNSW developed a technique for fabricating single-mode POF	Peng et al. (1995)
1995	Keio University and KAIST reported perfluorinated GI-POF	Polishuk (2006)
1997	Asahi Glass developed low loss perfluorinated GI-POF.	Polishuk (2006)
2001	Sydney University developed polymer PCF	Eijkelenborg et al. (2001)
2005	Chromis developed continuous extrusion process for low cost production of PF GI-POF	Polishuk (2006)
2015	Sydney University and UNSW developed a POF technique based on 3D printing	Cook et al. (2015)

considerations of POFs. Two main concerns for POFs are attenuation and bandwidth that directly determine the highest possible distance and data rate in transmission.

Poly(Methyl MethAcrylates) (PMMA) has so far been the primary thermoplastic material for POFs. PMMA is an amorphous vinyl polymer, made by free radical vinyl polymerization from the monomer: methyl methacrylate. In addition to its excellent optical quality and high flexibility, PMMA has other good properties, e.g., processability, good resistance to alkalis, aqueous inorganic salts and dilute acids, hydrolysis, and UV-induced aging. It is a widely used material in many industries under several different trade names such as Perspex, Plexigras, Crylux, and Acrylite.

The material attenuation is always the most challenging issue of POF for transmission purpose. Initial progress made by Schleinitz at DuPont demonstrated that PMMA-based POF could have an attenuation as low as 300 dB/km in mid-1970s. Mitsubishi developed a continuous extrusion process in manufacturing POF based on high purity acrylics. This has since led to the development of Eska, Mitsubishi POF with PMMA core, and Poly(fluoroalkyl methacrylate) cladding. Nowadays the PMMA-based POF typically has attenuation ranging from 80 dB/km

Table 2 Comparison of key features of POF, silica fiber, and copper cable

Feature	POF	Silica fiber	Copper cable
Transmission loss	✗	✓	✓
High bandwidth	✓	✓	✗
Low cost material and fabrication	✓	✓	✓
Low installation cost	✓	✗	✓
Light weight	✓	✓	✗
Operation in the visible	✓	✓	
Excellent elasticity, flexibility, and impact resilience	✓	✓	✓
Immunity to electromagnetic interference (EMI)	✓	✓	✗
Handling safe, as compared with silica fiber	✓	✗	✓
Compatibility with organic and biomedical materials	✓	✗	✗
Thermal stability	✗	✓	✓
Long term stability	✗	✓	✓

Table 3 Fundamental vibrations of typical bonds in polymers

Bond type	Fundamental vibration [μm]
O-H	~2.8
C-H	3.3 ~ 3.5
C-D	~4.5
C-F	7.6 ~ 10.0
C-C	7.6 ~ 10.0
C-O	7.9 ~ 10.0
C-Cl	11.7 ~ 18.2

to 120 dB/km in its transmission window around 650 nm. In addition to PMMA and PS, POFs were later made from a range of other materials, including polycarbonate (PC) (Tanaka et al. 1988) and perfluorinate (PF) (Giaretta et al. 2000). The IR absorption of POF is closely related to the fundamental vibration that is determined by the mass of the two atoms and determines the spectral attenuation of a particular bond. The fundamental vibrations of typical bonds in polymers are listed in Table 3. The tabled number is the resonant absorption wavelength corresponding to a fundamental vibration. Since hydrogen is the lightest atom, the fundamental vibrations of O-H (oxygen-hydrogen) and C-H (carbon-hydrogen) occur at relatively short wavelength. The high attenuation in the visible and IR wavelength range, which is of great concern for POFs, comes largely from the absorptions peaks linked to fundamental and higher harmonic vibrations of the C-H and O-H bonds.

Obviously the spectral attenuation of a material can be modified by atomic substitution. An important objective in research and development for POF materials is to replace H in C-H with heavier atoms such as D, F, and Cl. For example, in PMMA, the fundamental vibration of C-H bond corresponds to wavelength 3.3 μm. As a result, its second to eighth overtones are distributed in the range of 0.4 ~ 1.6 μm. These overtones give rise to large absorptions within the communications spectrum. By substituting hydrogen with a heavier atom, the wavelength of fundamental IR vibration will increase and consequently the wavelength of a certain overtone will also increase. This in turn reduces the absorption in the communications spectrum where lower order overtones are replaced by higher order overtones. The fundamental vibration of a C-D (carbon-deuterium) bond is at about 4.5 μm. The fundamental vibrations of C-F (carbon-fluorine) and C-Cl (carbon-chlorine) bonds are at even longer wavelengths, 7.6 ~ 10 μm and 11.7 ~ 18.2 μm, respectively, which are comparable to or better than that of the Si-O (silicon-oxygen) bond, typically 9 ~ 10 μm in silica glass. Because of the longer wavelengths of the molecular vibrations of C-D, C-F, and C-Cl bonds, deuteration, fluorination, or chlorination will significantly reduce attenuation in the visible and near infrared (600–1500 nm). The deuterated PMMA as a core material was proposed and demonstrated by Schleinitz to reduce the IR vibration absorption (Schleinitz 1977; Beasley et al. 1979). Kaino et al. at NTT, Japan, developed techniques to make POF with a deuterated PMMA core and the lowest loss they achieved was 20 dB/km at 680 nm (Kaino et al. 1983; Kaino 1987). However, deuteration of PMMA is very expensive and is not practical.

POFs based on perfluorinated polymers have been made with very low loss and high bandwidth (Ishigure et al. 1992; Tanio and Koike 2000; Koganezawa and Onishi 2000). The perfluorinated monomer, CYTOP, has all its hydrogen H replaced fluorine F, as shown in Fig. 1. CYTOP has the similar excellent chemical, thermal, electrical, and surface properties as conventional fluoropolymers such as teflon. The most attractive feature of this material is its very broad transmission window, ranging from 650 nm right up to 1300 nm. Within this window, a graded-index POF has been made with loss less than 100 dB/km; it bottomed out at 50 dB/km at 1300 nm (Ishigure et al. 1992). They have predicted that the material would have a theoretical attenuation limit of 0.3 dB/km, comparable to that of silica glass (Tanio and Koike 2000). In 2000, researchers in Asahi Glass Co. and Keio University in Japan reported that fluorinated polymer optical fiber achieved a loss as low as 16 dB/km (Koganezawa and Onishi 2000).

The attenuation of POFs is highly dependent on materials and fabrication techniques. The typical spectral attenuations of POFs made from different materials are summarized and shown in Fig. 2.

The long-term stability of POF's physical properties is important for practical applications. POFs may suffer undesirable changes in their optical, thermal, and mechanical properties when aging under high temperature, humidity, and UV exposure. Conventional PMMA POFs are typically working up to 85 °C. Their long-term performance remains a main concern. Under accelerated aging tests equivalent to environmental effects of typical temperature and humidity of about 10 years, considerable spectral changes in transmission have been observed (Daum et al.

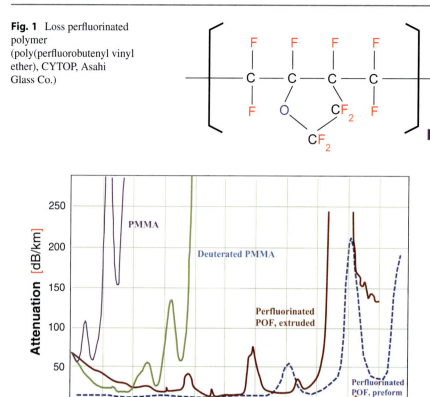

Fig. 1 Loss perfluorinated polymer (poly(perfluorobutenyl vinyl ether), CYTOP, Asahi Glass Co.)

Fig. 2 Characteristic attenuation spectra of polymer optical fibers based on various materials: PMMA, deuterated PMMA, extruded perfluorinated polymer fiber, and perfluorinated polymer fiber by preform

1997). POF doped with triphenyl phosphate (TPP) demonstrates higher thermal stability at high humidity (80 °C, 80% RH) (Ishigure et al. 1997).

For GI-POFs, the thermal stability in refractive index profiles could be stable at 85 °C over 5000 h. New dopants for GI-POF under investigation with Tg (glass transition temperatures) greater than 90 °C have been found to be stable after 600–700 h.

Types of POF

POF has been developed in many different designs and types for a great variety of purposes and applications. As silica optical fibers, POFs are classified into two types, step-index (SI) and graded-index (GI), according to their refractive index

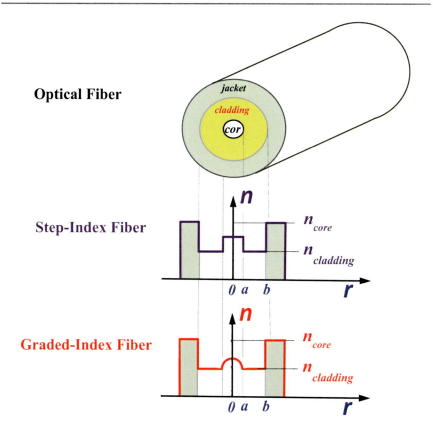

Fig. 3 Optical fibers with step and graded index profiles

profiles as shown in Fig. 3. Step-index POF (SI-POF) has a simple index profile and is the simplest to fabricate. POF for data communication purpose is multimode for simple and low cost implementation. Multimode SI-POF has very large modal dispersion that limits its bandwidth-distance product to about 5 MHz•km, which is too low for data-transmission purpose. GI-POF is preferred for its much higher bandwidth-distance product which are appropriate for local area networks or home networks.

SI-MM POF

Conventional commercial POFs are dominantly step-index multimode (SI-MM) fibers made from extrusion. These commercial POFs typically have 1 mm outer diameter with a core diameter of 980 μm. As an example, the specifications of a 1 mm diameter SI-MM POF (ESKA CK40) made by Mitsubishi Rayon are summarized in Table 4 (Kuang and Cantwell 2003a).

Table 4 Specification of a SI-MM POF (ESKA CK40) (Kuang and Cantwell 2003a)

	Core	Clad
Material	PMMA	Fluorinated polymer
Diameter (typical)	980 μm	1000 μm
Young's modulus	3.09 GPa	0.68 GPa
Poisson's ratio	0.3	0.3
Refractive index	1.492	1.405
Yield strength	82 MPa	
Transmission loss (@ 650 nm)	200 dB/km	
Maximum operating temperature	70 °C	
Approximate weight	1 g/m	

POFs with smaller outer and core diameters are also widely available. SI-MM POFs have long been used for sensor and illumination applications, in addition to transmission applications.

GI-MM POF

Graded-index multimode (GI-MM) POFs with both low loss and high bandwidth have been developed with well-tailored index profiles (Giaretta et al. 2000). Since early 1990s, intensive research has been carried out to produce a graded-index POF, which would have significantly larger bandwidth-length product. One significant advance in POF was made by Keio University when they developed a graded-index POF with a bandwidth-length product of 2 GHz km (Ishigure et al. 1992). This graded-index polymer optical fiber is produced using a photo copolymerization process. The original version of this process involved a glass tube being filled with a mixture of two or three monomers with a specific initiator and chain transfer. The tube was then rotated about its axis while under exposure to UV light. As a result, a copolymer phase was first formed on the inner wall. Since each constituent monomer has a different refractive index, the final polymer rod was solidified with a graded refractive index profile. This technique was subsequently refined by replacing the glass tube with a polymer tube. The mixture of monomers dissolves the polymer on the inner wall and a gel layer at the interface is formed. Since the polymerization in the gel phase is faster than that in the liquid phase due to the "gel effect," the polymerization progresses from the interface toward the center. It is reported that GI-POF achieved 40 Gbps transmission over 100 m of a double clad, 130 μm core graded index perfluorinated polymer optical fiber at 1.55 μm wavelength, achieving a very high bandwidth distance product >4000 MHz km (Nuccio et al. 2008). OFS laboratories now produce a series of commercial fluorinated GI-POFs (White et al. 2003). The main features of these fibers are summarized in Table 5 (White et al. 2003).

Table 5 Fluorinated GI-POFs produced by OFS Laboratories (White et al. 2003)

Cladding diameter [μm]	Core diameter [μm]	Attenuation @850 nm [dB/km]	Bandwidth distance [MHz km]
750	500	40	150–300
490	200	40	150–400
490	120	33	188–500
250	62.5	33	188–500

Fluorinated GI-MM POFs have low loss, high bandwidth, high thermal, and chemical stabilities that would make them very useful in many different applications.

SM POF

Single-mode POFs (SM POFs) are developed mainly for fiber sensing systems, especially for interferometer based (Bosc and Toinen 1992; Koike 1992; Tagaya et al. 1993; Peng et al. 1995) and for Fiber Bragg Grating (FBG) based (Peng et al. 1995; Xiong et al. 1999a; Schmitt 1999; Liu et al. 2003a).

Both silica and POF have been developed for many fiber sensor applications. POF could be very advantageous for some important sensing applications. For example, POF has very low Young's modulus in comparison to silica fiber. This property could be a significant advantage for strain-related sensing applications. Since a strain ε is related the applied stress σ by $\varepsilon = \frac{\Delta L}{L} = \frac{\sigma}{E}$, with E the Young's modulus of material, under a certain stress, a much lower Young's modulus of polymer fiber means much higher strain and thus much higher sensitivity. Also, the breakdown strain of POF is typically much larger than that of its silica counterpart. For strain sensing in civil engineering or composite structures, this larger breakdown strain could mean higher dynamic range. Moreover, it is possible to tailor the Young's modulus and elasticity of a polymer fiber with readily available synthesis techniques or to select appropriate materials with desirable Young's modulus or elasticity from a wide range of optical polymer materials. This is a remarkable feature that makes polymer optical fibers or polymer fiber Bragg gratings better candidates for sensing in various liquid and elastic material environments, duly covering a full range of strain-related sensor applications.

Table 6 briefly summarizes the relevant characteristics of typical silica and polymer fibers for sensing (Peng 2002a):

EO POF

Owing to these unique features of POF, POF may acquire special functionalities for many applications in photonics, material science, medicine, optical sensing, optical

Table 6 Typical parameters of silica and polymer optical fibers for sensing (Peng 2002a)

Property	Silica fiber	Polymer fiber
Attenuation (dB/km)	$0.2 \sim 3$	$10 \sim 100$
Young's modulus (GPa)	100	3
Breakdown strain (%)	$1 \sim 2$	$5 \sim 10$
Thermal optical coefficient (/°C)	10^{-5}	-10^{-4}
Thermal expansion coefficient (/°C)	5×10^{-7}	5×10^{-5}

spectroscopy, etc. In particular, POF is advantageous for optical device applications such as optical amplifiers, lasers, and EO modulators, since a great many organics, e.g., laser dyes, organic EO materials, could be directly incorporated into POF. Optical fiber is perhaps the most effective medium for devices based on nonlinear optical effects. Some nonlinear optical effects can be optimized with high intensity and long interaction length in optical fiber. It is well known that many optical organics are highly nonlinear and have a fast response. Incorporating these materials into an optical fiber would render it optically nonlinear. One unique advantage of POF is its relatively low process temperature (typically less than 250 °C). This allows a wide range of functional optical materials to be incorporated into the POF, which would be otherwise be impossible in silica-based fiber. The process temperature of silica fiber is so high (typically 1800 °C $\sim$ 2000 °C) that those useful organic materials would simply be destroyed. Another advantage of POF is its good compatibility with these functional polymer materials, and this means that various synthesis techniques and materials can be used to produce functional POFs for various applications.

Electro-optic (EO) POF has been developed by Kuzyk et al. for optical switching and modulation (Kuzuk et al. 1991; Welker et al. 1998). EO POFs could have great potential for applications such as voltage and electric field sensors (Peng et al. 2001a). Figure 4 shows two of our nonlinear POF designs (Peng et al. 2001b).

Segmented Cladding POF

Chiang et al. proposed a new type of segmented cladding fibers (Chiang and Rastogi 2002). A segmented cladding fiber is a novel fiber design that a core of high refractive index is surrounded by a cladding with alternate regions of high and low refractive indices in the radial direction. The fiber design was proposed as an alternative of the holey fiber structures for single-mode operation over an extended wavelength range. Due to the difficulties in silica fiber fabrication, initial attempt to make silica segmented cladding fibers was not successful.

Taking advantage of the flexibility in POF fabrication, Yeung et al. (2004) successfully fabricated segmented cladding POF by following the approach described in Peng et al. (1995). Both 4-segment and 8-segment POF have been fabricated, as shown in Figs. 5 and 6. These fibers have been shown to have a large range of single

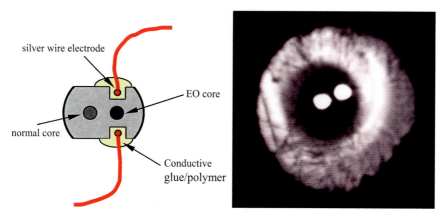

Fig. 4 Nonlinear POF designs. Left: a twin-core electro-optic POF; Right: the near-field of a doped twin-core nonlinear optical POF (Peng et al. 2001b)

Fig. 5 A 35-μm-core 8-segment POF (Yeung et al. 2004)

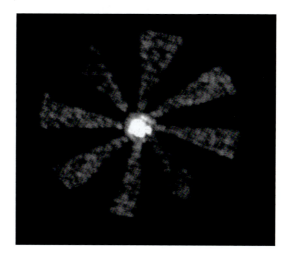

mode wavelength operation for large core diameters. For a 20-μm-core 4-segment POF, it is single-moded at 1.548 μm with a spot of ~15 μm in diameter, as shown in Fig. 6.

Scintillating POF

Scintillating POFs have been developed for radiation detection applications since 1980s (Thevenin et al. 1984; Blumenfeld et al. 1986). Scintillating POFs are doped with active materials such as fluorescent materials, laser-dyes, etc. These scintillating POFs have been used widely in detecting radiations and tracking

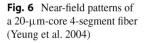

Fig. 6 Near-field patterns of a 20-μm-core 4-segment fiber (Yeung et al. 2004)

charged and high-energy particles in nuclear physics (Blumenfeld et al. 1987, 1991; Chiron 1991; Barni et al. 1996).

Oka et al. developed a radiation monitor by using scintillation POFs (Oka et al. 1998). A prototype monitor is built by combining scintillating POF for detection and silica fiber for optical propagation. They used a silica fiber of 100 m in length connected to the scintillating POF. The radiation monitor has a long detection length of 20 m. Even normal POF has been reported for radiation sensing. Naka et al. have reported a radiation distribution monitor using a normal POF. The monitor has a long detection length and can obtain continuous radiation distributions. The position sensing is based on a time-of-flight technique. The monitor is sensitive to beta rays or charged particles, gamma rays, and fast neutrons. The spatial resolutions for beta-rays, gamma-rays, and D-T neutrons are 30, 37, and 13 cm, respectively. The detection efficiencies for the beta-rays, gamma-rays, and D-T neutrons are 0.11%, $1.6 \times 10{-}5\%$ and $1.2 \times 10{-}4\%$, respectively.

More recently, various scintillating POF based on fiber-optic dosimeters have been developed and reported as promising candidates for dose information (Naka et al. 2001; Archambault et al. 2007; Yoo et al. 2013, 2015, n.d.). These fiber-optic dosimeters have many favorable dosimetric characteristics, such as water-equivalence, good flexibility, and immunity to ambient electromagnetic interference (EMI), over conventional dosimeters.

Dye-Doped POF

A number of different laser dye-doped POFs have been developed for fiber lasers and optical amplifiers since early 1990s. Tagaya et al. reported an active POF with laser dyes as dopants (Tagaya et al. 1995a, b). Peng et al. have also developed laser dye-doped POFs for active components applications (Peng et al. 1996a, b; Peng and Li 2001). Peng et al. have also reported a POF with β-carotene as the dopant for all-optical switching (Peng et al. 1994). Dye-doped POFs can be used for fiber amplifiers and lasers that operate at wavelengths other than the 1300 and 1550 nm used in silica-based fiber lasers. Laser dyes used to dye-dope POF included rhodamine 6G, rhodamine B, pyrromethane 650, and fluorescein.

Also dye-doped POFs, either doped in core or cladding, have been developed and reported for sensor applications. Laguesse developed a noncontact sensor using dye-doped POFs for detecting defects in thin-film sheet strips consisting of opacity variation, holes, cracks, and rendings or thickness variation of transparent film.

A wide range of fiber sensors can be built on dye-doped POFs. For example, Muto et al. reported dye-doped POFs for humidity sensing (Laguesse 1990, 1993; Destruel et al. 1992; Muto et al. 1989, 1990, 1994). A fault location system using a dye-doped POF for gas insulated switch gear (GIS) in high voltage electric environment was proposed by Kurosawa et al. (1997). GIS must be both maintenance free and inspection free, inhibiting the effects of the external environment by sealing the main circuit equipment with SF6 gas in a tank. Once an abnormality occurs inside it, however, it may develop into a major malfunction. Therefore, it is necessary to develop the fault locating technique. Suzuki et al. also proposed a dye-doped POF system to detect fault location in GIS (Suzuki et al. 1999). They investigated the system reliability using a full scale GIS. They conducted reliability and sensitivity tests of the POF against several factors such as the structure of the GIS, decomposed material of SF6 gas deposited on the fiber surface, and other environmental conditions. They found that they can practically locate a 100 A fault current, if the fault occurs somewhere in the unit of the GIS, even inside the current breaker. Sawada et al. built a simple fiber gas sensor to detect concentrations of alkalescent and acidic gases (Sawada et al. 1989). It was able to detect NH3 and HCl at concentrations of less than 10 ppm. They used a POF doped with fluorescent materials and demonstrated that a dye-doped POF can be a simple light source. This POF consists of a polycarbonate (PC) core with a glass transition at 150 °C, and a cladding of polymethylmethacrylate and polyvinylidene fluoride (PMMA/PVDF). The fiber converts incident white light to monochromatic light in less than 10 μs. These features make the fiber useful and highly efficient as 1 mW light sources.

A detailed survey on the development and applications of these laser-dye doped POFs can be found in a review paper by Bartlett et al. (2000). Here the core-doped POF is commercially available in a range of dopants, with either a circular or square cross section, and can be manufactured to specific requirements by various commercial and research sources such as Optectron, France; Tver Polymer Optical Fiber Scientific Production Association, USSR and Poly-Optical, USA (Bartlett et al. 2000).

The long-term stability of dye-doped POFs has always been an important issue for practical applications.

The long thermal stability of dye-doped POFs made by Tver Research Production Centre for POF in Russia was investigated by Klein et al. (1996). The POF samples were temperature cycled between 10 °C and 70 °C. They found the intensity of transmitted light at peak absorption or emission wavelengths changed by 5% but shifts in the peak absorption and emission wavelengths were not detected. However, when exposed to sunlight, the absorption and emission spectra of certain dye-doped FPOFs shifted to longer wavelengths.

Photorefractive POF

The development of photorefractive POF has been reported by Bian and Kuzyk of Washington State University (Bian and Kuzyk 2004). Holographic storage in photorefractive POF could have advantages over conventional recording media because of the enhanced sensitivity and diffraction efficiency and high angle access sensitivity. Dynamic holographic recording has been observed in POFs doped with Disperse Red 1, using a simple setup as shown in Fig. 7 (Bian and Kuzyk 2004). Their refractive-index grating is written by the polarized-light induced reorientation of the DR1 molecule. Two writing beams from a He-Ne laser 633 nm are launched into the POF.

The circles shown in Fig. 8a, b are the far field patterns on a screen perpendicular to the fiber axis for each of the two writing beams, respectively. Figure 13c, d shows the reconstructed images when each of the writing beams is successively blocked after 30 s recording. These results clearly indicate that holographic gratings are formed inside the fiber. They measured the diffraction efficiency, defined as the ratio of the power of the total diffracted light to the total output light from the fiber. The efficiency of about 60% has been achieved. The recorded gratings can be quickly erased with one laser beam.

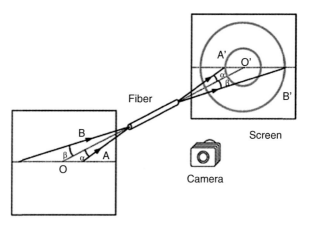

Fig. 7 Transmission-type holographic recording in POF (Bian and Kuzyk 2004)

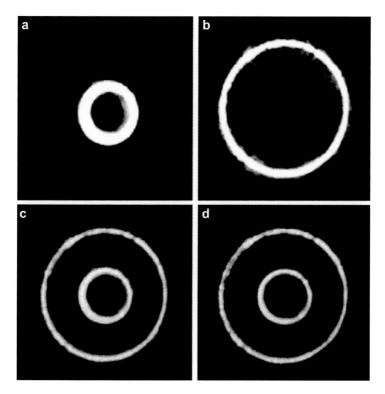

Fig. 8 Output images of writing beam (**a**) A and (**b**) B, respectively. Reconstructed images (**c**) and (**d**) when beam A and B are blocked, respectively (Bian and Kuzyk 2004)

Photosensitive POF

FBGs have many very important applications in optical communications and optical fiber sensing. POF FBGs are very attractive for many sensing applications such as strain, temperature, and humidity because of their desirable material properties (Peng 2002a; Peters 2010; Webb 2015).

Photosensitivity is one property essential for fabricating FBGs and hence developing photosensitivity POFs has been an important research area. There are a number of experimental studies on the photosensitivities of PMMA-based and fluorinated (CYTOP)-based POFs (Peng et al. 1999; Schmitt 1999; Yu et al. 2004; Luo et al. 2010; Sáez-Rodríguez et al. 2013).

Microstructured POF

A new class of special optical fibers has been investigated (Birks et al. 1997; Kubota et al. 2001; Ranka et al. 2000). These fibers can also be made in POFs and are

Fig. 9 A design of 4-layer holey microstructured POF

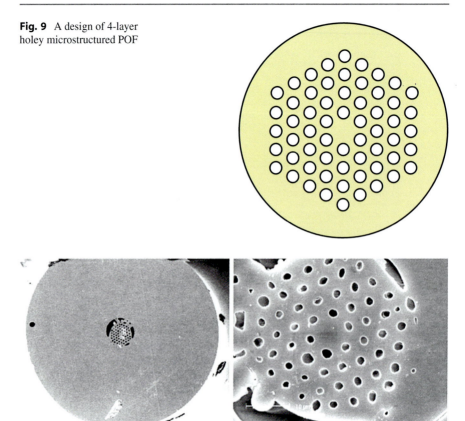

Fig. 10 A holey microstructured POF (Eijkelenborg 2001)

usually referred as microstructured POFs (Eijkelenborg 2001). The light guiding mechanism in microstructured POFs can be quite different from conventional POFs. In a holey microstructured POF, as shown in Fig. 9, a range of microscopic air holes produces an averaged refractive index distribution equivalent to the conventional refractive index distribution of a fiber.

Large et al. have made MPOFs from drilling patterned holes in commercial PMMA rods, as the electron microscope photos show in Fig. 10 (Eijkelenborg 2001). These commercially available extruded polymethyl methacrylate (PMMA) rods are not of good optical quality. They reported the attenuation of a raw fiber to be about 32 dB/m at a wavelength of 632.8 nm. From the photos, it can be seen that the air hole microstructure consists of four rings of holes in a hexagonal pattern embedded in an outer sleeve. Small deformations in the hole diameters and shapes are obvious.

The specially structured pattern of holes in a POF introduces novel properties similar to those of microstructured silica optical fibers. These novel properties include the possibility to realize single-mode operation with large effective core/mode areas or to guide the light in air rather in fiber materials. The novel properties certainly are significant for fiber sensor applications. For example, being able to have the guiding mechanism realized by means of an air microstructured band-gap POF is of great significance in gas or chemical sensor applications.

POF Fabrication Techniques

There are two conventional methods for POF fabrication currently: extrusion and preform method. The extrusion method is straightforward: extruders feed core and cladding materials separately through specially designed co-extrusion die to form a core-cladding structure and produce a desirable index profile required for POF. Although POF can be made by extruding ready-made polymer pellets, these POFs usually have high attenuation due to imperfections such as internal interfaces, microbubbles, and contaminations. For high quality low loss POFs, one would use normally a closed, continuous extrusion system which starts from monomers, through purification, initiation, and polymerization, and end-up with extrusion. The preform drawing method is also straightforward: firstly POF preforms are fabricated with desirable index profiles and core cladding structures and then draw into optical fibers. The process is similar to that in silica optical fiber fabrication.

One fully new method, 3D printing, for POF fabrication is proposed and demonstrated recently (Cook et al. 2015). This 3D printing optical fiber technology is now under intense development (Cook et al. 2016; Canning et al. 2016; Atakaramians et al. 2017; Toal et al. 2017; Zubel et al. 2016; Talataisong et al. 2018).

Extrusion Method

The more developed and widely used fabrication method for POFs, dominantly SI-POFs, is the extrusion method.

Extrusion of SI-POF

As mentioned above, early commercial SI-POFs were made by du Pont and Mitsubishi Rayon, and they used the continuous extrusion method for large production quantity and high optical quality (low optical attenuation). In this method a purified monomer (methyl methacrylate), an initiator and a chain transfer agent are fed into a polymerization reactor. The polymer that becomes the core of the SI-POF is then fed into an extruder by a gear pump. The core and cladding material are fed by separate extruders proceed into a spinning block where the concentric core-cladding structure of an SI-POF is formed. Although the description of this process may appear to be simple, numerous variables including the purity of the monomer, the reaction temperature, the degree of polymerization, the amount of initiator and chain transfer agent, and extrusion conditions play an important role in determining the optical and physical properties of the fiber. Eska products from Mitsubishi Rayon which currently dominate the SI-POF market are made by this process. One important advantage of continuous extrusion is high production rate.

Extrusion of SI-MPOF

Microstructured Polymer Optical Fiber (MPOF) has attracted great interest in recent years, not only for their novel features as compared to conventional POFs, but also for possible advantages over their glass counterparts.

Since the first demonstration (Eijkelenborg et al. 2001), MPOFs have attracted much attention, not only for their novel features as conventional microstructured optical fibers such as their "endlessly single-mode" properties (Birks et al. 1997), easy tailorability for dispersion and polarization control, the possibility of large core single moded fibers (Knight et al. 1999), and even the ability to make band-gap fibers in which the light is guided in air (Cregan et al. 1999), but also for many advantages over their glass counterparts, such as the much lower processing temperatures, the intrinsic tailorability of polymers, the cheap base materials processing, and more flexible even at large diameters compared with silica fiber (van Eijkelenborg et al. 2003).

So far, a range of different fabrication methods have been used to make MPOFs such as capillary stacking (Kondo et al. 2004), polymerization in a mold (Lyytikäinen et al. 2004), drilling (Barton et al. 2004), and (Zhang et al. 2006). But none of these are particularly and industrially suitable for mass production. In 2007, Mignanelli et al. (2007) fabricated at the first time the MPOFs via direct bicomponent extrusion, which is one of sophisticated techniques in textile industry. This opened the door to high-volume low-cost production of microstructured and photonic crystal fiber. Liu et al. had also applied the bicomponent extrusion technique to multicore POFs (Liu et al. 2009).

Using the bicomponent extrusion technique which has long been named as "islands in the sea" fiber by those in the textile trade, SI-MPOFs can be fabricated directly from melt polymers through a die with custom designed pattern of holes, structures or geometries. Such fibers are generally achieved through the coextrusion of two polymers using a double screw melt spinning extruder. The key advantage of the bicomponent extrusion techniques is flexibility in selecting the index difference and doping materials, and feasibility of making MPOFs with varying index differences, various structures, and various doping materials with mass low-cost production.

Because there is no air hole in the fiber from the bicomponent extrusion technique, the cross-sections can be protected well without hole distortion compared with the method used by Zagari et al. (2004). In addition, the technique is possible to achieve desirable smaller refractive index difference and larger effective mode area MPOFs, by properly choosing the "sea" and "island" polymer materials with smaller refractive index difference and to produce fibers of any cross sectional shape or geometry that can be imagined.

Two types of "sea-island" bicomponent MPOFs, with high refractive index "land" (glycol-modified-polyethylene-terephthalate (PETG), $n = 1.57$) and low index "sea" (poly(methyl methacrylate)(PMMA), $n = 1.49$), and low refractive index "sea" (PMMA) and high index "island" (PETG), were successfully fabricated by the extrusion method (Liu et al. 2009, 2010) as shown in Fig. 11.

The MPOFs are fabricated by combining both extrusion and draw techniques, extrusion with a purpose-built bicomponent extrusion system and drawing with a fiber spool, as shown in Fig. 12. The processing temperature and extrusion volume of each polymer component can be controlled separately. The fiber diameter can be controlled by varying the pickup speed of fiber spool.

Extrusion of GI-POF

As mentioned above, GI-POFs are perceived to be more preferable for the short-haul applications because of their high bandwidth, and high flexibility. High flexibility of polymers makes it possible for POFs to have a much larger core diameter in the order of 1 mm compared to about 10 μm of single-mode silica fibers. A large diameter makes implementation, coupling, and connection of these optical fibers much easier and economic.

GI-POFs required a continuously varying refractive index profile across their radial direction that enables a much higher bandwidth than SI-POFs. Especially when the refractive index profile is close to parabolic, the modal dispersion of GI-POF is minimized, making it possible to transmit data at a very high bandwidth distance product, e.g. 4000 MHz km (Nuccio et al. 2008). For these reasons, there have been considerable research efforts on developing fabrication techniques for POFs with graded index profiles, especially the optimum refractive index profile (Koike and Narutomi 1998; Sohn and Park 2001, 2002).

The bicomponent extrusion, mentioned above for SI-POF, has been used for GI-POF (Asai et al. 2007; Hirose et al. 2008). A schematic diagram of the bicomponent extrusion, also referred as coextrusion technique, for GI-POF is shown in Fig. 13.

Here a step-index cylindrical polymer preform is extruded from the two extruders, for core and cladding materials, respectively, through the spin die, and is fed into a diffusion heater. This polymer preform has uniform core and cladding and with step-index profile. The core material is composed of the base polymer and high refractive index dopant. The cladding material is composed of only the base polymer. Through the diffusion heater, the high refractive index dopant within the core material diffuses and redistributes central-symmetrically along the radial direction to form a dopant concentration profile. By proper control of the diffusion process, the step-index refractive profile of the polymer preform is converted into desirable graded-index profile for high bandwidth GI-POF. However, material diffusion could be a rather slow process that limits the production rate for commercial applications. Hence, further development of GI-POF from a multistep index core through thermal diffusion was carried out (Lee et al. 2003). In this method, the core of the preform is fabricated with multiple-step-like layers with

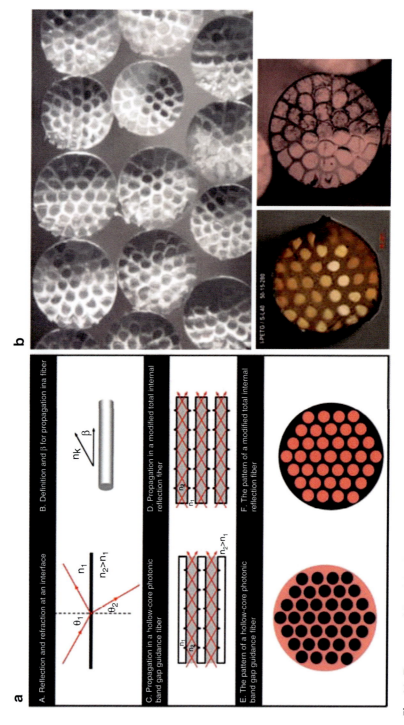

Fig. 11 Two types of "sea-island" MPOFs. (**a**) The MPOF design and operation principles; (**b**) micrographs of the fabricated MPOFs. Top: MPOFs with multiple "sea"; Bottom: MPOFs with multiple "island" (Liu et al. 2009, 2010)

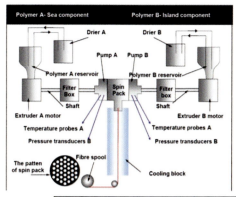

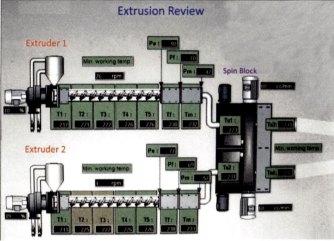

Fig. 12 The schematic diagram of POF fabrication using a purpose-built bicomponent extrusion system. Top left: The bicomponent POF fabrication schedule; Top right: a bundle of MPOFs out of the spin pack; Bottom: The operation overview diagram of the extrusion system

different indexes that roughly follow the optimum index profile and is then thermally diffused to produce a smooth graded-index profile (Lee et al. 2003).

Preform Method

Fabricating POF by the preform method consists of several steps: purification, initiation of monomers and additive materials, fabrication of POF preform, and fiber drawing. To meet high purity requirement, the preferred method for synthesizing polymer material for POF preform is bulk polymerization. Preform with the desired refractive index profile can be fabricated mainly by two ways: (1) direct in situ

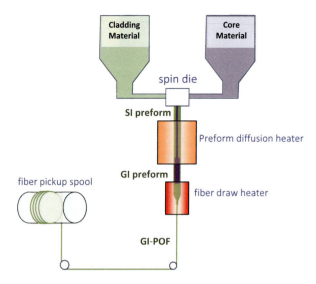

Fig. 13 A schematic diagram of the bicomponent extrusion and diffusion technique for GI-POF

polymerization and (2) extrusion and molding after polymerization. The latter is closely associated with the extrusion method above.

SI-POF Preform Fabrication

To make SI-POF, we may simply use the bicomponent extrusion technique to produce SI-POF preform and the process and draw it into fibers. Rod-in-tube technique can also be used. First polymer rods with higher refractive index and polymer tubes with lower refractive index are prepared for core and cladding, respectively. These rods can be inserted into the tubes to form preform sets and they can be heated and drawn into POFs straightforwardly.

GI-POF Preform Fabrication

To make GI-POF, a number of different techniques, including diffusion, interfacial-gel polymerization, and photopolymerization, have been developed for preform fabrication since 1970s (Ohtsuka and Yoshida 1976; Ohtsuka and Nakamoto 1976; Koike et al. 1989, 1995; Ishigure et al. 1995, 1996, 1998; Van Duijnhoven and Bastiaansen 1999; Liu et al. 1999; Park et al. 2000a; Sohn and Park 2001; Ferenets et al. 2004; Park 2004).

GI-POF by Diffusion: The diffusion method is the simplest technique that would convert a SI-POF preform into GI-POF preform though dopant diffusion under heat process, as discussion above in Fig. 13.

GI-POF by Interfacial-gel polymerization: The interfacial-gel polymerization technique starts with a polymeric tube (e.g., PMMA) which is to be filled with a mixture of at least two monomers of different refractive indices with initiators and chain transfer agents (see for example, Peng et al. 1995). One of the monomers with the lower refractive index is used as the host material (e.g., methyl methacrylate) and the other with higher refractive index as the guest, or dopant (e.g., vinyl benzoate). The filled tube is then polymerized thermally while the tube is rotated. During the polymerization process, the inner wall of the tube is swollen by the monomer mixture and forms a gel phase layer. The polymerization is faster inside the gel phase layer, referred as "gel effect," than that in the monomer bulk phase. As a result, the polymerization develops gradually from the inner surface of the tube toward to the tube center. By selecting proper reactivity ratio of the two monomers, the material composition and hence the refractive index of the copolymer change gradually in the radial direction. The achieving GI-POF with proper graded index profile, the host monomer with a lower refractive index should have higher reactivity than that of the guest/dopant monomer with higher refractive index (Ohtsuka and Yoshida 1976; Ohtsuka and Nakamoto 1976; Koike et al. 1989). The refractive index profile obtained by this method depends on the relative ratios of the monomers.

GI-POF by Photo polymerization: The procedure of making GI-POF preform by photopolymerization is similar to that of interfacial-gel technique (see for example, Ohtsuka and Yoshida 1976; Park 2004). It also starts with a polymeric tube (e.g., PMMA) or a glass tube which is to be filled with a mixture of at least two monomers of different refractive indices with initiators and chain transfer agents. The two monomers are used as the host and guest materials. The tube is then positioned vertically and rotating with respect to central axis while under UV irradiation. The radiation is to be applied to the tube locally using a shade and moved up from the tube bottom to top at a certain speed. The UV radiation is absorbed by initiators and monomers and this induces polymerization of the monomers. Because the UV is stronger near the inner wall of the tube, a gel phase layer is formed earlier around the inner wall of the tube. As the polymerization develops, the absorbing centers reduce, the UV penetrates deeper and deeper, moving the polymerization zone and the gel phase toward the tube center. The polymerization progressed from the bottom of the tube to the top as the irradiation moved upward. Usually preform fabrication is completed by thermal treatment.

POF preform can be drawn easily at relative low temperature. Most POFs are drawn between 200 °C and 300 °C. The traditional way of heating is to use an electric resistance furnace. For economic reasons, it is better to use large diameter preforms. However, the thermal conductivity of POF materials is low. This heating method is accompanied by an unavoidably large temperature gradient along the radial direction of the preform, and this limits the maximum possible size of POF preform. To overcome this difficulty, an IR furnace working in wavelength range of $1 \sim 2$ μm could be used. In this wavelength range, the thermal energy has good transmission and can penetration deeper into polymer materials. Hence, heat may reach inner region of the preform while the temperature gradient would be effectively reduced and controlled. A critical condition during fiber drawing is the draw tension. It must be maintained relatively low, normally between 50 and 100 g. If the draw tension is too high, POF could be over-stretched POF and may shrink considerably and cause fiber deformation and optical loss when heating at an elevated temperature, e.g., $\sim$80 °C.

MPOF by Drilling: Microstructured polymer optical fibers (MPOF) are an exciting new development in POF research, offering opportunities to develop fibers with new functionalities for specific applications (Eijkelenborg 2001). MPOF can be fabricated by the extrusion technique introduced above. The preform method for fabricating MPOF is drilling polymer rod with a required structure or pattern (van Eijkelenborg et al. 2003; Barton et al. 2004). Firstly the structure required in the final fiber, i.e., the hole pattern, is to be designed. Then the polymer rod is drilled to become a preform with the required structure using a well-controlled mill. The coated drill bits produce deep holes with minimal drill wander while leaving the inside of the holes with a smooth finish, the latter being of importance in that it minimizes the likelihood of surface roughness-induced scattering in the drawn fiber (van Eijkelenborg et al. 2003).

3D Printing Method

In recent years, Additive Manufacturing (AM) technologies, including photopolymerization, powder fusion/sintering, extrusion, 3D printing, direct writing/prototyping, are coming of age because of their ever rising importance for fabricating a huge variety of materials, structures, and components. Actually all these technologies can be closely related to future optical fiber fabrication techniques.

3D printing is one the key AM technologies that is finding new applications such as fiber fabrication (Cook et al. 2015, 2016; Canning et al. 2016; Atakaramians et al. 2017; Toal et al. 2017; Zubel et al. 2016; Talataisong et al. 2018). Fused deposition modeling (FDM) is the first and also the most commonly used 3D printing technique that a polymer is fed through a heated nozzle, which acts essentially as a mini extruder with programmable 3D positioning capability, to build custom-designed components and structures. Photopolymerization is also a 3D technique which uses UV light emitting diodes (LEDs) or lasers to carry out in situ fabrication and offer higher resolution and more flexibility. Selective Laser Sintering (SLS) is another 3D technique that a high power laser is used to sinter particles or powders of polymer, glass, metal, or ceramic materials into 3D objects (Cook et al. 2015). All these 3D printing technologies can be used for fabricating POFs. 3D printing will enable the fabrication of custom 3D printed POFs with unique structures and material compositions that are otherwise too difficult and expensive to achieve.

3D printing using FDM technique has been applied to print "light pipes" (Willis et al. 2012) and polymer optical fibers (Pereira et al. 2014). Willis et al. have shown that polymer optical components can be custom designed and printed with a 3D printer using VeroClear, a transparent material similar optical properties to PMMA with a refractive index of 1.47 at 650 nm wavelength. In the case of Pereira et al. (Pereira et al. 2014), printing POF is demonstrated using a modern multimaterial 3D printer with VeroClear as the core material.

The first 3D printed structured optical fibers were drawn from 3D printed POF preforms, as shown in Figs. 14 and 15. The preform used in their work was printed

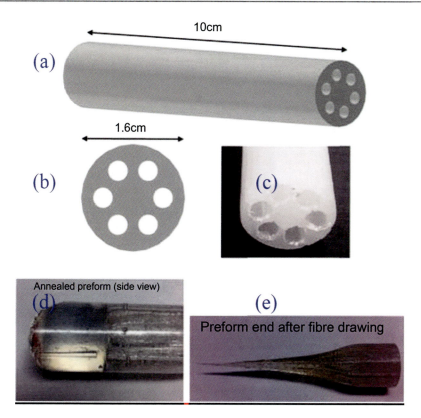

Fig. 14 3D printed structured POF preform (Cook et al. 2015). (**a**) and (**b**) CAD designed preform; (**c**) 3D printed preform; (**d**) preform after annealing and (**e**) preform after fiber drawing

laterally with the preform written by its side. A low-cost, commercially available 3D FDM printer was used to print preforms. Resolution of the printing on the xy-axes was set to $\Delta x = \Delta y = 0.4$ mm and $\Delta z = 0.1$ mm on the z-axis. It takes about 6 h for 3D printing a preform of 10 cm as shown in Fig. 14a.

More 3D printing works on POF fabrication have been reported since then (Cook et al. 2016; Canning et al. 2016; Atakaramians et al. 2017; Toal et al. 2017; Zubel et al. 2016; Talataisong et al. 2018). A first step index optical fiber drawn from a 3D-printed polymer preform was also reported (Cook et al. 2016). A first 3D-printed hollow-core POF is also drawn from a PMMA preform made using the FDM technique (Zubel et al. 2016). Recently a first hollow-core MPOF drawn from a 3D printed PETG preform for mid-IR application was also reported (Talataisong et al. 2018).

Comparing with conventional optical fiber fabrication, 3D printing allows optical fibers to be built on software-based design with arbitrary geometries and selected materials. Simply by changing software parameters, optical preform can be designed and fabricated by 3D printing. This can also be desirable for optical circuit design

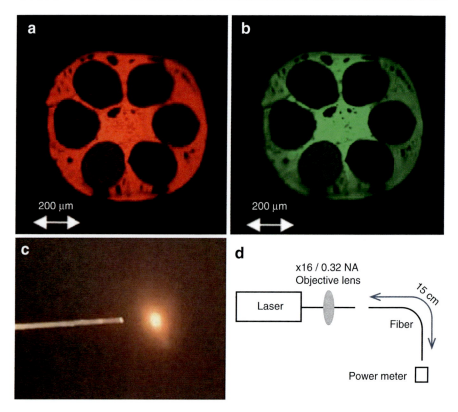

Fig. 15 Light guidance test of the drawn optical fiber (Cook et al. 2015): (**a**) 630 nm and (**b**) 515 nm guided light outputs at the fiber end face; (**c**) white light output projected onto a screen; (**d**) schematic of set-up for the guidance test

and fabrication. In contrast conventional manufacturing requires considerable effort for individual fiber optic strands to be mechanically assembled, fused/deformed with heat, or chemically bonded.

It is particularly attractive for research and development that require for rapid prototyping and manufacturing.

POF for Data Transmission

Terrestrial and transoceanic links-based silica fiber with substantial trunk communication capability has been installed worldwide. The bottleneck in telecommunication infrastructure is now shifting from metropolitan area networks toward the access networks, e.g., local area networks (LANs), customer access networks (CANs), and computer data links. The present access networks of metal cables and wires or wireless simply cannot meet the growing capacity and quality demand.

Table 7 Major progresses in developing high bandwidth POF transmission system

Year	Event	References
1994	Keio University reported 2.5 GHz over 100 m	Polishuk (2006)
1997	Fujitsu reports 2.5 Gb/s over 200 m with GI-POF, ~0.5 GHz-km	Polishuk (2006)
1999	Lucent reports 11 Gb/s over 100 m of Lucina (CYTOP GI-POF), ~1.1 GHz-km	Polishuk (2006)
2008	Georgia Tech reported 40 Gbps over 200 m PF GI-POF, demonstrated ~8 GHz-km	Polley et al. (2007)

Hence, future high-speed access networks will have to use more optical fiber and to choose more appropriate fiber from two most attractive candidates: silica optical fibers and POFs.

POFs, especially the low loss and high bandwidth PF GI-POFs, are very useful for data transmission (Lethien et al. 2011; Koike and Inoue 2016). They are easy and hazard, free to handle in comparing its glass counterpart silica fibers, since it allows more relaxed alignment and is much easier to terminate and connect POFs. A significant amount of efforts and progresses in research and development of POF data transmission have been made over these years, as seen in the summary in Table 7.

The development of POF has always been focusing on optical transmission applications, local access networks, and systems in particular. The digital revolution creates a great market for local networking or interconnection of internet, mobile, personal computer, consumer electronics, and video-audio equipment in vehicles, homes, offices or industrial sites. It is obvious that POF-based data transmission system has all the important attributes necessary for these applications: high data rate (up to 10Gbit/s), sufficient coverage (up to 100 m), electrical isolation, EMI free, flexibility, ease of installation, and most important of all, low cost. Currently commercial POFs can satisfy international standards such as IEEE 1394 and Ethernet and directly replace copper media or silica optical fibers in the applications. For example, Mitsubishi developed a home entertainment network technology (Optohome©) that uses its POF (ESKA™) to interconnect TV, audio/video, home movies, Internet, telephone and security panels for home, apartment, and business conference areas.

POF for Sensing

Optical fiber sensing has become one of the most important fields of modern science and technology and found wide applications in industrial manufacture, civil engineering, military technology, environmental protection, geophysical survey, oil exploration, medical and biological technologies, etc. The popular optical fiber sensors include rotation, temperature, strain, stress, vibration, acoustic, and pressure sensors, etc. Various sensor designs and system architectures, based on either silica optical fibers or polymer optical fibers, have been developed for these applications.

Due to the key POF features such as high flexibility and handling safety described above, POFs are used a wide range of areas including medical devices, pharmaceutical and chemical industries. POF sensors are particularly suitable for harsh industrial environments.

SI-MM POF: SI-MM POF is by far the most popular and widely used POF in sensor applications. In fact, polymer optical fiber sensors currently under research and development, as well as those already having been installed, dominantly use SI-MM POFs. Multimode fibers are mainly used for intensity-based sensor systems where a particular measurand affects the signal intensity directly. In comparison to interferometric-based sensors, intensity-based sensors are usually simpler, less expensive but also less sensitive. A great variety of POF-based sensors have been reported (Polishuk 2006; Bartlett et al. 2000; Chandy et al. 2000). POF sensors reported include humidity sensor (Hasseinibalim 1997; Muto et al. 2003), biosensor (Bansal et al. 2002; Philip-Chandy et al. 2000a), chemical sensor (Zhou et al. 1991; Ferguson and Walt 1997), gas sensor (Hecht and Kolling 2001; Morisawa et al. 2001), dew-point sensor (Hadjiloucas et al. 2000), flow sensor (Philip-Chandy et al. 1999, 2000b), PH sensor (Deboux et al. 1995; Netto et al. 1995), turbidity sensor (Kruszewski et al. 2003; Zhang et al. 1996; Chandy et al. 1998), structure health monitoring sensor (Kuang et al. 2001, 2003), etc. POF have been used for sensing and measuring a range of important physical parameters including liquid level (Vazquez et al. 2002; Borecki and Kruszewski 2001), electric discharge (Farenc et al. 1994; Kurosawa et al. 1991), magnetic field (Guerrero et al. 1993), refractive index (Ribeiro et al. 2002), temperature (Grattan and Kalymnios 1998; Alcala et al. 1995), wind speed (Zubia et al. 2000), rotation (Guerrero et al. 1994), vibration (Kamiya and Ikeda 1996; Kamiya et al. 1998), displacement (Ioannides et al. 1996), electric insulation (Takezawa et al. 1998), underwater sound (Capps et al. 1982), particle concentration (Zhang et al. 2000), etc. Nevertheless, all the developments are based on multimode POFs and these sensors are dominantly of the intensity-based type. Intensity-based fiber sensors are less complicated, but their applications are usually limited by their sensitivities and resolutions.

Single-Mode POF: Work on single-mode POF for sensor applications has been started only recently with the development of single-mode POFs (Bosc and Toinen 1992; Peng et al. 1995) as well as POF Bragg Gratings (Peng et al. 1999; Towrie et al. 2000). The application of single-mode polymer optical fiber in interferometric-based sensing could be advantageous (Peng 2002a). A main advantage is the low Young's modulus of polymer materials. The Young's modulus of polymer fiber materials is typically many times less than that of silica glass. Hence, for a strain related optical sensor such as fiber hydrophone, polymer optical fibers are intrinsically many times more sensitive to acoustic waves than silica glass fibers. For example, the use of polymer optical fiber in fiber hydrophones could revolutionize the sensor head design because of the much better acoustic compatibility between water and fiber. For the same reason, polymer optical fiber could be advantageous for sensing in environments of liquid and less rigid solid state materials.

The research and development of POF sensors has covered a wide spectrum of topics and applications and has used a wide range of POF types. For simplicity,

POF sensors can be roughly divided into two main categories: conventional POF based and special POF based. In the first category, we include any step-index multimode, graded-index multimode, and single-mode POFs made from ordinary polymer fiber materials and conventional fiber designs as they are counterparts of conventional silica glass fibers, although graded-index multimode POFs and single-mode POFs may not be considered as conventional POFs as yet. The other category includes POFs made with special functionalities such as laser-dye doped POFs (Peng et al. 1998), scintillating POFs, electro-optic POFs or with special designs such as microstructured POFs, twin-core POFs.

POFs are readily available at a low cost while exhibiting a good chemical resistance to many aqueous environments including water, lyes, dilute acids, petrol, mineral, and turpentine. While it is true that POF is potentially low cost, this is not very significant since the fiber length used in a sensor is typically short and its share of the sensor system cost is fairly small. Hence, we had better not over emphasize the low cost of POF as a major factor in POF's sensor applications.

Radiation Detection

Scintillating polymer fibers have been successfully used as radiation monitor and charged particle trackers in a number of nuclear physics experiments for many years (Thevenin et al. 1984; Blumenfeld et al. 1986, 1987, 1991; Chiron 1991; Barni et al. 1996; Oka et al. 1998). In recent years, scintillating POF detectors have been implemented in a number of important experimental facilities (Introduction to K2K Experiment n.d.; Kim et al. 2003; Ansorge et al. 1988; Suzuki et al. 2000).

To study the neutrino oscillation in relation to cosmic rays, an experimental program known as K2K -KEK to Super-K (Super-Kamiokande) has been established in Japan (Introduction to K2K Experiment n.d.). In the experimental facilities, a very large number of scintillating POF detectors are used for high tracking efficiency in reconstructing charged particles and their vertex positions when more than two particles are produced in neutrino interactions.

The scintillating POFs they used have a diameter of 0.7 mm. A scintillating fiber layer is used to provide the hit-position measurement in both horizontal and vertical directions. An individual scintillation sheet has a thickness of 1.3 mm, consisting of two staggered layers of scintillating fibers. The vertical and horizontal fiber sheets are separated by a 1.6-cm thick honeycomb panel. There are 20 scintillating fiber layers in total, and the adjacent layers are spaced 9 cm apart along the neutrino beam direction.

Biomedical and Chemical Sensor

The advantages of POF over its glass counterparts for biomedical applications are really remarkable: intrinsic good compatibility with biomedical materials or organics, a great selection of POF materials for desirable refractive indices, easiness in incorporating functional materials, etc. POFs have been used to construct a wide range of fiber sensors for biomedical applications (Kosa 1991; Macraith et al. 1995). These POF sensors are used for detecting chemical and biological agents (Yamakawa 1997; Merchant et al. 1998; Hawks et al. 2000), biofilm (Philip-Chandy et al. 2000a), micelle (Ogita et al. 1998, 2000), biological tissue (Dattamajumdar et al. 1998), and medical data (Raza and Augousti 1995; Scully et al. 1993).

Philip-Chandy developed an online sensor to determine the fouling properties of aqueous process fluids (Philip-Chandy et al. 2000a). The sensor detects material build-up at the core-cladding interface by means of refractive index modulation. This is achieved by evanescent field attenuation and intensity modulation. POF is sensitized by removing its cladding over a length. It is used to measure the growth of biofilms in a closed loop water process system.

Ogita et al. reported the use of POF for the detection of critical micelle concentration (CMC) surfactant solutions (Ogita et al. 1998, 2000). Their CMC detection is based on an adsorption effect in sample solutions consisting of sodium dodecylbenzenesulfonate. In their experiment, an incident beam was reflected at the interface between the fiber core and the solution, passing through the sensing region along the fiber with repeating reflections. The transmission of the light will change with the refractive index or absorption conditions near the core–solution interface. They tested two different types of fiber sensors. The first is a polymer coated (clad) silica fiber and the second is a POF sensor head. They have demonstrated that the output signals of both sensors worked well with significant changes at the CMC point as the surfactant concentration increases.

To detect patients who have Barrett's esophagus, a precursor to esophageal adenocarcinoma, a cost-efficient screening device developed using POF has been demonstrated (Dattamajumdar et al. 1998; Asai et al. 2007). A prototype system is built based on colorimetric assessment of esophageal lumen. The system consists of a small diameter (5 mm) fiber-optic probe, interfacing electronics, a probe-head position sensor, and a computer for display and analysis. The probe has a central (1 mm OD) plastic optical fiber through which white light (3200 K) is directed so as to be incident on the collapsed esophageal lumen via a conical mirror in the probe-head. The system performed well in tests using models of esophageal lumen which simulate patterns observed in Barrett's esophagus.

POF sensors for chemicals have also been developed (Zhou et al. 1991; Ferguson and Walt 1997). Chemical sensors have been made of novel porous POFs and selective chemical indicators (Zhou et al. 1991). Zhou et al. reported that, by careful selection of polymer materials and chemical indicators, the chemical reagents can be covalently bonded to the porous POFs. Their POF sensors could be used to detect a variety of chemical species and to measure various chemical parameters, both in vapor and solution. The sensors showed high sensitivity and stability. Sensor characteristics, including dynamic range, linearity, and response time, can be tailored to meet specific applications by altering the polymer composition and polymerization procedure.

Structural Health Monitoring

Optical fiber sensors have been targeted for structural health monitoring across a range of industries. Currently there are a number of nondestructive evaluation and testing techniques for engineering structure diagnostic purposes. These techniques employ a wide range of different technologies including electromagnetic methods (Shigeishi et al. 2001; Colla et al. 1998), acoustic emission method (Cheng and Sansalone 1993), electrical impact-echo (Philip-Chandy et al. 1999), ground penetrating radar systems (Hugenschmidt 2002), and corrosion monitoring with embedded reference electrodes (Castro et al. 1996). These technologies are susceptible to electromagnetic interference and/or limited in sensitivity or repeatability. In comparison, for civil engineering applications, fiber sensors are advantageous for their immunity to electromagnetic interference and disturbances, their resistance to humidity and corrosion, and their potential for long-term and high repeatability use.

POFs have been shown to be highly cost-effective sensors for structural health monitoring (Takeda et al. 1999; Takeda 2002; Kuang et al. 2002b; Kuang and Cantwell 2003a; Ishigure et al. 1998; Van Duijnhoven and Bastiaansen 1999). The sensing principle generally relies on monitoring the modulation of light intensity as a function of damage or deformation. Although, for long-term monitoring, intensity-based sensing techniques suffer from an inherent susceptibility to power fluctuations in the light source, this can be overcome using referencing techniques. Here, a strain-free reference fiber is coupled to the light source using an optical splitter, and any effects due to changes in the light intensity of the source can be removed. In addition, temperature-induced changes in the optical signal can be eliminated when the reference fiber is collocated with the sensing fiber.

A structural health testing or monitoring system is necessary for detecting and assessing localized fracture and ensuring the structural integrity of interest. There are quite a few traditional nondestructive evaluation techniques such as ultrasonic C-scanning, radiography, and infrared imaging for these purposes.

A number of new techniques are being developed in recent years for heath monitoring of composite material structures using optical fiber sensors (Okabe et al. 2000; Waite 1990). Other monitoring techniques detect the change in structural properties of the composite (e.g., stiffness and strength) which are related to and reflect the structure degradation or damage. Here structural degradation or damage information is obtained from monitoring dynamic characteristics of structures – viz. dynamic response in the frequency, mode shape, time, and impedance domain (Gadelrab 1996; Kessler et al. 2001; Whelan 1999; Valdes and Soutis 1999).

Cracking Detection

Cracking has a significant bearing on the structural integrity. Cracking is usually results from structural overloading, corrosion of reinforcing steel bars, and/or fatigue. It indicates the degree of structural deterioration and is closely linked to the safety of operation of the structure. If cracking is left unnoticed and unattended, it could end up with catastrophic structural failure and safety breakdown, because initial cracking in concrete structures could develop or propagate under the influence of mechanical and environmental factors.

Kuang et al. (2003b) of the University of Liverpool in UK have carried out a series of developments of a POF sensor for civil engineering applications in recent years. The application of POF sensors for damage or crack monitoring in civil engineering structures is especially advantageous over silica optical fiber sensors. The silica glass fiber sensors are vulnerable to fracture under huge stress concentrated at the fine crack points and could totally and immediately lose their sensing capability at the onset of initial cracking. Moreover, silica optical fiber sensors are susceptible to chemical damages caused by undesirable alkaline and aqueous conditions that are frequently encountered in civil engineering environments. In short, POF sensors could be much more chemically stable and potential advantageous for civil engineering applications.

In applications such as the detection of early cracking in civil engineering structures Fujimoto et al. 1996, accurate strain information is nonessential. Hence, rugged and low cost intensity-based optical fiber sensors are often preferred. Kuang et al. studied crack detection and vertical deflection monitoring in fiber reinforced composite structures using an intensity type of POF sensor system (Kosaka et al. 1999; Park et al. 2000b).

The method of sensitizing a POF removes only a small segment of the POF cross section (Kuang et al. 2003). It is different from other methods such as chemical tapering (Merchant et al. 1999), intermittent etching (Glossop 1989), and radial grooving (Kuang et al. 2002a) where the entire cladding of fiber cross-section is sensitized. This is why this type of sensor could be designed with directional sensitivity (Kuang and Cantwell 2003c). Under the fiber bending introduced by loading, the number of propagation modes that satisfy the total internal reflection condition could be either increased or decreased, depending on the direction of bending as well as loading.

They used the POF sensor to detect initial cracks, monitor postcrack vertical deflection, and detect failure cracks in concrete beams subjected to flexural loading conditions. A series of three- and four-point bend tests was carried out on a range of structures. In their experiment, the POF sensors were attached to both scalemodel concrete samples without reinforcement and life-size concrete beams containing reinforcing steel bars, to evaluate their ability to monitor beam deflection, detect cracks, and monitor crack development.

Full-scale reinforced concrete beams have also been tested by the group under similar flexural loading conditions and similar load-deflection and POF sensor output vs. deflection relations have been observed. These results further confirmed that POF sensors are capable of detecting initial hairline flexural cracks, monitoring crack propagation and damage development, as well as detecting ultimate failure in both scalemodel concrete specimens and fill-scale steel reinforced concrete beams. From these experiments, the distinctive advantage of the POF sensors is that it can operate over a wide range of loading – essentially the entire cracking dynamic regime has been demonstrated. Obviously, in comparison to silica fiber sensors, POF sensors could be more appropriate for real-time monitoring the cracking dynamics under loading – crack initiation, crack propagation, until ultimate failure of the structure of interest.

Impact Damage Assessment

POF sensors are useful not only for tests under quasi-static loading conditions but also for tests under dynamical loading conditions. A recent paper by Kuang et al. reported the use of a POF sensor for detecting impact damage in fiber reinforced composite structures (Kuang and Cantwell 2003c).

In this study, the POF sensor has been used was bonded to carbon fiber reinforced epoxy cantilever beams for monitoring their damping response under free vibration loading conditions, when they are impacted at varying impact energies up to 8 J. The postimpact strengths and damping characteristics of the damaged laminated structures were measured by the fiber sensor.

POF sensors were used to monitor the dynamic response of the undamaged and damaged composite specimens. Since the test durations were short (less than 5 s), the results obtained were not affected by any signal fluctuation usually associated with intensity-based sensing systems.

Their experiment has been carried out on plain carbon fiber epoxy beams by evaluating the postimpact flexural strength and stiffness. The results of these tests were then used as a reference to assess the damage sensitivity of the POF sensor technique.

The composite material system used in this study was based on a woven carbon fiber-reinforced epoxy resin. Beams with a POF sensor for the damping and free-vibration tests were produced in $240 \times 24 \times 2.4$ mm.

The POF is sensitized by removing the cladding layer by abrading the surface of POF over a length of approximately 70 mm. Initially, adhesive tapes were used to ensure the location and alignment of the POF. The fiber was then bonded to the specimen using a fast-curing cyanoacrylate-based adhesive and was left for several minutes to ensure complete curing of the adhesive prior to abrading the POF. An initial study showed that the adhesive did not adversely affect the properties of the POF. When light was passed through the POF, no change in the optical transmission efficiency was observed as a result of the bonding process. They simply used a razor blade to remove a segment of the POF cross-section. Care was taken to prevent the blade from cutting into the fiber by positioning the blade vertically or tilting it towards the direction of abrasion. Since the sensitization procedure was carried out after the POF had been bonded to the sample, the prepared region was free of any residual adhesive, which could have resulted in an optically nonuniform surface. This simple procedure has been found to be effective and offer a reasonably high level of repeatability.

The sensor was designed to be sensitive to flexural deformations. It operates simply on the principle that, when the fiber is bent or deformed, the number of guided modes in the sensitized POF section reduces, and the transmitted power through the POF is reduced accordingly.

All the above as well as a large amount of other research work reported (see for example Bartlett et al. 2000; Hasseinibalim 1997; Muto et al. 2003; Bansal et al. 2002; Papanicolaou et al. 1998) demonstrated the great potential of POF sensors in structure health monitoring.

Environment Monitoring

In a range of industrial environments and biomedical facilities, accurate measurement, monitoring, and control of humidity, dew-point, pH value, etc., are important.

Humidity Sensor

Low-cost, fast-response humidity sensors are very much needed. Over the years a number of fiber-optic humidity sensors using POF have been reported (Morisawa et al. 2001; Suzuki et al. 1999; Sawada et al. 1989; Klein et al. 1996; Brook et al. 1997). Some organic materials in the family of cellulose, such as a hydroxyethylcellulose (HEC) used by Muto et al. (Muto et al. 2003), swell in a humid atmosphere and show a decrease in refractive index due to attachment of water molecules. Based on this effect, simple optical humidity sensors with fast response and high sensitivity can be fabricated. The sensor makes use of the change of light transmission from leaking to guiding in POF that is induced by index change corresponding to humidity.

Brook et al. presented a new signal-processing method to extend the linear operational range of an optical-fiber humidity sensor (Brook et al. 1997). Their sensor is based on a Nafion-crystal violet complex immobilized on a glass substrate. POFs are employed as light guides to direct light from a tungsten halogen source to the sensor and from the sensor to a CCD-based spectrometer. They demonstrated that, by using artificial neural networks in analyzing generated spectra for varying relative-humidity levels, the linear response range of the fiber-optic relative-humidity sensor can be extended from the 40–55% humidity range previously reported to a nominal range of 40–82%.

Dew-Point Sensor

A dew-point sensor is needed in a number of applications including the monitoring of atmospheric and weather conditions, assessment of moisture content or water activity in crops, food products, or other agricultural products (Hadjiloucas et al. 1994, 2000; Harris and Andrews 1993). A dew-point hygrometer is often used as the reference instrument for the calibration of other types of hygrometers. Typical dew-point sensors consist of a mirror cooled by a small Peltier unit. The current fed to the cooler is adjusted until dew begins to form on the mirror; this is detected by a capacitive sensor (Harris and Andrews 1993) or optical sensor (Regtien 1982). The temperature of the mirror surface is then measured by a temperature transducer.

Most conventional condensation hygrometers have problems such as optical contamination by aerosol particles and relatively slow response times. In many industrial applications, the sensor must be capable of measuring dew points in environments where the air might be humid, static, and hot. Hence, it is required that opto-electronic components are insulated from the sensing head and a bifurcated fiber-optic cable with a random distribution of emitting and receiving fine glass fibers could be used. A proper cut thick POF could be used as a better alternative for this purpose.

A POF dew-point sensor that operates around the point of dew formation has been reported by Hadjiloucas et al. (2000). In their experiment, a POF reflectance sensor, which makes full use of the critical angle of the fibers, is implemented to monitor dew formation on a Peltier-cooled reflector surface. The dew sensor utilizes a novel dual optical fiber reflectance probe. They optimized the cutting of the fibers such that it permits better light coupling and reduces the stand-off distance between the reflector and the detecting fiber. Hence, this technique produced good signal related to the dew formation on the reflector and reduced the optical contamination by the use of a hydrophobic PMMA thin-film reflector and a feedback scheme around the point of dew formation. Under closed-loop operation, the sensor is capable of cycling around the point of dew formation at a frequency of 2.5 Hz.

Furthermore, new polymer optical fibers fabricated by Fujitsu Co. (Japan) using ARTON™ (Japan Synthetic Rubber Co. Ltd) with high thermal stability up to 150 °C, good transparency at the visible wavelength, high flexibility and high tensile strength properties may be implemented in the new humidity sensor. The minimum optical transmission loss spectrum is 0.8 dB/m at 680 nm and 1.2 dB/m at 780 nm (Hadjiloucas et al. 2000; Sukegawa et al. 1994). The thermal expansion of ARTON™ is much lower than that of conventional or low water absorption PMMA and can be considered negligible at high humidity. Hence, one particular advantage of this POF sensor is that it can operate at high humidity, where other sensors are often less reliable.

Oxygen Sensor

POF-based sensors have been developed for the continuous monitoring of gaseous oxygen and dissolved oxygen (Morisawa and Muto 1997; Morisawa et al. 1998b).

Liao et al. reported that the combined excited state phosphorescence lifetimes of alexandrite and platinum tetraphenylporphyrin Pt(TPP) were used to remotely monitor temperature and oxygen concentration simultaneously using a single-optical-fiber probe (Liao et al. 1997). In their experiment, temperature and oxygen concentration was successfully monitored in the physiological range from 15 °C to 45 °C and 0–50% O2 with a precision of 0.24 °C and 0.15% O2 and an accuracy of 0.28 °C and 0.2% O2. The particular sensor used a 750 μm core diameter POF with its core circumferentially coated with a 1 cm length of Pt(TPP). A luminescence signal was produced using a pulsed super bright blue LED light, and a frequency-domain representation of the time-domain phosphorescence decay was obtained by using a fast Fourier transform algorithm.

To improve the sensitivity and stability of the optical fiber sensor used for the continuous monitoring of gaseous oxygen, Vishnoi et al. utilized the quenching phenomena of cladding fluorescence (Vishnoi et al. 1997). In the case reported in Vishnoi et al. (1997), two polymers viz. Poly Cyclohexyl Methylacrylate (PCMA) and Poly (4-Methyl-1-Pentene) (PMP) were selected as the cladding and coating materials, respectively; both being doped with 9,10-Diphenyl Anthracene (DPA). The sensor head was prepared by dip-coating a polymer containing fluorescent dye on a 4.2 cm (Augousti et al. 1990). length of a specially designed ARTONTM POF with a 1 mm core diameter. The sensor was tested under various oxygen concentrations by mixing in nitrogen. The change in fluorescence, excited by a D2 lamp in their case, with oxygen content was recorded using a photo-multiplier tube at $\lambda = 430$ nm. The response was found to be fast, reversible, and reproducible with a recovery time of the order of a few seconds in both the cases (Vishnoi et al. 1997). They demonstrated that the sensor could operate in a wide range of oxygen concentrations, ranging from 0.5% up to 100% of O2. They reported that the fluorescence quenching rate, stability, and response time are dependent upon the cladding polymer used. Two additional polymers: poly(l-menthyl methacrylate) (PMtMA) and poly(4-methyl-1-pentene), having a relatively large oxygen permeability, have also been tested as cladding materials (Morisawa et al. 1998a). They observed better long-term stability with PCMA (Morisawa et al. 1998a).

The measurement of dissolved oxygen in water has been carried out using dye-doped POF by Morisawa et al. (Morisawa and Muto 1997; Morisawa et al. 1998b). This sensor is based on the enhancement of the fluorescence yield by an interaction between tetraphenylporphine (TPP) dye and oxygen molecules. This POF sensor head was fabricated by cladding the core of POF (ARTONTM; nD = 1.51) with poly-4-methyl-1-pentene (PMP; nD = 1.47) doped with TPP. Their experimental results indicated good enhancement of the fluorescence for both oxygen gas and dissolved oxygen in water. The response was found to be linear and stable in the range of 100 ppm-1% for oxygen gas and 1 mg/l-8 mg/l for dissolved oxygen.

Dangerous Gas Sensor

A number of volatile and flammable gases and vapor need to be able to be monitored for safety reasons. Researchers in University of Yamanashi in Japan have been developing various types of POF sensors for ammonia (Shadaram et al. 1997), gasoline (Uchiyama et al. 1997; Morisawa et al. 1999), alkane gases (Muto et al. 1998; Morisawa et al. 2000), and alcohol vapor (Morisawa et al. 2001).

Single-Mode POF Sensors and Applications

So far the POF sensors introduced use multimode POFs and are overwhelmingly intensity-based fiber sensors for simple and low-cost applications. The intensity-based fiber sensor systems are usually compromised in performance and thus their performance and applications are usually limited.

It is well known that single-mode fiber-based fiber sensors have always been a central part of silica optical fiber-based sensors. These are mostly interferometric-based sensors that can achieve very high sensitivity and resolution. These interferometric fiber sensors (often requiring single-mode fibers) provide much superior sensing performance, which intensity-based sensors (usually preferring multimode fibers) can hardly match. In recent years, fiber Bragg grating (FBG)-based sensors have become very popular (Kersey et al. 1997; Rao 1997; Guerrero et al. 1993; Ribeiro et al. 2002).

Optical fiber Bragg gratings are very useful fiber-optic components for optical fiber communications as well as optical fiber sensing. They have excellent wavelength selectivity in their spectral characteristics. They can be used as sensing elements or optical fiber reflectors for measuring temperature, pressure and strain, hydrosound, and industrial process (Othonos and Kalli 1999; Jin 1999; Peng and Chu 2001).

Due to the importance of fiber gratings for a great spectrum of applications, the fabrication of fiber gratings have been one of the main research topics in recent years. Work is dominantly concentrated on silica optical fiber gratings.

In the last few years, photosensitivities in various POFs, including doped and undoped, multimode and single-mode polymer fibers, have been experimentally characterized and evaluated for fiber grating applications (Peng et al. 1999; Scully et al. 2000). In particular, the wavelength dependence of material absorption and photosensitivity has been found to be essential to the fabrication of POF grating. With regard to the fabrication of polymer fiber gratings, remarkable progress has also been made by the researchers of the University of New South Wales in recent years (Peng and Chu 2000; Peng 2002b). They have successfully fabricated POF gratings and their grating quality and performance have gradually improved (Xiong et al. 1999a, b; Liu et al. 2002). The progress has largely benefited from better understanding from recent photosensitivity research and writing process study (Peng et al. 2002; Liu et al. 2003b, c).

POF has a low Young's modulus. The Young's modulus of polymer fiber materials is typically many times less than that of silica glass. For a strain-related optical sensor such as a fiber hydrophone and pressure sensor, POFs are intrinsically many times more sensitive than silica glass fibers. Also for better compatibility, POFs and POF gratings could be advantageous for sensing in liquids and less rigid solid state materials. Importantly, POF gratings have shown superior characteristics to silica FBGs in tenability, stress sensitivity, temperature sensitivity, and large breakdown yield for sensor applications (Xiong et al. 1999a). Hence, it is expected that POF gratings could be significantly advantageous over their better-known counterpart – silica FBGs for some sensing applications (Peng and Chu 2002).

Here we focus on sensor applications of single-mode polymer fiber. We discuss the recent work on photosensitivity in POFs, POF Bragg gratings, as well as related new opportunities for optical fiber sensor applications.

POF for Illumination

POFs are useful for simple light transmission and illumination purposes. A range of polymer materials such as polymethyl-metacrylate (PMMA) and polycarbonate (PC) can be produced by simply using single-screw and conical extrusion. These polymer materials can be readily made into lighting POFs, POF-based textile

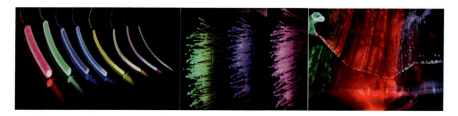

Fig. 16 LED-lit POFs and POF fabrics for illumination and decoration

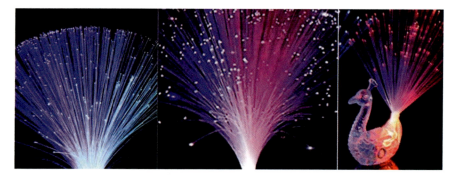

Fig. 17 Guided light using commercial POFs for illumination and decoration purposes

structures, or woven POF fabrics (Harlin et al. 2003; Selm et al. 2010; Huang et al. 2018; Ullah and Shin 2012). The woven POF fabrics can be fabricated as a flexible alternative to lighting elements where POFs are integrated in woven structures by means of handloom, narrow fabric weaving, and Jacquard technology (Harlin et al. 2003).

POFs have also found wide applications for illumination, lighting and decorating of commercial signage, consumer products, industrial, residential, and office areas. The rapid development of LEDs as light sources has provided many more lighting opportunities in critical sites, confined locations, medical procedures, scientific instruments, back-lit displays and signage, and so on. Figures 16 and 17 show a few examples of commercial POFs for illumination and decoration purposes.

There are many commercial companies offering POFs that emit light sideways periodically or uniformly across the entire length, in addition to the conventional way – end-emitting. POFs have been used for interior lighting in the automobile industry, for illuminating swimming pools, and for decorating commercial and residential areas. POFs also open up new possibilities in the textile industry and artistic work. In the form of individual, bundled, or cabled optical fibers, POFs can be either end-lit or side-lit to accommodate various illumination and artistic decorative needs, as shown in Fig. 18.

Fig. 18 Artistic works using commercial POFs

Final Remarks

The research, development, and commercialization of POFs have come through a long way and shown clearly attractive features of POFs that give them great potentials/advantages in niche applications, including (1) data transmission of office, home, automobile networks; (2) environmental and industrial sensing of physical parameters and conditions, gases, chemicals, radiation or high energy particles; (3) lighting and illumination of commercial displays and signage.

References

J.R. Alcala, S.C. Liao, J. Zheng, Real time frequency domain fiberoptic temperature sensor. IEEE Trans. Biomed. Eng. **42**(5), 471–476 (1995)

R.E. Ansorge et al., UA2 collaboration. Nucl. Instr. Meth. **A 265**, 33 (1988). P. Annis et al., CHORUS collaboration. Nucl. Instr. Meth. **A 412**, 19–23 (1998)

L. Archambault, A.S. Beddar, L. Gingras, F. Lacroix, R. Roy, L. Beaulieu, Water-equivalent dosimeter array for small-field external beam radiotherapy. Med. Phys. **34**(5), 1583–1592 (2007)

M. Asai, R. Hirose, A. Kondo, Y. Koike, High-bandwidth graded-index plastic optical fiber by the dopant diffusion coextrusion process. J. Lightwave Technol. **25**, 3062 (2007)

S. Atakaramians, A. Stefani, H. Li, M.S. Habib, J.G. Hayashi, A. Tuniz, X. Tang, et al., Fiber-drawn metamaterial for THz waveguiding and imaging. J. Infrared Millimeter Terahertz Waves **38**(9), 1162–1178 (2017)

A.T. Augousti, J. Mason, K.T.V. Grattan, A simple fibre optic level sensor using fluorescent fibre. Rev. Sci. Instrum. **61**, 3854–3858 (1990)

L. Bansal, S. Khalil, M.A. El-Sherif, Fiber optic neurotoxin sensor. Bioengineering, Proceedings of the Northeast Conference, pp. 221–222, 2002

E. Barni, G. Viscardi, C. D'Ambrosio, T. Gys, H. Leutz, D. Piedrossi, D. Puertolas, S. Tailhardat, U. Gensch, H. Gusten, P. Destruel, T. Shimzu, O. Shinij, M. Garg, A. Menchikov, Development of small diameter scintillating fibres detectors for particle tracking. POF'96, Fifth international conference on plastic optical fibres and applications, Paris, pp. 50–57, 22–24 Oct 1996

R.J. Bartlett, R. Philip-Chandy, P. Eldridge, D.F. Merchant, R. Morgan, P. Scully, Plastic optical fibre sensors and devices. Trans. Inst. Meas. Control. **22**(5), 431–457 (2000)

G. Barton, M.A. van Eijkelenborg, G. Henry, M.C.J. Large, J. Zagari, Fabrication of microstructured polymer optical fibres. Opt. Fiber Technol. **10**(4), 325–335 (2004)

J.K. Beasley, R. Beckerbauer, H.M. Schleinitz, F.C. Wilson, Low attenuation optical fiber of deuterated polymer. U.S. Patent 4,138,194, issued 6 Feb 1979

S. Bian, M.G. Kuzyk, Phase conjugation by low-power continuous-wave degenerate four-wave mixing in nonlinear optical polymer fibers. Appl. Phys. Lett. **84**(6), 858–860 (2004)

T.A. Birks, J.C. Knight, P.S.J. Russell, Endlessly single-mode photonic crystal fibre. Opt. Lett. **22**, 961–963 (1997)

H. Blumenfeld, M. Boudinaud, J.C. Thevenin, Scintillating plastic fibres for calorimetry and tracking devices. IEEE Trans. Nucl. Sci. **33**, 54–56 (1986)

H. Blumenfeld, M. Bourdinaud, J.C. Thevenin, Plastic fibres in high energy physics. Nucl. Inst. Methods **257**, 603–606 (1987)

H. Blumenfeld, M. Bourdinaud, J.C. Thevenin, Characterization of fluorescent plastic optical fibers for x-ray beam detection. Proc. SPIE Int. Soc. Opt. Eng. **1592**, 96–107 (1991)

M. Borecki, J. Kruszewski, Intelligent high resolution sensor for detecting of liquid mediums. Opt. Appl. **31**(4), 691–699 (2001)

D. Bosc, C. Toinen, Full polymer single-mode optical fiber. IEEE Photon. Technol. Lett. **4**(7), 749–750 (1992)

T.E. Brook, M.N. Taib, R. Narayanaswamy, Extending the range of a fibre-optic relative-humidity sensor. Sensors Actuators B Chem. **B39**(1–3 Part 2), 272–276 (1997)

J. Canning, M.A. Hossain, C. Han, L. Chartier, K. Cook, T. Athanaze, Drawing optical fibers from three-dimensional printers. Opt. Lett. **41**(23), 5551–5554 (2016)

R.N. Capps, I.J. Bush, S.T. Lieberman, S.E. Eveland, Evaluation of environmental effects on candidate polymeric materials for underwater optoacoustic sensors. Ind. Eng. Chem. Prod. Res. Dev. **21**(4), 540–545 (1982)

P. Castro, A.A. Sagüés, E.I. Moreno, L. Maldonado, J. Genesca, Characterization of activated titanium solid reference electrodes for corrosion testing of steel in concrete. Corrosion **52**, 609–617 (1996)

N. Cennamo, D. Massarotti, R. Galatus, L. Conte, L. Zeni, Performance comparison of two sensors based on surface plasmon resonance in a plastic optical fiber. Sensors **13**(1), 721–735 (2013)

R.P. Chandy, P.J. Scully, C. Whitworth, D. Fearnside, A novel fibre optic sensor for the determination of environmental turbidity (poster), Annual Meeting of the International Association for Great Lakes Research in Canada, Symposium on Biomarkers and Biomonitors as Indicators of Environmental Change, 18–22 May 1998

R.P. Chandy, P.J. Scully, C. Whitworth, Integrated, multi-angle, low turbidity measurement using fluorescent plastic optical fibre. Proc. SPIE **4185**, 374–377 (2000)

C. Cheng, M. Sansalone, The impact-echo response of concrete plates containing delaminations—numerical experimental and field studies. Mater. Struct. **26**, 274–285 (1993)

K.S. Chiang, V. Rastogi, Ultra-large-core single-mode fiber for optical communications: the segmented cladding fiber. Optical fiber communication conference OFC 2002, paper ThGG6, pp. 620–621, 2002

B. Chiron, Anamorphosor for scintillating plastic optical fiber applications. Proc. SPIE **1592**, 158–164 (1991)

C. Colla, P.C. Das, D.N. McCann, M.C. Forde, Sonic, electromagnetic and impulse radar investigation of stone masonry bridges. NDTE Int. **30**, 249–254 (1998)

K. Cook, J. Canning, S. Leon-Saval, Z. Reid, M.A. Hossain, J.-E. Comatti, Y. Luo, G.-D. Peng, Air-structured optical fiber drawn from a 3D-printed preform. Opt. Lett. **40**(17), 3966–3969 (2015)

K. Cook, G. Balle, J. Canning, L. Chartier, T. Athanaze, M.A. Hossain, C. Han, J.-E. Comatti, Y. Luo, G.-D. Peng, Step-index optical fiber drawn from 3D printed preforms. Opt. Lett. **41**(19), 4554–4557 (2016)

R.F. Cregan, B.J. Mangan, J.C. Knight, et al., Science **285**, 1537–1539 (1999)

A.K. Dattamajumdar, J.A. Myers, A.H. Proctor, D.S. Levine, P.L. Blount, B.J. Reid, R.W. Martin, Novel low-cost fiber-optic colorimetric instrument to rapidly screen pre-malignant esophageal tissue, in *Proc. SPIE*, vol. 3253, (1998), pp. 56–65

W. Daum, W. Hammer, K. Mader, Spectral transmittance of polymer optical fibres before and after accelerated aging. Proceedings POF Conference '97, Kauai, Hawaii, pp. 14–15, 1997

B.J. Deboux, E. Lewis, P.J. Scully, R. Edwards, A novel technique for optical fibre pH sensing based on methylene blue adsorption. J. Lightwave Technol. **13**, 1407–1414 (1995)

P. Destruel, J. Farenc, A. Saad, X. Liop, Luminescent plastic optical fibers in the field of active sensors. First plastic optical fibres and applications conference, Paris, 74–79, 22–23 June 1992

V. Eijkelenborg, A. Martijn, M.C.J. Large, A. Argyros, J. Zagari, S. Manos, N.A. Issa, I. Bassett, et al., Microstructured polymer optical fibre. Opt. Express **9**(7), 319–327 (2001)

C. Emslie, Review polymer optical fibres. J. Mat. Sci. **23**, 2281–2293 (1988)

J. Farenc, R. Mangeret, A. Boulanger, P. Destruel, M. Lescure, Fluorescent plastic optical fiber sensor for the detection of corona discharges in high voltage electrical equipment. Rev. Sci. Instrum. **65**(1), 155–160 (1994)

M. Ferenets, H. Myllymäki, K. Grahn, A. Sipilä, A. Harlin, Manufacturing methods for multi step index plastic optical fiber materials. Autex Research Journal **4**(4), 163–173 (2004)

J.A. Ferguson, D.R. Walt, Optical fibers make sense of chemicals. Photonics Spectra **31**(3) (1997)

Y. Fujimoto, E. Shintaku, S.C. Kim, Structural monitoring for fatigue crack detection and prediction. Proc. Int. Offshore Polar Eng. Conf **4**, 227–235 (1996)

R.M. Gadelrab, The effect of delamination on the natural frequencies of a laminated composite beam. J. Sound Vib. **197**, 283–292 (1996)

R. Galatus, P. Farago, J. Vallés, Optical data transmission with plastic scintillating fibers. In *Fiber Lasers and Glass Photonics: Materials Through Applications*, vol. 10683, p. 106832E. International Society for Optics and Photonics (2018)

G. Giaretta, W. White, M. Wegmuller, T. Onishi, High-speed (11 Gbit/s) data transmission using perfluorinated graded-index polymer optical fibers for short interconnects (<100 m). IEEE Photon. Technol. Lett. **12**(3), 347–349 (2000)

N.D.W. Glossop, An embedded fiber optic sensor for impact damage detection in compositematerials, Ph.D. Thesis, University of Toronto, Institute for Aerospace Studies, 1989

K.T.V. Grattan, D. Kalymnios, Fibre optic temperature measurement – the possibilities with POF. 7th International Plastic Optical Fibres Conference '98, Berlin, pp. 163–170, 5–8 Oct 1998

H. Guerrero, J.L. Escudero, E. Bernabeu, Magnetic-field sensor using plastic optical fiber and polycrystalline CdMnTe. Sensors Actuators A Phys **39**(1), 25–28 (1993)

H. Guerrero, J.L. Escudero, E. Bernabeu, Magneto-optical tachometer for anti-lock braking systems using plastic optical fibre. Meas. Sci. Technol. **5**(5), 607–610 (1994)

S. Hadjiloucas, L.S. Karatzas, D.A. Keating, M.J. Usher, A new plastic optical fibre displacement transducer, in *Trends in Optical Fibre Metrology and Standards*, NATO ASI Series, ed. by O. D. D. Soares (Ed), (1994), pp. 829–830

S. Hadjiloucas, J. Irvine, D.A. Keating, Feedback dew-point sensor utilizing optimally cut plastic optical fibres. Meas. Sci. Technol. **11**(1), 1–10 (2000)

A. Harlin, M. Makinen, A. Vuorivirta, Development of polymeric optical fibre fabrics as illumination elements and textile displays. Autex Res. J **3**(1), 8 (2003)

P.D. Harris, M.K. Andrews, A miniature dew-point hygrometer based on capacitance measurement, in *Sensors VI, Technology, Systems and Applications*, Sensors Series, ed. by K. T. V. Grattan, A. T. Augousti (Eds), (Institute of Physics Publishing, Bristol, 1993), p. 435

F. Hasseinibalim, Fluorescein coated plastic optical fibre humidity sensor, M.Sc. thesis, Liverpool John Moores University, 1997

M.R. Hawks, I. Dajani, C.A. Kutsche, F. Ghebremichael, Modeling and prototyping of polymer fiber based chemical and biological agent sensors. Proc. SPIE **4036**, 115–122 (2000)

H. Hecht, M. Kolling, A low-cost optode-array measuring system based on 1 mm plastic optical fibers – new technique for in situ detection and quantification of pyrite weathering processes. Sensors Actuators B Chem. **81**(1), 76–82 (2001)

R. Hirose, M. Asai, A. Kondo, Y. Koike, Graded-index plastic optical Fiber prepared by the coextrusion process. Appl. Opt. **47**, 4177 (2008)

J. Huang, D. Křemenáková, J. Militký, V. Lédl, Improvement and evenness of the side illuminating effect of side emitting optical fibers by fluorescent polyester fabric. Text. Res. J., 0040517518783344 (2018)

J. Hugenschmidt, Concrete bridge inspection with mobile GPR system. Constr. Build. Mater. **16**, 147–154 (2002)

Introduction to K2K Experiment. http://hep.bu.edu/~superk/k2k.html

N. Ioannides, D. Kalymnios, I.W. Rogers, An optimised plastic optical fibre (POF) displacement sensor. POF'96, Fifth International Conference on Plastic Optical Fibres and Applications, France, pp. 251–255, 22–24 Oct 1996

T. Ishigure, E. Nihei, Y. Koike, High bandwidth (2GHzkm) low loss (56dB/km) GI polymer optical fiber. SPIE **1799**, 67 (1992)

T. Ishigure, E. Nihei, Y. Koike, Graded-index polymer optical fiber for high-speed data communication. Appl. Opt. **33**(19), 4261–4266 (1994)

T. Ishigure, A. Horibe, E. Nihei, Y. Koike, High-bandwidth, high-numerical aperture graded-index polymer optical fiber. J. Lightwave Technol. **13**, 1686 (1995)

T. Ishigure, A. Horibe, E. Nihei, Y. Koike, Optimum refractive-index profile of the graded-index polymer optical fiber, toward gigabit data links. Appl. Opt. **35**, 2048 (1996)

T. Ishigure, M. Sato, E. Nihei, Y. Koike, Thermally stable GIPOF. Proceedings POF Conference '97, Hawaii, pp. 142–143, 22–25 Sept 1997

T. Ishigure, M. Sato, E. Nihei, Y. Koike, Graded-index polymer optical fiber with high thermal stability of bandwidth. Jpn. J. Appl. Phys. **37**, 3986 (1998)

W. Jin, Multiplexed FBG sensors and their applications. Proc. SPIE Int. Soc. Opt. Eng. **3897**, 468–479 (1999)

T. Kaino, Preparation of plastic optical fibers for near – IR region transmission. J. Polym. Sci. A Polym. Chem. **25**(1), 37–46 (1987)

T. Kaino, K. Jinguji, S. Nara, Low loss poly(methyl methacrylate-d8) core optical fibers. Appl. Phys. Lett. **42**(7), 567 (1983)

M. Kamiya, H. Ikeda, Simultaneous transmission of vibration sensor position control data and measured vibration data in opposite directions through single plastic optical fiber, in *IEEE Symposium on Emerging Technologies & Factory Automation, ETFA*, vol. 1, (1996), pp. 82–86

M. Kamiya, H. Ikeda, S. Shinohara, H. Yoshida, Data collection and transmission system for vibration test, in *Conference Record – IAS Annual Meeting (IEEE Industry Applications Society)*, vol. 3, (1998), pp. 1679–1685

A.D. Kersey, M.A. Davis, H.J. Patrick, Fiber Grating Sensors. J. Lightwave Technol. **15**(8), 1442–1462 (1997)

S.S. Kessler, S.M. Spearing, M.J. Atalla, C.E.S. Cesnik, C. Soutis, Structural health monitoring in composite materials using frequency response methods. Proc. SPIE **4336**, 1–11 (2001)

B.J. Kim et al., Tracking performance of the scintillating fiber detector in the K2K experiment. Nucl. Inst. Methods A **497**, 450–466 (2003)

K.F. Klein, S. Riesel, O. Schobert, L. Velte, Three colour sensor for VIS- and UV-A-region. POF '96, Fifth International Conference on Plastic Optical Fibres & Applications, Paris, pp. 213–219, 22–24 Oct 1996

J.C. Knight, T.A. Birks, R.F. Cregan, et al., Electron. Lett. **34**(13), 1347–1348 (1999)

K. Koganezawa, T. Onishi, Progress in perfluorinated GI-POF. Proceedings of the International POF Technical Conference, POF'2000, pp. 19–21, 2000

Y. Koike, Graded-index and single-mode polymer optical fibers. MRS Online Proceedings Library Archive 247, 1992

Y. Koike, A. Inoue, High-speed graded-index plastic optical fibers and their simple interconnects for 4K/8K video transmission. J. Lightwave Technol. **34**(6), 1551–1555 (2016)

Y. Koike, M. Narutomi, Graded-refractive-index-optical material and method for its production, U.S. Patent 5,783,636 (1998)

Y. Koike, N. Tanio, E. Nihei, Y. Ohtsuka, Gradient – index polymer materials and their optical devices. Polym. Eng. Sci. **29**(17), 1200–1204 (1989)

Y. Koike, T. Ishigure, E. Nihei, High-bandwidth graded-index polymer optical fiber. J. Lightwave Technol. **13**, 1475 (1995)

S. Kondo, T. Ishigure, Y. Koike, in *10th Microoptics Conference* (Jena, 2004), p. B–7

N.B. Kosa, Key issues in selecting plastic optical fibers used in novel medical sensors. Proc. SPIE **1592**, 114–121 (1991)

T. Kosaka, N. Takeda, T. Ichiyama, Detection of cracks in FRP by using embedded plastic optical fiber. Mater. Sci. Res. Int. **5**(3), 206–209 (1999)

J. Kruszewski, M. Beblowska, M. Borecki, Fibre optic nephelometer. Proc. SPIE **5064**, 128–131 (2003)

K.S.C. Kuang, W.J. Cantwell, The use of plastic optical fibre sensors for monitoring the dynamic response of fibre composite beams. Meas. Sci. Technol. **14**(6), 736–745 (2003a)

K.S.C. Kuang, W.J. Cantwell, Detection of impact damage in thermoplastic-based glass fibre composites using embedded optical fibre sensors. J Thermoplast. Compos. Mater. **56**, 213–229 (2003b)

K.S.C. Kuang, W.J. Cantwell, The use of plastic optical fibres and shape memory alloys for damage assessment and damping control in composite materials. Meas. Sci. Technol. **14**, 1305–1313 (2003c)

K.S.C. Kuang, R. Kenny, M.P. Whelan, W.J. Cantwell, P.R. Chalker, Residual strain measurement and impact response of optical fibre Bragg grating sensors in fibre metal laminates. Smart Mater. Struct. **10**, 338–346 (2001)

K.S.C. Kuang, W.J. Cantwell, P.J. Scully, Evaluation of novel plastic optical fibre sensor for axial strain and bend measurements. Meas. Sci. Technol. **13**, 1523–1534 (2002a)

K.S.C. Kuang, W.J. Cantwell, P.J. Scully, An evaluation of a novel plastic optical fibre sensor for axial strain and bend measurements. Meas. Sci. Technol. **13**(10), 1523–1534 (2002b)

K.S.C. Kuang, C.W.J. Akmaluddin, C. Thomas, Crack detection and vertical deflection monitoring in concrete beams using plastic optical fibre sensors. Meas. Sci. Technol. **14**(2), 205–216 (2003)

K.S.C. Kuang, S.T. Quek, C.G. Koh, W.J. Cantwell, P.J. Scully, Plastic optical fibre sensors for structural health monitoring: a review of recent progress. J. Sen. **2009** (2009)

H. Kubota, K. Suzuki, S. Kawanishi, M. Kakazawa, M. Tanaka, M. Fujita, Low-loss, 2 km-long photonic crystal fibre with zero GVD in the near IR suitable for picosecond pulse propagation at the 800 nm band, Postdeadline paper CPD3. Conference on Lasers and Electro-Optics CLEO 2001, Baltimore

K. Kurosawa, T. Sawa, H. Sawada, A. Tanaka, N. Wakatsuki, Diagnostic technique for electrical power equipment using fluorescent fiber. Proc. SPIE **1368**, 150–156 (1991)

K. Kurosawa, T. Sowa, K. Tanaka, Y. Yamada, Arc-discharge light detection using fluorescent plastic optical fiber in SF_6 gas. 12th International Conference on Optical Fiber Sensors. Technical Digest. Postconference Edition, pp. 249–252, 1997

M.G. Kuzuk, U.C. Paek, C.W. Dirk, Guest-host polymer fiber for non-linear optics. Appl. Phys. Lett. **59**, 902–904 (1991)

M.F. Laguesse, Optical detection and localization of holes in strips using a fluorescent fiber sensor. IEEE Trans. Instrum. Meas. **39**(1), 242–246 (1990)

M.F. Laguesse, Sensor applications of fluorescent plastic optical fibres. Proceedings of POF'93, pp. 14–19, 1993

A. Leal-Junior, A. Frizera-Neto, C. Marques, M.J. Pontes, Measurement of temperature and relative humidity with polymer optical fiber sensors based on the induced stress-optic effect. Sensors **18**(3), 916 (2018)

S. Lee, U.C. Paek, Y. Chung, Bandwidth enhancement of plastic optical fiber with multi-step core by thermal diffusion. Microw. Opt. Technol. Lett. **39**(2), 129–131 (2003)

C. Lethien, C. Loyez, J.-P. Vilcot, N. Rolland, P.A. Rolland, Exploit the bandwidth capacities of the perfluorinated graded index polymer optical fiber for multi-services distribution. Polymers **3**(3), 1006–1028 (2011)

S.-C. Liao, Z. Xu, J.A. Izatt, J.R. Alcala, Real-time frequency-domain combined temperature and oxygen sensor using a single optical fiber, in *Annual International Conference of the IEEE Engineering in Medicine and Biology – Proceedings*, vol. 5, (1997), pp. 2333–2336

S. Liehr, Polymer fiber sensors for structural and civil engineering applications, in *Handbook of Optical Fibers*, ed. by G. D. Peng (Ed), (Springer, 2018)

B.T. Liu, M.Y. Hsieh, W.C. Chen, J.P. Hsu, Gradient-index polymer optical fiber preparation through a co-extrusion process. Polym. J. **31**, 233 (1999)

H.Y. Liu, G.D. Peng, P.L. Chu, Polymer fiber Bragg gratings with 28dB transmission rejection. IEEE Photon. Technol. Lett. **14**, 935–937 (2002)

H.B. Liu, H.Y. Liu, G.D. Peng, P.L. Chu, Strain and temperature sensor using a combination of polymer and silica fibre Bragg gratings. Opt. Commun. **219**, 139–142 (2003a)

H.B. Liu, H.Y. Liu, G.D. Peng, Different types of polymer fiber Bragg gratings (FBGs) and their strain/thermal properties, in *Optical Memory and Neural Networks special issue on "Holographic Memory and Applications"*, vol. 12, (2003b), p. 147

H.Y. Liu, G.D. Peng, P.L. Chu, Observation of type I and type II Bragg grating behaviour in polymer optical fibre. Opt. Commun. **220**(4–6), 337 (2003c)

Y. Liu, C. Zhang, C. Brackley, S. Yang, Y. Lu, G.D. Peng, Extrusion fabrication of sea-island bicomponent microstructured polymer optical fibre. Proceedings of the 18th international conference on plastic optical fibers, 9–11 Sept 2009, Sydney, Paper 42 (2009)

Y. Liu, C. Zhang, C. Brackley, S. Yang, F. Yang, G.D. Peng, A new method to fabricate sea-island bicomponentmicrostructured polymer optical fibre, in Frontier Photonics and Electronics, ed. by G.D. Peng, J. Canning, Z. He, Proceedings of Joint Workshop on Frontier Photonics and Electronics (ISBN:[978-0-9807815-1-9]), (UNSW, Sydney, 2010), Paper [6–7], pp. 108–109

Y. Luo, Q. Zhang, H. Liu, G.-D. Peng, Gratings fabrication in benzildimethylketal doped photosensitive polymer optical fibers using 355 nm nanosecond pulsed laser. Opt. Lett. **35**(5), 751–753 (2010)

K. Lyytikäinen, J. Zagari, G. Barton, et al., Model. Simul. Mater. Sci. Eng **12**, S255–S265 (2004)

B.D. Macraith, C.M. McDonagh, G. O'Keefe, A.K. McEvoy, T. Butler, F.R. Sheridan, Sol-gel coatings for optical chemical sensors and biosensors. Sensors Actuators **B29**, 51–57 (1995)

C.A.F. Marques, G.D. Peng, D.J. Webb, Highly sensitive liquid level monitoring system utilizing polymer fiber Bragg gratings. Opt. Express **23**(5) (2015). https://doi.org/10.1364/OE.23.006058

D. Merchant, P.J. Scully, R. Edwards, J. Grabowski, Optical fibre fluorescence & toxicity sensor. Sensors Actuators B Chem. **B48**, 476–484 (1998)

D.F. Merchant, P.J. Scully, N.F. Schmitt, Chemical tapering of polymer optical fibre. Sensors Actuators A **76**, 365–371 (1999)

M. Mignanelli, K. Wani, J. Ballato, et al., Opt. Express **15**(10), 6183–6189 (2007)

Mitsubish Rayon Co. UK Patent 1,431,157 (1974); UK Patent 1,499,950 (1974)

M. Morisawa, S. Muto, POF sensors for detecting oxygen in air and water, 7th international plastic optical fibres conference '98, Berlin, pp. 243–44, 5–8 Oct 1997

M. Morisawa, G. Vishnoi, T. Hosaka, S. Muto, Comparative studies on sensitivity and stability of fiber-optic oxygen sensor using several cladding polymers. Jpn. J. Appl. Phys. **37**(8), 4620–4623 (1998a)

M. Morisawa, G. Vishnoi, S. Muto, Optical sensing of dissolved oxygen using dye-doped plastic optical fiber. Trans. Inst. Electr. Eng. Jpn. Part E **118-E**(12), 566–571 (1998b)

M. Morisawa, K. Uchiyama, G. Vishnoi, S. Muto, C.X. Liang, H. Machida, K. Kiso, Improvement of sensitivity in plastic optical fiber gasoline leakage sensors. Proc SPIE **3540**, 175–182 (1999)

M. Morisawa, H. Kozu, Y. Amemiya, S. Muto, Optical sensing of alkane and gasoline vapors using swelling plastic optical fiber. Trans. Inst. Electr. Eng. Jpn. Part E **120-E**(10), 452–457 (2000)

M. Morisawa, Y. Amemiya, H. Kohzu, C.X. Liang, S. Muto, Plastic optical fibre sensor for detecting vapour phase alcohol. Meas. Sci. Technol. **12**(7), 877–881 (2001)

S. Muto, A. Fukasawa, M. Kamimura, Fiber humidity sensor using fluorescent dye doped plastics. Jpn. J. Appl. Phys. **28**, L1065–L1066 (1989)

S. Muto, A. Fukasawa, T. Ogawa, M. Morisawa, H. Ito, Breathing monitor using dye-doped optical fibre. Jpn. J. Appl. Phys. **29**, 1618–1619 (1990)

S. Muto, H. Sato, T. Hosaka, Optical humidity sensor using fluorescent plastic fiber and its application to breathingcondition monitor. Jpn. J. Appl. Phys. **33**, 6060–6064 (1994)

S. Muto, K. Uchiyama, G. Vishnoi, M. Morisawa, C.X. Liang, H. Machida, K. Kiso, Plastic optical fiber sensors for detecting leakage of alkane gases and gasoline vapors. Proc. SPIE **3417**, 61–69 (1998)

S. Muto, O. Suzuki, T. Amano, M. Morisawa, A plastic optical fibre sensor for real-time humidity monitoring. Meas. Sci. Technol. **14**(6), 746–750 (2003)

R. Naka, K. Watanabe, J. Kawarabayashi, A. Uritani, T. Iguchi, N. Hayashi, N. Kojima, T. Yoshida, J. Kaneko, H. Takeuchi, Radiation distribution sensing with normal optical fiber. IEEE Trans. Nucl. Sci. **48**(6 II), 2348–2351 (2001)

E.J. Netto, J.I. Peterson, M. McShane, V. Hampshire, Fiber-optic broad-range pH sensor system for gastric measurements. Sensors Actuators B Chem. **B29**(1–3), 157–163 (1995)

S.R. Nuccio, L. Christen, X. Wu, S. Khaleghi, O. Yilmaz, A.E. Willner, Y. Koike, Transmission of 40 Gb/s DPSK and OOK at 1.55 μm through 100 m of plastic optical fiber, ECOC 2008. 34th European conference on, pp. 1–2. IEEE (2008)

M. Ogita, K. Yoshimura, M.A. Mehta, T. Fujinami, Detection of critical micelle concentration based on the adsorption effect using optical fibers. Jpn. J. Appl. Phys. Pt. 2 Lett. **37**(1 A-B, 15), L85–L87 (1998)

M. Ogita, Y. Nagai, M.A. Mehta, T. Fujinami, Application of the absorption effect of optical fibers for the determination of critical micelle concentration. Sensors Actuators B Chem. **64**(1), 147–151 (2000)

Y. Ohtsuka, I. Nakamoto, Light – focusing plastic rod prepared by photocopolymerization of methacrylic esters with vinyl benzoates. Appl. Phys. Lett. **29**(9), 559–561 (1976)

Y. Ohtsuka, M. Yoshida, Method of manufacturing a transparent light conducting element of synthetic resin having refractive index gradient. U.S. Patent 3,955,015, issued 4 May 1976

T. Oka, H. Fujiwara, K. Takashima, T. Usami, Y. Tsutaka, Development of fiber optic radiation monitor using plastic scintillation fibers. J. Nucl. Sci. Technol. **35**(12), 857–864 (1998)

Y. Okabe, S. Yashiro, T. Kosaka, N. Takeda, Detection of transverse cracks in CFRP composites using embedded fiber Bragg grating sensors. Smart Mater. Struct. **9**, 832–838 (2000)

A. Othonos, K. Kalli, Fiber Bragg Gratings, Fundamentals and applications in Telecommunications and Sensing, Artech House, 1999

G.C. Papanicolaou, A.M. Blanas, A.V. Pournaras, C.D. Stavropoulos, Impact damage and residual strength of FRP composites. Key Eng. Mater. **141–143**, 127–148 (1998)

C.W. Park, Fabrication techniques for plastic optical fibres, in *Polymer Optical Fibers*, ed. by H. S. Nalwa (Ed), (American Scientific Publishers, Los Angeles 2004)

C.W. Park, B.S. Lee, J.K. Walker, W.Y. Choi, A new processing method for the fabrication of cylindrical objects with radially varying properties. Ind. Eng. Chem. Res. **39**(1), 79–83 (2000a)

J.W. Park, C.Y. Ryu, H.K. Kang, C.S. Hong, Detection of buckling and crack growth in the delaminated composites using fiber optic sensor. J. Compos. Mater. **34**, 1602–1623 (2000b)

G.D. Peng, Polymer optical fibre systems and devices, in *A Book Chapter in Science and Technology-Advancing into the New Millennium*, ed. by J. Sun, L. Sun, J. Jin, A. Yu, Q. Zhang, P. Zhang (People's Education Press, Beijing, 1999), pp. 337–351 and 573–578, ISBN 710-7-13208-3

G.D. Peng, Prospects of POF and Grating for Sensing, invited paper, presented at the 15th International Conference of Optical Fiber Sensors, Portland, 6–10 May 2002a

G.D. Peng, Polymer Optical Fibre Bragg Gratings and Their Sensor Applications, invited paper, the Symposium on Optoelectronic Materials and Technology, in the Information Age (A-3) at the 104th Annual meeting of the American Ceramic Society, St. Louis, 28 April–1 May, 2002b

G.D. Peng, P.L. Chu, Polymer optical fiber photosensitivities and highly tunable fiber gratings. Fiber Integr. Opt. **19**(4), 277–293 (2000)

Peng G. D. and Chu P. L., Chapter 9: Optical fibre hydrophone systems, in *Fiber Optic Sensors*, ed. by F. Yu, S. Yin (Eds), (Marcel Dekker, Inc., New York, 2001), pp. 417–447

G.D. Peng, P.L. Chu, Polymer optical fiber sensing. Proc. SPIE **4929**, 303–311 (2002)

G. D. Peng and P. L. Chu, Chapter 4: Polymer optical fiber gratings, in *Polymer Optical Fibers*, ed. by H. S. Nalwa (Ed), (American Scientific Publishers, CA, USA, 2004), pp. 51–71. ISBN: 1-5888-3012-8

G.D. Peng, A.D. Li, Laser activity in polymer optical fibres doped with new organic materials, Proceedings of Progress in Electromagnetics Research Symposium, PIERS'2001, Osaka, 497, July 2001

G.D. Peng, A. Latif, P.K. Chu, R.A. Chaplin, Polymeric guest–host system for non-linear optical fiber. Proceedings, in *IEEE Non-Linear Optics, Material, Fundamentals and Applications*, (1994), pp. 86–88

G.D. Peng, P.L. Chu, L. Xia, R.A. Chaplin, Fabrication and characterisation of polymer optical fibres. J. IREEA **15**(3), 289–296 (1995)

G.D. Peng, P.L. Chu, Z. Xiong, T. Whitbread, R.P. Chaplin, Dye-doped polymer optical fibre for broadband optical amplification. IEEE/OSA J. Lightwave Technol. **14**(10), 2215–2223 (1996a)

G.D. Peng, P.L. Chu, Z. Xiong, T. Whitbread, R.P. Chaplin, Broadband tunable optical amplification in Rhodamine B-doped step-index polymer optical fibre. Opt. Commun. **129**, 353–357 (1996b)

G.D. Peng, Z. Xiong, P.L. Chu, Fluorescence decay and recovery in organic dye doped polymer optical fibres. J. Lightwave Technol. **16**, 2365–2371 (1998)

G.D. Peng, Z. Xiong, P.L. Chu, Photosensitivity and grating in dye-doped polymer optical fibres. Opt. Fiber Technol. **5**, 242–251 (1999)

G.D. Peng, P. Ji, P.L. Chu, Electro-optic and polarisation effects in polymer optical fibres, Proceedings of Progress in Electromagnetics Research Symposium, PIERS'2001, Osaka, p. 498, July 2001a

G.D. Peng, P. Ji, P.L. Chu, Electro-optic polymer optical fibers and their device applications. Proc. SPIE **4459**, 101–117 (2001b)

G.D. Peng, H.Y. Liu, P.L. Chu, Dynamics and threshold behaviour in polymer fibre Bragg grating creation. Proc SPIE **4803**, 164–178 (2002)

T. Pereira, S. Rusinkiewicz, W. Matusik, Computational Light Routing: 3D Printed Optical Fibers for Sensing and Display. ACM Trans. Graph. (TOG) TOG Homepage Arch. **33**(3), 24 (2014)

K. Peters, Polymer optical fiber sensors—a review. Smart Mater. Struct. **20**(1), 013002 (2010)

R. Philip-Chandy, R. Morgan, P.J. Scully, The measurement, instrumentation and sensors handbook, in *Drag Force Flowmeters Section 5,3,10*, (CRC Press LLC, Boca Raton, 1999)

R. Philip-Chandy, P.J. Scully, P. Eldridge, H.J. Kadim, M.G. Grapin, M.G. Jonca, M.G. D'Ambrosio, F. Colin, Optical fiber sensor for biofilm measurement using intensity modulation and image analysis. IEEE J. Sel. Top. Quantum Electron. **6**(5), 764–772 (2000a)

R. Philip-Chandy, P.J. Scully, R. Morgan, Design, development and performance characteristics of a fibre optic drag-force flow sensor. Meas. Sci. Technol. **11**(3), N31–N35 (2000b)

P. Polishuk, Plastic optical fibers branch out. IEEE Commun. Mag. **44**(9), 140–148 (2006)

A. Polley, K. Balemarthy, S.E. Ralph, Mode coupling: why POF supports 40 Gbps. Conference on Lasers and Electro-Optics, p. CWM5. Optical Society of America, 2007

A. Pospori, C.A.F. Marques, D. Sáez-Rodríguez, K. Nielsen, O. Bang, D.J. Webb, Annealing and etching effects on strain and stress sensitivity of polymer optical fibre Bragg grating sensors. 25th International Conference on Plastic Optical Fibres, pp. 260–263. University of Aston in Birmingham (2017)

J.K. Ranka, R.S. Windeler, A.J. Stentz, Visible continuum generation in air-silica microstructure optical fibers with anomalous dispersion at 800 nm. Opt. Lett. **25**, 25–27 (2000)

Y.J. Rao, In-fiber Bragg grating sensor. Meas. Sci. Technol. **8**, 355–375 (1997)

A. Raza, A.T. Augousti, Optical measurement of respiration rates. Proceedings of the Seventh Conference on Sensors and their Applications, Dublin, pp. 325–330 (1995)

P.P.L. Regtien, Solid state humidity sensors. Sensors Actuators **2**(85) (1982)

R.M. Ribeiro, J.L.P. Canedo, M.M. Werneck, L.R. Kawase, An evanescent-coupling plastic optical fibre refractometer and absorptionmeter based on surface light scattering. Sensors Actuators A Phys. **101**(1–2), 69–76 (2002)

D. Sáez-Rodríguez, K. Nielsen, H.K. Rasmussen, O. Bang, D.J. Webb, Highly photosensitive polymethyl methacrylate microstructured polymer optical fiber with doped core. Opt. Lett. **38**(19), 3769–3772 (2013)

H. Sawada, A. Tanaka, N. Wakatsuki, Plastic optical fiber doped with organic fluorescent materials. Fujitsu Sci. Tech. J. **25**(2), 163–116 (1989)

H.M. Schleinitz, Proceedings of the International Wire and Cable Symposium, p. 352, (1977)

N.F. Schmitt, UV photo-induced grating structures on polymer optical fibres. PhD thesis, Liverpool John Moores University, 1999

P.J. Scully, R. Holmes, G.R. Jones, Optical fibre-based goniometer for sensing patient position and movement within a magnetic resonance scanner using chromatic modulation. J. Med. Eng. Technol. **17**, 1–8 (1993)

P.J. Scully et al., UV laser photo-induced refractive index changes in poly-methyl-meth-acrylate and plastic optical fibres for application as sensors and devices. Proc. SPIE Int. Soc. Opt. Eng. **4185**, 854–857 (2000)

B. Selm, E.A. Gürel, M. Rothmaier, R.M. Rossi, L.J. Scherer, Polymeric optical fiber fabrics for illumination and sensorial applications in textiles. J. Intell. Mater. Syst. Struct. **21**(11), 1061–1071 (2010)

M. Shadaram, J. Martinez, F. Garcia, D. Tavares, Sensing ammonia with ferrocene-based polymer coated tapered optical fibers. Fiber Integr. Opt. **16**(1), 115–122 (1997)

M. Shigeishi, S. Colombo, K.J. Broughton, H. Rutledge, A.J. Batchelor, M.C. Forde, Acoustic emission to assess and monitor the integrity of bridges. Construction Building Mater **15**, 35–49 (2001)

I.-S. Sohn, C.-W. Park, Diffusion-assisted coextrusion process for the fabrication of graded-index plastic optical fibers. Ind. Eng. Chem. Res. **40**, 3740 (2001)

I.-S. Sohn, C.-W. Park, Preparation of graded-index plastic optical fibers by the diffusion-assisted coextrusion process. Ind. Eng. Chem. Res. **41**, 2418 (2002)

U. Steiger, Sensor properties and applications of POF. 7th International Plastic Optical Fibres Conference, Berlin, pp. 171–177, 1998

T. Sukegawa, M. Hirano, M. Tomatsu, T. Otsuki, H. Shinohara, Y. Hara, A. Tanaka, New polymer optical fiber for high temperature use, in *3rd International Conference on Plastic Optical Fibres and Applications (POF '94)*, (The European Institute of Communications and Networks, Yokohama, 1994), p. 92

H. Suzuki, Y. Yamada, K. Matsuo, T. Sowa, K. Yamashita, K. Kurosawa, K. Tanaka, Development of fault location system for GIS using fluorescent plastic optical fiber. Trans. Inst. Electr. Eng. Jpn Part B **119-B**(7), 840–846 (1999)

A. Suzuki et al., K2K collaboration. Nucl. Instr. Meth. A **453**, 165 (2000)

A. Tagaya, Y. Koike, T. Kinoshita, E. Nihei, T. Yamamoto, K. Sasaki, Polymer optical fiber amplifier. Appl. Phys. Lett. **63**(7), 883–884 (1993)

A. Tagaya, Y. Koike, E. Nikei, S. Teramoto, K. Fujii, T. Yamamoto, K. Sasaki, Basic performance of an organic dye-doped polymer optical fiber amplifier. Appl. Opt. **34**, 988–992 (1995a)

T.S. Tagaya, T. Yamamoto, K. Fujii, E. Nihei, K. Sasaki, Y. Koike, Theoretical and experimental investigation of Rhodamine B -doped polymer optical fibre amplifiers. IEEE J. Quantum Electron. **31**(12) (1995b)

N. Takeda, Characterization of microscopic damage in composite laminates and real-time monitoring by embedded optical fiber sensors. Int. J. Fatigue **24**(2–4), 281–289 (2002)

N. Takeda, T. Kosaka, T. Ichiyama, Detection of transverse cracks by embedded plastic optical fiber in FRP laminates. Proc. SPIE **3670**, 248–255 (1999)

Y. Takezawa, Y. Itoh, M. Shimodera, H. Miya, Development of a portable diagnostic apparatus for coil insulators in low-voltage induction motors. IEEE Trans. Dielectr. Electr. Insul. **5**(2), 290–295 (1998)

W. Talataisong, R. Ismaeel, T.H.R. Marques, S.A. Mousavi, M. Beresna, M.A. Gouveia, S.R. Sandoghchi, T. Lee, C.M.B. Cordeiro, G. Brambilla, Mid-IR hollow-core microstructured fiber drawn from a 3D printed PETG preform. Sci. Rep. **8**(1), 8113 (2018)

A. Tanaka, H. Sawada, T. Takoshima, N. Wakatsuki, New plastic optical fiber using polycarbonate core and fluorescence-doped fiber for high temperature use. Fiber Integr. Opt. **7**(2), 139–158 (1988)

N. Tanio, Y. Koike, What is the most transparent polymer? J. Polym **32**, 43–50 (2000)

J.C. Thevenin, L.R. Allemand, J. Calvet, J.C. Cavan, B. Chiron, F. Gauthier, Sintillating and fluorescent plastic optical fibers for sensors applications. Proc. SPIE **514**, 133–141 (1984)

P.M. Toal, L.J. Holmes, R.X. Rodriguez, E.D. Wetzel, Microstructured monofilament via thermal drawing of additively manufactured preforms. Addit. Manuf. **16**, 12–23 (2017)

M. Towrie et al., UV laser photo-induced refractive index changes in poly-methyl-meth-acrylate and plastic optical fibres for application as sensors and devices. Proc. SPIE **4185**, 854–857 (2000)

K. Uchiyama, G. Vishnoi, M. Moriswa, S. Muto, Plastic optical fibre gasoline leakage sensor. Proc. POF Conf. '97, Hawaii **22–25**, 140–141 (1997)

I. Ullah, S.Y. Shin, Development of optical fiber-based daylighting system with uniform illumination. J. Opt. Soc. Korea **16**(3), 247–255 (2012)

S.H.D. Valdes, C. Soutis, Delamination detection in composite laminates from variations of their modal characteristics. J. Sound Vib. **228**, 1–9 (1999)

F.G.H. Van Duijnhoven, C.W.M. Bastiaansen, Monomers and polymers in a centrifugal field, a new method to produce refractive-index gradients in polymers. Appl. Opt. **38**, 1008 (1999)

M.A. van Eijkelenborg, A. Argyros, G. Barton, I.M. Bassett, M. Fellew, G. Henry, N.A. Issa, et al., Recent progress in microstructured polymer optical fibre fabrication and characterisation. Opt. Fiber Technol. **9**(4), 199–209 (2003)

C. Vazquez, J. Garcinuno, J.M.S. Pena, A.B. Gonzalo, Multi-sensor system for level measurements with optical fibres. IECON Proc. Ind. Electron. Conf **4**, 2657–2662 (2002)

G. Vishnoi, M. Morisawa, T. Mizukami, K. Uchiyama, S. Muto, Studies on the improvement of sensitivity and stability of fiber optic oxygen sensing based on cladding fluorescence. Proc. SPIE **3105**, 114–121 (1997)

S.R. Waite, Use of embedded optical fibre for early fatigue damage detection in composite materials. Composites **21**, 148–154 (1990)

D.J. Webb, Fibre Bragg grating sensors in polymer optical fibres. Meas. Sci. Technol. **26**(9), 092004 (2015)

D.J. Welker, J. Tostenrude, D.W. Garvey, B.K. Canfield, M.G. Kuzyk, Fabrication and characterization of single-mode electro-optic polymer optical fiber. Opt. Lett. **23**(23), 1826–1828 (1998)

M. Whelan, Damage detection in vibrating composite panels using embedded fibre optic sensors and pulsed-DPSI key. Eng. Mater. **167**(/168), 122–131 (1999)

W.R. White, L.L. Blyer, R. Ratnagiri, M. Park, J.J. Refi, Perfluorinated POF, out of the lab, into the real world. Proceedings of the 12th International Conference on Polymer Optical Fiber (POF'2003), Seattle, pp. 16–19, Sept 2003

K. Willis, E. Brockmeyer, S. Hudson, I. Poupyrev, Printed optics: 3D printing of embedded optical elements for interactive devices, UIST '12. Proceedings of the 25th annual ACM symposium on User Interface Software and Technology, pp. 589–598, 2012

G. Woyessa, A. Fasano, C. Markos, Microstructured polymer optical fiber gratings and sensors, in *Handbook of Optical Fibers*, ed. by G. D. Peng (Ed), (Springer, Singapore, 2018)

Z. Xiong, G.D. Peng, B. Wu, P.L. Chu, Highly tunable Bragg gratings in single mode polymer optical fibers. IEEE Photon. Tech. Lett. **11**(3), 352–354 (1999a)

Z. Xiong, G.D. Peng, B. Wu, P.L. Chu, 73 nm wavelength tuning in polymer optical fiber Bragg gratings. The 24th Australian Conference on Optical Fibre Technology (ACOFT'99), Sydney, pp. 135–138, July 1999b

S. Yamakawa, Plastic optical fiber chemical sensor with pencil-shaped distal tip fluorescence probe. Proceedings POF '97 Conference, Kauai, pp. 109–110, 1997

A. Yeung, K.S. Chiang, V. Rastogi, P.L. Chu, G.D. Peng, Experimental demonstration of single-mode operation of large-core segmented cladding fiber. Optical Fiber Communication Conference in Los Angeles. 2004

W.J. Yoo, S.H. Shin, D. Jeon, S. Hong, S.G. Kim, H.I. Sim, K.W. Jang, S. Cho, B. Lee, Simultaneous measurements of pure scintillation and Cerenkov signals in an integrated fiber-optic dosimeter for electron beam therapy dosimetry. Opt. Express **21**, 27770–27779 (2013)

W.J. Yoo, S.H. Shin, D.E. Lee, K.W. Jang, S. Cho, B. Lee, Development of a small-sized, flexible, and insertable fiber-optic radiation sensor for gamma-ray spectroscopy. Sensors (Basel, Switzerland) **15**(9), 21265–21279 (2015)

W.J. Yoo, Sang Hun Shin, Dayeong Jeon, Seunghan Hong, Seon Geun Kim, Hyeok In Sim, Kyoung Won Jang, Seunghyun Cho, and Bongsoo Lee, Simultaneous measurements of pure scintillation and Cerenkov signals in an integrated fiber-optic dosimeter for electron beam therapy dosimetry, Optics Express **21**(23), 27770–27779 (2013)

J.M. Yu, X.M. Tao, H.Y. Tam, Trans-4-stilbenemethanol-doped photosensitive polymer fibers and gratings. Opt. Lett. **29**(2), 156–158 (2004)

J. Zagari, A. Argyros, N.A. Issa, et al., Opt. Lett. **29**(8), 818–820 (2004). A. Argyros, M.A. van Eijkelenborg, M.C.J. Large et al., Opt. Lett. **31**(2), 172–174 (2006)

F.H. Zhang, P.J. Scully, E. Lewis, A novel optical fibre sensor for turbidity measurement. 4th Divisional conference, Appl. Opt. Optoelectron. 370–373 (1996)

F.H. Zhang, E. Lewis, P.J. Scully, Optical fibre sensor for particle concentration measurement in water systems based on inter-fibre light coupling between polymer optical fibres. Trans. Inst. Meas. Control. **22**(5), 413–430 (2000)

Y. Zhang, K. Li, L. Wang, et al., Opt. Express **14**(12), 5541–5547 (2006)

W. Zhang, D.J. Webb, G.-D. Peng, Enhancing the sensitivity of poly(methyl methacrylate) based optical fiber Bragg grating temperature sensors. Opt. Lett. **40**(17), 4046–4049 (2015). https://doi.org/10.1364/OL.40.004046

Q. Zhou, M.B. Tabacco, K.W. Rosenblum, Development of chemical sensors using plastic optical fiber. Proc. SPIE Int. Soc. Opt. Eng. **1592**, 108–113 (1991)

O. Ziemann, J. Krauser, P.E. Zamzow, W. Daum, *POF Handbook* (Springer, 2008)

M.G. Zubel, A. Fasano, G. Woyessa, K. Sugden, H.K. Rasmussen, O. Bang. 3D-printed PMMA Preform for Hollow-core POF Drawing. The 25th international conference on plastic optical fibers 2016, 2016

J. Zubia, O. Aresti, J. Arrue, M. Lopez-Amo, Barrier sensor based on plastic optical fiber to determine the wind speed at a wind generator. IEEE J. Sel. Top. Quantum Electron. **6**(5), 773–779 (2000)

Optical Fibers in Terahertz Domain

26

Georges Humbert

Contents

Abstract

The terahertz (THz) frequency range spans between the microwave and the photonic domains. For more than 20 years, it is experiencing growing expansion justified by the new properties offered in telecommunication, spectroscopy, and imaging technologies, enabling numerous applications for today's society needs. Similarly as the optical fibers in the optical domain, THz fibers are key components for realizing complex, compact, and robust systems that are required by THz applications. Nevertheless, the developments of THz fibers are hindered by the strong degradations of material properties at THz frequencies. These constraints require to investigate and to develop THz fibers with innovative and

G. Humbert (✉)
XLIM Research Institute, UMR 7252 CNRS, University of Limoges, Limoges, France
e-mail: georges.humbert@xlim.fr

© Springer Nature Singapore Pte Ltd. 2019   1019
G.-D. Peng (ed.), *Handbook of Optical Fibers*,
https://doi.org/10.1007/978-981-10-7087-7_33

disruptive designs, which make the development of THz fibers challenging and very stimulating. Numerous strategies are inspired from the recent innovations in the field of specialty optical fibers. Since dry air is certainly the most favorable medium to propagate THz radiations. Two major approaches have been investigated. The first one is based on the propagation of THz waves into a fiber with a design that favors a large portion of evanescent field in air. The second way of beating these limits is by confining the THz waves in a hollow-core fiber with the help of reflectors in the fiber cladding. The main recent developments of THz fibers are presented in this chapter. The guiding mechanism of each THz fiber is detailed, in addition to a presentation of the recent experimental demonstrations and analyses of their drawbacks and advantageous properties.

Keywords

THz fiber · THz waveguide · Hollow-core fiber · specialty optical fibers · Photonic crystal fibers

Introduction

The terahertz domain refers to the electromagnetic wave spectrum form 100 GHz to 10 THz lying between microwave and optical spectra. It corresponds to wavelengths from 3 mm to 30 μm and energy levels from 0.4 to 41 meV. This domain is experiencing growing interest justified by the unique properties of THz wave – matter interactions. THz waves are weakly absorbed by nonmetallic and nonpolar materials (such as textiles, ceramics, plastics, semiconductors, etc.), while they are strongly absorbed by polar molecules (e.g., water molecule) and reflected by metallic materials. As a result, THz spectroscopy is a powerful technique for analyzing materials and biological or chemical substances and detecting gases and atmospheric pollutions. THz waves open new possibilities for imaging technologies since materials that are opaque in the optical domain are rather transparent. This offers numerous opportunities for nondestructive imaging and testing (mail checks, packaging inspections, plastic parts diagnosis, painting quality, etc.), in a large panel of applications such as in-line control of pharmaceutical products (drugs), detection of hidden weapons or explosives, material integrity control of composites materials or coatings, rapid fault detection in packaged electronic circuits, art restoration of paintings or manuscripts, etc. Furthermore, in contrast with X-rays that can ionize materials, the low energy of the THz waves makes them very attractive for noninvasive biomedical imaging without health issues (damages of living tissues, molecules). The penetration depth of this technique is nevertheless limited by the absorption of THz waves by water molecules and by the imaging resolution that is larger than the one obtain with X-rays, due to the larger size of THz waves. Besides, this frequency domain is envisioned as a key technology to satisfy the increasing demand for higher speed wireless communications, with potentially terabit-per-second link capacities.

The THz domain was known as the "THz gap," where technologies from microwaves or optical domains are not efficient, mainly due to the degradations of material properties. The last two decades have seen an impressive and unprecedented advance in the field of THz science and technology that was mainly driven by the development of new and higher-power sources and more efficient detectors. Even if THz devices, THz sensing, and imaging systems are commercially available, and if THz wireless communications have been demonstrated, intensive efforts are still focused on the development of the building blocks (sources, detectors, and components).

In this context, similarly as the optical fibers, THz fibers may play a significant role in the expansion of the THz domain by delivering THz waves in specific location, by enabling the realization of complex, compact, and robust THz systems, by improving sensing performances of THz systems, and by enabling advanced THz waves – matter interactions.

However, owing to the finite conductivity of metals yielding high ohmic losses at THz frequencies, the waveguide technologies developed in microwaves domain are not suitable (McGowan et al. 1999). The degradation of the transparency of most dielectric materials in the THz domain is another limiting factor of the realization of THz fibers based on standard designs of optical fibers. Numerous strategies for developing THz fibers are therefore inspired from the recent innovations in the field of specialty optical fibers for overcoming these constraints. After presenting the main constraints to develop THz fibers in section "Constraints and Challenges for Developing THz Fibers," different strategies of solid-core THz fibers will be detailed in section "Solid-Core THz Fibers." Then, the development of hollow-core THz fibers based on different fiber designs and guiding mechanisms will be presented in section "Hollow-Core THz Fibers."

Constraints and Challenges for Developing THz Fibers

Light is guided in optical fibers with an extremely low attenuation coefficient (<0.5 dB/km) due to the high transparency of silica glass in the optical domain. Silica glass has also excellent mechanical and thermo-optic properties, and it is largely used for fabricating optical devices and optical fibers with complex designs. However, in the THz domain, silica glass is no more the best material. Its large absorption coefficient limits its use for fabricating THz fibers or even THz devices for free-space propagations such as lenses. The absorption coefficient of silica increases from $\alpha = 155$ dB/m at 0.5 THz to 2850 dB/m at 2 THz (with $\alpha = 850$ dB/m at 1 THz and 1580 dB/m at 1.5 THz). These values are calculated from an averaging of the measures reported in Naftaly and Miles (2007). Furthermore as shown in Fig. 1a, the absorption coefficient of other common silicate glasses increases dramatically at higher frequencies (above 0.5 THz) with values much larger than the absorption coefficient of silica. It is worth to notify that the refractive index (real part) of these glasses is higher in the THz domain than in optical domain, and it is almost constant over the measured frequency span

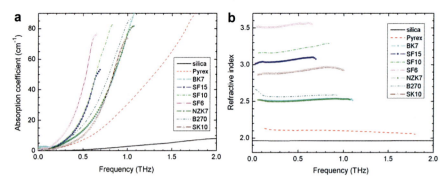

Fig. 1 (**a**) Absorption coefficient and (**b**) refractive index (real part) of different silicate glasses measured with THz time domain transmission spectroscopy method. (Reproduced from Naftaly and Miles (2007), with the permission of AIP Publishing)

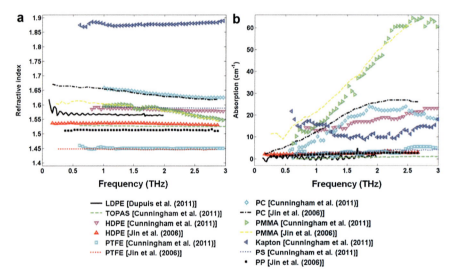

Fig. 2 (**a**) Refractive index and (**b**) absorption coefficient of common polymers. The different polymers are low-density polyethylene (LDPE), cyclic olefin copolymer (TOPAS®), high-density polyethylene (HDPE), polytetrafluoroethylene (PTFE), polycarbonate (PC), polymethyl methacrylate (PMMA), polyimide (Kapton®), polystyrene (PS), and polypropylene (PP). (Reproduced from Ung et al. (2011b), with the permission of OSA Publishing)

(Cf. Fig. 1b). For example, the refractive index of silica varies only from $n = 1.962$ at 0.5 THz to 1.965 at 2 THz (Naftaly and Miles 2007).

Polymer materials exhibit more interesting properties than silicate glasses for propagating THz waves. The real part of the refractive index of most common polymers is almost constant over the frequency span (0.1–3 THz, Cf. Fig. 2a), and the absorption coefficient is rather low (Cf. Fig. 2b). Among the different polymer materials, cyclic olefin copolymer (COC), call TOPAS (tradename), has the lowest material absorption (Nielsen et al. 2009; Cunningham et al. 2011). Due to the amorphous structure of the nonpolar COC, the absorption coefficient of TOPAS is approximately 100 times smaller than the one of PMMA, and the index of refraction has a constant value of n $= 1.5258 \pm 2 \cdot 10^{-4}$ in the 0.1–1.5 THz range (Nielsen et al. 2009).

Nevertheless, the absorption coefficients of polymers are still too large for developing long length THz fibers. Indeed, the absorption coefficient of TOPAS rises from about 35 dB/m at 0.5 THz up to 190 dB/m at 1.5 THz (with $\alpha_{mat} = 100$ dB/m at 1 THz) (Nielsen et al. 2009).

Therefore, the first challenge to develop THz fibers is finding a material with low absorption losses or a design that minimizes the confinement of THz waves in the material, by maximizing the propagation in dry air (or other dry gases). Hitherto, two major approaches are investigated. They consist in propagating a large portion of the evanescent waves in air or in confining and guiding THz waves within a hollow-core fiber. Both strategies require to develop new designs that are mostly inspired from innovative optical fiber designs.

The second challenge is inherent to the size of the wavelength at THz frequencies and the concept of THz fibers (that is associated with the notion of flexibility and bending capabilities). In THz domain, the size of the wavelength is about 300 times larger than in the optical domain. The translation of optical fiber designs to THz domain follows this order of magnitude yielding much large fiber diameters. For example, the translation of a standard telecom optical fiber (with a core diameter of 8 μm and an external diameter of 125 μm) to the frequency of 1 THz leads to a fiber core diameter and an external diameter of 2 and 25 mm, respectively. Even if this example is not possible due to the large absorption of silica glass at 1 THz, it clearly emphasizes the difficulty to develop THz fibers with mechanical flexibilities. The development of THz fibers requires therefore original designs with an external diameter limited to few millimeters for enabling fiber bending.

Furthermore, it is worth to notify that the necessity to use a polymer material for fabricating THz fibers is an additional difficulty. Even if polymer fibers with very complex designs have been realized in the optical domain, polymer materials are more difficult to use for fabricating fibers than the silica glass who requires less drastic fabrication processes (expected the high temperature requested for melting the glass). In comparison with polymers, silica has a very broad working-temperature range with a viscosity low enough for shaping it.

In this context, the development of THz fibers appears very stimulating because of the many underlying applications and of the challenges of designing and fabricating innovative fibers that fulfill the aforementioned constraints.

Solid-Core THz Fibers

Light is basically confined and guided in the core of an optical fiber by internal reflections at the interface between the core and the fiber cladding. These reflections are totals within a limited range of incident angles of light rays when the refractive index of the core is higher than the one of the cladding. In the THz domain, the large absorption coefficient of most dielectric materials requires to propagate THz waves mostly in air with unusual guiding regimes and to investigate and develop innovative fibers. This section is dedicated to THz fibers based on total internal reflection guiding mechanism. Two major approaches are investigated for developing THz fibers and associated components and systems. The first one is based on the propagation of a large portion of evanescent field in air, as the sub-wavelength diameter fibers presented in section "Sub-wavelength Diameter Fibers" and the porous fibers presented in section "Porous Fibers." The second strategy consists in exploiting the remarkable properties of photonic crystal fibers (PCF), initially introduced in the optical domain. The developments of THz PCF are presented in section "Solid-Core Photonic Crystal Fibers."

Sub-wavelength Diameter Fibers

A sub-wavelength diameter fiber is simply a rod or a fiber of a dielectric material with a diameter smaller than the wavelength of the propagated wave. The refractive index of the dielectric material is higher than the surrounding medium (usually air) for enabling wave guiding by total internal reflection (TIR) mechanism. The diameter is smaller than the wavelength for reducing the attenuation coefficient of the guided wave by propagating it mostly outside the fiber, in air, which one has a much lower absorption coefficient than the fiber material.

It is basically the simplest waveguide structure, i.e., a core with an infinite cladding. Theoretical analyses can be found in most handbooks on optical fibers (e.g., Snyder and Love 1983). Its properties are governed by the normalized frequency (V) following this expression:

$$V = \pi \frac{d}{\lambda} \sqrt{n_1^2 - n_2^2} \tag{1}$$

with d and n_1 the diameter and the refractive index of the fiber, λ the wavelength, and n_2 the refractive index of the surrounding medium. It is well known that single-mode condition is achieved for V < 2.405, below the cutoff frequency of the first higher-order mode (TE_{01} mode). In contrast with higher-order modes, the fundamental mode HE_{11} does not have a cutoff frequency. It is guided by TIR mechanism even for small value of the normalized frequency, as in the case of a very small diameter (d). In this weak guiding domain, most of the wave intensity is evanescent in air, and its effective refractive index (real part) is very close to unity (refractive index of air). The real part of the effective refractive index of a mode is defined as $n_{eff} = Re(\beta)$ c/ω, with β the propagation constant of the mode and $\omega = 2\pi f$. As shown in Fig. 3a for the case of a sub-wavelength diameter fiber (with refractive index of $n_1 = 1.5$) in air, for $d/\lambda < 0.48$ (V < 1.686), more than half the power is in air, and it is higher than 90% for $d/\lambda < 0.33$ (V < 1.159), 99% for $d/\lambda < 0.25$ (V < 0.878), and 99.9% for $d/\lambda < 0.22$ (V < 0.773). The augmentation of the fraction of evanescent power

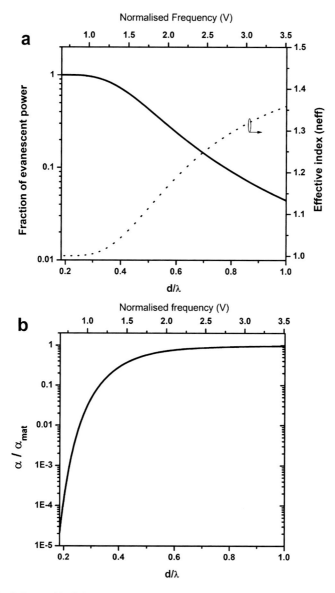

Fig. 3 Calculations with finite element method of the properties of the fundamental mode propagated in a rod (with a diameter (d) and a refractive index of 1.5) immersed in air. (**a**) Evolution of the fraction of the evanescent power and effective index of the fundamental mode. (**b**) Evolution of the attenuation coefficient (α) of the fundamental mode divided by the material absorption coefficient of the rod (α_{mat})

reduces the contribution of the absorption coefficient of the fiber material on the attenuation coefficient of the propagated wave, i.e., the ratio α/α_{mat}, which could be reduced by four orders of magnitude for $d/\lambda < 0.22$ (Cf. Fig. 3b).

These properties have been exploited for propagating THz waves on polyethylene (PE) fiber. An attenuation coefficient lower than 0.01 cm^{-1} ($\backslash 4$ dB/m) in the frequency range around 0.33 THz has been obtained with a fiber diameter of 200 μm (Chen et al. 2006). The measured free-space coupling coefficient from the source to the fiber is around 20%. The wave guiding is achieved by propagating more than 99% of the power in air, i.e., less than 1% of the power in the core ($d/\lambda = 0.22$ at this frequency).

However, the attenuation coefficient is strongly dependent on the wavelength of the propagated wave. Shorter wavelengths lead to larger fraction of the power in the fiber material (with a large absorption coefficient) yielding larger attenuation coefficient. As shown in Fig. 3b, the attenuation spectrum of sub-wavelength diameter fiber should show a decreasing trend toward larger wavelength (or lower frequency), until a minimum attenuation corresponding to the absorption coefficient of air medium.

However, this trend is for a perfect fiber without irregularity. In reality, the diameter of the fiber is not perfectly constant. Small variations of the diameter along the fiber length have a strong impact on the attenuation coefficient of the propagated wave when most of its power is in air. In this regime, the effective refractive index of the propagated wave approaches unity (Cf. Fig. 3a).

For example, this difference is only $\Delta n_{eff} \sim 1.3 \cdot 10^{-2}$ for $d/\lambda = 0.33$, $\Delta n_{eff} \sim 7.5 \cdot 10^{-4}$ for $d/\lambda = 0.25$, and $\Delta n_{eff} \sim 5.6 \cdot 10^{-5}$ for $d/\lambda = 0.22$. As a result, small irregularities and diameter variations along the fiber yield couplings of the guided wave to lossy radiation waves.

Radiation losses induced by fiber irregularities increase with longer wavelengths. Consequently, measured attenuation spectrum of sub-wavelength diameter fiber should show a decreasing trend toward larger wavelength (or lower frequency), until a minimum attenuation, and then an increasing trend. The wavelength of minimum attenuation corresponds at the balance between losses induced by material absorption and by couplings to radiation waves. Measured attenuation spectra of PE fibers with different diameter are plotted in Fig. 4a. These results obtained by Chen et al. (2007a) illustrate the existence of an attenuation minimum in a rather narrow transmission windows ($\Delta\lambda_{FHWM}/\lambda_{min} \sim 0.3$).

Radiation losses induced by fiber irregularities impose therefore a limitation on the smallest fiber diameter corresponding to the ratio d/λ at the minimum attenuation wavelength (λ_{min}).

These limitations have been observed and studied in tapered optical fiber with a waist diameter smaller than the propagated wavelength (typically few 100 nm). These fiber tapers, based on the same guiding mechanism, exploit the large fraction of evanescent power for sensing, analyzing the surrounding environment (Lou et al. 2005; Warken et al. 2007; Zhang et al. 2008). In this context, Sumetsky has developed an analytical solution, based on the theory of nonadiabatic transitions from quantum mechanics, for estimating the radiation loss threshold and therefore

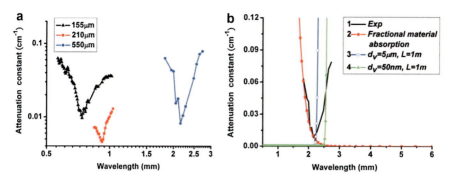

Fig. 4 (**a**) Measured attenuation spectrum of different sub-wavelength diameter fibers with a diameter of 155, 210, or 550 μm. (**b**) Measured attenuation spectrum of sub-wavelength diameter fibers with a diameter 550 μm (black curve), calculated attenuation coefficient from the fraction of wave absorbed by the material (red curve), and calculated radiation losses induced by diameter fluctuations of about dv = 5 μm (blue curve) or about dv = 50 nm (green curve) over a propagation length of 1 m. (Reproduced from Chen et al. (2007a), with the permission of OSA Publishing)

the smallest diameter for sub-micrometer diameter taper (d ≪ λ) following this expression for a silica taper (Sumetsky 2006):

$$\alpha \sim \exp\left[-\frac{0.687\,L}{\sqrt{d_w\,(d_0 - d_w)}}\exp\left(-\frac{0.27\,\lambda^2}{d_w^2}\right)\right] \tag{2}$$

This relation consider a taper shape with a diameter variation $d(z) = d_0 - (d_0 - d_w)/(1 - [(z - z_0)/L]^2)$, where d_0 and d_w are the diameter at both ends and at the waist of the taper, respectively. Chen et al. have translated this model to sub-wavelength diameter fibers by integrating the diameter variations in the term $d_v = d_0 - d_w$ (Chen et al. 2007a). The results plotted in Fig. 4b show a good agreement of the radiation loss threshold with the measured attenuation when diameter variations are only ~1%. It is noteworthy that smaller variations (<0.01%) do not strongly extend the minimum attenuation wavelength. The normalized wavelength shift (Δλ/λ) deduced from Fig. 4b is only around 0.13. This emphasizes that the waves are propagated at the edge of the TIR guiding mechanism requiring more and more drastic conditions when the fraction of power in the core decreased dramatically below 1% (i.e., d/λ < 0.2).

As for optical fibers, the chromatic dispersion of the group velocity of a THz fiber is a key property for evaluating the broadening of a pulse along the propagation in the fiber. It is typically quantified by the second-order term of the Taylor expansion of the propagation constant (β) of the propagated mode, which is calculated with the following expression:

$$\beta_2 = \frac{\partial^2 \beta}{\partial \omega^2} = \frac{2}{c}\frac{dn_{eff}}{d\omega} + \frac{\omega}{c}\frac{d^2 n_{eff}}{d\omega^2} \tag{3}$$

In optics, the group velocity dispersion (GVD) is expressed in ps²/km. In THz domain, frequency unit is preferred, and THz waves are rather propagated along metric than kilometric length. It is then more appropriate to express it in ps/(THz·m). The broadening of a pulse with a Gaussian shape is calculated with this equation:

$$T = T_0\sqrt{1 + \left(\frac{L\ |\beta_2|}{T_0^2}\right)^2} \sim \frac{L\ |\beta_2|}{T_0} \tag{4}$$

where L the propagation length and T_0 and T are the initial and final pulse width, respectively. The calculated GVD values of THz waves propagated (in the fundamental mode) along a PE sub-wavelength diameter fiber are presented in Fig. 5, for different fiber diameters. Since the refractive index of PE is almost constant at THz frequencies (Cf. Fig. 2a), chromatic dispersion from the material could be neglected in the calculation of the GVD. The GVD curves are characterized by a maximum value at a specific frequency, which both depend on the fiber diameter. The bold section in each curve corresponds to the transmission window of the sub-wavelength diameter fiber. It is limited by the radiation loss mechanism (induced by fiber irregularities) in low frequencies and by material absorption losses ($\alpha/\alpha_{mat} < 0.1$) in high frequencies. In this limited range, the GVD is not constant. As the fraction of the evanescent power in air increases (i.e., α/α_{mat} decreases), the

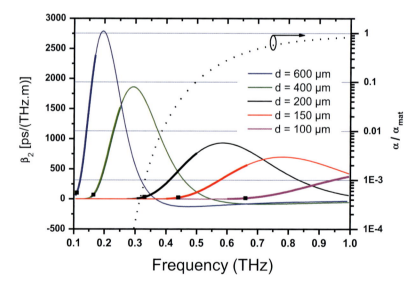

Fig. 5 Calculated group velocity dispersion values of THz waves propagated (in the fundamental mode) along a sub-wavelength diameter fiber (with a refractive index of 1.5), for different fiber diameters. The refractive index is considered frequency independent in the calculations. The sections in bold indicate the frequency range limited by the radiation loss threshold and the attenuation up to $\alpha/\alpha_{mat} < 0.1$. The squares highlight the frequency of minimal attenuation for each fiber diameter. The dashed curve corresponds to the ratio α/α_{mat} for a fiber diameter of 200 μm

GVD tends to zero (i.e., to the GVD of waves propagated in air). Nevertheless, at the frequency of minimal attenuation (indicated by a black square), the GVD is still rather larger. For example, it is around $\sim$34 ps/(THz·m) at 0.33 THz for a PE sub-wavelength diameter fiber with a diameter of 200 μm.

In conclusion, sub-wavelength diameter fibers enable low-loss propagation of THz waves. However, the narrowness of the transmission window associated with a rather larger GVD does not enable the propagation of broadband THz pulses, such as the ones delivered by photoconductive antenna in time domain THz spectrometer. Furthermore, this weak guiding regime does not allow fiber bends since the HE_{11} mode would not be guided along a fiber bend. The reduction of the effective index of HE_{11} mode induced by a bend would match the effective index values of radiative modes. Nevertheless, as for the optical fiber taper with sub-micrometer waist diameter (sub-wavelength scale), the very large fraction of evanescent power could be exploited for sensing the surrounding environment. For example, Borwen et al. have successfully detected tryptophan or polyethylene powders in the vicinity of a sub-wavelength diameter THz fiber (You et al. 2010b). This property could also be exploited for making directional fiber couplers (Chen et al. 2007b; Chang and Sun 2009). Unlike traditional fiber optic couplers that support symmetric and antisymmetric modes (arising from the coupling between the fundamental mode of both fibers), coupler based on sub-wavelength diameter fiber does not support the antisymmetric mode. In the guiding conditions of sub-wavelength diameter fiber, the antisymmetric mode that has a lower effective index than the symmetric mode is coupled to radiative modes. This yields losses of half the input power and a coupling ratio between both arms of the coupler that is weakly dependent on the coupling length and on the frequency (Chen et al. 2007b; Chang and Sun 2009). Besides, sub-wavelength diameter THz fibers and couplers have enabled the developments of some THz systems such as a THz fiber-scanning near-field microscope (Chiu et al. 2009; Chen et al. 2011) and a THz fiber endoscope (Lu et al. 2008).

Porous Fibers

Porous fibers are similar to sub-wavelength diameter fibers. They are also composed of a single rod, fiber in air, and THz waves are propagated in weak guiding conditions with a large fraction of evanescent power in air for reducing the attenuation coefficient of the waves. This guiding regime is characterized by a small value of the normalized frequency (V < 1.16), which one depends on the diameter and on the material refractive index of the fiber (Cf. Eq. 1). In contrast with sub-wavelength diameter fibers, the normalized frequency is reduced by introducing air holes into the fiber for decreasing the effective refractive index of the fiber material.

These fibers have been specifically developed in the THz domain. The first proposed fiber topology (Atakaramians et al. 2008; Hassani et al. 2008) is composed of a hexagonal array of air holes in a polymer fiber. As illustrated in Fig. 6a, the properties of THz waves guided by such fiber depend on the pitch (Λ), the hole diameter, the operating wavelength (λ), and the real and imaginary parts of the

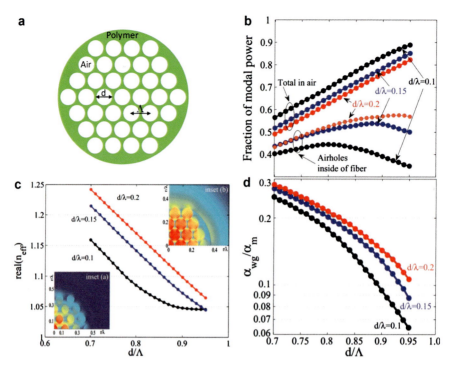

Fig. 6 (**a**) Cross-section schematic of a porous fiber composed of an array of air holes. (**b**) Evolution of versus the air-filling fraction (d/Λ), for different hole diameters (d/λ) of the fraction of the power in air and only in the air holes. (**c**) Effective refractive index of the porous fiber versus the air-filling fraction for different hole diameters. The 2D distribution of the wave intensity confined in the porous fiber (d/λ = 0.1) is shown in the insets (**a**) and (**b**) for d/Λ = 0.7 and 0.95, respectively. (**d**) Evolution of versus the air-filling fraction, for different hole diameters of the attenuation coefficient normalized by the material absorption coefficient. (Reproduced from Hassani et al. (2008), with the permission of AIP Publishing)

refractive index of the polymer (and their wavelength dependencies). For a given refractive index, the fraction of power guided in air is defined by the size of the air hole compared to the wavelength (d/λ) and the air-filling fraction (ratio d/Λ). As shown in Fig. 6c, the effective index of the guided wave decreases with higher d/λ and larger wavelength (over the size of the hexagonal array (d/Λ)). When the wavelength is comparable to the fiber diameter, the array of air holes could be seen by the wave as a homogenous material with an effective refractive index that depends on the air-filling fraction. Therefore, the propagations of THz waves could be achieved with a large fraction of power in the fiber, within the array of air holes (Cf. Fig. 6). This guiding condition enables a reduction of the contribution of the material absorption on the attenuation coefficient (α/α_{mat}) by more than one order of magnitude (Cf. Fig. 6d).

As shown in Fig. 6b, the fraction of power within the array of air holes increases with the ratio d/Λ until a maximum that depends on the ratio d/λ (i.e., diameter of the fiber over the wavelength). Above the maximal value, the fraction of power within the air holes decreases, while the fraction of power outside the fiber increases, leading to similar guiding conditions of a sub-wavelength diameter fiber. It is worth to emphasis that the larger fraction of power within the air holes is obtained with large d/Λ and large fiber diameter. This property of porous fibers enables the propagation of THz waves within the fiber (with a reduction by almost one order of magnitude of the material absorption), and an easier manipulation of the fiber with reduced parasitic coupling losses to the surrounding environment. However, these guiding conditions are limited toward larger wavelengths (i.e., lower frequencies) when the fraction of power guided outside the fiber becomes dominant. The porous fibers exhibit then similar limitations of sub-wavelength diameter fiber. As the wavelength increases, the attenuation coefficient strongly decreases due to larger fraction of power in air. Then, the propagation of THz waves is limited by radiation losses induced by fiber irregularities. Nevertheless, porous fibers enable broader transmission window compared to sub-wavelength diameter fiber of the same diameter and also a shift of the transmission windows toward shorter wavelength as the air-filling fraction increases.

The group velocity dispersion of the propagated wave is also modified by the introduction of the array of air holes in the fiber (Dupuis et al. 2010). In comparison with sub-wavelength diameter fibers, porous fibers exhibit a smaller GVD maximum, which depends on the air-filling fraction (Cf. Fig. 7). Larger air-filling fraction yields lower GVD values, which are reduced below 500 ps/(THz·m) for an air-filling fraction of 60% in the example of Fig. 7. It is worth to notify that the wavelength of maximum GVD shifts to shorter wavelengths as the air filling

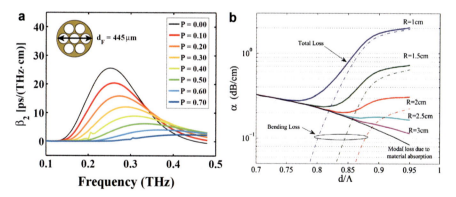

Fig. 7 (**a**) Calculated group velocity dispersion values of THz waves propagated (in the fundamental mode) in porous fiber with different porosity value. (Reproduced from Dupuis et al. (2010), with the permission of OSA Publishing). (**b**) Evolution of the attenuation coefficient versus the air-filling fraction (d/Λ) for different bending radius. The porous fiber is shown in Fig. 6a with d/λ = 0.1 and material absorption coefficient of 0.3 cm^{-1}. (Reproduced from Hassani et al. (2008), with the permission of AIP Publishing). The porosity is defined as the surface of air holes over the fiber surface

increases, which result from the augmentation of the fraction of evanescent power in the air holes.

The sensitivity to fiber bends is also dependent on the air-filling fraction (Hassani et al. 2008). As shown in the example of Fig. 7b, in the case of a constant fiber diameter, there exists a minimum of bending losses for an optimal value of d/Λ that depends on the bending radius. As the bend radius decreases, the delocalization of the wave outside the fiber needs to be compensated by a stronger confinement of the wave within the fiber material, which is achieved at a lower air-filling fraction. The minimum bend losses correspond then to the best compromise between bend losses and the losses from material absorption induced by light confinement within the material.

Numerous porous THz fibers have been realized mainly with low-density polyethylene (LDPE) or polymethyl methacrylate (PMMA) materials. The fiber has been fabricated following different fabrication processes such as (i) the stack-and-draw method that consists in stacking tubes and then drawing this structure down to an optical fiber (Dupuis et al. 2009); (ii) the extrusion method (heated bulk polymer billets are forced through an extrusion die, with a specific design, to make the preform that is then drawn down to a fiber) (Atakaramians et al. 2009); (iii) a subtraction technique that consists in fabricating a PE preform with PMMA rods, which ones are dissolved (to form the air holes) after drawing the preform down to a fiber (Dupuis et al. 2010); and (iv) mold-casting technique (melted PE billets are poured into a silica mold to cast the microstructured preform that is then drawn down to a fiber) (Dupuis et al. 2010).

Experimental comparisons between porous fibers and sub-wavelength diameter fibers have been reported by Dupuis et al. (2010). As shown in Fig. 8, the transmission spectra of porous fibers are broader and shifted to higher frequencies than the ones of sub-wavelength fibers with identic diameters. Furthermore, as expected from theoretical developments, higher air-filling fraction enables larger fiber diameter without degrading the transmission spectrum. The porous fiber with a diameter of 775 μm and a porosity of 86% has a transmission spectrum similar to the one of the porous fiber with smaller diameter and porosity (450 μm and 35%, respectively). These results demonstrate the interest of porous fiber for extending the transmission window toward higher THz frequencies that is rather complex to achieve with sub-wavelength diameter fiber due to the difficulty to handle the very small diameter fiber required for operating at higher frequencies. The minimum measured attenuation coefficient of these porous fibers is about 8 dB/m at 0.24 THz ($\sim$0.02 cm^{-1}) that correspond to a reduction by one order of magnitude of the material absorption ($\alpha_{PE} \sim 0.2$ cm^{-1}).

Numerous designs of porous fibers have been studied and fabricated as the examples shown in Fig. 9. A fiber with high birefringence of about 0.012 at 0.65 THz was obtained by organizing rectangular air holes following a rectangular array (Cf. Fig. 9a). This result contrasts with the symmetrical design from the spider-web topology (Cf. Fig. 9b) that does not induce birefringence.

The fiber shown in Fig. 9c is composed of an array of air holes with variable diameters and inter-hole separations that follow a gradient distribution of the

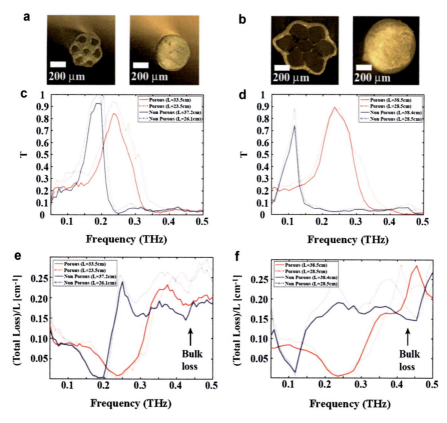

Fig. 8 Cross-section photographs of (**a**) small diameter and (**b**) large diameter porous fibers and (nonporous) sub-wavelength diameter fibers. (**c**) Normalized transmission spectra and (**e**) attenuation coefficient spectra of the small diameter porous and nonporous fibers for different fiber lengths. (**d**) Normalized transmission spectra and (**f**) attenuation coefficient spectra of the small diameter porous and nonporous fibers for different fiber lengths. (Reproduced from Dupuis et al. (2010), with the permission of OSA Publishing)

effective refractive index (Ma et al. 2015). In comparison with a porous fiber composed of an uniform array of air holes (Cf. Fig. 9d), this design leads to a better confinement of the waves at the center of the fiber and improves the propagation properties as lower GVD, larger transmission bandwidth, and higher wave-coupling efficiency to THz source (with Gaussian-shape).

The fibers shown in Fig. 9e–h are composed of a core suspended into an outer tube cladding with thin struts (Rozé et al. 2011). The first fiber has a core diameter of about 150 µm suspended in the middle of an outer cladding with a diameter of 5.1 mm. The second fiber has a porous core (with a diameter of 900 µm and a porosity of 10%) suspended inside an outer cladding with a diameter of 3 mm. The cladding protects the core from external perturbations that could degrade the guiding properties, such as dusts, core surface contamination, coupling to adjacent

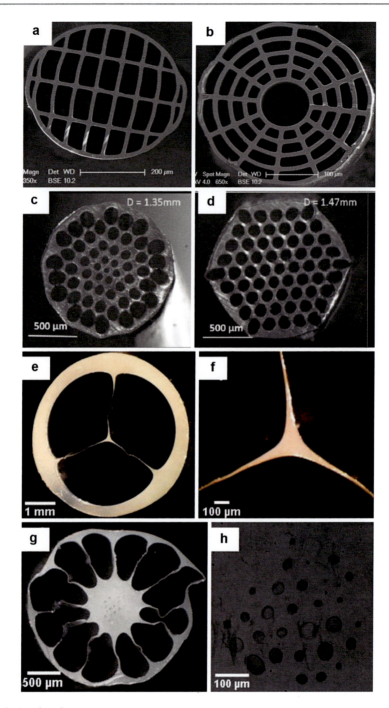

Fig. 9 (continued)

fibers, or fiber holders. Furthermore, it enables direct fiber manipulations and dry air encapsulation into the fiber for eliminating losses from water present in ambient air. The first fiber composed of a small core could be seen as a sub-wavelength diameter fiber in the middle of a cladding tube. Nevertheless, in the regime of very weak guidance when the propagation is limited by radiation losses, the outer cladding could improve the guiding properties. In this regime most of the waves are propagated outside the small core (i.e., $d/\lambda < 0.2$). The cladding could be considered as a tube that can confine a wave by antiresonant reflecting mechanism, as in hollow-core THz pipe (section "Hollow-Core Pipe Fibers"). Rozé et al. (2011) have demonstrated that this additional guiding mechanism results in the formation of a narrow transmission window in the low frequency side. It is worth to notify that this additional transmission window could be broader by optimizing the thickness of the cladding following the relation (11), which governs the guiding properties of hollow-core THz pipes.

A porous-core photonic bandgap (PBG) fiber has been developed on a similar concept (Nielsen et al. 2011; Bao et al. 2012). The fiber is composed of a porous core surrounded by a cladding with a honeycomb pattern of air holes (Cf. Fig. 10a). The cladding with a photonic crystal design exhibits PBG effects within specific frequency and effective index ranges that are governed by the optogeometrical parameters of the cladding (photonic crystal). As a result, allowed and forbidden photonic bands are formed within the mode spectrum of the photonic crystal cladding. The forbidden photonic bands (i.e., bandgaps) forbid the wave extension in the cladding leading to wave confinement and propagation in the core. As shown in Fig. 10b, these bands do not extend to effective index values below unity. PBG guiding is achieved with the help of the porosity that raises the effective refractive index of the core in the forbidden photonic bands.

Bao et al. (2012) have fabricated a fiber composed of porous core with a diameter of $Dc = 0.8$ mm surrounded by three rings of honeycomb cladding with hole diameter of $d = 165$ µm and a pitch of $\Lambda = 360$ µm (Cf. Fig. 10). The fiber is fabricated in TOPAS material that is the polymer with lowest absorption coefficient (Cf. section "Constraints and Challenges for Developing THz Fibers"). They have demonstrated PBG guidance in the porous core within a transmission band of 0.78–1.02 THz in which the attenuation coefficient (<1 dB/cm) is below the material absorption coefficient, with a minimum of 0.7 dB/cm at 0.88 THz (Cf. Fig. 10a).

Fig. 9 Cross-section photograph of polymer fibers composed of (**a**) an array of rectangular air holes, (**b**) an air of rectangular air holes following a spider-web design, (**c**) an array of air holes with a gradient distribution of air hole diameters and inter-hole separations, (**d**) a uniform array of air holes, (**e**) a small triangular core suspended in the middle of an outer cladding by 3 thin struts, and (**g**) a large porous core suspended in the middle of an outer cladding by 12 thin struts. Zoom-in photographs of the core of the fibers (**e**) and (**g**) are shown in (**f**) and (**h**), respectively. (Reproduced from (**a, b**) Atakaramians et al. (2009), (**c, d**) Ma et al. (2015), and (**e–h**) Rozé et al. (2011), with the permission of OSA Publishing)

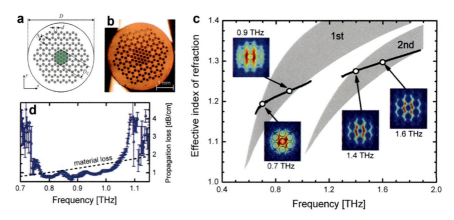

Fig. 10 (**a**) Design of a porous-core fiber with honeycomb cladding. (**b**) Photograph picture of the cross section of the fabricated fiber. (**c**) Dispersion diagram of effective indices vs. frequency of modes supported by the honeycomb cladding (white area) and by the porous core (bold curve). Gray areas show domains where the cladding forbids wave extensions (i.e., bandgaps where no mode is confined in the cladding). Inset shows the 2D distribution of the intensity of the fundamental mode confined in the porous core. (**d**) Measured attenuation spectrum (from cutback method) of the fabricated fiber. (Reproduced from Bao et al. (2012), with the permission of OSA Publishing)

It is worth to notify that the attenuation coefficient might be further reduced by improving the fiber design.

In conclusion, porous fibers enable a reduction of the material absorption coefficient by one order of magnitude with a minimum attenuation coefficient of about 8 dB/m around 0.2 THz (Dupuis et al. 2010; Rozé et al. 2011). However, this loss level limits the interest of these fibers for propagating THz waves along long length. Nevertheless, the versatility of the fiber designs offers numerous advantages (especially in comparison with sub-wavelength diameter THz fibers), such as a broader transmission window, higher compatibility for propagating THz waves at high frequencies, reduction and control of the GVD, extremely high birefringence, and efficient protection against external perturbations. Porous fiber is therefore an interesting platform for propagating THz waves in single-mode regime over few tens of centimeters, typically for applications that require the mechanical flexibility of a fiber.

Solid-Core Photonic Crystal Fibers

The development of solid-core photonic crystal fibers (PCF) in the optical domain has inspired numerous works on THz fibers for exploiting in the THz domain the remarkable properties of PCF. These fibers are composed of a core surrounded by an array of air holes that enables light confinement and propagation by TIR mechanism. The array of air holes could be seen as a cladding with a lower effective refractive index than the one of core that satisfies the TIR guiding conditions. The large index contrast between the air holes and the fiber material (glass or polymer) enables a strong control of the light confinement conditions and flexibilities in the fiber designs (core shape, topology of the air hole array) that offer a myriad of possibilities for developing optical fibers with original properties, for example, endlessly single-mode propagation, engineering of the chromatic dispersion curve, high birefringence, enhanced light confinement in the core, large fraction of evanescent light in the air holes for light liquid/gas interactions, and light propagation in multicores (Knight 2003; Russell 2003).

In the THz domain, THz PCF have been studied by stacking high-density polyethylene (HDPE) tubes with a diameter of 500 μm and a thickness of 50 μm to form a periodic array (a photonic crystal) of air holes (of 400 μm diameter (d)) with a pitch of $\Lambda = 500$ μm (Han et al. 2002). An HDPE rod (with a diameter of 500 μm) is inserted in the center to form the fiber core (Cf. Fig. 11a). However, the fiber length is limited to 2 cm due to the large attenuation coefficient of about 217 dB/m (including insertion losses) over the measured transmitted spectrum (0.1–3 THz). This large value comes mainly from the absorption coefficient of the HDPE material since THz waves are mostly confined in the fiber core. Nevertheless, this PCF exhibits a large GVD below 0.4 THz (with a maximum of 1400 ps/(THz·m) at 0.25), a zero GVD around 0.5 THz, and a small negative GVD of −30 ps/(THz·m) above 0.6 THz. This result demonstrates the possibility offered by the PCF designs for engineering the chromatic dispersion curve (i.e., stronger contribution of the waveguide dispersion than the material one). Furthermore, a high birefringent with a record value of 0.021 at 0.3 THz has been demonstrated by adding another HDPE rod on the side of the core in order to create two propagation axis (Cho et al. 2008), as shown on Fig. 11b.

A significant progress was realized by Nielsen et al. who have used TOPAS material for fabricating two PCFs (Nielsen et al. 2009). This material has a low absorption coefficient at THz frequencies (from 20 dB/m at 0.3 THz to 180 dB/m at 1.5 THz), which enables the realization of THz PCFs with an absorption coefficient

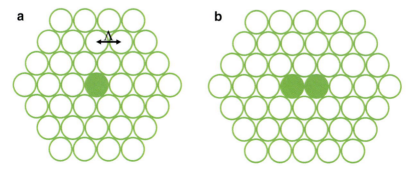

Fig. 11 Cross-section schematics of photonic crystal fibers realized by stacking thin-wall tubes around a core formed by a single rod (**a**) or two rods for inducing two propagation axis (**b**)

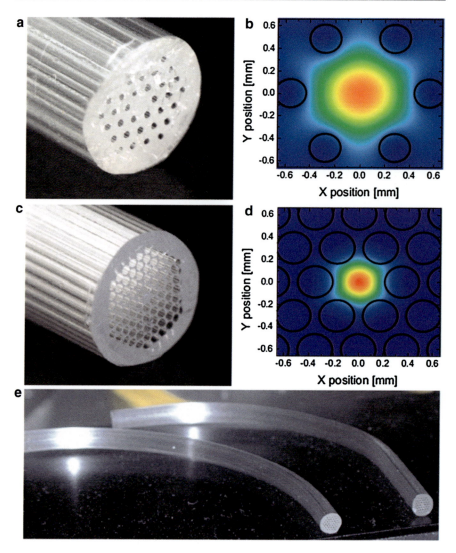

Fig. 12 Photographs of a large-core PCF (**a**) and a small-core PCF (**c**). Calculated 2D distribution of the guided mode intensity at 1 THz of the large-core (**b**) and small-core PCF (**d**). Photograph of two samples of the large-core PCF shaped into 90° bends. (Reproduced from Nielsen et al. (2009), with the permission of OSA Publishing)

of about <10 dB/m at 0.5 THz. This value has been measured with the cutback method on different fiber length from 6 to 92 mm. The fiber is composed of three rings of air holes around a solid core, with a pitch of $\Lambda = 560$ μm, a hole diameter of $d = 250$ μm ($d/\Lambda = 0.45$), and an external diameter of 6 mm (Cf. Fig. 12). THz wave propagation in the HE_{11} mode has been demonstrated from 0.2 to 1 THz, at least. Nevertheless, it is worth notifying that at low frequency (below 0.4 THz), the attenuation coefficient of the HE_{11} mode increases due to weaker wave confinements in the core. The wavelengths are much larger than the pitch ($\lambda/\Lambda > 1.3$). On the other side, at higher frequency (above 0.7 THz), the wavelengths are smaller than the pitch ($\lambda/\Lambda < 0.76$) leading to stronger confinement of the waves in the core and therefore to higher attenuation coefficient and scattering losses from surface roughness.

PCF technology offers the possibility to engineer the chromatic dispersion curve of the HE_{11} mode by adjusting the pitch and the hole diameter of the photonic crystal. The dispersion curve of this fiber, obtained from calculations, exhibits a zero-dispersion frequency (ZDF) at 0.55 THz for the HE_{11} mode with a rather low dispersion ($\beta_2 < 100$ ps/(THz·m)). By fabricating a PCF with a pitch of 350 μm and a hole diameter of 280 μm ($d/\Lambda = 0.8$), Nielsen et al. have demonstrated the possibility to shift the ZDF to higher frequency (0.6 THz) with an augmentation of the slope of β_2 in normal dispersion regime ($\beta_2 > 0$) at frequencies below the ZDF. The calculated results have been corroborated experimentally.

The fibers have been realized with the help of drawing tower used for fabricating polymer optical fibers. The holes of the PCF pattern are drilled in a preform formed by casting TOPAS granulates into a cylinder. The preform has been then heated and drawn down to fibers. It is worth to notify that bent THz fibers could be achieved by slightly heating and bending the fiber (Cf. Fig. 12e). This drill-and-draw fabrication process is very suitable for developing THz fibers with a good repeatability, high uniformity of the fiber cross section along its length, and a large robustness to external forces.

Following this work, THz PCFs have been fabricated by this process with Zeonex polymer that is a cycloolefin polymer (COP), similar to TOPAS polymer, with an absorption coefficient of 60 dB/m at 0.2 THz and 130 dB/m at 1.2 THz (Anthony et al. 2011a). Fibers with an external diameter of 3 or 4 mm are composed of a solid core of about 400 or 600 μm diameters surrounded by five large air holes ($d > 700$ μm). As expected, the attenuation coefficient of the propagated waves (from 43 dB/m at 0.2 THz to 173 dB/m at 1.2 THz) is mainly caused by the material losses.

Even if the absorption coefficient of TOPAS or Zeonex is still too large for enabling the development of low-loss fibers, THz PCFs might find some applications when short fiber length with managed chromatic dispersion is required or in the development of fiber-based components such as a fiber coupler. In this prospect, Bao et al. (2014) have proposed a design of a broadband directional coupler based on a dual-core THz PCF, in which the cores are down-doped by introducing an array of micrometer size air holes.

Hollow-Core THz Fibers

In contrast with solid-core THz fibers, hollow-core THz fibers enable the propagation of THz waves in a core filled with air that could even be dried for reducing further the losses. In hollow-core fibers, the waves are confined and propagated with the help of cladding reflectors. The developments of hollow-core fibers are still very intense in the optical domain, and the salient achievements are currently exploited and translated for realizing THz fibers. Hollow-core THz fibers based on a dielectric/metal reflector are presented in the following section. Then in section "Hollow-Core Bragg Fibers," the hollow-core Bragg fibers based on a PBG cladding are presented. Hollow-core fibers based on the antiresonance guiding mechanism are then presented in section "Hollow-Core Pipe Fibers" for the simplest design composed of simple dielectric pipe, and more sophisticated fibers designed as the Kagome hollow-core photonic crystal fibers and the tube lattice hollow-core fibers are presented in sections "Kagome Hollow-Core Photonic Crystal Fibers" and "Tube Lattice Hollow-Core Fibers," respectively.

Dielectric-/Metal-Coated Hollow-Core Fibers

Dielectric-/metal-coated hollow-core fibers have been translated from optical domain to THz frequencies. These fibers also named "hollow-glass fiber" are composed of a dielectric tube (glass or polymer) with its inner surface coated by a metallic layer and a dielectric layer (Cf. Fig. 13). Their large hollow-core makes them very attractive for the transmission of IR radiations, in particular for delivering the beam of high-power CO_2 laser or Er:YAG laser that is impossible to propagate in silica-based solid-core fibers (Abe et al. 1998).

Light waves are guided in the hollow-core by multiple reflections on the metallic layer. A thin dielectric layer is added for reducing propagation losses induced by ohmic losses and for changing the dominant mode structure. In hollow metallic waveguide, a portion of electric field that does not vanish entirely on the metallic surface interacts with electronic charges in the metal, leading to losses due to resistive heating. As a result, the mode with the lowest attenuation is the TE_{01} mode since its electric field is parallel to the circular metallic wall, while TM modes experience large losses.

The addition of a dielectric layer changes the boundary conditions of the hollow-core leading to the reduction of the attenuation of TM modes and the propagation of hybrid modes. The hybrid HE_{11} mode, for which the electric field at the boundary is reduced along the entire contour of the fiber wall, becomes dominant with a lower

Fig. 13 Schematic illustration of the cross section of a dielectric-coated hollow-core metallic fiber

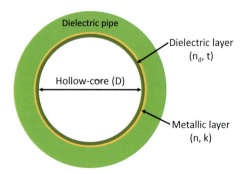

attenuation coefficient than the TE_{01} mode. The thickness of the dielectric layer has a strong effect on the attenuation coefficient of the HE_{11} mode. The optimum thickness (t) of the dielectric layer is given with the following relation (Bowden et al. 2007):

$$t = \frac{\lambda}{2\pi \sqrt{n_d^2 - 1}} \tan^{-1} \left(\frac{n_d}{\sqrt[4]{n_d^2 - 1}} \right) \tag{5}$$

where n_d is the refractive index of the dielectric layer and λ is the wavelength of the propagated wave. In consequence, the transmitted spectrum is composed of high-loss peaks corresponding to destructive interferences at the wavelengths (λ_m), which are calculated from this expression:

$$\lambda_m = \frac{4t \sqrt{n_d^2 - 1}}{m} \text{ or } f_m = \frac{m\,c}{4t \sqrt{n_d^2 - 1}} \tag{6}$$

with m the order of the interference peak (a positive integer). The bandwidth of the transmission windows delimited by the interference (high-loss) peaks is obtained from Eq. 6:

$$\Delta f = \frac{c}{4t \sqrt{n_d^2 - 1}} \tag{7}$$

For a lossless dielectric layer, the minimum of the attenuation coefficient of the HE_{11} mode (in the transmission windows) could be approximated with the following relation (Miyagi et al. 1984; Bowden et al. 2007):

$$\alpha = \frac{4u^2}{k_0^2 D^3} \frac{n}{n^2 + k^2} \left(1 + \frac{n_d^2}{\sqrt{n_d^2 - 1}} \right) \; [m^{-1}] \tag{8}$$

with D the diameter of the hollow-core, n and k the real and imaginary part of the complex refractive index of the metal, k_0 the wavenumber in vacuum, and u the first zero of the Bessel function $J_0(x)$ (i.e., the mode constant, $u = 2.4048$ for the HE_{11} mode).

For example, the attenuation spectrum of the HE_{11} mode propagated in the hollow-core of a dielectric-coated hollow metallic fiber is presented in Fig. 14a. This fiber is similar to the one theoretically studied by Tang et al. (2009). It is composed of a hollow-core (with a diameter of 2 mm) surrounded by a layer of polystyrene (PS) with a thickness of 25 μm and a thick gold layer (thicker than the penetration depth). A minimum attenuation of 0.21 dB/m at 1.82 THz and 0.045 dB/m at

3.65 THz is obtained for the first and second transmission windows, respectively. This simulation result demonstrates the interest of this kind of fiber for guiding THz waves. Nevertheless, these values are obtained for a lossless dielectric material and for a rather large core diameter (D/λ ~12 at 1.82 THz and D/λ ~24 at 3.65 THz). Larger core leads to lower attenuation coefficient but with a larger number of propagated modes. The second transmission window is thus more multimodes than the first one.

As already mentioned, the main challenge for developing THz waveguides is limiting the absorption of the propagated waves by the materials. As shown in the example in Fig. 14a, the minimum attenuation coefficient rises by almost a factor 3 (α_{min} = 0.61 dB/m at 1.71 THz) in the first windows and by a factor 6 in the second one (α_{min} = 0.266 dB/m at 3.65 THz), when the absorption of the PS layer is considered. Even if the absorption of the dielectric layer increases strongly the attenuation of the propagated waves, it does not modify the position and the bandwidth of the transmission windows, neither the effective index curve (as shown in Fig. 14b, both effective index curves are identical) and thus neither the group velocity dispersion that governs the propagation properties of short pulses. It is worth to notify that the group velocity dispersion of the waveguide (i.e., without the dispersion of the material) is extremely low compared to the one of the solid-core THz fibers. Furthermore, larger core diameter leads to flat curve with small GVD over a broad bandwidth (Cf. Fig. 14b, second transmission windows), but this is associated with an augmentation of the number of propagated modes that could increase the propagation losses due parasitic couplings to less confined modes.

The first fabrication of a dielectric-coated hollow metallic fibers for delivering THz waves has been reported by Bowden et al. (2007). The fiber is composed of a hollow-core of 2.2 mm diameter surrounded by a layer of PS with a thickness of 8.2 μm and a layer of silver (Ag, 1 μm thick) deposited on the inner surface of a glass tube. The lowest measured attenuation coefficient of the HE_{11} mode is 0.95 dB/m at 2.5 THz. This value has been measured with the cutback technique on a 90-cm-long length fiber. In comparison with the attenuation coefficient of 3.9 dB/m (at 1.89 THz) measured on a silver- or copper-coated tube with a hollow-core diameter of 3 mm (D/λ ~19) (Harrington et al. 2004), this result demonstrates the significant role of the dielectric layer on the propagation losses. As shown in Fig. 15a, the attenuation coefficient of the HE_{11} mode could be reduced by optimizing the thickness of the dielectric layer (Bowden et al. 2008). When the dielectric layer is too thin, the TE_{01} mode is dominant (with lower losses) over the HE_{11} mode leading to a large augmentation of the propagation losses. This behavior induces an effective cutoff frequency of the HE_{11} mode that depends on the thickness and the refractive index of the dielectric layer and on the operating wavelength. From Fig. 15b, we can observe that the HE_{11} mode is not anymore dominant at the wavelengths higher than at least 44 t, for a PS layer.

However, the large core diameter (D/λ > 14) enables the propagations of higher-order modes. This requires careful conditions for launching THz waves into the hollow-core to excite the HE_{11} mode only, which has the lowest attenuation coefficient and the fastest velocity. The propagation of multimodes has also an

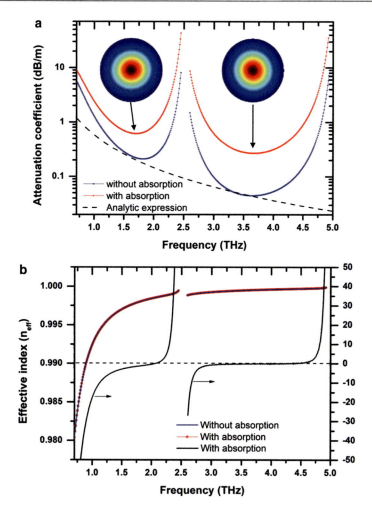

Fig. 14 (a) Attenuation spectra of the HE_{11} mode propagated within the hollow-core of a PS-/Au-coated hollow-core THz fiber, obtained from numerical calculations with a mode solver based on the finite element method (FEM) or from Eq. (8). The 2D distribution of the Poynting vector of the HE_{11} mode at 1.71 and 3.65 THz (with material absorption of the PS layer) is shown in the insets. (b) Evolution of the effective index and group velocity dispersion of the HE_{11} mode versus frequency. These curves are obtained from numerical calculations with a mode solver based on FEM. The hollow-core of a PS-/Au-coated hollow-core THz fiber is composed of a core diameter of D = 2 mm, a PS layer of t = 25 μm with a refractive index of $n_d = 1.58 - i*3.58\cdot10^{-3}$, and a gold layer with refractive index of n = 356 − i*444. For simplicity, these values are considered constant of this frequency range. FEM-based calculations are realized with or without the material absorption of the PS layer (i.e., imaginary part of n_d)

impact to the propagation of pulses by deteriorating and elongating the waveform (in time domain) due to different mode velocities. As shown in Fig. 16, the 2D electric field (Ex) distribution of different modes, at the fiber output, could be selectively measured by varying the launching conditions of THz waves into the

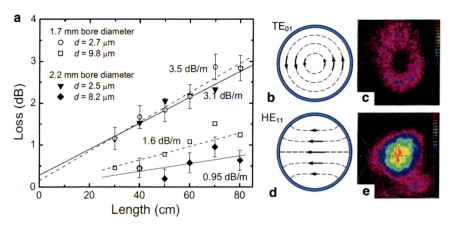

Fig. 15 (**a**) Attenuation coefficient measured with the cutback method for PS-/Ag-coated hollow-core THz fibers with a core diameter of 1.7 or 2.2 mm and a PS layer with different thickness. (**b, d**) Schematic illustrations of the electric field lines of (**b**) the TE_{01} mode and (**d**) the HE_{11} mode. (**c**) Measured 2D intensity distribution of THz waves (at 2.5 THz) propagated through 45-cm-long length of a PS-/Ag-coated hollow-core fiber composed of a core diameter of 1.6 mm and a PS layer thickness of 2 μm. Measured 2D intensity distribution of THz waves (at 2.5 THz) propagated through 90-cm-long length of a PS-/Ag-coated hollow-core fiber composed of a core diameter of 1.6 mm and a PS layer thickness of 10 μm. (Reproduced from Bowden et al. (2008), with the permission of AIP Publishing)

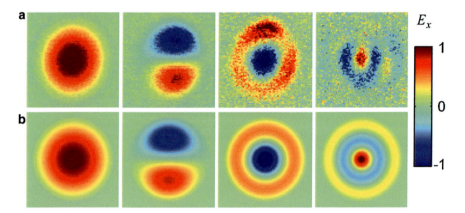

Fig. 16 (**a**) Measured spatial 2D distribution of the normalized electric field (Ex) of the first modes (HE_{11}, TE_{01}, HE_{12}, and HE_{13}, respectively) propagated through an PS-/Ag-coated hollow-core THz fiber ($D_{core} = 1.8$ mm, t = 14 μm) over a fiber length of 133.5 mm. The measures of different modes have been realized by varying the launching conditions of THz radiations into the hollow-core and by extracting the map (2 × 2 mm^2 area) at different times (t = 0, 1.5, 4.2, and 10.2 ps, respectively) in the waveforms recorded at each location of the map. (**b**) Calculated 2D distribution of the normalized electric field (Ex) of the corresponding modes. (Reproduced from Mitrofanov and Harrington (2010), with the permission of OSA Publishing)

hollow-core and by extracting a 2D distribution at different time position in the 2D waveform (Mitrofanov and Harrington 2010). Indeed, the different mode velocities yield separated pulse packets in the waveform, at the fiber output. The first one corresponds to the mode HE_{11}, followed by the TE_{01}, HE_{12}, and HE_{13} modes.

Besides, since the THz waves are guided in the hollow-core with a very low penetration in the polymer layer, propagated pulses experience a very low time broadening. The dispersion of the HE_{11} mode is estimated of about 0.08 ps/(μm·m) at 2.3 THz (i.e., β_2 ~6 ps/(THz·m)), for a PS-/Ag-coated hollow-core fiber (D_{core} = 1.8 mm, t = 14 μm) (Mitrofanov and Harrington 2010).

Fibers with mechanical flexibility could be realized by using a polymer tube instead of a glass tube. Doradla et al. (2012) have coated the inner surface of a polycarbonate tube with PS/Ag layers for measuring the bending losses of such fibers. They have reported bend losses of about 0.3 dB/m (at 1.4 THz) for fibers with a hollow-core diameter from 2 to 4.1 mm that are bent around an aluminum disc of 6.4 cm diameter. However, the fibers are bent along an angle of 20° only. This limitation is due to the degradation of the PS layer uniformity under the mechanical constraints created by the bend.

In order to improve the flexibility capability of such fibers, Navarro-Cía et al. (2013) have fabricated an PS/Ag hollow-core fiber. An Ag layer with a thickness of ~0.6 μm and a PS layer of 10 μm thick have been deposited on the inner surface of a silica capillary. The diameter of the hollow-core is only 1 mm for enabling a rather good mechanical flexibility of the fiber (up to a bending radius of 1.25 m). The first transmission window of the HE_{11} mode is from 2 to 5.5 THz. As shown in Fig. 17, THz radiations are propagated in the HE_{11} mode that is still the dominant mode when the fiber is bent, at least until a minimal radius of 1.25 m for which

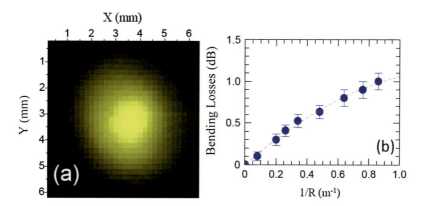

Fig. 17 (**a**) Measured 2D distribution of the intensity in far field at the fiber output of THz radiation propagated through 1-m-long length of the Ag/PS hollow-core fiber (D = 1 mm and t = 10 μm). (**b**) Bend losses measured at 2.85 THz for 1-m-long length fiber under different bending radius. (Reproduced from Navarro-Cía et al. (2013), with the permission of OSA Publishing)

the bending losses reach 1.0 dB/m at 2.85 THz. It is worth notifying that bend loss measurements have been realized with the polarization of the incoming THz radiations perpendicular to the plane of the bend, which is more favorable than the polarization parallel to the bending plane.

However, reducing the core diameter for gaining in flexibility yields a higher attenuation coefficient of the HE_{11}, which one is experimentally measured below 10 dB/m from 2 to 2.85 THz and estimated below 3 dB/m from 3 to 5 THz. Even if specific applications requiring a short fiber length and a relative mechanical flexibility, this loss level is too large for most of the applications. Furthermore, this fiber structure that give some mechanical flexibilities is not transposable to lower frequencies. The flexibility is obtained with a core diameter of 1 mm, and bend measurements have been realized at 2.85 THz. For example, in the case of an operating frequency of 0.6 THz, the same guiding properties are obtained for a core diameter of 4.7 mm (by applying the same ratio $D/\lambda = 9.5$) leading to much less mechanical flexibilities.

Despite its rather large attenuation coefficient, the mechanical flexibility of this fiber has been successfully exploited for delivering THz waves generated by a quantum cascade laser (QCL) over a distance of about 50 cm, within the HE_{11} mode (Vitiello et al. 2011). The coupling from the QCL to the fiber was realized through a copper-waveguide coupler that is robust to misalignment error, with a coupling efficient higher than 90% between the coupler and the fiber. This result demonstrates the advantages of such hollow-core THz fibers that exhibit broad transmission windows, a low GVD, waves guiding in the HE_{11} mode (Gaussian-like mode), and a relative mechanical flexibility (at high frequencies). Nevertheless, these fibers suffer from their large multimode guiding regime.

Hollow-Core Bragg Fibers

Hollow-core Bragg fibers are composed of a hollow-core surrounded by a periodic stack of alternating high- and low-refractive index layers, as illustrated in Fig. 18. These fibers have been developed in the optical domain (Fink et al. 1999; Ibanescu

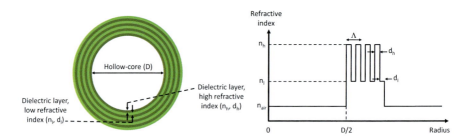

Fig. 18 Schematic illustration of the cross section and refractive index profile of a hollow-core Bragg fiber

et al. 2000) and then translated to THz frequencies. Light is confined and propagated in the hollow-core by multiple reflections in the periodic multilayer cladding that lead to constructive interferences of high reflectivity (Bragg reflections) in specific frequency bands (i.e., transmission windows). These bands exhibit photonic bandgaps properties (i.e., light propagation through the multilayer cladding is forbidden for any polarization) that reflect and confine the light in the hollow-core as a mirror (Fink et al. 1998, 1999; Temelkuran et al. 2002).

The properties of this Bragg reflector depend on the thickness and the refractive index of the two layers and the number of stacked bilayers (number of periods). When the diameter of the hollow-core is significantly larger than the wavelength of the guided wave, the wave propagates almost parallel to the core-cladding interface (the effective index of the fundamental mode is almost equal to unity, $n_{eff} \sim 1$) by consecutive reflections with grazing angles of incidence on the Bragg reflector cladding. In the idealistic case of an infinite number of bilayers, any polarization of light is completely reflected at the Bragg wavelength (λ_B) that is given by this relation (Skorobogatiy and Dupuis 2007):

$$\lambda_B = 2\,(d_h \tilde{n}_h + d_l \tilde{n}_l) \tag{9}$$

with $\tilde{n}_h = \sqrt{n_h + n_{eff}} \sim \sqrt{n_h + 1}$, $\tilde{n}_l = \sqrt{n_l + n_{eff}} \sim \sqrt{n_l + 1}$, and n_h, d_h, n_l, d_l, the refractive index and the thickness of the high- and low-refractive index layers, respectively. The bandwidth of the bandgap is proportional to the refractive index contrast ($n_h - n_l$), and it is maximized for the so-called quarter-wave reflector condition:

$$d_h \tilde{n}_h = d_l \tilde{n}_l = \frac{\lambda_B}{4} \tag{10}$$

In the realistic case of a finite size cladding, the light is not completely reflected by the Bragg reflector. The efficiency of the Bragg reflector increases with the refractive index contrast. Larger index contrast leads to larger reflections resulting to stronger light confinement in the hollow-core and thus to lower propagation losses. The size of the hollow-core has an effect on the propagation losses that decrease with a larger core diameter (D) according to the trend of λ^2/D^3.

Temelkuran et al. have fabricated a hollow-core Bragg fiber with remarkable properties (Temelkuran et al. 2002). The Bragg reflector is composed of alternating layers of arsenic triselenide (As_2Se_3) and poly(ether sulfone) (PES). They have reported an attenuation coefficient of about 0.95 dB/m at the wavelength 10.6 μm, for a hollow-core diameter of 700 μm. The large core diameter ($D/\lambda = 70$) reduces significantly the propagation losses, but it allows the propagation of an extremely large number of modes. It is worth to notify that this attenuation coefficient is much lower than the absorption coefficient of the materials of the Bragg reflector, which are 10 dB/m and 10^5 dB/m, respectively, for As_2Se_3 and PES materials. This result demonstrates the efficiency of the Bragg reflector that enables light guiding in a hollow-core with materials that are 10^5 more absorbing than the attenuation coefficient. This property of the hollow-core Bragg fibers makes them very suitable for propagating waves in the THz domain, where most of the materials are very absorbing.

The development of hollow-core Bragg fibers at THz frequencies is limited by the difficulties to find high- and low-refractive index materials with low absorption coefficients. Dupuis et al. have proposed to realize polymer/air bilayers or doped/undoped polymer bilayers (Dupuis et al. 2011).

The polymer/air bilayers were obtained by randomly dispersing polymer powder on the top of a polymer film. The fiber was then fabricated by rolling the polymer film around a mandrel that is then removed to form the hollow-core. The powder particles act as spacers to form the air layers between the polymer layers. PTFE tape was then wrapped around the fiber for holding it. The fabricated polymer/air Bragg fiber has a hollow-core diameter of 6.73 mm surrounded by five bilayers consisting of a 254 μm thick PTFE polymer and an air layer of about 150 μm thick.

As shown in Fig. 19a, the transmission spectrum of THz radiations propagated through 21.4 cm of the polymer/air hollow-core Bragg fiber is composed of clear transmission windows corresponding to the Bragg bandgaps. The measurements of the attenuation coefficient of this fiber are rather difficult. The large core enables the propagation of a large number of modes leading to transmission spectrum curves that depend on the modal interference between the excited modes. Slight fiber

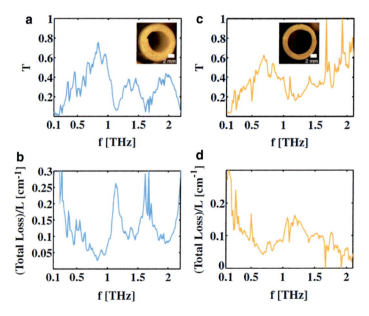

Fig. 19 Measured transmission properties of a polymer/air hollow-core Bragg fiber (left column) and of a doped/undoped polymer hollow-core Bragg fiber (right column). (**a, c**) Normalized transmission (amplitude) spectrum of the fiber, with a cross-section photograph shown in the inset. (**b, d**) Attenuation spectrum of the fiber. (Reproduced from Dupuis et al. (2011), with the permission of OSA Publishing)

misalignments lead to significant variations in the transmission spectrum making the attenuation coefficient measurements poorly accurate. In this context, Dupuis et al. have measured the overall attenuation spectrum (including coupling losses) with a minimum attenuation lower than 0.028 cm^{-1} ($\sim$12 dB/m) at 0.82 THz (Cf. Fig. 19b). This value is much larger than the calculated attenuation coefficient of $\sim$3·10^{-4} cm^{-1} ($\sim$0.13 dB/m) for the HE$_{11}$ mode (Dupuis et al. 2011). This discrepancy emphasis the difficulty to measure accurately the attenuation spectrum of THz waveguides and the strong effect of intermodal couplings on the attenuation spectrum due to irregularities along the fiber and if the THz radiations are not carefully coupled to the HE$_{11}$ only. For example, modes couplings to higher-order modes increase the attenuation coefficient by more than one order of magnitude, from 3·10^{-4} cm^{-1} (HE$_{11}$ mode) to 7·10^{-3} cm^{-1} (3 dB/m, HE$_{14}$ modes) (Dupuis et al. 2011).

The doped/undoped polymer bilayers were realized by doping polyethylene (PE) polymer with TiO$_2$ particles. The refractive index of PE was increased from 1.5 to 3 with a doping concentration of 80 wt.% of TiO$_2$. It is worth notifying that the absorption coefficient increases substantially and the thermomechanical properties of the polymer are also modified. The doped PE polymer has a higher viscosity and a much higher melting temperature than pure PE. A stack of doped/undoped PE films was hot-pressed together to consolidate the bilayer. The fiber is fabricated by rolling the stack around a mandrel that is then removed to form the hollow-core. The fabricated Bragg fiber has a hollow-core diameter of 6.63 mm surrounded by a six bilayers consisting of a 135 μm thick high-index layer (of 80 wt.% TiO$_2$ doped PE) and a 100 μm thick low-index layers of undoped PE.

The transmission spectrum of this fiber is less structured (Cf. Fig. 19c). Only one transmission band (in the low frequency side) is observable, for a fiber length of 22.5 cm. The overall attenuation coefficient spectrum is shown in Fig. 19d. The minimum attenuation coefficient is about 0.042 cm^{-1} (18 dB/m) at 0.69 THz. As for the polymer/air Bragg fiber, this value is larger than the calculated attenuation coefficient of 10^{-3} cm^{-1} (0.4 dB/m) for the TE$_{01}$ mode. Furthermore, the large absorption coefficient of the doped polymer layer limits the formation of Bragg bandgap for HE modes that have wider field penetration into the Bragg cladding than the TE modes. In contrast to the polymer/air fiber, the lowest attenuated mode is the TE$_{01}$ mode, which one requires specific conditions for properly coupling THz radiations into it. Furthermore, the fabrication of a TiO$_2$ doped polymer film and the Bragg fiber is rather difficult to realize without irregularities or film defects (density inhomogeneity, cracks, etc.) that are additional factors of losses.

Improvements of the fabrication processes and the use of less absorbing doping materials are two routes for improving the performances of this Bragg fiber that has the advantage to be more robust than the polymer/air fiber. In this prospect, Ung et al. have numerically study the compromise between the high refractive index contrast and material losses as function of TiO$_2$ doping concentrations in polymer (Ung et al. 2011a). They have demonstrated two situations of interest: (i) low doping concentration ($\sim$10 wt.% of TiO$_2$) that leads to the minimum attenuation coefficient of HE$_{11}$ mode with narrow transmission windows or (ii) very high doping concentration (~85 wt.% of TiO$_2$) with large transmission windows and relatively low attenuation coefficient. These two doping concentrations are more favorable than intermediate values (between 30 and 70 wt.% of TiO$_2$) where the attenuation coefficient and the transmission windows are both limited.

Single-mode propagation in a Bragg fiber has been proposed by drastically reducing the size of the hollow-core and by carefully designing the Bragg reflector for guiding only the mode TE$_{01}$ (Hong et al. 2017). The proposed Bragg fiber is composed of a hollow-core of 1.834 mm diameter surrounded by four bilayers TOPAS/air (with a thickness of about t$_h$ = 77.5 μm and t$_l$ = 1.154 mm, respectively) that are held together by thin TOPAS bridges (t = 15 μm). The core diameter and the Bragg reflector are designed for matching the position of the second bandgap with the TE$_{01}$ mode and for exhibiting bandpass properties (no confinement in the hollow-core) for the other modes, within the frequency range 0.8–1.2 THz. The confinement of the TE$_{01}$ mode is realized in the second bandgap for enabling loss discriminations between the TE$_{01}$ mode and the low order modes (HE$_{11}$, TM$_{01}$, HE$_{21}$) by positioning a bandpass (band between the first and second bandgap) within the guiding region of these modes, as shown in Fig. 20a. As a result, the attenuation coefficient of the TE$_{01}$ mode is lower than 1.2 dB/m from 0.85 to 1.15 THz (with a minimum of at 0.98 THz), which is more than one order of magnitude lower than the attenuation coefficient of the main competing mode HE$_{11}$ (Cf. Fig. 20b). It is worth notifying this relatively low attenuation is obtained for a small hollow-core diameter that is only six times the wavelength (D/λ ~6.1 at 1 THz). Nevertheless, the fabrication of this Bragg fiber is not demonstrated, probably due to the complexity due to complexity of the design (very thin bridges associated with thin TOPAS layers spaced by large air gaps).

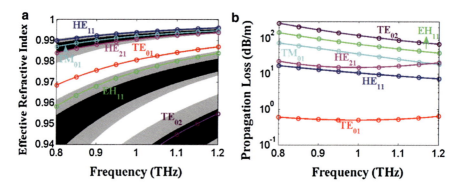

Fig. 20 (a) Dispersion diagram of effective indices vs. frequency of modes confined within the hollow-core and the modes supported by the Bragg cladding (white area). Gray and black areas show domains where the Bragg cladding forbids wave extensions (i.e., bandgaps where no mode is confined in the cladding) for TE/HE and TM/EH modes, respectively. (b) Attenuation spectra of the first six modes guided within the hollow-core Bragg fiber. The fiber is composed of a hollow-core of 1.834 mm diameter surrounded by four bilayers TOPAS/air (with a thickness of about t$_h$ = 77.5 μm and t$_l$ = 1.154 mm, respectively) that are held together by thin TOPAS bridges (t = 15 μm). (Reproduced from Hong et al. (2017), with the permission of IOP Publishing)

The recent progresses in the 3D printing technology might open new possibilities for fabricating THz fibers with complex design. Li et al. have fabricated a hollow-core Bragg fiber with a 3D printer (Li et al. 2017). The fiber is composed of a hollow-core surrounded by ten bilayers of high- and low-refractive index, namely, the printing resin ($n_h \sim 1.654$) and air that are held with additional micro-bridges. The thickness of each bilayer is about 512 μm, with a predicted fundamental bandgap centered at 0.18 THz. The diameter of the hollow-core is reduced to 4.5 mm ($D/\lambda = 2.6$) for enabling the propagation of the HE_{11} mode only (at the expense of a higher attenuation coefficient) and achieving an effectively single-mode operation at the fundamental bandgap. The measured attenuation coefficient is about ~ 0.12 cm^{-1} (~ 52 dB/m) within the narrow bandgap (45 GHz bandwidth centered at 0.18 THz), which is significantly smaller than the corresponding bulk absorption loss of the printing resin ($\alpha \sim 1$ cm^{-1}). It is worth notifying that this process is limited to the fabrication of 25-mm-long length fiber.

Nevertheless, this length is long enough for developing fiber-based sensors. Li et al. have realized a sensor by increasing the thickness of the first layer (high index) of the Bragg reflector in order to allow the confinement of a mode in this defect, by TIR mechanism at the interface hollow-core/high-index layer, and by the Bragg reflector (bandgap) (Li et al. 2017). This defect mode that exists within the bandgap frequencies is coupled with the HE_{11} mode at a specific frequency, which results as a sharp peak of losses in the transmission spectrum. The frequency of this peak is then sensitive to change in the hollow-core, especially around the defect layer. This property has been exploited for detecting added PMMA films on the inner core surface with a sensitivity of 0.1 GHz/μm and for sensing α-lactose monohydrate powder (deposited on the inner core surface) with a reliable detection of 3 μm change in the analyte layer thickness. This performance which is among the best ever reported in THz sensing demonstrates the large potential of this strategy for developing fiber-based sensors.

Hollow-Core Pipe Fibers

Hollow-core pipe fiber is simply a pipe composed of an air channel surrounded by a dielectric layer, as illustrated in Fig. 21. THz waves are confined and propagated in the air channel by the antiresonant reflecting mechanism. These THz fibers are analog to hollow-core photonic crystal fibers with a Kagomé lattice (Benabid et al. 2002; Pearce et al. 2007) and especially the simplified version composed of an air channel surrounded by a thin silica ring suspended in air by six silica struts (Gérôme et al. 2010).

Antiresonant reflecting mechanism has been first developed by Duguay et al. (1986) for guiding light in planar optical waveguides. It acts in wave structures composed of one or several dielectric layers with thicknesses in the same order of magnitude of the wavelength of the confined wave. For hollow-pipe fibers, the dielectric layer of the pipe acts as a Fabry-Perot resonator; at the resonant frequencies, the wave crosses the layer, while it is reflected otherwise (under the

Fig. 21 Schematic representations of a hollow-core pipe fiber, of the two-dimensional distribution of the Poynting vector of the fundamental mode (HE_{11}) confined into the air channel and of its transmission spectrum

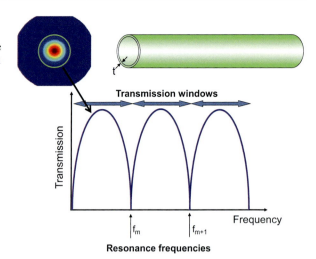

antiresonant condition) leading to wave confinement and propagation into the air channel. Transmission windows are delimited by the resonant frequencies that fulfill the following relation of the Fabry-Perot resonance conditions for a dielectric layer:

$$f_m = \frac{m\,c}{2\,t\,\sqrt{n^2 - 1}} \tag{11}$$

with m the resonance order; t and n the thickness and the refractive index (real part) of the dielectric layer, respectively; and c the speed of light in vacuum. As illustrated in Fig. 21, the waves are guided in the air channel of the pipe within frequency-delimited transmission windows.

Lai et al. (2009, 2010) have demonstrated the propagation of THz waves in the air channel of Teflon or polymethacrylate of methyl pipes. These commercially available pipes are composed of a hollow-core diameter (D) of 9 mm and a dielectric layer thickness (t) of 1 or 0.5 mm. The measured free-space coupling coefficient from the source to the fiber is around 40% in average with a maximum of 84%. The attenuation coefficient measured by the cutback method with a pipe length up to 3 m is as low as 0.0008 cm^{-1} ($\alpha = 0.35$ dB/m) below 1 THz. The attenuation coefficient is proportional to the hollow-core diameter following a $1/D^4$ dependency. It is worth notifying that this trend is different than the $1/D^3$ dependence of a hollow-core surrounded by an infinite thick dielectric layer or dielectric pipe with a thick layer (t $\gg \lambda$). Therefore, lower attenuation coefficient of the fundamental mode HE_{11} is obtainable with large core size.

Nevertheless, high-order modes are also propagated into the air channel. The modes propagated in the air channel pipe by the antiresonant reflecting guiding mechanism are not pure modes. They are leaky modes. In contrast with the modes of classical optical fibers (based on TIR guiding mechanism), the modes exhibit propagation losses, and they do not have a cutoff frequency. The distributions

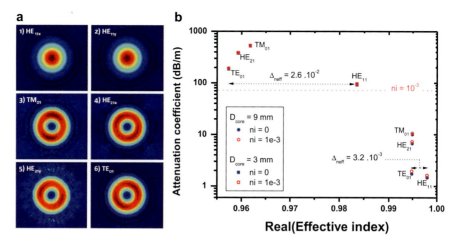

Fig. 22 (**a**) 2D distribution of the Poynting vector of the fundamental mode (HE_{11}) and first higher-order modes. (**b**) Attenuation coefficient and effective index (real part) of the fundamental mode (HE_{11}) and first higher-order modes, for dielectric pipes (n = 1.4) of different core diameters, with or without material absorption

(in two dimensions) of the Poynting vector of the first six modes (with the highest effective index) confined in the air channel are shown in Fig. 22a. Their attenuation coefficients are plotted in Fig. 22b versus their effective indices for a core diameter of D = 9 or 3 mm. The leaky nature of the modes is highlighted by their large attenuation coefficients in the idealistic case of a dielectric layer without material absorption. As already mentioned, the size of the core has a strong effect on the attenuation coefficient of the modes. For example, the attenuation coefficient of the fundamental mode (HE_{11}) is reduced by almost two orders of magnitudes (from 92 to 1.5 dB/m) when the core diameter is increased from $D/\lambda = 4$ to 12. This trend is similar for higher-order modes. Therefore, increasing the core size is a good strategy for reducing the propagation losses, but without improving the discrimination between the fundamental and high-order modes. For the TE_{01} mode, the mode with the closest attenuation coefficient to HE_{11} mode, the ratio $\alpha_{TE01}/\alpha_{HE11}$ decreases from ~2 to ~1.2, for a core diameter of 3 and 9 mm, respectively. Furthermore, larger core size increases the effective index of all modes toward unity ($n_{air} = 1$) pushing the higher-order modes closer to the fundamental modes. For example, the difference between the effective index of the HE_{11} and TE_{01} modes ($\Delta neff = neff_{HE11} - neff_{TE01}$) decreases from $2.6 \cdot 10^{-2}$ to $3.2 \cdot 10^{-3}$ for a core diameter of 3 and 9 mm, respectively. In consequence, couplings between the fundamental mode and higher order ones are more likely to happen with larger core size leading to multimode propagations and additional losses from intermodal couplings.

The effect of material absorption from the dielectric layer is rather small due to the confinement of the waves in the air channel with low penetration in the dielectric layer. As shown in Fig. 22, the material absorption is reduced by 45 times for a core diameter of 9 mm. The ratio $\alpha_{mat}/\alpha_{HE11}$ is mainly governed by the attenuation coefficient of the pipe without material absorption. Indeed, adding a material absorption of $\alpha_{mat} = 72.76$ dB/m at 400 GHz (corresponding to an imaginary part of refractive index $n_i = 10^{-3}$) increases the attenuation coefficient of the mode HE_{11} by only 1.04 or 1.10 for a core diameter of 3 and 9 mm, respectively. For a material absorption of $n_i = 10^{-2}$, this ratio increases only to 1.19 and 1.35 for a core diameter of 3 and 9 mm, respectively. This leads to a reduction of the material absorption by 260 times for a core diameter of 9 mm. It is worth notifying that the effect of material absorption is slightly lower for smaller core diameter. But this has to be balanced by the large augmentation of the attenuation coefficient of the modes (even with $\alpha_{mat} = 0$), which ones could be higher than the material absorption coefficient, as it is the case for a diameter of 3 mm.

For numerous optical fibers, bends generate additional transmission losses. In the case of hollow-core pipe fiber, bend losses of about 2.6 dB/m (0.006 cm^{-1}, at 420 GHz) have been reported for a bend radius of 60 cm applied on a pipe fiber (D = 9 mm and t = 0.5 mm) (Lu et al. 2010). In these fibers, the bend losses are function of the spectrum of the transmission bands. They are larger at frequencies close to the resonant frequencies and smaller in the middle of the transmission bands where the wave confinements are stronger (Lu et al. 2010). Furthermore, the bend losses increase with the thickness of the dielectric layer (t) and decrease with larger core size.

Nevertheless, Lai et al. (2014) have demonstrated that for a given bend radius, there is a critical core diameter above which bend loss reductions are less efficient. Chih-Hsien Lai et al. explain this behavior with the picture of ray propagations. At a given frequency and bend radius, the waves are bounced off the wall of the air channel at both inner and outer core-wall interfaces (in the plan of the bend). When the core diameter increases, the distance between two consecutive reflections also increases. As a result, the number of reflection inside the waveguide reduces, and thus the loss resulting from the reflection decreases. Above the critical core diameter, the waves are reflected only at the outer core-wall interface. This regime is similar to whispering gallery modes that are confined by multiple reflections at the outer boundary of a sphere or a toroid. This regime is more likely to appear with small bend radius, large air core diameter, or higher operating frequency.

Since the transmission windows are delimited by the resonant frequencies, the refractive index and the thickness of the dielectric layer define their bandwidth ($\Delta\lambda$) with the following relation:

$$\Delta f = \frac{c}{2\,t\,\sqrt{n^2 - 1}} \qquad (12)$$

A lower refractive index and a thinner wall lead therefore to broader transmission windows. Broad transmission windows of 0.36 and 0.58 THz have been demonstrated with silica pipes composed of a layer thickness of 250 and 155 μm, respectively, and an outer diameter of 5 mm (Nguema et al. 2011). Transmission

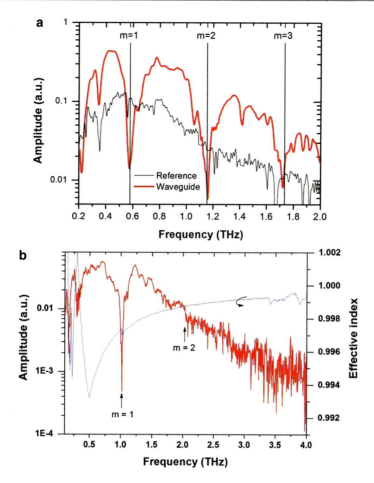

Fig. 23 (**a**) Fourier amplitude spectra of measured THz pulses after propagation in free space (reference) or through a 40-cm-long length silica pipe (t = 155 μm, D = 5 mm). (**b**) Measured THz pulses after propagation through a 50-cm-long length "commercial drinking straw": (red curve) Fourier amplitude spectrum, (blue curve) measured effective index of the propagated THz fields

spectrum of the silica pipe with a layer thickness of 155 μm is shown in Fig. 23a. Much wider transmission windows could be obtained with lower refractive index materials than silica (n ~1.95) such as polyethylene (PE, n ~1.5). As shown in Fig. 23b, a transmission window of 1.02 THz is obtained with a simple "commercial drinking straw" composed of a 120 μm thick PE layer and an outer diameter of 6 mm.

THz wave guidance through the air channel of the straw is confirmed by the evolution of the measured effective index of the electromagnetic field (Cf. Fig. 23b). The values are below unity, and the curve presents the characteristic shape within the transmission window and with variations closed to the resonance frequencies (at least for the first one).

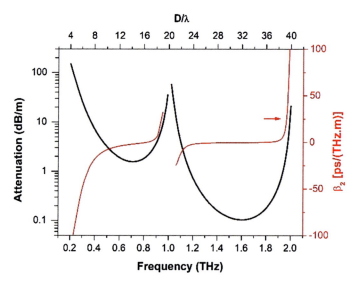

Fig. 24 Simulated attenuation coefficient and group velocity dispersion of the HE$_{11}$ mode propagated into the air channel of a thin PE pipe (t = 143 μm, n = 1.45, D = 6 mm)

It is worth notifying that the effective index is tangent to unity for larger frequency leading to lower GVD values. As shown in Fig. 24, the GVD of the HE$_{11}$ mode propagated into the air channel of a thin PE pipe (t = 143 μm, n = 1.45, D = 6 mm) is strongly affected by the degradations of the guiding conditions close to the resonant frequencies leading to sharp variations of the GVD from negative to positive values. The GVD is smaller in the second transmission windows. At larger frequencies, the size of the core compared to the wavelength size is larger. As a result, the HE$_{11}$ mode could propagate more parallel (to the propagation axis) with a larger distance between two consecutive reflections at the dielectric layer interface, leading to properties closer to the ones of a wave propagating in air. The ratio D/λ is two times larger in the second windows than in the first one. As a consequence, the GVD is below ±2 ps/(THz·m) in a broader frequency range, in the second window (Δλ = 700 GHz, from 1.2 to 1.9 THz) than in the first one (Δλ = 160 GHz, from 0.7 to 0.86 THz).

In comparison to other THz fibers, their broad transmission windows, relatively low attenuation coefficient, and very low GVD make these fibers very attractive especially for pulse propagations. However, these performances are obtained for a large core diameter (D > 20λ) that induces favorable conditions for multimode propagations and additional losses from intermodal couplings (as shown in Fig. 22), leading to drastic conditions for coupling THz radiations to only the HE$_{11}$ mode of the air channel.

Even if most of the work reported on hollow-core THz pipe fibers used commercial polymer pipes, some THz pipes have been especially developed by drawing down thin-wall silica capillaries (from a silica tube) with optical fiber fabrication facilities (Nguema et al. 2011), by fabricating thin-wall chalcogenide-glass capillaries with a double crucible glass drawing technique (Mazhorova et al. 2012), or also by drawing down thin-wall PMMA capillaries (from a PMMA tube) with low-temperature fiber drawing tower (Xiao et al. 2013a). Xiao et al. have also fabricated a self-supporting PMMA pipe by drawing down a PMMA tube filled with seven PMMA capillaries (Xiao et al. 2013b). The fiber is composed of a thin hexagonal inner pipe delimiting a hollow-core that is supported by six thin walls attached on the outer pipe. By this process, they have been able to reduce the thickness of the inner pipe down to $\sim$20 μm leading to broad transmission windows in the highest THz frequency range ($\Delta\lambda$ $\sim$6.5 THz). Finally, square or rectangular polymer pipes have been fabricated by sticking four PE or PMMA strips (Lu et al. 2011a). Rectangular pipes have been developed for studying and demonstrating some polarization sensitivity of the attenuation coefficient that could be controlled by adjusting the structure of the rectangle. Furthermore, directional coupler composed of two squares or rectangular pipes have been demonstrated by simply placing both pipes in contact side by side (Lu et al. 2011b).

A hollow-core THz pipe fiber has been successfully integrated into a reflective THz fiber-scan imaging system for in situ and dynamically monitoring the chemical reaction of hydrochloric acid (HCl) and ammonia (NH_3) vapors to generate ammonium chloride (NH_4Cl) aerosols in a sealed chamber (You and Lu 2016). A simple Teflon pipe has been used as a sensing and image-scanning arm. The bending capability of the reflective THz pipe fiber-based scan system has been demonstrated by imaging and identifying an array of tablets. Applications of hollow-core THz pipe fibers have been also demonstrated for high-performance sensing by tracking the shift of the resonant frequencies, when a sub-wavelength-thick molecular overlayer (with different Carbopol aqueous solutions) is adhered to the inner core surface of a tube (You et al. 2010a), when various powders are loaded on the outer surface of the pipe (composed of an absorptive layer), or when different vapors are inserted into the air channel of the pipe (You et al. 2012).

The guiding mechanism and properties of hollow-core THz pipe fibers are the basement of others hollow-core THz fibers, even if they have been developed independently. These THz fibers are presented in the following sections.

Kagome Hollow-Core Photonic Crystal Fibers

Kagome hollow-core PCF have been intensively developed in the optical domain for exploiting their broadband transmissions. These fibers are analog to hollow-core pipe fibers; they are composed of concentric thin hexagonal inner pipes supported by thin walls forming a Kagome lattice. Light is confined and propagated in the hollow-core by antiresonance guiding mechanism with inhibited coupling to cladding modes, if the wall connections are free of apex (where parasitic couplings could appear due to cladding mode confinement) (Couny et al. 2007).

The properties of Kagome hollow-core PCF are therefore similar to the ones of the hollow-core pipe fibers. As shown in Fig. 25a, the attenuation spectrum

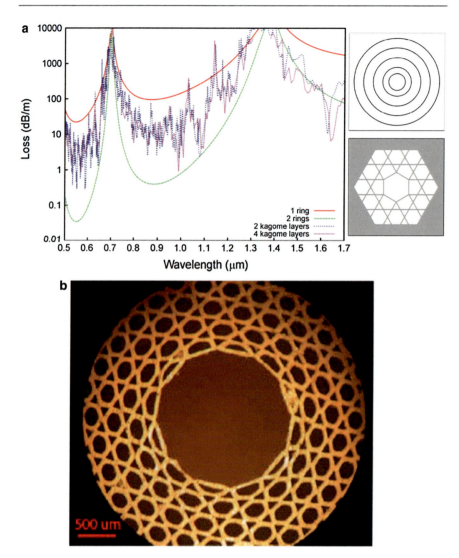

Fig. 25 (a) Numerically calculated attenuation spectra of the fundamental mode of a Kagome hollow-core PCF composed of two or four layers, together with that of a hollow-core pipe fibers composed of one or two layers. Schematics of the fibers structures are shown in the insets. Black and gray areas represent glass and white areas represent air. (Reproduced from Pearce et al. (2007), with the permission of OSA Publishing). (**b**) Optical micrograph of the cross section of a PMMA Kagome fiber showing the hollow-core and the innermost four of the six rings of the cladding. (Reproduced from Anthony et al. (2011b), with the permission of OSA Publishing)

is smaller than the one of a hollow-core pipe, due to the addition of at least one antiresonant layer (i.e., a thin hexagonal layer) (Pearce et al. 2007). Nevertheless, the supporting thin walls add some parasitic couplings between the core mode and cladding modes that perturb the attenuation spectrum with high attenuation peaks.

Anthony et al. have fabricated different Kagome hollow-core PCF in PMMA by drawing a stack of PMMA tubes (filled within a larger PMMA tube) down to fibers with core and outer diameters around 2 and 6 mm, respectively (Cf. Fig. 25b) (Anthony et al. 2011b). The hollow-core was formed by removing seven tubes from the stack composed of six hexagonal rings of tubes (in a triangular lattice) to form the cladding. Even if the cladding is not a perfectly Kagome lattice, they have demonstrated the propagation of THz waves with an attenuation coefficient about 0.6 cm^{-1} (260 dB/m) in the frequency range from 0.65 to 1.0 THz, corresponding to a reduction of about 20 times of the absorption coefficient of PMMA material.

Yang et al. have fabricated a Kagome hollow-core PCF in polymer material by 3D printing (Yang et al. 2016). The fiber is composed of a hollow-core of 9 mm diameter surrounded by two layers of air holes with a diameter of 3 mm and a wall thickness of 0.35 mm. The maximum fiber length was 30 cm. The cladding design is more a triangular lattice of connected capillaries than a Kagome lattice. Nevertheless, they have demonstrated the propagation of THz waves with an average attenuation coefficient of 0.02 cm^{-1} (8.7 dB/m) within the frequency range from 0.2 to 1.0 THz with the minimum attenuation about 0.002 cm^{-1} (0.87 dB/m) at 0.75 THz.

Even if the Kagome-lattice cladding protects the waves guided in the core to external perturbations, the large outer diameter of such fiber could be reduced since most of the confinement is realized by the two first layers as shown on Fig. 25a. In this prospect, Wu et al. have proposed a numerical study on the reduction of the overall fiber diameter by eliminating the outermost cladding layers for the resulting fibers to be practical and flexible (Wu et al. 2011).

Tube Lattice Hollow-Core Fibers

Tube lattice hollow-core THz fibers are a new family of hollow-core fibers based on antiresonance guiding mechanism with inhibited coupling to cladding modes (Vincetti 2010). These fibers are composed of a hollow-core surrounded by a cladding formed by a periodic arrangement of tubes in a triangular lattice (Cf. Fig. 26a, b). The properties of these fibers are similar to the ones of the hollow-core pipe fiber or Kagome hollow-core PCF that are characterized by broadband transmission windows. The wave confinement and propagation in the hollow-core is obtained by inhibiting the couplings between the core modes and the cladding modes. The field overlap between the core and the cladding modes is reduced by the tube lattice cladding that is free of node, avoiding localization of cladding modes.

As in pipe or Kagome fibers, the transmission windows are limited by sharp propagation losses at the cutoff frequencies of cladding modes where core modes are coupled to leaky cladding modes. These frequencies are well approximated by those of a single tube from the cladding that is defined by its refractive index, wall thickness, and external diameter (Vincetti and Setti 2010). Wider

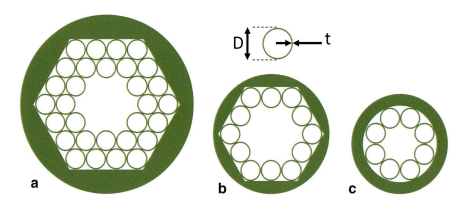

Fig. 26 Schematics of the tube lattice hollow-core THz fiber composed of (**a**) two layers of tube lattice, (**b**) one layer of tube lattice, and (**c**) one layer of tube lattice with an octagonal symmetry

transmission windows are obtained for a tube with a thinner wall and a large diameter (compared to the wavelength). In this condition, the cutoff frequencies are similar to those of an infinite slab that are following the resonance conditions of a Fabry-Perot (Eq. 11). As for a pipe or a Kagome fiber, the transmission windows depend therefore on the wall thickness of the tubes that form the cladding. This property enables the fabrication of THz fibers with broader transmission windows than that of pipe fiber, since it is easier to fabricate several very thin-wall tubes with a relative small diameter ($\sim$1 mm) than one with a large diameter (several millimeters). Furthermore, a large thin-wall pipe is less flexible than a bundle of small diameter thin-wall tubes, making the tube lattice hollow-core fiber an interesting design for developing low-loss flexible THz fibers. The tube lattice cladding presents another interest for reducing the propagation losses. In this configuration, a part of the external boundary of each tube in the first ring of the cladding delimits the hollow-core that could be represented as a large circle composed of negative curvatures (half perimeter) from each tube. These negative curvatures minimize the spatial overlap between the core modes and the cladding modes (in the tube lattice) leading to a better confinement and lower attenuation coefficient of the core modes (Wang et al. 2011; Pryamikov et al. 2011).

An attenuation coefficient of the HE_{11} mode as low as 0.07 dB/m at the frequency of 1.9 THz (and lower than 0.1 dB/m over a range of about 800 GHz) has been numerically predicted for a fiber composed of 12 Teflon tubes (of 1 mm diameter and wall thickness of 44 μm) surrounding an hollow-core (Vincetti 2010). The core shape corresponds to seven missing tubes packed in a triangular lattice, leading to a core diameter smaller than 3 mm.

The use of the triangular lattice in the fiber design comes from the fabrication process of PCF that is based on stacking capillaries or tubes in a triangular lattice. The tube lattice fibers composed of only one ring are not restricted to this design. A tube lattice fiber with octagonal symmetry (Cf. Fig. 26c), i.e., 8 tubes surrounding the hollow-core instead of 12, could improve the loss discrimination between the HE_{11} mode and TE_{01} mode, by a factor of about 2.6 (Vincetti et al. 2010). This is attributed to better phase-matching conditions between the high-order core modes and cladding modes.

Following this development, Setti et al. (2013) have fabricated a hollow-core tube lattice THz fiber composed of height PMMA tubes with an external diameter of about 1.99 mm and a wall thickness around 252 μm, leading to a core size around 3.42 mm (Cf. Fig. 27a). Small pieces (L ~5 mm) of jacket PMMA tubes have been added each 10 cm on the fiber structure for ensuring its mechanical stability. They have measured the propagation of THz waves within two transmission windows (0.3–0.5 THz and 0.6–0.95 THz) with an attenuation coefficient of 0.3 dB/cm at 0.375 THz and 0.16 dB/cm at 0.828 THz, corresponding to a reduction to the bulk PMMA material by 31 and 272 times, respectively. These measurements and the simulated attenuation spectra of the fundamental core mode for different material absorption coefficient of PMMA (imaginary part of the refractive index) are shown in Fig. 27c.

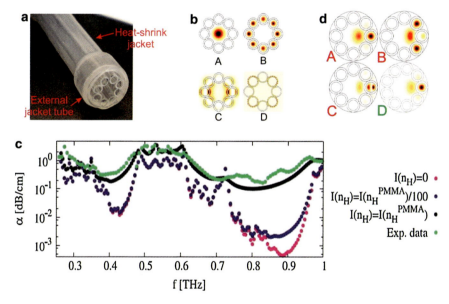

Fig. 27 (**a**) Picture of the tube lattice hollow-core THz fiber fabricated by Setti et al. (**b**) Simulated Poynting vector distribution of the fundamental core mode (A) and different cladding modes confined in tube holes (B) or in tube walls (C, D). (**c**) Measured attenuation coefficient of the THz waves propagated through the THz fiber (green dots) and numerical results of attenuation coefficient of the fundamental core mode with different values of the imaginary part of the refractive index of the tube material. (**d**) Simulated Poynting vector distribution of the fundamental core mode for different fiber bending radius (A, R_b ~21 cm; B, R_b ~15 cm; C, R_b ~7 cm; D, R_b ~12 cm) at frequency where couplings to cladding modes appear, leading to extra losses. (Reproduced from Setti et al. (2013), with the permission of OSA Publishing)

As previously mentioned, this fiber structure is favorable for broadband transmissions and bending. Setti et al. have experimentally demonstrated a clear reduction in the transmission spectrum of the first or second window for a bend radius of 10 or 30 cm, respectively, showing that the lower frequency transmission window is the most robust against bends (Setti et al. 2013). Bends induce a reduction of the transmission windows by a shift of the higher frequency edges to lower frequency. This shift is induced by a variation of the cutoff frequencies of the cladding modes. The bend could be considered following the conformal mapping technique, where the bent fiber is analyzed as a straight fiber with negative and positive variations (in the bend direction) of the refractive index, respectively, in the inner and outer bend curvature, and with amplitude depending on the bend radius. In this picture, smaller bend radius leads to higher refractive index in the outer bent curvature and therefore to smaller cutoff frequencies of the cladding modes that delimit the transmission windows. Furthermore, when the bend radius is small, extra losses due to the resonances between the core mode and hole modes appear (Cf. Fig. 27d). The refractive indices inside the cladding holes change according to their relative position along the bend direction and the bend radius. In particular, the effective indices of the hole modes of the tubes increase approaching the one of the core mode until the phase matching condition is reached, causing the extra losses (Setti et al. 2013). The mechanical flexibility of this kind of fiber has been successfully exploited as a probe in an endoscope geometry and used to guide radiation to and from a sample for enabling THz hyperspectral imaging by scanning the end face of the probe in front of the sample (Lu and Argyros 2014).

The reduction of the attenuation coefficient was realized by replacing the PMMA tubes with Zeonex polymer tubes (Lu et al. 2016). The absorption coefficient of Zeonex cycloolefin polymer is about of $\alpha = 0.14$–0.3 cm^{-1} for frequencies between 0.2 and 1.2 THz. Lu et al. have measured an attenuation coefficient of about 5 dB/m around 0.275 THz. They have also estimated from simulation an attenuation coefficient below 0.1 dB/m in the frequency range 0.8–1.2 THz for a fiber composed of a hollow-core of about 3.3 mm diameter surrounded by ten Zeonex tubes with a wall thickness of 92 μm (Lu et al. 2016). These results are among the lowest losses reported in this frequency range, making the tube lattice hollow-core THz fibers one of the best THz fibers.

Further reduction of the attenuation coefficient might be achievable by exploiting advanced design of tube lattice hollow-core fibers developed in the optical domains. Belardi and Knight (2014) have demonstrated from a simulation study the reduction of the attenuation coefficient by several orders (more than two orders) of magnitudes by adding one or two tubes within each tube of the cladding (composed of only a single layer), with the same thickness and attached to the cladding jacket at the same azimuthal position. The attenuation coefficient could be further reduced by at least one order of magnitude by adding some air gap between the main tubes (with or without added smaller tubes) of the cladding (Poletti 2014). Nevertheless, the interest of these improved designs has to be evaluated in THz domain by considering the large absorption coefficient of the tube material.

Conclusion

The degradation of the conductivity of metals and the low transparency of most dielectric materials in the THz domain call for innovative fiber designs that are mostly inspired from designs of specialty optical fibers. Porous fibers and tube lattice hollow-core fibers are part of fiber designs that originates from THz domain. They have been developed by taking advantages of the larger size of fiber design features at THz frequencies (than at optical ones). It is worth to notify that the tube lattice hollow-core fibers have been then successfully developed in the optical domain. This achievement emphasizes another interest of THz fibers for developing and testing new fiber designs that could be benefit for specialty optical fibers in the optical and infrared domains. Numerous fiber desings that have been investigated open been investigated that open various strategy for improving the performances of THz fibers. Even if the attenuation coefficient of THz fibers does not enable to deliver THz radiations over long fiber lengths, the performances of THz fibers become satisfactory for developing components and systems. These emerging realizations exploit the properties of specific THz fibers depending on the targeted applications.

Solid-core fibers have the advantages of (i) single-mode propagation (HE_{11} mode, with a Gaussian intensity distribution), (ii) a large fraction of evanescent power, and (iii) a GVD curve that depends on the fiber design. On the other side, they are sensitive to external perturbations, have a limited transmission windows (expected for the PCF), their properties and designs strongly depend on the refractive index and absorption coefficient of the fiber material (especially for the PCF), and, the mean GVD value is rather high (limiting pulse propagations to few tens of centimeters).

The confinement of the THz waves in a hollow-core yields a low GVD, a rather low attenuation coefficient, and a low sensitivity to external perturbations (except for hollow-core pipe fiber). Nevertheless, hollow-core THz fibers have limited transmission windows, and they could guide a large number of higher-order modes (in addition to the fundamental one) if THz radiations are not carefully coupled to the fundamental mode. It is difficult to fairly compare the performances of each hollow-core THz fiber, since the reported properties depend on the core size and the operating wavelength (i.e., D/λ), on the materials used, on the fabrication processes employed, and on the measurement methods. Even if some fiber designs require complex fabrication process that could affect the performances of the fibers, a comparison of the hollow-core THz fibers should be realized by simulating the different fiber designs with the same core size, refractive index (complex), and operating wavelength. Besides, the hollow-core fibers could lead to a very low attenuation coefficient by increasing the core size, but at the expense of stronger intermodal couplings to higher-order modes. Future designs of hollow-core fibers might therefore decrease the intermodal couplings by increasing the losses and phase differences between the fundamental and high-order modes, in order to enable an effective single-mode propagation.

In order to sustain the intense efforts on the development of THz devices, upcoming developments of THz fibers should investigate novel designs (to reduce the attenuation coefficient and to improve bending capabilities) and new materials (with reduced absorption coefficient). The developments of fabrication processes dedicated to THz fiber specifications (materials and size), and of rigorous characterization protocols of THz fibers (compatible with low attenuation coefficient measurements) are also required. These underlying challenges make the development of THz fibers a very stimulating research area with myriad of potential applications.

References

Y. Abe, Y. Matsuura, Y.-W. Shi, Y. Wang, H. Uyama, M. Miyagi, Polymer-coated hollow fiber for CO_2 laser delivery. Opt. Lett. 23, 89–90 (1998)

J. Anthony, R. Leonhardt, A. Argyros, M.C.J. Large, Characterization of a microstructured Zeonex terahertz fiber. J. Opt. Soc. Am. B 28(5), 1013–1018 (2011a)

J. Anthony, R. Leonhardt, S.G. Leon-Saval, A. Argyros, THz propagation in Kagome hollow-core microstructured fibers. Opt. Express 19, 18470–18478 (2011b)

S. Atakaramians, S. Afshar V, B.M. Fischer, D. Abbott, T.M. Monro, Porous fibers: a novel approach to low loss THz waveguides. Opt. Express 16(12), 8845–8854 (2008)

S. Atakaramians, S. Afshar V, H. Ebendorff-Heidepriem, M. Nagel, B.M. Fischer, D. Abbott, T.M. Monro, THz porous fibers: design, fabrication and experimental characterization. Opt. Express 17(16), 14053–15062 (2009)

H. Bao, K. Nielsen, H.K. Rasmussen, P.U. Jepsen, O. Bang, Fabrication and characterization of porous-core honeycomb bandgap THz fibers. Opt. Express 20, 29507–29517 (2012)

H. Bao, K. Nielsen, H.K. Rasmussen, P.U. Jepsen, O. Bang, Design and optimization of mechanically down-doped terahertz fiber directional couplers. Opt. Express 22, 9486–9497 (2014)

W. Belardi, J.C. Knight, Hollow antiresonant fibers with reduced attenuation. Opt. Lett. 39, 1853–1856 (2014)

F. Benabid, J.C. Knight, G. Antonopoulos, P. St, J. Russell, Stimulated Raman scattering in hydrogen-filled hollow-core photonic crystal fiber. Science 298, 399–402 (2002)

B. Bowden, J.A. Harrington, O. Mitrofanov, Silver/polystyrene-coated hollow glass waveguides for the transmission of terahertz radiation. Opt. Lett. 32, 2945–2947 (2007)

B. Bowden, J.A. Harrington, O. Mitrofanov, Low-loss modes in hollow metallic terahertz waveguides with dielectric coatings. Appl. Phys. Lett. 93(18), 181104 (2008)

H.-C. Chang, C.-K. Sun, Subwavelength dielectric-fiber-based THz coupler. J. Lightwave Technol. 27(11), 1489–1495 (2009)

L.-J. Chen, H.-W. Chen, T.-F. Kao, J.-Y. Lu, C.-K. Sun, Low-loss subwavelength plastic fiber for terahertz waveguiding. Opt. Lett. 31, 308–310 (2006)

H.-W. Chen, Y.-T. Li, C.-L. Pan, J.-L. Kuo, J.-Y. Lu, L.-J. Chen, C.-K. Sun, Investigation on spectral loss characteristics of subwavelength terahertz fibers. Opt. Lett. 32, 1017–1019 (2007a)

H.-W. Chen, J.-Y. Lu, L.-J. Chen, P.-J. Chiang, H.-C. Chang, Y.-T. Li, C.-L. Pan, C.-K. Sun, in *THz Fiber Directional Coupler. Proceedings of CLEO/QELS'2007, Baltimore* (2007b)

H. Chen, W.-J. Lee, H.-Y. Huang, C.-M. Chiu, Y.-F. Tsai, T.-F. Tseng, J.-T. Lu, W.-L. Lai, C.-K. Sun, Performance of THz fiber-scanning near-field microscopy to diagnose breast tumors. Opt. Express 19, 19523–19531 (2011)

C.-M. Chiu, H.-W. Chen, Y.-R. Huang, Y.-J. Hwang, W.-J. Lee, H.-Y. Huang, C.-K. Sun, All-terahertz fiber-scanning near-field microscopy. Opt. Lett. 34, 1084–1086 (2009)

M. Cho, J. Kim, H. Park, Y. Han, K. Moon, E. Jung, H. Han, Highly birefringent terahertz polarization maintaining plastic photonic crystal fibers. Opt. Express **16**, 7–12 (2008)

F. Couny, F. Benabid, P. Roberts, P. Light, M. Raymer, Generation and photonic guidance of multi-octave optical-frequency combs. Science **318**(5853), 1118–1121 (2007)

P.D. Cunningham, N.N. Valdes, F.A. Vallejo, L.M. Hayden, B. Polishak, X.-H. Zhou, J. Luo, A.K. Jen, J.C. Williams, R.J. Twieg, Broadband terahertz characterization of the refractive index and absorption of some important polymeric and organic electro-optic materials. J. Appl. Phys. **109**(4), 043505 (2011)

P. Doradla, C.S. Joseph, J. Kumar, R.H. Giles, Characterization of bending loss in hollow flexible terahertz waveguides. Opt. Express **20**, 19176–19184 (2012)

M.A. Duguay, Y. Kokubun, T.L. Koch, L. Pfeiffer, Antiresonant reflecting optical waveguides in SiO_2-Si multilayer structures. Appl. Phys. Lett. **49**, 13 (1986)

A. Dupuis, J.-F. Allard, D. Morris, K. Stoeffler, C. Dubois, M. Skorobogatiy, Fabrication and THz loss measurements of porous subwavelength fibers using a directional coupler method. Opt. Express **17**(10), 8012–8028 (2009)

A. Dupuis, A. Mazhorova, F. Désévédavy, M. Rozé, M. Skorobogatiy, Spectral characterization of porous dielectric subwavelength THz fibers fabricated using a microstructured molding technique. Opt. Express **18**(13), 13813–13828 (2010)

A. Dupuis, K. Stoeffler, B. Ung, C. Dubois, M. Skorobogatiy, Transmission measurements of hollow-core THz Bragg fibers. J. Opt. Soc. Am. B **28**, 896–907 (2011)

Y. Fink, J.N. Winn, S. Fan, C. Chen, J. Michel, J.D. Joannopoulos, E.L. Thomas, A dielectric omnidirectional reflector. Science **282**, 1679–1682 (1998)

Y. Fink, D.J. Ripin, S. Fan, C. Chen, J.D. Joannopoulos, E.L. Thomas, Guiding optical light in air using an all-dielectric structure. J. Lightwave Technol. **17**, 2039–2041 (1999)

F. Gérôme, R. Jamier, J.-L. Auguste, G. Humbert, J.-M. Blondy, Simplified hollow-core photonic crystal fiber. Opt. Lett. **35**, 1157–1159 (2010)

H. Han, H. Park, M. Cho, J. Kim, Terahertz pulse propagation in a plastic photonic crystal fiber. Appl. Phys. Lett. **80**, 2634–2636 (2002)

J.A. Harrington, R. George, P. Pedersen, E. Mueller, Hollow polycarbonate waveguides with inner Cu coatings for delivery of terahertz radiation. Opt. Express **12**, 5263–5268 (2004)

A. Hassani, A. Dupuis, M. Skorobogatiy, Low loss porous terahertz fibers containing multiple subwavelength holes. Appl. Phys. Lett. **92**(7), 071101 (2008)

B. Hong, M. Swithenbank, N. Somjit, J. Cunningham, I. Robertson, Asymptotically single-mode small-core terahertz Bragg fibre with low loss and low dispersion. J. Phys. D. Appl. Phys. **50**(4), 045104 (2017)

M. Ibanescu, Y. Fink, S. Fan, E.L. Thomas, J.D. Joannopoulos, An all-dielectric coaxial waveguide. Science **289**, 415–419 (2000)

J.C. Knight, Photonic crystal fibers. Nature **424**, 847–851 (2003)

C.H. Lai, Y.C. Hsueh, H.W. Chen, Y.J. Huang, H.C. Chang, C.K. Sun, Low-index terahertz pipe waveguides. Opt. Lett. **34**(21), 3457–3459 (2009)

C.H. Lai, B. You, J.Y. Lu, T.A. Liu, J.L. Peng, C.K. Sun, H.C. Chang, Modal characteristics of antiresonant reflecting pipe waveguides for terahertz waveguiding. Opt. Express **18**(1), 309–322 (2010)

C.-H. Lai, T. Chang, Y.-S. Yeh, Characteristics of bent terahertz antiresonant reflecting pipe waveguides. Opt. Express **22**, 8460–8472 (2014)

J. Li, K. Nallappan, H. Guerboukha, M. Skorobogatiy, 3D printed hollow core terahertz Bragg waveguides with defect layers for surface sensing applications. Opt. Express **25**, 4126–4144 (2017)

J. Lou, L. Tong, Z. Ye, Modeling of silica nanowires for optical sensing. Opt. Express **13**, 2135–2140 (2005)

W. Lu, A. Argyros, Terahertz spectroscopy and imaging with flexible tube-lattice fiber probe. J. Lightwave Technol. **32**(23), 4019–4025 (2014)

J.-Y. Lu, C.-C. Kuo, C.-M. Chiu, H.-W. Chen, Y.-J. Hwang, C.-L. Pan, C.-K. Sun, THz interferometric imaging using subwavelength plastic fiber based THz endoscopes. Opt. Express **16**, 2494–2501 (2008)

J.-T. Lu, Y.-C. Hsueh, Y.-R. Huang, Y.-J. Hwang, C.-K. Sun, Bending loss of terahertz pipe waveguides. Opt. Express **18**, 26332–26338 (2010)

J.-T. Lu, C.-H. Lai, T.-F. Tseng, H. Chen, Y.-F. Tsai, I.-J. Chen, Y.-J. Hwang, H.-c. Chang, C.-K. Sun, Terahertz polarization-sensitive rectangular pipe waveguides. Opt. Express **19**, 21532–21539 (2011a)

J.-T. Lu, C.-H. Lai, T.-F. Tseng, H. Chen, Y.-F. Tsai, Y.-J. Hwang, H.-c. Chang, C.-K. Sun, Terahertz pipe-waveguide-based directional couplers. Opt. Express **19**, 26883–26890 (2011b)

W. Lu, S. Lou, A. Argyros, Investigation of flexible low-loss hollow-core fibres with tube-lattice cladding for terahertz radiation. IEEE J. Sel. Top. Quantum Electron. **22**(2), 214–220 (2016)

T. Ma, A. Markov, L. Wang, M. Skorobogatiy, Graded index porous optical fibers – dispersion management in terahertz range. Opt. Express **23**, 7856–7869 (2015)

A. Mazhorova, A. Markov, B. Ung, M. Rozé, S. Gorgutsa, M. Skorobogatiy, Thin chalcogenide capillaries as efficient waveguides from mid-infrared to terahertz. J. Opt. Soc. Am. B **29**, 2116–2123 (2012)

R.W. McGowan, G. Gallot, D. Grischkowsky, Propagation of ultrawideband short pulses of terahertz radiation through submillimeter-diameter circular waveguides. Opt. Lett. **24**, 1431–1433 (1999)

O. Mitrofanov, J.A. Harrington, Dielectric-lined cylindrical metallic THz waveguides: mode structure and dispersion. Opt. Express **18**(3), 1898–1903 (2010)

M. Miyagi, A. Hongo, S. Kawakami, Design theory of dielectric coated circular metallic waveguides for infrared transmission. J. Lightwave Technol. **LT-2**, 116–126 (1984)

M. Naftaly, R.E. Miles, Terahertz time-domain spectroscopy of silicate glasses and the relationship to material properties. J. Appl. Phys. **102**(4), 043517 (2007)

M. Navarro-Cía, M.S. Vitiello, C.M. Bledt, J.E. Melzer, J.A. Harrington, O. Mitrofanov, Terahertz wave transmission in flexible polystyrene-lined hollow metallic waveguides for the 2.5–5 THz band. Opt. Express **21**, 23748–23755 (2013)

E. Nguema, D. Férachou, G. Humbert, J.-L. Auguste, J.-M. Blondy, Broadband terahertz transmission within the air channel of thin-wall pipe. Opt. Lett. **36**, 1782–1784 (2011)

K. Nielsen, H.K. Rasmussen, A.J.L. Adam, P.C.M. Planken, O. Bang, P.U. Jepsen, Bendable, low-loss Topas fibers for the terahertz frequency range. Opt. Express **17**(10), 8592–8601 (2009)

K. Nielsen, H.K. Rasmussen, P.U. Jepsen, O. Bang, Porous-core honeycomb bandgap THz fiber. Opt. Lett. **36**(5), 666–668 (2011)

G.J. Pearce, G.S. Wiederhecker, C.G. Poulton, S. Burger, P. St, J. Russell, Models for guidance in Kagome-structured hollow-core photonic crystal fibers. Opt. Express **15**, 12680–12685 (2007)

F. Poletti, Nested antiresonant nodeless hollow core fiber. Opt. Express **22**, 23807–23828 (2014)

A.D. Pryamikov, A.S. Biriukov, A.F. Kosolapov, V.G. Plotnichenko, S.L. Semjonov, E.M. Dianov, Demonstration of a waveguide regime for a silica hollow-core microstructured optical fiber with a negative curvature of the core boundary in the spectral region >3.5 μm. Opt. Express **19**(2), 1441–1448 (2011)

M. Rozé, B. Ung, A. Mazhorova, M. Walther, M. Skorobogatiy, Suspended core subwavelength fibers: towards practical designs for low-loss terahertz guidance. Opt. Express **19**, 9127–9138 (2011)

P.S.J. Russell, Photonic crystal fibers. Science **299**(5605), 358–362 (2003)

V. Setti, L. Vincetti, A. Argyros, Flexible tube lattice fibers for terahertz applications. Opt. Express **21**, 3388–3399 (2013)

M. Skorobogatiy, A. Dupuis, Ferroelectric all-polymer hollow Bragg fibers for terahertz guidance. Appl. Phys. Lett. **90**, 113514 (2007)

A.W. Snyder, J. Love, *Optical Waveguide Theory* (Springer, Norwell, 1983). ISBN 978-0-412-09950-2, 738. Hardcover

M. Sumetsky, How thin can a microfiber be and still guide light? Opt. Lett. **31**, 870–872 (2006)

X.-L. Tang, Y.-W. Shi, Y. Matsuura, K. Iwai, M. Miyagi, Transmission characteristics of terahertz hollow fiber with an absorptive dielectric inner-coating film. Opt. Lett. **34**, 2231–2233 (2009)

B. Temelkuran, S.D. Hart, G. Benoit, J.D. Joannopoulos, Y. Fink, Wavelength-scalable hollow optical fibres with large photonic bandgaps for CO_2 laser transmission. Nature **420**, 650–653 (2002)

B. Ung, A. Dupuis, K. Stoeffler, C. Dubois, M. Skorobogatiy, High-refractive-index composite materials for terahertz waveguides: trade-off between index contrast and absorption loss. J. Opt. Soc. Am. B **28**, 917–921 (2011a)

B. Ung, A. Mazhorova, A. Dupuis, M. Rozé, M. Skorobogatiy, Polymer microstructured optical fibers for terahertz wave guiding. Opt. Express **19**, B848–B861 (2011b)

L. Vincetti, Single-mode propagation in triangular tube lattice hollow-core terahertz fiber. Opt. Commun. **283**, 979–984 (2010)

L. Vincetti, V. Setti, Waveguiding mechanism in tube lattice fibers. Opt. Express **18**(22), 23133–23146 (2010)

L. Vincetti, V. Setti, M. Zoboli, Terahertz tube lattice fibers with octagonal symmetry. IEEE Photon. Technol. Lett. **22**, 972–974 (2010)

M.S. Vitiello, J.-H. Xu, F. Beltram, A. Tredicucci, O. Mitrofanov, J. Harrington, H.E. Beere, D.A. Ritchie, Guiding a terahertz quantum cascade laser into a flexible silver-coated waveguide. J. Appl. Phys. **110**, 063112 (2011)

Y.Y. Wang, N.V. Wheeler, F. Couny, P.J. Roberts, F. Benabid, Low loss broadband transmission in hypocycloid-core Kagome hollow-core photonic crystal fiber. Opt. Lett. **36**, 669–671 (2011)

F. Warken, E. Vetsch, D. Meschede, M. Sokolowski, A. Rauschenbeutel, Ultra-sensitive surface absorption spectroscopy using sub-wavelength diameter optical fibers. Opt. Express **15**, 11952–11958 (2007)

D.S. Wu, A. Argyros, S.G. Leon-Saval, Reducing the size of hollow terahertz waveguides. J. Lightwave Technol. **29**(1), 97–103 (2011)

M. Xiao, J. Liu, W. Zhang, J. Shen, Y. Huang, THz wave transmission in thin-wall PMMA pipes fabricated by fiber drawing technique. Opt. Commun. **298–299**, 101–105 (2013a)

M. Xiao, J. Liu, W. Zhang, J. Shen, Y. Huang, Self-supporting polymer pipes for low loss single-mode THz transmission. Opt. Express **21**, 19808–19815 (2013b)

J. Yang, J. Zhao, C. Gong, H. Tian, L. Sun, P. Chen, L. Lin, W. Liu, 3D printed low-loss THz waveguide based on Kagome photonic crystal structure. Opt. Express **24**, 22454–22460 (2016)

B. You, J.-Y. Lu, Remote and in situ sensing products in chemical reaction using a flexible terahertz pipe waveguide. Opt. Express **24**, 18013–18023 (2016)

B. You, J.-Y. Lu, J.-H. Liou, C.-P. Yu, H.-Z. Chen, T.-A. Liu, J.-L. Peng, Subwavelength film sensing based on terahertz anti-resonant reflecting hollow waveguides. Opt. Express **18**, 19353–19360 (2010a)

B. You, J.-Y. Lu, T.-A. Liu, J.-L. Peng, C.-L. Pan, Subwavelength plastic wire terahertz time-domain spectroscopy. Appl. Phys. Lett. **96**, 051105 (2010b)

B. You, J.-Y. Lu, C.-P. Yu, T.-A. Liu, J.-L. Peng, Terahertz refractive index sensors using dielectric pipe waveguides. Opt. Express **20**, 5858–5866 (2012)

L. Zhang, F. Gu, J. Lou, X. Yin, L. Tong, Fast detection of humidity with a subwavelength-diameter fiber taper coated with gelatin film. Opt. Express **16**, 13349–13353 (2008)

Optical Fibers for Biomedical Applications 27

Gerd Keiser

Contents

Abstract

This chapter describes the various types of optical fibers that are used for precise delivery of light to specific biological tissue areas and for collection of reflected, scattered, or fluorescing light resulting from the light-tissue interaction. First, the Introduction gives some background information on biophotonics as it applies to the human body. Next, section "Basic Concepts of Optical Fibers" discusses

G. Keiser (✉)
Boston University, Boston, MA, USA
e-mail: gkeiser@photonicscomm.com

© Springer Nature Singapore Pte Ltd. 2019
G.-D. Peng (ed.), *Handbook of Optical Fibers*,
https://doi.org/10.1007/978-981-10-7087-7_35

the fundamental principles of how conventional fibers guide light along the fiber. Here the term "conventional" refers to the structure of optical fibers that are used widely in telecom networks. This discussion will be the basis for describing how light propagates in other optical fiber structures. In addition, section "Basic Concepts of Optical Fibers" describes the necessary performance characteristics of optical fibers for applications in specific spectral bands. Based on this background information, section "Optical Fibers Used in Biophotonics" then describes categories of optical fiber structures that are appropriate for use in different biophotonic applications.

Introduction

The category of photonics that deals with the interaction between light and biological material, such as human tissue, is known as *biophotonics* or *biomedical optics* (Popp et al. 2011; Keiser 2016; Ho et al. 2016). The result of this light-tissue interaction includes reflection, absorption, fluorescence, and scattering manifestations of light by tissue samples. These interactions are widely used in facilities such as research laboratories, pathology departments, hospitals, and health clinics to analyze the characteristics and health conditions of biological tissues and to carry out therapy on diseased or injured tissue. Note that commonly the word *human tissue*, or simply *tissue*, is used to designate all categories of human biological material including components such as flesh, bones, blood and lymph vessels, and body fluids.

From a general viewpoint, biophotonics involves the detection, reflection, emission, modification, absorption, creation, and manipulation of photons as they interact with biological tissue components. The biomedical areas of interest include (a) imaging techniques of biological elements ranging from cells to organs, (b) noninvasive measurements of biometric parameters such as blood oxygen and glucose levels, (c) light-based treatment of injured or diseased tissue, (d) detection of injured or diseased cells and tissue, (e) monitoring of wound healing progress, and (f) surgical procedures such as laser cutting, tissue ablation, and removal of cells and tissue.

The use of various types of optical fibers is attractive for such biophotonic applications because fibers allow pinpoint illumination of tissue areas in order to investigate the structural, functional, mechanical, biological, and chemical properties of biological material and systems. In addition, optical fibers play a key role in biophotonic methodologies that are being used extensively to investigate and monitor the health and well-being of humans. Among the numerous diverse biophotonic applications of optical fibers are imaging, spectroscopy, endoscopy, tissue pathology, blood flow monitoring, light therapy, biosensing, biostimulation, laser surgery, dentistry, dermatology, and health status monitoring.

As shown in Fig. 1, the spectral bands of interest for biophotonics range from the mid-ultraviolet (about 190 nm) to the mid-infrared (about 10.6 μm) regions. Numerous applications are in the visible 400–700 nm spectrum, because of the

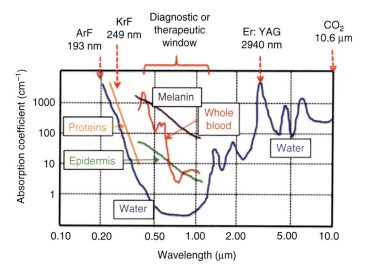

Fig. 1 Absorption coefficients of several major tissue components as a function of wavelength with example light source emission peaks (After Keiser 2016)

relatively low absorption of light in this band compared to other spectral regions. As Fig. 1 shows, quite different levels of photon absorption occur in the spectral regions that are of interest to biophotonics. Thus, a broad range of diverse photonic tools and both standard commercially available and custom-made optical fibers that operate efficiently in designated lightwave spectral bands are employed in biophotonics. For example, spectral regions with low absorption (such as the visible band) are ideal for imaging relatively deep into tissue. The ultraviolet and infrared spectral bands exhibit strong absorption and thus are suitable for cutting and removal of tissue material.

The unique physical and light transmission properties of optical fibers enable them to help resolve challenging biomedical implementation issues. These challenges include:

(a) Collecting emitted low-power light from a tissue specimen and transmitting it to a photon detection and analysis instrument
(b) Delivering a wide span of optical power levels to a tissue region during different categories of therapeutic healthcare sessions
(c) Accessing a diagnostic or treatment area within a living being with an optical detection probe or a radiant energy source in a manner that is the least invasive to the tissue host

Consequently, diverse types of optical fibers are finding widespread use in biophotonics instrumentation for clinical and biomedical research applications. In terms of transmission characteristics, flexibility, strength, and size, each optical fiber structure has certain advantages and limitations for specific uses in different spectral bands (Keiser et al. 2014). Therefore, it is essential that biophotonics researchers and designers of clinical biomedical optical fiber-based tools know what fiber type is best suited for a certain application.

Basic Concepts of Optical Fibers

This section discusses the fundamental principles for light guiding in two categories of conventional solid-core fibers. These discussions will be a guide for understanding light-guiding mechanisms in other optical fiber structures.

Several categories of optical fiber structures consisting of various materials are appropriate for use at different wavelengths in biomedical research and clinical practice. These include conventional and specialty solid-core fibers, double-clad fibers, hard-clad silica fibers, internally coated hollow-core fibers, photonic crystal fibers, polymer fibers, side-emitting and side-firing fibers, middle-infrared fibers, and fiber bundles. The fiber materials that are employed at different wavelengths include standard silica, UV-resistant silica, halide glasses, and polymer materials. Table 1 summarizes the characteristics of these optical fibers. The biophotonic applications in this table have been designated by the following three general categories with some basic examples:

1. Light care: healthcare monitoring, laser surfacing or photorejuvenation
2. Light diagnosis: biosensing, endoscopy, imaging, microscopy, spectroscopy
3. Light therapy: ablation, photobiomodulation, dentistry, laser surgery, oncology

Light-Guiding Principles in Conventional Fibers

A conventional solid-core fiber is a dielectric waveguide that operates at optical frequencies (Keiser 2015). This fiber waveguide is normally cylindrical in form. It confines and guides electromagnetic energy at optical wavelengths within its surfaces. The propagation of light along an optical waveguide can be described by means of a set of guided electromagnetic waves called the *modes* of the waveguide. Each guided mode is a pattern of electric and magnetic field distributions that is repeated along the fiber at periodic intervals. Only a certain discrete number of modes can propagate along the optical fiber. These modes are those electromagnetic waves that satisfy the homogeneous wave equation in the fiber and the boundary condition at the waveguide surfaces.

Figure 2 shows a schematic of a conventional optical fiber. This waveguide structure is a cylindrical silica-based glass *core* surrounded by a glass *cladding* that has a slightly different composition. The core has a refractive index n_1 and the cladding has a slightly lower refractive index n_2. Encapsulating these two layers is a polymer buffer coating that protects the fiber from adverse mechanical and environmental effects. The refractive index of pure silica varies with wavelength

Table 1 Major categories of optical fibers and their applications to biomedical research and clinical practice (After Keiser 2016)

Optical fiber types		Characteristics	Biophotonic applications
Conventional solid-core silica fibers	Multimode	Multimode propagation; carry more optical power	Light diagnosis; light therapy
Specialty solid-core fibers	Single-mode	Single-mode propagation	Light diagnosis
	Photosensitive	High photosensitivity to UV radiation; FBG fabrication	Light care; light therapy
	UV-resistant	Low UV sensitivity and reduced attenuation below 300 nm	Light diagnosis
	Bend-loss insensitive	High NA and low bend-loss sensitivity	Light therapy
	Polarization-maintaining	High birefringence and preserve the state of polarization	Light diagnosis
Double-clad fibers		Single-mode core and multimode inner cladding	Light diagnosis
Hard-clad silica fibers		Silica glass core with thin plastic cladding; increased fiber strength; high-power transmission	Light diagnosis; light therapy
Coated hollow-core fibers		Low absorption for mid-IR and high optical damage threshold	Light therapy
Photonic crystal fibers		Low loss; transmit high optical power without nonlinear effects	Light diagnosis; light therapy
Plastic optical fibers or polymer optical fibers		Low cost; fracture resistance; biocompatibility	Light diagnosis
Side-emitting fibers and side-firing fibers		Emit light along the fiber or perpendicular to the fiber axis	Light therapy
Mid-infrared fibers		Efficient IR delivery; large refractive index and thermal expansion	Light diagnosis; light therapy
Optical fiber bundles		Consist of multiple individual fibers	Light diagnosis

ranging from 1.453 at 850 nm to 1.445 at 1550 nm. By adding certain impurities such as germanium oxide to the silica during the fiber manufacturing process, the index can be changed slightly.

Design variations in the optical fiber material and the diameter of the conventional solid-core fiber structure dictate how a light signal is transmitted along a fiber. These variations also influence how the lightwaves in the fiber respond to

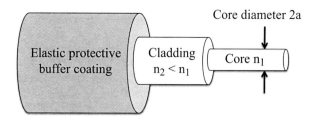

Fig. 2 Schematic of a conventional silica fiber structure

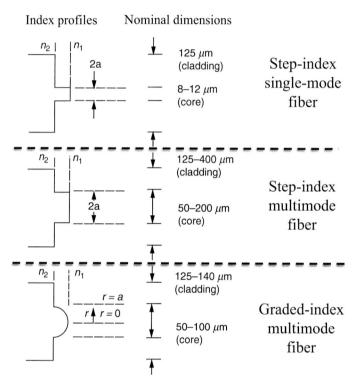

Fig. 3 Comparison of conventional single-mode and multimode step-index and graded-index optical fibers (After Keiser 2016)

environmental perturbations, such as stress, bending, and temperature fluctuations. Changing the material composition of the core gives rise to two commonly used fiber types, as shown in Fig. 3. In the first case, the refractive index of the core is uniform throughout and undergoes an abrupt change (or step) at the cladding boundary. This is called a *step-index fiber*. In the *graded-index fiber*, the core refractive index varies as a function of the radial distance from the center of the fiber.

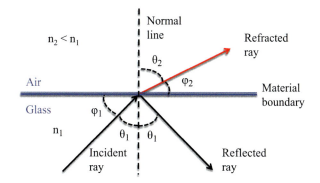

Fig. 4 Reflection and refraction of a light ray at a material boundary (After Keiser 2016)

Both the step-index and the graded-index fibers can be further divided into single-mode and multimode classes. As the name implies, a *single-mode fiber* (SMF) has only one propagating mode, whereas a *multimode fiber* (MMF) contains many hundreds of modes. Figure 3 gives examples of a few representative sizes of single- and multimode fibers to provide an idea of the dimensional scale. Multimode fibers have several advantages compared with single-mode fibers. First, the larger core radii of multimode fibers make it easier to launch optical power into the fiber and to collect light emitted or reflected from a biological sample. An advantage of multimode graded-index fibers is that they have larger data rate transmission capabilities than a comparably sized multimode step-index fiber. Single-mode fibers are more advantageous when delivering a narrow light beam to a specific tissue area and also are needed for applications that deal with coherence effects between propagating light beams.

In standard conventional fibers, the core of radius a has a refractive index n_1, which for silica-based fibers is typically equal to 1.48. The core is encapsulated by a cladding of slightly lower index n_2, where $n_2 = n_1(1 - \Delta)$. The parameter Δ is the *core-cladding index difference* or simply the *index difference*. Typical values of Δ range from 1% to 3% for multimode fibers and from 0.2% to 1.0% for single-mode fibers.

Ray Optics Concepts

For a conceptual illustration of how light travels along a fiber, consider the situation when the core diameter is much larger than the wavelength of the light. In this picture a basic geometric optics approach based on the concept of light rays can be used. To start, first consider Snell's law, which describes what happens when a light ray is incident on the interface between two different materials. When a light ray encounters a smooth interface that separates two different dielectric media, part of the light is reflected back into the first medium, and the remainder is bent (or refracted) as it enters the second material. This is illustrated in Fig. 4 for the interface between two materials with refractive indices n_1 and n_2, where $n_2 < n_1$.

Fig. 5 Ray optics picture of the propagation mechanism in an optical fiber (After Keiser 2016)

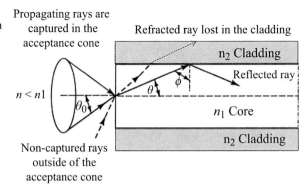

The relationship between the incident, reflected, and refracted rays at the interface is known as *Snell's law* and is given by

$$n_1 \sin \theta_1 = n_2 \sin \theta_2 \tag{1}$$

or, equivalently, as

$$n_1 \cos \varphi_1 = n_2 \cos \varphi_2 \tag{2}$$

where the angles are defined in Fig. 4. The angle θ_1 between the incident ray and the normal to the surface is known as the *angle of incidence*.

The application of Snell's law to an optical fiber can be illustrated by means of Fig. 5. Here a light ray enters a fiber core from a medium of refractive index n at an angle θ_0 with respect to the fiber axis. Inside the fiber the light ray strikes the core-cladding interface at an angle ϕ relative to the normal of the interface. If this angle is such that the light ray is totally internally reflected at the core-cladding interface, then the ray follows a zigzag path along the fiber core.

Total internal reflection occurs when the refracted angle is $\theta_2 = 90°$. Thus, from Snell's law, the minimum or critical angle ϕ_c that supports total internal reflection at the core-cladding interface is given by

$$\sin \phi_c = \frac{n_2}{n_1} \tag{3}$$

Incident rays meeting the interface at angles less than ϕ_c will refract out of the core and be lost in the cladding, as the dashed line in Fig. 5 shows. By applying Snell's law to the air-fiber face boundary, the condition of Eq. 3 can be related to the maximum entrance angle $\theta_{0, \max}$, which is called the *acceptance angle* θ_A, through the relationship:

$$n \sin \theta_{0,\max} = n \sin \theta_A = n_1 \sin \theta_c = \left(n_1^2 - n_2^2\right)^{1/2} \tag{4}$$

where $\theta_c = \pi/2 - \phi_c$. Those rays having entrance angles θ_0 less than θ_A will be totally internally reflected at the core-cladding interface. Thus, θ_A defines an *acceptance cone* for an optical fiber. Rays outside of the acceptance cone, such as the ray shown by the dashed line in Fig. 5, will refract out of the core and be lost in the cladding.

Equation 4 also defines the *numerical aperture* (NA) of a step-index fiber:

$$NA = n \sin\theta_A = \left(n_1^2 - n_2^2\right)^{1/2} \approx n_1 \sqrt{2\Delta} \tag{5}$$

The approximation on the right-hand side of the above equation holds because the parameter Δ is much less than 1. Because it is related to the acceptance angle, the NA is used commonly to describe the light acceptance or gathering capability of a multimode fiber and to calculate the source-to-optical fiber power coupling efficiencies. The NA value can be found on vendor data sheets for fibers.

As an example, consider a multimode step-index silica fiber that has a core refractive index $n_1 = 1.480$ and a cladding index $n_2 = 1.460$. From Eq. 3, the critical angle is given by

$$\varphi_c = \sin^{-1}\frac{n_2}{n_1} = \sin^{-1}\frac{1.460}{1.480} = 80.5^\circ$$

From Eq. 5 the numerical aperture is

$$NA = \left(n_1^2 - n_2^2\right)^{1/2} = 0.242$$

From Eq. 5 the acceptance angle in air ($n = 1.00$) is

$$\theta_A = \sin^{-1} NA = \sin^{-1} 0.242 = 14.0^\circ$$

Modal Concepts

The light ray picture gives a general concept of how light propagates along a fiber. However, mode theory is necessary to get a more detailed understanding of concepts such as mode coupling, dispersion, coherence or interference phenomena, and light propagation in single-mode and few-mode fibers. Figure 6 shows a longitudinal cross-sectional view of an optical fiber, which illustrates the field patterns of some of the lower-order transverse electric (TE) modes. The order of a mode is equal to the number of field zeros across the guiding core. The plots illustrate that the electric fields of the guided modes are not completely confined to the core but extend partially into the cladding. These cladding-mode components are important for understanding the operation of certain types of optical fiber sensors. Inside the core region of refractive index n_1, the fields vary harmonically, and they decay exponentially in the cladding of refractive index n_2. The exponentially decaying

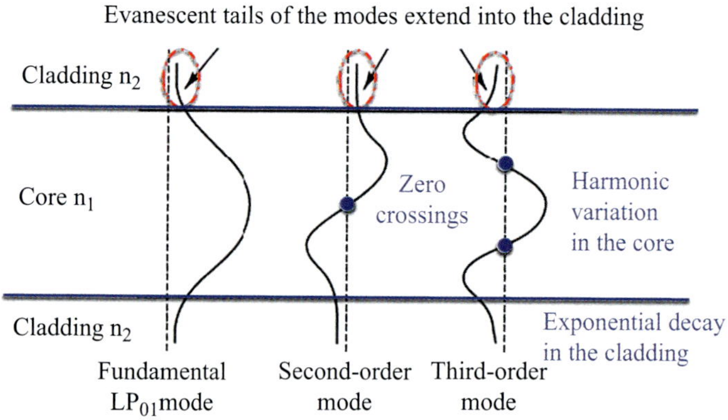

Evanescent tails of the modes extend into the cladding

Fig. 6 Electric field distributions of lower-order guided modes in an optical fiber (After Keiser 2016)

field is referred to as an *evanescent field*. For low-order modes, the fields are tightly concentrated near the optical fiber axis and penetrate very little into the cladding region. Higher-order mode fields are distributed more toward the edges of the core and penetrate farther into the cladding.

As the optical fiber core radius is made progressively smaller, all modes except the *fundamental mode* (the zeroth-order linearly polarized mode designated by LP_{01}) shown in Fig. 6 will start getting cut off and will not propagate in the fiber. A *single-mode fiber* results when only the fundamental mode can propagate along the fiber axis. An important parameter related to the cutoff condition is the V number, which is defined by

$$V = \frac{2\pi a}{\lambda}\left(n_1^2 - n_2^2\right)^{1/2} = \frac{2\pi a}{\lambda}NA \approx \frac{2\pi a}{\lambda}n_1\sqrt{2\Delta} \tag{6}$$

where the approximation on the right-hand side comes from Eq. 5. The V number is a dimensionless parameter that determines how many modes a fiber can support. Except for the lowest-order fundamental mode, each mode can exist only for values of V that exceed the limiting value V = 2.405 (with each mode having a different V limit). The wavelength at which all higher-order modes are cut off is called the *cutoff wavelength* λ_c. The fundamental mode has no cutoff and ceases to exist only when the core diameter is zero.

The V number also can be used to express how many modes M are allowed in a multimode step-index fiber when V is large. For this case, an estimate of the total number of modes supported in such a fiber is

$$M = \frac{1}{2}\left(\frac{2\pi a}{\lambda}\right)^2\left(n_1^2 - n_2^2\right) = \frac{V^2}{2} \tag{7}$$

Because the field of a guided mode extends partly into the cladding, as shown in Fig. 6, another important parameter is the fractional power flow in the core and cladding for a given mode. As the V number approaches cutoff for any particular mode, more of the power of that mode is in the cladding. At the cutoff point, all the optical power of the mode resides in the cladding. For large values of V that are far from the cutoff condition, the fractional optical power residing in the cladding can be estimated by

$$\frac{P_{clad}}{P} \approx \frac{4}{3\sqrt{M}} \tag{8}$$

where P is the total optical power in the fiber.

Graded-Index Optical Fibers

Core Index Structure
In contrast to a step-index fiber in which the core index is constant, in a graded-index fiber, the refractive index of the core decreases with increasing radial distance r from the fiber axis and is constant in the cladding. A commonly used refractive-index variation is the power law relationship:

$$n\left(r\right) = n_1\left[1 - 2\Delta\left(\frac{r}{a}\right)^{\alpha}\right]^{1/2} \text{ for } 0 \leq r \leq a$$
$$= n_1(1 - 2\Delta)^{1/2} \approx n_1\left(1 - \Delta\right) = n_2 \text{ for } r \geq a \tag{9}$$

Here, r is the radial distance from the fiber axis, a is the core radius, n_1 is the refractive index at the core axis, n_2 is the refractive index of the cladding, and the dimensionless parameter α defines the shape of the index profile. For example, a value of $\alpha = 2.0$ describes a parabolic index profile, whereas when $\alpha = \infty$ the fiber has a step-index profile. The index difference Δ for the graded-index fiber is given by

$$\Delta = \frac{n_1^2 - n_2^2}{2n_1^2} \approx \frac{n_1 - n_2}{n_1} \tag{10}$$

The approximation on the right-hand side reduces this expression for Δ to that of the step-index fiber. Thus, the same symbol Δ is used in both cases.

Graded-Index Numerical Aperture
Whereas for a step-index fiber the NA is constant across the core, for graded-index fibers, the NA is a function of position across the core end face. Geometrical optics analyses show that light incident on the fiber core at position r will propagate as a

guided mode only if it is within the local numerical aperture NA(r) at that point. The local numerical aperture is defined as

$$
\begin{aligned}
\text{NA}(r) &= \left[n^2(r) - n_2^2 \right]^{1/2} \approx \text{NA}(0) \sqrt{1 - (r/a)^\alpha} \text{ for } r \leq a \\
&= 0 \text{ for } r > a
\end{aligned}
\tag{11}
$$

where the axial numerical aperture is defined as

$$
\text{NA}(0) = \left[n^2(0) - n_2^2 \right]^{1/2} = \left(n_1^2 - n_2^2 \right)^{1/2} \approx n_1 \sqrt{2\Delta}
\tag{12}
$$

Thus, the NA of a graded-index fiber decreases from NA(0) to zero as r moves from the fiber axis to the core-cladding boundary. The number of bound modes M_g in a graded-index fiber is

$$
M_g = \frac{\alpha}{\alpha + 2} \left(\frac{2\pi a}{\lambda} \right)^2 n_1^2 \Delta \approx \frac{\alpha}{\alpha + 2} \frac{V^2}{2}
\tag{13}
$$

Typically, a parabolic refractive index profile given by $\alpha = 2.0$ is used for a graded-index fiber. In this case, the number of modes is $M_g = V^2/4$, which is half the number of modes supported by a step-index fiber (for which $\alpha = \infty$) that has the same V value.

Cutoff Wavelength in Graded-Index Fibers

Similar to step-index fibers, graded-index fibers can be designed as single-mode fibers in which only the fundamental mode propagates at a specific operational wavelength. An empirical expression of the V number at which the second lowest-order mode is cut off for graded-index fibers has been shown to be

$$
V_{cutoff} = 2.405 \sqrt{1 + \frac{2}{\alpha}}
\tag{14}
$$

Equation 14 shows that in general for a graded-index fiber, the value of V_{cutoff} decreases as the profile parameter α increases. This also shows that in a parabolic graded-index fibers ($\alpha = 2$), the critical value of V for the cutoff condition is a factor of $\sqrt{2}$ larger than for a similar-sized step-index fiber. In addition, from the definition of V given by Eq. 6, the numerical aperture of a graded-index fiber is larger than that of a step-index fiber of comparable size.

Performance Characteristics of Generic Optical Fibers

When selecting an optical fiber to use in a particular biophotonics system application, various performance characteristics need to be considered. These include optical signal attenuation as a function of wavelength, optical power-handling capability, the degree of signal loss as the fiber is bent, and mechanical properties of the optical fiber.

Attenuation Versus Wavelength

Attenuation (i.e., decrease in optical power) in a fiber is due to absorption, scattering, and radiative losses of optical energy as light propagates along a fiber. This parameter normally is measured in units of decibels per kilometer (dB/km) or decibels per meter (dB/m). A variety of materials are used to make different types of optical fibers for biophotonic applications. The basic reason for this selection is that each material type exhibits different light-attenuation characteristics in various spectral bands. For example, silica (SiO_2) glass is the common material used for solid-core fibers in telecom applications. This material has low losses in the 800–1600 nm telecom operating region, but the loss is significantly higher for ultraviolet and mid-infrared wavelengths. Thus, other fiber types and/or materials are needed for biophotonic applications that use wavelengths outside of the telecom spectral band.

Bend-Loss Insensitivity

When conventional and certain other types of optical fibers are bent, light escapes from the core of the curved fiber section. This bending loss increases exponentially as the bending radius decreases. The loss is unobservable for slight bends, but at a certain critical radius, the loss becomes observable. If the bend radius is made smaller than the critical radius, the bend losses quickly become extremely large.

Specially designed fibers that are less sensitive to bending loss are available commercially to provide optimum low bending loss performance at specific operating wavelengths, such as 820 or 1550 nm. These fibers usually have an 80 μm cladding diameter. Such a smaller outer diameter yields a reduced coil volume compared with a standard 125 μm cladding diameter when a length of this low-bend-loss fiber is coiled up within a miniature optoelectronic device package or in a compact biophotonics instrument.

Mechanical Properties

Several unique mechanical properties of optical fibers make them appealing for biomedical applications. For example, the fact that optical fibers are a thin highly flexible medium allows minimally invasive medical treatment or diagnostic procedures to take place in a living body. Such applications include endoscopic procedures, cardiovascular surgery, and microsurgery.

Another mechanical-related characteristic is that by monitoring the signal change resulting from some intrinsic physical variation of an optical fiber (e.g., elongation or refractive index changes), one can create fiber sensors to measure external physical parameter changes. For example, if a varying external parameter, such as temperature or pressure fluctuations, elongates the fiber or induces refractive index difference changes at the outer cladding boundary, this effect can modulate the intensity, phase, polarization, wavelength, or transit time of light in the fiber.

The degree of light modulation then can be used to directly measure changes in the external physical parameter. For biophotonic applications, the external physical parameters of interest include fluctuations in pressure, temperature, stress and strain in tissue and bones, and the molecular composition of a liquid or gas surrounding the fiber.

Optical Power-Handling Capability

In biomedical photonics applications such as imaging and fluorescence spectroscopy, the fibers transmit optical power levels of less than 1 μW. In other applications the fibers need to transmit optical power levels of 10 W and higher. A principal application is laser surgery, which includes bone ablation, cardiovascular procedures, dentistry, dermatology, ophthalmology, and oncological treatments. Optical fibers that can transmit high-power levels include hard-clad silica fibers with fused silica cores that contain very low contaminant levels, coated hollow-core fibers, photonic crystal fibers, and germinate glass fibers.

Optical Fibers Used in Biophotonics

This section describes various categories of optical fiber structures that are appropriate for use in different biophotonic spectral bands. The fiber categories include conventional and specialty multimode and single-mode solid-core fibers, double-clad fibers, hard-clad silica fibers, conventional coated hollow-core fibers, photonic crystal fibers, polymer optical fibers, side-emitting and side-firing fibers, middle-infrared fibers, and optical fiber bundles.

Conventional Solid-Core Fibers

Extensive development work by telecom companies has resulted in highly reliable and widely available conventional solid-core silica-based optical fibers. These fibers come in a variety of core sizes and are used throughout the world in telecom networks and also in many biophotonic applications. Figure 7 shows the optical signal attenuation per kilometer as a function of wavelength. The shape of the attenuation curve is due to the following three factors:

(a) Intrinsic absorption resulting from electronic absorption bands causes high attenuations in the ultraviolet region for wavelengths less than about 500 nm.
(b) The Rayleigh scattering effect starts to dominate the attenuation for wavelengths above 500 nm. However, this effect decreases rapidly with increasing wavelength because of its $1/\lambda^4$ behavior.
(c) Atomic vibration bands in the optical fiber material produce intrinsic absorption in the infrared spectrum. This is the dominant attenuation mechanism in the infrared region for wavelengths above 1500 nm.

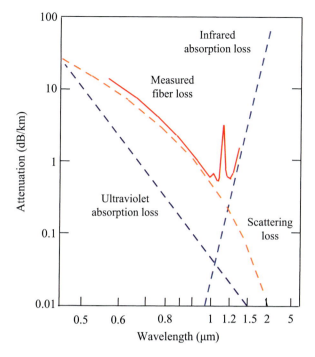

Fig. 7 Typical attenuation curve of a silica fiber as a function of wavelength (After Keiser et al. 2014)

Figure 7 illustrates that in silica fibers, these attenuation mechanisms result in a low-loss region in the spectral range of 700–1600 nm, which matches the low-absorption biophotonics window illustrated in Fig. 1.

Absorption by residual water ions in the silica material gives rise to an attenuation spike around 1400 nm. Removing a large percentage of the water ions during fiber manufacturing can reduce this attenuation spike significantly. The end product is called a *low-water-content fiber* or *low-water-peak fiber*.

Multimode fibers are commercially available with standard core diameters of 50, 62.5, 100, 200 μm, or larger. Applications of such multimode fibers include laser or LED light delivery, photobiomodulation, optical fiber probes, and oncological therapy. Single-mode fibers have core diameters around 10 μm and are used in clinical fiber sensors, in endoscopes or catheters, and in imaging systems.

Specialty Solid-Core Fibers

Specialty solid-core fibers can be custom-designed through either material or structural variations. Such fibers enable functions such as lightwave signal manipulation for optical signal-processing functions, extending the spectral operating

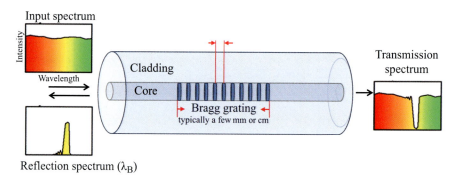

Fig. 8 A periodic index variation in the core of a single-mode fiber creates a fiber Bragg grating (After Keiser et al. 2014)

range of the fiber, sensing fluctuations in a physical parameter such as temperature or pressure, or analyzing the contents of biomedical fluids. The main specialty solid-core fiber types for biophotonic applications are photosensitive fibers, fibers resistant to darkening from ultraviolet light, bend-loss-insensitive fibers for use along circuitous paths inside bodies, and polarization-preserving optical fibers for imaging. Descriptions and utilizations of such fibers are given in the following subsections.

Photosensitive Optical Fiber

In a photosensitive fiber, the refractive index changes when the fiber material is exposed to ultraviolet light. This feature can be employed to fabricate a short *fiber Bragg grating* (FBG) in the fiber core, which is a periodic variation of the refractive index along the fiber axis (Kashyap 2010; Al-Fakih et al. 2012).

This index variation is illustrated in Fig. 8, where n_1 is the core refractive index, n_2 is the cladding index, and Λ is the period of the grating, that is, the spacing between the maxima of the index variations. If a specific incident lightwave at a wavelength λ_B (which is known as the *Bragg wavelength*) encounters a periodic refractive index variation, λ_B will be reflected back if the following condition is met:

$$\lambda_B = 2n_{eff}\Lambda \tag{15}$$

Here n_{eff} is the effective refractive index, which has a value falling between the refractive indices n_1 of the core and n_2 of the cladding. The grating reflects the Bragg wavelength and all others will pass through. Fiber Bragg gratings are available in a selection of Bragg wavelengths with the width of the reflection bands at specific wavelengths varying from a few picometers to tens of nanometers.

A common biophotonics utilization of a FBG is to sense a variation in a physical parameter such as temperature or pressure. For example, an external strain will slightly stretch the fiber, which causes the period Λ of the FBG to lengthen and thus will change the characteristic Bragg wavelength value. Similarly, rises or drops in temperature will lengthen or shorten the value of Λ. The resuting change in the Bragg wavelength $_B$ then can be related to the change in temperature.

Fibers Resistant to UV-Induced Darkening

Conventional solid-core silica fibers darken when exposed to ultraviolet (UV) light. The UV-induced darkening losses are known as *solarization* and occur strongly at wavelengths less than about 260 nm. Consequently, standard silica optical fibers can only be implemented above approximately 300 nm.

However, recently special glass processing techniques have allowed the fabrication of fibers with moderate (around 50% attenuation) to minimal (a few percent additional loss) UV sensitivity below 260 nm (Khalilov et al. 2014; Gebert et al. 2014). When these solarization-resistant fibers are exposed to UV light, the transmittance decreases rapidly and then stabilizes to an asymptotic value. The exact asymptotic value and the time duration needed to reach the plateau depend on the specific fiber type and on the UV wavelength. Shorter wavelengths induce larger attenuation changes, and a longer time is needed to reach the asymptotic value.

Solarization-resistant fibers for use in the 180–850 nm range are commercially available with core diameters ranging from 50 to 1000 μm and a numerical aperture of 0.22. Manufacturers recommend that prior to use, these fibers should first be exposed to UV radiation for approximately 5 min or more (depending on the operational wavelength) to establish loss equilibrium.

Bend-Insensitive Fiber

Often when employing optical fibers within a living body for medical applications, the fibers follow a winding path with sharp bends through arteries that snake around bones and organs. As noted in section "Bend-Loss Insensitivity," radiative losses can occur whenever the fiber undergoes a bend with a finite radius of curvature. This factor is negligible for slight bends. However, the bending loss effects increase exponentially as the radius of curvature decreases until at a certain critical radius the losses become extremely large. The bending loss is more sensitive at longer wavelengths as Fig. 9 shows. For example, suppose a conventional fiber has a 1 cm bending radius as indicated by the dashed vertical line in Fig. 9. At 1310 nm this results in an additional loss of 1 dB. However, for this 1 cm bend radius, there will be an additional loss of about 100 dB at 1550 nm.

Optical fiber research activities in the telecom industry led to bend-loss-insensitive fibers that can tolerate numerous sharp bends. These same fibers also can be applied in the medical field (Matsui et al. 2011; Kusakari et al. 2013). The fibers are available with either an 80 μm or a 125 μm cladding diameter as standard products. The 80 μm reduced-cladding fiber results in a much smaller volume compared with a 125 μm cladding diameter when a fiber length is coiled up within a miniature optoelectronic device package or in a compact biophotonics instrument.

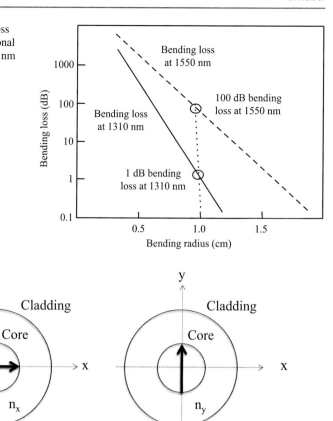

Fig. 9 Generic bend-loss sensitivity for conventional fibers at 1310 and 1550 nm (After Keiser 2016)

Fig. 10 Two polarization states of the fundamental mode in a single-mode fiber (After Keiser 2016)

Polarization-Maintaining Fibers

The fundamental mode in a single-mode fiber can be viewed as consisting of two orthogonal polarization modes. These modes can be designated as horizontal and vertical polarizations in the x direction and y direction, respectively, as shown in Fig. 10. In general, the electromagnetic field of the light traveling along the fiber is a linear superposition of these two orthogonal modes and depends on the polarization state of the light at the input point of the fiber.

In ideal fiber with perfect rotational symmetry, the polarization state of any light injected into the fiber will propagate unchanged along the fiber. In actual fibers any small imperfections (such as asymmetric lateral stresses, noncircular cores, and variations in refractive-index profiles) disturb the circular symmetry of the ideal fiber. The two orthogonal polarization modes then travel along the fiber with

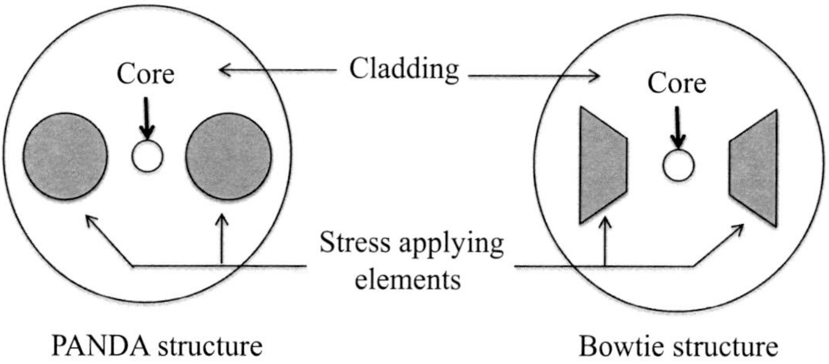

Fig. 11 Cross-sectional geometry of two different polarization-maintaining fibers (After Keiser 2016)

different phase velocities. This effect causes the state of polarization to fluctuate as a lightwave travels through the fiber.

Specially designed *polarization-maintaining fibers* have been created that preserve the state of polarization along the fiber with little or no coupling between the two modes. Figure 11 illustrates the cross-sectional geometry of two commonly used polarization-maintaining fibers. The light circles represent the usual core and the cladding material. The dark areas are stress elements made from a different type of glass that are embedded in the cladding. The purpose of the stress-applying components is to create slow and fast axes in the core that will guide light in each mode at a different velocity. Thereby when polarized light is launched into the fiber it will maintain its state of polarization as it travels along the fiber. These fibers are used in special biophotonic applications such as fiber-optic sensing and interferometry where polarization preservation is essential (Tuchin et al. 2006).

Double-Clad Fibers

A double-clad fiber (DCF) is being used widely in the medical field for imaging systems (Lemire-Renaud et al. 2011; Liang et al. 2012; Beaudette et al. 2015). As shown in Fig. 12, a DCF structure contains a core region, an inner cladding, and an outer cladding arranged concentrically. Typical dimensions are a 9 μm core diameter, a 105 μm inner cladding diameter, and an outer cladding with a 125 μm diameter. Light in the core region is single-mode and is multimode in the inner cladding. The combination of a single-mode and multimode structure allows using just one optical fiber for the delivery and collection of probing light. The illumination light is transmitted via the single-mode core, and the multimode inner cladding is employed for the collection of light coming from the tissue.

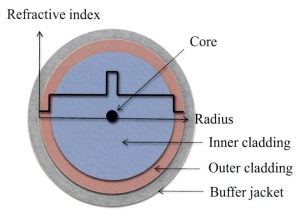

Fig. 12 Cross-sectional representation of a DCF and its index profile (After Keiser 2016)

Table 2 General specifications of selected HCS fibers

	HCS fiber 1	HCS fiber 2	HCS fiber 3
Core diameter	$200 \pm 5\,\mu m$	$600 \pm 10\,\mu m$	$1500 \pm 30\,\mu m$
Cladding diameter	$225 \pm 5\,\mu m$	$630 \pm 10\,\mu m$	$1550 \pm 30\,\mu m$
Wavelength range (high OH content)	300–1200 nm	300–1200 nm	300–1200 nm
Wavelength range (low OH content)	400–2200 nm	400–2200 nm	400–2200 nm
Maximum power capability (CW)	0.2 kW	1.8 kW	11.3 kW
Maximum power capability (pulsed)	1.0 MW	9.0 MW	56.6 MW
Maximum long-term bend radius	40 mm	60 mm	150 mm
Max attenuation at 850 nm	10 dB/km (0.010 dB/m)	12 dB/km (0.012 dB/m)	18 dB/km (0.018 dB/m)

Hard-Clad Silica Fibers

A multimode hard-clad silica (HCS) optical fiber is useful in applications such as laser light delivery, endoscopic procedures, oncological therapy, and biosensing systems. The HCS structure consists of a silica glass core that is encapsulated by a thin plastic cladding. The hard cladding gives greater fiber strength and reduces static fatigue effects in humid environments. Other features of these HCS fibers include bend insensitivity, long-term reliability, ease of handling, and resistance to harsh chemicals. Core diameters of commercially available HCS fibers range from 200 to 1500 μm. Table 2 lists some performance parameters of three selected HCS fibers.

A common HCS fiber structure for medical applications comprises a 200 μm core diameter and a cladding diameter of 230 μm. Such fibers are very strong and have a low attenuation (<10 dB/km or 0.01 dB/m at 820 nm), a numerical aperture of 0.39, and negligible bending-induced loss for 40 mm bend diameters.

Coated Hollow-Core Fibers

Owing to the fact that a conventional solid-core silica fiber exhibits extremely high light signal absorption above about 2 μm, internally coated hollow-core fibers were developed. Such fibers offer one alternative for sending mid-infrared (2–10 μm) light to a localized site with a low transmission loss. Optical sources that emit at wavelengths above 2 μm include Er:YAG (2.94 μm) and CO_2 (10.6 μm) lasers. Such lasers are used in urology, dentistry, otorhinolaryngology, and cosmetic surgery. In addition, because silica-based fibers exhibit high transmission losses in the ultraviolet region, this is another application area for hollow-core fibers.

As shown in Fig. 13, hollow-core fibers are made from a glass tube with metal and dielectric layers deposited at the inner surface plus a protection jacket on the outside (Yu et al. 2012; Monti and Gradoni 2014). To fabricate these fibers, first a layer of silver (Ag) is deposited on the inside of a glass tube, and then a thin dielectric film, such as silver iodide, is added. Light travels along the fiber through mirror-type reflections from the inner metallic layer. The dielectric layer is normally less than 1 μm thick and is chosen to yield a high reflectivity at a particular wavelength or a band of wavelengths. Although capillaries of other materials, such as plastic and metal, could be used as hollow-core waveguides, a glass-based hollow-core fiber provides more flexibility and better performance. The diameters of the fiber hole, which are called the *bore sizes*, can range from 50 to 1200 μm. However, since the loss of all hollow-core fibers varies as $1/r^3$, where r is the bore radius, the more flexible smaller-bore hollow-core fibers have higher losses.

In contrast to solid-core fibers, the damage threshold from intense optical powers is higher in hollow-core fibers, there are no cladding modes to give unwanted signal interference, and there is no need for angle cleaving or anti-reflection coating at the

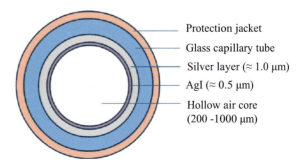

Fig. 13 The cross section of a typical hollow-core fiber (After Keiser 2016)

Protection jacket

Glass capillary tube

Silver layer ($\approx$ 1.0 μm)

AgI ($\approx$ 0.5 μm)

Hollow air core (200 -1000 μm)

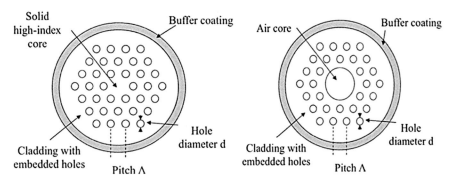

Fig. 14 Sample structural arrangements of air holes in solid-core (left) and hollow-core (right) PCF (After Keiser et al. 2014)

fiber end to mitigate laser feedback effects. To direct the light emerging from the hollow-core fiber into a certain direction, sealing caps with different shapes (e.g., cone and slanted end) at the light exit end of the fiber can be employed.

Photonic Crystal Fibers

The structure of a *photonic crystal fiber* (PCF) consists of a geometric arrangement of cylindrical air holes that run along the entire length of the fiber (St. John Russell 2006; Poli et al. 2007; Gong et al. 2015). The PCF core can be solid or hollow as shown in Fig. 14 by the cross-sectional images of two typical PCF structures. This air hole arrangement in a PCF offers another dimension of light control in the fiber compared to a conventional solid-core fiber. The arrangement, size, and spacing (known as the *pitch*) of the holes and the refractive index of the PCF material determine the fiber transmission characteristics.

A core consisting of pure silica gives the PCF a number of operational advantages over conventional fibers that typically have a Ge-doped silica core. These advantages include very low optical power losses, the ability to transmit high optical power levels without encountering nonlinear effects, and a strong resistance to darkening effects from ionizing radiation. Single-mode photonic crystal fibers can be made to operate over wavelengths ranging from 300 to 2000 nm.

For biophotonic applications, certain hollow-core PCFs can be employed for delivering mid-infrared light with a broadened transmission window, reduced attenuation, and low bending loss. In addition, a wide selection of optical fiber-based biosensors are employing the unique properties of photonic crystal fibers.

Plastic Fibers

A *plastic optical fiber* or *polymer optical fiber* (both designated by POF) is an alternative to glass optical fibers for areas such as biomedical sensors (Ziemann et al.

2008; Zhou et al. 2010; Bilro et al. 2012). The advantages of a POF include low cost, inherent fracture resistance, a low Young's modulus, and biocompatibility. Most POFs are fabricated from polymethyl methacrylate (PMMA). This material has a refractive index of around 1.492, which is a bit higher than the 1.48 index of silica. A POF can have core diameters ranging up to 0.5 mm. Both multimode and single-mode plastic optical fibers are available commercially. The standard 50 μm and 62.5 μm core diameters of multimode POFs are compatible with the core diameters of conventional multimode glass telecom fibers.

Single-mode POF structures enable the fabrication of fiber Bragg gratings inside of plastic fibers, which provides an extended range of possibilities for POF-based biosensing. In addition to PMMA POFs, single-mode perfluorinated POFs with a 1.34 refractive index have been fabricated. This lower index allows for an enhanced performance improvement in biosensors, because such a POF produces a stronger optical coupling between a light signal in the fiber and the material surrounding the biosensor that is being analyzed.

Side-Emitting or Glowing Fibers

For signal-transmission functions, the operational goal of optical fibers is to transport light to a recipient with as little optical power loss and signal distortion as possible. In contrast, for certain biomedical applications, a category of optical fibers that continuously emit light radially along the fiber length is of high interest (Shen et al. 2013; Krehel et al. 2014; George and Walsh 2009). Such a *side-emitting fiber* or *glowing fiber* acts as an extended radially emitting optical source. These side-emitting fibers are being used in many popular nonmedical applications such as decorative lamps, submersible lighting, event lighting, and lighted signs.

Another variation for localized lateral light emission is the *side-firing fiber*. This structure consists of either micro-optic prism elements or an angled end face at the exit end of the fiber. This enables the light to be emitted almost perpendicular to the fiber axis onto a localized spot. Biomedical procedures that employ side-emitting and side-firing fibers include oncological procedures, therapy for atrial fibrillation, treatment of prostate enlargements, and orthodontic treatments.

Both plastic and silica materials have been used to make side-emitting fibers available. One method is to add impurity materials into either the core or cladding of the fiber, which creates a side-emitting effect through the scattering of light from the fiber core into the cladding and then into the medium outside of the fiber. As Fig. 15 shows, one implementation technique is to attach a short length of side-emitting fiber to a longer delivery fiber. This fiber length can vary from 10 to 70 mm for short diffusers or 10–20 cm for long diffusers. The delivery fiber nominally is from 2 to 2.5 m long and runs from the laser to the side-emitting unit.

One consideration in a side-emitting fiber is that the scattering effects cause the light intensity to decrease exponentially along the fiber length. Under the assumption that the scattering process is much stronger than light absorption and other losses in the fiber, then the radiation intensity I_S that is emitted radially in any

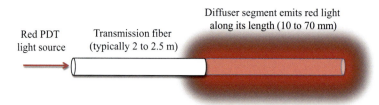

Fig. 15 The red input light is scattered radially along the length of the side-emitting fiber (After Keiser 2016)

direction per steradian at a distance x from the light input end is

$$I_S = \frac{I_0}{4\pi} \exp(-kx) \tag{16}$$

where I_0 is the input light intensity and k is the side-scattering efficiency coefficient. Typical values of k range from 0.010 to 0.025 m^{-1}.

This exponential decrease in radially emitted light intensity normally is not a major problem in biomedical applications where the glowing fiber lengths are on the order of 10–20 cm. However, if a more uniform light distribution is desired along a fiber, a micro-optic reflector element can be attached to the far end of the fiber. Thus, a relatively uniform radially emitted light distribution can result from the combined transmitted and reflected light.

Middle-Infrared Fibers

The high losses in silica-based fibers at wavelengths greater than 2 μm generated interest in a *middle-infrared fiber*, or simply an *infrared fiber* (IR fiber), which transmits light efficiently in the spectral band above 2 μm (Eggleton et al. 2011; Damin and Sommer 2013; Israeli and Katzir 2014). Depending on the constituent material and the specific structure, IR fibers can be categorized as glass, crystalline, and photonic crystal fibers.

Glass materials for IR fibers include heavy metal fluorides, chalcogenides, and heavy metal germanates. Fibers made from *heavy metal fluoride glasses* (HMFG) are useful for the 1.5-to-4.0 μm spectral region, which is increasingly used in medical applications. Two available HMFG fibers are made from InF_3 and ZrF_4, which have attenuations of <0.25 dB/m between 1.8 and 4.7 μm.

Chalcogenide glasses consist of the chalcogen elements sulfur (S), selenium (Se), and tellurium (Te) together with the addition of other elements such as Ge, As, and Sb. These glasses are quite stable, durable, and insensitive to moisture and can transmit at wavelengths up to 10 μm. Some generic loss values are less than 0.1 dB/m at the commonly used 2.7 μm and 4.8 μm biomedical wavelengths. The three common chalcogenide fibers have the following maximum losses:

(a) Sulfides: <1 dB/m over 2–6 μm
(b) Selenides: <2 dB/m over 5–10 μm
(c) Tellurides: <2 dB/m over 5.5–10 μm

Germanate (GeO_2) glass fibers contain heavy metal oxides such as PbO, Na_2O, and La_2O_3 in order to move the infrared absorption edge from around 2 μm to longer wavelengths. The advantages of GeO_2 glass fibers include a high laser damage threshold (e.g., they can handle up to 20 watts of power) and a high glass transition temperature. Thus, GeO_2 glass fibers offer high mechanical and thermal stability. The attenuation of GeO_2 glass is <1 dB/m in the spectrum ranging from 1.0 to 3.5 μm.

Crystalline IR fibers can transmit light wavelengths up to 18 μm. These fibers typically are *polycrystalline silver-halide fibers* with AgBr cores and AgCl claddings. The attenuation of Ag-halide fibers usually is <1 dB/m in the 5-to-12 μm spectral band and slightly higher in the 12-to-18 μm spectral band. For 10.6 μm operation, the loss nominally is 0.3–0.5 dB/m.

Optical Fiber Bundles

Greater throughput of optical power can be achieved with multiple flexible glass or plastic fibers that are packaged in a bundle (Coté et al. 2012). Each fiber acts as an independent waveguide that allows light to be transmitted over long distances with minimal attenuation. Large bundles of fibers can contain a few thousand to 100,000 individual fibers that are between 2 and 20 μm in diameter, thereby yielding a bundle diameter of less than 1 mm. Typically the large fiber bundles are employed for illumination of tissue areas. Some advantages of fiber bundles in illumination systems include the following:

1. Multiple locations can be illuminated with a single optical source by splitting the bundle into two or more branches.
2. Several light sources can be merged into a single output.
3. Different types of specialty fibers can be integrated into one bundle.

Individual fibers inside large bundles normally are arranged randomly in the bundle during the manufacturing process. However, for certain biomedical applications, individual or a small group of fibers can be arranged in specific patterns within the bundle. These groupings can be bundled together and aligned in such a fashion that the orientations of the individual or grouped fibers are identical at both ends of the bundle. Such an arrangement is called a *coherent fiber bundle* or an *ordered fiber bundle*. In this case, the term "coherent" refers to the correlation between the spatial arrangements of the fibers at both ends of the fiber bundle. Because the fiber arrangements are matched at both ends of the bundle, any incident illumination pattern at the input end of the bundle is duplicated when it emerges

Fig. 16 (**a**) Randomly arranged fibers in a bundle containing many fibers. (**b**) Coherent bundle cable with identical arrangements of fibers on both ends

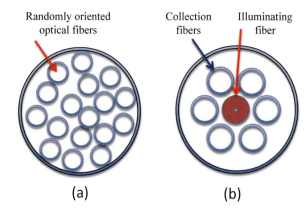

from the output end. Figure 16 shows examples of illumination and coherent fiber bundle configurations.

References

E. Al-Fakih, N.A. Abu Osman, F.R.M. Adikan, Sensors **12**, 12890 (2012)

K. Beaudette, H.W. Bac, W.-J. Madore, M. Villiger, N. Gadbout, B.E. Bouma, C. Boudoux, Biomed. Opt. Exp. **6**, 1293 (2015)

L. Bilro, N. Alberto, J.L. Pinto, R. Nogueira, Sensors **12**, 12184 (2012)

G.L. Coté, L.V. Wang, S. Rastigar, in *Introduction to Biomedical Engineering*, ed. by J.D. Enderle, J.D. Bronzino (Academic, New York, 2012), p. 1111

C.A. Damin, A.J. Sommer, Appl. Spectros. **67**, 1252 (2013)

B.J. Eggleton, B. Luther-Davies, K. Richardson, Nat. Photonics **5**, 141 (2011)

F. Gebert, M.H. Frosz, T. Weiss, Y. Wan, A. Ermolov, N.Y. Joly, P.O. Schmidt, P.S.J. Russell, Opt. Express **22**, 15388 (2014)

R. George, L.J. Walsh, Lasers Surg. Med. **41**, 214 (2009)

T. Gong, N. Zhang, K.V. Kong, D. Goh, C. Ying, J.-L. Auguste, P. P. Shum, L. Wei, G. Humbert, K.-T. Yong, M. Olivo, J. Biophotonics. (2015). https://doi.org/10.1002/jbio.201500168

A. H.-P. Ho, D. Kim, M. G. Somekh (eds.), *Handbook of Photonics for Biomedical Engineering* (Springer, Singapore, 2016)

S. Israeli, A. Katzir, J. Appl. Phys. **115**, 023104 (2014)

R. Kashyap, *Fiber Bragg Gratings*, 2nd edn. (Academic, New York, 2010)

G. Keiser, *Optical Fiber Communications*, 5th edn. (McGraw-Hill, Singapore, 2015)

G. Keiser, *Biophotonics: Concepts to Applications* (Springer, Singapore, 2016)

G. Keiser, F. Xiong, Y. Cui, P.P. Shum, J. Biomed. Optics **19**, 080902 (2014)

V. Khalilov, J.H. Shannon, R.J. Timmerman, Proc. SPIE **8938**, 89380A (2014)

M. Krehel, M. Wolf, L.F. Boesel, R.M. Rossi, G.-L. Bona, L.J. Scherer, Biomed. Opt. Express **5**, 2537 (2014)

D. Kusakari, H. Hazama, R. Kawaguchi, K. Ishii, K. Awazu, Optics Photon. J. **3**, 14 (2013)

S. Lemire-Renaud, M. Strupler, F. Benboujja, N. Godbout, C. Boudoux, Biomed. Opt. Exp. **2**, 2961 (2011)

S. Liang, A. Saidi, J. Jing, G. Liu, J. Li, J. Zhang, C. Sun, J. Narula, Z. Chen, J. Biomed. Opt. **17**, 070501 (2012)

T. Matsui, K. Nakajima, Y. Goto, T. Shimizu, T. Kurashima, J. Lightw. Technol. **29**, 2499 (2011)

T. Monti, G. Gradoni, J. Sel. Topics Quantum Electron. **20**, 6900409 (2014)

F. Poli, A. Cucinotta, S. Selleri, *Photonic Crystal Fibers* (Springer, New York, 2007)

J. Popp, V.V. Tuchin, A. Chiou, S.H. Heinemann, *Handbook of Biophotonics: Vol. 1: Basics and Techniques* (Wiley, Berlin, 2011)

J. Shen, C. Chui, X. Tao, Biomed. Opt. Express **4**, 2925 (2013)

P. St. John Russell, J. Lightw. Technol. **24**, 4729 (2006)

V.V. Tuchin, L.V. Wang, D.A. Zimnyakov, *Optical Polarization in Biomedical Applications* (Springer, New York, 2006)

F. Yu, W.J. Wadsworth, J.C. Knight, Opt. Express **20**, 11153 (2012)

G. Zhou, C.-F.J. Pun, H.-Y. Tam, A.C.L. Wong, C. Lu, P.K.A. Wai, IEEE Photon. Technol. Lett. **22**, 106 (2010)

O. Ziemann, J. Krauser, P.E. Zamzow, W. Daum, *POF Handbook*, 2nd edn. (Springer, Berlin, 2008)

Part VI
Optical Fiber Measurement

Basics of Optical Fiber Measurements

28

Mingjie Ding, Desheng Fan, Wenyu Wang, Yanhua Luo,
and Gang-Ding Peng

Contents

M. Ding · D. Fan · W. Wang · G.-D. Peng (✉)
Photonics and Optical Communications, School of Electrical Engineering and
Telecommunications, University of New South Wales, Sydney, NSW, Australia
e-mail: m.ding@student.unsw.edu.au; desheng.fan@student.unsw.edu.au;
wenyu.wang@student.unsw.edu.au; g.peng@unsw.edu.au

Y. Luo (✉)
Photonics and Optical Communications, School of Electrical Engineering and
Telecommunications, University of New South Wales, Sydney, NSW, Australia

Key Laboratory of Optoelectronic Devices and Systems of Ministry of Education and Guangdong
Province, Shenzhen University, Shenzhen, China
e-mail: yanhua.luo1@unsw.edu.au

© Springer Nature Singapore Pte Ltd. 2019   1099
G.-D. Peng (ed.), *Handbook of Optical Fibers*,
https://doi.org/10.1007/978-981-10-7087-7_57

Abstract

This chapter is devoted to introducing fundamental properties of optical fibers and related measurement techniques. The basics are firstly introduced to give a clear working principle of an optical fiber as a light waveguide. Then the definitions of the related parameters are described, which include acceptance angle, numerical aperture, refractive index, cut-off wavelength, mode field diameter, spot size etc. For measurement of these parameters, the common optical components, instruments, as well as fiber handling are briefed. Then, the measurement techniques are presented along with the geometry specification of optical fibers. Each of the introduced measurement technique will be provided with a practical example for a better understanding.

Introduction

In 1960s, Sir Charles K. Kao firstly calculated the necessary condition of a glassy fiber for practically guiding the optical wave with high information capacity (Kao and Hockham 1966). In particular, it has been concluded that two factors are significant for an optical fiber to be an effective light waveguide: the ratio of core diameter to cladding size and the difference of the refractive index between core and cladding. With the development of fiber fabrication technology, the loss at 1550 nm region has been successfully reduced to as low as 0.2 dB/km (Miya et al. 1979), which boosted the development of the optical fiber communications and finally brought us from the electronic era into the photonic era. Nowadays, application of the optical fiber has already extended to the sensor, imaging, medical, etc. In order to understand the optical fiber and its application in such a wide range, it is necessary to have more knowledge about the fundamental properties of the optical fiber. This chapter will focus on the basics of the optical fiber and related measurement techniques. Fundamental properties of the optical fiber including acceptance angle, numerical aperture, refractive index, cut-off wavelength, mode field diameter, spot size, and attenuation coefficient are discussed. Then, the mostly used optical components, optical equipment, and the fiber handling are introduced. At last, the related measurement techniques are described along with the geometry specification of the optical fibers. For the sake of better understanding, more than one measurement techniques are presented for each property with practical examples.

Optical Fiber Basics

Basics of Optical Fiber

A typical structure of a step index optical fiber consists of two coaxial layers of homogeneous glasses as shown in Fig. 1. The central part is the core and the outer part is the cladding of the fiber. In order to confine the light in the core, the refractive

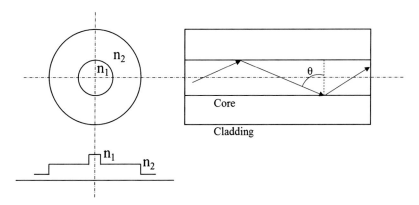

Fig. 1 Structure of a step index optical fiber and refractive index distribution

index of core is designed to be higher than that of the cladding. As shown in Fig. 1, the refractive index distribution in the range of core and cladding is a step function. Thus, optical fiber with such structure is called step index optical fiber (Gloge 1971).

In order to have total internal reflection at the interface of core and cladding without leaking to cladding, the reflection angle θ at the interface of core and cladding is required to be larger or equal to the critical angle θ_c, which is calculated as:

$$\sin\theta_c = -\frac{n_2}{n_1},$$ (1)

where n_1 and n_2 are the refractive indices of core and cladding. One common issue of the step index optical fiber is the modal dispersion. As propagation paths of different modes in the step index optical fiber vary, the time that each mode spent on passing the same axial length of the fiber differs. The advent of the graded-index (GI) fiber successfully solved this problem. A GI fiber is a fiber whose core material is inhomogeneous (Gloge 1971). The refractive index of core decreases as a function of the distance from the core center as shown in Fig. 2. Graded-index fiber is named for such special refractive index distribution.

The distribution of the index along the radial direction can be represented by Gloge and Marcatili (1973):

$$n(r) = n_1 \sqrt{1 - 2(r/a)^\alpha \Delta}, r \leq a,$$ (2)

$$n(r) = n_1 \sqrt{1 - 2\Delta} = n_2 \, r > a,$$ (3)

$$\Delta = \frac{n_1^2 - n_2^2}{2n_1^2},$$ (4)

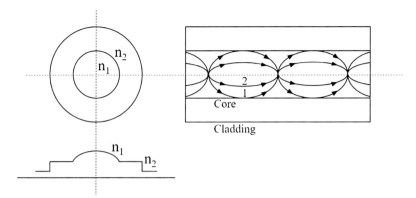

Fig. 2 Structure of a graded-index optical fiber and refractive index distribution

where n_1 is the refractive index along the fiber axis in the core, n_2 is the refractive index in cladding part, a is the core radius, and α is used to describe the refractive-index variation. Because of the variation of the refractive index in the core, the direction of the light changes toward to axis continuously during the propagation along the waveguide as shown in Fig. 2. Let us discuss the travel speed and travel distances of propagation trace 1 and 2 in Fig. 2. Trace 1 travels longer distance compared with trace 2 between two focus points. At the same time, trace 1 travels faster than trace 2 does because the refractive index in the path of trace 1 is lower. This feature makes both of the lights take nearly the same travel time. Therefore, one significant advantage of GI fiber is the suppression of model dispersion.

Besides the layer of core and cladding, most fibers have a polymer coating surrounding the cladding layer. The purpose of the coating is to increase the tensile strength and prevent the fiber surface from the physical damage. Furthermore, another jacket layer will be applied to the optical fiber, which can have higher tensile strength and be used for harsh conditions. Especially, specific color code of the jacket is widely implemented in practical, so that fiber types can be easily distinguished.

Basic Parameters and Definitions

Acceptance Angle and Numerical Aperture

To realize the light transmission over the optical fiber, the launching angle of the incident beam should satisfy certain condition as shown in Fig. 3. Considering the situation that the beam enters the optical fiber with a launching angle of θ_α as the beam 1 shown in Fig. 3, the reflection angle of the beam at the interface of the core and cladding just equals to the critical angle θ_c, so that no light leaks to the cladding in this circumstance. However, in the case of beam 2 (dash line), the launching angle is increased so that the reflection angle surpass the critical angle at the interface of

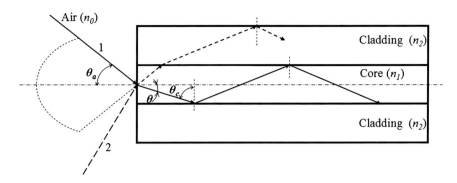

Fig. 3 Beam path from the air entering the fiber. Beam 1 enters with the entrance angle equals to acceptance angle θ_α. Beam 2 (dash line) enters with the entrance angle greater than the acceptance angle

core and cladding, then the beam escapes from the core to the cladding. Therefore, the launching angle θ_α ensures that any beam whose launching angle is smaller than θ_α can propagate inside the core part of the optical fiber is known as acceptance angle. In other word, if the launching angle is greater than θ_α, the beam cannot experience a total internal reflection at the interface of the core and cladding.

In order to have a more general expression of the acceptance angle when taking the refractive index of air, core, and cladding into account, the term numerical aperture (NA) is introduced. Taking the beam 1 as the example again, the beam 1 enters the fiber from the air with refractive index of n_0 with launching angle θ_α. With the Snell's law, the entrance angle θ_α at the core side is given by:

$$n_0\sin\theta_\alpha = n_1 sin\theta \tag{5}$$

As $\theta_c = 90° - \theta$ and the critical angle is given by Eq. 1, NA is referred to as:

$$NA = n_0\sin\theta_\alpha = \left(n_1^2 - n_2^2\right)^{1/2}, \tag{6}$$

where n_1 and n_2 are the refractive indices of core and cladding. Since NA is a metric and independent of the fiber size, it is very useful in a practical view.

Attenuation Coefficient

The minimum detection level of the receiver limits the distance of the optical transmission system, thus transmission attenuation in optical fiber is another significant specification needed to be considered when integrating fibers into a communication system. The attenuation coefficient α at wavelength λ is defined as:

$$P_1(z) = P_0 e^{-\alpha(\lambda)z} \tag{7}$$

where P_0 is the input light power and $P_1(z)$ is the power at the position z away from the input end of the fiber.

Two types of transmission attenuation are discussed in this section. One is the fiber attenuation related with fiber material. The other one occurs at fiber-fiber connection point, where core concentricity error and non-circularity lead to the connection loss of splicing or physical connector.

Attenuation in optical fiber is caused by several mechanisms including absorption, scattering, and geometric effects. Light absorption in silica glass optical fiber is usually caused by the material impurity. In the case of glass fiber, OH absorption is one of the main absorption origins (Lines 1984). The OH in optical fiber are generally created during the preform fabrication or fiber drawing. Multiple absorption peaks of OH are observed in the wavelength range from visible band to infrared band. The fundamental absorption peak locates at 2.7 μm, and the first, second overtones are found to be at 1.4 μm and 0.95 μm. Besides, silica or dopant-associated resonance frequencies also induce infrared absorption peaks. In addition, photon energy induced electronic transitions between quantum energy level in transition-metals is another contributor of light absorption in glass optical fiber. The resonance absorption peak wavelength is decided by the energy band gap between the initial and final energy state of the involved ions, which can be defined as:

$$\lambda = \frac{hc}{E_2 - E_1} \tag{8}$$

where h and c are the Planck constant and the light velocity in vacuum, and E_2 and E_1 are the final and initial energy level. Some of the metals such as Fe, Cu, V, Co, Ni, Mn, and Cr has strong absorption coefficient in the range of communication band (Lines 1984). Thus, the concentrations of these materials are not allowed to surpass level of a few ppb. With the refinement of the fiber fabrication technique, such kind of material impurity loss is hardly observed in high purity silica fibers.

Rayleigh scattering is mainly responsible for the scattering loss in glass optical fiber. As the result of fiber fabrication process, microscopic variations of fiber material component density, randomly distributed material defects, and inhomogeneous material structure all contribute to variation of the refractive index in the scale of much smaller than the wavelength of interest. By the analogy with such variation and small objects, the Rayleigh scattering is the process of energy scattering when the traveling light interacts with the small objects. The metric of Rayleigh scattering is proportional to λ^{-4}, therefore the significance of the Rayleigh scattering reduced with the increase of wavelength as shown in Fig. 4.

Geometric effect-induced optical loss is caused when bending a fiber. Such bending loss is categorized to into two types: macroscopic and microscopic bending loss. The former one is produced whenever apply curvatures to the fiber. The practical examples are wrapping in a spool and fiber installation at the corner. Macro bending loss can be explained from the view point of changing of the angle when traveling light interacting with interface the core and cladding. Such variation of

Fig. 4 Attenuation spectrum in a silica optical fiber at infrared band

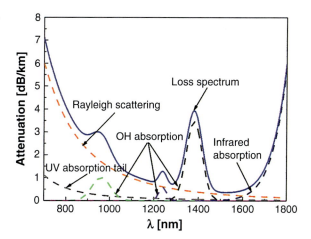

angle tends to become smaller than the critical angle, where the total reflection does not occur anymore, and portion of the light launches into the fiber cladding. In particular, the high mode light with traveling angle close to critical angle is more sensitive to the macro bending loss. Microscopic bending loss is much weaker. It is caused by the strain or stresses distributed along the fiber, which is subjected to temperature variation along the length or setup in the cabling process. The amount of microscopic bending loss depends on the how well the fiber is coated or packed.

Cut-Off Wavelength

The cut-off wavelength for the first high-order mode (LP_{11}) is an important parameter for single mode fiber, which separates the fundamental mode LP_{01} from other higher order modes. Figure 5 shows the Bessel functions J_0 and J_1 against the normalized frequency (also called V-number) in a step index optical fiber (Gloge 1971). The cutoff point for separating each mode is located at when J_0 or J_1 crosses the zero. Thus, as indicated from Fig. 5, single-mode operation exists when the normalized frequency is below 2.405. This leads to a theoretical cut-off wavelength λ_c given by:

$$\lambda_c = \frac{2\pi a}{V_c}\left(n_1^2 - n_2^2\right)^{1/2}, \qquad (9)$$

beyond which only the LP_{01} mode propagates in the fiber.

Generally, a multimode fiber has a number of cut-off wavelengths as there are more bound propagating modes. Usually the cut-off wavelength refers to the case of single-mode fiber. The theoretical cut-off wavelength can also be calculated from the fiber refractive index profile (Marcuse 2013). However, the cut-off wavelength measured is always smaller than the theoretical value by as much as 100–200 nm because the attenuation of the LP_{11} mode is strongly influenced by fiber bend, length, and cabling near cut-off wavelength (Kalish and Cohen 1987; Neumann 2013).

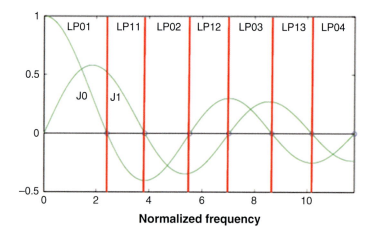

Fig. 5 The allowed regions for the various LP modes against V-number for a step index optical fiber

Mode Field Diameter and Spot Size

In a single mode optical fiber, as a portion of the light travels in the cladding, the transverse range of the fundamental model field is larger than the size of the core. The intensity distribution of the energy in the transverse direction can be described by a Gaussian profile, and the diameter when the intensity is reduced to 1/e is the mode field diameter (MFD) as depicted in Fig. 6. The value of the MFD can be numerically calculated as (Ghatak and Thyagarajan 1998):

$$d_{MFD} = 2\omega_{NF} = 2\left[\frac{2\int_0^\infty r^2 \Psi^2(r)r dr}{\int_0^\infty \Psi^2(r)r dr}\right]^{1/2}, \qquad (10)$$

where $\Psi(r)$ is modal intensity at a distance r away from the fiber center in the transverse direction and ω_{NF} is defined as near-field spot size. Both of the MFD and spot size are very important parameters, because they are usually used to evaluate the characterization of the light traveling, such as coupling efficiency, bending loss, and even waveguide dispersion. The measurement of the spot size will be discussed later in this chapter.

Components and Test Equipment

Components and Handling Techniques

Generally, an optical fiber measurement link consists of four major components as depicted in Fig. 7. The light source is the device generating optical signal. The fiber connector/coupler is the connection components to link the fiber or changing the

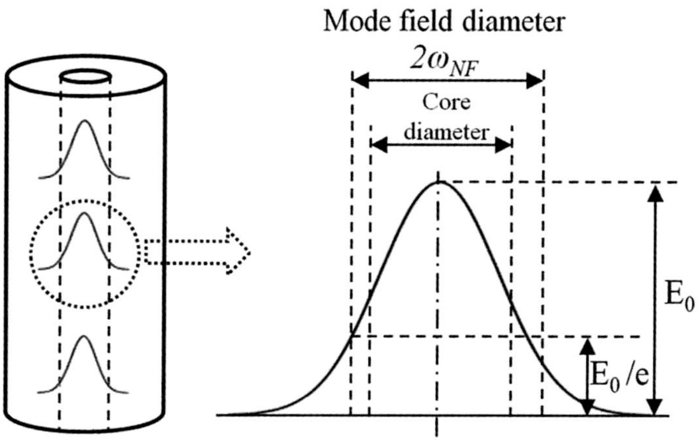

Mode field diameter

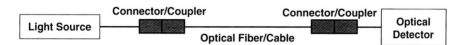

Fig. 6 Illustration of the mode field diameter

Fig. 7 Schematic diagram of a simplified fiber optic test system

light path for specific purpose, such as energy division/combination, wavelength division/combination, etc. The optical detector is an optical component to convert the optical signal to the electrical signal so that the optical signal can be processed by the electrical signal analysis hardware. In this section, the functions and various types of these mentioned optical components will be introduced.

Light Source
The optical source in an optical fiber measurement system is used to provide optical signal (light) that can be coupled into the optical fiber. Three kinds of optical light sources are commonly used for optical fiber test and measurement. These sources include: broadband light sources (tungsten or mercury lamp), monochromatic incoherent sources (light emitting diodes (LED), and monochromatic coherent sources (lasers).

LED
LED is a P-N junction diode that emits light with different wavelengths depending on the semiconductor materials used. When LED is forward biased, electrons cross the junction from the n- to the p-type material, recombine with holes and produce light. Most of the semiconductor materials consist of III–V ternary & quaternary compounds. Typical emitting wavelengths of LEDs with different materials are summarized and shown in Fig. 8. LED should have a high radiance to ensure

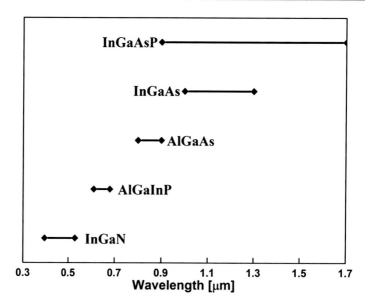

Fig. 8 Typical wavelength ranges of different LEDs

sufficient coupling efficiency, fast response time to obtain broad bandwidth, and high quantum efficiency for fiber optic system.

Laser

Laser is an acronym coming from Light Amplification by Stimulated Emission of Radiation. Lasers can produce coherent light with high power and narrow spectral width. The light emitting mechanism of laser involve the processes of stimulated emission in a gain medium.

To obtain stimulated emission and hence lasing, two conditions must be satisfied to ensure stimulated emission overwhelm the spontaneous emission. One condition is known as population inversion of gain medium; the other condition is requiring an optical cavity to provide feedback. Based on the active medium, laser systems can be categorized as shown in Table 1 and the operation wavelength is mainly dependent upon the gain medium. The gain medium can be either a solid-state medium or a gas-phase medium. Even the same gain medium, the operation wavelength can further be selected or optimized by the cavity.

Photodetector

The photodetector (PD) in an optical fiber measurement system is used to convert the received optical signal into electrical signals prior to further amplification and processing. Requirements for PDs include high sensitivity, short response time for a desirable bandwidth, low noise contribution, small size for effective coupling, wide linear response range, stability, low cost, and low bias voltage. Among different types of PDs are photodiode, avalanche photodiode (APD), photomultiplier tube,

Table 1 Types and operation wavelengths of available lasers

Laser type	Laser gain medium	Operation wavelength(s) [nm]
Gas laser	Helium-neon	632.8
	Argon	488.0, 454.6, 514.5
	Nitrogen	337.1
	Carbon dioxide	10,600
	Excimer	193(ArF), 248(KrF), 308(XeCl), 353(XeF)
Chemical laser	Hydrogen fluoride	2700–2900
	Deuterium fluoride	$\sim$3800 (3600–4200)
	COIL(Chemical oxygen-iodine laser)	1315
Dye laser	Stilbene	390–435
	Coumarin 102	460–515
	Rhodamine 6G	570–640
Metal-vapor laser	Helium-cadmium	325, 441.563
	Helium-mercury	567, 615
	Helium-silver	224.3
	Neon-copper	248.6
Solid-state laser	Ruby	694.3
	Nd:YAG	1064
	Er:YAG	2940
	Ti: sapphire	650–1100
	Tm:YAG	2000
	Yb:YAG	1030
	Er: erbium-ytterbium doped glass	1530–1560
	$Sm:CaF_2$	708.5
Semiconductor laser	GaN	400
	InGaN	400–500
	AlGaInP, AlGaAs	630–900
	InGaAsP	1000–2100
Other type	Free electron laser	0.1 nm – several mm
	Raman laser	1–2 μm (fiber)

and infrared sensor. Although all these PDs can be used to detect optical signal, they have distinct sensing mechanisms.

Internal Photoelectric Effect

For photodiode, including Positive-Negative (PN) junction photodiode, Positive-Intrinsic-Negative (PIN) photodiode, and avalanche photodiode, the sensing principle is based on internal photoelectric effect as shown in Fig. 9. The PN junction diode operation principle is just the opposite of an LED. In semiconductors, absorption of light with photon energy hv excites electron from valence band to

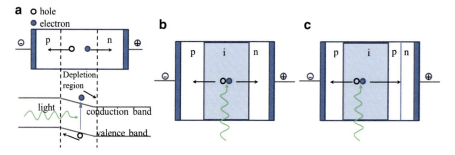

Fig. 9 Internal photoelectric effect for (**a**) PN junction photodiode, (**b**) PIN photodiode, and (**c**) avalanche photodiode

conduction band, leading to the generation of electron-hole pair. Under the electric field of a reverse bias voltage, these free charge carriers flow and create a current in the external circuit. For PIN photodiode as shown in Fig. 9b, such structure increases the width of the depletion region which improves the quantum coefficient and increases speed. Avalanche photodiode is specially designed for an internal gain by impact ionization. Such gain is achieved through generation of many secondary electrons and holes by absorption of single primary electron/hole as shown in Fig. 9c.

Characteristic parameters for photodiode include quantum efficiency, responsivity, and dark current. Quantum efficiency is the capability of converting light signal to electrical signal. Responsivity of photodiode is the ratio of the photocurrent output to that of radiant energy incident on the detector. Dark current is the output from the photodiode without incident light.

Currently, photodiodes using Silicon, Germanium and Indium Gallium Arsenide (InGaAs) as the fabricating material are for different applications. Features including quantum efficiency, detection wavelength range, and dark current are summarized in Table 2. The primary advantage of an APD over a PN or PIN photodiode is high gain and therefore a much higher sensitivity. However, the noise is also amplified the same as the signal so that an APD usually has higher dark current as compared with PN or PIN photodiode.

External Photoelectric Effect

Photomultiplier tubes (PMTs) are extremely sensitive optical detectors that can detect light signal with very low intensity. As shown in Fig. 10, the incident photon interacts with the photocathode and ejects photo-electrons as a consequence of the photoelectric effect. These primary photo-electrons are reflected to the focusing electrode, where electrons are amplified by a series of secondary emission process. Therefore, the PMT can achieve high sensitivity and generate high SNR signal due to multiple level amplification. Because of its high sensitivity, PMT is suitable for the optical detection requiring high sensitivity and low noise, e.g., photon-counting task.

Table 2 Typical characteristic parameters of photodiodes

Photodiode	Wavelength range [nm]	Peak [nm]	Responsivity [A/W]	Quantum efficiency (%)	Dark current [nA]
Silicon PN	500–1000	900	0.41–0.7	0.75	1–5
Silicon PIN	400–1100	900	0.6	65–90	1–10
Silicon APD	400–1100	830	77–130	77	0.1–1.0
Germanium PIN	800–1600	1550	0.65–0.7	50–55	50–500
Germanium APD	800–1600	1300	3–28	55–75	10–500
InGaAs PIN	900–1700	1550	0.75–0.97	60–70	1–20
InGaAs APD	900–1700	1550		60–70	1–5

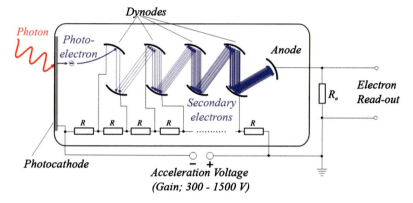

Fig. 10 Schema of a PMT

Others

Infrared sensor such as thermopile is a heat sensitive detector that is used to measure temperature change caused by absorbed infrared energy. When infrared light strikes the "sensor face," it generates energy after receiving the infrared heat. Since such detectors are sensitive to ambient environment, they are usually sealed in vacuum to prevent it from heat transfer except by radiation.

Cables and Connectors

Fiber optic links require connectors to connect the light source with the optical fiber or the fiber with the detector. Optical couplers are usually used to distribute light from one port to another or more fibers.

Fiber Connectors

Physical contact connectors as shown in Table 3 are the most general type of connectors. Such connectors are rugged, easy to clean, and cost-effective. Moreover,

Table 3 Common fiber connectors

Standard Connector (SC)		Simple, rugged, and low cost; available in single-mode and multimode; offer a push on/pull off operation
Ferrule Core Connector (FC) or FC/APC (Angled Physical Contact)		Low back reflection; exact alignment of the fiber in relation to the detector and the optical source with the help of a threaded container and a position locatable notchAngled FC connector known as FC/APC minimizes the reflected signal due to the angle design
Lucent Connector (LC) or LC/APC (Angled Physical Contact)		Pull-proof design; small size; available in simplex or duplex versionsAngled LC connector known as LC/APC minimizes the reflected signal due to the angle design
Straight Tip Connector (ST)		Utilize a bayonet lock maintaining the spring-loaded force between the two fiber cores; available in single mode and multimode
SMA Connector		Forerunner to the ST connector; replaced by the ST and SC connectors

the insertion loss for these connectors is usually low. There are many different types of fiber connectors as summarized in Table 3.

Fiber Coupler

The mechanism of the optical fiber coupler is classified into two categories based on the way the power transfer takes place. One is the light transfers through the fiber core cross section by the interaction between the fiber butts. The other method is sharing the light power by changing the core light modes to cladding and refracted mode through the fiber surface along the light-guiding path (Senior 2004).Various types of optical fiber couplers are available and generally classified into the following types:

(a) Three port coupler and four port coupler that are used to split, distribute, and combine signal.
(b) Star coupler for transmitting signal to/from a single port from/to multiple outputs. M and N are the number of the input and output port respectively.
(c) Wavelength division multiplexing (WDM) are special couplers designed to transfer different permitted wavelength signal to one single fiber. In contrast, wavelength de-multiplexer separates the different wavelength signals from one single port to multiple outputs.

Fiber Handling
Since dirt, stress, and flaws on the surface of bared fiber may influence the functionality of the system and even result in the failure of the fiber, attention should be stated in the fiber handling, including operation and processing processes. For example, do not overbend the fiber; avoid placing the fiber on top of hard items or in touch with sharp edges; always clean the bared fiber with ethanol instead of abrasive or organic solvents. In addition, for safety operation, always wear protective safety goggles and never look into the fiber end when light is launched. Moreover, remember to dispose all cleaved fiber part into the sharp container since it is sharp enough to penetrate your skin and extremely dangerous.

Test Tools and Equipment

Splicer and Cleaver

Splicer
Fusion splicing is a technique for permanent splicing optical fibers with the low connection loss. In fusion splicing, two fibers are fused together by electric-arc welding when they are correctly aligned using a fiber splicer. The splicing loss can be as low as 0.05 dB equaling to less than 1% power loss. A splicer is usually equipped with a microscope or a charge coupled device (CCD) camera and a liquid crystal display (LCD) to observe the alignment condition of the fiber to be spliced. The automatic alignment of the fiber is carried out by monitoring the positions through the objective lens and CCD. Most splicers, as shown in Fig. 11, can hold fibers of various sizes including single mode fibers and multimode fibers. However, high loss can be induced by different type or size fibers. Some splicers can even automatically splice multicore and ribbon cables up to 12 fibers at a time.

Fig. 11 Fiber fusion splicer

Cleaver

Fiber cleaving is a process to get a clean and flat fiber end surface. In particular, in fiber cleaving, the fiber is scored or scratched, and stress is applied to break the fiber in a smooth manner. Then, the fiber can be cleaved with a clean surface perpendicular to the length of the fiber, with no protruding glass on either end if the cleaving has been done properly. A cleaver is the tool used to cleave the fiber with low tension. It scores the fiber surface at the proper location and applies larger tension until the fiber is broken. Some cleavers, as shown in Fig. 12, can automatically produce consistent results so that users only need to clamp the fiber into the cleaver and operate its controls. Other cleavers are less automation requiring. As users exert force manually to break the fiber, it makes them more dependent on operator technique.

Optical Power Meter

The optical power meter is the device to measure the power of an optical signal. One typical optical power meter consists of a PD, amplifier, and display. The PD is selected according to the required measurement wavelength and intensity level. In order to be clearly seen when wearing the laser safety goggle, optical power meter is usually equipped with a vacuum fluorescent display (VFD) as shown in Fig. 13. For the accurate power measurement, the optical power meter should provide wavelength dependence individual correction, temperature stabilization, wide power region with good linearity, good spatial homogeneity, low polarization

Fig. 12 Components of a
typical fiber cleaver

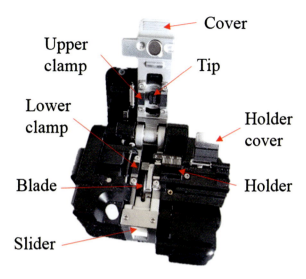

Fig. 13 Optical power meter
with an optical head

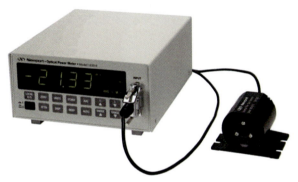

dependence, low reflections, and compatibility with different types of fibers. For the high power level measurement, an attenuator is used to decrease the incident power so that the max peak power will not saturate the power meter avoiding a wrong reading. Furthermore, a class of thermal detection power meter has been developed for the laser with high peak power, such as femtosecond pulse laser.

Optical Spectrum Analyzer

An optical spectrum analyzer (OSA) is an optical instrument to analyze the optical spectrum information. A typical front panel of the OSA is shown in Fig. 14. The display is used to real time tracing the optical power in the vertical scale and the wavelength in the horizontal scale. The OSA receives an optical signal (usually via an optical fiber) and transfers it to an optical spectrum, which shows the optical power as a function of the wavelength. For the application, the OSA can be employed to measure the amplifier gain, obtain ASE spectral shape, calculate optical signal to noise ratio, and characterize light source.

Fig. 14 Front panel of an
OSA

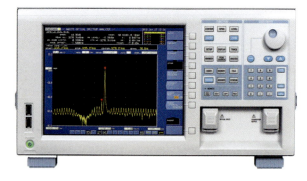

Fig. 15 Illustration of the
cut-back attenuation
measurement

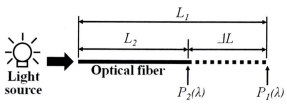

Attenuation Measurements

There are three typical methods for optical fiber attenuation measurement: cut-back technique, insertion loss technique, and Optical Time Domain Reflectometer (OTDR) backscattering technique. Each method has its own advantage and disadvantage with respect to the setup complexity and measurement accuracy.

Spectral Attenuation Measurement

In the process of cut-back spectral attenuation measurement, two transmission spectra are measured. As shown in Fig. 15, the transmission spectrum of the fiber with L_1 length $P_1(\lambda)$ is firstly measured. Then the fiber is cut for length of ΔL, and the transmission spectrum $P_2(\lambda)$ is measured again. As a result, the attenuation spectrum $\alpha(\lambda)$ of the fiber in the unit of length is calculated to be:

$$\alpha(\lambda) = \frac{10}{\Delta L} \log_{10} \frac{P_2(\lambda)}{P_1(\lambda)} \tag{11}$$

A typical setup for the measurement is shown in Fig. 16. A halogen lamp can be a good candidate of light source for the attenuation measurement ranging at visible and infrared band. The incident light from lamp is focused via a lens and launched into the optical fiber after being chopped by a chopper. Then the transmitted light is split by the monochromator (Mono) and the optical signal is converted to electronic

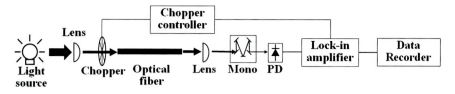

Fig. 16 Set up for cut back spectral loss measurement

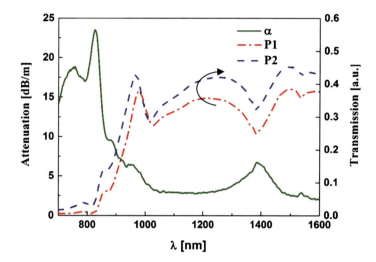

Fig. 17 Attenuation measurement result using cut back technique

signal by the PD. Finally, the signal is extracted and enhanced by the lock-in amplifier. The PD is selected according to different measurement wavelength range. An InGaAs PD has higher responsivity at the wavelength above 1.0 μm, while a Si PD is suitable for the signal in the range between 0.5 and 1.0 μm. The stability of the system is important for accurate measurement using this system, which may include the launching emission intensity stability and interruption of any movement in the system. Besides, a suitable launching technique and usage of index matching oil can help to reduce the influence of the light traveling in cladding mode.

An illustration of the spectral attenuation measurement result by cut-back technique of a bismuth erbium co-doped optical fiber is shown in Fig. 17. First, two transmission spectra P_1 and P_2 are measured. The optical signal of P_2 is stronger than that of P_1 due to the difference length of the sample fiber. Then the spectral attenuation spectrum is calculated according to Eq. 11 as shown in Fig. 17. Since several elements, including bismuth, erbium, are doped in this fiber, multiple absorption peaks are observed: 1535 nm, 1420 nm, 830 nm. Meanwhile, the loss tends to increase in the short wavelength range due to the enhancement of Rayleigh scattering loss.

Insertion Loss Measurement

Insertion loss is important in evaluating fiber systems and components. The measurement system and operation technique used in determining insertion loss is similar to in the cut-back method. Two transmission spectra are measured: one is the transmission spectrum excluding the fiber under test in the system, the other one is measured when inserting the fiber under test into the system. In order to complete this insertion measurement, a stable fiber-fiber coupling method is acquired. An adjustable device like V-groove or directly splicing the fiber with low splicing loss is alternative. Thus, by introducing the uncertainty during the insertion operation into the measurement, cut-back method is more precise than the insertion loss method. Meanwhile, non-destructive insertion loss method is more suitable when the integrity of the fiber needs to be ensured.

Optical Time Domain Reflectometer Loss Measurement

The basic idea of OTDR is employing the backward scattering light in optical fiber. Detection of such light signal provides not only the level of scattering loss but also its position information in the optical fiber. As the amount of the scattered light varies with the microscopic fluctuations of the refractive index and flaws in the fiber, not only the absorption but also splice loss, micro-bending loss, and fault detection can be implemented using OTDR scheme. The operation principle can be described in two steps: periodically launching a pulse light into the one end of the fiber and detecting the backward scattered light from the same end. However, since the signal of the scattered light is usually very low, special detection setup is necessary to identify and extract the periodic signal from the noise. One detection setup is illustrated in Fig. 18.

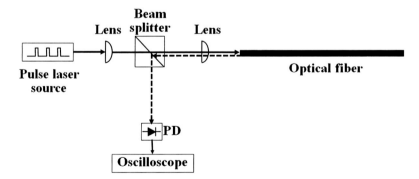

Fig. 18 Setup for OTDR attenuation measurement

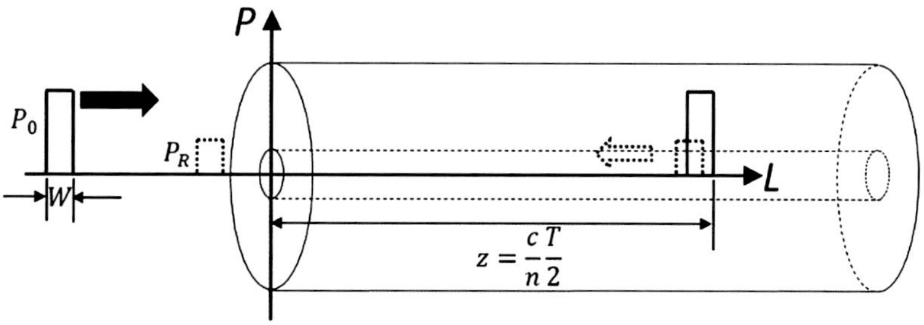

Fig. 19 Illustration of OTDR loss measurement

A pulse laser source launches short pulses periodically. The pulse laser source can be a Q-switched Nd:YAG laser, Er^{3+} lasers, or AlGaAs lasers for different wavelengths. The generated pulse is split into two pulses, and one is injected into the fiber to yield the backscattering light. In a typical OTDR system for measuring single mode fibers, the backward Rayleigh scattering light P_R at the input end of the fiber (as shown in Fig. 19) was firstly numerically calculated by Personick (1977), where the P_R as a function of time T can be expressed as:

$$P_R(T) = \frac{P_0 \alpha_s S W v_g}{2} e^{-2\alpha \frac{v_g T}{2}}, \qquad (12)$$

where α_s is the Rayleigh scattering constant, S is the fraction of backscattered Rayleigh light, W is the pulse width launched into the fiber, v_g is the light group velocity in the fiber, n is the material index, α is the fiber attenuation coefficient. It is necessary to be noted that W should be smaller enough than $2n/c$ (Bernard and Depresles 1988). In practical, Rayleigh scattering light signal is six orders of magnitude lower than the input optical signal. Therefore, a lot of efforts have been made to enhance the detection sensitivity. One method is to employ an optical amplifier (Mears et al. 1987; Sato and Aoyama 1991). A semiconductor laser was used to amplify the backscattered signal by 4.5 dB (Suzuki et al. 1984). High sensitivity detection devices, such as single photon detectors, were also introduced to OTDR system. Besides, the advanced detection techniques, including optical frequency domain reflectometry (Eickhoff and Ulrich 1981) and heterodyne detection, have been developed for OTDR system as well (Tateda and Horiguchi 1989).

Figure 19 illustrates the reflection signal in a single mode optical fiber measured by a typical OTDR system. The signal decreases along the distance at a constant speed, whose slope is decided by the strength of Rayleigh scattering. When the incident light experiences the impurity of the fiber, splicing point, and Fresnel reflection at the end of fiber, strong reflections are detected by the OTDR system as shown in Fig. 20. Therefore, OTDR system is a useful tool to locate the events described above in the fiber.

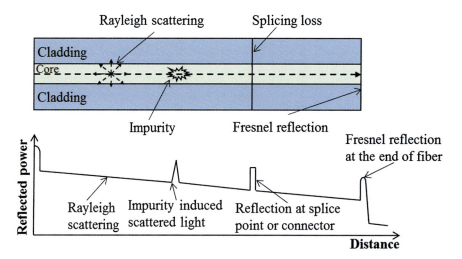

Fig. 20 Illustration of OTDR measurement result in an single mode optical fiber

Index Profile and Geometry Measurement

Fiber Design Parameters

Cut-Off Wavelength Measurement

The cut-off wavelength determined experimentally is referred to as the effective cut-off wavelength, because the real cut-off wavelength tends to vary with the measurement technique due to the dependence of the LP_{11} mode attenuation on the fiber length and bending radius. Several widely utilized measurement methods are proposed and adopted to study the cut-off wavelength and will be discussed. Among them, three measurement methods are recommended by the International Telephone and Telegraph Consultative Committee (CCITT), including bending-reference technique, power step method, and spot size technique. All these methods will be included in the following and other alternatives will be introduced briefly. Measurement configuration for the cut-off wavelength by bending-reference technique and power step method is illustrated in Fig. 21. It consists of a quasi-straight 2 m fiber with one single loop of 28 cm diameter (I. T. Union 2009; T. I. Association 2003). For bending-reference technique and power step method, including bending-reference technique and power step method, the wavelength at which the total transmitted power decreased by 0.1 dB is defined as the effective cut-off wavelength. In bending-reference method, the total transmitted power, including fundamental mode and higher order modes power, is decreased by introducing a small loop. For power step method, the relative attenuation is calculated between the total transmitted power of fiber sample and that of a multimode fiber. Detailed descriptions about these two methods are as follows.

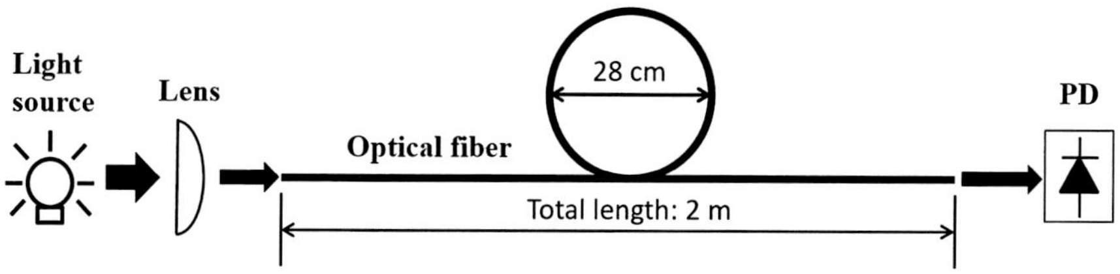

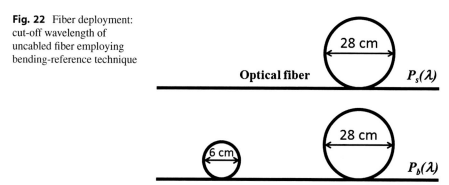

Fig. 21 Configuration for the measurement of cut-off wavelength by bending reference technique and power step method (I. T. Union 2009; T. I. Association 2003)

Fig. 22 Fiber deployment: cut-off wavelength of uncabled fiber employing bending-reference technique

Bending-Reference Technique

In the bending-reference method (known as single bend attenuation method), the launching power is kept fixed and the transmitted power through the fiber in Fig. 22 is recorded as a function of wavelength. Then at least one loop with small diameter (6 cm or less) is introduced into the fiber to filter higher modes and keep fundamental mode. In this case, the transmitted power $P_b(\lambda)$ is measured and the bend attenuation $\alpha_b(\lambda)$ is calculated as:

$$\alpha_b (\lambda) = 10\log_{10} \frac{P_s (\lambda)}{P_b (\lambda)} \qquad (13)$$

The bend attenuation exhibits a peak in such wavelength region that the bending loss for the LP_{11} mode is much higher than that for the LP_{01} mode, as can be seen from Fig. 23. According to the CCITT definition of effective cut-off wavelength λ_{ce}, it is determined as the longest wavelength at which the bend attenuation equals to 0.1 dB (T. I. Association 2003).

Power Step Method

In power step method, the transmitted power for the sample fiber $P_s(\lambda)$ and a two meters long multimode fiber $P_m(\lambda)$ are measured and the relative attenuation is calculated as:

Fig. 23 Bend attenuation spectrum in the bending-reference technique for the cut-off wavelength measurement (T. I. Association 2003)

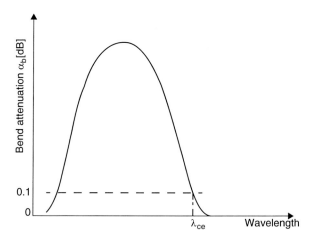

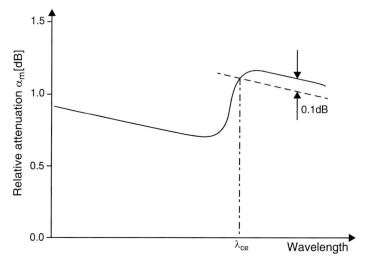

Fig. 24 Relative attenuation spectrum in the power step method for cut-off wavelength measurement (Senior 2004)

$$\alpha_m\left(\lambda\right) = 10\log_{10}\frac{P_m\left(\lambda\right)}{P_s\left(\lambda\right)} \tag{14}$$

To determine the effective cut-off wavelength using this method, the long-wavelength region is firstly fitted to a straight line which then dropped by 0.1 dB (dash line) as depicted in Fig. 24. Its intersection with the relative attenuation spectrum then produces the effective cut-off wavelength.

A variant of the power step method, known as multiple bend method, introduces different curvature R into the fiber and determines the cut-off wavelength for each curvature $\lambda_{ce}(R)$ using a method similar with power step method. By fitting $\lambda_{ce}(R)$

Fig. 25 Mode field diameter vs. wavelength in the spot size technique for the cut-off wavelength measurement (Suzuki et al. 1984; Eickhoff and Ulrich 1981)

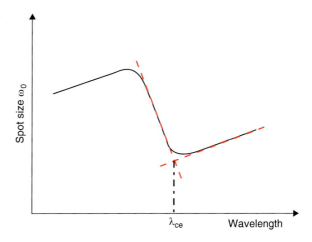

dependence on the curvature using the following power series expansion, one can obtain effective cut-off wavelength (Franzen 1985):

$$\lambda_{ce}(R) = \lambda_{ce0} - A/R + B/R^2 \tag{15}$$

where λ_{ce0}, A, and B are constants.

Alternative Test Method

Alternative methods for determining the effective cut-off wavelength include spot size technique, near-field pattern technique (Franzen 1985; Murakami et al. 1979), and multiple bend method (Murakami et al. 1979). In the spot size technique, the curve of spot size versus wavelength determines the effective cut-off wavelength, which is based on the theory that the spot size increases linearly with wavelength in the single-mode region. For multiple bend method, the cut-off wavelengths under different fiber curvature are measured with method similar to power step method and fitted using power series expansion to give the effective cut-off wavelength. The effective cut-off wavelength for the near-field pattern technique is determined by monitoring the spectral near-field intensity profile with wavelength based on the difference of radial intensity profile across the fiber core between fundamental mode and higher order modes.

Among these methods, the CCITT recommend the spot size technique as an alternative test method for the determination of the effective cut-off wavelength. In this method, a mode field diameter as a function of wavelength is measured using the transverse offset method. Theoretically, the fundamental mode field diameter increases linearly with wavelength when the fiber is operating in the single-mode region. At the cut-off wavelength, there is a large variation because of the contribution of the LP_{11} mode. Mode field diameter variation is linearly extrapolated. The intersection of linear extrapolations determines the effective cut-off wavelength as shown in Fig. 25 (Lines 1984; Galtarossa et al. 1993).

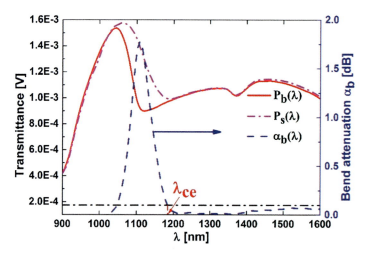

Fig. 26 Cut-off wavelength measurement result using bending-reference technique

Practical Measurement Demonstration

Two of aforementioned methods, bending-reference technique and power step method, are implemented and compared. The experimental setup shown in Fig. 16 can be applied to this cut-off wavelength measurement. The white light is coupled into the fiber under test through a lens and the transmitted light through Mono is detected by a PD and lock-in amplifier.

For bending-reference technique, the measurement steps are as follows: (1) the transmitted light spectrum through tested sample as illustrated in Fig. 22 is measured as $P_s(\lambda)$; (2) the transmission spectrum after introducing another loop of diameter of 6 cm is recorded and indicated as $P_b(\lambda)$. The transmission spectra and bend attenuation $\alpha_b(\lambda)$ calculated using Eq. 13 are displayed in Fig. 26. The obtained effective cut-off wavelength for this case is 1185 nm.

Similarly, for power step method, the measurement steps are as follows: (1) the transmitted light spectrum through fiber under test as illustrated in Fig. 27 is measured as $P_s(\lambda)$; (2) the transmission spectrum after a multimode fiber is recorded and indicated as $P_m(\lambda)$. The transmission spectra and relative attenuation $\alpha_m(\lambda)$ calculated using Eq. 14 are plotted in Fig. 27. The obtained effective cut-off wavelength for this case is 1180 nm, which is quite close to that determined using bending-reference technique.

Spot Size Measurement

Since the cross section of a single mode fiber is only several micro meter, it is difficult to accurately measure the field distribution in the fiber core directly to determine the near-field spot size. Therefore, in order to obtain the near-field spot size, it is recommended to firstly measure the far-field spot size, i.e., the light from the fiber end is projected onto a screen at some distance away. The far-field intensity

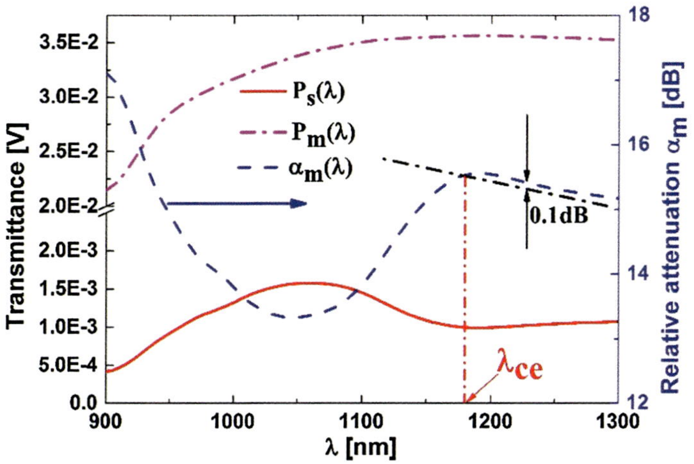

Fig. 27 Cut-off wavelength measurement result using power step method

spot size ω_{FF} is defined in Eq. 16, which can be measured by a far-field pattern.

$$(\omega_{FF})^2 = \frac{\int_0^{\pi/2} F_0^2(\theta)\,\theta^3 d\theta}{\int_0^{\pi/2} F_0^2(\theta)\,\theta d\theta}, \tag{16}$$

where $F_0^2(\theta)$ is the intensity of the far-field, as a function of the angle θ. Then the near-field spot size ω_{NF} can be calculated as follow:

$$\omega_{NF} = \frac{\lambda}{2\pi \tan \omega_{FF}} \tag{17}$$

One measurement system for far-field intensity $F_0^2(\theta)$ is demonstrated in Fig. 28a. A visible light source from a laser is launched into the single mode fiber through an objective lens, and chopped by a chopper. A PD is mounted on a rotating table to measure the intensity distribution of the light coming from the fiber end. The angle related to the far-field intensity is changed by a step motor controller. Figure 28b shows a typical far-field light intensity distribution from a 600 nm single mode fiber by the setup in Fig. 28a with 632.8 nm He–Ne laser and silicon detector. With the measured far-field pattern, the far-field spot size and the near-field spot size can be calculated using Eqs. 16 and 17, respectively.

Geometry

Fiber Diameter
Fiber outer diameter is one of the most important parameters to be controlled in the fabrication process. It determines the splice loss, connector design, etc.

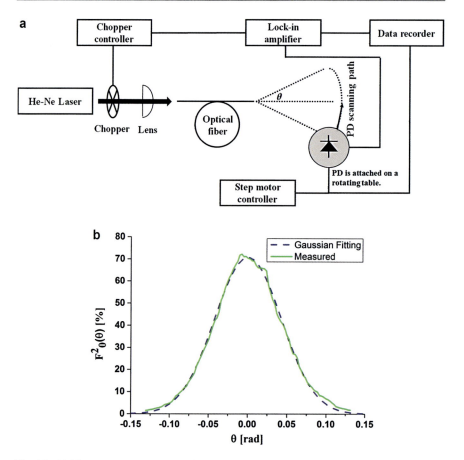

Fig. 28 (**a**) Measurement system setup for far-field intensity. (**b**) Measured far-field intensity as a function of PD rotation angle. The fiber sample is a 630 nm single mode fiber

Therefore, the measurement of fiber diameter is necessary for fiber fabrication quality control. In this section, two techniques of fiber diameter measurement are discussed, including microscope technique and interferometric technique.

Microscope Technique

A short fiber sample with several millimeters is prepared for the dimension measurement and both ends of the fiber have to be cleaved well. A microscope is used for observing and measuring the fiber diameter (Marcuse and Presby 1979). As shown in Fig. 29a, the fiber is attached onto a fiber holder, which is located on the stage of the microscope. The fiber has to be kept vertical, because one end needs to align with the objective lens and the other end faces to the light bulb on the bottom of the microscope. The bulb on the bottom of the microscope provides a beam of light to illuminate the fiber and the light beam transmits through the fiber directly.

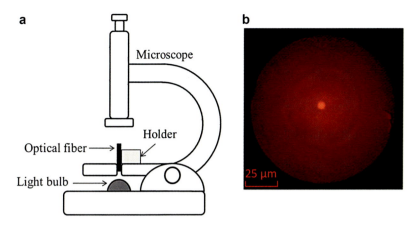

Fig. 29 (**a**) Fiber outer diameter measurement by microscope (**b**) An image taken for the fiber outer diameter measurement by microscope

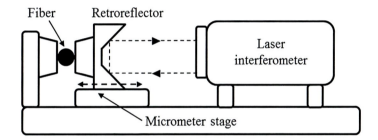

Fig. 30 Schema of fiber diameter measurement by interferometer

Then the image of the fiber cross section can be observed by the eyepiece of the microscope. Generally, there are two methods to read the fiber diameter. One is to use a filar eyepiece to read the fiber diameter with the calibrated scales. Another method is measuring the diameter through the cross section image captured by an integrated camera system. Figure 29b is a typical cross section image of single mode fiber taken by the integrated camera system. By measuring the size of the fiber on the image, the fiber outer diameter can be easily calculated via the scale, e.g., 125 mm.

Interferometric Technique
In the interferometric technique, an interferometric micrometer is used for fiber diameter measurement. The fiber is located between two fused silica probes which have to be plane and parallel, as shown in the Fig. 30. The distance between these two silica probes, which is exactly the fiber diameter, are measured by an interferometric micrometer (Smithgall and Schroeder 1980).

Interferometric technique and light scattering technique provide alternate static measurement. They are the most accurate methods of fiber diameter measurement. The interferometric method is easy to implement. Light scattering technique could

Fig. 31 Definition of CCE

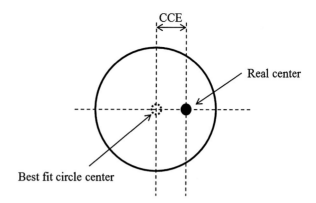

be taken as a fiber diameter monitor during fiber drawing. Microscopy technique is the most widely used method. However, from the view of the measurement consistency, the microscopy technique has the highest error of ±0.5 μm among these techniques. Meanwhile, interferometric and light scattering methods have only ±0.1 and ± 0.2 μm error (Smithgall and Schroeder 1980). Therefore, an average value is usually taken after multiple measurements when the microscopy technique is employed.

Core Concentricity

Core concentricity is the deviation of the core center from the best fit circle center, which is the center axis of the cladding and often described with the core concentricity error (CCE). The definition of CCE is shown in Fig. 31. CCE is mainly related to splicing and connection loss, which is essential for overall system loss budget and fiber specification. The core alignment should be controlled well during the fiber splicing to reduce the CCE, which imposes extra difficult and expenses from installation.

Image Processing Measurement Method

The image processing method is based on an image processing system (Warnes and Millar 1984) as shown in Fig. 32. The fiber is illuminated by a white light through two lens and a rotatable image is formed to be recorded by a TV camera. In the camera, the image is linearized, calibrated, and quantized. The line scans through the core center and the cladding center is compared and analyzed in the computer. The operation processing is independent when the system is set up.

Image Shearing Method

This method was fully discussed before (Moore 1983a). An image shearing microscope produces two images, which are the main image and sheared image, as shown in Fig. 33. The CCE measurement is carried out by bringing the cladding of sheared image with the core edge of main image. The CCE value can be regarded

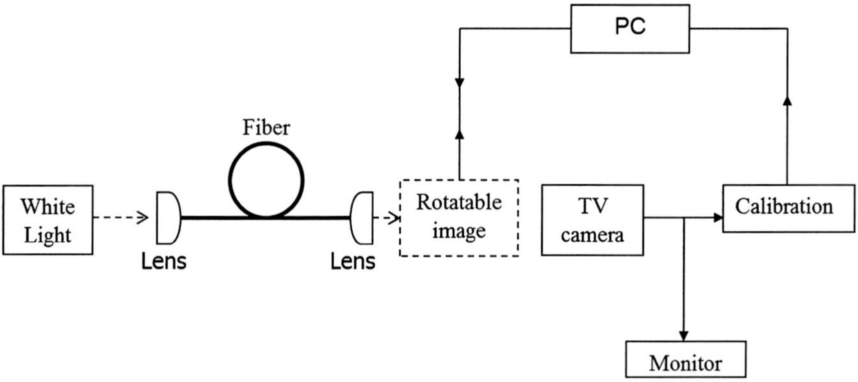

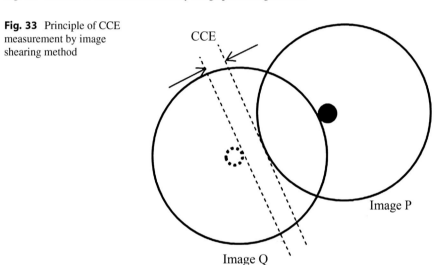

Fig. 32 Schema of CCE measurement by image processing method

Fig. 33 Principle of CCE measurement by image shearing method

as half of the distance between the core edge of sheared image and cladding edge of the main image. Compared with many other measurement methods that are based on operator adjustment of a field coincidence, image shearing method may be criticized as rather subjective. Because this method greatly reduces the measurement error induced by operator objectivity, the condition of fiber end, and illumination setting.

Non-Circularity

Non-circularity of optical fiber is another important parameter to be controlled in the fabrication procedure. It determines the coupling efficiency. The non-circularity measurement techniques have been developed for both core and cladding. Generally, it is defined by the difference between the core/cladding admissible error and the average measured diameter. Some of the measurement methods involve observing and analyzing the image of the fiber cross section. In microscope method, a microscope is used for fiber cross section observation. For image cutting method (Moore 1983b), a microscope is utilized together with an optoelectronic imaging system and a computer to take the image of the fiber cross section. Another method is pulse counting method. In this method, fiber cross section image is taken by a microscope as well, and the edge of fiber section curve is obtained from the comparison of the threshold value and the corresponded gray level of the fiber image. All these methods are complex as the microscope is utilized, and poor to adapt the environment.

In order to improve the adaptability to the surrounding and make the set up easier, an advanced method which is named far-field method is proposed for cladding non-circularity measurement (Xiao et al. 2008). The measurement system setup is shown in Fig. 34. The laser beam goes into the beam splitter prism through the fiber. And then the output beam from the fiber is reflected to the display screen by the beam splitter prism. A spot is formed on the screen, which is taken by a digital camera. The image taken is shown on the observation screen and processed by the computer. N pieces of the images are taken before the calculation of the non-circularity β, which is expressed as:

$$\beta = \frac{R_{max} - R_{min}}{\frac{1}{N}\sum_{i=1}^{N} R_i}, \qquad (18)$$

where R_{max} and R_{min} are the largest and the smallest diameter among the measured results, and R_i is the diameter measured at the ith time. The color image got by CCD is translated into gray image, and then contrast is enhanced for retaining the information. Next, a contour line is picked up by selecting a group of gray level every 20 units with a Prewitt operator. Finally, the non-circularity of fiber is calculated with Eq. 18. A circle is fitted by picking up three different points on each of the contour line, and then the diameter of the circle is calculated. The average diameter result can be obtained from them for the non-circularity calculation.

A single mode optical fiber is measured in Xiao et al. (2008), of which the measurement setup is the same as Fig. 34. When the fiber is attached with the measurement system, the spot image of the fiber is obtained. Then the color image is translated to gray image and the contrast of it is enhanced after that. A median filter is used to remove the noise of the environment. The contour line is generated by selecting the steps of gray level every 20 units. Finally, the result is calculated by the Eq. 18.

A demonstration result showing the non-circularity dependence on the gray level and sampling number is presented in Fig. 35. The gray level has strong effect on the non-circularity other than sampling numbers. With the increase of gray level, the non-circularity changes from 3% to over 10%. On the other hand, the non-circularity has less dependence regarding with the selection of sampling number 90 or 180.

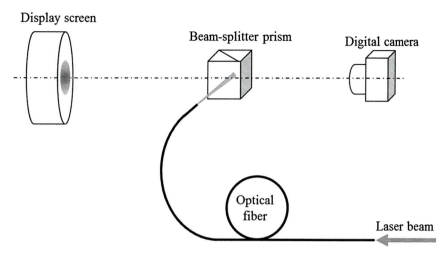

Fig. 34 Measurement schema for fiber non-circularity

Fig. 35 Demonstration of
one measurement result of
non-circularity

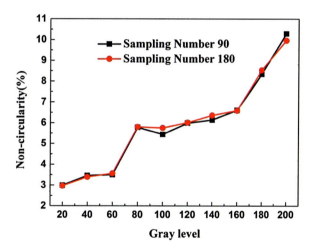

Refractive Index Profiling

Refractive index profile in the optical fiber is one of the parameters to determine the transmission properties of the fibers. In this section, several methods of refractive index measurement are discussed.

Interference Microscopy Method

Interference microscopy method can be used as a standard method due to its high accuracy. There are two interferometric methods which are slab method and transverse method. Slab method is time consuming since nearly 1 day is needed for sample preparation. The test sample is a fiber slab, which is obtained by polishing the fiber to make the surface flat and parallel. The setup needs an interference microscope to observe the sample. The problem of long-time operation of slab method can be eliminated by the transverse method. In this method, a light illuminates the fiber sample which is immersed in the matching oil. The refractive index profile is achieved from the measured fringe shift by solving an integral equation. Circular symmetry is assumed in this technique. Therefore, geometric variations have effect on refractive index profile practically. As a result, several measurements based on different rotational positions of the fiber are necessary for the accurate results.

Focusing Method and Ray Tracing Method

Both of ray tracing method (Liu 2004) and focusing method (Marcuse 1979) compute the index profile through the deflection suffered from the transverse ray passing through the core. The deflection can be measured directly in ray tracing method. In focusing method, the deflection is deduced from the distribution of focusing light intensity. Ray tracing method has the advantage of high resolution. Focusing method is accurate and fast. Circular symmetry is assumed in this method as well.

Light Scattering Method and Reflection Method

Backward and forward light scattering methods are two main methods in the category of light scattering refractive index measurement. In both methods, lights are scattered transversely to the fiber to build the refractive index distribution. But the scattering directions of them are different, i.e., forward and backward to the scattering axis. The backward scattering is more popular compared to the forward one due to its simple implementation. Moreover, it can be applied to both fiber and perform, whereas forward scattering method can only work on the fiber. The reflection method takes the advantage of the reflectivity from the fiber end to determine the refractive index profile, which is easy to implement.

There are many other techniques for measuring refractive index profile. Near-field method (Morishita 1986) deduces the profile from the intensity distribution of incoherent light passing a short fiber. Meanwhile refracted near-field method makes use of the light refracted out of the fiber. Immersion technique (Eickhoff and Weidel 1975) finds the refractive index profile by comparing index matching fluids. Near-field method and immersion technique are easy to implement, but immersion technique is limited in accuracy, and leaky mode corrections are necessary for the near-field method. Refracted near-field does not need leaky mode corrections and has good resolution. The refractive index can also be achieved with X-ray microprobes (Chevallier et al. 1996). However, it has disadvantages of sophisticated instrumentation, limited resolution, and difficult calibration.

Digital Holographic Microscopy Method

Digital holographic microscopy method is a new technology which is developed from the principle of digital holography. In this technology, the object light wave is amplified by micro objective and then interferes with the reference light on the CCD

Fig. 36 Multilayer fiber
model

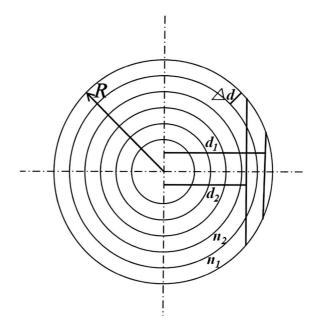

surface to form a hologram. The single phase diagram obtained by the traditional
digital holographic microscopy technology is two-dimensional distribution. The
phase distribution is the path integral of refractive index and light, which follows
as (Linglong et al. 2013):

$$\varphi\left(x_i, y_i\right) = \frac{2\pi}{\lambda} \int_l \left[n\left(x, y\right) - n_0\right] dl \tag{19}$$

In the formula, $\varphi(x_i,y_i)$ is the refractive index distribution of fiber sample, n_0
is the refractive index of surrounding (e.g., refractive index matching oil), l is the
geometric path of sample, and λ is the wavelength. Therefore, in order to obtain the
refractive index distribution of fiber, it is necessary to combine the fiber structure
and digital holographic microscopy technology (Wang et al. 2012; Ting-Ting et al.
2015). A multilayer fiber structure model is used to calculate the refractive index as
shown in Fig. 36. In this model, the fiber with radius R is divided into N layers with
concentric rings with equal thickness $\Delta d = R/N$. In particular, the refractive index
of ith layer corresponds to $n_{i,}$ and in each layer, the refractive index is assumed to
have little difference.

During the measurement, the sample fiber is immersed in the refractive index
matching oil whose refractive index is the same as that of cladding. Then the optical
path of the probe light going through the fiber can be considered as a straight line
in this multilayer structure fiber. The distance between the light path and the central
axis of the fiber is d_i and the optical path h_Q of light passing through the Q layer are
calculated as (Ting-Ting et al. 2015; Xiaoman et al. 2004):

$$d_i = (N - i - 0.5)\,\Delta d, \tag{20}$$

$$
h_Q =
$$
$$
\sum_{i=1}^{Q-1} 2n_i \left[\sqrt{\left(R - (i-1)^2 \Delta d\right)^2 - d_Q^2} - \sqrt{(R - i\,\Delta d)^2 - d_Q^2} \right] \tag{21}
$$
$$
+ 2n_Q \left[\sqrt{(R - (Q-1)\,\Delta d)^2 - d_Q^2} \right].
$$

In order to have the refractive index of each layer n, the matrix of n, h, φ, are defined as:

$$
n = \begin{bmatrix} n_1 \\ n_2 \\ n_3 \\ \dots \\ n_N \end{bmatrix}, h = \begin{bmatrix} h_1 \\ h_2 \\ h_3 \\ \dots \\ h_N \end{bmatrix}, \varphi = \begin{bmatrix} \varphi_1 \\ \varphi_2 \\ \varphi_3 \\ \dots \\ \varphi_N \end{bmatrix}, \tag{22}
$$

where the relationship between n and h can be expressed as:

$$h = \varphi \cdot [\lambda / 2\pi] \tag{23}$$

From Eqs. 19 and 21, h can be presented as:

$$h = M \cdot [n - n_0], \tag{24}$$

where M is the coefficient matrix derived from Eqs. 21 and 23:

$$M =$$

$$
\begin{bmatrix}
2\sqrt{R^2 - d_1^2} & 0 & \dots & \dots & 0 \\
2\left[\sqrt{R^2 - d_2^2} - \sqrt{(R - \Delta d)^2 - d_2^2} \right] & 2\sqrt{(R - \Delta d)^2 - d_2^2} & 0 & \dots & 0 \\
2\left[\sqrt{R^2 - d_3^2} - \sqrt{(R - \Delta d)^2 - d_3^2} \right] & 2\left[\sqrt{(R - \Delta d)^2 - d_3^2} - \sqrt{(R - 2\Delta d)^2 - d_3^2} \right] & 2\sqrt{(R - 2\Delta d)^2 d_3^2} & \dots & 0 \\
\dots & \dots & \dots & \dots & 0 \\
2\left[\sqrt{R^2 - d_N^2} - \sqrt{(R - \Delta d)^2 - d_N^2} \right] & 2\left[\sqrt{(R - \Delta d)^2 - d_N^2} - \sqrt{(R - 2\Delta d)^2 - d_N^2} \right] & 2\left[\sqrt{(R - 2\Delta d)^2 - d_N^2} - \sqrt{(R - 3\Delta d)^2 - d_N^2} \right] & \dots & 2\sqrt{(R - (N-1)\Delta d)^2 - d_N^2}
\end{bmatrix} \tag{25}
$$

Then, from Eq. 24, the refractive index matrix n can be calculated by:

$$n = M^{-1} \cdot h + n_0 \tag{26}$$

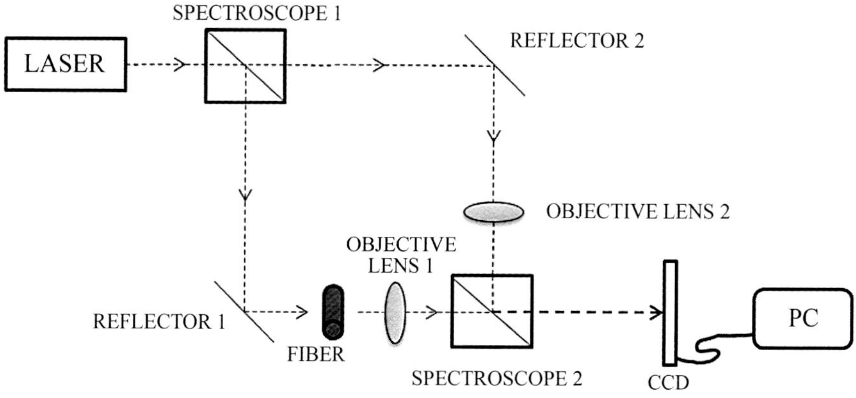

Fig. 37 Refractive index measurement schema by digital holographic microscopy

Since the refractive index distribution of fiber is unique due to cylindrical symmetry structure, the index profile at the different angle could be considered identical. As a result, two-dimensional refractive index distribution can be obtained from the combination of the phase distribution of objective light wave extracted by a single digital hologram and the multilayer model of the optical fiber. It can be seen that accurate extraction of phase distribution from the digital hologram is important for obtaining refractive index profile. The first step of refractive index profile measurement is the pre-processing for digital holograms by using digital image processing technology to eliminate the zero order items and conjugate term. Then the complex amplitude distribution of objective light wave is reproduced by the numerical reconstruction. The common numerical reconstruction methods are finel construction method, convolution construction method, and angular spectrum reconstruction method.

In Fig. 37, one refractive index measurement setup using digital holographic microscopy method is illustrated. The light from the laser generator is separated into two beams by a spectroscope (or a fiber coupler). One of them is taken as the reference light. The other one passes through the fiber sample to propose the phase delay. The light goes through the fiber interferes with the reference light at the surface of the CCD, which is the holography plane. The hologram base on digital holography is recorded by CCD. It shows the inner structure information of the optical fiber. As a result, the phase distribution can be constructed from the hologram. Then the refractive index distribution is produced. The experiment should be done carefully to avoid the error generated by tilt (Hu et al. 2009; Jianglei et al. 2008).

By using the setup shown in Fig. 37, a result of a common single mode fiber refractive index profile is plotted in Fig. 38. The axis X is related with fiber diameter and the axis Y is corresponded with the refractive index difference between the measured fiber and the refractive index oil. From the result, it is found that the fiber

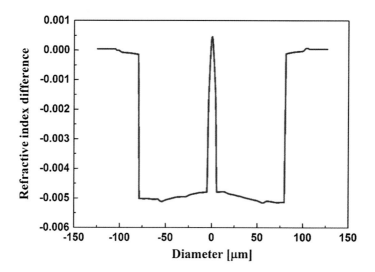

Fig. 38 Refractive index measurement result by digital holographic microscopy

outer diameter is around 125 μm. As the refractive index of the oil used is 1.452, the refractive indices of the fiber cladding and core are close to 1.447 and 1.452.

Acknowledgements Authors are thankful for the support of National Natural Science Foundation of China (61520106014, 61405014 and 61377096), Key Lab of In-fiber Integrated Optics, Ministry Education of China, State Key Laboratory of Information Photonics and Optical Communications (Beijing University of Posts and Telecommunications) (IPOC2016ZT07), Key Laboratory of Optical Fiber Sensing & Communications (Education Ministry of China), Key Laboratory of Opto-electronic Devices and Systems of Ministry of Education and Guangdong Province (GD201702) and Science and Technology Commission of Shanghai Municipality, China (SKLSFO2015-01 and 15220721500).

References

T. I. Association, (2003)

J. Bernard, E. Depresles, in *Proceedings of EFOC/LAN-88,* (Marcoussis 1988), p. 127

P. Chevallier, P. Dhez, A. Erko, A. Firsov, F. Legrand, P. Populus, X-ray microprobes Nucl. Instrum. Methods Phys. Res., Sect. B: Beam Interactions with Materials and Atoms **113**, 122 (1996)

W. Eickhoff, R. Ulrich, Appl. Phys. Lett. **39**, 693 (1981)

W. Eickhoff, E. Weidel, Opt. Quant. Electron. **7**, 109 (1975)

D. Franzen, J. Lightwave Technol. **3**, 128 (1985)

A. Galtarossa, E. Nava, G. Valentini, in *Single-mode Optical Fiber Measurement: Characterization and Sensing*, ed. by G. Cancellieri (Artech House Publishers, Norwood, 1993)

A. Ghatak, K. Thyagarajan, *An Introduction to Fiber Optics* (Cambridge University Press, New York, 1998)

D. Gloge, Appl. Opt. **10**, 2442 (1971)

D. Gloge, E. Marcatili, Bell Labs Tech. J. **52**, 1563 (1973)

C. Hu, J. Zhong, Y. Gao, J. Weng, Acta Opt. Sin. **29**, 3317 (2009)

I. T. Union, **G.652** (2009)
D. Jianglei, Z. Jianlin, F. Qi, Acta Opt. Sin. **28**, 56 (2008)
D. Kalish, L.G. Cohen, Bell Labs Tech. J. **66**, 19 (1987)
K. Kao, G. A. Hockham, in *Proceedings of the Institution of Electrical Engineers* (IET, 1966), p. 1151
M. Lines, Science **226**, 663 (1984)
S. Linglong, M. Lihong, W. Hui, L. Yong, Chin. J. Lasers **10**, 032 (2013)
L. Liu, J. Quant. Spectrosc. Radiat. Transf. **83**, 223 (2004)
D. Marcuse, Appl. Opt. **18**, 9 (1979)
D. Marcuse, *Theory of Dielectric Optical Waveguides* (Elsevier, London, 2013)
D. Marcuse, H.M. Presby, Appl. Opt. **18**, 402 (1979)
R.J. Mears, L. Reekie, I. Jauncey, D.N. Payne, Electron. Lett. **23**, 1026 (1987)
T. Miya, Y. Terunuma, T. Hosaka, T. Miyashita, Electron. Lett. **15**, 106 (1979)
D.S. Moore, Fibre Opt. **83** (1983a)
D. Moore, in *Proceedings of SPIE,* 1983b, p. 125
K. Morishita, J. Lightwave Technol. **4**, 1120 (1986)
Y. Murakami, A. Kawana, H. Tsuchiya, Appl. Opt. **18**, 1101 (1979)
E.G. Neumann, *Single-Mode Fibers: Fundamentals*, vol 57 (Springer, Berlin, 2013)
S. Personick, Bell Labs Techn. J. **56**, 355 (1977)
Y. Sato, K. Aoyama, IEEE Photon. Technol. Lett. **3**, 1001 (1991)
J. M. Senior, (Englewood Cliffs, 2004)
D. Smithgall, C. Schroeder, in *Techical Digest. Symposium on Optical Fiber Measurements (SOFM 80),* (Boulder, Colorado, 1980), p. 41
K. Suzuki, T. Horiguchi, S. Seikai, Electron. Lett. **20**, 714 (1984)
M. Tateda, T. Horiguchi, J. Lightwave Technol. **7**, 1217 (1989)
G. Ting-Ting, H. Su-Juan, Y. Cheng, M. Zhuang, C. Zheng, W. Ting-Yun, Acta Phys. Sin. **64**, 10 (2015)
D. Wang, Z. Chang, S. Huang, in *Proceedings of SPIE*, (Beijing, 2012), p. 85561K
G. Warnes, C. Millar, in *Fibre Optics' 84* (International Society for Optics and Photonics, Bellingham, 1984), p. 138
S. Xiao, H. Hu, X. Mao, B. Zhang, in *International Conference of Optical Instrument and Technology* (International Society for Optics and Photonics, Bellingham, 2008), p. 716006
D. Xiaoman, Z. Liu, Y. Chen, Y. Zheng, C. Yin, X. Xu, Opt. Tech. **5** (2004)

Measurement of Active Optical Fibers

29

Gui Xiao, Ghazal Fallah Tafti, Amirhassan Zareanborji,
Anahita Ghaznavi, and Qiancheng Zhao

Contents

Abstract

Active optical fiber owns its special optical properties to laser-active dopants in fiber. This chapter presents key properties and their characterization, including measurement principles, experimental techniques, as well as test results, of active optical fiber. Firstly, the fundamental optical properties and relations between light and matter in active fiber are introduced, including Einstein relation, the absorption and emission cross sections, energy transfer, as well as up-conversion. Then the measurements of the absorption, emission, and gain are described.

G. Xiao (✉) · G. Fallah Tafti · A. Zareanborji · A. Ghaznavi · Q. Zhao
Photonics and Optical Communications, School of Electrical Engineering and
Telecommunications, UNSW, Sydney, NSW, Australia
e-mail: xgsheen@gmail.com; ghazal.tafti@gmail.com; zarean@msn.com;
Anahitaghaznavi7@gmail.com; qiancheng.zhao@unsw.edu.au

© Springer Nature Singapore Pte Ltd. 2019   1139
G.-D. Peng (ed.), *Handbook of Optical Fibers*,
https://doi.org/10.1007/978-981-10-7087-7_56

Introduction

With the recent development of water-free fiber technology, silica-based optical fiber has great transmission capability cross the whole spectrum from 1200 to 1700 nm region with a loss lower than 0.2 dB/km (Nagel 1987; Poole et al. 1985). It enhances the efficiency of transmission and accelerates the growing of long distance telecommunication. To compensate the signal loss in long distance transmission, fiber amplifier and laser based on active optical fiber are essential. Different from standard transmission fiber, active fiber is one type of specialty optical fibers, whose core is doped with laser-active ions, such as Er^{3+}, Yb^{3+}, Bi ions. According to the doped active centers, there is erbium doped fiber (EDF), ytterbium doped fiber (YDF), thulium doped fiber (TDF), praseodymium doped fiber (PDF), as well as recently developed bismuth doped fiber (BDF). These doped active centers give them different natures from other types of fibers, which can be used to amplify an optical signal. Thanks to its unique geometry structure and exceeding transparency, high intensity and long interaction length can be provided rendering active fiber a superior device to its counterparts (bulk glass or crystal) in applications. Based on such features, they have widely been used for fiber amplifiers and lasers in optical communication and laser physics.

It is known that the interaction of light with matter has many forms, where the photons can be annihilated or created. The photon annihilation involves absorption processes and the photon generation relates to emission processes, which can be spontaneous emission or stimulated emission as shown in Fig. 1 for an atom of two energy levels/states. In the absorption process (Fig. 1a), the atom at a lower level absorbs a photon of the frequency v and is excited to its upper level. Here hv is the photon energy with h the Plank constant. In the spontaneous emission process (Fig. 1b), the atom at an upper level transits spontaneously to its lower level and, if the transition between E_2 and E_1 is radiative, a photon of energy hv is emitted in a random direction and with a random phase. In the stimulated emission process (Fig. 1c), an incident photon causes the atom at upper level back to its lower level and emitting a "stimulated" photon with its properties identical to those of the incident photon. The term "stimulated" underlines the fact that this kind of radiation only occurs if an incident photon is present. This process achieves optical amplification, or optical gain, since two photons are produced by one incident photon in the process.

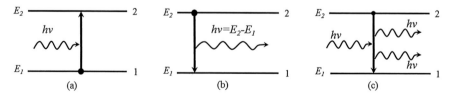

Fig. 1 Two level diagrams of (**a**) absorption, (**b**) spontaneous emission, and (**c**) stimulated emission

The processes in Fig. 1 commonly exist in the applications of active fibers. Due to the difference in the doping ions and host glasses, the absorption, emission, and gain properties of active doped fiber will be different. Therefore, it is necessary to have suitable or standard measurement method to quantify and evaluate these properties for their application and development. In this chapter, key parameters and properties of active optical fibers are firstly introduced and then their measurements are described in detail.

Basics of Active Optical Fibers

Here the basic physics of laser and the features of rare earth ions are introduced. Firstly, the Einstein relation which is fundamental to lasing and amplification is explained. Then the absorption and emission processes and the inherent parameters to quantify them are presented. Some relevant transition processes such as energy transfer, excited state absorption (ESA), and up-conversion are also introduced.

Interaction Between Light and Matter

Einstein Relation

Absorption
As shown in Fig. 1a, in a light field with a frequency of $v = (E_2 - E_1)/h$, the atoms (or ions and molecules) on level E_1 can transit to level E_2 in a certain probability and every excited atom absorbs an energy of $E_2 - E_1$ or a photon hv, which is called excited absorption or absorption for short. Absorption is a process where particles interact with the external field and transit from a low energy level to a higher one. The absorption rate is defined as

$$\left(\frac{dn_1}{dt}\right) = -W_{12}n_2 \tag{1}$$

where n_1 is the number density on level 1 at the given moment, W_{12} is the absorption rate with the unit of s^{-1}. W_{12} depends on not only the level system, but also the intensity of light, which is

$$W_{12} = B_{12}\rho(v) \tag{2}$$

where $\rho(v)$ is energy density and B_{12} is Einstein stimulated absorption coefficient.

Spontaneous Emission
As schemed in Fig. 1b, when the electrons of ions in level 2 transit to level 1 and release energy $E_2 - E_1$, if this energy is released in a form of light/photon, the process is called spontaneous emission. The frequency of spontaneous emission is described as $v = (E_2 - E_1)/h$, and the photon energy $hv = E_2 - E_1$ is often used for

characterization of the spontaneous emission. The other spontaneous transition is called nonradiative transition where energy is released in other forms like kinetic or thermal energy.

For an individual ion, spontaneous emission is random in possibility and uncertain in time. But for a plenty of ions, statistically, the possibility of it in a certain time can be assured. If in a unit volume, the number of ions on level 1 and 2 is n_1 and n_2, respectively, in a unit of time the number transiting from E_2 to E_1 shall be proportional to n_2. If the ratio is A_{21}, the transition rate is

$$\left(\frac{dn_2}{dt}\right)_{sp} = -A_{21}n_2 \tag{3}$$

"−" represents a decrease of the number. The ratio A_{21} is the Einstein coefficient of spontaneous emission. It represents the ratio between the number of ions generating spontaneous emission and the total number n_2, so the physical meaning of this ratio is the spontaneous emission probability of each ion in a unit time. Integrating Eq. (3), we can get

$$n_2(t) = n_{20}e^{-A_{21}t} = n_{20}e^{-\frac{t}{\tau_s}} \tag{4}$$

where $\tau_s = 1/A_{21}$ is called spontaneous emission lifetime or fluorescence lifetime, which will be described in the sections "Lifetime" and "Measurement of Fluorescent Lifetime." Physically, it indicates that after a period time of τ_s, the number of ions on E_2 decreases to $1/e$ of initial number n_{20}. τ_s implies the time of ions staying on the upper level. For a certain energy level, if $\tau_s \to \infty$, it is a stable level; if τ_s is relatively larger (often reach microseconds or even milliseconds), it is called a metastable level. The spontaneous emission probability or Einstein spontaneous emission coefficient only depends on the two levels involved in the system.

Stimulated Emission

If the frequency of the light field is the same with the transition frequency between two energy levels $v = (E_2 - E_1)/h$, the ions on level 2 can be induced to transit to level 1 and release photons with the same frequency (or energy as $E_2 - E_1$). Different from spontaneously emitted photons whose phase, direction, and polarization are random, these photons have the same quantum state or optical mode as external light field and good coherent property. The stimulated emission rate can be given as

$$\left(\frac{dn_2}{dt}\right)_{st} = -W_{21}n_2 \tag{5}$$

where W_{21} is given as

$$W_{21} = B_{21}\rho(v) \tag{6}$$

B_{21} is the Einstein stimulated emission coefficient.

Einstein coefficients are introduced above to indicate the probability of the emission and absorption of the ions. The Einstein A coefficient is related to the rate of spontaneous emission of light and the Einstein B coefficients are related to the absorption and stimulated emission of light. Einstein coefficients A_{21}, B_{21}, and B_{12} are all determined by the nature of energy level system, instead of the external light field. In a given energy system, they have an intrinsic relation called the Einstein relation. In a two-level system, if $n_1{}^e$ and $n_2{}^e$ are the numbers of ions on each level, in thermal equilibrium condition, they follow the Boltzmann distribution as

$$\frac{n_2^e}{n_1^e} = \exp\left(-\frac{E_2 - E_1}{k_B T}\right) = \exp\left(-\frac{h\nu}{k_B T}\right) \tag{7}$$

where k_B is Boltzmann constant. If three processes all happen, the energy change between ions and external light field can reach a balance as

$$A_{21} n_2 + B_{21}\rho(\nu) n_2 = B_{12}\rho(\nu) n_1 \tag{8}$$

or

$$\frac{n_2}{n_1} = \frac{B_{12}\rho(\nu)}{A_{21} + B_{21}\rho(\nu)} \tag{9}$$

If both thermal equilibrium Eq. (7) and interaction balance between light and material ions Eq. (9) are achieved, there is $n_2{}^e/n_1{}^e = n_2/n_1$, but $\rho(\nu)$ is

$$\rho(\nu) = \frac{A_{21}}{B_{21}} \frac{1}{(B_{12}/B_{21}) \exp\left[h\nu/(k_B T)\right] - 1} \tag{10}$$

Compare it with Planck's Law at temperature T, if refractive index is n,

$$\rho(\nu) = \frac{8\pi h\nu^3}{c^3} \frac{1}{\exp\left[h\nu/(k_B T)\right] - 1} \tag{11}$$

so

$$B_{12} = B_{21} \tag{12}$$

$$\frac{A_{21}}{B_{21}} = \frac{8\pi h\nu^3 n^3}{c^3} \tag{13}$$

Equations (12 and 13) are called the Einstein relation. As Einstein A and B coefficient are intrinsic to the energy system, so it can be applied to a laser system and won't be affected by the nonequilibrium condition.

If degeneracies of energy level are taken into consideration (Fernicola and Rosso 2000) as shown in Fig. 2, the degeneracy of E_1 is g_1 and that of E_2 is g_2. Define n_{1i}

Fig. 2 A typical degenerated
two-level system

j } $E_2 g_2 n_2$

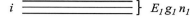

i } $E_1 g_1 n_1$

Fig. 3 Absorption of the
light in a medium

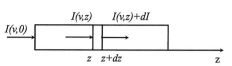

and n_{2j} are the numbers of each manifold of two levels. n_2 and n_1 are the populations in each level. Equation (12) in Einstein relation will be changed into a common form as:

$$B_{21} g_2 = B_{12} g_1 \tag{14}$$

Light Absorption and Gain

In a homogeneous medium, the states of the ions will be distributed as the thermal equilibrium condition in Eq. (7). When a light of given frequency v propagates in this medium, it will be absorbed by the material. As shown in Fig. 3, the launched light intensity is $I(v,0)$. At position z, the intensity is $I(v,z)$, and change to $I(v,z + dz)$ after propagating a distance of dz, so the absorbed intensity is $dI(z)$. Accordingly, the absorption coefficient is

$$\alpha(v) = -\frac{1}{I(v,z)} \frac{dI(z)}{dz} \tag{15}$$

where $dz = (c/n)dt$, as the velocity of light is c and the refractive index of the medium is n.

To build the relation of the absorption coefficient, the number of ions, and the transition rate, assume a two-level system and the number density on each level is n_2 and n_1, respectively. The light attenuation from z to $z + dz$ should be ascribed to the summary of stimulated absorption and stimulated emission. In time $t \sim t + dt$, the absorbed photon number per unit volume should be

$$dN_{12} = n_1 B_{12} g(v, v_0) \rho(v, z) dt \tag{16}$$

where $g(v,v_0)$ is the line function of light field. In a certain period, energy absorbed by unit volume medium is $dN_{12}hv$ and should equal to the absorbed amount of incident light, which is

$$-d\rho_1(v, z) = dN_{12}hv = n_1 \rho(v, z) B_{12} g(v, v_0) hv dt \tag{17}$$

Using the relation $I(v,z) \propto \rho(v,z)$,

$$dI(v,z) = -\left(n_1 - \frac{g_1}{g_2}n_2\right) I(v,z) B_{12} hv g(v,v_0) dt \qquad (18)$$

Substituting Eq. (18) into Eq. (15),

$$\alpha(v) = -\frac{1}{I}\frac{dI}{dz} = \left(n_1 - \frac{g_1}{g_2}n_2\right) B_{12} hv \frac{n}{c} g(v,v_0) \qquad (19)$$

If the medium is in a state of population inversion (e.g., under pumping), the change of light through propagation distance is $dI(v,z)/dz$. The gain coefficient is defined as

$$G(v,z) = -\frac{1}{I(v,z)}\frac{dI(z)}{dz} \qquad (20)$$

The physical meaning of G is the increasing ratio of light intensity in unit length by population inversion. Substitute Eq. (18) to (19),

$$G(v,z) = \left(n_2 - \frac{g_2}{g_1}n_1\right) B_{21} hv \frac{n}{c} g(v,v_0) \qquad (21)$$

The Absorption and the Emission Cross Sections

To further characterize the inherent ability of absorption and gain for a certain material, the absorption cross section and the emission cross section are introduced. The absorption coefficient in Eq. (19) can be expressed as

$$\alpha(v) = \sigma_{12}\left(n_1 - \frac{g_1}{g_2}n_2\right) \qquad (22)$$

where the absorption cross section (σ_{12}) is given by:

$$\sigma_{12} = \frac{hv}{\upsilon} B_{12} g(v,v_0) \qquad (23)$$

where h is the Planck constant and υ is the velocity of light. It has the unit of area (m^2 or cm^2) implying the absorption of light is a result from absorption center. Every center has an area of σ_{12}.

Similarly, the gain coefficient in Eq. (20) can be expressed by

$$G(v) = \sigma_{21}\left(n_2 - \frac{g_2}{g_1}n_1\right) \qquad (24)$$

where emission cross section (σ_{21}) is given by

$$\sigma_{21} = \frac{h\nu}{\nu} B_{21} g\,(\nu, \nu_0) \tag{25}$$

Equation (24) indicates that the gain coefficient is proportional to the inversion population. The value of σ_{21} is determined by line function $g(\nu, \nu_0)$ and spontaneous emission rate A_{21} (according to the Einstein relation).

Füchtbauer–Ladenburg Theory

According to the Einstein relation, the absorption and the emission cross section can be expressed as (Grattan and Zhang 1994)

$$\sigma_{12} = \frac{g_2}{g_1} \frac{\lambda^2}{8\pi n^2} A_{21} g\,(\nu, \nu_0) \tag{26}$$

$$\sigma_{21} = \frac{\lambda^2}{8\pi n^2} A_{21} g\,(\nu, \nu_0) \tag{27}$$

where λ is the center wavelength of emission/absorption spectrum corresponding to center frequency ν_0. Spontaneous emission rate A_{21} can be evaluated by $1/\tau_2$, in which τ_2 is the lifetime of level 2 when nonradiative relaxation is neglected. The effective linewidth is defined as

$$I_{peak}\Delta\lambda_{eff} = \int_0^\infty I\,(\lambda)\,d\lambda \tag{28}$$

Label the effective linewidth of absorption as $\Delta\lambda_A$ and that of emission as $\Delta\lambda_E$. The line shape function is

$$g\,(\nu) = \frac{I_{peak}}{\int I d\nu} \tag{29}$$

Substitute Eq. (29) to (28), we can get

$$g\,(\nu) = \frac{\lambda^2}{c} \frac{1}{\Delta\lambda_{eff}} \tag{30}$$

So the Füchtbauer-Ladenberg Eqs. (26 and 27) can be expressed as

$$\sigma_{12} = \frac{g_2}{g_1} \frac{\lambda^4}{8\pi n^2 c} \frac{1}{\tau_2 \Delta\lambda_A} \tag{31}$$

$$\sigma_{12} = \frac{g_2}{g_1} \frac{\lambda^4}{8\pi n^2 c} \frac{1}{\tau_2 \Delta\lambda_E} \tag{32}$$

Obviously, there is a certain ratio between absorption and the emission cross section in a given system, which is

$$\frac{\sigma_{12}}{\sigma_{21}} = \frac{g_2}{g_1} \frac{\Delta\lambda_E}{\Delta\lambda_A} \tag{33}$$

If the emission spectrum is known, emission and absorption cross section can be obtained through Eqs. (31) and (32). It should be noticed that several assumptions are adopted: (a) the quantum efficiency is nearly unity, (b) all the stark components contribute the same to the transition process or they are equally populated.

McCumber Theory

Taken the thermal distribution of each manifold into consideration, McCumber theory (Miniscalco and Quimby 1991) is derived from Einstein relation, and emission (σ_e) and absorption (σ_a) cross section is given by

$$\sigma_e(v) = \sigma_a(v) \exp\left[(\varepsilon - hv)/kT\right] \tag{34}$$

where ε is the temperature dependent excitation energy. In the Er^{3+} energy level system, the physical interpretation of ε is the net free energy required to pump an Er^{3+} ion from the $^4I_{15/2}$ to the $^4I_{15/3}$ state at temperature T. As shown in Fig. 4, the emission cross sections measured and calculated by McCumber theory have a small offset with the absorption cross sections. According to Eq. (34), only when the wave frequency is $v_c = \varepsilon/h$ defined as the crossing wavelength where the absorption and emission cross section are equal.

Lifetime

Fluorescence lifetime is an essential parameter for evaluation and application of active optical fibers (Zareanborji et al. 2013, 2014, 2015; Pengc). It is one of the most important properties of active fiber for optical gain media. The variation of fluorescence lifetime results in the variation of the fraction of spontaneous and stimulated emission, hence leads to various pump efficiency and gain. Moreover, the fluorescence lifetime is also essential for the determination of the emission cross section, the absorption cross section, the cooperative up-conversion emission, and other process of energy transfer. It is an absolute parameter of emission processes, unlike the emission and absorption parameters, which are relative to the external conditions; therefore, it is known as an intrinsic property of active centers in active fibers.

As mentioned above, when short pulse of excitation light passes through a piece of active fiber, electrons of atoms, ions, or molecules in the ground state or a low energy state are excited to a higher energy state by absorbing the excitation photons. These electrons will stay for a short time at the excited state and then transfer to a lower energy state. The energy of these electrons can be transferred by radiative transition (spontaneous emission) or by nonradiative transitions (without any emission). The fluorescence lifetime is the average time that electrons spend

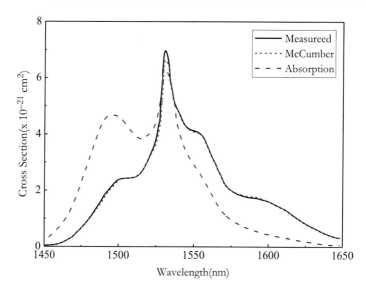

Fig. 4 The absorption cross section, the emission cross section measured and calculated by McCumber theory

in the excited state before returning to the ground state or the lower energy state by spontaneous emission. Radiative lifetime depends on dopants, host materials, energy levels, and environmental parameters. Nonradiative lifetime depends on the host materials and the coupling between the vibrations of the lattice ions and the states of the rare earth ions.

Generally, the decay of fluorescence is the first order kinetics; however, in many cases, the decay is more complex. Influences such as quenching, energy transfer, multiphonon transitions, inhomogeneous environment can change the decay profile to multi- or nonexponential decays. In the case of quenching, the nonradiative emission competes with the radiative emission, which results in a decrement of the radiative emission. In the case of energy transfer, the population of electrons at a metastable level is reduced due to the transfer of electrons to another active center (Fig. 5). In the case of multiple emission processes, the fluorescence decay will follow the multiexponential function.

The fluorescence decay of a single emission process is often exponential with one decay constant. This decay constant is defined as the fluorescence lifetime or upper-state lifetime. In some cases, due to the existence of more than one emission processes in the emission band of interest, like up-conversion, energy transfer, or quenching, the florescence decay is nonexponential. The lifetime of electrons in the excited state can range from nanoseconds to milliseconds. Generally, the lifetime is defined as the time when the number of excited electron decays to $1/e \sim 0.3678\%$ of the original excited population, as illustrated in Fig. 6. In fluorescence lifetime measurements, the aim is to characterize the temporal decays of the fluorescence signals, which generally are exponential (Ehn et al. 2012).

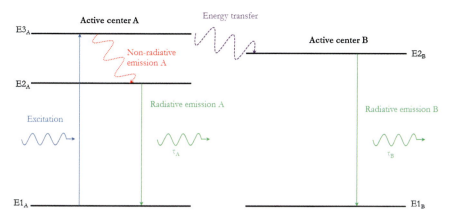

Fig. 5 Excitation, emission, and energy transfer processes

Fig. 6 Excitation,
florescence decay, and
lifetime

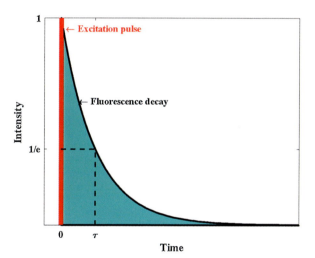

Typical Active Ions and Properties

The main difference between active and passive fiber is that passive fiber can only receive/guide the light from outside environment so extra light is needed to boost the optical signal in long distance transmission as signal decays during the propagation. So active fiber is developed for light generation and amplification as having additional laser-active dopants. The rare earth elements are a group of natural candidates as laser-active dopants in glass fiber by providing sharp fluorescence covering visible or infrared band. The partially filled 4f shell of rare earth elements can generate luminescence at multiple wavelengths between different orbits/energy levels and the sharpness is due to the shielding by 5 s and 5p outer shells. Concomitantly, rare earth ions emit and absorb light at narrow bands barely

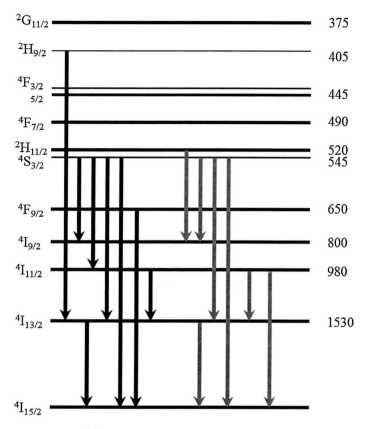

Fig. 7 Energy levels of Er^{3+} and reported lasing transitions (Miniscalco and Digonnet 1993)

affected by the host material; their lifetimes of metastable states are long, and the quantum efficiencies are high.

Among the rare earth elements, **Neodymium** (Nd^{3+}) is the first and still one of the most important trivalent rare earth ions used in a laser. Stimulated emission can be generated at 0.9, 1.06, and 1.35 μm (from $^4F_{3/2} \rightarrow ^4I_{11/2}, ^4I_{13/2}$, respectively) (Koechner 2013). So far, **Erbium** (Er^{3+}) has become the dominant dopant in active fiber when it is applied to the modern optical communication. The energy diagram of Er^{3+} is plotted in Fig. 7. 1.5 μm luminescence by $^4I_{13/2} \rightarrow ^4I_{15/2}$ of Er^{3+} in EDF spans the third optical communication window and gets the most extensively studied. At room temperature $^4I_{15/2}$ is the ground state and it forms a three-level laser system with a relatively high threshold so **ytterbium** (Yb^{3+}) often co-doped with Er^{3+} as a sensitizer (Fig. 9). In addition, **Holmium** (Ho^{3+}) and **Thulium** (Tm^{3+}) are also used and co-doped for 2.0 μm lasing and amplification.

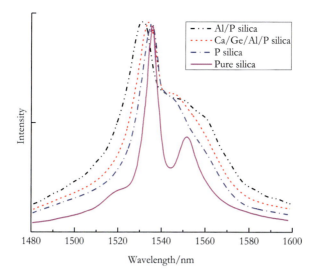

Fig. 8 Emission spectrum of erbium doped active fiber (Miniscalco and Digonnet 1993)

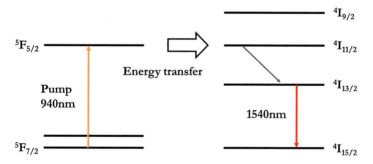

Fig. 9 Energy level diagram in Er/Yb co-doped glass

Energy Level System of Er^{3+}

The energy levels of Er^{3+} and lasing transitions reported in crystalline hosts and glass host are shown in Fig. 7. The black arrows in the left side are emissions reported in crystalline host and the grey arrows in the right side indicate those in glass hosts. It is worth notice that the properties of host materials, such as macroscopic mechanism, thermal and optical properties, and their unique microscopic lattice, can affect the position of electron orbits and manifolds of active ions' energy levels. The emission spectra of Er^{3+} in silica host with different host composition are shown in Fig. 8. It demonstrates the compositions of the hosts affect the spectral profile slightly on the bandwidth and position of peaks.

Ion-Ion Interactions

Energy transfer can happen in one isolated ion (Fig. 7) or between two ions (Fig. 9). In this section, several important energy transfer processes between Er^{3+}/Yb^{3+} or Er^{3+}-Er^{3+}, also known as ion-ion interactions, will be discussed.

Er^{3+}/Yb^{3+} Interaction-Sensitized Fluorescence In an Er/Yb co-doped fiber applying to 1.5 μm generation (Fig. 9), the radiation at 940 nm can pump Yb^{3+} effectively through transition $^5F_{7/2} \rightarrow {}^5F_{5/2}$. There exists a good overlap between $^2F_{5/2}$ of Yb^{3+} and the upper state $^4I_{11/2}$ of Er^{3+}. The transfer rate between these two states is 10 times greater than the $^2F_{5/2}$ relaxation rate of Yb^{3+}, so this energy transfer can occur in a good chance. In addition, it reduces the probability of transfer from Er^{3+} to Yb^{3+} that relaxation from $^2I_{11/2}$ of Er^{3+} is relatively quick (<10 μs) for 1.5 μm lasing. It is also a typical kind of energy transfer, called sensitized fluorescence.

Er^{3+}-Er^{3+} Interaction-Cooperative Up-Conversion Energy transfer can also happen between the same types of ions, for example, when the Er^{3+} concentration is reasonably high enough and the distance between ions is close relatively, energy transfer can happen between Er^{3+}-Er^{3+} in a way, called cooperative up-conversion mechanism as shown in Fig. 10.

The excited states $^4I_{13/2}$ of Er^{3+} has a long fluorescent lifetime of 10 ms providing ample time for interaction between its neighbors. The excited ion 2 (Fig. 10) can be prompt to a higher energy state through accepting energy from ion 1 and 980 nm up-conversion emission from $^4I_{11/2} \rightarrow {}^4I_{15/2}$ can occur in this case. In this process, deactivation of ion 1 back to the ground state provides another alternative pathway of population loss of $^4I_{13/2}$, which decreases the lifetime.

Two Er^{3+} ions as an ion pair can stay in three possible states: none, one, or both ions excited. When the ions distribute evenly in fiber host, all three states can exist. However, when two ions in Fig. 10 are close enough and form a smallest cluster-an ion pair, things could be a bit more complicated. As they are too close to each other, once both ions are excited, one ion will transfer the energy to the other and the cooperative up-conversion process is highly efficient (<50 μs), so

Fig. 10 Cooperative up-conversion process between two erbium ions excited to their respective $^4I_{13/2}$

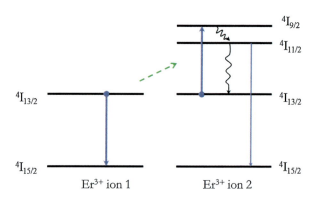

$^4I_{9/2}$

$^4I_{11/2}$

$^4I_{13/2}$

$^4I_{13/2}$

$^4I_{15/2}$

$^4I_{15/2}$

Er^{3+} ion 1

Er^{3+} ion 2

state $^4I_{13/2}$ is quenched. The result is no pairs can stay in the state of both ions excited. Consequently, a pair of ions functions as one ion by either nonexcited or one ion excited. This concentration-induced quenching can change the pump absorption coefficient (section "Pump Absorption") when pump power changes.

Excited State Absorption (ESA)

Excited state absorption is only possible when an electron has already been excited from the ground state to a lower excited state. Hence, the further transition to a higher state by absorbing some photon energy is called excited state absorption. The ESA is usually an undesired effect when it usually reduces the gain of amplifiers, but it can be useful in up-conversion. The excited state absorption measurement is often done by pump-probe techniques. It is not as easy as the ground-state absorption test due to its dependence on pump condition. In some cases, complete bleaching of the ground state is required. The ESA measurement setup is usually the same to that of on/off gain measurement which will be discussed in section "Principle of Gain Measurement." Figure 11 shows Er^{3+} energy levels with possible pump and signal ESA.

Up-conversion Emission

The up-conversion emission occurs when an excited ion is further excited to an even higher energy level and emits photons whose energy is higher than that of the exciting photon (Zhou et al. 2015). Up-conversion can happen inter- and intra-ions. The mechanisms giving rise to up-conversion including energy transfer (inter-ions) and ESA (intra-ions) are discussed above. Normal emission follows the ground state absorption (GSA) of pump light with lower photon energy (from lower energy level and with longer wavelength) than pump light. When up-conversion happens and generates emission in the following relaxation, the wavelength of emission is possibly shorter than the pump light, which is called *up-conversion emission* (Scheps 1996). It can be applied in up-conversion lasers, where the laser wavelength is shorter than the pump wavelength. However, it is often detrimental for laser or amplifier applications because it sacrifices the pump power without contributing to the amplification of the desired wavelength (Philipps et al. 2002).

The energy transfer diagram in Er^{3+} doped silicate fiber is given in Fig. 12 under 980 nm pumping. Both the sequential ESA1 and the energy transfer up-conversion (ETU1) at $^4I_{11/2}$ lead to a population increase at level $^4F_{7/2}$. Due to the small energy gap among $^4F_{7/2}$, $^2H_{11/2}$ $^4S_{3/2}$, and $^4F_{9/2}$ level, the ions of $^4F_{7/2}$ level can decay to the $^2H_{11/2}$ $^4S_{3/2}$ and $^4F_{9/2}$ by multiphonon relaxation process (MPR). Afterwards, emission centered at 524 nm, 547 nm, and 648 nm occurs corresponding to Er^{3+}: $^2H_{11/2} \rightarrow {}^4I_{15/2}$, $^4S_{3/2} \rightarrow {}^4I_{15/2}$, and $^4F_{9/2} \rightarrow {}^4I_{15/2}$ radiative transitions, respectively. Furthermore, a red up-conversion emission from $^4F_{9/2}$ can be obtained by ESA2 process ($^4I_{13/2} \rightarrow {}^4F_{9/2}$).

The relationship between the up-conversion intensity for different bands and the launched pump power is illustrated in the inset of Fig. 13. A slope of ~ 2 indicates

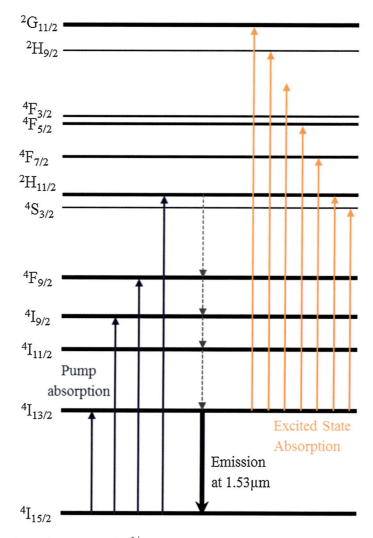

Fig. 11 Energy level diagram of Er^{3+} and possible ESA (Laming et al. 1988)

that the visible up-conversion is dominantly due to two-photon absorption process, namely, the pump ESA.

Measurement of Absorption

Absorption is uniform. The same amount of the same length of fiber always absorbs the same fraction of light at the same wavelength. If there are three fibers of the same type of glass, each 1-centimeter thick, all three will absorb the same fraction

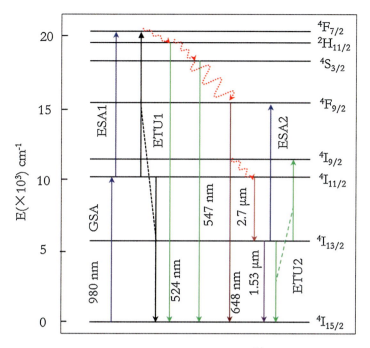

Fig. 12 Energy level diagram and energy transfer sketch of Er^{3+} under 980 nm pumping (Wei et al. 2014)

of the light passing through them. In addition, absorption also is cumulative, so it depends on the total length of fiber the light passes through. If the absorption is 1% per centimeter, it absorbs 1% of the light in the first centimeter, and 1% of the remaining light the next centimeter, and so on (Optical Fiber Loss and Attenuation). Especially, in optical fiber, the absorption coefficient is defined as a measure of the extent to which the united length of fiber absorbs the photon energy.

A fiber absorption spectrum is the fraction of incident radiation absorbed by the fiber over a range of frequencies. The absorption spectrum is primarily determined by the composition of the material. Radiation is more likely to be absorbed at frequencies that match the energy difference between two quantum mechanical states of the ions/molecules as shown in Fig. 1a. The absorption that occurs due to a transition between two states is referred as an absorption line and a spectrum is typically composed of many lines (Kalita et al. 2008). The frequencies where absorption lines occur, as well as their relative intensities, primarily depend on the electronic and molecular structure of the sample. The frequencies will also depend on the interactions between molecules in the sample, the crystal structure in solids, and several environmental factors (e.g., temperature, pressure, electromagnetic field). The lines will also have a width and shape that are primarily determined by the spectral density or the density of states of the system (Kalita et al. 2008).

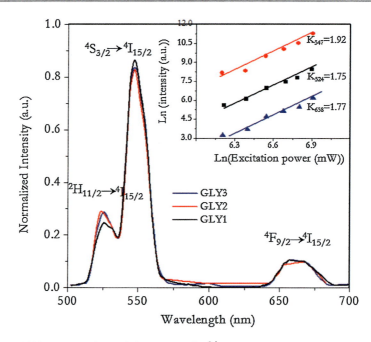

Fig. 13 Visible up-conversion emission spectra of Er^{3+} in germanate glass. The inset is the power dependence of up-conversion in Ln-Ln scale (Wei et al. 2014)

GSA and ESA In quantum mechanics, the ground state is the lowest-energy state, whose energy is known as the zero-point energy of the system, while the excited state is any states with energy greater than the ground state. As illustrated in Fig. 1a, the absorption or excited absorption is a process, where every excited particle (molecule, ion, defect, etc.) absorbs energy E_2-E_1 or a photon hv if particles on E_1 transit to E_2 in certain probability under a field with a frequency of $v = (E_2-E_1)/h$. According to the type of state of E_1 in Fig. 1a, the absorption can be classified as *ground state absorption* (GSA) and *excited state absorption* (ESA). When E_1 is the ground state, the excitation of a system (an atom or molecule) from ground state (E_1) to any excited state (E_2) with the absorption of a photon is called *ground state absorption* (GSA), while when E_1 is the low-energy excited state, the excitation of a system (an atom or molecule) from low-energy excited state (E_1) to a high-energy excited state (E_2) with the absorption of a photon is called *excited state absorption* (ESA). Both GSA and ESA are very common processes in active doped fiber.

Signal Absorption and Pump Absorption Active doped fibers are mostly used for fiber lasers and amplifiers. In these devices, the gain medium of active doped fiber can be powered by the energy from both signal and pump. According to the incident power of the light, absorption will often be classified as *signal absorption* and *pump absorption*. Especially, the signal absorption is usually for the absorption of relative weaker light. It can result in the simple excitation from a lower energy level to a higher one and be included into the description of both GSA and ESA. However, *pump absorption* is more complex and important for power efficiency in the application of these devices which usually consists of multiple mechanisms.

Intrinsic Absorption and Extrinsic Absorption The intrinsic absorption is caused by interaction with one or more of the components of the glass, while extrinsic absorption is caused by impurities within the glass. Intrinsic absorption in the ultraviolet region is caused by electronic absorption bands. Basically, absorption occurs when a photon interacts with an electron of the fiber material and is excited to a higher energy level. The main cause of intrinsic absorption in the infrared region is the characteristic vibration frequency of atomic bonds. In silica glass, absorption is caused by the vibration of silicon-oxygen (Si-O) bonds. The interaction between the vibrating bond and the electromagnetic field of the optical signal causes intrinsic absorption. Extrinsic impurity absorption is mainly caused by the presence of minute quantity of metallic ions (such as Fe^{2+}, Cu^{2+}, Cr^{3+}) and the OH^- from water dissolved in glass. In pure silica fiber, intrinsic absorption is less significant than extrinsic absorption, where a low loss window exists between 800 and 1600 nm (http://opti500.cian-erc.org/opti500/pdf/sm/Module3%20Optical%20Attenuation.pdf). However, in active doped fiber, the intrinsic absorption by the active dopants such as Er^{3+}, Yb^{3+}, and Bi ions will be more evident than other extrinsic absorption.

Ground State Absorption

Fiber absorption involves the direct transfer of energy from the propagating light to the fiber material, resulting in the excitation, especially dopant active center to a higher energy state. In this concept, the dopant was described in terms of a collection of oscillators, each available to absorb energy from an optical field oscillating at a frequency that is equal (or nearly equal) to the fundamental resonance frequency of given oscillator. The energy absorption promotes the oscillator to a higher energy level. Such concept is equivalent to GSA in active fiber. This resonance condition is directly observed in the dispersion (frequency dependent) behavior of the complex dielectric function for the material which, in turn, gives rise to the complex refractive index at optical frequencies (Thipparapu et al. 2015).

More specifically, the propagation of light within a material can be described in terms of a complex refractive index (n^*) (Thipparapu et al. 2015):

$$n^* = n(\omega) + i\kappa(\omega) \tag{35}$$

where: $n(\omega)$ = real portion of the refractive index and $\kappa(\omega)$ = extinction coefficient (Thipparapu et al. 2015)

$$\text{and} : \quad \alpha(\omega) = \frac{2\omega\kappa}{c} \tag{36}$$

where $\alpha(\omega)$ = absorption coefficient and c = speed of light. In terms of the dopant in the active fiber, different structural dopants will each contribute to the overall magnitude, and frequency dependence, of the optical absorption. Thus, the material composition and specific structural characteristics will directly impact the optical absorption observed at some wavelengths.

Measurement

According to the Beer–Lambert law, the overall optical throughput (transmission) of an optical fiber can be quantified in terms of the input optical power, P(0), and the output power, P(z) observed after light propagates a distance, z, along the fiber length (Thipparapu et al. 2015):

$$P(z) = P(0)e^{-\alpha_{total}z} \tag{37}$$

and

$$T = P(z)/P(0) \times 100\% \tag{38}$$

where α_{total} = the total attenuation coefficient, involving all the contributions by absorption, scattering as well as bending; T is the percentage optical power transmission.

In an optical fiber transmission, the attenuation coefficient above is often expressed in base-10 form (Thipparapu et al. 2015):

$$\alpha_{total}\,(dB/km) = \frac{10}{z}\lg\left[\frac{P(0)}{P(z)}\right] = 4.343\alpha_{total}\left(km^{-1}\right) \tag{39}$$

which is often referred as the fiber loss.

The fiber loss mainly includes absorption loss (α_{abs}) caused by the fiber itself or by impurities in the fiber, scattering loss caused by the interaction of the photons with the glass itself ($\alpha_{scattering}$) and bending loss ($\alpha_{bending}$) induced by physical stress on the fiber, which can be given by:

$$\alpha_{total} = \alpha_{abs} + \alpha_{scattering} + \alpha_{bending} \tag{40}$$

Material absorption is caused by absorption of photons within the fiber. Two main contributions to optical absorption in fiber optic glasses are intrinsic absorption of base glass and extrinsic absorption from the impurities within the glass. Scattering is a process whereby all or some of the optical power in a mode is transferred into a leaky or radiation mode. Two basic types of scattering exist linear scattering (Rayleigh and Mie scattering) and nonlinear scattering (Stimulated Brillouin and Stimulated Raman). Rayleigh scattering is the dominant loss mechanism in the low loss silica window between 800 and 1700 nm. Raman scattering is an important issue in dense WDM systems. Especially, the Rayleigh loss falls off as a function of the fourth power of wavelength, given by:

$$\alpha_r = \frac{A}{\lambda^4} \tag{41}$$

where λ in this empirical formula is expressed in microns (μm), and the Rayleigh scattering coefficient A is a constant for a given material (http://opti500.cian-erc.org/opti500/pdf/sm/Module3%20Optical%20Attenuation.pdf).

Cut-Back Method

As schemed in ▶ Chap. 28, "Basics of Optical Fiber Measurements", the fiber loss is often measured by the cut-back method. For the measurement, white light is often used as the broadband source. For weak signal, fiber absorption is for GSA, resulting in the transition from the ground state to any excited state. In addition, the fiber sample will keep without any bending to avoid the bending loss introduced. In general, Rayleigh scattering is the main type of the linear scattering, caused by small scale inhomogeneities (small compared with the light wavelength). Therefore, Eq. (40) will often be simplified as:

$$\alpha_{total} = \alpha_{abs} + \alpha_r \tag{42}$$

As Rayleigh loss will follow with Eq. (41), for the wavelength without absorption, the value of fiber loss can be taken as Rayleigh scattering loss and the Rayleigh scattering coefficient A will be calculated from that value. Through Eq. (42), the absorption will be given by

$$\alpha_{abs} = \alpha_{total} - \alpha_r = \alpha_{total} - \frac{A}{\lambda^4} \tag{43}$$

For good fiber fabrication (preform development and drawing conditions), even the Rayleigh scattering can be neglected. Thus, via Eq. (43), the absorption coefficient almost contributes to the output power observed through its participation in the attenuation coefficient, α_{total}. Figure 14 shows a typical absorption spectrum of erbium and ytterbium co-doped optical fiber by the cut-back method. The absorption spectrum in Fig. 14 shows some typical absorption band around 912 nm for Yb^{3+}, 968 nm for both Er^{3+}, and Yb^{3+}, 1380 nm for OH^- and 1530 nm for Er^{3+}.

Pump Absorption

Definitions

Due to the existence of pump ESA, cooperative up-conversion (in section "Typical Active Ions and Properties"), etc., the pump absorption often varies with the pump power as shown in Fig. 15. Basically, the pump absorption of active fiber is comprised of two parts: the saturable absorption α_s and the unsaturable absorption α_{us}. In the small pump regime, the pump light can be fully absorbed and contribute to the saturable absorption. As the pump power increases, all the active ions in the

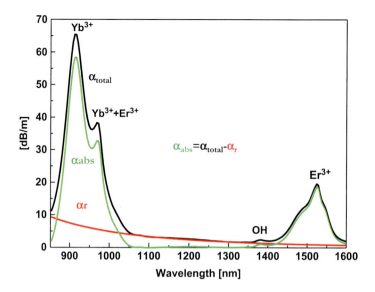

Fig. 14 Typical absorption spectrum of erbium and ytterbium co-doped optical fiber

ground state are excited to the higher-lying levels. In other words, the density of absorbing ions is depleted. While other mechanisms of attenuation can still exist such as exited state absorption, energy transfer, and scattering loss, they are not able to be saturated. As a result, the unsaturable absorption appears and becomes dominant gradually as shown in Fig. 15. Especially the maximal pump absorption α_m has a relation as shown in Eq. (44)

$$\alpha_s(\lambda) = \alpha_m(\lambda) - \alpha_{us}(\lambda) \tag{44}$$

Measurement

Pump absorption is measured by Fig. 16 and unsaturable absorption α_{us} will be the plateau value when pump power increased, according to Eq. (45):

$$\alpha_p = \frac{10}{\Delta L} \times \lg\left(\frac{P_2}{P_1}\right) \tag{45}$$

where P_2 is the pump power at the output end of active fiber, P_1 is the output end of active fiber after cutting the fiber with a length of ΔL. For better accuracy, the cut of ΔL can be repeated for several times for error reduction.

Figure 17 demonstrates that the gain of bismuth doped aluminosilicate active fiber has great dependence upon the unsaturable absorption (Sathi 2015). Seen from Fig. 17 (a), the ratio of unsaturable loss to the small signal loss is 35% at 1120 nm and 65% at 1047 nm. However, the gain coefficient has increased from 4.4 to

Fig. 15 Typical saturable and unsaturable absorption in pump absorption (Sathi 2015)

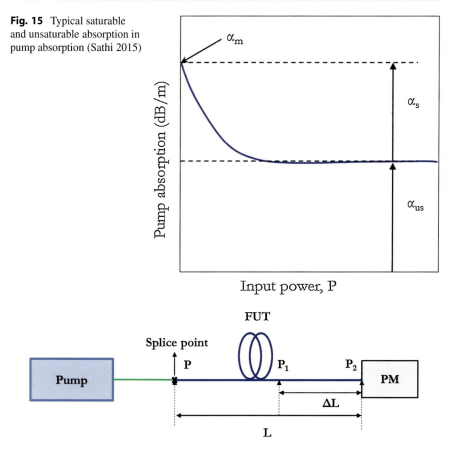

Fig. 16 Pump absorption scheme to determine the unsaturable loss

8 dB when the pump wavelength is changed from 1047 to 1120 nm as shown in Fig. 17 (b). Such comparison indicates that the unsaturable absorption has a significant influence on the gain performance of BDF. It can further deduce that the unsaturable absorption can be reduced and suppressed by optimizing the pump wavelength, host material, or other external conditions.

Measurement of Emission

Measurement of Spectral Emission

Emission is one of the key properties of active optical fibers. The analysis of the experimental spectral emission based on bandwidth and intensity is a suitable assessment to quantify and improve their performance. The experimental techniques to measure the spectral emission of active fiber mainly divided into four configures

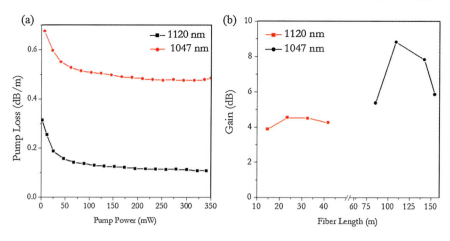

Fig. 17 (**a**) Pump absorption for 1120 and 1047 nm pump. (**b**) Gain variation with fiber length at 1180 nm for two different pump wavelengths (Pump power: 350 mW) (Sathi 2015)

according to the pump and detection position. It is namely forward emission (forward pump and forward detection), side emission (forward pump and side detection), backward emission (forward pump and backward detection), axial emission (side pump and axial detection).

The forward pump and forward detection configure is suitable when there is no overlapping between pump and emission spectra. In addition, the residual pump should be removed by filter if possible as the residual pump power could be significant, resulting in a strong background noise. A forward pump and side detection configure is used for the intrinsic emission measurement of active fiber, where the waveguide and loss effect can be ignored (Peng et al. 1996). Forward pump and backward detection configure shows better results due to neglecting the effect of ASE; however, a suitable WDM or another broad coupler is required. Side pump and forward detection configure is suitable for the active fiber with strong luminescence and low loss (Zhang et al. 2012) which led to a simple and nondestructive method to simultaneous measurement of emission and absorption. In additional, a powerful tool to identify the active centers in active fibers through luminescence intensity-dependency of the contour graphs is introduced.

Forward Emission

Figure 18 shows the experimental setup for forward emission measurement with forward pump and forward detection. Seen from Fig. 18, it consists of a MONO, a lock-in amplifier, a pump source, a chopper, a photodetector (PD), optical filters, lenses, and a computer (PC) to record the transmitted spectral from the PD. The emission from FUT pumped by laser source is launched into the MONO passing through chopper plate (and possibly filters). The emission signal spectrum is scanned by MONO and converted into electrical signal by PD, which is further amplified by lock-in amplifier and recorded by the PC.

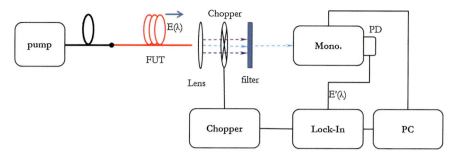

Fig. 18 Forward emission measurement setup

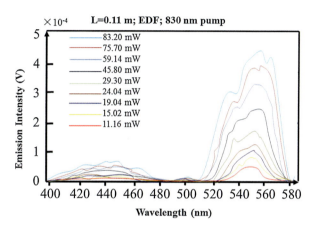

Fig. 19 The up-conversion spectrum of EDF in the visible band measured by mono and lock-in system

It is noted that to prevent re-absorption of emitted light or ASE effect from FUT, shorter fiber length is desirable. The original emission spectrum ($E(\lambda)$) is extracted from the recorded emission spectrum ($E'(\lambda)$) by compensating the effect of filters' ($F(\lambda)$) and the detector's ($D(\lambda)$) (Peng et al. 2013). The real forward emission should be given as

$$E(\lambda) = \frac{E'(\lambda)}{F(\lambda) \times D(\lambda)} \tag{46}$$

This setup can also be used for the measurement of up-conversion emission (short as up-conversion emission). Due to the high loss in a shorter wavelength band, a short piece of active fiber is often used for the up-conversion experiment. As the up-conversion wavelength is shorter than the pump wavelength, a short pass filter (i.e., KG5) is often used to remove the influence of the residual pump at the longer wavelength. Figure 19 shows an example of up-conversion spectra of EDF under 830 nm pumping under different pump power. Two up-conversion bands are clearly demonstrated with peaks around 450 nm ($^4F_{5/2} \rightarrow {}^4I_{15/2}$) and 550 nm ($^4S_{3/2} \rightarrow I_{15/2}$).

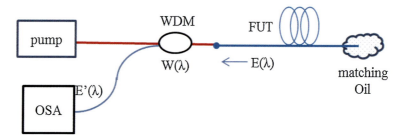

Fig. 20 Backward emission measurement setup with the forward pump and backward detection configure

Backward Emission

Backward emission is measured with the forward pump and backward detection configure. The typical experimental setup for the backward emission measurement is shown in Fig. 20. It consists of a WDM (generally broadband type), a white light source (WLS), and an OSA to record the spectral result. The emission is pumped by the light from pump laser passing through a WDM and the backward emission is transmitted through the WDM and measured by OSA. It is noted that the selection of the WDM depends upon both the pump wavelength and the emission wavelength. It is proved that the true emission spectra ($E(\lambda)$) is extracted from recorded emission spectra ($E'(\lambda)$) by removing the effect of the WDM loss. It should be noticed that both mono and lock-in system as well as OSA can be used in this configure. The mono system is capable of eliminate noise signal while OSA usually has higher sensitivity. Depending on the requirement of experiments, these two systems can exchange in the test. The fiber end in this setup is dipped in the matching oil to remove unwanted reflection and noise from outside.

Axial Emission by Side Pumping

Side pump and axial detection configure in Fig. 21 can be used for a nondecon-structive evaluation of active optical fibers for both absorption and emission (Zhang et al. 2012). It is free from fiber cutting, intrinsically low pump background, and simultaneous accurate measurement of emission and absorption. Seen from Fig. 21, the absorption of BEDF can be measured based on the emission spectra (Channel 1: I_{1A}, I_{1B}, Channel 2: I_{2A}, I_{2B}) that obtained from two measurements at two different positions Z_A and Z_B, which can be given by (Zhang et al. 2012):

$$\alpha\left(\lambda\right) = \frac{\ln\left[\frac{I_{1A}(\lambda)\times I_{2B}(\lambda)}{I_{1B}(\lambda)\times I_{2A}(\lambda)}\right]}{2\left(Z_B - Z_A\right)} \tag{47}$$

It could be used as a method to measure the uniformity of the doping concentration along the active fiber just by measuring the spectra at a few different pumping

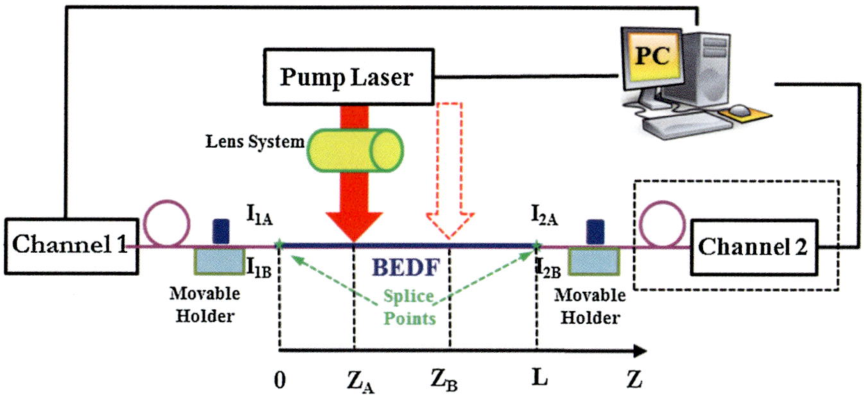

Fig. 21 The experimental setup for the simultaneous emission and absorption measurement of active optical fiber by side pumping

positions. That is needed and expected to be applied in the active waveguide area as well. There are several points to be noted for it (Zhang et al. 2012):

- The absorption measurement is limited to the emission band. For the very weak emission, the lock-in amplifier technique is expected to apply to obtain the spectra, instead of using OSAs.
- The higher pump power is needed for the side pumping scheme compared to the backward emission based on a WDM coupler.
- The resonance pump should be applied by using the appropriate laser to realize the full inversion if the emission cross section of active fiber is needed.

Combined Excitation-Emission Spectroscopy

The detailed measurements of the luminescence intensity depend on both emission and excitation wavelengths, which often varies in a wide spectral range. The contour graphs of the dependence of luminescence intensity, λ_{em}, and λ_{ex}, is a useful tool for the identification of the luminescence center in active fiber, especially the recent developed BDF (Firstov et al. 2011). For the complicate doped active fiber, multiactive center co-existed. The fluorescence profiles are highly pumped wavelength dependent. Therefore, it is necessary to have more information by the simultaneous emission and excitation measurement through combined excitation-emission spectroscopy. For the measurement of combined excitation-emission spectroscopy, broad tuned light source, i.e., tunable laser source (TLS), is often used as the pump source and OSA as the detection system. Figure 22 shows a typical excitation-emission spectrogram of a typical BEDF (Hang et al. 2013). Seen from the spectrogram, there are two obvious emission bands with the central wavelengths at 1420 and 1530 nm, which are related to BAC-Si and Er^{3+}, respectively.

Fig. 22 The excitation-emission spectrogram of a typical BEDF

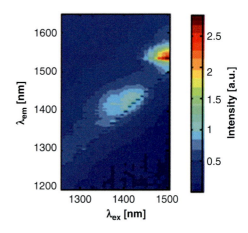

Measurement of Fluorescent Lifetime

Varieties of fluorescence lifetime measurement techniques and setups have been developed for different applications. For instance, streak camera for the biological and physical-chemistry applications, Time-Correlated Single-Photon Counting (TCSPC) instrument for biological and clinical applications, AMD technique for fluorescence lifetime imaging microscopy (FLIM), stroboscopic optical boxcar for chemistry applications, and monochromator spectroscopic for fiber laser, amplifier, and sensor applications are existed.

In the case of active fiber, the fluorescence lifetime measurement setup is quite different from other applications. Time domain and frequency domain analyses are two methods of fluorescence lifetime measurement. For system design, two important parameters are excitation wavelengths and emission wavelengths. These two parameters should be designed with regard to the characteristics of active centers of interest.

Time Domain

For the time-domain method, the fiber sample is excited with a short pulse of light, and the statistical decay curve of emission (the impulse response) is measured by a fast photodetector. The transient response of fluorescence substances to excitation pulse can be captured in the time domain by sampling enough data points within the time span of the decay (Jo et al. 2004). This falling curve demonstrates the lifetime. The time domain function of the fluorescence intensity is given by:

$$I(t) = \alpha e^{-\frac{t}{\tau}} \tag{48}$$

where $I(t)$ is the intensity at time t, α is a normalization term (the pre-exponential factor), and τ is the fluorescence lifetime. Time domain methods are important especially for complex emission process with multi or non-exponential decays, such as co-doped fiber with several fluorescence processes with overlapping emission and absorption spectra. In such cases, the fluorescence properties can be extracted from the decay profile separately.

If the measurement scheme is applied in broadband and complex emission processes, it is essential to (1) cover a relatively wide range of lifetimes (emission wavelengths and excitation wavelength), (2) have a wavelength selection capability, and (3) be able to measure the lifetimes of emission processes, with either single or multiple lifetimes. Time domain methods can cover these properties better than the frequency domain parameters.

In the time domain lifetime measurement, the following parameters need to be designed according to the fluorescence lifetime and active centers of interest:

• Excitation pulse parameters
• Detection section parameters
• Processing technique

For the excitation pulse, the effective parameters are wavelength, power, pulse width, rising, and falling time of excitation pulse. To measure a lifetime in the time domain, the excitation impulse can be generated using several methods, such as ultrafast lasers (ultrashort pulse lasers, gain-switching, and Q-switching, tunable titanium- sapphire laser, optical modulator, and mechanical choppers (Zareanborji et al. 2016).

Three essential properties for excitation pulse design include: (1) pulse width has to be short enough compared with the lifetime of interest to prevent saturation, resonance absorption, and stimulated emission; (2) the pulse power has to be long enough to provide measurable emission power; (3) duty cycle of excitation pulse needs to be short enough to cover a complete emission process; and (4) rising time and falling time of excitation pulse need to be short, to minimize the effects of measurement setup on the measured signal. The short duty cycle would guarantee that most electrons in an excited state return the lower levels of energy and subsequently, the emission process is completed. In the case of the high duty cycle for excitation pulse, the excitation pulse could occur, while some electrons are still in the excited state. This would change the normal excitation, which could possibly lead to measurement uncertainty.

Detection may consist of a monochromator, fast detector, data acquisition, and amplifiers. If the measurement scheme is applied in broadband and complex emission processes, the existence of monochromator for wavelength selection is necessary. The detector speed, sensitivity, and wavelength need to be selected based on emission process of interest. Data acquisition system can be used for recording measured data as well as processing such as averaging for SNR improvement. In the case of the low power emission signal, amplifier can be utilized.

Finally, the processing algorithms are an essential stage to extract the fluorescence lifetime from the experimental data. In the simple case with a single exponential emission, the lifetime can extract directly from experimental data.

Fig. 23 $I_1(t)$ and $I_2(t)$ are two emission processes with single exponential decays, I(t) is the fluorescence decays of these emission processes in the same wavelength

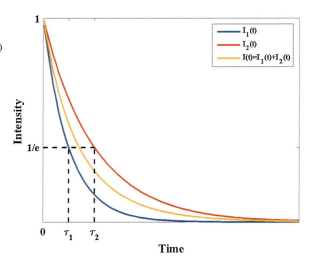

In contrast, in the case of complex emission process or ultra-short lifetime with multi or nonexponential decay, utilizing processing algorithm is an essential. Especially, as shown in Fig. 23 in the case of two single processes in an active fiber with two different lifetimes, in the same emission band (same wavelength) utilizing processing algorithm is necessary. The $I_1(t)$ and $I_2(t)$ are two emission processes with different lifetimes in the same wavelength:

$$
\begin{aligned}
I_1(t) &= \alpha_1 e^{\frac{-t}{\tau_1}} \\
I_2(t) &= \alpha_2 e^{\frac{-t}{\tau_2}}
\end{aligned}
\tag{49}
$$

The fluorescence intensity of the emission is given by:

$$
I(t) = \alpha_1 e^{\frac{-t}{\tau_1}} + \alpha_2 e^{\frac{-t}{\tau_{21}}}
\tag{50}
$$

where $I(t)$ is the intensity of dual emission processes at time t, α_1 and α_2 are a normalization terms of emission process 1 and 2, τ_1 and τ_2 are the lifetimes of emission process 1 and 2, respectively.

Various algorithms for fluorescence lifetime determination such as the nonlinear least-square iterative reconvolution, Fourier transformation, Laplace transformation, exponential series, recursive convolution integral, stretched exponential, direct deconvolution, plausible convolution, and trial convolution have been developed (Zareanborji et al. 2016). Convolution and deconvolution based are two general algorithms to measure fluorescence lifetime from the experimental data. The convolution and deconvolution are mathematically identical but different in applications to calculate lifetime(s). The convolution-based algorithm has superior noise tolerance and accuracy compared with the direct deconvolution, Fourier transformation, and Laplace transformation.

Generally, to apply DSP (Digital Signal Processing) algorithm for data processing, the measurement setup should be linear time invariant (LTI). One of the nonlinear effects, which may cause uncertainty in the measurement, is saturation. There are two main types of saturation effects in fluorescence lifetime measurement: detector saturation and emission saturation. Detector saturation is caused by the nonlinear response of the detector mostly due to the high power of laser or emission signals.

In the time domain, following the end of an excitation pulse, the fluorescence decays exponentially with time and the analysis of the signal can be quite straightforward. In practice, however, the observed decaying exponential waveform is often superimposed to baseline offset, leakage of excitation light, and other sources of optical and electronic noise (Fernicola and Rosso 2000).

To process the experimental data with low SNR, averaging algorithm is an essential. By using data acquisition system and averaging over multiple emission signals, the SNR will improve. Figure 24 demonstrates the effect of averaging and saturation on the measured signals. The emission is increased linearly with pump power increment, until saturation. The linear and unclipped rising edge of emission and reference signals indicate nonsaturation of both emission and detector.

Frequency Domain

In the frequency domain, the modulated excitation signal at an appropriate frequency induces the emission of active fiber. The effects of the fluorescence dynamics manifest themselves in the phase shift and change in modulation depth between the excitation and emission. Due to the relaxation of fluorescence, the fluorescence signal is delayed in time relative to the excitation, inducing a phase-shift, which is used to calculate the fluorescence lifetime. A sequence of measurements at different frequencies provides the full transfer function. For mono-exponential fluorescence decays, however, the decay time constant (often referred to simply as the fluorescence lifetime) is readily obtained from a measurement at a single frequency. Generally, by utilizing a sinusoidal excitation light source, the lifetime is determined as the phase shift and demodulation depth of the fluorescence signal (Fernicola and Rosso 2000; Grattan and Zhang 1994; Jo et al. 2004).

Figure 25 demonstrates the excitation and emission signal parameters in frequency domain lifetime measurement, where Ex_{DC} and Em_{DC} are the DC value excitation and emission, respectively, Ex_{AC} and Em_{AC} are the AC value of excitation and emission, respectively, and $\Delta\varphi$ is the phase lag of emission signal compare with the excitation signal. Excitation intensity is:

$$E(t) = E_0 \left[1 + M_E \sin \omega t\right] \tag{51}$$

where E(t) and E_0 are the intensity at time t and 0, M_E is the modulation factor, and the angular frequency of signal is $\omega = 2\pi f$. The result of this test is phase shift and demodulation as follows:

Fig. 24 (**a**) averaging algorithm for SNR increment, (**b**) saturation effect of fluorescence

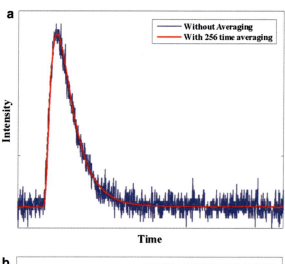

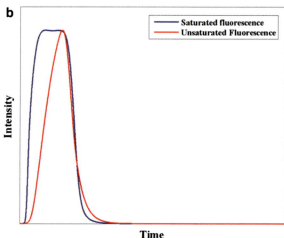

$$M_E = \frac{\frac{Em_{AC}}{Em_{DC}}}{\frac{Ex_{AC}}{Ex_{DC}}} = \frac{1}{\sqrt{1+(\omega\tau)^2}}$$

$$\Delta\phi = tg^{-1}(\omega\tau) \tag{52}$$

In the frequency domain, either in single or multiple frequency cross-correlation phase measurements, the contributions to the experimental errors come from the leakage of excitation light into the detection channel, the nonlinearity of the electronic filters, as well as the phase noise associated with the fluorescence response. In both cases, a careful design of the data acquisition and signal averaging instrumentation and a suitable choice of the algorithms used to analyze and process the data is required (Fernicola and Rosso 2000).

Fig. 25 Illustration of frequency-domain fluorescence lifetime measurement. The excitation signal (red) is modulated in amplitude at a frequency, while the fluorescence signal (blue) is emitted with the same modulation frequency but with a phase shift in time $\Delta\varphi$

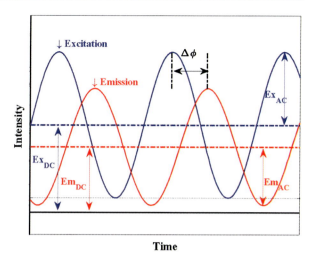

Measurement of Gain

Principle of Gain Measurement

Discussion in section "Interaction Between Light and Matter," gain in rate equations or propagation equations is usually using unit m^{-1}, while in application and measurement dB is the dominant unit. In this section, all measured data are designated as dB mathematically. The definition of net gain and on/off gain needs to be distinguished firstly.

As shown in Fig. 26a, when the input power of signal is P_0, it will be attenuated (absorbed through GSA) by the fiber in the absence of pump light and the output in signal power will be P_1. The attenuation/absorption coefficient will be

$$\alpha = 10\lg\frac{P_0}{P_1} \tag{53}$$

When a signal incident into the fiber together with a proper pump as shown in Fig. 26b, the signal can be amplified, the output signal P_2 is a combination after the amplification and attenuation. If ESA is absent, the signal is attenuated in a rate of α and amplified simultaneously. It is easy to predict that P_2 is stronger than P_1 while it depends if it is stronger than P_0. Define the on/off gain as

$$G_{on/off} = 10\lg\frac{P_2}{P_1}\ (P_2 > P_1) \tag{54}$$

However, when ESA of signal exists, which means the signal is attenuated not only by GSA but also ESA, P_2 is possible to be lower than P_1. In this case, the ESA can be defined as

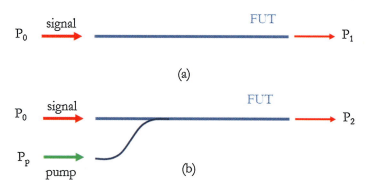

Fig. 26 Signal attenuation and amplification in fiber (**a**) signal attenuation; (**b**) signal amplification

$$\alpha_{ESA} = 10 \lg \frac{P_1}{P_2} \ (P_1 > P_2) \tag{55}$$

So it is easy to understand that on/off gain and the signal ESA test can share the same setup. As for the pump ESA, it affects the gain performance by reducing the pump efficiency.

In the application of devices, the relation between P_2 and P_0 counts more when the devices' efficiency, energy consumption, and economy need to be taken into consideration. Define the net gain as

$$G_{net} = 10 \lg \frac{P_2}{P_0} \ (P_2 > P_0) \tag{56}$$

If the attenuation is stronger than the amplification, the output P_2 will be lower than the input signal; $G_{net} < 0$, which means in a real amplification system the input signal is not amplified. The ideal case is when $P_2 > P_1$ and $G_{net} > 0$ which means the attenuation is overwhelmed by amplification. Physically or mathematically, it is obvious that the net gain is the difference between on/off gain and loss as following:

$$G_{net} = G_{on/off} - \alpha = 10 \lg \frac{P_2}{P_1} - 10 \lg \lg \frac{P_0}{P_1} = 10 \lg \frac{P_2}{P_0} \tag{57}$$

Gain Measurement

As defined above, the net gain is the amplification to input signal, while the on/off gain is the amplification to output signal without pump. Gain is a comprehensive performance of system and is wavelength dependent, and the gain coefficient is sometimes tested at only the interested wavelength (single wavelength) and other times among the whole spectrum (at multiple wavelengths). Two measurement

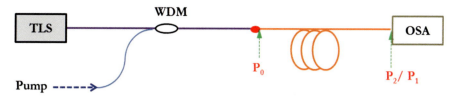

Fig. 27 The experimental setup for gain measurement at single wavelength

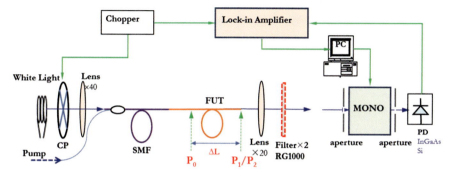

Fig. 28 Measurement setup of gain spectra

setups are explained in this case with different light sources and data acquisition devices which can be chosen according to the requirements and feasibilities of equipment.

A typical gain measurement setup is shown in Fig. 27. A tunable laser source (TLS) is adopted as the signal source. A WDM should be chosen for coupling the pump and signal in the FUT. The output end of the fiber is connected to OSA (or power meter) to record the output power. The pump and signal wavelength are selected based on the type of the fiber, i.e., for EDF, the pump wavelength is 980 nm and the signal wavelength is 1530 nm. OSA collects the output power at the target wavelengths. The input signal power is P_0; when the pump is off, the output power is P_1; when the pump is on, the output power is P_2. Thus, the on/off gain or ESA and net gain can be calibrated by Eqs. (53, 54, and 55).

To obtain the gain spectrum, two conditions must be satisfied. One is TLS that can emit wavelengths as needed and the output signal can be collected and measured at multiple wavelengths to draw the curve of spectrum. The setup in Fig. 28 provides an alternative way for spectral test as WLS emits broadband and the mono and lock-in system connected with PC allow a handier operation to split and sweep the light signal.

As shown in Fig. 28, the white light source acts as a signal source with broad spectrum. White light is chopped by the chopper, passes through the lens, and launches into the fiber. Similar to the setup above, signal and pump light are coupled into the active fiber by a selected WDM. A mono and lock-in system are adopted as the spectrum collector which can be replaced by OSA but will have

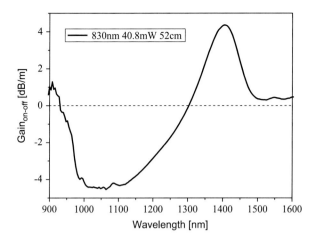

Fig. 29 On/off gain spectrum of BEDF fiber

lower signal-noise ratio. The signal is chopped before launching into fiber and only chopped signal can be amplified in the lock-in amplifier, so the noise is effectively reduced. Take 830 nm pumped BDF as an example, two RG1000 filters are applied to remove the pump light before data acquisition. If the other pump is chosen, the filter should be changed correspondingly. For BDF, the infrared gain from 1000 to 1600 nm is of most interest so an InGaAs photo detector will be used. WDM, filter, and photo detector can all be swapped accordingly. All data collected by the system will be recorded by PC and the gain and ESA can be calibrated by Eq. (57). This setup can also be changed into a backward one where the pump incident is from opposite direction of signal for further reducing the pump influence.

Equation (54) is used to calculate the on/off gain (ESA). In the calculated spectrum, the positive area is on/off gain, while the negative area is ESA. Figure 29 shows a typical on/off gain measurement for our home-made Bi/Er co-doped fiber. Negative part is attributed to the ESA and positive part indicates on/off gain.

Furthermore, combining the cut-back method, P_0 as shown in Fig. 28 can be measured after fiber cut. The $\alpha(\lambda)$ and $G_{net}(\lambda)$ can be obtained and the on/off gain (or ESA) can be calibrated (Riumkin et al. 2014). As shown in Fig. 30, it should be noticed that in this figure the positive area is loss. $\alpha(\lambda)$ is the loss spectra obtained as the cut-back method. $G_{net}(\lambda)$ is the net gain spectra and the negative part of this curve indicated net gain. $G_{on/off}$ is the difference of two curves, and if the blue dash and dot curve show positive area, it indicates ESA.

Without adopting cutback method, the above two systems can still test the on/off gain directly. The net gain can be obtained by Eq. (57) if the absorption coefficient/spectrum is known. But in this case, the ESA/on-off gain spectrum and absorption spectrum are tested separately in different setups and with different samples. Additionally, to measure the absorption spectrum, an operation of cut-back without any pumping is still needed. By combining cut-back method, the net

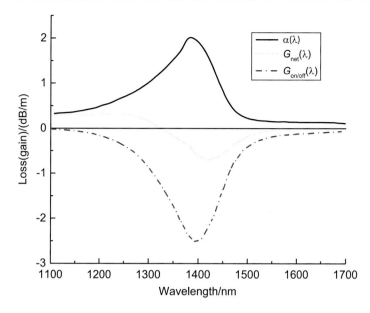

Fig. 30 Optical loss and transmission change in bismuth-doped fiber (Riumkin et al. 2014)

gain spectrum can be obtained more precisely because the absorption spectrum and ESA/on-off gain spectrum are tested simultaneously in an integrated set-up without changing the sample.

References

A. Ehn et al., Opt. Express **20**(3) (2012)

V. Fernicola, L. Rosso, Time-and frequency-domain analysis of fluorescence lifetime for temperature sensors, in *2000 Conference on Precision Electromagnetic Measurements Digest*, IEEE, 2000

S.V. Firstov et al., Opt. Express **19**(20) (2011)

K. Grattan, Z. Zhang, *Fiber Optic Fluorescence Thermometry, in Topics in Fluorescence Spectroscopy* (Springer, London, 1994)

J. Hang et al., Opt. Express **21**(6) (2013)

J.A. Jo et al., J. Biomed. Opt. **9**(4) (2004)

M.P. Kalita, S. Yoo, J. Sahu, Opt. Express **16**(25) (2008)

W. Koechner, *Solid-State Laser Engineering* (Springer, New York, 2013)

R.I. Laming, S.B. Poole, E. Tarbox, Opt. Lett. **13**(12) (1988)

W. Miniscalco, M.J. Digonnet, *Rare Earth Doped Fiber Lasers and Amplifiers* (Marcel Dekker, New York, 1993)

W.J. Miniscalco, R.S. Quimby, Opt. Lett. **16**(4) (1991)

S. Nagel, IEEE Commun. Mag. **25**(4) (1987)

Optical Fiber Loss and Attenuation https://www.fiberoptics4sale.com/blogs/archive-posts/95048006

G.D. Peng et al., J. Lightwave Technol. **14**(10) (1996)

G.D. Peng, J. Zhang, Y. Luo, Z. Sathi, A. Zareanborji, J. Canning, Developing new active optical fibers with broadband emissions, in *Fourth Asia Pacific Optical Sensors Conference* 2013 Oct 15, International Society for Optics and Photonics, p. 89240E

G.-D. Pengc, High temperature assessment of an Er^{3+}/Yb^{3+} co-doped phosphosilicate optical fiber for lasers, amplifiers and sensors, in *Proceedings of SPIE*. https://doi.org/10.1117/12.2194612

J.F. Philipps et al., Appl. Phys. B **74**(3) (2002)

S. Poole, D.N. Payne, M.E. Fermann, Electron. Lett. **21**(17) (1985)

K. Riumkin et al., Opt. Lett. **39**(8) (2014)

E.J.Z. Sathi, *Bismuth, Erbium and Ytterbium Co-doped Fibers for Broadband Applications*. University of New South Wales, School of Electrical Engineering & Telecommunications. (2015)

R. Scheps, Prog. Quantum Electron. **20**(4) (1996)

N.K. Thipparapu et al., Opt. Lett. **40**(10) (2015)

T. Wei et al., Opt. Mater. Express **4**(10) (2014)

A. Zareanborji et al., Time-resolved emission characteristics of Bi/Er codoped fiber for ultra-broadband applications, in *Workshop on Specialty Optical Fibers and their Applications*, Optical Society of America, 2013

A. Zareanborji et al., Time-resolved fluorescence measurement based on spectroscopy and DSP techniques for Bi/Er codoped fiber characterisation, in *2014 OptoElectronics and Communication Conference and Australian Conference on Optical Fiber Technology*, IEEE, 2014

A. Zareanborji, Y. Luo, G.-D. Peng, Characterization and assessment of multiple bismuth active centres in Bi/Er doped fiber, in *2015 2nd International Conference on Opto-Electronics and Applied Optics (IEM OPTRONIX)*, IEEE, 2015

A. Zareanborji et al., J. Lightwave Technol. **34**(21) (2016)

J. Zhang et al., Opt. Express **20**(18) (2012)

B. Zhou et al., Nat Nano **10**(11) (2015)

Characterization of Specialty Fibers

30

Quan Chai, Yushi Chu, and Jianzhong Zhang

Contents

Q. Chai
Key Laboratory of In-Fiber Integrated Optics, Ministry Education of China, Harbin Engineering University, Harbin, China

Y. Chu
Key Laboratory of In-Fiber Integrated Optics, Ministry Education of China, Harbin Engineering University, Harbin, China

Photonics and Optical Communications, School of Electrical Engineering and Telecommunications, UNSW, Sydney, NSW, Australia

interdisciplinary Photonics Laboratories (iPL), Global Big Data Technologies Centre (GBDTC), Tech Lab, School of Electrical and Data Engineering, University of Technology Sydney, Sydney, NSW, Australia

J. Zhang
Key Lab of In-fiber Integrated Optics, Ministry of Education, Harbin Engineering University, Harbin, China
e-mail: zhangjianzhong@hrbeu.edu.cn

© Springer Nature Singapore Pte Ltd. 2019
G.-D. Peng (ed.), *Handbook of Optical Fibers*,
https://doi.org/10.1007/978-981-10-7087-7_59

Abstract

Specialty fibers play an important role both in scientific research and industrial applications. The past decades have also witnessed a significant benefit from specialty fibers. Behind these successes is a constant understanding of the performance of these fibers. In this chapter, characteristics of specialty fibers and their measurement technologies are discussed in detail, including dispersion characterization, polarization characterization, and other special characterization techniques.

Keywords

Dispersion characterization · Polarization characterization · Polarization mode dispersion · Polarization-dependent loss · Material characterization · Spectral characterization

Introduction

Single-mode fibers are the most common fibers widely used in our world. There are also some other types of optical fibers with uniform characteristics, such as multimode fibers, high-birefringence fibers, and active fibers. In this chapter, some main characteristics of specialty fiber and their measurement technologies are introduced.

Multimode fiber (https://en.wikipedia.org/wiki/Multi-mode_optical_fiber), shown in Fig. 1, due to its large core, allows for the transmission of light using different paths (multiple modes) along the link. The primary advantages of multimode fiber are easily coupling to light sources or other fibers, low-cost light sources, simplified connectorization, and splicing processes. However, its relatively high attenuation and low bandwidth limit the transmission of light over multimode fiber to short distances.

Multimode fibers are usually categorized into step-index multimode fiber and graded-index multimode fibers, which are shown in Fig. 2. Step-index (SI) multimode fiber guides light through total reflection on the boundary between the core and cladding. The refractive index is a constant in the core. SI multimode fibers always have a minimum core diameter of 50 μm or 62.5 μm, a cladding diameter

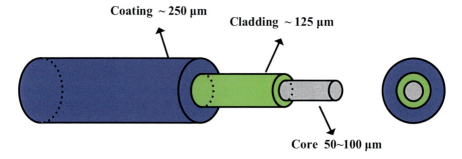

Fig. 1 Structure of multimode fiber

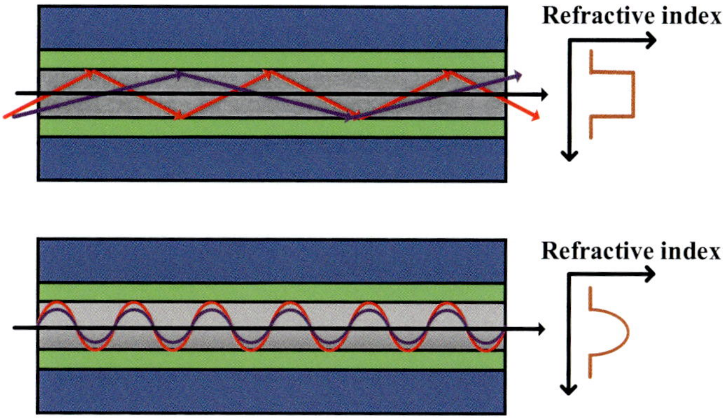

Fig. 2 Light propagation through the SI and GI multimode fiber

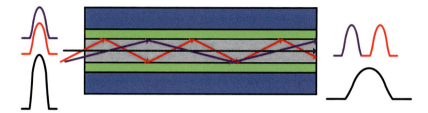

Fig. 3 Modal dispersion in step-index multimode fiber

between 100 and 140 μm, and a numerical aperture between 0.2 and 0.5. Step-index multimode fibers are surrounded by a plastic coating and used mostly for short-distance links that can accommodate high attenuations. The core of graded-index (GI) multimode fibers possesses a nonuniform refractive index, decreasing gradually from the central axis to the cladding (ITU-T G.651). This index variation of the core forces the rays of light to progress through the fiber in a sinusoidal manner.

The highest order modes have a longer path to travel, but outside of the central axis in areas of low index, their speeds increase. In addition, the difference in speed between the highest order modes and lowest order modes is smaller for graded-index multimode fibers than for step-index multimode fibers.

Multimode fibers are quite sensitive to modal dispersion, which reduce the effective bandwidth available for transmission. When a very short light pulse is injected into the fiber within the numeral aperture, all of the energy do not reach the end of the fiber at the same time. Figure 3 shows the modal dispersion in a step-index multimode fiber. Different modes of oscillation carry the energy go through the fiber, experiencing different paths and lengths. It limits the bandwidth of optical fibers. The details are discussed in this chapter.

Polarization-maintaining (PM) fibers work by intentionally introducing a systematic linear birefringence in the fiber, so that there are two well-defined polarization modes which propagate along the fiber with very distinct phase velocities. The beat length of a PM fiber (for a particular wavelength) is the distance (typically a few millimeters) over which the wave in one mode will experience an additional delay of one wavelength compared to the other polarization mode. A PM fiber does not polarize light as a polarizer. However, PM fiber maintains the linear polarization of linearly polarized light, which means it is launched into the fiber aligned with one polarization mode of the fiber. Launching linearly polarized light into the fiber at a different angle will excite both polarization modes, conducting the same wave at slightly different phase velocities. PM fibers can be classified into high-birefringence (HB) fibers and low-birefringence (LB) fibers.

The birefringence of conventional single-mode fibers is in the range $B_F = 10^{-6}$ to 10^{-5} (Payne et al. 1982). An HB fiber requires $B_F > 10^{-5}$ and a value better than 10^{-4} is a minimum for polarization maintenance (Stolen and Paula 1987). HB fibers can be separated into two types which are generally referred to two-polarization fibers and single-polarization fibers. In the latter case, in order to allow only one polarization mode to propagate through the fiber, a cutoff condition is imposed on the other mode by utilizing the difference in bending loss between the two polarization modes.

A selection of the most common structures of PM fiber is illustrated in Fig. 4. The fiber types (Noda et al. 1986) illustrated in Fig. 4a employ geometrical shape bire-fringence, while Fig. 4b–d utilizes various stress effects. Geometrical birefringence is a somewhat weak effect, and a large relative refractive index difference between the fiber core and cladding is required to produce high birefringence. The elliptical core fiber generally has high doping levels which tend to increase the optical losses as well as the polarization cross-coupling, which is shown in Fig. 4a (Eickhoff and Brinkmeyer 1984).

Stress birefringence may be induced using an elliptical cladding with a high thermal expansion coefficient. For example, borosilicate glass with some added germanium or phosphorus to provide index compensation can be utilized (Kaminow and Ramaswamy 1979). The HB fibers shown in Fig. 4b, c employ two distinct stress regions and are often referred to as the PANDA (Hosaka et al. 1985) and bow-tie (Birch et al. 1982) fibers because of the shape of these regions. Alternatively, the flat cladding fiber design illustrated in Fig. 4d has the outer edge of its elliptical cladding, touching the fiber core which therefore divides the stressed cladding into two separate regions (Stolen et al. 1984).

\looseness-1{}PM optical fibers are widely used in special areas, such as optical fiber sensors, interferometry, and telecommunications. PM fibers are rarely used for

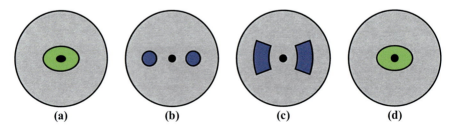

Fig. 4 Some structures of PM fibers: (**a**) elliptical core; (**b**) PANDA; (**c**) bow-tie; (**d**) elliptical stress cladding

long-distance transmission, due to the more expensive and higher attenuation than the single-mode fibers.

Active fibers, always doped rare earth element, are one of the driving forces developing guided wave optical fibers. When the optical signal travels in a waveguide, it usually suffers from attenuation due to the absorption, scattering, dispersion, and so on. To exploit the long path length provided by waveguide, the rare earth doped fibers play an important role in the optical communications and laser physics. Active elements, such as Er, Yb, Tm, Bi, Ho, Pr, Cr, etc. (Becker et al. 1999; Chen et al. 2007; Ohishi et al. 1991; Kasamatsu et al. 1999; Moulton et al. 2009; Dvoyrin et al. 2006, 2008; Bufetov and Dianov 2009; Yeh et al. 2012; Liang et al. 2004; Hu et al. 2011), have been doped in glasses (Chen et al. 2007), silica-based fibers (Ohishi et al. 1991; Kasamatsu et al. 1999; Moulton et al. 2009; Dvoyrin et al. 2006, 2008; Bufetov and Dianov 2009; Yeh et al. 2012), polymer fibers (Liang et al. 2004; Hu et al. 2011), or waveguides (Yliniemi et al. 2006; Della Valle et al. 2008), in order to realize the optical amplifiers and lasers for optical fiber communication and sensing. According to the doped ions and its related active centers, there are erbium-doped fiber amplifier (EDFA), ytterbium-doped fiber amplifier (YDFA), thulium-doped fiber amplifier (TDFA), praseodymium-doped fiber amplifier (PDFA), and so on.

EDFA is one of the most popular optical devices in optical communication systems. The optical pumping process in an Er-doped fiber is usually by a simplified three-level energy system (Tünnermann et al. 2010; Desurvire 2002; Quimby et al. 1994; Jarabo and Álvarez 1998). Pump photons at 980 nm are able to excite ground state carriers to the excited state and create population inversion (Quimby et al. 1994). The carriers stay in the excited state for only about 1 μs and decay into a metastable state through a nonradiative transition. Then radiative recombination happens when carriers step down from the bottom of the metastable state to the ground state and emit photons in the vicinity of 1550 nm. Er-doped fiber can also be pumped at 1480 nm, which corresponds to the bandgap between the top of the metastable state and the ground state (Jarabo and Álvarez 1998). Typically, 1480 nm pumping is more efficient than 980 nm pumping because it does not involve the nonradiative transition from the 980 nm to the 1480 nm band. Therefore, 1480 nm pumping is often used for high-power optical amplifiers. However, amplifiers with 980 nm pumps usually have lower noise figures.

Ytterbium doping provides the most efficient fiber amplifier systems, as essentially no competing absorption and emission mechanism exist (Paschotta et al. 1997). This is because in ytterbium, there are only two energy states that are resonant at the wavelengths of interest, viz., the ground state $^2F_{7/2}$ and the excited state $^2F_{5/2}$. There appear a strong absorption at wavelengths in the vicinity of 920 nm and an emission between 1000 and 1100 nm. Pump absorption is very efficient, which makes side pumping geometries practical. Because of the extremely high gain that is possible, Yb-doped fibers are a candidate for power amplifier of 1060 nm and have been employed in fiber laser configurations. YDFA has also been proven attractive as super fluorescent source, in which the output is simply amplified spontaneous emission and there is no signal input.

Ytterbium ions (Yb^{3+}) are often co-doped with erbium ions (Er^{3+}). There is a strong absorption of Yb^{3+} at the conventional 980 nm Er^{3+} pump wavelength (Yahel and Hardy 2003). The excited state of Yb^{3+} transition energy to Er^{3+}, and gain between 1530 and 1560 nm, is formed. With high pump absorption, side pumping is possible and high gain can be established over a shorter propagation distance in the fiber than EDFA. As a result, shorter length amplifiers having lower ASE noise can be constructed.

The strongest gain occurs in the praseodymium (Pr)-doped fiber that is in the vicinity of 1300 nm, with the pump wavelength at 1020 nm (Ohishi et al. 1991; Nishida et al. 1998). Gain formation is described by a basic three-level model, in which pump light excites the system from the ground state $^3H_4$ to the metastable excited state $^1G_4$. Gain for 1300 nm is associated with the emission from $^1G_4 \rightarrow ^3H_5$, which peaks in the range of 1320–1340 nm. The 1300 nm emission of Pr^{3+} ions has been extensively studied because of its suitability for optical amplification in the fiber telecommunication technology.

Holmium (Ho)-doped fiber amplifiers exhibit oscillations at the longest wavelengths for lasers based on silica-based fiber. Spectral range of lasing is centered near 2100 nm (Jackson 2003; Kurkov et al. 2009), that determined by $^5I_7 \rightarrow ^5I_8$ transition.

Because the doped elements are optical active, the doped fibers are widely used in optical fiber laser and amplifiers. The characteristics of active fiber are very important for evaluating and utilizing fiber lasers and amplifiers for various applications. We will describe the material and spectral characteristics of active fibers in detail in this chapter.

Dispersion Characterization of Optical Fibers

In optical fiber, dispersion is the phenomenon in which the phase velocity of a wave depends on its frequency. The optical signal with different frequency propagates along the optical fiber in different speeds. Fiber dispersion can be categorized into chromatic dispersion and intermodal dispersion. Chromatic dispersion is originated from the fact that different frequency components in each propagation mode may travel at slightly different speeds and therefore it is the dominant dispersion source in single-mode fibers. Intermodal dispersion is caused by the fact that different propagation modes in a fiber travel at different speeds. It is the major source of dispersion in multimode fibers.

Dispersion Characteristics

A single-frequency plane optical wave propagating in z-direction is usually expressed as

$$E(z, t) = E_0 \exp\left[-j \Phi(t, z)\right] = E_0 \exp\left[-j(\omega_0 t - \beta_0 z)\right] \tag{1}$$

where $\Phi(t, z) = \omega_0 t - \beta_0 z$ is the phase, ω_0 is the optical frequency, and $\beta_0 = 2\pi n/\lambda = n\omega_0/c$ is the propagation constant.

As we have already known, when the optical phase is a constant, the phase front of the light is a plane wave, and the propagation speed of the phase front is called phase velocity, which is defined as

$$v_p = \frac{dz}{dt} = \frac{\omega_0}{\beta_0} = \frac{c}{n} \tag{2}$$

Assume this single-frequency optical wave is modulated by a sinusoidal signal of frequency. Then, Eq. (2-1) can be expressed as

$$E(z, t) = E_0 \exp[-j(\omega_0 t - \beta_0 z)] \cos(\Delta\omega t) \tag{3}$$

Using Euler's formula, it can be expressed as

$$E(z, t) = \frac{1}{2} E_0 \left\{ e^{-j[(\omega_0 + \Delta\omega)t - (\beta_0 - \Delta\beta)z]} + e^{-j[(\omega_0 - \Delta\omega)t - (\beta_0 + \Delta\beta)z]} \right\}$$

$$= E_0 \exp[-j(\omega_0 t - \beta_0 z)] \cos(\Delta\omega t - \Delta\beta z) \tag{4}$$

where $E_0 \exp[-j(\omega_0 t - \beta_0 z)]$ represents an optical carrier and $\cos(\Delta\omega t - \Delta\beta z)$ is an envelope that is carried by the optical carrier, which represents the information that is modulated onto the optical carrier. Group velocity is defined as the propagation speed of this information-carrying envelope. Similar to the derivation of phase velocity, group velocity can be expressed as

$$v_g = \frac{dz}{dt} = \frac{\Delta\omega}{\Delta\beta} = \frac{d\omega}{d\beta} \tag{5}$$

If the refractive index n is a constant, the group velocity is the same as the phase velocity. However, in most optical materials, refractive index n is a function of the optical frequency (or optical wavelength). In these optical materials, group velocity is not equal to the phase velocity, and the propagation group delay is defined as the inverse of the group velocity:

$$\tau_g = \frac{1}{v_g} = \frac{d\beta}{d\omega} \tag{6}$$

Consider that two sinusoids with the frequencies $\Delta\omega \pm \delta\omega/2$ are modulated onto an optical carrier of frequency ω_0. When propagating along a fiber, each modulating frequency will have its own group velocity; then over a unit fiber length, the group delay difference between these two frequency components can be found as

$$\delta\tau_g = \frac{d\tau_g}{d\omega}\delta\omega = \frac{d}{d\omega}\left(\frac{d\beta}{d\omega}\right)\delta\omega = \frac{d^2\beta}{d\omega^2}\delta\omega \tag{7}$$

This group delay difference is introduced by the frequency dependency of the propagation constant. In general, the frequency-dependent propagation constant β_0 can be expended in a Taylor series around a central frequency ω_0:

$$\begin{aligned} \beta(\omega) &= \beta(\omega_0) + \frac{d\beta}{d\omega}\Big|_{\omega=\omega_0}(\omega-\omega_0) + \frac{1}{2}\frac{d^2\beta}{d\omega^2}\Big|_{\omega=\omega_0}(\omega-\omega_0)^2 + \ldots \\ &= \beta(\omega_0) + \beta_1(\omega-\omega_0) + \frac{1}{2}\beta_2(\omega-\omega_0)^2 + \ldots \end{aligned} \tag{8}$$

Here, $\beta_1 = \frac{d\beta}{d\omega}$ represents the group delay and $\beta_2 = \frac{d^2\beta}{d\omega^2}$ is the group delay parameter.

We can also express the relative delay in wavelength separation as following

$$\delta\tau_g = \frac{d\tau_g}{d\lambda}\delta\lambda = D\delta\lambda \tag{9}$$

where $D = d\tau/d\lambda$ is another parameter to describe group delay dispersion. The relationship between the two dispersion parameters D and β_2 can be found as

$$D = \frac{d\tau_g}{d\lambda} = \frac{d\omega}{d\lambda}\frac{d\tau_g}{d\omega} = -\frac{2\pi c}{\lambda^2}\beta_2 \tag{10}$$

In practical fiber-optic systems, the relative delay between different wavelength is usually picosecond orders; wavelength separation is usually in nanometers and fiber length is usually in kilometers. Therefore, the most commonly used units for β_1, β_2, and D are s/m, ps^2/nm, and ps/(nm·km), respectively.

Chromatic Dispersion

Chromatic dispersion is caused by different frequency signals that propagate in different speeds in optical fiber. Both the optical material property and waveguide structure may cause the propagation constant to be a function of wavelength. Therefore, chromatic dispersion can be categorized into material dispersion and waveguide dispersion.

Material Dispersion

Material dispersion is due to material refractive index that is a function of wavelength, and the wavelength-dependent propagation constant is $\beta(\lambda) = 2\pi n(\lambda)/\lambda$. For a unit fiber length, the wavelength-dependent group delay is

$$\delta\tau_g = \frac{d\beta(\lambda)}{d\omega} = -\frac{\lambda^2}{2\pi}\frac{d\beta(\lambda)}{d\lambda} = \frac{1}{c}\left[n(\lambda) - \lambda\frac{dn(\lambda)}{d\lambda}\right] \tag{11}$$

The group delay dispersion between two wavelength components separated by $\delta\lambda$ is then

$$\delta \tau_g = \frac{d \tau_g}{d \lambda} \delta \lambda = -\frac{1}{c} \left[\lambda \frac{d^2 n(\lambda)}{d \lambda^2} \right] \delta \lambda \tag{12}$$

From the equation, we can see that the material dispersion is proportional to the second derivative of the refractive index $n(\lambda)$.

Waveguide Dispersion

The waveguiding of the fiber may also create chromatic dispersion. It is because that the group velocity of a mode will change with wavelength.

Waveguide dispersion can be explained as the wavelength-dependent angle of the light ray propagating inside the fiber core (Fig. 5).

In the guided mode, the projection of the propagation constant β_z in z-direction has to satisfy $\beta_2^2 < \beta_Z^2 < \beta_1^2$, where $\beta_1 = kn_1$, $\beta_2 = kn_2$, and $k = 2\pi/\lambda$; it is equivalent to:

$$0 < \frac{(\beta_z/k)^2 - n_2^2}{n_1^2 - n_2^2} < 1 \tag{13}$$

If we define

$$b = \frac{(\beta_z/k)^2 - n_2^2}{n_1^2 - n_2^2} \approx \frac{(\beta_z/k) - n_2}{n_1 - n_2} \tag{14}$$

as a normalized propagation constant, the actual propagation constant in the z direction, β_z can be expressed as a function of b as

$$\beta_z(\lambda) = kn_2(b\Delta + 1) \tag{15}$$

where $\Delta = (n_1 - n_2)/n_2$ is the normalized index difference between the core and the cladding. Then the group delay can be found as

$$\tau_g = \frac{d\beta_z(\lambda)}{d\omega} = \frac{n_2}{c}\left(1 - b\Delta - k\frac{db}{dk}\Delta\right) \tag{16}$$

Fig. 5 Ray propagating inside the core of fiber

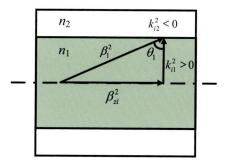

Generally speaking, due to the doping into silica introduce additional attenuation, material dispersion is difficult to modify. On the other hand, waveguide dispersion can be modified by designing suitable index profile.

The overall chromatic dispersion in an optical fiber is the combination of material dispersion and waveguide dispersion. Different types of fiber have different dispersion characteristics. For a standard single-mode fiber, the dispersion parameter D can usually be described by a Sellmeier equation (Neumann 1988):

$$D(\lambda) = \frac{S_0}{4} \left(\lambda - \frac{\lambda_0^4}{\lambda^3} \right) \tag{17}$$

where S_0 is the dispersion slope, which ranges from 0.08 to 0.09 ps/(nm^2·km), and λ_0 is the zero-dispersion wavelength, which is around 1315 nm.

By the definition of dispersion $D = d\tau_g/d\lambda$, the wavelength-dependent group delay $\tau_g(\lambda)$ can be obtained by integrating $D(\lambda)$ over wavelength; the group delay can be derived as

$$\tau_g(\lambda) = \int D(\lambda) d\lambda = \tau_0 + \frac{S_0}{8} \left(\lambda - \frac{\lambda_0^2}{\lambda} \right)^2 \tag{18}$$

Figure 6a shows the dispersion parameter D versus wavelength for standard single-mode fiber, which has dispersion slope $S_0 = 0.09$ ps/(nm^2·km) and zero-dispersion wavelength $\lambda_0 = 1315$ nm. $D(\lambda)$ is nonlinear; however, if we are only interested in a relatively narrow wavelength window, it is often convenient to linearize this parameter.

Figure 6b shows the relative group delay $\Delta\tau_g(\lambda) = \tau_g(\lambda) - \tau_0$ versus wavelength. The group delay is not sensitive to wavelength change around $\lambda = \lambda_0$ where the dispersion is zero.

Intermodal Dispersion

Chromatic dispersion specifies wavelength-dependent group velocity within one optical mode. In a multimode fiber, more than one mode exist. Different modes also have different propagation speeds, which induce intermodal dispersion.

Intermodal dispersion depends on the number of propagation modes that exist in the fiber, which is determined by the fiber core size and the index difference between the core and the cladding. Due to the ray optics theory, we can find that the fastest mode is the one that travels along the fiber longitudinal axis, whereas the ray trace of the slowest mode has the largest angle with respect to the fiber longitudinal axis. Since the group velocity of the fastest ray trace is c/n_1 (assume n_1 is a constant), the group velocity of the slowest ray trace should be $c \sin(\theta_i)/n_1 = cn_2/n_1^2$. Therefore, the maximum group delay difference in a fiber of with length L is approximately

$$\delta T_{\max} = \frac{n_1 L}{c} \left(\frac{n_1 - n_2}{n_2} \right) \tag{19}$$

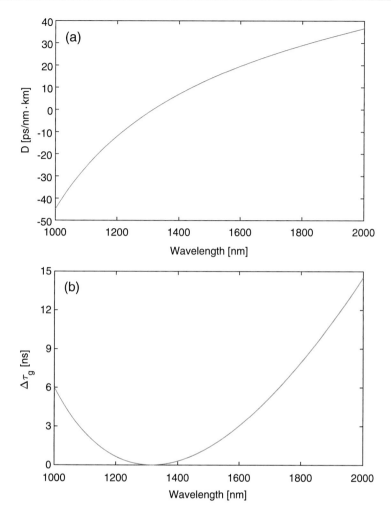

Fig. 6 (**a**) Chromatic dispersion D versus wavelength λ and (**b**) relative group delay versus wavelength

    Pulse broadening due to intermodal dispersion results from the propagation delay differences between modes within a multimode fiber. As the different modes which constitute a pulse in a multimode fiber travel along the channel at different group velocities, the pulse width at the output is dependent upon the transmission times of the slowest and fastest modes.

    Multimode step-index fibers exhibit a large amount of intermodal dispersion which gives the greatest pulse broadening. However, intermodal dispersion in multimode fibers may be reduced by adoption of an optimum refractive index profile which is provided by the near-parabolic profile of most graded-index fibers. Hence, the overall pulse broadening in multimode graded-index fibers is far less than that obtained in multimode step-index fibers (typically by a factor of 100). The graded-index fibers used with a multimode source give a tremendous bandwidth advantage over multimode step-index fibers.

Polarization Mode Dispersion

Polarization mode dispersion (PMD) is a special type of modal dispersion that exists in single-mode fibers. It is a random effect due to both intrinsic (caused by noncircular fiber core geometry and residual stresses in the glass material near the core region) and extrinsic (caused by stress from mechanical loading, bending, or twisting of the fiber) factors which in actual manufactured fibers result in group velocity variation with polarization state. In modern high-speed optical communications, polarization mode dispersion has become one of the most important reasons for the degradation of transmission performance. Since the perturbation causing birefringence is random, polarization mode dispersion in an optical fiber is also a random process. Its characteristics and measurement are described in detail in the next section.

Dispersion Measurement

Chromatic Dispersion Measurement

Chromatic dispersion is caused by different frequency signals that propagate in different speeds in optical fiber, which is the most important dispersion effect in single-mode fibers. Since the chromatic dispersion value is relatively small, the dispersion-limited information bandwidth is usually very wide, so it is difficult to measure this equivalent bandwidth directly. For example, for a standard single-mode fiber operating in the 1550 nm wavelength window, the chromatic dispersion is approximately 17 ps/(nm·km). If we want to directly measure the pulse broadening in a 1-km fiber, a picosecond-level pulse width light source and a signal receiver with a bandwidth larger than 100 GHz are required. Therefore, some alternative techniques are used to characterize chromatic dispersion in single-mode fibers.

Phase Shift Method

As illustrated in Fig. 7, modulation phase shift method is operated in time domain to characterize fiber chromatic dispersion. A tunable laser is used as the light source. To meet the measurement requirement, the adjusted wavelength range of the tunable laser should be greater than the wavelength window where the chromatic dispersion needs to be measured. An electro-optic intensity modulator is used to convert a sinusoidal electrical signal at frequency f_m into optical domain. A photodiode is used to detect the modulated optical signal that passes through the optical fiber to be tested and change it into electrical signal. An oscilloscope is used to receive the sinusoidal modulated electrical signal. The waveform of the same sinusoid source at frequency f_m is also directly measured by the same oscilloscope for phase comparison. The wavelength-dependent propagation delay caused by chromatic dispersion can be evaluated by the relative phase retardation of the received RF

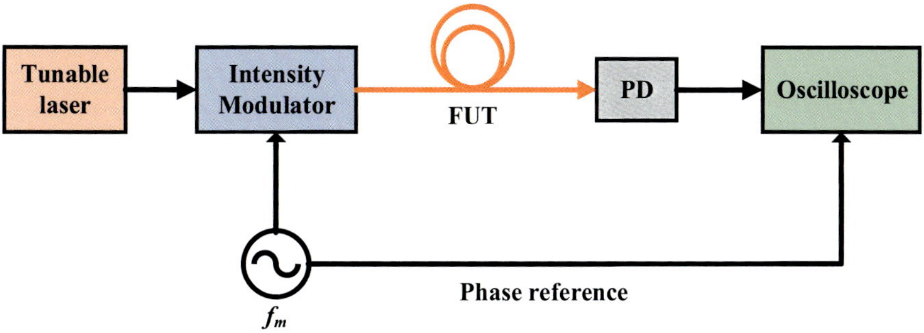

Fig. 7 Scheme of the phase shift method to measure fiber chromatic dispersion

signal. By adjusting the wavelength of the tunable laser, the RF phase delay as a function of the source wavelength $\phi = \phi(\lambda)$ can be obtained (Costa et al. 2007; Cohen 1985).

When the relative phase delay is smaller than 2π, the group delay versus optical signal wavelength can be expressed as

$$\Delta\tau_g(\lambda) = \frac{\phi(\lambda) - \phi(\lambda_r)}{2\pi f_m} \tag{20}$$

where $\phi(\lambda) - \phi(\lambda_r)$ is the RF phase difference measured between wavelength λ and a reference wavelength λ_r. As the definition of the chromatic dispersion coefficient, we can obtain

$$D(\lambda) = \frac{1}{L}\frac{d(\Delta\tau_g(\lambda))}{d\lambda} = \frac{1}{2\pi L f_m}\frac{d\phi(\lambda)}{d\lambda} \tag{21}$$

where L is the length of the fiber under test. This equation indicates that for a given fiber dispersion, the phase delay per unit wavelength change is linearly proportional to the modulation frequency $d\phi(\lambda)/d\lambda = 2\pi D(\lambda)Lf_m$. It also indicates that the value of $D(\lambda)$ is determined by the derivative of the group delay. The reference wavelength and the corresponding phase delay $\phi(\lambda_r)$ are not important. In practical measurement, a high modulation frequency f_m could increase the measurement accuracy. However, if the modulation frequency f_m is too high, the phase delay may be larger than 2π. In this case, the measurement system has to track the number of full cycles of 2π phase shift. To ensure enough measurement points in each 2π period, the tunable laser should be adjusted wavelength small enough.

AM Response Method

Due to the chromatic dispersion, different modulation sidebands may introduce different phase delays. The receiver can obtain their interference signal and measure the chromatic dispersion, which we called the baseband AM response method (Devaux et al. 1993; Christensen et al. 2002).

Figure 8 shows the scheme of the AM response method to measure fiber chromatic dispersion. A tunable laser is used as the light source. An electro-optic intensity modulator is used to modulate the output intensity of the tunable laser. An RF network analyzer is worked in S_{21} mode and provides a frequency-swept RF signal to modulate the output optical signal. The optical signal is detected by a wideband photodiode after propagate the test fiber. Then the weak RF signal is amplified and received by the network analyzer.

According to the AM modulation response, the photocurrent in the receiver is

$$I(f) \propto \cos\left[\frac{\pi\lambda_s^2 D(\lambda_s) f^2 L}{c} + \text{atan}(\alpha_{lw})\right] \qquad (22)$$

where α_{lw} is the modulator chirp parameter and f is the modulation frequency. The resonance zeroes at frequencies can be obtained by

$$f_k = \sqrt{\frac{c}{2D(\lambda_s) L \lambda_s^2}\left[1 + 2k - \frac{2}{\pi}\text{atan}(a_{lw})\right]} \qquad (23)$$

We need to eliminate the impact of modulator chirp of the transmitter to measure fiber chromatic dispersion. From Eq. 23, we can find that two consecutive zeroes in the AM response, f_k and f_{k+1}, are related only to the accumulated chromatic dispersion:

$$D(\lambda_s) = \frac{1}{L}\frac{c}{(f_{k+1}^2 - f_k^2)\lambda_s^2} \qquad (24)$$

Eqs. (2-24) indicate that the dispersion parameter can be measured directly by baseband AM response method, which is different from the phase shift method. It also has the advantage of high measurement accuracy because the AM response spectra are sharp. However, it needs an RF network analyzer which increases the cost.

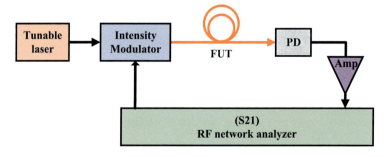

Fig. 8 Scheme of AM response method to measure fiber chromatic dispersion

Intermodal Dispersion Measurement

Compared to the chromatic dispersion in single-mode fiber, the intermodal dispersion in multimode fiber is much larger (typically >100 ps/km). Therefore, the measurement accuracy is usually not as straight as chromatic dispersion measurement. However, it is to be noted that the measurement value of the intermodal dispersion is just a statistic value because the mode coupling in multimode fibers is a random process. Intermodal dispersion of a multimode fiber can be measured in both time domain and frequency domain (Hernday 1998).

Time Domain Measurement

The output pulses will be broader than the input pulses due to the intermodal dispersion of the multimode fiber, and the amount of pulse broadening is linearly proportional to the intermodal dispersion of the multimode fiber. Therefore, the intermodal dispersion can be directly measured in time domain.

Figure 9 shows the scheme of time domain method to measure modal dispersion. An optical pulse that is generated by a laser diode is modulated by an electrical short pulse generated by a pulse signal source (Hernday 1998; IEC 60793-1-4; Hui and O'Sullivan 2009). A mode scrambler is used to provide a modal distribution that is independent of the optical source, which can improve the reproducibility of the measurements. Then the optical pulses are launched into the multimode fiber under test. A photodiode is used to convert the optical pulse into the electrical signal. An oscilloscope is used to receive the electrical signal.

It is to be noted that the output of the laser diode should not be directly launched into the multimode fiber under test. The primary purpose of a mode scrambler is to create a uniform, overfilled launch condition, which ensures that the measurement system has substantially the same emission conditions with different laser sources.

When the optical signal power is low enough, the measurement system works in linear regime. The relationship between the input waveform $A(t)$ and the output waveform $B(t)$ can be expressed by $B(t) = A(t) \otimes H(t)$, where $H(t)$ is the time domain transfer function of the fiber and $\otimes$ is the convolution operator. In multimode fiber, intermodal dispersion is much larger than the chromatic dispersion, so $H(t)$ can represent the intermodal dispersion of the multimode fiber. In realistic measurement, we usually use Gaussian function to represent the input and output signal. The

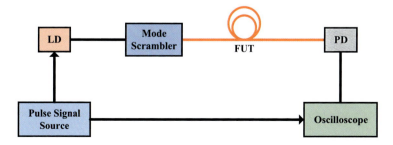

Fig. 9 Scheme of time domain method to measure modal dispersion

bandwidth of the multimode fiber can be calculated by the pulse width difference between the input signal $A(t)$ and the output signal $B(t)$. If the input pulse width is $\Delta\tau_i$ and the output pulse width is $\Delta\tau_0$, the response time of the fiber will be $\Delta\tau_H = \Delta\tau_0 - \Delta\tau_i$.

Frequency Domain Measurement

Although the time domain method is easily method to measure modal dispersion, a more straightforward way to find the transfer function of the fiber is in frequency domain (Hernday 1998; IEC 60793-1-4; Hui and O'Sullivan 2009). Converting the time domain waveforms $A(t)$ and $B(t)$ into frequency domains $A(\omega)$ and $B(\omega)$ by Fourier transform, the transform function of the fiber in frequency domain can be obtained by:

$$H(\omega) = \frac{B(\omega)}{A(\omega)} \qquad (25)$$

Figure 10 shows the scheme of frequency domain method to measure intermodal dispersion. A laser diode is used as the light source. An RF network analyzer is worked in S_{12} mode and provides a frequency-swept RF signal to modulate the intensity of the output optical signal. A mode scrambler is used to provide a modal distribution that is independent of the optical source, which can improve the reproducibility of the measurements. The optical signal is detected by a wideband photodiode after propagating the test multimode fiber. Then the optical signal converts into electrical domain and is received by the RF network analyzer. Since fiber dispersion can be thought of as a low-pass filter, the bandwidth limitation caused by the fiber under test can be obtained by comparing the difference in transfer functions of the system.

Both the time domain method and the frequency domain method are standard methods for bandwidth measurement. The time domain method is often used to characterize large-signal response, while the frequency domain method is only valid for small-signal characterization. As the nonlinear effect is usually negligible in the

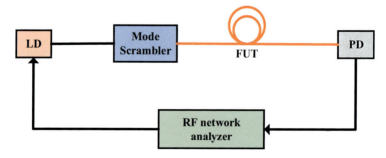

Fig. 10 Scheme of frequency domain method to measure modal dispersion

multimode fiber dispersion measurement, the measured result of the time domain method and the frequency domain method should be the same.

Polarization Characterization of Optical Fibers

State of Polarization (SOP)

Polarization is an important parameter in optics, especially in lasers and optical fiber telecommunications.

Most sources of light are classified as incoherent and unpolarized (or only partially polarized) because they consist of a random mixture of waves having different spatial characteristics, frequencies (wavelengths), phases, and polarization states. Light from many sources, such as the sun, flames, and incandescent lamps, is usually unpolarized light. It consists of shortwave trains with an equal mixture of polarizations. However, polarization is best understood by initially considering only pure polarization states and only a coherent sinusoidal wave at some optical frequency.

In free space, light is a transverse wave, where the wave motion is perpendicular to the direction of propagation. A single-frequency plane optical wave propagating in z direction is usually expressed as $E(z,t) = E\exp[-j(\omega_0 t - \beta_0 z)]$; here t is the time, ω_0 is the optical frequency, and β_0 is the propagation constant. The field oscillates in xoy plane. Polarization is a vector quantity describing the behavior of the electric field. When the phase difference between the horizontal and vertical components, E_x and E_y, is 0 or an integer multiple of π, the light wave is linearly polarized and the plane of oscillation is constant, which is shown in Fig. 11a, b. When the horizontal and vertical components have equal amplitude and their phase difference is an odd integer multiple of $\pi/2$, the light is circularly polarized, which is shown in Fig. 11c, d. When the phase shift between horizontal and vertical polarization components is not 0, $\pi/2$ and π, the light is elliptical polarized, which is shown in Fig. 11e. It is to be noted that circular or elliptical polarization can involve either a clockwise or counterclockwise rotation of the field. These correspond to distinct polarization states.

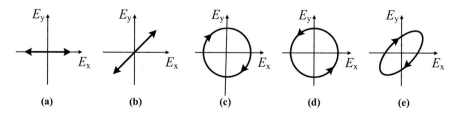

Fig. 11 Illustration of the polarization states

The orientation of the x and y axes used in description polarization state is arbitrary. One would typically choose axes to suit a particular problem such as x being in the plane of incidence. Since there are separate reflection coefficients for the linear polarizations in and orthogonal to the plane of incidence (p and s polarizations), that choice greatly simplifies the calculation of a wave's reflection from a surface. Moreover, one can use as basic functions any pair of orthogonal polarization states, not just linear polarizations. For instance, choosing right and left circular polarizations as basic functions simplifies the solution of problems involving circular birefringence.

Birefringence and Beat Length

In an ideal single-mode fiber, the cross section of the fiber is symmetry, which causes the degenerate modes in x and y directions to propagate at the same speed. However, in the manufacturing process of optical fibers, there are some birefringence resulting from differences in the core geometry from variations in the internal and external stresses and fiber bending. Then the fiber has an asymmetry cross section and introduces an effective refractive index difference in x and y axes, which caused two orthogonally polarized modes to propagate with different speeds. Assume β_x and β_y are the propagation constants for the slow mode and the fast mode, respectively. When the fiber cross section is independent of the fiber length L in z direction, the modal birefringence BF for the fiber is given by (Kaminow 1981):

$$B_F = \frac{\beta_x - \beta_y}{(2\pi/\lambda)} \tag{26}$$

where λ is the optical wavelength.

Due to the different phase velocities in the two orthogonal directions, there is a linear delay in z direction given by (Kaminow 1981): $\Phi(z) = (\beta_x - \beta_y)L$, assuming that the phase coherence of the two mode components is maintained. The phase coherence of the two mode components is achieved when the delay between the two transit times is less than the coherence time of the source.

However, when phase coherence is maintained, the state of polarization (SOP) of the optical signal will rotate when propagating in the fiber. The incident linear polarization which is at $\Phi = \pi/4$ with respect to the x axis becomes circular polarization at $\Phi = \pi/2$ and linear again at $\Phi = \pi$, then the above process continues through another circular polarization at $\Phi = 3\pi/2$ before returning to the initial linear polarization at $\Phi = 2\pi$. The characteristic length LB for this process corresponding to the propagation distance for which a 2π phase difference accumulates between the two modes is known as the beat length. It is defined as

$$L_B = \frac{\lambda}{B_F} = \frac{2\pi}{\beta_x - \beta_y} \tag{27}$$

In short fibers, birefringence is relatively simple where the refractive indices are slightly different on two orthogonal axes of the cross section of the fiber. If the fiber is long enough, the birefringence become more complex because bending, twisting, or nonuniformity of the fiber may rotate the birefringence axes. In addition, the two orthogonally polarized propagation modes couple to each other in the fiber. Both of the processes are random and unpredictable, which make polarization mode dispersion a complex problem to understand and to solve (Poole and Nagel 1997).

Polarization Mode Dispersion (PMD)

Polarization mode dispersion (PMD) is a special type of modal dispersion where two different polarizations of light in a waveguide, which normally travel at the same speed, travel at different speeds due to random imperfections and asymmetries, causing random spreading of optical pulses.

In an ideal optical fiber, the core has a perfectly circular cross section. In this case, the fundamental mode has two orthogonal polarizations that travel at the same speed. The signal that is transmitted over the fiber is randomly polarized.

In a realistic fiber, however, due to both intrinsic (caused by noncircular fiber core geometry and residual stresses in the glass material near the core region) and extrinsic (caused by stress from mechanical loading, bending, or twisting of the fiber) factors which in actual manufactured fibers break the circular symmetry cause the two polarizations to propagate with different speeds. In this case, the two polarization components of a signal will slowly separate, causing pulses to spread and overlap. Because the imperfections are random, the pulse spreading effects correspond to a random walk and thus have a mean polarization-dependent time-differential $\Delta\tau$, which is commonly referred to as differential group delay (DGD) of the fiber:

$$\Delta\tau = \frac{n_x - n_y}{c}L = \Delta n_{\text{eff}}L/c \tag{28}$$

where n_x and n_y are the effective refractive indices of the two polarization modes, L is the length of the fiber, and c is the speed of light. $\Delta n_{\text{eff}} = n_x - n_y$ is the effective differential refractive index. Δn_{eff} is typically on the order of 10^{-5} to 10^{-7} in coiled standard single-mode fibers.

The following are a few important parameters related to fiber PMD:

• Principal state of polarization (PSP)

DSP indicates two orthogonal polarization states corresponding to the fast and slow axes of the fiber. Under this definition, if the polarization state of the input optical signal is aligned with one of the two PSPs of the fiber, the output optical signal will keep the same SOP. In this case, the PMD has no impact in the optical signal, and the fiber only provides a single propagation delay. It is important to note that PSPs exist not only in "short" fibers but also in "long" fibers. In a long fiber, if higher-order PMD is negligible, although the birefringence along the fiber is random and there is energy coupling between the two polarization modes, an equivalent set of PSPs can always be found. Again, PMD has no impact on the optical signal if its polarization state is aligned to one of the two PSPs of the fiber. However, in practical fibers the orientation of the PSPs usually change with time, especially when the fiber is long. The change of PSP orientation over time is originated from the random changes in temperature and mechanical perturbations along the fiber.

Using Stokes parameter representation on the Poincare sphere, the two PSPs are represented as two vectors, which start from the origin and point to the two opposite extremes on the Poincare sphere as shown Fig. 12. In a short fiber, the PSP of the fiber is stable and independent of optical signal frequency. When the wavelength or frequency of the optical signal is changed, the SOP vector will rotate on the Poincare sphere in a regular circle around the PSP, which is shown in Fig. 12. Otherwise, the PSP is no longer stable in a long fiber because of the random mode coupling, and the regular circles shown in Fig. 12 will no longer exist.

- Differential group delay (DGD)

DGD indicates the propagation delay difference between the signals carried by the two PSPs. In general, DGD is a random process due to the random nature of the birefringence in the fiber. The probability density function (PDF) of DGD in an optical fiber follows a Maxwellian distribution:

$$P(\Delta\tau) = \sqrt{\frac{2}{7T}} \frac{\Delta\tau^2}{\alpha^3} \exp{-\left(\frac{\Delta\tau^2}{2\alpha^2}\right)} \tag{29}$$

where $P(\Delta\tau)$ is the probability that the DGD of the fiber $\Delta\tau$ and α is a parameter related to the mean DGD.

- Mean DGD

Fig. 12 Definition of principal state of polarization on Poincare sphere

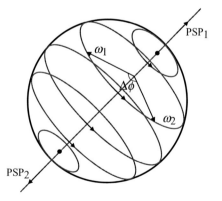

It is the average DGD; $\langle \Delta r \rangle$ is defined as

$$\langle \Delta r \rangle = \alpha \sqrt{\frac{8}{\pi}} \qquad (30)$$

The unit of mean DGD is in picoseconds; mean DGD in a fiber is usually counted as the average value of DGD over certain ranges of wavelength, time, and temperature. For short fibers, mean DGD is proportional to the fiber length L: $\langle \Delta \tau \rangle \propto L$, whereas for long fibers, the mean DGD is proportional to the square root of fiber length $\sqrt{L}$: $\langle \Delta \tau \rangle \propto \sqrt{L}$. This is due to the random mode coupling and the rotation of birefringence axes along the fiber. The mean DGD is an accumulated effect over fiber; therefore random mode coupling makes the average effect of PMD smaller. However, due to the random nature of mode coupling, the instantaneous DGD can still be quite high at the event when the coupling is weak.

• PMD parameter

It is defined as the mean DGD over a unit length of fiber. For the reasons we've discussed, for short fibers, the unit of PMD parameter is ps/km , whereas for long fibers, the unit of PMD parameter becomes ps/$\sqrt{km}$. Because of the time-varying nature of DGD, the measurement of PMD is not trivial.

In the following sections, we discuss several techniques that are often used to measure PMD in fibers.

PMD Measurement

Pulse Delay Method

The simplest DGD measurement method is to measure the differential delay of short pulses that are simultaneously carried by two polarization modes (Poole and Giles 1988). Figure 13 shows the scheme of pulse delay method to measure PMD. A mode-locked laser is used as a light source to provide short optical pulses. A polarization controller is used to adjust the polarization state of the optical pulses. Then the optical pulses were launched into the fiber under test and changed into electrical signal by a photodiode. An oscilloscope is used to receive the electrical signal. As mentioned above, DGD indicates the pulse propagation delay difference between the fast axis and the slow axis of the fiber. We can measure the fastest and the slowest arrival times of the optical pulses at the oscilloscope.

The pulse delay method is a time-domain technique that directly measures the DGD in a fiber, which is easy to understand. By adjusting the wavelength of the laser, the DGD versus wavelength can be measured. However, the accuracy of this measurement method depends on the temporal width of the optical pulses used. As the chromatic dispersion in the fiber will broaden and distort the optical pulses, the pulse delay method is suitable for the fibers with low chromatic dispersion and relatively high levels of DGD.

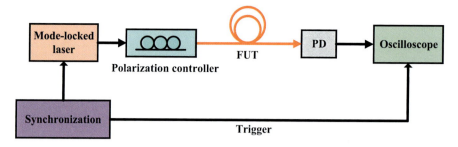

Fig. 13 Scheme of pulse delay method to measure PMD

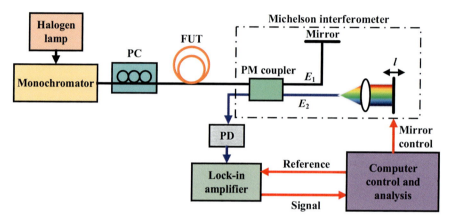

Fig. 14 Scheme of interferometric method to measure DGD

Interferometric Method

Because the pulse delay method requires short optical pulses and ultrafast detection, its application is limited. The interferometric method based on the low coherent interferometric technique can overcome this limitation (Gisin et al. 1991; Weid et al. 1987).

Figure 14 shows the scheme of interferometric method to measure PMD. Different from pulse delay method, a wideband light source is used instead of the laser. A monochromator is used to select the wavelength and the spectral bandwidth of the wideband light source, which determined the coherence length of the measurement system. The coherence length can be approximated as

$$\Delta l = \frac{\lambda^2}{2n\Delta\lambda} \tag{31}$$

where λ is the center wavelength, $\Delta\lambda$ is the spectral bandwidth, and n is the refractive index. When the spectral bandwidth $\Delta\lambda$ is large, the coherence length Δl is limited, which causes the interference only if the arm length difference is around zero.

When the light is passed through a birefringent optical fiber, the optical signal will be separated into the fast axis and the slow axis; thus, the optical field can be expressed as

$$E = E_x \exp(j\beta_x L) + E_y \exp(j\beta_y L) \qquad (32)$$

where x and y are the two orthogonal PSPs of the fiber, β_x and β_y are the propagation constants of the fast and the slow axis, E_x and E_y are the amplitude of the optical signals carried by these two orthogonal modes, and L is the fiber length. A photodiode is used to receive the interference signal from the Michelson interferometer. This interference signal can be expressed as

$$E_0 = \frac{1}{2}\left(E_x e^{j\beta_x L} + E_y e^{j\beta_y L}\right)\left(e^{j\beta l} + 1\right) \qquad (33)$$

where l is the differential delay between the two interferometer arms and β is the propagation constant in the interferometer arm. Then the intensity of the interference signal can be obtained

$$I = \eta|E_0|^2 = \frac{\eta}{4}\left|E_x\left(e^{j\beta_x L} + e^{j(\beta_x L + \beta l)}\right) + E_y\left(e^{j\beta_y L} + e^{j(\beta_y L + \beta l)}\right)\right|^2 \qquad (34)$$

If $l = 0$, there is an interference peak, which is caused by the self-mixing terms of $|E_x|^2$ and $|E_y|^2$. On the other hand, if one meets the requirement of $(\beta_y - \beta_x)L \pm \beta L = 0$ or approximately $l = \pm \Delta\tau c$, there are two secondary interference peaks. Here, $\Delta\tau$ is the DGD of the fiber under test, which can be measured by the location of the two secondary interference peaks.

There may be random mode coupling between the two polarization modes when the fiber is long, which caused the interference to have randomly scattered sidebands. In this case, the DGD value is generally a function of wavelength, and the measurement result is the mean DGD of the spectral bandwidth. Since the overall envelope of the energy distribution is usually represented by a Gaussian function, the mean DGD can be calculated as $\langle\Delta\tau\rangle = \sigma\sqrt{2/\pi}$, whereas the RMS value of DGD is $\langle\Delta\tau^2\rangle^{1/2} = \sqrt{3}\sigma/2$, where σ is the standard deviation of this Gaussian fitting (Hernday 1998).

Poincare Arc Method

PMD can also be measured by the frequency dependency of polarization rotation. If we use Stokes parameters to represent the SOP, the polarized optical vector $\vec{S}$ will rotate around the PSP on the Poincare sphere when the wavelength or frequency changed because of the birefringence of the fiber. In a short fiber, the PSP of the fiber is stable and independent of optical signal frequency. When the wavelength or frequency of the optical signal is changed, the SOP vector will rotate on the Poincare sphere in a regular circle around the PSP, which is shown in Fig. 15a. Otherwise, the PSP is no longer stable in a long fiber because of the random mode coupling. As

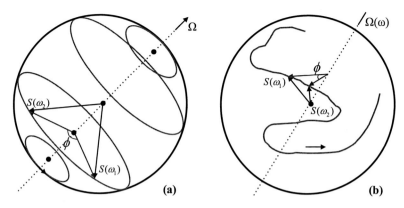

Fig. 15 Traces of signal polarization vector on Poincare spheres when optical frequency is varied (**a**) for a short fiber and (**b**) for a long fiber

shown in Fig. 15b, each frequency has its own PSP and SOP vector on the Poincare sphere, and the SOP vector rotated on the Poincare sphere in a random path when the frequency of the optical signal is changed.

For an infinitesimal change of the signal optical frequency, the SOP vector $\vec{S}$ versus optical signal frequency ω can be obtained by $\vec{S}$ versus ω. Obviously, the amount of this change is directly proportional to fiber birefringence (or PMD). For an infinitesimal change of the signal optical frequency, the relationship is

$$\frac{d\vec{S}}{d\omega} = \Omega \times \vec{S} \tag{35}$$

where Ω is the PMD vector originated from the center of the Poincare sphere and points toward the PSP. It is also convenient to use the scalar relationship (Hernday 1998):

$$\phi = \Delta\tau\Delta\omega \tag{36}$$

where ϕ is the angular change of the polarization vector $\vec{S}$ in radians on the plane perpendicular to the PSP, as shown in Fig. 15, and $\Delta\tau$ is the DGD between the two PSP components. Then the DGD of the fiber can be obtained by measuring the angular change ϕ.

Figure 16 shows the scheme of the DGD measurement using Poincare arc method. A tunable laser is used as a light source. A polarization controller is used to change the SOPs injected into the fiber under test. Adjust the tunable laser step by $\Delta\omega$. The polarimeter measured the Stokes parameters corresponding to each frequency of the light source $\vec{S}(\omega) = \vec{a}_x S_1(\omega) + \vec{a}_y S_2(\omega) + \vec{a}_z S_3(\omega)$, where

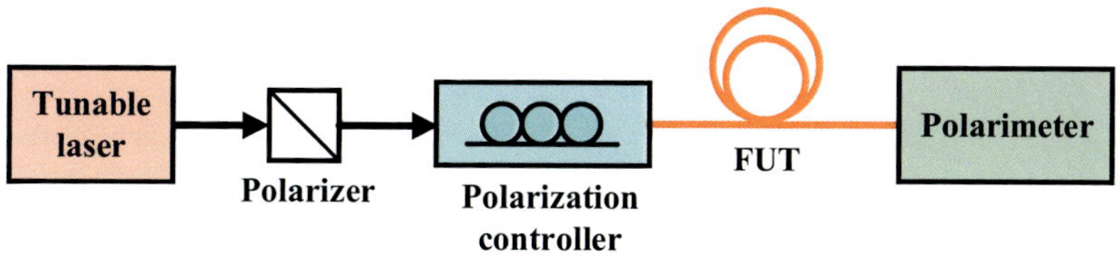

Fig. 16 Scheme of DGD measurement using a polarimeter

$\vec{a}_x$, $\vec{a}_y$, and $\vec{a}_z$ are unit vectors. The angular rotation ϕ can be evaluated by

$$\phi = \mathrm{acos}\left\{ \frac{\left[\vec{S}(\omega_1) \times \vec{\Omega}(\omega)\right]\left[\vec{S}(\omega_2) \times \vec{\Omega}(\omega)\right]}{\left|\vec{S}(\omega_1) \times \vec{\Omega}(\omega)\right\|\vec{S}(\omega_2) \times \vec{\Omega}(\omega)\right|} \right\} \tag{37}$$

Then the DGD value at this frequency can be obtained by:

$$\Delta\tau(\omega) = \frac{\phi}{\Delta\omega} \tag{38}$$

where $\omega = (\omega_1 + \omega_2)/2$ is the average frequency.

The Poincare arc method is easy to understand. However, it requires a polarimeter. To make this method easier to measure PMD in practical applications, a power meter or an OSA is used instead of the polarimeter (Poole and Favin 1994), as shown in Fig. 17.

As illustrated in Fig. 17a, another polarizer with a fixed polarization state is used to convert the SOP change into an optical power change; then a power meter is used to detect the optical power change. On the other hand, in Fig. 17b, a wideband source provides a broad optical spectrum, and an OSA is used to detect the output spectrum.

In the Poincare sphere representation, the power transfer function through a perfect polarizer can be expressed as

$$T(\omega) = \frac{P_{\mathrm{out}}}{P_{\mathrm{in}}} = \frac{1}{2}[1 + \hat{s}(\omega) \cdot \hat{p}] \tag{39}$$

where $\hat{s}(\omega)$ is the unit vector representing the polarization state of the input optical signal into the polarizer and $\hat{p}$ is the unit vector representing the high transmission state of the polarizer.

Due to the birefringence in the fiber, the power transmission efficiency $T(\omega)$ through the fixed polarization analyzer should be in range of [0,1]. If the birefringence orientation is random along the fiber, the transfer function $T(\omega)$ will have a uniform probability distribution between 0 and 1 with the average value

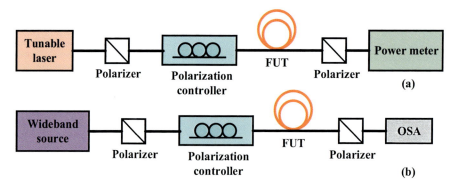

Fig. 17 Scheme of PMD measurement using a fixed analyzer method: (**a**) tunable laser and power meter combination; (**b**) wideband source and OSA combination

of 0.5. Then, we have

$$T'(\omega) = \frac{dT(\omega)}{d\omega} = \frac{1}{2}\left[\frac{d\hat{s}(\omega)}{d\omega} \cdot \hat{p}\right] \tag{40}$$

Using Eq. 35, it can be expressed as

$$T'(\omega) = \frac{1}{2}[(\Omega \times \hat{s}) \cdot \hat{p}] = \frac{1}{2}[(\hat{s} \times \hat{p}) \cdot \Omega] \tag{41}$$

According to Eq. 41, at the mean value of $T(\omega) = 0.5, \hat{s}(\omega) \cdot \hat{p}$ has to be equal to zero, which is equivalent to $\hat{s}(\omega) \times \hat{p} = \hat{1}$, where $\hat{1}$ is a unit vector and, therefore, $T'(\omega) = 0.5\left[\hat{1} \cdot \Omega\right]$.

We define a parameter γ_m to specify how often the random variable $T(\omega)$ crosses through its average value per unit frequency interval:

$$\gamma_m = \frac{1}{2}\left\langle\left|\hat{1} \cdot \Omega\right|\right\rangle = \frac{1}{2}\langle|\Omega|\rangle\langle|\cos\theta|\rangle \tag{42}$$

In a long fiber with truly random mode coupling, $\cos\theta$ is uniformly distributed in range of [1, 1]. Also, by definition, the magnitude of the PMD vector Ω is equal to the DGD of the fiber, $\Delta\tau = |\Omega|$, so $\gamma_m = 0.25\langle\Delta\tau\rangle$. In a long fiber with truly random mode coupling, $\cos\theta$ is uniformly distributed in range of [1, 1]. Also, by definition, the magnitude of the PMD vector Ω is equal to the DGD of the fiber, $\Delta\tau = |\Omega|$, so $\gamma_m = 0.25\langle\Delta\tau\rangle$. Within a frequency interval $\Delta\omega$, if the average number of crossovers is $\langle N_m\rangle$, we will have $\gamma_m = \langle N_m\rangle/\Delta\omega$, that is:

$$\langle\Delta\tau\rangle = 4\langle N_m\rangle/\Delta\omega \tag{43}$$

Sweep the optical frequency and count the number of transmission crossovers through its average value within a certain frequency interval to evaluate the average DGD. Another way to evaluate average DGD is to count the average number of extrema (maximum + minimum) within a frequency interval. Without further derivation, the equation to calculate the DGD can be found as (Poole and Favin 1994)

$$\langle \Delta\tau \rangle = 0.824\pi \frac{\langle N_e \rangle}{\Delta\omega} \tag{44}$$

where $\langle N_e \rangle$ is the number of extrema within a frequency interval $\Delta\omega$.

Jones Matrix Method

As we know, Jones Matrix is widely used in polarization optics, which represents the polarization state of a light or the transfer matrix of an optical device. Based on Jones Matrix, we can also measure the PMD parameters. As illustrated in Fig. 18, the setup of the Jones Matrix method is similar to the Poincare arc method. DGD of the fiber under test can be measured by adjusting the SOP of the input optical signal and measuring the corresponding response of the output optical signal by a polarimeter (Heffner 1992, 1993).

The transfer function of a passive optical device can be written as a four-element matrix as

$$M = \begin{bmatrix} M_{11} & M_{12} \\ M_{21} & M_{22} \end{bmatrix} \tag{45}$$

if the input and output Jones vector of the optical signal are $\vec{E}_{\text{in}} = \begin{bmatrix} E_{x,\text{in}} \\ E_{y,\text{in}} \end{bmatrix}$ and $\vec{E}_{\text{out}} = \begin{bmatrix} E_{x,\text{out}} \\ E_{y,\text{out}} \end{bmatrix}$, respectively. The Jones Matrix of the optical device must be:

$$\begin{bmatrix} E_{x,\text{out}} \\ E_{y,\text{out}} \end{bmatrix} = \begin{bmatrix} M_{11} & M_{12} \\ M_{21} & M_{22} \end{bmatrix} \begin{bmatrix} E_{x,\text{in}} \\ E_{y,\text{in}} \end{bmatrix} \tag{46}$$

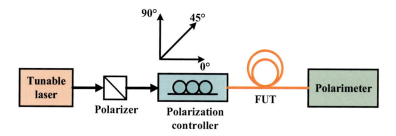

Fig. 18 Scheme of the Jones Matrix method to measure fiber PMD

If the light passed through a fiber, the SOP of output optical signal is a function of the optical frequency ω because of the birefringence in the fiber. The Jones Matrix of the fiber should be a function of the optical frequency ω, then we have

$$M(\omega + \Delta\omega) M^{-1}(\omega) = \begin{bmatrix} M_{11} & M_{12} \\ M_{21} & M_{22} \end{bmatrix} \tag{47}$$

The fiber DGD can be expressed as the group delay difference as the

$$\Delta\tau(\omega) = |\tau_{g1} - \tau_{g2}| = \left| \frac{\text{Arg}(\rho_1/\rho_2)}{\Delta\omega} \right| \tag{48}$$

Here, $\text{Arg}(\rho_1/\rho_2)$ denotes the phase angle of (ρ_1/ρ_2). ρ_1 and ρ_2 are two eigenvalues that can be calculated with

$$\rho_{1,2} = \frac{(m_{11} + m_{22}) \pm \sqrt{(m_{11} + m_{22})^2 + 4(m_{12}m_{21} - m_{11}m_{22})}}{2} \tag{49}$$

Equation 48 emphasizes the principle of Jones Matrix method for fiber DGD measurement. We can measure DGD in the following steps:

1. Measure Jones Matrix elements at different optical signal frequencies: $M(\omega_1)$, $M(\omega_2)$, $M(\omega_3)$, and so on.
2. Convert the results of step 1 into a series of new matrices shown in Eq. 47 for different wavelengths.
3. Find the fiber DGD value at each optical frequency using Eqs. 48 and 49.

It is to be noted that a suitable frequency step $\Delta\omega$ is needed for the measurement. For a fiber system, it is better to choose $\Delta\omega < \pi/(4\Delta\tau)$ (the corresponding wavelength step $\Delta\lambda < \lambda^2/(8c\Delta\tau)$). However, if $\Delta\omega$ is too small, the measurement time will be increased, and the result will be strongly affected by any fiber path instability, wavelength inaccuracy of laser source, or polarimeter accuracy, which are not expected.

Mueller Matrix Method

The polarization characteristics of an optical device can also be expressed by a Mueller Matrix. If the input polarization vector is $\vec{S}_{\text{in}}$ and the corresponding output polarization vector is $\vec{S}_{\text{out}}$, we have

$$\begin{bmatrix} S_{\text{out}1} \\ S_{\text{out}2} \\ S_{\text{out}3} \end{bmatrix} = \begin{bmatrix} m_{11} & m_{12} & m_{13} \\ m_{21} & m_{22} & m_{23} \\ m_{31} & m_{32} & m_{33} \end{bmatrix} \begin{bmatrix} S_{\text{in}1} \\ S_{\text{in}2} \\ S_{\text{in}3} \end{bmatrix} \tag{50}$$

Here, $M = [mij]$ is the Mueller Matrix. Based on Mueller Matrix, we can also measure the PMD parameters. Figure 19 shows the scheme of the Mueller Matrix to measure PMD. As illustrated in Fig. 19, the difference between the Jones Matrix method and Mueller Matrix method is only two independent linear input polarization states (not need two orthogonal polarization states) are required.

Assume the measurements at two frequencies ω_1 and $\omega_2 = \omega_1 + \Delta\omega$, their Mueller matrices are $[M(\omega_1)]$ and $[M(\omega_2)]$, and they should be satisfied as

$$[M_\Delta] = [M(\omega_2)][M(\omega_1)]^T \tag{51}$$

The rotational matrix, $[M_\Delta]$, can be determined through the measurement using the setup shown in Fig. 19. In fact, the rotational matrix $[M_\Delta]$ can be defined as a spatial vector, which is described by a rotation angle ϕ around an axis $\vec{r}$. The explicit forms relating $[M_\Delta]$ with the rotation angle ϕ around an axis $\vec{r}$ are

$$
\begin{aligned}
\cos\phi &= \frac{1}{2}(Tr([M_\Delta]) - 1) \\
r_1 &= \frac{m_{\Delta 23} - m_{\Delta 32}}{2 \sin\phi} \\
r_2 &= \frac{m_{\Delta 31} - m_{\Delta 13}}{2 \sin\phi} \\
r_3 &= \frac{m_{\Delta 12} - m_{\Delta 21}}{2 \sin\phi}
\end{aligned}
\tag{52}
$$

where $Tr([M_\Delta])$ is the trace of the matrix $[M_\Delta]$.

The procedure of the Mueller Matrix technique can be summarized in the following steps:

1. Measure Mueller Matrix elements at different optical signal frequencies $M(\omega_1)$ and $M(\omega_2)$.
2. Derive the rotation matrix $[M_\Delta]$ using Eq. 51, and then obtain the rotation angle ϕ and the orientation of the PMD vector $\vec{r}$ based on Eq. 52.
3. Find the fiber DGD based on $\Delta\tau = \phi/\Delta\omega$.

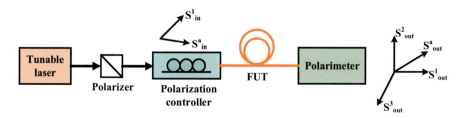

Fig. 19 Scheme of the Mueller Matrix method to measure fiber PMD

Compared to the Jones Matrix method, Mueller Matrix method measures the frequency-dependent orientation of the PMD vector $\vec{r}$, and the frequency step size is not very strict. In addition, Mueller Matrix method can also be used to measure polarization-dependent loss (PDL) in an optical fiber, which will be described in detail in the next section.

Polarization-Dependent Loss (PDL)

A PMD-related effect is polarization-dependent loss (PDL), in which two polarizations suffer different rates of loss in the fiber due to asymmetries.

The power transfer function of an optical device depends on the SOP of the optical signal. Generally speaking, the PDL of the optical fiber itself is very small, and the PDL of the whole optical system depends on the PDLs of all the optical devices in the system. The PDL of each device in the system is easily determined. However, the overall system PDL cannot be obtained by simply adding up the PDL of each device. It is because the two orthogonal polarization directions of each device is randomly and the presence of PMD in the fiber.

Fig. 20 shows the scheme of PDL measurement of an optical fiber system. A polarization controller is used to adjust the SOP of the optical signal injected into the device under test (DUT). It is to be noted that the polarization controller should be repeatable very well, or another polarimeter is required to monitor the SOPs of the optical signal. Then a photodetector is used to measure the output optical power. With the scanning of the signal SOP across the Poincare sphere, the difference in the maximum and the minimum value indicates the PDL of the DUT, which can be expressed as

$$\text{PDL} = \frac{T_{\max} - T_{\min}}{T_{\max} + T_{\min}} \tag{53}$$

where $T_{\max}$ and $T_{\min}$ are the maximum and the minimum power transmission efficiencies of the DUT. If the signal optical power P_{in}, entering the DUT, is polarization-independent, it can be written as

$$\text{PDL} = \frac{P_{\max}/P_{\text{in}} - P_{\min}/P_{\text{in}}}{P_{\max}/P_{\text{in}} + P_{\min}/P_{\text{in}}} = \frac{P_{\max} - P_{\min}}{P_{\max} + P_{\min}} \tag{54}$$

where $P_{\max}$ and $P_{\min}$ are the maximum and the minimum optical powers received by the photodetector, respectively. However, it is impossible to scan the SOP to cover

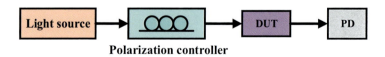

Fig. 20 Scheme of polarization-dependent loss (PDL) measurement

the Poincare sphere in real measurement. Therefore, we use a 4×4 Mueller Matrix to describe PDL as following:

$$\begin{bmatrix} s_{out0} \\ s_{out1} \\ s_{out2} \\ s_{out3} \end{bmatrix} = \begin{bmatrix} m_{11} & m_{12} & m_{13} & m_{14} \\ m_{21} & m_{22} & m_{23} & m_{24} \\ m_{31} & m_{32} & m_{33} & m_{34} \\ m_{41} & m_{42} & m_{43} & m_{44} \end{bmatrix} \begin{bmatrix} s_{in0} \\ s_{in1} \\ s_{in2} \\ s_{in3} \end{bmatrix} \tag{55}$$

where $S_{in} = [s_{in0} \ s_{in1} \ s_{in2} \ s_{in3}]^{T}$ and $S_{out} = [s_{out0} \ s_{out1} \ s_{out2} \ s_{out3}]^{T}$ are the Stokes parameter of the input and output optical signal, respectively. If the signal is polarized, the input and output optical power are $s_{in0} = \sqrt{s_{in1}^2 + s_{in2}^2 + s_{in3}^2}$ and $s_{out0} = \sqrt{s_{out1}^2 + s_{out2}^2 + s_{out3}^2}$, respectively.

The system PDL with the Mueller Matrix elements is (Hui and O'Sullivan 2009; Hentschel and Schmidt)

$$PDL = \frac{\sqrt{m_{12}^2 + m_{13}^2 + m_{14}^2}}{m_{11}} \tag{56}$$

and the elements of the Mueller Matrix are

$$m_{11} = \frac{1}{2}\left(\frac{P_1}{P_a} + \frac{P_2}{P_b}\right)$$

$$m_{12} = \frac{1}{2}\left(\frac{P_1}{P_a} - \frac{P_2}{P_b}\right)$$

$$m_{13} = \frac{P_3}{P_c} - \frac{1}{2}\left(\frac{P_1}{P_a} - \frac{P_2}{P_b}\right) \tag{57}$$

$$m_{14} = \frac{P_4}{P_d} - \frac{1}{2}\left(\frac{P_1}{P_a} - \frac{P_2}{P_b}\right)$$

The meaning of P_i is shown in Table 1. Using the experimental setup shown in Figure and four well-defined SOPs of the input optical signal, we can measure the output optical power by the photodetector (Hentschel and Schmidt).

It is worthwhile to note that both PMD and PDL are system properties that should be independent of the optical signal that is launched into the system. However, the

Table 1 Input signal SOP setup for PDL measurement (Hentschel and Schmidt)

Input signal SOP	Input Stokes vector	Measured output power
Linear horizontal (0°)	$S_{in,a} = [P_a P_a \ 0 \ 0]$	$P_1 = m_{11}P_a + m_{12}P_a$
Linear vertical (90°)	$S_{in,b} = [P_b - P_b \ 0 \ 0]$	$P_2 = m_{11}P_b - m_{12}P_b$
Linear diagonal (45°)	$S_{in,c} = [P_c \ 0 \ P_c \ 0]$	$P_3 = m_{11}P_c + m_{13}P_c$
Right-hand circular	$S_{in,d} = [P_d \ 0 \ 0 \ P_d]$	$P_4 = m_{11}P_d + m_{14}P_d$

impact of PMD and PDL on the performance of an optical system may depend on the polarization state of the input optical signal.

Special Characterization Techniques

Material Characterization

Before some properties of active fibers will be measured more in detail, microstructures of the material should be first characterized, because on the basis of understanding of the relationship between the organizational structure and the performance of the material, the microstructure formation conditions can be controlled by a certain processing method, so that it forms the desired organizational structure to achieve the desired performance. For the active fibers, the properties mightily depend on their condition, such as whether the fiber is crystallized and the distribution of elements is uniform. These characteristics can be described by some material analysis methods, shown in Fig. 21.

Physical Analysis of Active Fiber

For the physical analysis, cross-sectional view and refractive index profile are usually used to evaluate the quality of almost every kind of fibers, such as single-mode fibers, multi-core fibers, hollow-core fibers, photonic crystal fibers, doped fibers, etc. The cross-sectional images are usually obtained by a scanning electron microscope (SEM) or even a transmission electron microscope (TEM). Figure 22a shows the structural block diagram of SEM. It is consisted by three parts, electronic optics system (electron gun, electromagnetic lens, scanning coil, sample room), signal processing system (signal detector, signal amplification, kinescope, sweep generator), and vacuum system. It uses the various physical signals stimulated by the fine focused electron beam scanning in the sample surface to image. Figure 22b demonstrates the SEM image of a well-cut bismuth-doped holey-fiber cross section (Zlenko et al. 2011). The structure can be clearly observed.

TEM will play a protagonist if more specified information is needed. It is a high-resolution, high-magnification electronic optical instrument that used very short electron beam as a lighting source, with electromagnetic lens focus imaging. It is consisted by electronic optics system, power supply and control system, and vacuum system. Electronic optics system is the kernel, shown in Fig. 23a; the electron beam emitted by the electron gun passes through the condenser and converges into a thin, bright, and uniform spot and irradiates on the sample in the sample stage; then electron beam carries the structural information of sample projects on the phosphor screen after zooming. The quality of samples influences results strongly; high-quality samples are particularly important. There are two methods for sample preparation of optical fibers. First, a region of an optical fiber is thinned to less than 100 nm using ion beam milling. Second, optical fiber is grinded to nanoscale, then the nanopowder is put in alcohol and vibrated for at least an hour by ultrasonic, and the supernate is dropped to a copper grid and put on the sample stage of TEM.

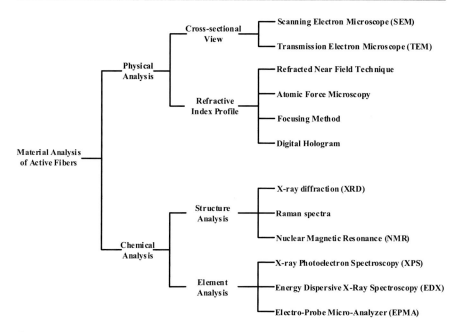

Fig. 21 Material analysis of active fibers

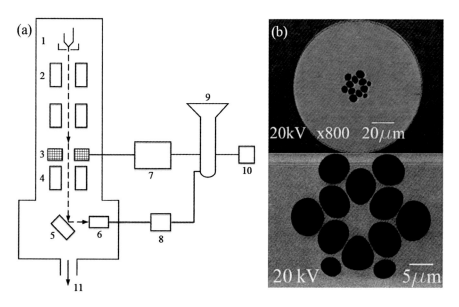

Fig. 22 (**a**) Structure schematic diagram of SEM: 1, electron gun; 2, 4, electromagnetic lens; 3, scanning coil; 5, sample room; 6, signal detector; 7, 8, signal amplification and processing; 9, kinescope; 10, sweep generator; 11, vacuum system. (**b**) Cross-sectional images of bismuth-doped holey fiber

For example, the size of quantum dots (QDs) must be known in the QD-doped glass fibers; Fig. 23b–d demonstrates TEM images and size distribution of these glass fibers. Quasi-spherical PbS QDs dispersed in the glass fiber with average size of ~2.61 nm for fiber FB (as the obtained fibers were about 350 m long in total, three parts were selected as the fiber samples corresponding to different fiber length abscissas: beginning of drawing for 0–15 m, middle for 150–165 m, and end for 300–315 m, and each was named as FB, FM, and FE for beginning, middle, and end of drawing, respectively), ~2.55 nm for fiber FM, and ~2.36 nm for fiber FE. The size distribution of QDs became smaller and narrower as the fiber samples were collected along different fiber abscissas: beginning, middle, and end of drawing (Huang et al. 2017).

Refractive index profile (RIP) is another factor to assess the quality of fibers. It contacts some parameters which reveal the properties of light transition, such as mode, fiber bandwidth, loss, and dispersion. Based on different principles, there are variety of different refractive index measurement methods, such as refraction near-field method (White 1979), atomic force etching method (Huntington et al. 1997), focusing method (Saekeang et al. 1980), and holographic measurement method (Wahba and Kreis 2009). One of the most reliable methods is the refraction near-field method, and the famous S14 refractive index analyzer is based on this principle.

Chemical Analysis of Active Fiber

As for the chemical analysis of active fibers, it is significant to know the structure and state of doped elements. X-ray diffraction (XRD), Raman spectra, and nuclear magnetic resonance (NMR) are usually described the properties of crystallization, molecular radicals, and chemical bond, respectively.

XRD is usually consisted by X-ray source, goniometer, radiation detector, and recording system; goniometer is the core and shown in Fig. 24a. Convergent beam will be obtained when a divergent X-ray is irradiated onto the sample satisfying the Bragg relationship; a receiving slit mounted on the holder rotates around the O at the same time as the counter tube; a reflected light can be received when the counter tube is rotated to the proper position read from the scale. Both optical fibers and fiber powders can be used as test samples. Figure 24 shows the XRD pattern of bismuth- and erbium-doped optical fibers under different heating temperatures for 2 h (Hao et al. 2018). The crystallizing phase of SiO_2 is clearly observed.

Raman spectroscopy is a scattering spectrum. The information of molecular vibration and rotation is obtained by comparing the different frequencies of incident and scattered light. Raman spectrometer is usually consisted by source (laser), outside the light road (sample room, lens), dispersion system (monochromator), receiving system (detector), and signal processing and display system (signal transformation and output), shown in Fig. 25a. The Raman signal excited by a laser is processed by monochromator and then converted to the data recorded by a computer. Both optical fibers and fiber powders can be used as test samples. For example, there are five major peaks located at ~140, 367, 721, 914, and 1324 cm^{-1} in the Raman spectra of bismuth borosilicate glasses, respectively, shown in Fig. 25b. The strongest peak (130 cm^{-1}) attributes to the vibrational

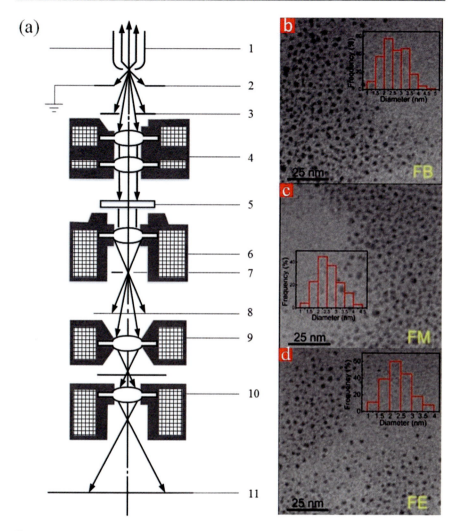

Fig. 23 (**a**) Structure schematic diagram of TEM: 1, electron gun; 2, anode; 3, 7, 8, aperture; 4, 6, 9, 10, lens; 5, sample stage; 11, fluorescent screen. (**b–d**) TEM images of the PbS QD-doped glass fiber FB, FM, and FE

mode of Bi^{3+} in [BiO_6] octahedron. The 367 cm^{-1} peak is caused by the distorted vibration of Bi-O-Bi and Si-O-Si in [BiO_6] and [SiO_4]. The 721 cm^{-1} band is due to the distorted vibration of B-O-B in [BO_3] and [BO_4]. The vibration of Bi-O-Si accounts for the peak located at 914 cm^{-1}. The 1324 cm^{-1} peak is induced by the asymmetric telescopic vibration of B-O$^-$ in [BO_3] triangle and [BO_4] tetrahedron (Chu et al. 2016).

The NMR is originated from the movement of angular momentum of the nucleus under an additional magnetic field. It is widely used in structural confirmation, thermodynamics, dynamics, and reaction mechanism. Figure 26a demonstrates structure schematic diagram of NMR; it is consisted by magnet, radio-frequency generator, probe detector, scanning unit, signal processor, and computer system. The NMR data is got by processing absorption and dispersion in frequency domain obtained by the samples under a single radio-frequency field. Optical fiber powders can be used as test samples. Figure 26b demonstrates the ^{31}P NMR analyses of polyphosphate glasses. For [ZnO] of 10–35 mol.%, Q^2 and Q^3 phosphate species are dominating, indicating a network of interconnected chains. For even higher ZnO, the further appearance and later dominance of Q^0 species indicate rapid depolymerization of the phosphate network into an assembly of phosphate and zinc oxide tetrahedrons with less and less localized interconnectivity (Tan et al. 2016).

X-ray photoelectron spectroscopy (XPS), energy dispersive X-ray spectroscopy (EDX), and electron probe microanalysis (EPMA) are always used for element analysis; XPS is one of the most widely used surface analysis methods and aims to identify the valence state of element, EDX and EPMA, often with a SEM or TEM, and focus on the component and distribution. Schematic diagram of XPS is shown in Fig. 27a. A well-cut optical fiber is used as test sample. When a tested sample irradia by X-ray, photoelectrons produced by photoionization are transported from the place where they are produced to the surface, then offsets work function and emits. This is the three-step process of X-ray photoelectron emission. Analysis of photoelectron function by energy analyzer is the X-ray photoelectron spectroscopy. The XPS measurement clearly shows both Ce^{3+} and Ce^{4+} can coexist in bismuth borosilicate glasses, shown in Fig. 27b; the XPS spectra demonstrate that trivalent Ce^{3+} dominates in the samples, as confirmed by the resemblance of the XPS spectra to that of the standard reference CeF_3 crystal and the absence of the characteristic doublet related to Ce^{4+} at ~916 eV (Chu et al. 2016).

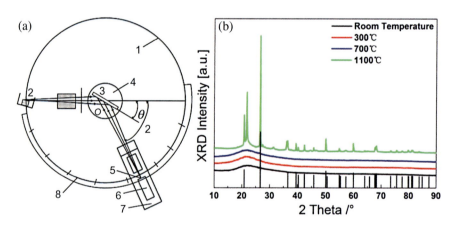

Fig. 24 (**a**) Structure schematic diagram of XRD: 1, goniometer round; 2, X-ray source; 3, samples; 4, sample stage; 5, receiving slit; 6, counter tube; 7, holder; 8, calibrated scale. (**b**) XRD patterns of bismuth- and erbium-doped optical fibers. The standard JCPDF card no. 46-1045 for quartz

Fig. 25 (**a**) Structure schematic diagram of Raman spectrograph: 1, laser; 2, samples room; 3, lens; 4, monochromator; 5, detector; 6, signal transformation; 7, output. (**b**) Raman spectra of bismuth borosilicate glasses doped with different concentrations of CeO_2

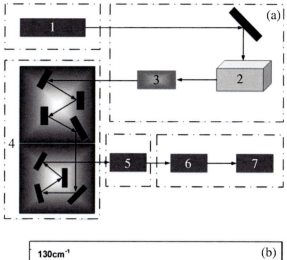

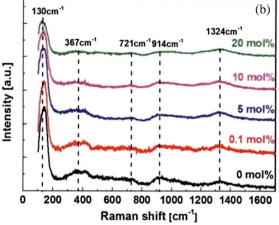

EDX is achieved by analyzing the elemental characteristic X-ray wavelength and intensity of the sample, and the relative content of the element is determined according to the strength of the element contained in the sample. Commonly used EDX detector is silicon lithium detector shown in Fig. 28a. When the characteristic X-ray photon enters the silicon lithium detector, the silicon atoms are ionized to produce several electron-hole pairs, which are proportional to the energy of the photon. These electron-hole pairs are collected by bias and then passed through a series of converters and then turned into voltage pulses to supply the multi-pulse height analyzer and count the number of pulses per energy band in the energy spectrum. A well-cut optical fiber is used as test samples. Figure 28b–d provides the EDX compositional profile in elemental atom percentages, across the bismuth- and erbium-doped fiber (BEDF) core, corresponding to the point shown in this figure. The result shows that the fiber core contains Bi, Al, Ge, and Er (Peng et al. 2013).

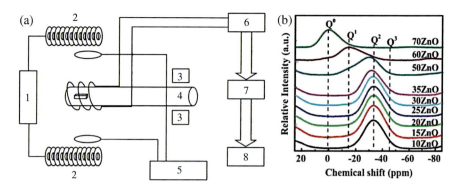

Fig. 26 (**a**) Structure schematic diagram of NMR: 1, scanning element; 2, magnet pole; 3, rotors; 4, sample tube; 5, radio-frequency generator; 6, multiprobe detector; 7, signal processor; 8, output. (**b**) 31P NMR spectra of polyphosphate samples

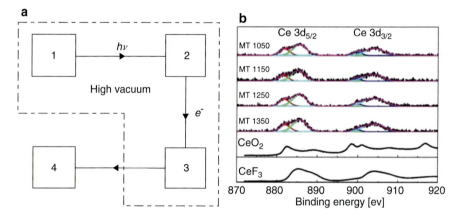

Fig. 27 (**a**) Structure schematic diagram of XPS: 1, X-ray source; 2, samples; 3, energy analyzer; 4, micro-signal retrieval and data processing. (**b**) XPS of Ce 3d spin orbit doublet of bismuth borosilicate glasses doped with 5 mol. % of CeO_2 but melted at different melting temperatures (MT)

The function of EPMA is mainly for microanal component analysis. It is a high-efficiency analysis equipment based on the electronic optics and X-ray spectroscopy. The structure schematic diagram of EPMA is shown in Fig. 29a, characteristic X-ray generated by electron beam focusing on the surface of samples, analyzing the wavelength (wavelength dispersive X-ray spectroscopy) and intensity (energy dispersive X-ray spectroscopy) of characteristic X-ray to know the variety and content of elements, respectively. A well-cut optical fiber can be used as test samples. Elemental analysis was conducted by EPMA mapping to present the spatial element distribution in the cross section of fiber FB, shown in Fig. 29b–g. As can be seen in the backscattered electron image of fiber FB (Fig. 29b), the fiber and fiber core were both fine circles with diameter about 125.5 and 18.1 μm, respectively,

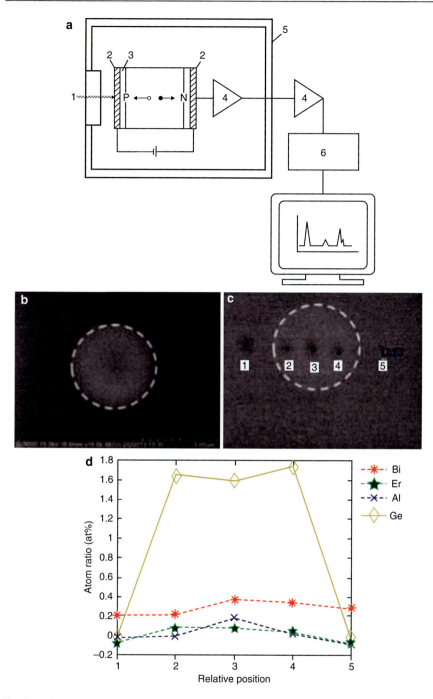

Fig. 28 (**a**) Structure schematic diagram of EDX: 1, X-ray source; 2, gold electrode layer; 3, silicon electrode layer; 4, amplifier; 5, liquid-nitrogen cooling system; 6, multi-channel analyzer. (**b**) SEM image of the core region of BEDF; (**c**) SEM image and EDX sampled points of the BEDF; (**d**) Radial elemental composition profile of the BEDF

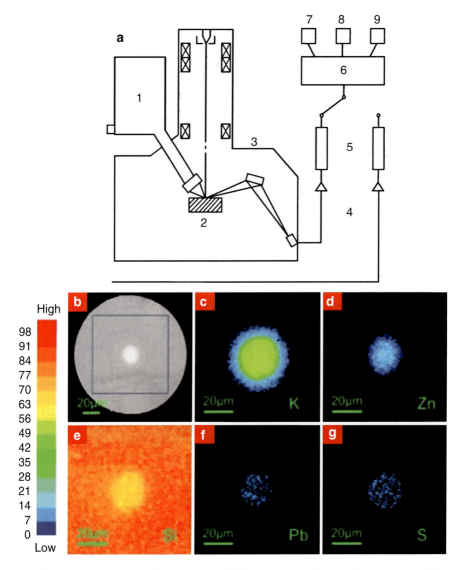

Fig. 29 (**a**) Structure schematic diagram of EPMA: 1, energy dispersive X-ray spectroscopy; 2, sample stage; 3, wavelength dispersive X-ray spectroscopy; 4, 5, amplifier; 6, multi-channel analyzer; 7, 8, 9, output. (**b**) Backscattered electron image of cross section of fiber FB. (**c–g**) The EPMA images of different elements from the fiber FB cross section

which matched well with the original proportion of core-cladding in preform. As a result, the fiber dimensions including the fiber and core diameter can be well preserved along the fiber length. Figure 29c–g shows images of distribution in fiber FB cross section of K, Zn, Si, Pb, and S elements. It can be seen that all the elements exhibited diffusion to some extent (Huang et al. 2017).

Spectral Characterization

Absorption and Emission Measurement

Fiber loss, which mainly contains absorption loss and scattering loss, is one of the most important characteristics of optical fibers. The absorption loss is caused by the absorbing light of the optical fiber's materials and impurities, while the scattering loss is caused by the scattering effect and changing transmission direction internal the fiber. In terms of active fiber, the doped rare earth elements have a high absorption at the specific wavelength, so the absorption loss can be considered as the total loss of the optical fiber. The absorption coefficient of active fiber is calculated by measuring the absorption loss at the wavelength of pump light. Usually, we measure it by the cutback method.

For spectral emission measurement, usually the co- and counter-pumping schemes are employed. The counter-pumping is preferred since that, in the co-pumping scheme, the residual pump power could be significant and produces a background noise mixed with the emission spectra to be tested. However, when the emission spectra are broad, e.g., over 200 nm (Dvoyrin et al. 2006) or even wider, it is difficult to find a very broadband WDM coupler needed in the counter-pumping scheme. In addition, the selection of the active fiber length is always an issue because the emission spectra could be changed with fiber lengths due to the spectral absorption (Becker et al. 1999). So a short section of active fiber is normally used to minimize the effect in order to observe the original emission spectra.

Absorption Spectrum Measurement Based

Cutback Method

The cutback technique is a direct application of the definition in which the power levels $P_1(\lambda)$ and $P_2(\lambda)$ are measured at two points of the fiber without change of input conditions. $P_2(\lambda) = P_0 e^{-\alpha z}$ is the power emerging from the far end of the fiber, and $P_1(\lambda) = P_0 e^{-\alpha(z - L)}$ is the power emerging from a point near the input after cutting the fiber.

The scheme of the cutback method is shown in Fig. 30. A lamp source is usually used in the measurement for the absorption spectrum of the active fiber. Chopped white light from a lamp was launched into one end of a length of active fiber by passing a mode filter, and the output power $P_2(\lambda)$ was analyzed with a monochrometer (FWHM −2 nm). Keeping the launching conditions fixed, the fiber is cut to the cutback length L. The cladding mode filter is refitted and the output power $P_1(\lambda)$ from the cutback length is recorded. The absorption spectrum of the active fiber can be evaluated by:

$$\alpha(\lambda) = \frac{\ln\left[P_2(\lambda)/P_1(\lambda)\right]}{L} \qquad (58)$$

In this measurement, the light source should be stable in position, intensity, and wavelength over a time period sufficiently long to complete the measurement. The

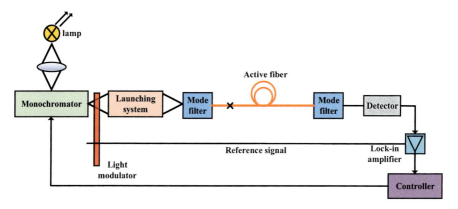

Fig. 30 Scheme of absorption spectrum measurement based on cutback method

spectral linewidth (FWHM) should be specified such that the linewidth is narrow compared with any features of the fiber spectral absorption. A lamp source is usually used in the measurement for the absorption spectrum of the active fiber. A chopper is used to modulate the light source in order to improve the signal/noise ratio (SNR) at the receiver.

If such a procedure is adopted, the detector should be linked to a signal processing system synchronous with the source modulation frequency. The detecting system should be substantially linear in sensitivity. A cladding mode filter encourages the conversion of cladding modes to radiation modes. The spectral response should be compatible with spectral characteristics of the source. The detector must be uniform and have linear sensitivity characteristics.

The cutback method is a simple, direct, and accurate measurement method, which is usually employed for absorption spectrum measurement. However, it is a destructive measurement method because of cutting the fiber. On the other hand, its measurement accuracy, determined by the repeatability of power transmitted or coupled before and after the fiber cutting, can be easily affected by the uncertainty in cutting, re-splicing, or realigning fibers – especially research-purpose active optical fibers with nonconventional materials and small core geometries.

Backward Method

Figure 31 shows the experimental system using an optical spectral analyzer (OSA) to record the backward luminescence spectra of our Bi/Er co-doped fiber (Zhang et al. 2013). An 830 nm laser diode, connected to an 810/1310 nm WDM coupler, offers a maximum ~60 mW power launched into the Bi/Er fiber. The Bi/Er co-doped fiber is spliced with lead-in single-mode fiber with a splice loss of ~1 dB because their mode fields are not matched. A power meter is used to monitor the left pump power and useful for the luminescence analysis later.

The luminescence spectrum emitted from a 3 m-long Bi/Er co-doped fiber ($[Er_2O_3]$ ~0.01, $[Al_2O_3]$ ~0.15, $[Bi_2O_3]$ ~0.02, $[P_2O_5]$ ~0.94, and $[GeO_2]$

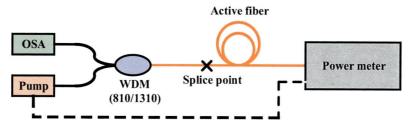

Fig. 31 Experimental scheme for luminescence measurement

~12.9 mol.%) under ~60 mW pump power and measured by the OSA with a 5 nm bandwidth is shown in Fig. 32a. The red and green curves correspond to the directly measured and the true (corrected) emission spectra, respectively. The true spectrum is determined correctly from the directly measured by compensating the spectral transmission of the 810/1310 nm WDM coupler. The true spectral intensity is over – 45 dBm from 1100 to 1570 nm and over −50 dBm from 900 to 1100 nm. This intensity is over 10 dB stronger than that of the normal white light source coupled in single-mode fiber, such as xenon lamps and some commercialized white light source with a single-mode fiber output (AQ4305). The emission covers all optical fiber communication bands and is reasonably good for the spectral measurement of single-mode fiber-based devices, although the intensity remains significantly lower than the ASE and super luminescence sources that have the narrower bandwidth. Since the OSA spectral measurement limitation of ~−85 dBm at the resolution of 5 nm, we will have ~40 dB dynamic measurement range for the broadband spectral measurement of kinds of fiber devices. With a lock-in-based optical spectrum measurement system, a much better dynamic range can be achieved.

Figure 32b shows the relationship between the pump power and total emission power over the whole spectrum in the range of 900–1600 nm. We measured the spectrum every 5 min over an hour period and found that the standard deviation in the emission power of the broadband spectrum is <0.2 dB as shown in Fig.~32c. This shows its good stability, which is important for the spectral measurement application. The careful observation and analysis of such broadband spectra is needed, and this is to be carried out in the following with more detail emission observation.

Absorption and Emission Spectra Measurement Based on Side Pumping Method

The direct side pumping scheme is a nondestructive measurement method that can overcome some disadvantages of the cutback method. This scheme is free from fiber cutting, intrinsically low pump background, and simultaneous accurate measurement of absorption and emission (Zhang et al. 2012). Moreover, it can be a simple probe method with appropriate spatial resolution that provides local absorption and emission information in the range of millimeters to centimeters – determined by the pump beam size.

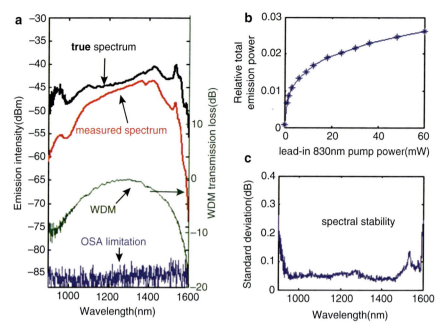

Fig. 32 (**a**) The broadband emission of Bi/Er co-doped fiber, (**b**) the relationship between the total emission power and the pump power, (**c**) the stability of the whole emission spectrum (Zhang et al. 2013)

To illustrate the side pumping scheme, we measured the emission and absorption of a commercial L-band Er-doped fiber against the known results from the supplier. We also successfully demonstrated the scheme's application over a very broad spectral range of a Bi and Er co-doped fiber we made ourselves.

The scheme of side pumping scheme is shown in Fig. 33. A test sample can be simply prepared by sandwiching a section of the active optical fiber to be tested (fiber under test, FUT) between two half standard SMF connectors. Then the absorption of the FUT can be tested by side pumping a short section of FUT and by connecting one to two optical spectrum analyzers (OSAs). In the experiment, the whole FUT is hold by two fiber holders sat on two movable stages, and the FUT is in between, which could realize the parallel shift of FUT easily. The pump laser light is compressed as a line shape and focused perpendicularly on FUT based on a cylinder lens system to realize the side pump. Then the FUT begins to emit the light spectra. Therefore, the emission spectra is got because a part of them would be attracted by the fiber and guided to the spectrometer.

The FUT absorption in the emission band can be realized based on the spectra that are obtained from two measurements at two different positions Z_A and Z_B, and their coordinate system is given as Fig. 33b. The FUT is along Z axis and their left and right connection points are 0 and Z_0. The spectra, $I_1(\lambda)$ and $I_2(\lambda)$, are got

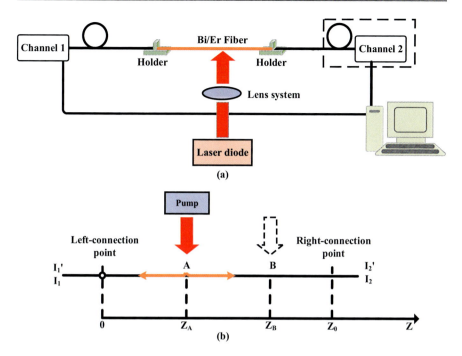

Fig. 33 (a) Scheme of absorption spectrum measurement based on side pumping method, (b) the coordinate system of the active fiber

from the channel 1 and channel 2 when the pump is in the position Z_A. They can be expressed as

$$
\begin{aligned}
I_1(\lambda) &= C_L(\lambda)S_{A-L}(\lambda)e^{-\alpha(\lambda)Z_A} \\
I_2(\lambda) &= C_R(\lambda)S_{A-R}(\lambda)e^{-\alpha(\lambda)(Z_0-Z_A)}
\end{aligned}
\tag{59}
$$

where $S_{A-L}(\lambda)$ is the part of the left-transmitted emission spectra attracted and guided by the optical fiber and it equals to that of the right-transmitted emission spectra $S_{A-R}(\lambda)$ because of the symmetry. $C_L(\lambda)$ is the total loss of the left splice point and the left SMF connector, and $C_R(\lambda)$ is that of the right splice point and the right SMF connector. $\alpha(\lambda)$ is the absorption of the active fiber per unit length. $I_1'(\lambda)$ and $I_2'(\lambda)$ are the spectra obtained when the side pump is at the position BZ. They are

$$
\begin{aligned}
I_1'(\lambda) &= C_L(\lambda)\,S_{B-L}(\lambda)\,e^{-\alpha(\lambda)Z_B} \\
I_2'(\lambda) &= C_R(\lambda)\,S_{B-R}(\lambda)\,e^{-\alpha(\lambda)(Z_0-Z_B)}
\end{aligned}
\tag{60}
$$

where $S_{B-L}(\lambda)$ is the part of the left-transmitted emission spectra attracted and guided by the active fiber core and it equals to that of the right-transmitted emission

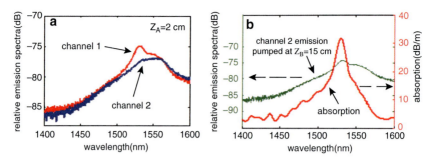

Fig. 34 The side pumping experimental results of Er-doped fiber (Costa et al. 2007)

spectra $S_{B-R}(\lambda)$ because of the symmetry. When the FUT is uniform and side pumping laser is stable, $S_{A-L}(\lambda)$ equals to $S_{B-L}(\lambda)$ approximately, and one channel method could be realized based on the spectra from channel 1. The FUT absorption is expressed as

$$\alpha(\lambda) = \frac{\ln\left[I_1(\lambda)/I_1'(\lambda)\right]}{Z_B - Z_A} \tag{61}$$

If the FUT is not uniform or the power and pattern of the pump source are not stable, the four spectra are needed to obtain the absorption as

$$\alpha(\lambda) = \frac{\ln\left[I_1(\lambda)\,I_2'(\lambda)/I_1'(\lambda)\,I_2(\lambda)\right]}{2\,(Z_B - Z_A)} \tag{62}$$

called two-channel method.

We measure a commercial Er-doped fiber (EDL001) of 15 cm in length based on the methods proposed above. The pump laser is a multimode laser diode of 810 nm in central wavelength and 600 mW in total power. The laser light is compressed as a line shape, smaller than $\sim$125 um wide and $\sim$7 mm long, and projected on the FUT (EDL001). Two-channel spectrometer with a 5 nm resolution, based on two synchronized spectral analyzers (Aglent 86143B), is used to collect the emission spectra. In order to observe the emission spectra with the minimum absorption effect, we pumped the FUT nearby the two connection points. The two spectra, when the pump is projected near the left connection point, are shown in Fig. 34a. Their profiles are obviously different because the spectrum at channel 2 experienced the absorption of the $\sim$13 cm-long Er-doped fiber and the other at channel 1 didn't. In order to observe the emission spectrum without the FUT reabsorption, we pump the fiber at the right connection point, and the emission spectra is recorded by the nearby OSA (channel 2), shown in Fig. 34b.

We also show the measurement of our lab-made Bi/Er co-doped, Al-activated silica optical fiber sample based on the same setup. Our fiber is fabricated by in situ doping of [Er$_2$O$_3$] $\sim$0.01, [Al$_2$O$_3$] $\sim$0.15, [Bi$_2$O$_3$] $\sim$0.16, [P$_2$O$_5$] $\sim$0.94, and

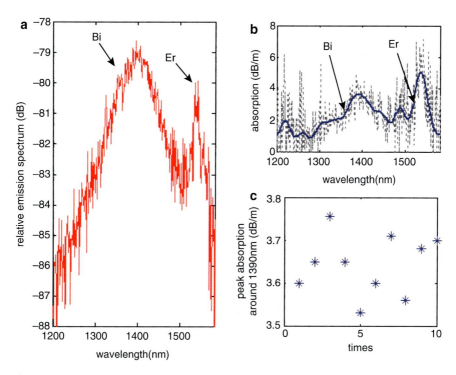

Fig. 35 (**a**) The emission spectra of Bi/Er co-doped fiber, (**b**) the calculated absorption spectra based on a two-channel method, and (**c**) the measurement stability of the peak absorption at around 1390 nm (Costa et al. 2007)

[GeO_2] ~12.9 mol.%, respectively. The core of our nonstandard fiber sample is off axis of about 5 um, which is a big trouble for the normal cutback scheme because the coupling with other components is quite different every time, while it is convenient to use the proposed side pumping-based scheme. The emission and absorption spectra can be readily obtained, as shown in Fig. 35a, b, respectively. The bandwidth of the emission spectrum is as wide as over 300 nm. There are two obvious emission and absorption bands at 1400 nm for ES1→GS of Bi center (Dvoyrin et al. 2008) and 1530 nm for $^4I_{13/2} \rightarrow {}^4I_{15/2}$ of Er ions (Becker et al. 1999). The stability of the peak absorption at ~1390 nm is shown in Fig. 35c.

The side pumping scheme is a simple and nondestructive scheme to measure the uniformity of the doping concentration along the active fiber just by measuring the spectra at a few different pumping positions. The absorption measurement based on the side pumping scheme is limited to the emission band. For the very weak emission, the lock-in amplifier technique is expected to apply to obtain the spectra, instead of using OSAs simply. It could be used as a calibration method to calibrate the results of the cutback measurement when the absorption spectrum in the non-emission band is needed.

It needs to be mentioned that the higher pump power is needed for the side pumping scheme compared to the case that pump based on a WDM coupler. In addition, the resonance pump should be applied by using the appropriate laser to realize the full inversion if the emission cross section of active fiber is needed.

Conclusion

In summary, characteristics of specialty fiber and their measurement technologies are introduced in this chapter. Dispersion is the phenomenon in which the phase velocity of a wave depends on its frequency. It can be categorized into material dispersion, waveguide dispersion, intermodal dispersion, and polarization mode dispersion. Their corresponding measurement technologies are described in detail. Some characteristics of polarization-maintaining (PM) fibers and the polarization-dependent loss (PDL) measurement technologies are also introduced. Finally, the main characteristics and measurement technologies of active fibers, including material analysis, absorption and emission spectra measurement, luminescence measurement, are described in detail.

References

P.M. Becker, A.A. Olsson, J.R. Simpson, *Erbium-Doped Fiber Amplifiers: Fundamentals and Technology* (Academic Press, San Diego, 1999)

R.D. Birch, D.N. Payne, M.P. Varnham, Fabrication of polarisation-maintaining fibres using gas-phase etching. Electron. Lett. **18**(24), 1036–1038 (1982)

I.A. Bufetov, E.M. Dianov, Bi-doped fiber lasers. Laser Phys. Lett. **6**(7), 487 (2009)

J.C. Chen, Y.S. Lin, C.N. Tsai, et al., 400-nm-bandwidth emission from a Cr-doped glass fiber. IEEE Photon. Technol. Lett. **19**(8), 595–597 (2007)

B. Christensen, J. Mark, G. Jacobsen, et al., Simple dispersion measurement technique with high resolution. Electron. Lett. **29**(1), 132 (2002)

Y. Chu, J. Ren, J. Zhang, et al., $Ce^{3+}/Yb^{3+}/Er^{3+}$ triply doped bismuth borosilicate glass: a potential fiber material for broadband near-infrared fiber amplifiers. Sci. Rep. **6**, 33865 (2016)

Y. Chu, J. Hao, J. Zhang et al., Temperature properties and potential temperature sensor based on the Bismuth/Erbium co-doped optical fibers[C]//Optical Fiber Sensors Conference (OFS), 2017 25th. IEEE, 1–4 (2017)

L.G. Cohen, Comparison of single-mode fiber dispersion measurement techniques. J. Lightwave Technol. **3**(5), 958–966 (1985)

B. Costa, M. Puleo, E. Vezzoni, Phase-shift technique for the measurement of chromatic dispersion in single-mode optical fibres using LEDs. Electron. Lett. **19**(25), 1074–1076 (2007)

G. Della Valle, A. Festa, G. Sorbello, et al., Single-mode and high power waveguide lasers fabricated by ion-exchange. Opt. Express **16**(16), 12334–12341 (2008)

E. Desurvire, *Erbium-Doped Fiber Amplifiers: Principles and Applications* (Wiley-Interscience, Hoboken, 2002)

F. Devaux, Y. Sorel, J.F. Kerdiles, Simple measurement of fiber dispersion and of chirp parameter of intensity modulated light emitter. J. Lightwave Technol. **11**(12), 1937–1940 (1993)

V.V. Dvoyrin, V.M. Mashinsky, L.I. Bulatov, et al., Bismuth-doped-glass optical fibers – a new active medium for lasers and amplifiers. Opt. Lett. **31**(20), 2966–2968 (2006)

V.V. Dvoyrin, O.I. Medvedkov, V.M. Mashinsky, et al., Optical amplification in 1430–1495 nm range and laser action in Bi-doped fibers. Opt. Express **16**(21), 16971–16976 (2008)

W. Eickhoff, E. Brinkmeyer, Scattering loss vs polarization holding ability of single-mode fibers. Appl. Opt. **23**(8), 1131–1132 (1984)

N. Gisin, J.P.V.D. Weid, J. Pellaux, Polarization mode dispersion of short and long single-mode fibers. J. Lightwave Technol. **9**(7), 821–827 (1991)

J. Hao, Y. Chu, Z. Ma et al., Effects of thermal treatment on photoluminescence properties of bismuth/erbium co-doped optical fibers. Opt. Fiber Technol 46, 141–146 (2018)

B.L. Heffner, Automated measurement of polarization mode dispersion using Jones matrix eigenanalysis. IEEE Photon. Technol. Lett. **4**(9), 1066–1069 (1992)

B.L. Heffner, Accurate, automated measurement of differential group delay dispersion and principal state variation using Jones matrix eigenanalysis. IEEE Photon. Technol. Lett. **5**(7), 814–817 (1993)

C. Hentschel, S. Schmidt, PDL Measurements Using the Agilent 8169A Polarization Controller, Product Note, Agilent Technologies.

P. Hernday, in *Fiber-Optic Test and Measurement*, ed. by D. Derickson. Dispersion measurement (Prentice Hall, Upper Saddle River, 1998)

T. Hosaka, Y. Sasaki, J. Noda, et al., Low-loss and low-crosstalk polarisation-maintaining optical fibres. Electron. Lett. **21**(20), 920–921 (1985)

Z. Hu, W. Qiu, X. Cheng, et al., Optical amplification of Eu (TTA) 3 Phensolution-filled hollow optical fiber. Opt. Lett. **36**(10), 1902–1904 (2011)

X. Huang, Z. Fang, Z. Peng, et al., Formation, element-migration and broadband luminescence in quantum dot-doped glass fibers. Opt. Express **25**(17), 19691–19700 (2017)

R. Hui, M. O'Sullivan, in *Fiber Optic Measurement Techniques*. Optical fiber measurement (Elsevier/Academic Press, Amsterdam/London, 2009), p. 365–479

S.T. Huntington, P. Mulvaney, A. Roberts, et al., Atomic force microscopy for the determination of refractive index profiles of optical fibers and waveguides: a quantitative study. J. Appl. Phys. **82**(6), 2730–2734 (1997)

S.D. Jackson, 2.7-W Ho^{3+}-doped silica fibre laser pumped at 1100 nm and operating at 2.1 μm. Appl. Phys. B **76**(7), 793–795 (2003)

S. Jarabo, J.M. Álvarez, Experimental cross sections of erbium-doped silica fibers pumped at 1480 nm. Appl. Opt. **37**(12), 2288–2295 (1998)

I. Kaminow, Polarization in optical fibers. IEEE J. Quantum Electron. **17**(1), 15–22 (1981)

I.P. Kaminow, V. Ramaswamy, Single-polarization optical fibers: slab model. Appl. Phys. Lett. **34**(4), 268–270 (1979)

T. Kasamatsu, Y. Yano, H. Sekita, 1.50-μm-band gain-shifted thulium-doped fiber amplifier with 1.05-and 1.56-μm dual-wavelength pumping. Opt. Lett. **24**(23), 1684–1686 (1999)

A.S. Kurkov, E.M. Sholokhov, O.I. Medvedkov, et al., Holmium fiber laser based on the heavily doped active fiber. Laser Phys. Lett. **6**(9), 661 (2009)

H. Liang, Q. Zhang, Z. Zheng, et al., Optical amplification of Eu (DBM) 3 Phen-doped polymer optical fiber. Opt. Lett. **29**(5), 477–479 (2004)

P.F. Moulton, G.A. Rines, E.V. Slobodtchikov, et al., Tm-doped fiber lasers: fundamentals and power scaling. IEEE J. Sel. Top. Quantum Electron **15**(1), 85–92 (2009)

E.G. Neumann, *Single-Mode Fibers Fundamentals*, vol 57(4) (Springer, Tokyo, 1988), pp. 201–203

Y. Nishida, M. Yamada, T. Kanamori, et al., Development of an efficient praseodymium-doped fiber amplifier. IEEE J. Quantum Electron. **34**(8), 1332–1339 (1998)

J. Noda, K. Okamoto, Y. Sasaki, Polarization-maintaining fibers and their applications. J. Lightwave Technol. **4**(8), 1071–1089 (1986)

Y. Ohishi, E. Snitzer, G.H. Sigel, et al., Pr^{3+}-doped fluoride fiber amplifier operating at 1.31 μm. Opt. Lett. **16**(22), 1747–1749 (1991)

R. Paschotta, J. Nilsson, A.C. Tropper, et al., Ytterbium-doped fiber amplifiers. IEEE J. Quantum Electron. **33**(7), 1049–1056 (1997)

D.N. Payne, A. Barlow, J.J. Ramskov Hansen, Development of low- and high-birefringence optical fibers. IEEE J. Quantum Electron. **18**(4), 477–488 (1982)

G.D. Peng, Y. Luo, J. Zhang, et al., Recent development of new active optical fibres for broadband photonic applications. Photonics (ICP), 2013 IEEE 4th International Conference on. IEEE, 2013, pp. 5–9.

C.D. Poole, D.L. Favin, Polarization-mode dispersion measurements based on transmission spectra through a polarizer. J. Lightwave Technol. **12**(6), 917–929 (1994)

C.D. Poole, C.R. Giles, Polarization-dependent pulse compression and broadening due to polarization dispersion in dispersion-shifted fiber. Opt. Lett. **13**(2), 155–157 (1988)

C.D. Poole, J. Nagel, Polarization effects in lightwave systems. Opt. Fiber Telecommun. **IIIA**, 114–161 (1997)

R.S. Quimby, W.J. Miniscalco, B. Thompson, Clustering in erbium-doped silica glass fibers analyzed using 980 nm excited-state absorption. J. Appl. Physiol. **76**(8), 4472–4478 (1994)

C. Saekeang, P.L. Chu, T.W. Whitbread, Nondestructive measurement of refractive-index profile and cross-sectional geometry of optical fiber preforms. Appl. Opt. **19**(12), 2025–2030 (1980)

R.H. Stolen, R.P. De Paula, Single-mode fiber components. Proc. IEEE **75**(11), 1498–1511 (1987)

R.H. Stolen, W. Pleibel, J.R. Simpson, High-birefringence optical fibers by preform deformation. J. Lightwave Technol. **2**(5), 639–641 (1984)

L. Tan, S. Kang, Z. Pan, et al., Topo-chemical tailoring of tellurium quantum dot precipitation from supercooled polyphosphates for broadband optical amplification. Advanced Optical Materials **4**(10), 1624–1634 (2016)

A. Tünnermann, T. Schreiber, J. Limpert, Fiber lasers and amplifiers: an ultrafast performance evolution. Appl. Opt. **49**(25), F71–F78 (2010)

H.H. Wahba, T. Kreis, Characterization of graded index optical fibers by digital holographic interferometry. Appl. Opt. **48**(8), 1573–1582 (2009)

J.P. Weid, L. Thevenaz, J.P. Pellaux, Interferometric measurements of chromatic and polarisation mode dispersion in highly birefringent single-mode fibres. Electron. Lett. **23**(4), 151–152 (1987)

K.I. White, Practical application of the refracted near-field technique for the measurement of optical fibre refractive index profiles. Opt. Quant. Electron. **11**(2), 185–196 (1979)

E. Yahel, A. Hardy, Modeling high-power Er3+-Yb3+ codoped fiber lasers. J. Lightwave Technol. **21**(9), 2044 (2003)

S.M. Yeh, S.L. Huang, Y.J. Chiu, et al., Broadband chromium-doped fiber amplifiers for next-generation optical communication systems. J. Lightwave Technol. **30**(6), 921–927 (2012)

S. Yliniemi, J. Albert, Q. Wang, et al., UV-exposed Bragg gratings for laser applications in silver-sodium ion-exchanged phosphate glass waveguides. Opt. Express **14**(7), 2898–2903 (2006)

J. Zhang, Y. Luo, Z.M. Sathi, et al., Test of spectral emission and absorption characteristics of active optical fibers by direct side pumping. Opt. Express **20**(18), 20623–20628 (2012)

J. Zhang, Z.M. Sathi, Y. Luo, et al., Toward an ultra-broadband emission source based on the bismuth and erbium co-doped optical fiber and a single 830 nm laser diode pump. Opt. Express **21**(6), 7786–7792 (2013)

A.S. Zlenko, V.V. Dvoyrin, V.M. Mashinsky, et al., Furnace chemical vapor deposition bismuth-doped silica-core holey fiber. Opt. Lett. **36**(13), 2599–2601 (2011)

Characterization of Distributed Birefringence in Optical Fibers

31

Yongkang Dong, Lei Teng, Hongying Zhang, Taofei Jiang, and Dengwang Zhou

Contents

Abstract

Birefringence is the fundamental physical parameter of optical fibers which characterizes their polarization properties, and it can be classified into phase birefringence and group birefringence. Phase birefringence is the difference in effective index between the two orthogonal linear polarization modes of an optical fiber, while the group birefringence is related to group index representing the polarization mode dispersion. In this chapter, we introduce a distributed phase birefringence measurement method based on Brillouin dynamic grating

Y. Dong (✉) · L. Teng · T. Jiang · D. Zhou
National Key Laboratory of Science and Technology on Tunable Laser, Harbin Institute of Technology, Harbin, China
e-mail: aldendong@163.com; aldendong@gmail.com; tengl_hit@163.com; jiangtaofei390@126.com; cishixitie@163.com

H. Zhang
Institute of Photonics and Optical Fiber Technology, Harbin University of Science and Technology, Harbin, China
e-mail: zhy_hit@163.com

© Springer Nature Singapore Pte Ltd. 2019
G.-D. Peng (ed.), *Handbook of Optical Fibers*,
https://doi.org/10.1007/978-981-10-7087-7_60

(BDG), which creates a new horizon for optical fiber evaluation. When two parallel polarized pump waves, with a frequency offset equal to the fiber Brillouin frequency shift, counter-propagate along the fiber, a BDG can be excited through simulated Brillouin scattering (SBS), and another orthogonally polarized probe wave injected into fiber is used to detect the BDG. When the frequency difference between the probe wave and the co-propagating pump wave meets the phase-matching condition, the maximum reflection on probe wave from the BDG can be obtained. The interaction of the excitation and the probing of a BDG involves four optical waves, and the Brillouin-enhanced four-wave mixing model completely describes this coupling process. In the following sections of this chapter, the theoretical operation principles, numerical simulations, and experimental implementation of distributed phase birefringence measurement with BDG are described; some sensing applications of distributed birefringence measurement with BDG are also given including simultaneous distributed temperature and strain measurement, distributed transverse pressure sensing, and distributed hydrostatic pressure sensing.

Keywords

Birefringence · Polarization-maintaining fiber · Distributed measurement · Stimulated Brillouin scattering · Brillouin dynamic grating

Introduction

This chapter includes an introduction to and review of the principles of distributed phase birefringence measurement in the case of a polarization-maintaining fiber (PMF) based on the Brillouin dynamic grating (BDG). The primary motivations of the technique are introduced, a description provided the operation principles upon which a BDG is generated and detected, and a detailed examination of the particular manner in which the distributed phase birefringence measurement based on the BDG is undertaken. These are followed by a review of selected real sensing applications investigated for particular sensing areas, together with details of a range of experimental systems and data.

The applications of single-mode fibers (SMFs) to coherent optical fiber communication systems and fiber sensor systems require a definite state of polarization along the fiber length (Keiser 2003). It is well known that in a conventional SMF with imperfect circular cross-section geometry and asymmetrical material distribution, the polarization state of the output signal is usually random; the mode coupling and the output polarization state are highly sensitive to external perturbations. PMFs such as shape-induced PMF, with an elliptical core, and stress-induced PMF, e.g., Panda fiber, have a high internal birefringence that exceeds external perturbing birefringence and maintains linear polarization along the fiber (Kaminow 1981; Noda et al. 1986). PMFs are birefringent in principle: as the two polarization states propagate with slightly different velocities, and when a broadband source is used, the spurious cross-state waves lose coherence with the main primary signal. The mode coupling between the two orthogonal polarization modes is minimized in a PMF, because the difference between their propagation constants is large (Eugene 2002). Among the fundamental parameters characterizing the polarization-maintaining properties of a PMF, birefringence is a critical one.

There are two definitions of birefringence to describe the polarization properties of a high-birefringence fiber: one is the phase birefringence Δn, i.e., the difference in effective refractive indexes n_x and n_y between the two orthogonal linear polarization modes (Eugene 2002):

$$\Delta n (\lambda) = n_x (\lambda) - n_y (\lambda) \tag{1}$$

which is associated with the polarization beat length $L_B(\lambda)$ by

$$L_B (\lambda) = \lambda / \Delta n (\lambda) \tag{2}$$

and the other is the group birefringence Δn^g

$$\Delta n^g (\lambda) = n_x^g (\lambda) - n_y^g (\lambda) = \Delta n (\lambda) - \lambda \frac{d \Delta n (\lambda)}{d (\lambda)}. \tag{3}$$

The group birefringence is closely related to polarization mode dispersion (PMD) $\tau = \Delta n^g(\lambda)/c$, which limits the fiber transmission system (Wai and Menyak 1996). Here, n_i and n_i^g denote the phase and group refractive indexes which are related to the phase velocity and group velocity, respectively, and c is the velocity of light in a vacuum (Eugene 2002). In telecommunications, it is extremely important to limit the impacts of PMD, while in other applications, such as polarimetric optical sensors, a high value of the phase birefringence is required.

The development of the fiber-optic gyroscope has meant that it has become the main application area for PMFs. To solve the problems of polarization-induced signal fading and lack of the polarization rejection, the development of the high-quality PMF is an important step toward a compact and practical device. In a fiber gyro, the polarization conservation maintains most of the power in the primary reciprocal wave, avoiding signal fading.

A cross-polarization coupling point could occur due to the external perturbation, and the main and crossed wave trains lose their overlap. At this point, a phase mismatch exists because of the different velocities, which leads to the statistical decorrelation between both crossed polarizations (Lefevre 2014).

From the above analyses, it can be found that the polarization-maintaining characters of a PMF play a significant role in practical applications, and the measurement of birefringence is very important for characterizing the polarization properties of the PMF, especially in fiber gyro monitoring and evaluation.

The value of the phase birefringence is often expressed by the beat length L_B between the two orthogonal polarization modes. Several methods exist to measure the beat length (Kikuchi and Okoshi 1983; Takada et al. 1985; Szczurowski et al. 2011; Huang and Lin 1985; Sikka et al. 1998), and the classic schemes are based on spectral interferometry and the method of a lateral point-like force applied on the fiber (Kaczmarek 2012; Hlubina and Ciprian 2007).

In the experimental setup of the periodic lateral force method for measuring the beat length shown in Fig. 1, the monochromatic light output from the laser diode (LD) is coupled into a principal polarization mode. At a certain location, a periodic lateral force F is applied over a very short length of the fiber, which can be regarded as a concentrated (point-like) perturbation. The force generates the polarization cross talk, and a fraction of the light is coupled into the polarization mode which is not excited at the input end of the fiber. The force sweeps along the fiber to cause the interference oscillatory intensity changes when observed through a polarization analyzer (Pol.). The interference signal is recorded by photoelectric detector (PD) and data acquisition device (DAQ). And the beat length is measured through dividing the fiber length by the number of oscillations. And the phase birefringence can be obtained from Eq. 2.

The high-birefringence fiber Sagnac loop operates like a wavelength-division-multiplexing filter (Fang and Claus 1995), which follows from the dependence of the phase difference between the two orthogonal polarization modes propagating in the high-birefringence fiber.

Figure 2 shows the experimental setup of a typical Sagnac-based birefringence measurement system. The output light from the broadband light source (BBS) is separated by a 3 dB coupler into two beams that counter-propagate in the opposite directions, which recombine in the coupler after traveling through the loop. The two orthogonal polarization modes in the PMF create an interference pattern in the coupler, which is dependent on the phase difference. The practical measurement accuracy depends on the measurement accuracy of fiber length and spectral resolution of optical spectrum analyzer (OSA) (Kaczmarek 2012).

The typical Sagnac interference coefficient $T(\lambda)$ approximates well to a periodic function of wavelength, which can be described by the formula (Xu et al. 2016; Kim and Kang 2004; Hlubina and Ciprian 2007; Fang and Claus 1995):

$$T(\lambda) = 1 - \cos(\varphi)/2 \qquad (4)$$

where φ is the phase difference between the two polarization modes in a PMF with a length of L, defined by the formula of $\varphi = 2\pi \Delta n L/\lambda$.

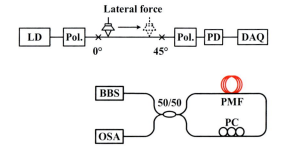

Fig. 1 The typical scheme of the periodic lateral force method for measuring the L_B of the PMF

Fig. 2 Experimental setup of birefringence measurement method based on Sagnac

The relationship between the spectrum oscillation and the wavelength changes can be retrieved by calculating the derivative of the phase difference with respect to the wavelength λ, where

$$\frac{d\varphi}{d\lambda} = \frac{2\pi L}{\lambda^2} \left[\frac{d\Delta n}{d\lambda}\lambda - \Delta n \right] = \frac{2\pi L}{\lambda^2} \left(-\Delta n^g \right). \tag{5}$$

Equation 5 indicates that wavelength scanning of broadband light source yields the group birefringence. And the separation between two adjacent maxima ($\Delta\lambda$) is related to a phase change of 2π and to the group birefringence by

$$\Delta n^g = \frac{\lambda^2}{\Delta\lambda L} \tag{6}$$

which shows an approximate solution for the Sagnac-based method measuring the group birefringence.

Although the existing birefringence measurement methods have many advantages, such as simple structures and quick responses, they all belong to the point measurement systems which cannot realize the distributed birefringence measurement. Owing to the nonuniformity of the materials and changes of environmental conditions during the fabrication of the fibers, birefringence variation along the fiber is inevitable. Therefore, it is insufficient to obtain the average birefringence of the PMF for practical applications. Distributed birefringence measurement methods are needed for characterizing the birefringence variation along the fiber (Suzuki et al. 2001).

In 2010, the first truly distributed birefringence measurement of a PMF based on a BDG was reported (Dong et al. 2010a). The results represent the birefringence distribution of two types of widely used PMFs, bow tie fiber and Panda fiber with a length of 8 m and in a spatial resolution of 20 cm. The experimental results confirm that the birefringence features a periodic variation along the fiber. This work was a significant achievement toward the realization of birefringence measurement using BDG. The measurement principles and experimental results of the investigation of the present work, being the primary contents for this chapter, are included in Section III. In addition to the distributed birefringence measurements, the BDG has two unique features when compared with conventional fiber Bragg gratings (FBG): one is that it is a moving grating, which can produce a Brillouin frequency shift (BFS, $\Delta\nu_B$) to the reflected wave with respect to the probe wave, and the other is that a lifetime is associated with it ($\sim$10 ns for silica fiber) for its existence after removing the pump waves (Zhou et al. 2011; Song et al. 2008; Dong et al. 2010b).

To date, the BDG has been generated in a frequency correlation domain (Zou et al. 2009) and time domain (Zhou et al. 2011; Song et al. 2008; Dong et al. 2010a). In the case of the correlation-based technique, a BDG is generated using two synchronized and frequency-modulated continuous pump waves and detected using a separate orthogonally sinusoidal frequency-modulated probe wave (Zou et al. 2009). In the time domain, a BDG can be generated using two frequency-locked pump waves through stimulated Brillouin scattering (SBS) and can also be generated by a strong pump pulse through spontaneous Brillouin scattering. Moreover, the BDG has been successfully realized in many other platforms besides a PMF (Dong et al. 2009; Song 2012), such as conventional SMF (Song 2011; Dong et al. 2014), dispersion-shifted fibers (Zou and Chen 2013), few-mode fibers (Li and Li 2012), and polarization-maintaining photonic crystal fiber (Dong et al. 2009), and in a photonic chip (Pant et al. 2013), where different acoustic and spatial modes are used in certain materials corresponding to diverse phase relationships.

As well as the basic studies of the theoretical properties of BDG, a wide range of potential applications have been explored, including all-optical information processing (Santagiustina et al. 2013), light storage (Winful 2013), microwave photonics (Sancho et al. 2012), optical delay lines (Chin and Thévenaz 2012), optical spectrum analysis (Dong et al. 2014), and distributed sensing (Dong et al. 2009; Song 2012).

The remaining sections in this chapter are arranged as follows: in the second section, the basic theories of BDG are covered, including the phase-matching condition, the coupled wave equations of the analytical model, and the characteristics of the reflection spectrum; in the third section, distributed birefringence measurement is described in theory and using experiments, and the sensing applications, e.g., distributed temperature, strain, transverse pressure, and hydrostatic pressure measurements, are summarized; and in the last section, the conclusions are presented.

Operation Principle of BDG

The basic operation principles for generating and probing a BDG, the analytical model of the BDG with coupled wave equations of Brillouin-enhanced four-wave mixing (FWM), and some reflection characteristics such as reflectivity and bandwidth are introduced in this section. The contents focus on the discussion of the phase-matching condition and the FWM model explanation for the case of a BDG in PMF.

Theoretical Analysis of BDG

Principle of Generation and Detection of BDG

The Brillouin dynamic grating is similar to a conventional FBG in that they are both examples of local refractive index-modulated grating structures. A brief introduction about the BDG was provided in Section I. In this part, the BDG generation and detection processes in a PMF for the case without depletion are theoretically investigated. The interference of the excitation and probing of the BDG involves four waves interacting with each other through localized density variations of the material, and the resulting reflected wave exhibits a Brillouin frequency difference with respect to the probe wave (Dong et al. 2010b). There are two cases for the interaction between the probe wave and the BDG: one is that they propagate in the same direction, in which case the reflected wave frequency is down-converted by a fiber BFS through the process of coherent Stokes Brillouin scattering; the other is that they propagate in opposite directions, producing a reflected wave with an up-converted frequency through a fiber BFS in the coherent anti-Stokes Brillouin scattering process.

In the case of coherent Stokes Brillouin scattering, there are two different configurations to perform the birefringence measurement by exchanging the roles of the slow and fast axes, as shown in Fig. 3 (Dong et al. 2010b).

There exists a similar situation for the case of coherent anti-Stokes Brillouin scattering. Therefore, there are a total of four different configurations for generation and detection of a BDG within one PMF by exchanging the options of the slow and fast axes or reversing propagation direction of the probe wave. The following discussions are confined to the case of coherent anti-Stokes Brillouin scattering.

As shown in Fig. 4, a BDG is an optically local refractive index-modulated grating as it is in the case of a FBG. Two counter-propagating pump waves, with a frequency offset of BFS of the PMF, are launched into one axis of the PMF, and SBS occurs when the two pump waves beat together giving rise to density variations associated with acoustic wave propagation within the fiber through the electrostriction effect. The property of the BDG can be detected by monitoring the reflected wave from a probe wave launched into the other axis. The maximum reflection from the BDG can be obtained when the frequency difference between the probe and the co-propagating pump wave satisfies the phase-matching conditions (Song et al. 2008; Dong et al. 2010b).

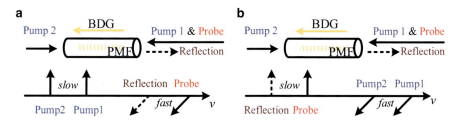

Fig. 3 The schematic representation of two different configurations based on coherent Stokes Brillouin scattering

Fig. 4 Schematic diagram of generation and detection of a BDG in PMF

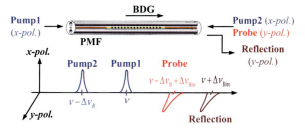

Fig. 5 Wave vector mismatch of the four optical waves

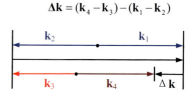

Phase Matching Condition

Due to conservation of momentum during the interaction process of the excitation and the probing of the BDG (Dong et al. 2010a), the four optical waves should meet the criterion of wave vector relationship shown in Fig. 5, where

$$\Delta \mathbf{k} = (\mathbf{k}_4 - \mathbf{k}_3) - (\mathbf{k}_1 - \mathbf{k}_2) \tag{7}$$

and Eq. 7 can also be expressed as

$$\frac{2\pi \cdot \Delta v_B}{V} = \left(\frac{2\pi \cdot n_y (v_4) v_4}{c} + \frac{2\pi \cdot n_y (v_3) v_3}{c} \right) - \left(\frac{2\pi \cdot n_x (v_2) v_2}{c} + \frac{2\pi \cdot n_x (v_1) v_1}{c} \right) \tag{8}$$

where V and v are the acoustic wave velocity and light frequency, respectively, with the former representing the velocity of the BDG.

The phase-matching condition can be derived from the Taylor expansion in Eq. 8 around the refractive indexes of the two axes omitting the higher-order terms, which is

$$\frac{n_y (v_3) v_3}{c} + \frac{n_y (v_4) v_4}{c} = \frac{n_x (v_3) v_3}{c} + (v_2 - v_3) \frac{d (n_x (v_3) v_3)}{c d v_3} + \frac{n_x (v_4) v_4}{c}$$

$$+ (v_2 - v_3) \frac{d (n_x (v_4) v_4)}{c d v_4} \tag{9}$$

where $n_x^g (v_i) = d (n_x (v_i) v_i) / d v_i$, where i can be 3 or 4, is denoted as the group refractive index of the x axis, and the Eq. 9 can be simplified as

$$(v_3 - v_2) \left(n_x^g (v_3) + n_x^g (v_4) \right) = \left(n_x (v_3) - n_y (v_3) \right) v_3$$

$$+ \left(n_x (v_4) - n_y (v_4) \right) v_4. \tag{10}$$

When ignoring the group modal dispersion, the group refractive index satisfies the relation of $n_x^g = n_x^g (v_3) = n_x^g (v_4)$. Because that the optical frequency v (~hundreds of terahertz) is much larger than the BFS (~tens of gigahertz) of a fiber, it can be approximately presented that $v_3 \approx v_3 + \Delta v_B$, and then

$$\Delta n\,(v_3) = n_x\,(v_3) - n_y\,(v_3) = n_x\,(v_4) - n_y\,(v_4)\,. \tag{11}$$

Finally, the phase-matching condition can be obtained as

$$\Delta v_{\text{Bire}} = \Delta n\,(v)\,v/n_x^g \tag{12}$$

where $\Delta v_{\text{Bire}} = (v_3 - v_2)$ is the birefringence-induced frequency shift (BireFS, Δv_{Bire}) and it has a proportional relationship with the phase birefringence $\Delta n(v) = n_x - n_y$.

Coupled Wave Equations of the Brillouin-Enhanced FWM Process

As previously mentioned, the process for the generation and detection of a BDG involves four light waves, two pump waves, a probe waves, and a reflected wave, and the interaction is characterized as Brillouin-enhanced FWM in which the longitudinal acoustic wave couples those four optical waves with different polarization states together (Agrawal 2007; Zhou et al. 2009). Characterization of the BDG spectrum reveals that the shape of the BDG spectrum is similar to that of a weak FBG (Dong et al. 2010b; Erdogan 1997; Song and Yoon 2010).

In the discussions below, the mathematical model for the case of coherent anti-Stokes Brillouin scattering is formulated, with the frequency of the BDG reflection being higher than that of the probe wave. The spatial representation of the four optical waves $\tilde{E}_j$ ($j = 1, 2, 3,$ and 4) and the acoustic wave are illustrated in Fig. 6.

A section of PMF with a horizontal axial coordinate between $z = 0$ and $z = L$ is considered together with the two orthogonal axes of the x and y directions. Two pump waves (pump 1, $\tilde{E}_1$, and pump 2, $\tilde{E}_2$), with Electric field frequencies of v_1 and v_2, respectively, and a frequency difference equal to the BFS of the fiber of $\Delta v_B = v_1 - v_2$ are injected from opposite directions with the same polarization state (x pol.). An acoustic wave with an amplitude of ρ, frequency of v_o, and wave vector of k_5 is then excited by the two pump waves through SBS, which can be expressed in consideration of the medium density modulation, as

$$\tilde{\rho}\,(z,t) = \rho_0 + \{\rho\,(z,t)\exp\left[i\,(k_5 z - v_o t)\right] + c.c.\} \tag{13}$$

where ρ_0 denotes the mean density of the medium.

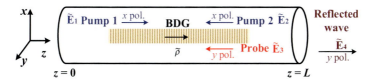

Fig. 6 Scheme of interaction between the optical and acoustic waves in a Brillouin-enhanced FWM process associated with a BDG

The electric fields can be expressed as

$$\tilde{E}_j\,(z,t) = E_j\,(z,t)\exp\left[i\left(k_j z - v_j t\right)\right] + c.c. \tag{14}$$

where $E_j\,(z,\,t)$ $(j = 1,\,2,\,3,$ and $4)$ represents the amplitude function of position z and time t. And k_j is the wave vector.

The interaction of the four optical waves can be represented as

$$\tilde{E}\,(z,t) = \tilde{E}_1\,(z,t) + \tilde{E}_2\,(z,t) + \tilde{E}_3\,(z,t) + \tilde{E}_4\,(z,t). \tag{15}$$

Considering the electrostriction effect, the equation of medium density modulation can be described as follows:

$$\frac{\partial^2 \tilde{\rho}}{\partial t^2} - \Gamma'\nabla^2\frac{\partial \tilde{\rho}}{\partial t} - V^2\nabla^2\tilde{\rho} = \nabla\cdot\mathbf{f} \tag{16}$$

where Γ' and V denote the damping parameter and the velocity of the acoustic wave related to material characteristics, respectively.

Regarded as the source term of Eq. 16, the divergence of the force $\mathbf{f}$ at per unit volume relates to the electrostriction term, by $\mathbf{f} = -\frac{1}{2}\varepsilon_0\gamma_\varepsilon\nabla\left\langle\tilde{E}^2\right\rangle$.

Substituting Eq. 13 into Eq. 16 and considering the slowly varying amplitude approximation, the further material density equation can be obtained as

$$-2i\,v_0\frac{\delta\rho}{\delta t} - i\,v_0\Gamma_\mathrm{B}\rho - 2i\,v^2\frac{\delta\rho}{\delta z} = \varepsilon_0\gamma_e k_5\left(E_1 E_2^* + E_3^* E_4 e^{i\Delta kz}\right) \tag{17}$$

where $\Gamma_\mathrm{B} = q^2\Gamma'$ represents the Brillouin linewidth connected with the phonon lifetime $\tau_\mathrm{p} = 1/\Gamma_\mathrm{B}$ and $\Delta k = (k_4 + k_3) - (k_2 + k_1)$ is the phase mismatch of the four optical waves. Due to the orthogonal polarization states of the two pump waves and the probe wave, there are only two driven terms of the practical acoustic field remained in Eq. 16.

The spatial evolution for the interactions can be described using the wave equation

$$\frac{\partial^2 \tilde{E}_j}{\partial z^2} - \left(\frac{n}{c}\right)^2\frac{\partial^2 \tilde{E}_j}{\partial t^2} = \frac{1}{\varepsilon_0 c^2}\frac{\partial^2 \tilde{I}_j}{\partial t^2} \tag{18}$$

where the source term $\tilde{I} = \varepsilon_0\Delta\chi\tilde{E} = \varepsilon_0\rho_0^{-1}\gamma_e\tilde{\rho}\tilde{E}$ indicates the nonlinear polarization.

The coupled wave equations for the interaction of the optical waves and the acoustic wave in the Brillouin-enhanced FWM process can be obtained by combining Eqs. 14 and 18 and making the slowly varying amplitude approximation:

$$\left(\frac{\partial}{\partial z} + \frac{n_x}{c}\frac{\partial}{\partial t}\right)E_1 = ig_o\rho E_2 \tag{19a}$$

$$\left(-\frac{\partial}{\partial z} + \frac{n_x}{c}\frac{\partial}{\partial t}\right) E_2 = ig_o\rho^* E_1 \tag{19b}$$

$$\left(-\frac{\partial}{\partial z} + \frac{n_y}{c}\frac{\partial}{\partial t}\right) E_3 = ig_o\rho^* E_4 e^{i\Delta kz} \tag{19c}$$

$$\left(\frac{\partial}{\partial z} + \frac{n_y}{c}\frac{\partial}{\partial t}\right) E_4 = ig_o\rho E_3 e^{-i\Delta kz} \tag{19d}$$

$$\left(\frac{\partial}{\partial t} + \frac{\Gamma_{\mathrm{B}}}{2}\right) \rho = ig_a\left(E_1 E_2{}^* + E_3{}^* E_4 e^{i\Delta kz}\right) \tag{19e}$$

where the effective refractive indexes of the two polarization axes of the PMF n_x and n_y are related to the optical frequency; g_o and g_a are coupling coefficients of the optical wave and acoustic wave, respectively; and $g_{\mathrm{B}} = 4g_o g_a/\Gamma_{\mathrm{B}}$ is the Brillouin gain factor.

Characteristics of the BDG Reflection Spectrum

Under the steady-state condition with a uniform BDG generated by two continuous pump waves and considering the slowly varying envelope approximation, the analytic expressions of the probe wave E_3 and the reflection wave E_4 can be obtained, and the coupled equations for the latter two optical waves can be simplified as follows:

$$\begin{aligned} \frac{\partial E_3}{\partial z} &= -ig_o\rho * E_4 e^{-i\Delta kz} \\ \frac{\partial E_4}{\partial z} &= ig_o\rho E_3 e^{i\Delta kz} \end{aligned} \tag{20}$$

The analytical expression of the reflection wave E_4 can be expressed as follows:

$$E_3(z) = E_3(L) \cdot \frac{2g\cosh(gz) - i\,\Delta k\,\sinh(gz)}{2g\cosh(gL) - i\,\Delta k\,\sinh(gL)} e^{-i\Delta k(z-L)/2} \tag{21a}$$

$$E_4(z) = E_3(L) \cdot \frac{2K_2\sinh(gz)}{2g\cosh(gL) - i\,\Delta k\,\sinh(gL)} e^{-i\Delta k(z+L)/2} \tag{21b}$$

where $K_1 = -\frac{1}{2}g_{\mathrm{B}}E_1^* E_2$, $K_2 = -\frac{1}{2}g_{\mathrm{B}}E_1 E_2^*$ and g is related to $g^2 = K_1 K_2 - (\Delta k)^2/4$.

The reflectivity of the BDG in steady state can be expressed as

$$R = \frac{|E_4(L)|^2}{|E_3(L)|^2} = \frac{\sinh^2(gL)}{\cosh^2(gL) - \Delta k^2/(4K_1 K_2)} \tag{22}$$

In most of the BDG experiments, the pump 1 power is much higher than the pump 2 power for a strong BDG distribution. In order to achieve a strong reflected signal from the BDG, the probe power is also chosen to be on the same order as the pump 1 power. Therefore, a more reasonable approximation compared to a moving FBG model is that the pump 1 and probe powers are undepleted, since the length of PMF used in the experiments is short, usually less than tens of meters. Moreover, since the probe power is also high in most experiments, the reflected signal also experiences an SBS amplification. With these assumptions, the numerical and the analytical solutions are compared and analyzed with the simulation parameters shown in Table 1, with the results shown graphically in Fig. 7.

From Fig. 7a, it can be observed that the analytical solutions (the black line for 0.5 m and the blue one for 1 m) agree well with the numerical solutions (the red dots for 0.5 m and the pink ones for 1 m). The full width at half-maximum (FWHM) of

Table 1 The simulation parameters

Parameter	Value
Power of pump 1	0.1 W
Power of pump 2	0.01 W
Power of probe	0.1 W
Fiber length (L)	1.0/0.5 m
Brillouin gain spectrum linewidth	30 MHz
Brillouin gain factor (g_B)	2.5×10^{-11} m/W
Phonon lifetime (τ_p)	5.3 ns
Effective modal area (A_{eff})	50 μm^2
Effective refractive index of slow axis (n_x)	1.4686
Effective refractive index of fast axis (n_y)	1.4683

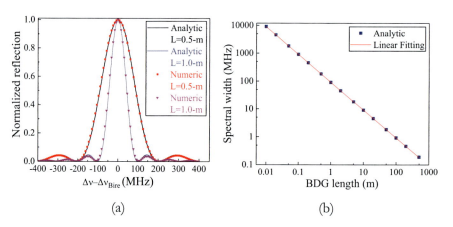

(a) (b)

Fig. 7 (**a**) The normalized BDG spectra; (**b**) the spectrum width (FWHM) versus BDG length

the BDG reflection spectrum is inversely proportional to the BDG length, which is shown in Fig. 7b, indicating that the spectral width becomes narrower as the BDG is lengthened.

Distributed Birefringence Measurement of PMFs and Its Sensing Applications

As stated above, the polarization characteristic of a PMF can be described using the phase birefringence and the group birefringence. The former represents the effective refractive index difference of the two polarization axes, which affects the polarization-maintaining capacities; and the latter is connected with the fiber group velocity, which is as a consequence of the fiber dispersion. In Section I, the Sagnac interferometer was introduced as well as its potential use for the measurement of the average group birefringence. For PMFs dominated by stress-induced birefringence with a small dispersion, it is not necessary to distinguish between phase birefringence and group birefringence, and hence the Sagnac interferometer can be used to evaluate the polarization-maintaining properties of the Panda fiber. However, it should be clarified that for the shape-induced PMFs, it is common that phase birefringence and group birefringence differ quite significantly, and thus they need to be measured separately using different measurement methods in order to characterize the PMF.

In this section, the distributed measurement of phase birefringence of a PMF using a BDG is theoretically analyzed and experimentally investigated. The phase birefringence of PMFs is verified by using the two-dimensional finite element numerical calculations and experimental measurements based on BDG. In addition, some sensing applications of birefringence measurement using a BDG are introduced, including simultaneously distributed temperature and strain measurement and distributed measurement of transverse pressure and hydrostatic pressure.

Numerical Calculations of the Birefringence

From Eq. 11, it is clear that the phase birefringence is proportional to the frequency shift between the probe and the co-propagating pump, defined as BireFS, indicating that the BDG can be used to evaluate the phase birefringence. The following content will provide further discussion for the phase birefringence measurement with BDG by comparing the numerical and experimental results.

First, the theoretical values of the phase birefringence and group birefringence are calculated using a two-dimensional finite element analysis method. Four different kinds of commercial PMFs as shown photographically in Fig. 8, including a Panda fiber (PMF-a), an elliptical core (ECORE) fiber (PMF-b), and

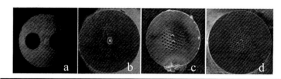

	PMF-a	PMF-b	PMF-c	PMF-d
Type	Panda	ECORE	PCF	PCF
Dia. (μm, cladding)	125	125	99	125
v_B (GHz, 25 ℃)	10.871	9.894	11.084	11.063

Fig. 8 The cross sections (upper) and the parameters (lower) of the FUT

Table 2 Calculated phase birefringence and group birefringence of the FUTs

	PMF-a	PMF-b	PMF-c	PMF-d
λ (nm)	1550			
Phase birefringence (10^{-4})	3.748	1.12	3.61	4.86
Group birefringence (10^{-4})	3.806	−2.02	−6.81	−8.71

two polarization-maintaining photonic crystal fibers (PM-PCFs) (PMF-c, PMF-d) are investigated as the fibers under test (FUT).

Among the four PMFs, the birefringence of the Panda fiber is induced by anisotropic stress surrounding the core, which is classified as the stress-induced birefringence PMF, and those of the ECORE fiber and the PM-PCFs, already used in fiber sensing applications for the temperature stability and high-birefringence characteristics, are induced by the imperfect symmetry geometrical effect, which are classified as the shape-induced birefringence PMF.

The phase birefringence is calculated for a set of incident wavelength, and the nonlinear relationship between the phase birefringence and the incident wavelength is numerically approximated with a quartic polynomial. Consequently the dispersion term $d\Delta n/d\lambda$ can be obtained and Δn^g calculated using Eq. 3. The calculated results for the four different PMFs are shown in Fig. 9, which show that the phase birefringence and group birefringence are different. For the shape-induced birefringence PMF, the values of phase birefringence and group birefringence not only differ in magnitude but have opposite signs. For stress-induced birefringence PMF, PMA-a, the absolute values of the phase birefringence and group birefringence exhibit very small difference at the long wavelength range.

The calculated phase birefringence and group birefringence of the four FUTs under the wavelength of 1550 nm are summarized in Table 2

The results show that there are large differences between the phase birefringence and group birefringence for PMF-b, PMF-c, and PMF-d, not only in opposite signs but also the absolute value. One of the major reasons for a large group birefringence is the dispersion, which is caused by the geometric asymmetry in structure, out of concentricity (or airhole distribution for PM-PCFs) and optical modal diffusion.

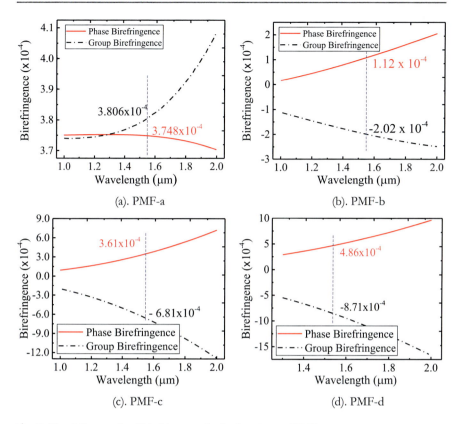

Fig. 9 Simulation results of birefringence for the four types of PMFs

Distributed Phase Birefringence Measurement of the PMFs with BDGs

Experimental Measurement for Different PMFs

Although several effective methods have been proposed for birefringence measurement as referred to in Section I, to date all of them have concentrated on point sensors that can measure only the average birefringence of the test fiber and cannot characterize the birefringence variation along the fiber. As verified earlier in this chapter, under the phase-matching condition expressed by Eq. 12, the phase birefringence can be obtained with the BireFS, which can be obtained through the reflected spectrum of the BDG, and can be measured to evaluate the polarization property with the BDG technique. Using a pulsed probe wave, position-dependent time delays occur for the grating reflection signals at different fiber axial positions, which provide the possibility of distributed phase birefringence measurement. In this section, the distributed phase birefringence measurement of PMF using BDG is discussed, and typical measurement results are provided.

Fig. 10 Experimental setup
for generating and probing a
BDG

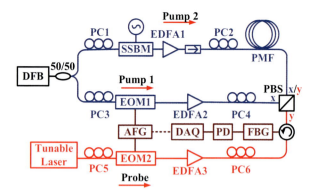

In the birefringence measurement scheme presented, a relatively long pump 1 pulse, a continuous-wave pump 2, and a short probe pulse are chosen to generate and detect the BDG in a PMF, where the long pump 1 pulse can effectively excite a BDG with a relatively low power to avoid other nonlinear effects, i.e., self-phase modulation or modulation instability, thus extending the measurement range in a long-length fiber, and the short probe pulse is used to provide good spatial resolution.

In this case, the configuration presented in Fig. 3a was adopted, and the experimental setup for generating and probing a BDG is shown in Fig. 10. A narrow linewidth distributed feedback (DFB) laser diode with a center wavelength at ∼1550 nm and a tunable laser were used as light sources of the pumps and the probe, respectively, with a frequency difference of several tens of GHz according to the BireFS of the FUT. The original frequency wave output from the DFB is regarded as the Brillouin pump wave (pump 1), while the Brillouin Stokes wave (pump 2) is generated by a single-sideband modulator (SSBM) and microwave generator. Two high extinction ratio electrooptic modulators (EOMs) were used to generate the pump 1 and the probe pulses. The two pump waves were polarized to the x polarization state, and the probe wave was polarized to y polarization state through the use of polarization controllers (PCs) and a polarization beam splitter (PBS). The BDG is generated following the pump 1 pulse along the fiber through the SBS process between pulsed pump 1 and continuous pump 2 and detected by the short probe pulse injected immediately following pump 1. The reflected wave is recorded using a photodiode (PD) and photo-amplifier whose output is connected to a data acquisition (DAQ). A reflection spectrum is obtained by sweeping the frequency of the tunable laser.

For this scheme, the spatial resolution of the measurement is determined by the duration of the probe pulse, and hence a high spatial resolution can be obtained using a short probe pulse (Song et al. 2016). The duration of the pump 1 pulse should be larger than that of the acoustic wave (∼10 ns) so that the latter can grow to its full extent through the SBS process realizing efficient excitation of the BDG. This setup has proved to be an effective scheme for long-range and high spatial resolution

Table 3 Group birefringence of the FUTs detected by a Sagnac interferometer

	PMF-a	PMF-b	PMF-c	PMF-d
$\Delta\lambda$ (nm)	1.46	1.09	0.612	0.406
Length (m)	4.06	9.83	5.7	7.5
λ (nm)	1550			
Δn_g (10^{-4})	4.05	2.25	6.89	7.89

distributed birefringence measurement without unwanted pump power depletions and nonlinear effects, i.e., self-phase modulation or modulation instability (Dong et al. 2013).

A comparison of the group birefringence as detected by the Sagnac interferometer and calculated according to Eq. 6 is shown in Table 3. The results agree well with the calculated group birefringence shown in Fig. 9 which vindicates the use of the simulation model.

Figure 11 shows the typical measurement results of phase birefringence distributions of the four PMFs using a BDG. The measurement results exhibit good agreement with the corresponding calculated results shown in Fig. 9, which confirms the validity of the phase birefringence measurement using a BDG in both theory and practice.

The data summarized in Table 4 lists the simulation and experiment results of birefringence for the four FUTs. The variance between the simulation and experiment results is quantified using the ratio of phase birefringence to group birefringence (PGR) which has been introduced as a reference parameter. The distributed phase birefringence is obtained in the case of a large value fluctuation. The average phase birefringence value along the FUTs in the case of simulation and experiments was selected to be comparable (as close as possible) in each case (Table 4).

From the comparison of the results and analysis above, the BDG clearly measures the phase birefringence, and for most PMFs, the phase and group birefringence values are different, especially for the shape-induced PMF such as PM-PCFs which have relatively large dispersion value. Thus, for evaluation of the polarization characteristics of certain kinds of PMFs, the selection of the birefringence detection methods should be carefully considered: the Sagnac method is a suitable group birefringence measurement method, while the BDG can be an effective method for distributed phase birefringence measurement.

Extension of the Measurement Range

It has already been stated that the BDG can be generated using two counter-propagating pump waves through the SBS interaction, and when the BDG is generated in the slow axis of a PMF, there are two possible probing processes which are defined according to the relative movement directions of the probe wave and the BDG. One is the coherent Stokes Brillouin scattering with the two propagating in the same direction, where the reflected wave frequency is down-converted by a fiber

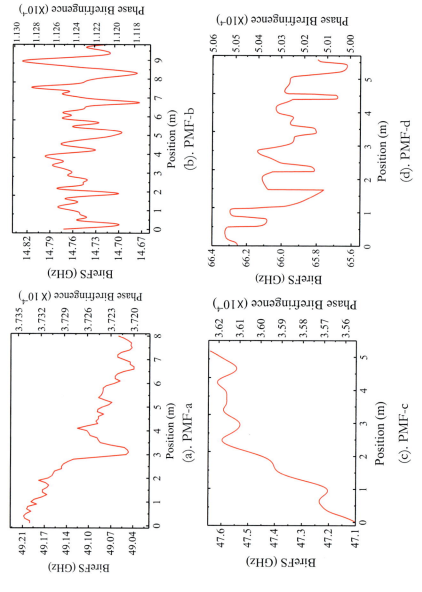

Fig. 11 Typical examples of measured phase birefringence distributions of the four PMFs

Table 4 Birefringence of different PMF samples ($\lambda = 1550$ nm)

	Simulation results			Experiment results		
	Phase (10^{-4})	Group (10^{-4})	PGR	BDG phase (10^{-4})	Sagnac group (10^{-4})	PGR
PMF-a	3.748	3.806	0.98	3.73	4.05	0.92
PMF-b	1.12	−2.02	−0.55	1.12	2.25	0.50
PMF-c	3.61	−6.81	−0.53	3.60	6.89	0.52
PMF-d	4.866	−8.71	−0.56	5.02	7.89	0.64

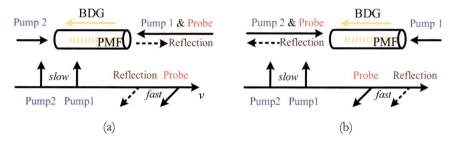

Fig. 12 BDG probing schemes of (**a**) coherent Stokes Brillouin scattering and (**b**) coherent anti-Stokes Brillouin scattering

BFS as shown in Fig. 12a; the other is the coherent anti-Stokes Brillouin scattering process for which the probe wave and the BDG propagating in opposite directions, resulting in a frequency up-converted reflected wave, as shown in Fig. 12b (Dong et al. 2010b). If two continuous-wave (C.W.) light sources are used to excite the BDG, a serious pump depletion occurs that greatly limits the measurement range.

In the investigation of this article, a scheme using a pulsed pump and a continuous pump is adopted in order to extend the measurement distance. Through optimizing the pumping and probing scheme, long-range and high spatial resolution distributed birefringence measurement can be achieved. There are two schemes of BDG generation according to the energy transfer direction between the two pumps: the pump pulse attenuated scheme and the pump pulse amplified scheme depending on whether the energy is transferred to or from the pulsed pump (Dong et al. 2013).

Pump Pulse Attenuated Scheme

In order to minimize the depletion effect between the two pumps, the pump pulse attenuated scheme shown in Fig. 13 was adopted, involving a C.W. pump 2, a relative long pump 1 pulse, where the latter can effectively excite a BDG with a relatively low power to avoid additional nonlinear effects in a long fiber, and the short probe pulse was used to improve the spatial resolution. The frequency of pump 1 is higher than that of pump 2 with a BFS, such that the energy of pulsed pump 1 is transferred to C.W. pump 2. The energy of pump 1 is therefore attenuated following

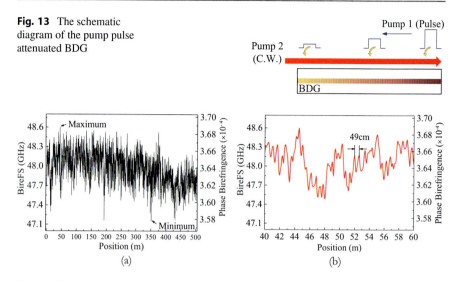

Fig. 13 The schematic diagram of the pump pulse attenuated BDG

Fig. 14 The measured birefringence over (**a**) the entire 500 m FUT and (**b**) the segment of 40–60 m

a long transmission through the fiber. The duration of pulsed pump 1 limits the interaction length between the two pumps, so as to minimize the depletion of the higher-frequency pump 1 power.

Figure 14a shows the distributed phase birefringence measurement results of a 500 m length of Panda fiber with a 20 cm spatial resolution (Dong et al. 2013) using the pump pulse attenuated BDG shown in Fig. 13. The maximum phase birefringence achieved was 3.6869×10^{-4} located at the position of 44.55 m, and the minimum value was 3.5772×10^{-4} located at the position of 350.65 m, resulting in a fluctuation of ~3% over the entire length of the fiber. The detailed birefringence distribution for the fiber segment of 40–60 m is shown in Fig. 14b. The results show a periodic variation of the birefringence with a spatial period of 49 cm, which corresponds to the spatial period of the fiber spool. It indicates that the fiber is subject to uneven stress when wound to the spool, and the birefringence variation could result from both uneven axial stress and uneven transverse pressure between different fiber layers, and it can be observed that the change of BireFS caused by the uneven stress is in the range of a few hundreds of megahertz (the difference between maximum and minimum BireFS values in the trace).

In addition to the peaks with a spatial period of 49 cm, there exist several other peaks with longer spatial periods, which could be caused by the residual stress induced by various disturbance factors, e.g., during the fiber drawing and coating processes or the nonuniformity in the fiber preform.

Pump Pulse Amplified Scheme

For the pump pulse attenuated scheme, although the interaction length between the two pumps is limited and the pump depletion is minimized, the continuous power

Fig. 15 The schematic diagram of the pump pulse amplified BDG.

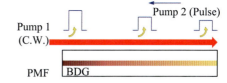

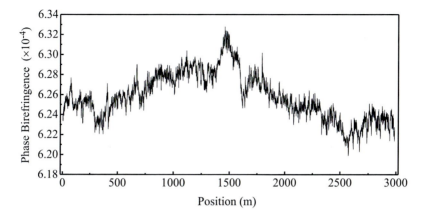

Fig. 16 The measured birefringence over the entire 3 km FUT

attenuation of the pulsed pump will still weaken the signal intensity, which limits the measurement range.

The pump pulse amplified BDG with a C.W. pump 1 and a pulsed pump 2 is shown in Fig. 15, where the frequency of the former is higher than that of the latter by a BFS. Because of the time invariance C.W. pump 1, the energy transferred from pump 1 to pump 2 remains constant, and the latter is continuously amplified, enabling stable BDG excitation at all axial positions along the entire fiber.

Figure 16 shows the distributed phase birefringence measurement results of a 3 km Panda fiber with a 20 cm spatial resolution using the pump pulse amplified BDG. It is worth noting that the distribution symmetry of the phase birefringence is caused by winding, and the periodic variation of birefringence is caused by the nonuniform distribution of stress or transverse pressure when winding the fiber onto the spool. Such a nonuniform stress distribution measurement is of great use in the field of online monitoring of fiber fabrication and fiber gyro evaluation.

Sensing Applications

In recent years, distributed optical fiber sensors based on Brillouin scattering have gained much interests owing to their inherent advantages of long-range measurement, high spatial resolution, and high accuracy. However, the existing Brillouin-based distributed fiber sensors, such as Brillouin optical time-domain analysis (BOTDA) and Brillouin optical correlation-domain analysis (BOCDA), are limited by a narrow measurement range, only being able to measure distributed temperature and strain (Martynkien et al. 2010; Bao and Chen 2011), which has restricted the growth of such distributed fiber sensors. In this section, the sensing applications of the BDG birefringence measurement are discussed. The sensing mechanism of a BDG derives from the sensitivity of the fiber birefringence to externally applied environmental parameters. The strain, temperature, and pressure fields affect the response of the fiber birefringence directly through expansion and compression of the fiber size and through the electrooptical effect, which results in the external perturbation-induced modification of the birefringence. On the basis of distributed birefringence measurement, a wider range of sensing applications can be realized, which provide numerous new and excellent opportunities for distributed fiber sensing.

Distributed Temperature and Strain Measurement

The existing Brillouin scattering-based distributed fiber sensor can realize distributed temperature and strain measurement, separately, through measuring the BFS distribution. However, since both strain and temperature can produce a BFS change, these distributed sensors are inevitably subject to cross sensitivity to these parameters. This is a regularly encountered problem when attempting discrimination of temperature and strain. Formerly, an additional measurable parameter, which has an independent relationship with strain and temperature, has been induced to allow separation of temperature and strain measurement. The BireFS associated with a BDG can be employed as the supplementary parameter to realize simultaneous and independent temperature and strain measurement.

It has been proved that phase birefringence can be introduced as the second parameter to simultaneously and completely discriminate the strain and temperature using a single length of Panda PMF (Zou et al. 2009). Two independent parameters in a fiber, the BFS and phase birefringence, were measured to obtain two independent responses to strain and temperature. Compared with the positive strain and temperature coefficients of BFS, the phase birefringence has a positive strain coefficient and negative temperature coefficient, which ensures a high discrimination accuracy.

The BireFS and BFS distributions obtained by the BDG and typical BOTDA can be used to determine the temperature and strain, which were described given by Dong et al. (2010c) as follows:

$$\begin{bmatrix} \Delta\varepsilon \\ \Delta T \end{bmatrix} = \frac{1}{C_B^\varepsilon C_{\mathrm{Bire}}^T - C_B^T C_{\mathrm{Bire}}^\varepsilon} \begin{bmatrix} C_{\mathrm{Bire}}^T & -C_B^T \\ -C_{\mathrm{Bire}}^\varepsilon & C_B^\varepsilon \end{bmatrix} \begin{bmatrix} \Delta v_B \\ \Delta v_{\mathrm{Bire}} \end{bmatrix} \tag{23}$$

where ΔT and $\Delta\varepsilon$ are changes in strain and temperature, respectively, and C_B^ε, C_B^T, $C_{\mathrm{Bire}}^\varepsilon$, and C_{Bire}^T are coefficients of strain and temperature related to BFS and BireFS, respectively. Here, C_B^ε, C_B^T, and $C_{\mathrm{Bire}}^\varepsilon$ have a positive sign, while C_{Bire}^T has a negative sign, and hence $C_B^\varepsilon C_{\text{Bire}}^T - C_B^T C_{\text{Bire}}^\varepsilon$ is a large value, which ensures a high discrimination accuracy (Zou et al. 2009).

The individual temperature and strain dependences of BFS of the Panda fiber were measured using a standard BOTDA system, with $C_B^T = 1.12$ MHz/°C and $C_B^\varepsilon = 0.0482$ MHz/$\mu\varepsilon$, respectively. The dependence of the BireFS on temperature and strain were measured, and temperature and strain coefficients of BireFS were $C_{\text{Bire}}^T = -54.38$ MHz/°C and $C_{\text{Bire}}^\varepsilon = 1.13$ MHz/$\mu\varepsilon$ as shown in Fig. 17, which indicates that the BireFS provides a higher sensitivity for temperature and strain measurement compared with the BFS.

A 6 m Panda fiber is used to conduct the temperature and strain simultaneous measurement experiment, with a 1 m segment stressed under a strain of 670 $\mu\varepsilon$ and 1 m segment heated under the temperature higher than room temperature by 30 °C, as shown in Fig. 18.

The BFS was obtained using a one-peak Lorentz fitting, and the BFS difference (subtracting the BFS at room temperature and loose state) is shown in Fig. 19a. The BireFS difference (subtracting the BireFS at room temperature and loose state) is also shown in Fig. 19a. The measured strain and temperature coefficients and distributions of BireFS and BFS along the sensing fiber are shown in Fig. 19a, and these in turn allow the distributions of the strain and temperature to be simultaneously calculated as shown in Fig. 19b. The fitting uncertainty of BFS is 0.4 MHz, corresponding to the temperature and strain accuracy of the discrimination of $\delta T = 0.36$ °C and $\delta\varepsilon = 8.3$ $\mu\varepsilon$, and the fitting uncertainty of BireFS is 3 MHz, corresponding to $\delta T = 0.06$ °C and $\delta\varepsilon = 2.7$ $\mu\varepsilon$. Therefore the BireFS provides higher accuracy compared to BFS, as the discrimination accuracy is limited by the

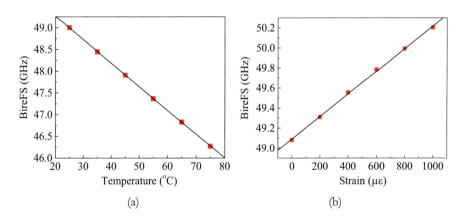

Fig. 17 The dependence of BireFS on (**a**) temperature and (**b**) strain of the Panda fiber

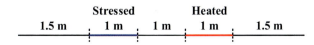

Fig. 18 Layout of the 6 m sensing fiber with a 1 m stressed segment and 1 m heated segment

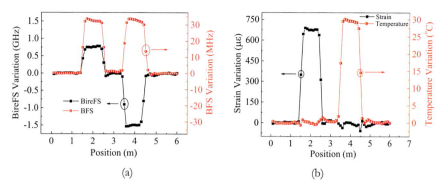

Fig. 19 (a) Measured BireFS and BFS and (b) calculated temperature and strain

uncertainty of BFS. From the results in Fig. 19, it can be seen that the temperature and strain can be discriminated from each other except at the transition regions, at which the uncertainty in BFS and BireFS due to the spatial resolution causes the error.

Distributed Transverse Pressure Measurement

In recent years, transverse pressure sensing techniques have gained much attention mainly due to the increasing demands on high-sensitivity non-axisymmetric pressure detection in fields of structural health monitoring and civil engineering (Kringlebotn et al. 1996; Shao et al. 2010). The reliable and effective measurement of transverse pressure is needed in actual manufacture. The traditional fiber transverse pressure sensor has largely been based on a FBG with a high-sensitivity and a simple structure (Jewart et al. 2006; Zu et al. 2011); however, it cannot realize distributed measurement. Although Brillouin scattering-based fiber sensors have existed for two decades and are widely used in external physical parameter distributed sensing areas, such as temperature and strain, the traditional Brillouin distributed fiber sensing systems are insensitive to transverse pressure.

Fiber birefringence changes because of the elastic-optical effect when subjected to the transverse pressure. Distributed transverse pressure measurement can be realized by measuring the birefringence changes using a BDG. Figure 20 shows the self-designed pressure applying platform with a 10-m-long test fiber and a support fiber embedded between a 20 cm glass plate and a controllable metal support platform. An ECORE fiber is used as FUT due to its lower sensitivity to temperature. The temperature sensitivity of ECORE fiber is almost an order of magnitude lower than that of Panda fiber, hence resulting in weaker temperature cross talk. A SMF is used as the supporting fiber; since the SMF has the same structural mechanical properties as the test fiber, including the geometric size, the Young's modulus, and the Poisson's ratio, when applying the pressure in the middle of the two fibers, each fiber carries half of the pressure. The sensing fiber is maintained with a low axial strain to prevent unwanted twist by the two rotary mounts. By rotating the mounts,

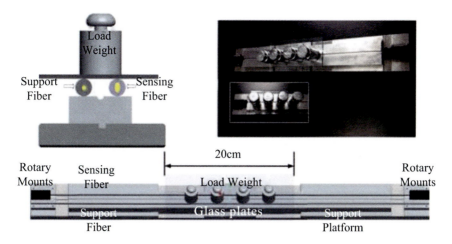

Fig. 20 Setup of pressure weight applied with two rotary mounts

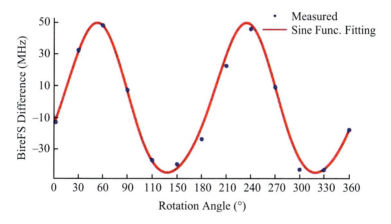

Fig. 21 Change in birefringence as a function of pressure direction

the angle between the direction of application of the pressure and the fiber axis can be adjusted (Dong et al. 2015).

Figure 21 indicates the birefringence change as a function of the pressure direction. The measured results (blue dots) agree well with a sinusoidal fitting curve, where the maximum increment of birefringence appears at the slow-axis direction (50° and 230°) while the maximum decrement at the fast-axis direction (140° and 320°)

Figure 22 shows the transverse pressure dependences of BireFS of the ECORE fiber and Panda fiber. The results for ECORE fiber in Fig. 22a illustrate the transverse pressure sensitivities of 6.217 GHz/Nmm^{-1} for the fast-axis direction and 6.28 GHz/Nmm^{-1} for the slow-axis direction. And for the Panda fiber shown in

Fig. 22b, the measured sensitivities are 4.097 GHz/Nmm^{-1} and 3.439 GHz/Nmm^{-1} for the slow and fast axes, respectively (Dong et al. 2015).

To verify the distributed measurement and the direction-sensitive abilities of transverse pressure sensing, the simultaneous measurement of two transverse pressures applied on two separate fiber locations is performed, with one pressure along the fast axis and the other along the slow. The results are shown in Fig. 23, where the red and green lines represent the measured BireFS distributions with and without pressure, respectively, and the blue line represents the BireFS difference and the corresponding transverse pressure. The position, magnitude, and direction of the pressure can be distinguished in Fig. 23.

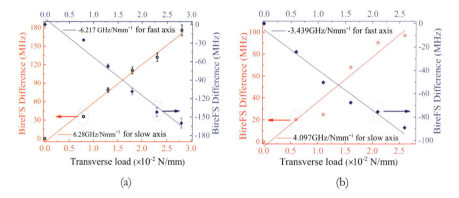

Fig. 22 BireFS differences versus transverse pressures for (**a**) ECORE fiber and (**b**) Panda fiber

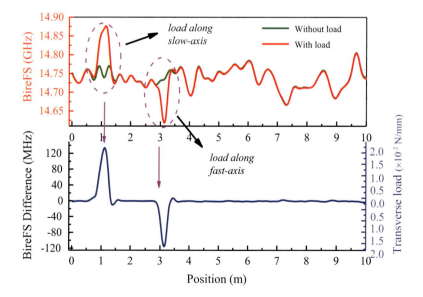

Fig. 23 Results of the distributed transverse pressure measurements

Distributed Hydrostatic Pressure Measurement

Hydrostatic pressure measurement is widely needed for a variety of areas, such as automotive production, aerospace engineering, and harsh environment detection. Currently available optical fiber hydrostatic pressure sensors are mainly based on the FBGs and interferometric sensors (Tarnowski et al. 2013; Bhowmik et al. 2015) which are point sensors that cannot realize distributed measurement. Distributed hydrostatic pressure sensing is needed in several industrial application areas, e.g., for measurement of oil well pressure with high temperature and where the pressure range is ultrahigh, being above 100 MPa. It would be highly significant for the oil-gas production to be able to successfully realize distributed hydrostatic pressure measurement.

When the fiber experiences the hydrostatic pressure, its birefringence will change because of the elastic-optical effect, based on which, hydrostatic pressure along the fiber can be measured by detecting the birefringence change of the fiber using a BDG.

A PM-PCF is adopted as the test fiber to conduct the distributed hydrostatic pressure measurement experiment. This kind of fiber features a porous structure, a facility to deform, and a pure silica material. The thermal expansion coefficients between the solid core and the porous cladding of the pure silica PM-PCF are approximately equal; therefore, there is a lower-temperature birefringence coefficient, and the temperature cross talk can be more accurately compensated. These characteristics make it a highly suitable choice for increasing the measurement sensitivity and decreasing the temperature cross talk effect (Teng et al. 2016).

Figure 24 shows the layout of the PM-PCF which is divided into three parts with two 20 cm pressurized parts in a self-designed pressure vessel and a heated part in a controllable oven, where the inset shows the cross section of the PM-PCF.

The temperature and hydrostatic pressure dependences of BFS of the PM-PCF are investigated using the typical BOTDA system, with the results shown in Fig. 25, which indicates the temperature dependence of the BFS, with a coefficient of 1.07 MHz/°C. And it can be seen that the BFS has low sensitivity to hydrostatic pressure, with a maximum BFS fluctuation of only 1.88 MHz within the pressure range of 0–1.1 MPa, which can be neglected.

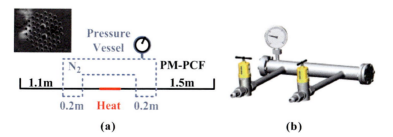

Fig. 24 (a) The cross section and layout of the FUT and (b) the self-designed pressure vessel

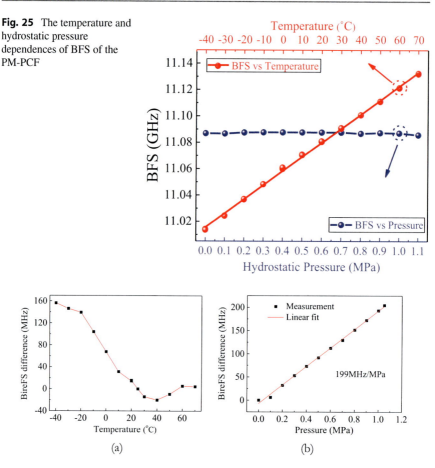

Fig. 25 The temperature and hydrostatic pressure dependences of BFS of the PM-PCF

Fig. 26 Measured BireFS differences of PM-PCF versus the (**a**) temperature and (**b**) hydrostatic pressure

In order to eliminate temperature cross talk effects on the birefringence changes, the first step is to interrogate BFS changes by obtaining the temperature distribution over the FUT and then obtain the BireFS changes of the FUT induced when both temperature and pressures are interrogated. Next, the effect on the fiber birefringence from the ambient temperature can be subtracted according to the measured temperature distribution. Consequently, the hydrostatic pressure-induced birefringence changes without temperature disturbance can be isolated.

The temperature and hydrostatic pressure dependences of the BireFS of the PM-PCF have been investigated with the results shown in Fig. 26.

Figure 26a shows the measured BireFS with respect to temperature, with a maximum difference of 176 MHz for the temperature range of −40 °C to 40 °C. The characteristic of the nonlinear temperature dependence of birefringence is mainly due to the effect of coating material. The standard deviation of temperature measurement of BOTDA is about ±0.5 °C, which corresponds to temperature-induced

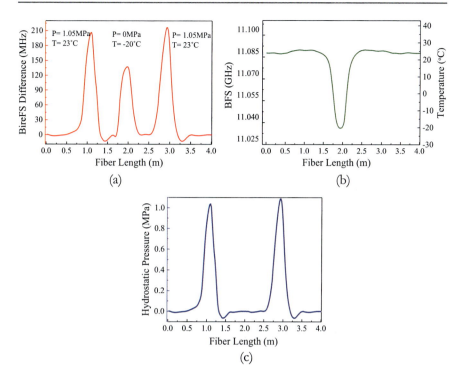

Fig. 27 Results of temperature-compensated distributed hydrostatic pressure sensing, (**a**) the measured BireFS difference of the FUT, (**b**) the measured BFS of the FUT, and (**c**) the measured hydrostatic pressure with temperature compensation

BireFS changes of ± 1.1 MHz for the PM-PCF and BireFS changes of ± 27.5 MHz for Panda. From Fig. 26, it can be calculated that the BireFS changes at ± 1.1 MHz are equivalent to a hydrostatic pressure change of 0.005 MPa; by contrast, the BireFS changes at ± 27.5 MHz are equal to a hydrostatic pressure change of ± 0.14 MPa. Therefore, the PM-PCF is preferred when compensating for the temperature changes and reducing the cross talk to pressure measurement.

Figure 26b shows the linear dependence of BireFS difference on hydrostatic pressure, with a sensitivity of 199 MHz/MPa. When considering the measurement standard deviation of $\delta\, v_{\mathrm{Bire}} = \pm 5$ MHz from the measured distributed BireFS results, the accuracy is as high as 0.025 MPa. When taking the error propagation from the temperature measurement error of 0.5 °C of BOTDA into consideration, the maximum measurement accuracy should be modified to less than 0.03 MPa.

The BFS difference, Δv_{B}, and BireFS difference, Δv_{Bire}, caused by temperature and pressure can be expressed as

$$\begin{aligned} \Delta v_{\mathrm{B}} &= C_{\mathrm{B}}^{T}\Delta T + \Delta v_{\mathrm{B}}^{P} \\ \Delta v_{\mathrm{Bire}} &= \Delta v_{\mathrm{Bire}}^{T} + C_{\mathrm{Bire}}^{P}\Delta P \end{aligned} \tag{24}$$

where C_{B}^{T} and C_{Bire}^{P} are coefficients to the temperature related to BFS and to hydrostatic pressure related to BireFS, respectively. It has been demonstrated that the pressure-induced BFS changes, $\Delta v_{\mathrm{B}}^{P}$, can be ignored. And the temperature-induced BFS changes, $C_{\mathrm{B}}^{T}\Delta T$, can be obtained from the BOTDA data, and the temperature changes ΔT can be computed. Therefore, the temperature-induced BireFS difference, $\Delta v_{\mathrm{Bire}}^{T}$, can be calculated according to Fig. 26a. Finally, $\Delta v_{\mathrm{Bire}}^{T}$ can be subtracted, and the hydrostatic pressure-induced birefringence changes, $C_{\mathrm{Bire}}^{P}\Delta P$, and the hydrostatic pressure changes ΔP without temperature disturbance can be obtained.

Figure 27 shows the process of extraction of the hydrostatic pressure distributed measurement and the temperature compensation. In Fig. 27a, the hydrostatic pressure (1.05 MPa) and temperature (-20 °C) are simultaneously measured using the BDG. Figure 27b shows the temperature measurement result using a standard BOTDA, which clearly distinguishes the temperature. The hydrostatic pressure measured after temperature compensation is shown in Fig. 27c. The results shown in Fig. 27 are consistent with the results in Fig. 26.

Conclusion

Phase birefringence is an important optical parameter for defining the polarization property of a PMF. Distributed phase birefringence measurement can be realized using the BDG technique, creating a new view of PMF and fiber gyro quality test. In this chapter, the basic principles of distributed birefringence measurement of PMF based on BDG have been introduced. The BDG is an acoustic-modulated weak optical grating obeying the fundamental laws of the standard FBG. The generation and probing of a BDG involve four optical waves, and the FWM model clearly demonstrates this coupling process. The BireFS has a proportional relationship with the phase birefringence, which provides the foundation of birefringence measurement with BDG. The theoretical analysis and numerical simulation of birefringence measurement with BDG have been demonstrated. The chapter is concluded with examples of real sensing applications of distributed phase birefringence measurement with BDG: simultaneous temperature and strain measurement, distributed transverse pressure sensing, and hydrostatic pressure sensing.

References

P. G. Agrawal (ed.), *Nonlinear Fiber Optics* (Academic Press, San Diego, 2007)
X. Bao, L. Chen, Sensors **11**, 4 (2011)
K. Bhowmik, G.D. Peng, Y. Luo, J. Lightwave Technol. **33**, 12 (2015)
S. Chin, L. Thévenaz, Laser Photonics Rev. **6**, 724 (2012)
Y. Dong, X. Bao, L. Chen, Opt. Lett. **34**, 17 (2009)
Y. Dong, L. Chen, X. Bao, Opt. Lett. **35**, 2 (2010a)
Y. Dong, L. Chen, X. Bao, Opt. Express **18**, 18 (2010b)
Y. Dong, L. Chen, X. Bao, IEEE Photon. Technol. Lett. **22**, 18 (2010c)

Y. Dong, H. Zhang, Z. Lu, J. Lightwave Technol. **31**, 16 (2013)
Y. Dong, T. Jiang, L. Teng, Opt. Lett. **39**, 10 (2014)
Y. Dong, L. Teng, P. Tong, Opt. Lett. **40**, 21 (2015)
T. Erdogan, J. Lightwave Technol. **15**, 8 (1997)
H. Eugene (ed.), *Optics* (Pearson Education, San Francisco/Boston/New York, 2002)
X. Fang, R.O. Claus, Opt. Lett. **20**, 20 (1995)
P. Hlubina, D. Ciprian, Opt. Express **15**, 25 (2007)
S. Huang, Z. Lin, Appl. Opt. **24**, 15 (1985)
C. Jewart, K.P. Chen, B. McMillen, Opt. Lett. **31**, 15 (2006)
C. Kaczmarek, Opt. Appl. **42**, 4 (2012)
I. Kaminow, IEEE J. Quantum Electron. **17**, 1 (1981)
G. Keiser (ed.), *Optical fibers communications* (Wiley, New York, 2003)
K. Kikuchi, T. Okoshi, Opt. Lett. **8**, 2 (1983)
D.H. Kim, J.U. Kang, Opt. Express **12**, 19 (2004)
J.T. Kringlebotn, W.H. Loh, R.I. Laming, Opt. Lett. **21**, 22 (1996)
H.C. Lefevre, *The fiber-optic gyroscope* (Artech House, Boston/London, 2014)
S. Li, M.J. Li, Opt. Lett. **37**, 22 (2012)
T. Martynkien, G. Statkiewicz-Barabach, J. Olszewski, Opt. Express **18**, 14 (2010)
J. Noda, K. Okamoto, Y. Sasaki, J. Lightwave Technol. **4**, 8 (1986)
R. Pant, E. Li, C.G. Poulton, Opt. Lett. **38**, 3 (2013)
J. Sancho, N. Primerov, S. Chin, Opt. Express **20**, 6 (2012)
M. Santagiustina, S. Chin, N. Primerov, L. Ursini, L. Thévenaz, Sci. Rep. **3**, 1594 (2013)
L.Y. Shao, Q. Jiang, J. Albert, Appl. Opt. **49**, 36 (2010)
V. Sikka, S. Balasubramanian, A. Viswanath, Appl. Opt. **37**, 2 (1998)
K.Y. Song, Opt. Lett. **36**, 23 (2011)
K.Y. Song, Opt. Express **20**, 25 (2012)
K.Y. Song, J.H. Yoon, Opt. Lett. **35**, 17 (2010)
K.Y. Song, W. Zou, Z. He, Opt. Lett. **33**, 9 (2008)
K.Y. Song, K. Hotate, W. Zou, J. Lightwave Technol. **35**, 16 (2016)
K. Suzuki, H. Kubota, S. Kawanishi, Opt. Express **9**, 13 (2001)
M. Szczurowski, W. Urbanczyk, M. Napiorkowski, Appl. Opt. **50**, 17 (2011)
K. Takada, J. Noda, R. Ulrich, Appl. Opt. **24**, 24 (1985)
K. Tarnowski, A. Anuszkiewicz, J. Olszewski, Opt. Lett. **38**, 24 (2013)
L. Teng, H. Zhang, Y. Dong, Opt. Lett. **41**, 18 (2016)
P.K.A. Wai, C.R. Menyak, J. Lightwave Technol. **14**, 2 (1996)
H.G. Winful, Opt. Express **21**, 8 (2013)
B. Xu, C.L. Zhao, F. Yang, Opt. Lett. **41**, 7 (2016)
W. Zhou, Z. He, K. Hotate, Opt. Express **17**, 3 (2009)
D.P. Zhou, Y. Dong, L. Chen, Opt. Express **19**, 21 (2011)
W. Zou, J. Chen, Opt. Express **21**, 12 (2013)
W. Zou, Z. He, K.Y. Song, Opt. Lett. **34**, 7 (2009)
P. Zu, C.C. Chan, Y. Jin, Meas. Sci. Technol. **22**, 2 (2011)

Characterization of Distributed Polarization-Mode Coupling for Fiber Coils

32

Jun Yang, Zhangjun Yu, and Libo Yuan

Contents

J. Yang · Z. Yu · L. Yuan (✉)
Key Lab of In-Fiber Integrated Optics, Ministry Education of China, Harbin Engineering University, Harbin, China

College of Science, Harbin Engineering University, Harbin, China
e-mail: yangjun@hrbeu.edu.cn; yuzhangjun@yeah.net; lbyuan@vip.sina.com

© Springer Nature Singapore Pte Ltd. 2019
G.-D. Peng (ed.), *Handbook of Optical Fibers*,
https://doi.org/10.1007/978-981-10-7087-7_58

Abstract

Fiber coil wound by polarization-maintaining fiber is an important part of fiber optic gyroscope. Its distributed polarization-mode coupling degrades the performance of fiber optic gyroscope on drift bias. Characterization of distributed polarization-mode coupling coefficient could evaluate the quality of fiber coils. In addition, it can help improve the wound technique, and then suppress the polarization-mode coupling. The optical coherence domain polarimetry system based on white-light interferometer is a suitable instrument for characterizing the distributed polarization-mode coupling of polarization-maintaining fibers. This chapter contains the measurement and analysis of polarization-mode coupling, system performance improvement of the optical coherence domain system, dispersion effect suppression of fiber coil under test, and the diagnosis for fiber coil.

Introduction

In the late 1980s, to accurately positioning the fault location in fiber and waveguide components, a novel measurement technique referred to as optical coherence domain reflectometry (OCDR) or optical low coherence reflectometry (OLCR) is presented based on the white-light interferometry (Danielson and Whittenberg 1987; Takada et al. 1987b; Youngquist et al. 1987). OLCR is a noninvasive measurement technique with high spatial resolution of micron dimension and high sensitivity for Rayleigh scattering measurement. The rapid development of OLCR is because that it fills a vacancy on the performance and measurement field relative to the optical time domain reflectometry (OTDR) (Barnoski et al. 1977) and optical frequency domain reflectometry (OFDR) (Eickhoff and Ulrich 1981).

OLCR is a typical distributed measurement system for reflected (or scattered) parameter testing. It consists of a Michelson interferometer or Mach-Zehnder interferometer that excited by a wideband spectrum light source. It can be widely used for measuring the optical fiber components and waveguide. There are two types of OLCR techniques: amplitude-sensitive (AS) and phase-sensitive (PS) OLCR. The AS-OLCR technique detects the envelop peak intensity of white-light interferometric signal and corresponding optical path difference. It is mainly used for positioning the fault points in fiber optical components and devices. The PS-OLCR technique could measure the distributed complex refractive index, dispersion, spectral response function with high accuracy. It requires precisely recovering the phase information from the white-light interferometric signal. To enhance the accuracy of the scanning optical path, a reference laser should be added to the OLCR system. Accordingly, the system complexity, control, and signal processing difficulty of PS-OLCR is quite higher than the AS-OLCR.

Table 1 Comparison of different distributed measurement techniques

	AS-OLCR	PS-OLCR	OCDP	OFDR	OTDR
Measured quantity	Reflectivity	Complex refractive index, dispersion, spectral response function	Polarization crosstalk, PMD	Fault points, dispersion	Fault points
Spatial resolution	<2 μm	<2 μm	~5 cm	~1 mm	~0.5 m
Measurement range	<5 m	<5 m	<5 km	~2 km	~100 km
Sensitivity	−162 dB	−162 dB	−95 dB	−130 dB	−50 dB
Dynamic range	>120 dB	>120 dB	>100 dB	~70 dB	~50 dB
Time to measure	~100 s	~100 s	~100 s	~3 s	~100 s
Light source	White light	White light	White light	Tunable laser	Pulse laser
Detection method	Coherence	Coherence	Coherence	Coherence	Direct

AS-OLCR amplitude-sensitive optical low coherence reflectometry, *PS-OLCR* phase-sensitive optical low coherence reflectometry, *OCDP* optical coherence domain polarimetry, *OFDR* optical frequency domain reflectometry, *OTDR* optical time domain reflectometry

Optical coherence domain polarimetry (OCDP) is a distributed polarization-mode coupling (PMC) measurement technique with high accuracy. It is like the OLCR technique, but aim at polarization features of components. It is also based on the white-light interferometry. Interference between different polarized modes is achieved by a scanning interferometer to optical path compensation. OCDP could accurately measure the position of PMC occurred, the intensity of PMC, and the polarization extinction ratio (PER) of components. Thus, it is widely used in polarization-maintaining fiber (PMF) fabricating (Hotate and Kamatani 1993), axis alignment of PMF (Takada et al. 1987a), and PER measurement (Choi and Jo 2009). It directly and truly describes the transmitting behavior of light in the optical circuit. Therefore, it is suitable to test and evaluate the optical fiber components, devices, and ultrahigh accuracy fiber optic gyroscope. The comparison of different distributed measurement techniques is shown in Table 1.

This chapter deals with the distributed polarization crosstalk, i.e., polarization-mode coupling, of PMF, and begins with its measurement and analysis methods. The succeeding three sections discuss the measurement range and accuracy. The measurement accuracy consists of two aspects, optical delay line-induced optical power fluctuation and dispersion-induced amplitude fading. In the end, the diagnosis method for PMF coil is discussed.

Measurement and Analysis for Distributed Polarization Crosstalk

Distributed Polarization Crosstalk

Polarization-maintaining fiber (PMF), just as its name implies, could hold the linear polarization state of light propagating through it. As is well known, there are two categories of PMFs, low-birefringent and high-birefringent fibers. The high-birefringent fiber could propagate two polarization modes, HE_{11x} and HE_{11y}, with low polarization-mode coupling. This enables PMF to fulfill its potential in distributed sensing.

Polarization crosstalk, also known as polarization-mode coupling, is the optical power coupling between the two polarized modes of PMF, which occurs at position of inner structural imperfection or external perturbation of the PMF. Polarization crosstalk degenerates the performance of device based on the polarization-maintaining feature of PMF. On the contrary, polarization crosstalk leads the PMF to be widely used in distributed sensing. In both the aspects, the distributed polarization crosstalk of PMF is worth investigating. In the case of fiber optic gyroscope coil wound with PMF, for instance, measuring and analysis, the distributed polarization crosstalk contributes to improve the winding technology and suppress the polarization crosstalk.

Optical Coherence Domain Polarimetry System

The optical coherence domain polarimetry (OCDP) system based on white-light interferometry (WLI) for measuring the distributed polarization crosstalk of PMF is shown in Fig. 1 (Li et al. 2016). The white light from a superluminescent light-emitting diode (SLD) is divided into two beams through a 98:2 fiber coupler. Two percent of the light is for monitoring the output power of the light source, and the remaining light is polarized by a $0°$-rotated polarizer 1. Then the linearly polarized light is launched into the slow-axis of PMF under test. The PMF under test with multiple perturbation points (Points X_1, X_2, ..., X_J) is spliced to Polarizers 1 and 2 at Points X_{in} and X_{out}, respectively. A part of linearly polarized light along the slow-axis will be coupled into the orthogonal axis at a perturbation point of PMF. Then it will generate two optical paths (OPs) with orthogonally eigenmodes, which will induce OPD due to the birefringence Δn of the PMF. Afterward, the first-order coupling interferograms are detected with photodiodes (PD) by the scanning Mach–Zehnder interferometer (MZI) that will compensate the OPD. The large dynamic range of system achieved in the previous works is improved from many aspects: Firstly, a differential detection is completed by adopting two PDs (Yuan et al. 2015). Secondly, the dispersion of fiber-based WLI is compensated by inserting a segment of dispersion-shifted fiber (DSF) into one arm of MZI. Thirdly, we utilize the differential scanning MZI with two lenses to suppress the optical power fluctuation (Li et al. 2015).

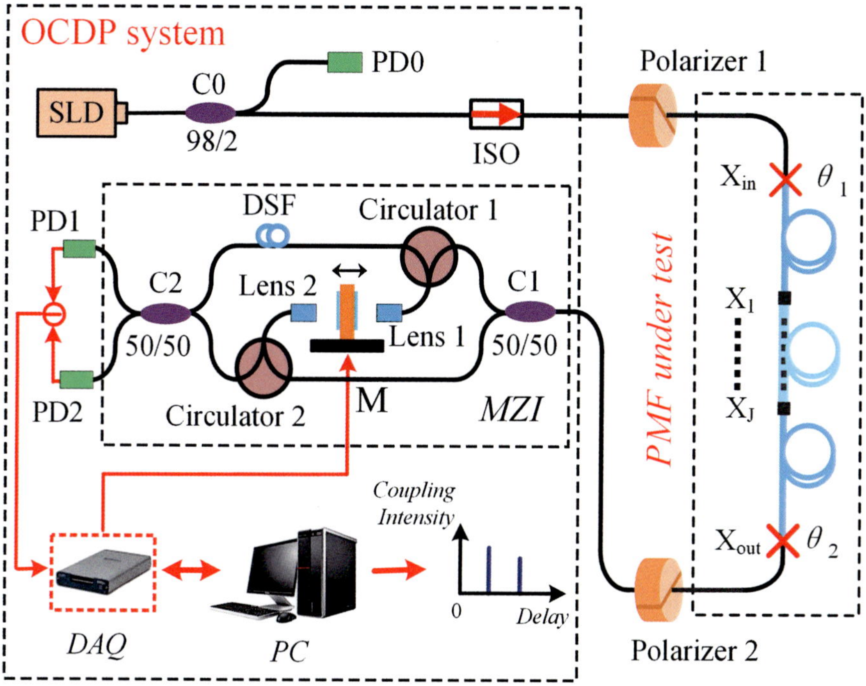

Fig. 1 Optical coherence domain polarimetry system based on WLI for measuring the distributed polarization crosstalk of PMF (C: coupler, PD: photodiode, ISO: Isolator, M: motor, MZI: Mach–Zehnder interferometer, DSF: dispersion-shifted fiber, DAQ: data acquisition)

Fig. 2 The schematic diagram of Jones matrix model for a long PMF

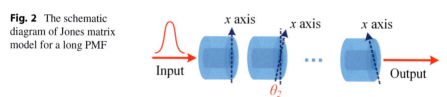

Jones Matrix Method

Jones calculus approach is a common tool to analyze polarized light. It represents the state of polarization of light by a two-component vector, which is known as a Jones vector. In addition, it uses a 2×2 matrix, namely Jones matrix, to calculate the effect of transmission medium on a given state of polarization of light. The more detailed contents of Jones calculus approach could be found in many books about polarized light (Kumar and Ghatak 2011; Rogers 2008). Here, we use the Jones matrix method to model and analyze the polarization crosstalk of PMF (Fig. 2).

Considering a long PMF that is spliced by n segments short PMF, every segment can be regarded as a birefringent element without perturbation point. The Jones matrix of the ith birefringent element can be expressed as:

$$J_i = A_i \cdot P_i \cdot C_i = \begin{bmatrix} \alpha_{xi} & 0 \\ 0 & \alpha_{yi} \end{bmatrix} \begin{bmatrix} e^{\mathrm{i}\beta_{xi}l_i} & 0 \\ 0 & e^{\mathrm{i}\beta_{yi}l_i} \end{bmatrix} \begin{bmatrix} \cos\theta_i & \sin\theta_i \\ -\sin\theta_i & \cos\theta_i \end{bmatrix} \tag{1}$$

where matrices A_i, P_i, C_i is the transmission loss, phase delay, and power coupling matrix, respectively; α_{xi} and α_{yi} denote the transmission loss of x polarized light and y polarized light, respectively; β_{xi} and β_{yi} denote the propagation constants of the x and y axis, respectively, and l_i is the length of this element; the nonitalic i is the imaginary unit; θ_i is the axis alignment angle between the x axes of this element and the $(i-1)$th element. When $i = 1$, θ_i denotes the angle between the x axis of this element and the input polarized light. Accordingly, for this PMF with n birefringent element, i.e., $n-1$ perturbation points, the Jones matrix could be expressed as:

$$J = \prod_{i=1}^{n} (A_i P_i C_i) \tag{2}$$

The power coupling matrix can also be expressed as:

$$C_i = \begin{bmatrix} \sqrt{1-\rho_i^2} & \rho_i \\ -\rho_i & \sqrt{1-\rho_i^2} \end{bmatrix} \tag{3}$$

where $\rho_i = \sin\theta_i$ is the amplitude coupling coefficient of this splicing point. Thus, other types of perturbation point with amplitude coupling coefficient ρ, for instance, external force-induced perturbation, can be regarded as a splicing point with axis alignment angle θ.

Optical Path Tracking Method

In addition to the Jones matrix method, optical path tracking (OPT) method (Li et al. 2016) can be used to analyze the polarization crosstalk of PMF. It has been recognized that a pair of OPs with an OPD less than the coherence length will suffer interference at the output of MZI and lead to an interferogram. For an identical scanning OPD in the spatial domain, there will be numerous possible pairs of OPs introduced by multiple perturbation points along PMF. The interferograms corresponding to the same scanning OPD with distinct OP pairs will give rise to the superposition of interference intensity. Therefore, the direct analysis of polarization crosstalk for the entire PMF with multiple perturbation points, such as Jones matrix (Choi and Jo 2009), will be rather complicated and cannot obtain the general formulas due to complex superposition phenomenon. Here, the OPT method based on the enumeration method and graphic method is presented to simplify the analysis of polarization crosstalk.

The steps of OPT method can be briefly described as follows: (1) Stable unit – we divide an entire PMF into stable units based on the corresponding OPD conditions and list all the OP pairs with graphic method; (2) Recursion formula – then we obtain the recursion formula between arbitrary adjacent stable units; (3) General formulas – finally we extend the recursion formulas to the entire PMF under test and derive the general formulas of interference intensity.

Stable Unit and Recursion Formula

We define the segment $(X_{j-p}, X_j]$ ($p \geq 1$) of PMF as a stable unit with the following three characteristics: (a) the pair of OPs merely occurs once coupling between the orthogonal axes of PMF at the right end (Point X_j) of segment $(X_{j-p}, X_j]$; (b) The position of X_{j-p} satisfies that if we move it right until to Point X_j, the OPD of segment $(X_{j-p}, X_j]$ is always invariable; (c) Point X_{j-p} is chosen as the leftmost point that satisfies condition (b) in order to guarantee that all the stable units are linked end-to-end.

Then stable unit can be classified into two categories based on the corresponding OPD introduced by the OP pairs in the segment; for simplicity, we denote stable unit by $B_{(i,0)}$ with OPD $= 0$ and $B_{(i,+)}$ with OPD $\neq 0$, respectively. Obviously, the OPDs of arbitrary adjacent stable units are different, so that we might set the sequence of the ith adjacent units to $B_{(i,0)} \cup B_{(i,+)}$. As shown in Fig. 3, the only four kinds of connections of adjacent units can be diagramed by enumeration method. The output intensity of the PMF segment $(X_{j-p}, X_{j+q}]$ from fast-axis and slow-axis at Point X_{j+q} are denoted by $P_{X_{j+q},F}$ and $P_{X_{j+q},S}$, respectively, which can be evaluated as:

$$
\left\{
\begin{aligned}
P_{X_{j+q},S} &= P_{X_{j-p},S}\left(\rho_j \sqrt{1-\rho_j^2}\right)\left(-\rho_{j+q}\sqrt{1-\rho_{j+q}^2}\right) \\
&\quad + P_{X_{j-p},F}\left(-\rho_j \sqrt{1-\rho_j^2}\right)\left(-\rho_{j+q}\sqrt{1-\rho_{j+q}^2}\right) \\
P_{X_{j+q},F} &= P_{X_{j-p},S}\left(-\rho_j \sqrt{1-\rho_j^2}\right)\left(\rho_{j+q}\sqrt{1-\rho_{j+q}^2}\right) \\
&\quad + P_{X_{j-p},F}\left(\rho_j \sqrt{1-\rho_j^2}\right)\left(\rho_{j+q}\sqrt{1-\rho_{j+q}^2}\right)
\end{aligned}
\right. , p,q \geq 1 \qquad (4)
$$

where, ρ_j and ρ_{j+q} are the coupling coefficients of the Point X_j and X_{j+q}, respectively. The sign of ρ_j changes only for coupling from the fast- to the slow-axis as shown in (Szafraniec et al. 1995). In most cases, it has the relation $\rho_j \ll 1$ in the detection for distributed polarization couplings along PMF (Tsubokawa et al. 1989). Here, we are reasonable to neglect the slight errors introduced by the approximation $\sqrt{1-\rho_j^2} \approx 1$, which can be used to simplify the analysis. For any two adjacent units, Eq. 1 can be rewritten as:

$$
\left\{
\begin{aligned}
P_{i,S} &= -(P_{i-1,S} - P_{i-1,F})\,\rho_{i,j}\rho_{i,j+q} \\
P_{i,F} &= (P_{i-1,S} - P_{i-1,F})\,\rho_{i,j}\rho_{i,j+q}
\end{aligned}
\right. , i,q \geq 1 \qquad (5)
$$

where, $P_{i,S}$ and $P_{i,F}$ represent light intensities from the slow-axis and fast-axis after passing through the ith adjacent units, respectively, $\rho_{i,j}$ and $\rho_{i,j+q}$ are the coupling coefficients of the point at the right end of segment $(X_{j-p}, X_j]$ and $(X_j, X_{j+q}]$, respectively. From Eq. 5, the stable units linked end-to-end can be expressed as:

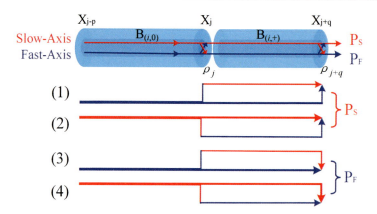

Fig. 3 The graphics of any two adjacent units of PMF. Segment $(X_{j-p}, X_{j+q}]$ are denoted by $B_{(i,0)} \cup B_{(i,+)}$, where the subscript i represents the ith adjacent unit combination of PMF; the subscripts (0) and (+) represent the corresponding OPD $= 0$ and OPD $\neq 0$, respectively; X_{j-p}, X_j, and X_{j+q} ($p, q \geq 1$) are the perturbation points of PMF, respectively; ρ_j is the coupling coefficient of the corresponding Point X_j; P_F and P_S are the light intensities out of the fast-axis and slow-axis of PMF, respectively

$$
\begin{cases}
P_{i,S} = -(P_{In,S} - P_{In,F})\, 2^{i-1} \displaystyle\prod_{i=1}^{\max\{i\}} \rho_{i,j}\,\rho_{i,j+1} \\[4mm]
P_{i,F} = (P_{In,S} - P_{In,F})\, 2^{i-1} \displaystyle\prod_{i=1}^{\max\{i\}} \rho_{i,j}\,\rho_{i,j+1}
\end{cases}
,\, i, q \geq 1 \qquad (6)
$$

where, $P_{In,S}$ and $P_{In,F}$ are the initial intensities that launch into the slow-axis and fast-axis of the first stable unit along PMF under test, respectively.

Classifications and General Formulas

In this section, we consider the pair of OPs of the first and last segments of the PMF under test. As mentioned above, we set the sequence of adjacent units as $B_{(i,0)} \cup B_{(i,+)}$ to simplify the analysis. However, the two end segments of the entire PMF under test might not be always satisfied the sequence. The OPD of the first and last segments could also conform to the sequence of $\{B_{(in,+)} \cup (B_{(1,0)} \cup B_{(1,+)}) \cup \cdots\}$ and $\{\cdots \cup (B_{(last,+)} \cup B_{(last,+)}) \cup B_{(out,0)}\}$, respectively, where the first segment $B_{(in,+)}$ and the last segment $B_{(out,0)}$ satisfy the features of $B_{(i,+)}$ and $B_{(i,0)}$, respectively. Therefore, the scanning OPDs of the entire PMF can be categorized into four classifications based on the possible end segments conditions. As shown in Fig. 4, the scanning OPDs of the entire PMF, for simplicity, are denoted by (A) $\{B_{(1,0)}, B_{(out,+)}\}$, (B) $\{B_{(in,+)}, B_{(out,0)}\}$, (C) $\{B(1,0), B_{(out,0)}\}$, and (D) $\{B_{(in,+)}, B_{(out,+)}\}$, respectively.

The initial intensities ($P_{In,S}$ and $P_{In,F}$) and terminal intensities ($P_{Out,S}$ and $P_{Out,F}$) for the four conditions in Fig. 4 are expressed as:

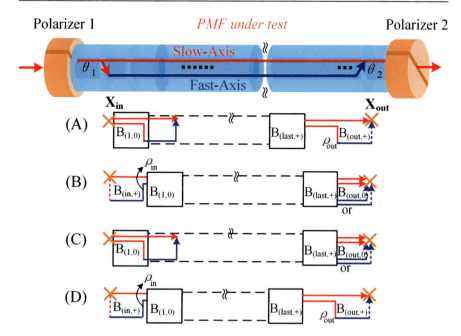

Fig. 4 Depending on the two end unit types ($B_{(i,0)}$ or $B_{(i,+)}$), the scanning OPDs of the entire PMF under test are categorized into four kinds, which are notated by (A) $\{B_{(1,0)}, B_{(out,+)}\}$, (B) $\{B_{(in,+)}, B_{(out,0)}\}$, (C) $\{B_{(1,0)}, B_{(out,0)}\}$, and (D) $\{B_{(in,+)}, B_{(out,+)}\}$, respectively. The consecutive units between the two black boxes in each kind conform with sequence of $B_{(i,0)} \cup B_{(i,+)}$. Besides, ρ_{in} and ρ_{out} represent the coupling coefficients of the points before the first unit $B_{(1,0)}$ and after the last unit $B_{(last,+)}$, respectively

$$\begin{cases} P_{In,S} = \cos^2\theta_1, & P_{In,F} = \sin^2\theta_1, & \text{first segment} \in B_{(i,0)} \\ P_{In,S} = \sin\theta_1 \cos\theta_1 (-\rho_{in}), & P_{In,F} = \sin\theta_1 \cos\theta_1 \rho_{in}, & \text{first segment} \in B_{(i,+)} \end{cases}$$

(7)

$$\begin{cases} \begin{aligned} P_{Out,S} &= P_{i,S}\cos^2\theta_2, \\ P_{Out,F} &= P_{i,F}\sin^2\theta_2, \end{aligned} & \text{last segment} \in B_{(i,0)} \\ \\ \begin{aligned} P_{Out,S} &= P_{i,S}\rho_{out} (-\sin\theta_2) \cos\theta_2, \\ P_{Out,F} &= P_{i,F} (-\rho_{out}) (-\sin\theta_2) \cos\theta_2, \end{aligned} & \text{last segment} \in B_{(i,0)} \end{cases}$$

(8)

where ρ_{in} and ρ_{out} are the coupling coefficients of the points before the first unit $B_{(1,0)}$ and after the last unit $B_{(last,+)}$, respectively, $P_{Out,S}$ and $\underline{P}_{Out,F}$ represent the output intensity from slow-axis and fast-axis at spliced point X_{out}, respectively. Because the polarizer is aligned to the slow-axis of its PM pigtail, the amplitude changing of polarized light that launched into slow-axis of PMF at point X_{in} is $\cos\theta_1$, and that coupled into fast-axis is $\sin\theta_1$. It is similar at the spliced point X_{out}.

Therefore, the final interference intensity with a given OPD based on Eqs. 6, 7, and 8 can be expressed as:

$$|P| = |P_{Out,S} + P_{Out,F}|$$

$$= \begin{cases} 2^{i-1} T_i \rho_{out} \cos 2\theta_1 \sin 2\theta_2, & OPD \in \{B_{(1,0)}, B_{(out,+)}\} \\ 2^{i-1} T_i \rho_{in} \sin 2\theta_1 \cos 2\theta_2, & OPD \in \{B_{(in,+)}, B_{(out,0)}\} \\ 2^{i-1} T_i \cos 2\theta_1 \cos 2\theta_2, & OPD \in \{B_{(1,0)}, B_{(out,0)}\} \\ 2^{i-2} T_i \rho_{in}\rho_{out} \sin 2\theta_1 \sin 2\theta_2, & OPD \in \{B_{(in,+)}, B_{(out,+)}\} \end{cases}$$

$$T_i = \begin{cases} \prod_{i=1}^{\max\{i\}} \left(\rho_{i,j}\rho_{i,j+q}\right), & i \geq 1 \\ 1, & i = 0 \end{cases}$$
(9)

where $i = 0$ represents there is no stable unit $B_{(i)}$. In addition, the central interferogram intensity is calculated as $|P_{central}| = \cos^2\theta_1\cos^2\theta_2 + \sin^2\theta_1\sin^2\theta_2$.

In reality, it might occur negative stable unit denoted by $B_{(i,-)}$ while there exist a positive term $B_{(i,+)}$. Here, the connection conditions of adjacent units can be classified to (a) $B_{(i,+)} \cup B_{(i,0)} \cup B_{(i,-)}$ and (b) $B_{(i,0)} \cup B_{(i,+)} \cup B_{(i,-)}$. Like the above analysis, we generalize the results as follows. In case of (a), the interference intensities with given OPD situations are unchanged. In case of (b), the interference intensities only should be multiplied by ρ_i^2 instead of the corresponding ρ_i, which is produced at the corresponding kink point between $B_{(i,+)}$ and $B_{(i,-)}$, and the other terms are remained the same.

Some summaries can be acquired by the above analysis, if we define the interference-order as $N = N_1 + N_2 + \cdots + N_i$ that can be found in the coupling coefficients term $\rho_1^{N1} \cdot \rho_2^{N2} \cdots \rho_i^{Ni} (Ni = 0,1,2)$ in Eq. 9. Because there are obviously even-number times couplings in arbitrary adjacent two units, the interference-order N of the four conditions in Fig. 4 can be summarized as $N \in$ odd-order when OPD$\in$ case (A) or (B), and $N \in$ even-order when OPD$\in$ case (C) or (D). Note that the intensities of every interferogram are related to the inject angle θ_1 at polarizer 1 and the output angle θ_2 at polarizer 2 in Eq. 9. Especially, 45° and 0° for θ_1 or θ_2 would introduce interesting results. The intensity of odd-order interferences has the maximum and even-order interferences are reduced to zero when $\theta_1 - \theta_2$ are 0°–45°, or 45°–0°, respectively. However, the variation trend of intensities are the exactly opposite results when $\theta_1 - \theta_2$ are 0°–0°, or 45°–45°, respectively.

Range Extension of Optical Delay Line for OCDP System

Principle of Operation

The pivotal issue of the range-extension of the optical delay line is broadening the continuous-scanning optical path difference between the two arms of the reference interferometer as shown in Fig. 5. The extension of the continuous-scanning optical path difference between two arms relies on the design of the path imbalance of

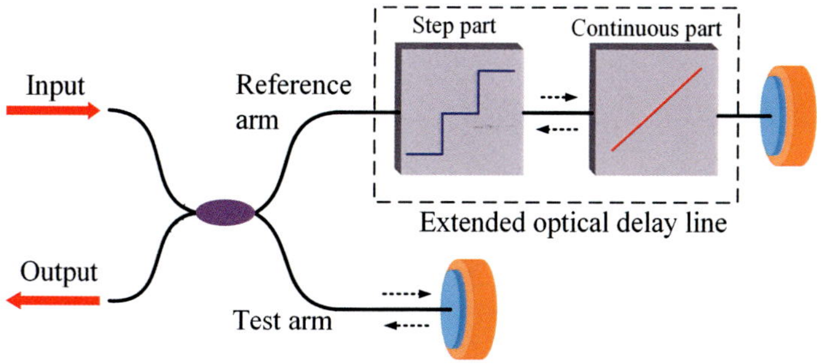

Fig. 5 The diagram of the WLI with the extended optical delay line

the two arms of the interferometer. The extension method we proposed consists of two parts placed in one arm (reference arm): the step part and the continuous part. The step part can cause the step change in optical path and provide delayed replicas of the original continuous part, while the continuous part obtains the delay changed continuously. The step part can be constructed by fibers with certain length connected to the optical switches or multiport couplers. Each port of the optical switch or coupler chooses its own fixed delay fibers represented as $L_1, L_2, \ldots, L_{n-1}, L_n$ after converting into optical path, respectively, and the delay at the port is fixed, then. The continuous part can be focused on the mechanically scanning stage.

The principle of the range extension method is illustrated in Fig. 6 in detail. Figure 6a represents the step part of the range extension method where the delay is step-change as the port number k of the optical switches or multiport couplers. Figure 6b shows the continuous part of the range extension method where the delay is continuously changed as the displacement of the reflector with a range of X.

In the reference arm, each port number adds a delay equal to a multiple of the continuous scanning range. Through the process of the copying and jointing, the large continuous tuning range constructed as the result of the combination of Fig. 6a the step part and Fig. 6b the continuous part as the red solid line shown in Fig. 6c shown can be expressed as:

$$L_m = nX \tag{10}$$

However, in the case of the step error ε of the L_{k+1} for the system error or environmental perturbation, the discontinuities of the delay can be introduced in the process of jointing between the adjacent delays as the green dashed line shown in Fig. 6c. To eliminate the influences of delay missing in the jointing procedure, i.e., discontinuous regions, we should induce overlaps ΔX_k between the adjacent delays, making the fiber length difference of the step part between two adjacent ports just slightly less than the original measurement range X as red-dashed line shown in Fig. 6c. The optical paths brought by the switchable delay fibers are revised as:

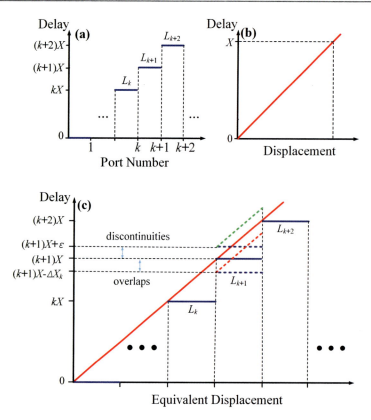

Fig. 6 The schematic diagram of the range extension method. The step part (**a**) and the continuous part (**b**) constitute the range-extension delay (the red solid line as shown in (**c**)). The different fiber optic delays L_k, L_{k+1}, L_{k+2} of the step part can be selected by activating the corresponding port number k. In (**c**), overlap delay ΔX_k (the red-dashed line) from the step part is constructed for the improvement of discontinuous delay (the green-dashed line) from the step error ε

$$\begin{cases} L_1 > 0 \\ L_{k+1} = L_k + X - \Delta X_k \end{cases} \tag{11}$$

where ΔX_k is the overlap scanning optical path between the kth port and the $(k+1)$th port as shown in Fig. 6.

The combination of step part and continuous part laid in one arm of the interferometer can extend the continuous-scanning OPD between two arms through setting the ports of the optical switches or multiport couplers in sequence. However, the lengths of delay fibers L_k and L_{k+1} cannot be accurately determined to obtain the values of ΔX_k unless a specialized measuring instrument with high precision is adopted. We add a self-calibration part in the other arm to quantitate the optical path of the overlap regions ΔX_k we configure, avoiding the large measuring error. The calibration part is constructed by a fiber ring with a constant length of S to obtain the

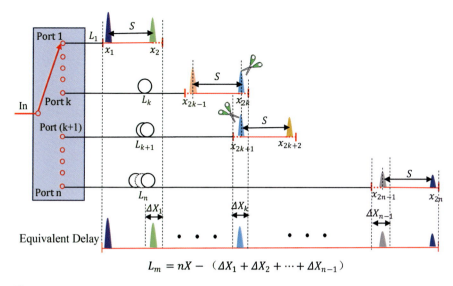

$$L_m = nX - (\Delta X_1 + \Delta X_2 + \cdots + \Delta X_{n-1})$$

Fig. 7 The principle of the self-calibration in the extension method. (The red segments represent the continuous part. X: the continuous-scanning range of the continuous part. The black segments with length L_k: the optical path of the step part at the kth port. Peaks in the same color represent the same paths in the fiber ring. S: the optical path of the fiber ring. $x_1, x_2, \ldots, x_{2k-1}, \ldots, x_{2k-1}, x_{2n}$: the positions of the calibration peaks in the continuous scanning part. $\Delta X_1, \Delta X_2, \ldots, \Delta X_k, \ldots,$ $\Delta X_{n-2}, \Delta X_{n-1}$: the overlap joints of the adjacent ports)

self-localization signals for the jointing. The fiber optic ring is intended to increase the optical path in the test arm through adding the times of loops traveling along the fiber ring. One principle of the design of S is to recover the path length imbalance twice during only one scanning of the scanning stage, namely to create two profiles of the interference fringes as shown in Fig. 7. The other principle is the path of the first peak has one more fiber loop than the second one at kth port, however, has the same optical path in test arm as the latter peak at $(k-1)$th port, making these peaks flags for self-localization. The relations at adjacent ports are summarized as:

$$
\begin{aligned}
L_{\text{reference}} + x_{2k-1} + L_k &= (k-1)S + L_{\text{test}} \\
L_{\text{reference}} + x_{2k} + L_k &= kS + L_{\text{test}} \\
L_{\text{reference}} + x_{2k+1} + L_{k+1} &= kS + L_{\text{test}} \\
L_{\text{reference}} + x_{2k+2} + L_{k+1} &= (k+1)S + L_{\text{test}}
\end{aligned}
\tag{12}
$$

where k is a valid integer greater than 0. $L_{\text{reference}}$ and L_{test} are the optical paths brought by the fibers in the reference and test arm, respectively. $x_{2k-1}, x_{2k}, x_{2k+1},$ x_{2k+2} are the positions of the calibration peaks during the continuous scanning part. k is the order of the traveling circle in the fiber-ring cavity. S is the optical path of the fiber ring. n is the total number of the ports.

The introduction of the fiber ring can locate and analyze the profiles of the interference fringes depending on the order of the ports of the optical switch or multiport coupler. The peaks for calibration change the measurement for the fiber lengths of the delay fibers into the identifications of locations, which is to acquire the locations to cut and piece for extension that expressed as:

$$\Delta X_k = L_k + X - L_{k+1} = X - x_{2k} + x_{2k+1} \tag{13}$$

After removing the overlap scanning regions and piecing all the pieces, Eq. 10 is revised and we can obtain a large continuous tuning range that expressed as:

$$L_m = nX - \sum_{k=1}^{n-1} \Delta X_k \tag{14}$$

In the measuring mode, the fiber ring is off state and the testing results are cut and pieced based on the locations acquired in calibration procedure.

Device, Implementation, and Performance

The schematic of the WLI system with the expanded optical delay line is illustrated in Fig. 8. One arm (reference arm) of the WLI we proposed comprises a continuous-scanning stage, switchable delay fibers, and a faraday rotator mirror (FRM) for doubling the optical paths. The continuous-scanning stage we adopt is a differential optical delay line (Li et al. 2015) consisting of a driving motor and a movable mirror, which can be equivalent to only one arm and where the delay varies from the positions of the movable mirror. It can provide continuous tuning but over only a relatively narrow range ($0\sim 2.4$ m) after converting into optical path. The scanning speed and stability of the driving motor and the quality of alignment between the collimation lens and mirrors used will directly affect the accuracy of measurement (Guo et al. 2011; Ning et al. 1996). The switchable delay fibers are constructed through the optical switches with low insertion loss. The other arm (test arm) of the WLI comprises an FRM and a fiber-ring cavity in which light can travel for multiple trips before interfering with the light from the reference arm of the WLI. The fiber ring with the length of S in air is constructed by an 80:20 fiber coupler. With the F1 connected, the WLI system is in calibration mode. While the F1 is off state, the WLI system is in measuring mode. The two faraday rotator mirrors at the end are used to eliminate the polarization fading in reflective Michelson interferometer. As schematically demonstrated in Fig. 8, a broadband light source of superluminescent diode (SLD) with flatten spectrum at wavelength of 1550 nm is split into the two separate ends of the differential optical delay line, which functions as the continuous-scanning stage in the interferometer, through a 3-dB single-mode coupler C1 after a circulator. The two beams from the outputs of the differential delay line are reflected by the FRMs at the end, after passing through the switchable

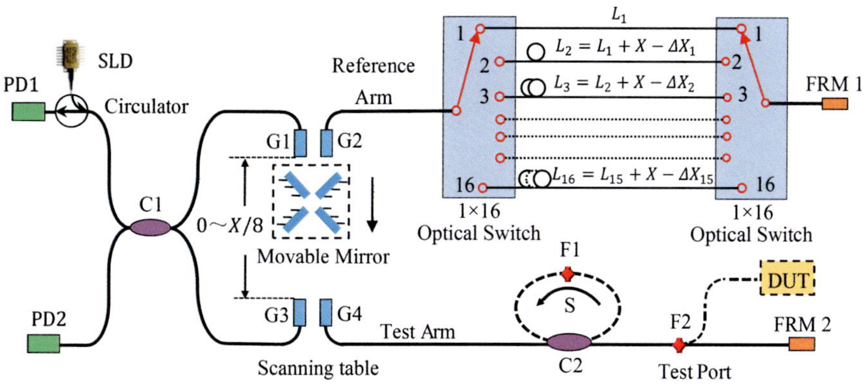

Fig. 8 Diagram of WLI system with the expanded optical delay line. (SLD: superluminescent diode, PD: photodiode, C1 and C2: couplers, 0~X: the scanning range of the scanning stage, G: gradient index lens, F1 and F2: connectors, FRM1 and FRM2: Faraday rotator mirrors, DUT: device under test)

paths, respectively. The reflected beams return to C1 along the original paths and differentially detected by the two photodetectors (PD) (Yang et al. 2014).

In the calibration mode, two calibration peaks can be obtained at the PDs when the ports of the optical switches are set to be the same and expressed as S_{2k-1}, S_{2k} in terms of the port number k as illustrated in Fig. 9. The signals at port 1 and port 2 are set as an example of the jointing. The right side from the middle of S_2 and the left side from that of S_3 are removed and the rest of the two parts are joined together a whole. The differences between the experimental amplitude changes and the expectations are believed to be due to polarization effects, which were not adjusted during the measurement, for the heights have no effect on the calibration results. Obviously, the dispersion in the setup in Fig. 9 decreases the contrast of interference fringes. The contrast decrease will increase the possibilities that the calibration peaks produced in the calibration mode of the experimental setup be submerged in the noise. However, it is true that the dispersion in the setup can be measured and compensated in our system with certain techniques, such as the algorithm in (Yu et al. 2016) and (Jin et al. 2013).

ΔX_k is quantitated from the location of the calibration peaks and we can obtain a large continuous tuning range as expressed in Eq. 14 after removing the overlap scanning regions and piecing all the pieces.

$$L_m = 16X - \sum_{k=1}^{15} \Delta X_k = 37.2 \text{ m} \tag{15}$$

As we have recognized in the previous paper (Li et al. 2015), the loss coefficient fluctuation of the extended delay line can introduce the differences between the measured and actual values. The loss coefficient fluctuation of the optical delay

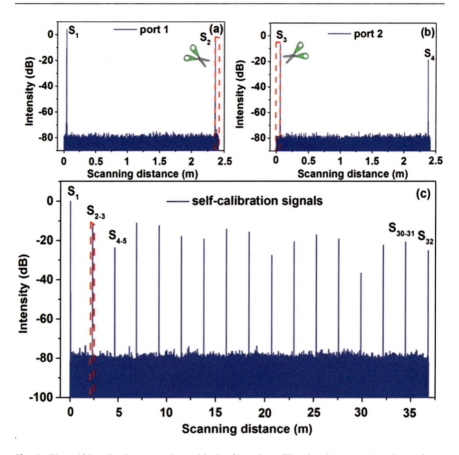

Fig. 9 The self-localization procedure with the fiber ring. (The signals at port 1 and port 2 are set as the example for jointing. S_1, S_2, ..., S_{31}, S_{32}: the peaks for self-localization. S_{2-3}, S_{4-5}, ..., S_{30-31}: the jointing results of the two corresponding peaks)

line can bring a modulation coefficient of the light intensity in the reference arm, introducing great errors in the measurement, especially when we employ one of the peaks for calibration of amplitude. The insertion loss is caused by optical switches, the scanning stage, and the connectors or fusion splices. However, the loss fluctuation mainly results from the scanning of the stage and the different ports of the optical switches. The loss and loss coefficient fluctuation can be measured by 10 points equally divided along the optical delay line at each port. After we piece the results at each port as the calibration mode illustrated, we obtain Fig. 10. The average loss of the optical delay line we proposed is 1.45 dB. With the standard deviation calculated, the loss of the optical delay line can be written as:

$$\text{Loss} = (1.45 \pm 3\sigma) \text{ dB} = (1.45 \pm 0.15) \text{ dB} \qquad (16)$$

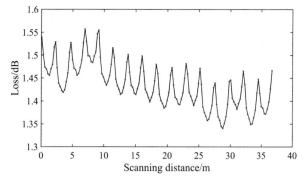

Fig. 10 The loss distribution of the extended optical delay line

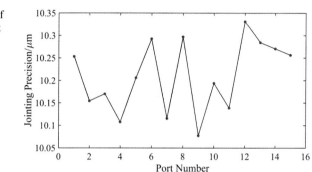

Fig. 11 The distribution of the jointing errors $3\sigma_{\Delta Xk}$ at the kth port

The delay precision of the extended optical delay line mainly depends on the precise quantitation of the ΔX_k produced in the jointing procedure. As demonstrated in Eq. 13, the delay precision of the extended delay line falls on the position uncertainties of the peaks for self-calibration. In the calibration mode, the two peaks for calibration at adjacent ports work together to affect the jointing errors. Then we can calculate the jointing error based on Eq. 13 as follows:

$$\sigma_{\Delta X_k} = \sqrt{\sigma_{2k}^2 + \sigma_{(2k+1)}^2} \tag{17}$$

where k is the number of the port. $\sigma_{\Delta Xk}$ is the jointing error when we joint the delay at the kth port and $(k + 1)$th port. σ_{2k} is the standard deviation of the right peak for calibration at the kth port, σ_{2k+1} is the standard deviation of the left peak for calibration at the $(k + 1)$th port.

The σ_{2k} or σ_{2k+1} is mainly influenced by the identification precision of the zero-order peak fringe, environmental disturbances, and the measurement accuracy of the moving distance that the scanning stage travels (Guo et al. 2011). With the three factors above working together, the statistical jointing errors $3\sigma_{\Delta Xk}$ at each port are distributed as summarized in Fig. 11. The average and the standard deviation of the jointing precision of the extended are 10.21 μm and 0.08 μm, respectively. The influences of the factors are detailed in the next section.

Measurement Uncertainties

In many respects, the sources of measurement uncertainties of the proposed structure are like those of any white-light interferometer, but unique and mainly influential aspects of the proposed system must be discussed. The delay precision of the extended optical delay line mainly depends on the jointing uncertainties produced in the jointing procedure. The influential factors of the jointing precision will be detailed in this section.

The Precise Identification σ_A of the Zero-Order Fringe

As is well known, the normalized output signal of the central interference packet of an interferometer that is illuminated by an SLD has a cosine function modified by an approximately Gaussian visibility profile (Gerges et al. 1990). When the coherence length of the SLD is relatively large, the autocorrelation function (or visibility profile) of the fringe pattern is flat near the central fringe; hence, in the presence of noise, it is difficult to identify the central fringe. Thus, there is a high probability of large errors occurring in the measurement of the measured quantity that is unacceptable in a practical system. When the central fringe position is being identified, the relative amplitude difference between the central fringe and the adjacent fringes determines the minimum system SNR requirement (Rao et al. 1993) as summarized below:

$$SNR_{\min} = -20\lg\left\{1 - \exp\left[-\left(\frac{2\lambda}{L_c}\right)^2\right]\right\} \quad (18)$$

where the central wavelength $\lambda = 1550$ nm and the coherence length of the SLD $L_c = 50$ nm. The calculated minimum system SNR requirement is -48.32 dB. The improved dynamic range over 80 dB in our system can guarantee the identification error of the central fringe is within $\pm\lambda\cdot(3\sigma)$, namely, 1.55 μm.

Temperature Instability σ_B

Temperature is another important influential factor affecting the precision in the experiment because of its effects on the fiber length and refractive index. Since both the reference and test arms have almost the same fiber types, the influence of temperature fluctuation only originates from the optical path difference between the two arms. Considering the temperature coefficient, Eq. 19 is expressed as follows (Yuan and Yang 2003):

$$S_{\text{fiber}} = n_0\left(\alpha_l + \alpha_n\right) \quad (19)$$

where S_{fiber} is the temperature sensitivity coefficient for silica-based optical fibers, n_0 is the original refractive index, and α_l and α_n are the thermal expansion coefficient and the refractive index temperature coefficient of the optical fiber, respectively. For the standard commercial communication single-mode fiber at wavelength $\lambda = 1550$ nm, the parameters are $n_0 = 1.4675$, $\alpha_n = 8.11 \times 10^{-6}/°C$, $\alpha_l = 5.5 \times 10^{-7}/°C$. And we finally calculate and obtain $S_{\text{fiber}} = 12.71$ μm/(m°C). The position errors of peaks for self-calibration introduced by the temperature changes are less than 30.504 μm/°C, locating among the scanning stage with the range of 2.4 m. The change in the optical path difference indicates that the temperature influence is less than 3 μm if the environmental temperature fluctuation is controlled within 0.1 °C.

The Measurement Accuracy σ_C of the Scanning Stage

The uncertainty due, for instance, to imperfection of the delay line or the mechanical vibrations of the measurement system introduces a spurious (frequency) modulation of the acquired data, causing a distortion of the time–domain interference pattern and a frequency–domain spectral broadening (Canavesi et al. 2009). The position error caused by the mechanical vibration can be recovered by using an additional highly coherent light source, whose interference pattern is employed to control, measure, and compensate for the interferometer unbalance inaccuracies. A laser interferometer (Renishaw RLE 10 fiber optic laser interferometer) is employed to measure the displacement of the reflector in the continuous-scanning stage, and the repeated positioning accuracies of the stepping motor near the start and the end of the mechanical device are 5.52 μm(3σ) and 6.96 μm(3σ), respectively.

We take the jointing precision $\sigma_{\Delta X1}$ at the first port for example. Based on Eq. 17

$$\sigma_{\Delta X_1} = \sqrt{\sigma_2^2 + \sigma_3^2} \qquad (20)$$

In terms of the positions of the second and third calibration peaks x_2, x_3, obtained in the calibration mode, Eq. 20 can be rewritten as:

$$
\begin{aligned}
\sigma_{\Delta X_1} &= \sqrt{\sigma_{A2}^2 + \sigma_{B2}^2 + \sigma_{C2}^2 + \sigma_{A3}^2 + \sigma_{B3}^2 + \sigma_{C3}^2} \\
&= \sqrt{\sigma_{A2}^2 + \left(S_{fiber}x_2\sigma_T\right)^2 + \sigma_{C2}^2 + \sigma_{A3}^2 + \left(S_{fiber}x_3\sigma_T\right)^2 + \sigma_{C3}^2}
\end{aligned}
\qquad (21)
$$

where σ_{A2}^2, σ_{B2}^2, σ_{C2}^2, σ_{A3}^2, σ_{B3}^2, σ_{C3}^2 are the three kinds of precisions of the second and third calibration peaks, respectively, and $\sigma_T = 0.02$ °C is standard deviation the of the temperature changes during 10 min. Substituting all the related parameters into Eq. 21, we get $3\sigma_{\Delta X1} = 9.19$ μm. The theoretical predictions for systematic and statistical errors are compared with experimental results, which show that the calculated jointing precision is quite close to the measured results as shown in Fig. 11. According to Eq. 12 and Fig. 7, we know that the measurement accuracy of $S, \cdot \sigma_S$, is another parameter that can be used to characterize the measurement error in the system. The theoretical predictions of $3\sigma_S = 9.31$ μm for σ_S and the experimental results of 10.15 μm(3σ) are obtained in a similar way to $\sigma_{\Delta Xk}$.

Greater delay precision can be achieved; however, the price is increased cost. Three suggested techniques are as follows: (1) The identification precision of the zero-order peak fringe can be greatly improved by use of a broader-band, shorter-wavelength source, such as a 830-nm SLD, or by the technique of synthesized source for WLI by using two multimode laser diodes, as Rao et al. suggested (Rao et al. 1993). (2) A dedicated continuous-scanning stage with high quality could be added to locate the calibration interference fringe patterns with higher accuracy. Alternatively, we can use an additional highly coherent light source, whose interference pattern is employed to control, measure, and compensate for the interferometer inaccuracies (Palavicini et al. 2005). (3) It is necessary to insert the whole extended optical delay line into a temperature controller. We can place the system in homothermic medium, like water, and then the fluencies resulting from environmental disturbances can be erased.

Accuracy Improvement of Optical Delay Line for OCDP System

Distributed Polarization Crosstalk Measurement with Loss Coefficient

The schematic of white-light interferometric OCDP based on single GRIN lens delay line is shown in Fig. 12a. Polarized light with power of I_0 travels along one principal axis of a high-birefringence single-mode fiber. If there is a polarization coupling point (coupling strength $\rho \ll 1$), a part of the light along the principal axis will be coupled to the orthogonal principal axis. The amplitudes of the excitation mode E_x and that of the coupling mode E_y after the coupling point are expressed as (Takada et al. 1985):

$$\begin{aligned} E_x &\propto E_0\sqrt{1-\rho} \\ E_y &\propto E_0\sqrt{\rho} \end{aligned} \tag{22}$$

where E_0 is the field amplitude of light power. The two wave trains transmitted along the orthogonal axes are sent to a scanning Michelson interferometer through a polarization analyzer. The output signal of the Michelson interferometer is detected by a photodetector (PD). When the optical path difference (OPD) between the two arms of the Michelson interferometer is less than the coherence length of the light source, a principal fringe (fringe B in Fig. 12b) corresponding to the interference between the same kind of polarization modes can be observed. When the OPD of the interferometer is equal to the OPD between the exciting and coupling modes in the device under test (DUT), coupling fringes (fringes A and C in Fig. 12b) corresponding to interference between the different types of polarization modes can be detected. Therefore, the AC components of the principal and coupling interference fringes can be written as (Jing et al. 2002):

$$I_{\text{main}} \propto \left(E_x E_x^* + E_y E_y^*\right)\sqrt{p(Z_0)} = I_0\sqrt{p(Z_0)} \tag{23}$$

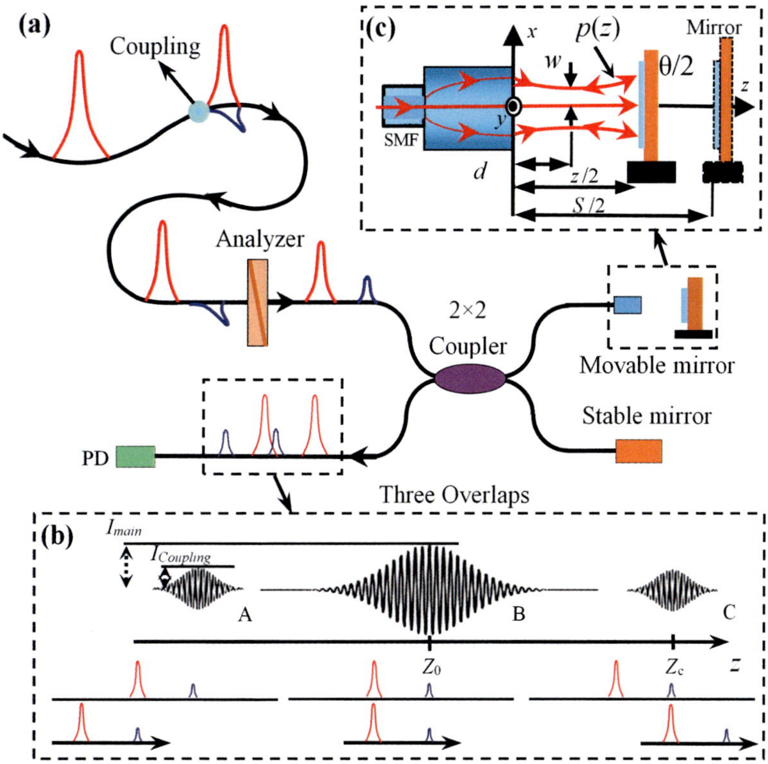

Fig. 12 (**a**) The schematic of OCDP with white-light interferometry. (**b**) Principle of polarization coupling measurement. (**c**) The GRIN lens coupling process of the delay line

$$I_{\text{coupling}} \propto \left(E_x E_y^*\right) \sqrt{p\left(Z_c\right)} = I_0 \sqrt{\rho\left(1-p\right)} \sqrt{p\left(Z_c\right)} \approx I_0 \sqrt{\rho p\left(Z_c\right)} \quad (24)$$

where Z_0 and Z_c are double distance of the scanning mirror when the principal and coupling fringes are observed, respectively, and $p(z)$ is the power attenuation coefficient of the delay line which is determined by the interval between the scanning mirror and the GRIN lens. By dividing Eq. 2 by Eq. 3, the measured coupling strength ρ can be expressed as (Takada et al. 1985):

$$\rho = \left[\frac{p\left(Z_0\right)}{p\left(Z_c\right)}\right]\left[\frac{I_{\text{coupling}}}{I_{\text{main}}}\right]^2 = H\rho_0 \quad (25)$$

where $\rho_0 = (I_{\text{coupling}}/I_{\text{main}})^2$ is the actual coupling strength and $H = p(Z_0)/p(Z_c)$ is a modulation coefficient of the coupling strength. From Eq. 4, we can see that there is a relative difference H between the measured and actual coupling strengths. The existence of H is mainly due to the light propagating in free space between the GRIN lens and the scanning mirror in the delay line. In other words, the measurement accuracy of coupling strength is directly related to performance of the delay line, especially the GRIN lens.

Generally, the light beam output from a GRIN lens has a Gaussian distribution (Sakamoto 1986) (see Fig. 12c). If we establish a coordinate system at center of the end face of the GRIN lens, the relative power loss of the GRIN lens L can be written as:

$$L = -10 \lg \left[\frac{4 \exp(2J)}{W^2 (z - d) + W^2(d)|F|^2} \right] = -10 \lg [p(Z)] \qquad (26)$$

where

$$
\begin{aligned}
F &= F_r + \mathrm{i} F_i = \tfrac{1}{W^2(z-d)} + \tfrac{1}{W^2(d)} + \mathrm{i} \tfrac{k}{2} \left[\tfrac{1}{R(z-d)} + \tfrac{1}{R(d)} \right], \\
J &= \frac{-F_r(k\theta)^2}{4|F|^2}, \; W^2(z) = w^2 \left[1 + \left(\frac{\lambda z}{\pi n w^2} \right)^2 \right], \\
k &= \frac{2\pi n}{\lambda}, \; R(z) = z \left[1 + \left(\frac{\pi n w^2}{\lambda z} \right)^2 \right]
\end{aligned}
\qquad (27)
$$

where z is the double distance between the scanning mirror and the exit face of the GRIN lens, d is the distance between the exit face of the GRIN lens and the location of the Gaussian beam waist, θ is double the angle between the planes of the mirror and the end face of the GRIN lens, w is the waist radius of the Gaussian beam, λ is the wavelength of light source, and n is the refractive index of the transmission medium. F, J, W, and R represent the relational parameters. F_r and F_i represent the real and imaginary values of F, respectively.

Simulations are done based on Eqs. 26 and 27 to show the impacts of the waist radii w, the distances d, and the angle θ on the loss coefficient L. In the calculations, the double moving range of the scanning mirror is set to $S = 800$ mm and other parameters are set to $\lambda = 1550$ nm and $n = 1$. The simulated results are shown in Fig. 13, in which (a) presents the loss coefficient under different w, and is obtained with $d = 230$ mm, $\theta = 0$ mrad, and $w = 440$ μm, 400 μm, and 360 μm; (b) shows the loss coefficient under different d, and is obtained with $w = 440$ μm, $\theta = 0$ mrad, and $d = 230$ mm, 200 mm, and 150 mm; (c) is the loss coefficient under different θ, and is obtained with $w = 440$ μm, $d = 230$ mm, and $\theta = 0$ mrad, 0.2 mrad, and 0.4 mrad.

For ease of comparing the responses of L to different w, d, θ, the fluctuation of loss coefficient is defined as:

$$\Delta L = L_{\max} - L_{\min} \qquad (28)$$

where $L_{\max}$ and $L_{\min}$ are the maximum and minimum of power loss within the scanning range, respectively. Higher loss represents the lower utilization of light

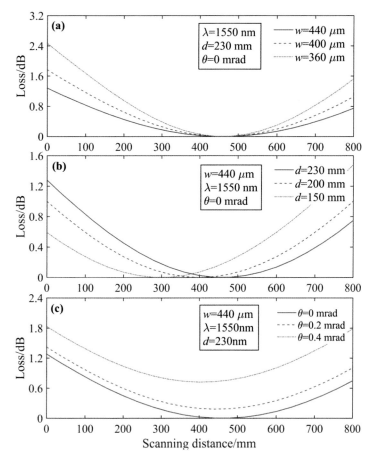

Fig. 13 Change of the loss coefficient with the changing of scanning distance under different waist radii w (**a**), the distances d (**b**), and the angle θ (**c**)

Table 2 The minimum and the fluctuation of loss coefficient

w/μm	440	400	360
L_{min}/dB	0	0	0
ΔL/dB	1.3	1.8	2.5
d/mm	**230**	**200**	**150**
L_{min}/dB	0	0	0
ΔL/dB	1.3	1.0	1.5
θ/mrad	**0**	**0.2**	**0.4**
L_{min}/dB	0	0.2	0.7
ΔL/dB	1.3	1.2	1.1

source. L_{min} also represents the utilization of light source. ΔL and L_{min} of each curve in Fig. 13 are listed in Table 2.

From Fig. 13 and Table 2, we can see that the loss coefficient L is not only a function of the distance z, but also relates to the waist radius w, the distance d, and the angle θ. The power attenuation coefficient $p(z)$ can be written as $p(z, w, d, \theta)$. The fluctuation ΔL decreases with the increase of w and gets its minimum value as $d = S/4$ and $\theta = 0$ mrad.

To reduce the effect of delay line on OCDP measurement, we should reduce the fluctuation of the loss ΔL to get stable power during the scanning of delay line. According to the above analysis, for a single GRIN lens, the delay line can reach its best performance when $d = S/4$ and $\theta = 0$, and the beam waist radius is the larger the better. However, in a practical situation, such a delay line is not appropriate for large scanning range applications due to three reasons. Firstly, w cannot be too large because a larger w will induce a larger insertion loss and more stray light to the delay line. Secondly, it is difficult to adjust the plane of the mirror parallel to the end face of the GRIN lens. Lastly, fluctuation compensation by using software is complicated because the compensation algorithm should be accurately reset each time owing to the variation of environmental factors. Thus, the fluctuation ΔL as well as the modulation coefficient H cannot be reduced sufficiently in the single-GRIN-lens based delay line.

Differential Delay Line Structure

A differential delay line structure is proposed to further reduce the fluctuation of the modulation coefficient H on coupling strength ρ. The schematic structure of the differential delay line is shown in Fig. 14a. It is composed of two GRIN lenses, which are fixed on the two ends of a moving stage, and a scanning platform with a double-faced mirror. According to the symmetry of the differential structure, the two GRIN lenses 1 and 2 have the same parameters and complementary displacements. Some relationships shown in Fig. 14 are

$$\begin{cases} z_1 = z, z_2 = l - z_1, z_d = 2z_1, \\ S = z_{max} - z_0, S_d = 2S, \\ w_1 = w_2 = w, d_1 = d_2 = d \end{cases} \tag{29}$$

where $z_i (i = 1, 2)$ are the double distance between the scanning mirror and the ith GRIN lens, l is the double distance between the two GRIN lenses, S is double the moving range of the scanning mirror, z_0 is double the original distance between exit face of the GRIN lens and the original location of the platform where the principal fringe is observed (see Fig. 14b), z_{max} is double the largest distance of platform (see Fig. 14c), w_i is the ith ($i = 1, 2$) GRIN lens waist radius of the Gaussian beam, and d_i is the distance between the exit face of the ith ($i = 1, 2$) GRIN lens and the location of the Gaussian beam waist. Owing to the differential characteristics, in case the distance z_i of one GRIN lens increases, the other one z_i decreases by the same length accordingly. The OPD between the two GRIN lenses will be extended

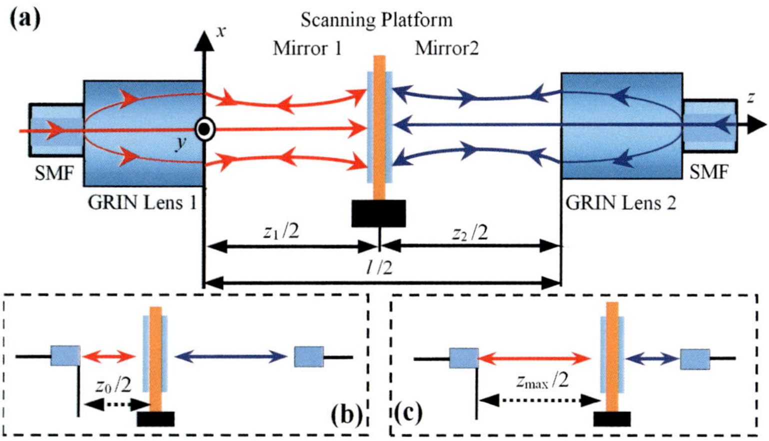

Fig. 14 (a) Differential delay line with two identical GRIN lenses and a double-faced mirror. (b) and (c) are original and furthest distance of the mirror, relatively

twice. Therefore, the equivalent distance z_d and the moving range of the scanning mirror S_d of the differential delay line are twice those of the single GRIN lens.

We assumed that $p_i(z_i, w_i, d_i, \theta_i)$ $(i = 1, 2)$ are the power attenuation coefficients of ith GRIN lens. According to optical interference theory, the interference light power is proportional to the product between the complex amplitude and its complex conjugate. Therefore, neglecting the DC term of the interference signal, the AC term of composition power attenuation coefficient $f(z_d, w, d, \theta)$ of differential scanning structure can be written as:

$$f(z_d, w, d, \theta) \propto \sqrt{p_1(z_1, w_1, d_1, \theta_1)} \sqrt{p_2(z_2, w_2, d_2, \theta_2)} \qquad (30)$$

Simulations were done based on Eqs. 26, 27, 29, and 30. The parameters were set to $S = 400$ mm, $S_d = 800$ mm, $z_0 = 0$, $w = 440$ μm, $\lambda = 1550$ nm, and $d = 230$ mm. The calculated power loss of the differential delay line is shown in Fig. 15a as expressed in Eq. 30. The patterns of GRIN lens 1 and 2 expressed as Eqs. 26 and 27 retain the same meaning with the relationship of Eq. 29. The red- and green-dashed curves present the losses of GRIN lenses 1 and 2, respectively, and the solid line shows the loss of the differential delay line. To intuitively present the power loss difference between the differential structure and the optimal single-GRIN lens-based structure, the losses of the two structures are plotted in Fig. 15b. In the calculation, we take $z_0/2 = 30$ mm for the single-GRIN-lens-based delay line. From Fig. 15, we can see that ΔL of differential structure (blue solid curve) is greatly reduced relative to the optimal performance of a single lens (red-dashed curve), even though the minimum loss coefficient L_{min} becomes a little larger for $d < 250$ mm.

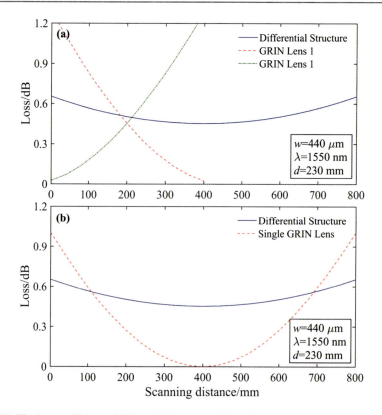

Fig. 15 The loss coefficients of differential structure (blue solid curve) obtained from GRIN lens 1 (red-dashed curve) and 2 (green-dashed curve) are shown in (**a**). The comparison between the loss coefficients of the differential structure and the optimal performance of the single-GRIN lens structure is shown in (**b**)

According to the curves in Fig. 13b, the distance d of these parameters plays an important role in the loss coefficient of the delay line. Different d determines different proper work distance. Therefore, the minimum loss coefficient L_{min} and the fluctuation of loss coefficient ΔL under different d are further calculated for the differential delay line. The calculated results are shown in Fig. 16.

Figure 16a demonstrates that the differential structure has the lower utilization of light power. Generally, it is feasible if the minimum loss coefficient is less than 3 dB. In this case, the distance d should be less than about 500 mm with $S_d = 800$ mm. From Fig. 16b, we can see ΔL_d of the differential structure is small and flat, while ΔL_s of the single GRIN lens is much larger and sharp, and the minimum loss fluctuation occurs at the point of $d = S_d/4 = 200$ mm. At this point, substituting Eqs. 26 and 27 into Eq. 30, ΔL_s of the single lens and ΔL_d of the differential structures in Eq. 28 can be expressed as:

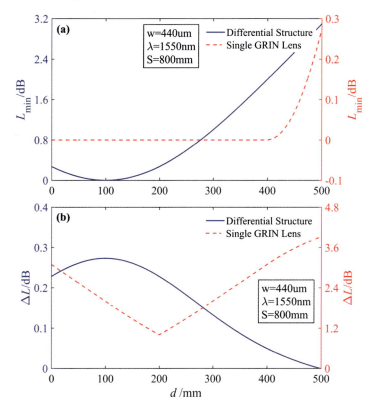

Fig. 16 The minimum loss coefficient (**a**) and the fluctuation of loss coefficient (**b**) change with the changing of different distance d. The vertical axis on the left and right sides of the figure shows the performances of differential structure and single GRIN lens, respectively

$$
\begin{cases}
\Delta L_s = -10\lg\left(\dfrac{4n^2\pi^2w^4}{4n^2\pi^2w^4 + S_d^2\lambda^2}\right) \\[2ex]
\Delta L_d = -10\lg\left(\dfrac{16n^2\pi^2w^4 + S_d^2\lambda^2}{\sqrt{4n^2\pi^2w^4 + S_d^2\lambda^2}} \cdot \dfrac{1}{8n^2\pi^2w^4}\right) \\[2ex]
\Delta = \Delta L_d - \Delta L_s \\[2ex]
\quad = -10\lg\left[\sqrt{1 + \dfrac{S_d^2\lambda^2}{4n^2\pi^2w^4}} \cdot \left(1 + \dfrac{S_d^2\lambda^2}{16n^2\pi^2w^4}\right)\right] < 0
\end{cases}
\tag{31}
$$

where Δ is the difference of fluctuation between the two structures. Eq. 31 shows that the fluctuation of loss coefficient with differential structure is less than that of the single-lens based structure in the whole scanning range. In other words, the fluctuation of $f(z_d, w, d, \theta)$ is smaller than $p(z)$ within the valid range of $L_{\min}$. Substituting Eq. 30 into Eq. 25, the coupling strength ρ_d of the differential structure can be written as:

$$\rho_d = \left[\frac{\sqrt{p\,(z_0)} \cdot \sqrt{p\,(l - z_0)}}{\sqrt{p_1\,(z_1)} \cdot \sqrt{p_2\,(l - z_1)}} \right] \left[\frac{I_{\text{coupling}}}{I_{\text{main}}} \right]^2 = H_d \rho_0 \qquad (32)$$

where H_d is the modulation coefficient of differential structure. From the two modulation coefficients of Eqs. 25 and 32, the numerators of H_d and H are constant while the changing of denominators of H_d is flatter than H. Therefore, the normalized coupling strength ρ_d is closer to the real value ρ_0 in comparison to ρ.

Measurement with Differential Structure Delay Line

To verify the loss fluctuation reducing of the proposed differential delay line, we constructed a white-light Mach–Zehnder interferometer (MZI) each arm of which includes a single GRIN-lens delay line. The configuration of the interferometer is shown in Fig. 17. As shown in Fig. 17a, the light from the superluminescent light-emitting diode (SLD) was split into two beams by a 3-dB coupler. The two beams were reflected respectively by two independent scanning mirrors and then combined by the second 3-dB coupler. The output signals of the MZI were detected by two PDs. The SLD has central wavelength of 1550 nm, bandwidth of >50 nm, and output power of >2 mW. The moving ranges of the scanning mirrors M1 and M2 are 200 mm and 300 mm, respectively. The effective d and w of the GRIN lenses are 230 mm and 440 μm, respectively.

To get an equivalent state of the differential delay line with the configuration as shown in Fig. 17a, the distance between one GRIN-mirror pair should be in section S_A and the distance of the other GRIN-lens pair should be in section S_C which is symmetric with S_A, as illustrated in Fig. 17c. From Fig. 15b, the loss curve of the single-GRIN-lens based delay line with $d = 200$ mm is symmetric around $S = 400$ mm. If we set the delay line M1 working in section S_A, the working range of M1 is 0–400 mm due to the 200-mm moving range of mirror 1. To make the delay line M2 work at the range symmetric with that of M1, the working range of M2 should be 520–920 mm. Therefore, the original distance $z_0/2$ of M2 is 260 mm. In the experiment, the fiber lengths in the two arms of the interferometer were adjusted firstly to make sure that zero OPD occurred when the two delay lines worked at their original distances. Then the mirror of delay line M1 was moved away from the GRIN lens with steps of 10 mm, and the mirror of delay line 2 was moved accordingly. Therefore, a series of interference fringes were generated at the optical path matching point. The intensity of the interference signals is plotted as triangle points in Fig. 18 which includes the own error of the motorized linear stage. To show the difference between the differential delay line and the single-GRIN-lens delay line, the mirror of one delay line was covered to detect the light only from one delay line. The detected data were shown as diamond points and box points in Fig. 18. The data in each curve were processed by subtracting the smallest ones. From the figure, we can see that the loss of delay line M1 decreases from 1.2 dB to 0 dB

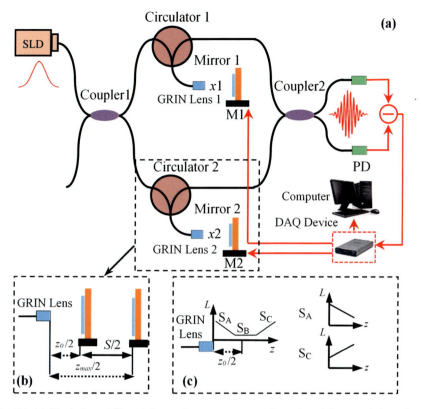

Fig. 17 (**a**) Equivalent differential scanning structures are investigated by two independent scanning platforms. (**b**) The detail of scanning delay line. (**c**) The loss of delay line is divided into three sections simply based on the basic changing trend

Fig. 18 Equivalent differential delay line. Dots and solid curves are experimental and theoretical data, respectively

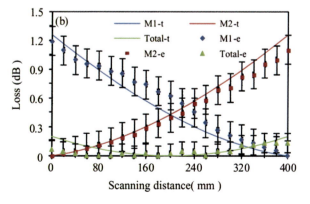

and that of M2 increases from 0 dB to 1.1 dB, which corresponds to a fluctuation of about ± 0.6 dB. As a contrast, the loss of the equivalent differential delay line is much flatter and has a fluctuation of ± 0.1 dB. Therefore, the loss coefficient fluctuation of the differential structure drops to 4.5% (0.2 dB) from 24.1% (1.2 dB), which means an improved accuracy of less than ± 0.1 dB with scanning range of 0.8 m (corresponding to a sensing range ability of 1.6 km, assuming birefringence of 5×10^{-4}). The relative accuracy error of the differential delay line is 4.5%, which corresponds to an accuracy improvement of more than 80% relative to the error of 20% of the typical single GRIN-lens delay line (± 0.5 dB accuracy).

A 1100-m FOG PMF coil was tested by using a differential delay line-based OCDP to verify the vitality of the differential delay line. In the experiment, to match the double scanning distance, an equivalent state of a single GRIN lens based on the differential structure and the standard differential delay line, respectively, were investigated for comparison. The original distances of M1 and M2 in standard differential delay line are both 0, while those in the equivalent state of a single GRIN lens are about 0 mm and 260 mm, respectively. Zero OPD occurs when all delay lines work at their original distances. In the case of standard differential delay line, this structure happens to work in the section S_A. One of the distances between the GRIN-lens and movable mirror is positive changing and the other is reverse. In the equivalent state of single GRIN lens, it makes one GRIN lens pair be in section S_A and makes the other be in section S_C. When the mirrors are scanning, the GRIN-lens pairs changed inversely happen to have the same change trend of loss. The detected signals of the differential structure and the equivalent single-GRIN-lens delay line are shown in Fig. 19a. From the figure, it is difficult to distinguish the difference of the two delay lines; 10 peak points (followed by A, B . . .) are selected from the peaks (corresponding to the strains at coil conversion layers). Therefore, the coupling intensities (peak values) of each peak of the equivalent single-GRIN-lens delay line and the standard differential delay line, respectively, are expressed as triangle and square points in Fig. 19c. The diamond points denote the coupling intensity difference between the two kinds of delay lines, the dashed line is the fitted line, and the solid line is the difference between the green and blue lines in Fig. 18 with an error of 0.2 dB which comes from the system. From Fig. 19c and Fig. 18, the actual difference between the two delay lines matches well with the calculated theoretical difference, which indicates the validity of the proposed differential delay line.

Some errors of loss might be caused by aberrations in the GRIN lens, and the possible small undesired angular tilt in both experiments. The delay line is added an additional GRIN lens, yet the adjustment process become easier. The differential structure could compensate defects with each other even though one or both could not achieve their own best state. Although the equivalent state of a single GRIN lens based on the differential structure is not the best state to be equivalent to the single GRIN lens delay line, the contrast of the two experiments is obvious and the results of ideal differential scanning structure are at accuracy of ± 0.1 dB with scanning range of 0.8 m. Although the moving

Fig. 19 (a) Polarization crosstalk curves of an FOG PMF coil with differential scanning structure. (b) A larger version of the dotted square in (a) with an offset of 5 dB between these two curves. (c) Typical 10 points of crosstalk intensity analysis

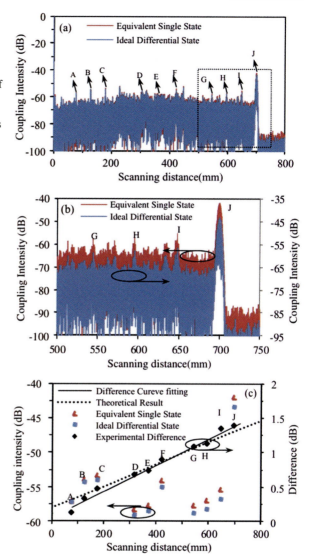

range of the scanning platform is just 200 mm in this paper, the delay line has been extended to 800 mm. We also can link two pairs of GRIN lenses to increase the length of delay line to 1.6 m (sensing range of 3.2 km, assuming birefringence of 5×10^{-4}). In addition, the optical delay line using differential structure has a positive effect on temperature and dispersion compensation. However, the stability of loss fluctuation is at the expense of part of the optical power. Despite consuming more energy, pursuing high amplitude accuracy is still significant.

Iterative Dispersion Compensation for Measuring PMF

Birefringence Dispersion of PMF

Birefringence dispersion (BD) is the chromatic dispersion difference between the two polarized modes of PMF. Moreover, the product of the length matching to the location of polarization crosstalk and the BD of PMF is the distributed BD. Because it is accumulated along with the fiber length, we call it accumulated dispersion (AD) to avoid being confused with the BD. Furthermore, the product of the fiber length and its BD is referred to as length-dispersion product (LDP) of a PMF, i.e., the maximum of AD. On the other hand, the AD is proportional to the width of interference peak, but inversely proportional to its intensity squared. Accordingly, it dynamically degrades the spatial resolution and accuracy of the polarization crosstalk measurement along with the fiber length.

Iterative Dispersion Compensation Method

Considering a large-length gyroscope coil wound by PMF, the interferometric signal generated by its two orthogonal polarized modes can be expressed as (Takada and Mitachi 1998):

$$I_0(x) = \Re \left(i \int_0^L \Gamma_0(\eta) \left\{ \int_0^{+\infty} G(\omega) e^{i[\omega x + \Delta\beta(\omega)(L-\eta)]} d\omega \right\} d\eta \right) \tag{33}$$

where x is the optical path difference; the nonitalic i is the imaginary unit; L is the length of PMF; $\Gamma_0(\eta)$ is the polarization crosstalk at position η that indicates the distance from the output end of PMF; $G(\omega)$ is the optical source spectrum with respect to the optical angular frequency ω; and $\Delta\beta(\omega)$ is the propagating constant difference of the two polarized modes. The Taylor-expansion of the $\Delta\beta(\omega)$ is taken around the optical angular frequency ω_0 that is corresponding to the central wavelength of the optical source spectrum

$$\Delta\beta(\omega) = \beta_0 + DGD(\omega - \omega_0) + \frac{SBD}{2!}(\omega - \omega_0)^2 + \frac{TBD}{3!}(\omega - \omega_0)^3 + \cdots \tag{34}$$

where DGD is the differential group delay of the two polarized modes; SBD is the second-order BD and TBD is the third-order BD.

For second-order AD, $SAD = SBD(L-l)$, and third-order AD, $TAD = TBD(L-l)$, the corresponding phase package is

$$\varphi_c = \frac{SAD}{2!}(\omega - \omega_0)^2 + \frac{TAD}{3!}(\omega - \omega_0)^3 \tag{35}$$

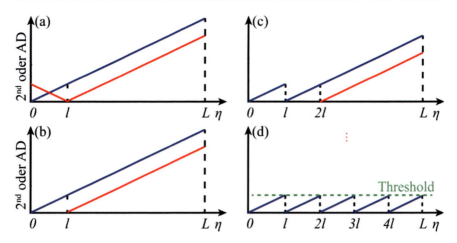

Fig. 20 (**a**) Second-order AD before (blue line) and after (red line) using the PP method. (**b**, **c**) Evolution of the second-order AD during using the IPP method. (**d**) The second-order AD after using the IPP method

where l is a certain distance. Using the phase package (PP) method (Jin et al. 2013) on the interferometric signal expressed by Eq. 33 with φ_c, we have

$$I_0(x) = \Re \left(i \int_0^L \Gamma_0(\eta) \left\{ \int_0^{+\infty} G(\omega)\, e^{i[\omega x + \Delta\beta(\omega)(L-\eta)]} \cdot e^{-i\varphi_c}\, d\omega \right\} d\eta \right)$$

$$= \Re \left\{ i\Gamma_0(l) \left[\int_0^{+\infty} G(\omega)\, e^{i\{\omega x + [\beta_0 + DGD(\omega - \omega_0)](L-l)\}}\, d\omega \right] \right\}$$

$$+ \Re \left\{ i \int_{\eta \neq l} \Gamma_0(\eta) \left[\int_0^{+\infty} G(\omega) \exp\left(i\{\omega x \right. \right. \right.$$ (36)

$$+ [\beta_0 + DGD(\omega - \omega_0)](L - \eta)$$

$$+ \left[\frac{SBD}{2!}(\omega - \omega_0)^2 + \frac{TBD}{3!}(\omega - \omega_0)^3 \right](l - \eta)\right\} \right) d\omega \right] d\eta \right\}$$

where the first equation indicates the process of PP method. The first term in the second equation signifies that the AD of peak at distance of l is removed totally, whereas the second term shows that the residual AD of a peak anywhere else is proportional to $l-\eta$. Figure 20a schematically shows the distributed SAD before and after using the PP method, and the case of third-order AD is similar.

Figure 20b–d indirectly show the process of iterative phase package (IPP) method from the perspective of AD evolution. We maintain the data segment with AD less than the threshold, such as the $0{\sim}l$ segment in Fig. 20b, and use the PP method on the rest data. Similarly, we maintain the $0{\sim}2l$ segment in Fig. 20c, and use the PP

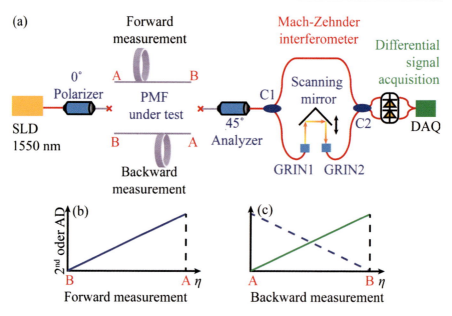

Fig. 21 (**a**) Setup for distributed polarization crosstalk measurement. (**b**) Second-order AD of the PMF under forward measurement. (**c**) Second-order AD of the PMF under backward measurement (green line) and forward measurement (blue-dashed line)

method on the rest data. Repeat the above operation until the AD of all data are less than the threshold, as shown in Fig. 20d. Every iterative step uses the identical phase package corresponding to the threshold.

Figure 21a shows the setup for distributed polarization crosstalk measurement. Broadband light with FWHM of 50 nm and centered at 1550 nm is injected into a 0° polarizer. After transmitting through the PMF under test, the light output to a 45° analyzer. Both the excited light and the coupled light are injected into a Mach–Zehnder interferometer with a scanning mirror in one arm. In the end, a differential signal acquisition system is used to obtain the interferometric signal. The PMF coil under test is a quadrupole fiber gyro coil. The length of PMF is $\sim$ 3 km, and the coil diameter is 10$\sim$20 cm (more details can be seen in section "Periodicity of PMF Coil"). Moreover, the BD of the PMF is SBD $= -301.4$ fs^2/m ($\sim$0.236 ps/nm/km), TBD $= 1.81 \times 10^3$ fs^3/m.

In Fig. 21a, we depicted the forward measurement and the backward measurement for the same PMF. It depends on that the port A or port B is splicing to the polarizer. Even if both directions can measure the polarization crosstalk, the output interferometric signal is quite different due to the BD. Figure 21b schematically shows the second-order AD of the PMF under forward measurement. The port B is located at the zero position if ignoring the pigtail of the analyzer. Figure 21c schematically shows the second-order AD of the PMF under backward measurement (green line), but the port A is located at the zero position. Inverting the interferometric signal output from the forward measurement, the corresponding second-order AD is shown in Fig. 21c with the blue-dashed line. Accordingly, comparing the forward and backward interferometric signals is a good evaluation of IPP method.

High-Resolution Measurement Cancelling Dispersion

Figure 22 demonstrates the performance of the IPP method via the overall interferometric signal of the ~3 km-length PMF. The spatial resolution of the original data in Fig. 22a is degrading. The FWHM of the peak rightmost is 12.36 m in the inset (ii). Besides, the inset (i) could be regarded as blurring. The distributed polarization crosstalk after using IPP method is shown in Fig. 22b, from which we can see the resolution is nearly identical anywhere. The counterpart of inset (ii) is shown in the inset (iv), and the FWHM of the same peak is recovered to 0.09 m (more than two orders of magnitude enhancement). In addition, the counterpart of inset (i) is shown in the inset (iii), from which we can see the comb-like peaks.

As stated in the previous section, the comparison of different measurement directions is a good evaluation means for a dispersion compensation method. We used the IPP method to process the data from forward and backward measurement. The results are demonstrated in Fig. 23, from which we can see the two sets of data are matched well. The minor difference in the inset (i) and (ii) may be the small residual AD.

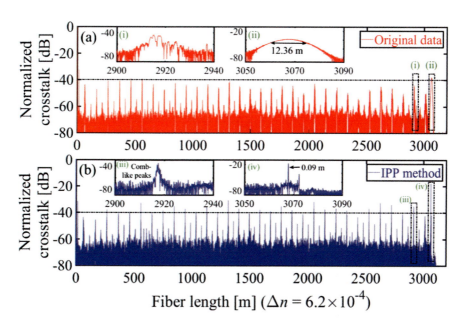

Fig. 22 Comparison between the (**a**) original data and (**b**) the counterpart after using the IPP method. Inset: close-up view of the labeled dashed boxes, (i)~(iv)

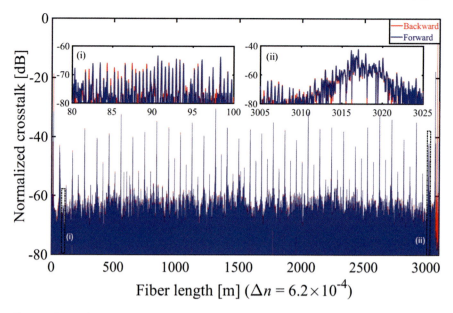

Fig. 23 Comparison between different measurement directions. Inset: close-up view of the labeled dashed boxes, (i)~(ii)

PMF Coil Diagnosis

Analysis Method of PMF Coil Data

Benefit from the dispersion compensation technology, we could obtain the distributed polarization crosstalk of the PMF coil with sufficient spatial resolution. Nevertheless, the figure of measurement result alone demonstrates little information of the PMF coil. Sometimes, it will mislead us when evaluating a PMF coil. To diagnosis a PMF coil, we need more analysis method, the PMF coil floor analysis, and the periodicity analysis, for instance (Fig. 24).

Firstly, we provide a reasonable interpretation for the comb-like peaks in the distributed polarization crosstalk of PMF coil. The high birefringence of PMF is induced by the two symmetric stress bars, and the principle axes of PMF are along this direction and perpendicular to this direction. When a segment of PMF is pressed by a force that has different direction to the principle axes of the PMF, its principle axes will rotate by an angle. Thus, this segment of PMF can be regarded as spliced with the adjacent PMFs. Therefore, a uniform stress on a segment of PMF will generate two polarization crosstalk peaks at the boundary of this stress. The comb-like peaks perhaps indicate that the stress of every circle of PMF coil is different.

PMF coil floor is the distributed polarization crosstalk without comb-like peaks. It is the distributed polarization crosstalk of the PMF itself. The comb-like peaks

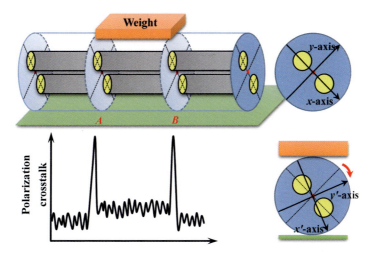

Fig. 24 Schematic diagram of the generation of the comb-like peaks

provide information of the PMF coil from another aspect. PMF coil floor and the periodicity information can be divided in the Fourier domain.

Diagnosis Results of PMF Coil

PMF Coil Floor

Figure 25 illustrates the PMF coil floor of a 3051-m-length fiber optic gyroscope coil made by PMF. We could intuitively observe from the PMF coil floor (black line) that the PMF coil floor at central section is quite higher than anywhere else, yet there is nothing noteworthy in the distributed polarization crosstalk (red line). In addition, the amplitude difference between the distributed polarization crosstalk and the PMF coil floor demonstrates the amplitude of the peaks. According to the quadrupolar winding pattern of PMF coil (Li et al. 2013), the central section of PMF coil is wound on the skeleton, instead of fiber. It should be the reason of higher PMF coil floor.

Periodicity of PMF Coil

The comb-like peaks in the distributed polarization crosstalk seem to be uniformly spaced. The fact is that the spacing between peaks changes gradually according to the diameter of this circle of PMF. The periodicity analysis of the PMF coil is illustrated in Fig. 26. The left two labeled peaks indicate the length of PMF per layer is approximately 48 m. The right two labels indicate that the perimeter of a circle of PMF is from 0.399 m to 0.456 m. It means that the diameter of the PMF coil is from 0.127 m to 0.145 m. The inset demonstrates the variation of amplitude at different diameter of the PMF coil.

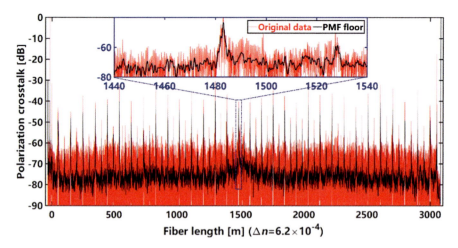

Fig. 25 Comparison of the distributed polarization crosstalk of a 3051-m-length PMF coil and its PMF coil floor

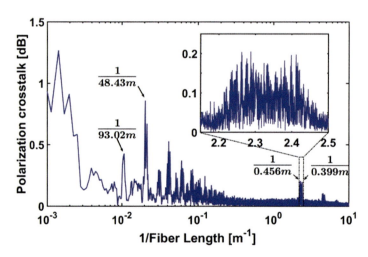

Fig. 26 Periodicity analysis of the 3051-m-length PMF coil

Diagnosis Results at Different Temperatures

Considering the extreme environment that fiber optic gyroscope perhaps faced, the PMF coil should be tested in different temperatures. Figure 27 demonstrates the PMF coil floor three different temperatures. The extreme low temperature, −45 °C notably raises the PMF coil floor, while the extreme high temperature, 60 °C also raises the PMF coil floor a little. Figure 28 shows the periodicity analysis of the PMF coil at three different temperatures. Unexpectedly, the extreme low temperature (red line) has the lowest amplitude for comb-like peaks from the inset of Fig. 28.

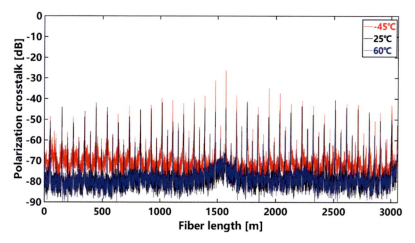

Fig. 27 The PMF coil floor at different temperatures

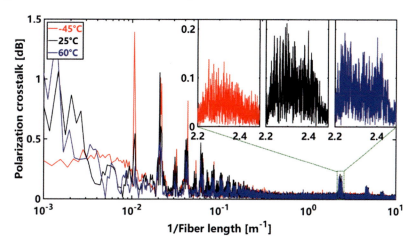

Fig. 28 The periodicity analysis at different temperatures

Conclusion

This chapter gives a brief review of Jones matrix method and a detailed introduction of optical path track method for analyzing the distributed polarization crosstalk of PMF. The instrument, OCDP system, for measuring the distributed polarization crosstalk is discussed, including its measurement range and accuracy improvement. The problem of amplitude fading induced by birefringence dispersion of PMF under test is basically solved. Finally, the high-performance OCDP system is used to measure the distributed polarization crosstalk of PMF coil, which is widely used in optic fiber gyroscope. Corresponding data analysis methods for PMF coil diagnosis are discussed in detail.

References

M. Barnoski, M. Rourke, S. Jensen, R. Melville, Appl. Opt. **16**, 9 (1977)
C. Canavesi, F. Morichetti, A. Canciamilla, F. Persia, A. Melloni, J. Lightwave Technol. **27**, 15 (2009)
W.-S. Choi, M.-S. Jo, J. Opt. Soc. Korea **13**, 4 (2009)
B. Danielson, C. Whittenberg, Appl. Opt. **26**, 14 (1987)
W. Eickhoff, R. Ulrich, Appl. Phys. Lett. **39**, 9 (1981)
A. Gerges, T. Newson, D. Jackson, Appl. Opt. **29**, 30 (1990)
Z. Guo, G. Zhang, X. Chen, D. Jia, T. Liu, Appl. Opt. **50**, 20 (2011)
K. Hotate, O. Kamatani, J. Lightwave Technol. **11**, 10 (1993)
J. Jin, S. Wang, J. Song, N. Song, Z. Sun, M. Jiang, Opt. Fiber Technol. **19**, 5 (2013)
W. Jing, Y. Zhang, G. Zhou, H. Zhang, Z. Li, X. Man, Opt. Express **10**, 18 (2002)
A. Kumar, A. Ghatak, *Polarization of light with applications in optical fibers* (SPIE, Bellingham, 2011), p. 75
Z. Li, Z. Meng, T. Liu, X. Yao, Opt. Express **21**, 2 (2013)
C. Li et al., Meas. Sci. Technol. **26**, 4 (2015)
C. Li et al., Sensors **16**, 3 (2016)
Y. Ning, K. Grattan, A. Palmer, Meas. Sci. Technol. **7**, 4 (1996)
C. Palavicini et al., Opt. Lett. **30**, 4 (2005)
Y. Rao, Y. Ning, D. Jackson, Opt. Lett. **18**, 6 (1993)
A. Rogers, *Polarization in optical fibers* (Artech House, Norwood, 2008), p. 101
T. Sakamoto, Appl. Opt. **25**, 15 (1986)
B. Szafraniec, J. Feth, R. Bergh, J. Blake, in *SPIE proceedings 2510*, 1995
K. Takada, S. Mitachi, J. Lightwave Technol. **16**, 8 (1998)
K. Takada, K. Okamoto, J. Noda, JOSA A **2**, 5 (1985)
K. Takada, K. Chida, J. Noda, Appl. Opt. **26**, 15 (1987a)
K. Takada, I. Yokohama, K. Chida, J. Noda, Appl. Opt. **26**, 9 (1987b)
M. Tsubokawa, T. Higashi, Y. Sasaki, J. Lightwave Technol. **7**, 1 (1989)
J. Yang et al., J. Lightwave Technol. **32**, 22 (2014)
R. Youngquist, S. Carr, D. Davies, Opt. Lett. **12**, 3 (1987)
Z. Yu et al., Opt. Express **24**, 2 (2016)
L. Yuan, J. Yang, Sens. Actuators A Phys. **105**, 1 (2003)
Y. Yuan et al., Photonics Res. **3**, 4 (2015)

Part VII
Optical Fiber Devices

Materials Development for Advanced Optical Fiber Sensors and Lasers

Peter Dragic and John Ballato

Contents

Abstract

Beyond their utility for all modern communications, glass optical fibers are of significant additional present and future value for defense, sensor, and manufacturing systems. However, the extreme commercial scale of communication optical fibers has relegated these special applications to a niche dual-use industry. Accordingly, optical fibers are variations on a commercial theme and, generally, do not adequately address more extreme performance defense and

P. Dragic (✉)
Department of Electrical and Computer Engineering, University of Illinois at Urbana-Champaign, Urbana, IL, USA
e-mail: p-dragic@illinois.edu

J. Ballato
Center for Optical Materials Science and Engineering Technologies (COMSET) and the Department of Materials Science and Engineering, Clemson University, Clemson, SC, USA
e-mail: jballat@clemson.edu

© Springer Nature Singapore Pte Ltd. 2019
G.-D. Peng (ed.), *Handbook of Optical Fibers*,
https://doi.org/10.1007/978-981-10-7087-7_21

sensor demands. Making matters worse, the majority of present global research into optical fibers has focused on geometric microstructuring in order to force the light to behave counter its natural inclinations. As a result, today's "highest-performance" microstructured optical fibers (MOFs) and photonic crystal fibers (PCFs) are remarkably complex in their structures. Accordingly, their costs are significant and their availability is limited.

This chapter reviews the state of knowledge of next-generation optical fibers whose properties arise from attacking performance limitations at their fundamental origin: the interaction of the light with the material through which it propagates. The chapter is divided into two general themes: (1) intrinsically low optical nonlinearities of critical need for high-power and narrow line-width laser applications and (2) novel material effects of interest to sensing. In each case, the material considerations are first introduced, followed by a literature survey of some best results to date. A discussion then is provided that relates to the impact of such materially engineered fibers on specific laser and sensor applications followed by opportunities for further research.

Introduction

Low-loss optical fibers made long-distance high-capacity communications cost-effective and opportune. Nearly 40 years and two billion kilometers later (enough to connect the Earth with Saturn), optical fibers enable all modern communications; play central roles in medicine, energy, manufacturing, and sensing; and serve as unique tools for scientific inquiry.

Be it more Internet traffic, higher-volume manufacturing, or greater directed energy, the optical power propagating through optical fibers today has grown markedly. Indeed in only 20 years, fiber lasers have scaled from just a few W to over 30 kW (Ballato and Dragic 2016). The confined mode, often of small diameter, and long propagation lengths lead to a bouquet of nonlinear optical phenomena, which can either be useful or parasitic, depending on the application. In order to enhance or diminish such nonlinearities, the optical fiber community has developed fibers possessing periodic regions of high and/or low refractive indices to form optical bandgaps that control the properties of light. These microstructured optical fibers (MOFs) or photonic crystal fibers (PCFs) are very complex and have yielded incremental improvements in managing nonlinearities. Their complexities pose significant manufacturing barriers, making such fibers expensive and insufficiently available.

However, it is the authors' supposition that ever more complicated fiber structures, using the same basic materials, miss the most foundational opportunity to study and purposefully tailor nonlinearities. At the most elementary level, it is the interaction of the light with the material that induces these nonlinearities.

The optical nonlinearities of greatest application consequence – good or bad – are stimulated Brillouin scattering (SBS), stimulated Raman scattering (SRS), self-phase modulation (SPM), and four-wave mixing (FWM). SBS is detrimental in high-energy laser (HEL) systems but useful for infrastructure sensing. SRS also is detrimental to HELs yet can be an efficient amplifier for communications. Wave mixing is parasitic in HELs and communications, but central to super-continuum generation for broadband light sources, particularly for IRCMs and chemical sensors. Additional effects include transverse mode instabilities (TMI) and the intrinsic coupling of thermal and strain effects in sensors. These phenomena all share a common origin: the interaction of the optical wave with the fiber material.

Due to growing demands for higher or lower nonlinearities, the fiber community has tailored fiber performance using MOFs and PCFs made from conventional telecomm or IR glasses. In the best cases, these complex structures yield several dB of suppression or up to about 15 dB of enhancement. To be truly transformative, nonlinearities need to be at least 20–30 dB greater or lower than the state of the art. The approach advocated herein is to tailor optical nonlinearities to a degree hereto-fore never considered possible, based on the materials from which the optical fiber is made, rather than its geometry. Central to this approach is balancing fiber compositions through novel core/clad interaction products, which take on positive and negative property values. An exemplar is SBS, which originates from the (p_{12}) photoelasticity. Simply put, materials exist where p_{12} is positive (e.g., SiO_2) and where p_{12} is negative (e.g., Al_2O_3, BaO) such that a combination exists exhibiting zero photoelasticity, hence zero Brillouin scattering and potential for gain. Such zeroing of a fundamental physical property has not previously been considered, yet it will be immensely useful for communication and high-energy laser systems, for example.

The Materials Science of Optical Nonlinearities

The Nonlinearities

In this section, the optical nonlinearities are discussed from the perspective of the driving light-matter interaction. In the next section, a modeling approach to some of the relevant glass and fiber physical characteristics is presented.

Stimulated Brillouin Scattering (SBS)

SBS is an interaction between hypersonic (thermally excited) acoustic waves and the optical signal in a fiber. Simply put, the acoustic wave produces a periodic longitudi-nal pressure and, therefore, density variation. The spatially modulated density of the material then corresponds to a spatially modulated refractive index. The forward-going optical wave will interact with the Bragg-matched acoustic waves, which require that $\lambda_a = \lambda_o/2n_{eff}$ where λ_a is the acoustic wavelength, λ_o is the vacuum optical wavelength, and n_{eff} is the effective index of the optical mode, resulting in a back-reflected Stokes-shifted wave. Via electrostriction, the interference between the forward-propagating signal and backscattered light feeds the acoustic (pressure) wave. This "positive feedback" process increases in efficiency as the power is increased until "threshold" is reached wherein the acoustic wave becomes a highly efficient reflector to the optical signal. In general, SBS limits the amount of light per unit bandwidth that can be transmitted down or generated in an optical fiber. As such, it typically has the lowest threshold of all the nonlinear processes in narrow line-width systems and is a major limitation in the scaling to higher powers in high-energy laser (HEL) systems. Satisfying the classification "narrow line width" largely depends on the power spectral density of a laser source relative to the Brillouin gain bandwidth. This gain bandwidth is usually on the order of 20–200 MHz, and therefore sources with much broader spectra are typically not limited by SBS.

The Brillouin gain coefficient, g_B, is given by

$$g_B = \frac{2\pi n^7 p_{12}^2}{c \lambda_o^2 \rho V_a \Delta \nu_B} \qquad (1)$$

where n is the refractive index, p_{12} is the photoelastic coefficient, c is the speed of light, λ_o is the free space wavelength, ρ is the density, V_a is the acoustic velocity, and $\Delta \nu_B$ is the Brillouin line width. Present methods to suppress SBS have focused on fiber engineering and generally involve the broadening of the Brillouin gain spectrum (BGS) via tailoring of the fiber acoustic properties, namely, V_a, either in the radial or longitudinal directions. In the radial direction, this can include the excitation of multiple acoustic modes or the introduction of acoustic waveguide propagation losses (anti-wave guidance). Active spectral broadening may also be realized in a conventional fiber by applying a thermal or strain gradient along the fiber length (Ballato and Dragic 2013).

Looking to Eq. 1 from a materials perspective, it stands to reason that a low refractive index, low p_{12}, large density, large acoustic velocity, and large spectral width (acoustic attenuation) are desirable properties for intrinsically low Brillouin gain (Ballato and Dragic 2013). It is particularly important to point out that, of the materials coefficients that determine g_B, only the Pockels p_{12} coefficient has the physical potential to take on a zero value. For an optical wave polarized transverse to the propagation direction, p_{12} describes the change in the refractive index "seen" by that wave due to an applied longitudinal pressure (or strain), i.e., due to the presence of an acoustic wave. Photoelasticity is driven by a combination of a change in density and a change in electronic polarizability of a material in the presence of that strain, with the former contributing a positive-valued contribution to p_{12} and the latter a negative contribution. Simply put, materials with positive p_{12} are dominated by the change in material density, and those with negative p_{12} are dominated by the change in electronic polarizability (Ryan et al. 2015). More succinctly, for some materials, a strain, therefore, leads to a decrease in the refractive index (positive p_{12}), and for some it leads to an increase (negative p_{12}). For a material where g_B is zero, these effects balance each other, and the presence of a pressure wave imparts no net change to the refractive index as seen by the optical wave. Hence there is no spatially modulated refractive index variation from which the light wave will scatter, despite the presence of the acoustic wave. This can be found illustrated in Fig. 1.

Even if g_B is not zero, a significant reduction in the value of the Brillouin gain can be realized with a significant reduction in p_{12}. While the photoelastic

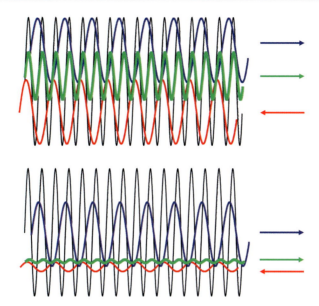

Fig. 1 Illustrated depiction of Brillouin gain suppression by reducing glass photoelasticity. *Top*: conventional fiber. The *black line* represents the acoustic wave, and the *green line* represents the concomitant refractive index distribution. The *blue line* represents the forward-going optical signal and the *red line* the backscattered Stokes wave. *Bottom*: low-p_{12} materials. The presence of the same acoustic wave results in a reduced-amplitude refractive index variation, thus lowering the scattering intensity. The SBS process, in this case, requires much more optical signal power to build. For the case of $g_B = 0$, scattering (red curve) diminishes to zero

constant of silica is well-known to be positive, for other materials, such as sapphire (alumina) and baria (BaO), it is negative (Dragic et al. 2012, 2013a, b). Thus, when such materials as the latter ($p_{12} < 0$) are mixed with those where $p_{12} > 0$ (silica), a binary composition exists where there can be significant cancelation of the photoelastic constant (Ballato and Dragic 2013; Dragic and Ballato 2014). Figure 2 provides a representative comparison of selected materials systems, whereby marked reductions in g_B are realized through this materials approach. Table 1 provides a more thorough compilation of the materials and their deduced Brillouin-related properties; note the richness of compounds that possess negative p_{12} values (Dragic et al. 2013a, b, 2010, 2014; Mangognia et al. 2013; Cavillon et al. 2016). Further inspection of Fig. 2 shows that, while there is only one composition for a given system whereby $g_B = 0$, the binaries have wide compositional ranges where g_B is reduced by 10 dB or more.

Stimulated Raman Scattering (SRS)

SRS is an interaction between an optical wave and optical phonons and can be considered a parasitic effect in high-peak-power fiber laser systems where wavelength control is mandatory (such as in spectrally beam-combined systems).

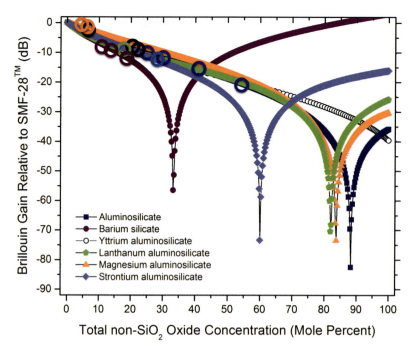

Fig. 2 Comparison of relative Brillouin gain from fibers fabricated using the molten core method; see Table 1 for additional details

Table 1 Deduced Brillouin-related physical parameters for oxide compounds

	$V_a (m/s)$	ρ (kg/m^3)	$\Delta_{\nu B}$[b] (MHz)	n	p_{12}	References
SiO$_2$	5970	2200	21	1.444	0.226	Dragic et al. (2013c)
GeO$_2$	3510	3650	124	1.571	0.268	Dragic and Ward (2010)
P$_2$O$_5$	3936	2390	177	1.488	0.255	Law et al. (2012)
B$_2$O$_3$	3315	1820	428	1.41	0.298	Dragic (2011)
Y$_3$Al$_5$O$_{12}$	7649	3848	253	1.868	0.022	Dragic et al. (2010)
Al$_2$O$_3$	9790	3350	274	1.653	−0.027	Dragic et al. (2013c)
MgO	8731	3322	[a]	1.81	[a]	Mangognia et al. (2013)
SrO	3785	4015	187	1.81	−0.245	Cavillon et al. (2016)
BaO	3131	4688	178	1.792	−0.33	Dragic et al. (2013b)
Yb$_2$O$_3$	4110	8102	1375	1.881	−0.123	Dragic et al. (2013c)
La$_2$O$_3$	3979	5676	181	1.877	−0.027	Dragic et al. (2014)
Lu$_2$O$_3$	3660	7928	145	1.66	−0.043	Dragic et al. (2016a)
Li$_2$O	6500	3150	[a]	1.97	−0.01	Dragic et al. (2015)

[a]Not yet characterized
[b]$n_{ref} = 11$ GH$_z$

The excitation of SRS from spontaneous scattering can lead to wavelength shifts and power instabilities that degrade laser performance, thus presenting the need for its suppression. However, unlike Brillouin scattering, the dependence of the Raman gain coefficient (g_R) is on material properties (Raman cross section, refractive index) that do not permit a zeroing of its value. Instead, an intrinsically low Raman gain material must be selected such that one or more of the following conditions are met: (a) the material is highly disordered so as to broaden the Raman gain spectrum thereby reducing the peak value; (b) high concentrations of materials with low g_R are utilized; and (c) materials are utilized that have Raman spectra with minimal overlap. For (c), a mixture of two materials with similar strength but not overlapping Raman gain spectra could cut the peak Raman gain coefficient in half (Dragic and Ballato 2013).

In compositions possessing large yttria (Y_2O_3) and alumina concentrations, significant Raman gain reduction can be realized (with 3 dB reduction relative to silica demonstrated experimentally in a fabricated fiber), without the use of complicated fiber filtering schemes. Figure 3 shows the relative g_R for yttrio-aluminosilicate glass versus $Y_2O_3 + Al_2O_3$ content (in mole%) (Dragic and Ballato 2013). These reductions only partly relate to the weaker g_R in yttria and

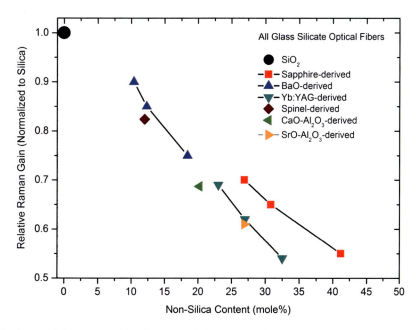

Fig. 3 g_R relative to pure silica for crystal derived optical fiber as a function of total non-silica content. The *numbers* denote specific fibers described in Dragic and Ballato (2013) and Ballato and Dragic (2014)

alumina (which has replaced the much higher-gain silica). The reduction also is due, in part, to the highly non-equilibrium nature of the molten core method employed to fabricate these fibers. In this case, the glass is realized directly from the melt such that the distribution of molecules comprising the glass is more disordered, further lowering the peak g_R values for each of the contributing vibrations.

Transverse Mode Instability (TMI)

Thermally induced longitudinal modulations of the refractive index associated with stimulated Rayleigh scattering result in modal instabilities (mode coupling) in "effectively single-mode" fiber lasers. These higher-order transverse mode instabilities (TMIs) limit power scaling in fiber laser systems by dynamically randomizing the beam modal distribution.

The link between TMI and stimulated Rayleigh scattering yields a materials solution to TMI. The TMI threshold is proportional to $\rho C/Q$ where ρ is the mass density, C is the specific heat, and Q is the thermo-optic coefficient (dn/dT) (Dong 2013; Smith and Smith 2016). Here, in an analogous manner to Brillouin scattering, a material with dn/dT = 0 would completely obviate TMI. Accordingly, combining materials with thermo-optic coefficients of opposite sign can give rise to a significant reduction in dn/dT and possibly even yielding a value of zero. Materials such as SiO_2, GeO_2 (dn/dT larger than silica), and Al_2O_3 (dn/dT similar to silica) have positive dn/dT, but this value is also negative for several materials such as P_2O_5 and B_2O_3 among many others. As an illustrative example based on B_2O_3-doped optical fibers, the thermo-optic coefficient for borosilicate glass is shown in Fig. 4. The origins of positive and negative values of dn/dT are fully analogous to those for positive and negative p_{12}: a heated material thermally expands, and the resulting decrease in density leads to a decrease in refractive index via the Clausius–Mossotti relation. However, this expansion can also tend to increase the electronic polarizability (i.e., increase in refractive index) of the material. As such negative dn/dT materials are dominated by thermal expansion and positive dn/dT materials by the increase in polarizability. The additive materials model, discussed in detail below, was employed to generate the solid curve shown.

n_2-Related Effects

Relative to Brillouin and Raman gain suppression through the enabling materials, less work has been conducted on compositions to reduce the nonlinear refractive index, n_2. The general effect of n_2-related nonlinear optical processes is to broaden and modify the optical spectrum and is undesirable in high-peak-power laser systems (Ballato and Dragic, 2014). Accordingly, suppression of parametric nonlinear phenomena requires a minimization of n_2. It is further worth noting that the Raman gain is proportional to the imaginary part of $\chi^{(3)}$, and so low n_2 materials should further aid in the reduction in SRS.

One approach to n_2 reduction would be to incorporate greater amounts of fluorine into the glass structure, since compounds containing fluorine are known to possess reduced n_2 values relative to silica (Boling et al. 1978; Nakajima and

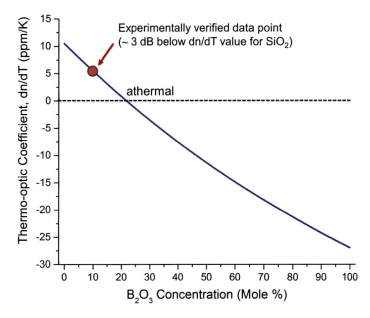

Fig. 4 Thermo-optic coefficient of borosilicate glass with increasing B_2O_3 content. A measured datum on a fabricated fiber where the thermo-optic coefficient, dn/dT, is reduced by 3 dB is provided

Ohashi 2002). However, there are two principal issues with fluorination. First, due both to the glass structure and also to the very high processing temperatures associated with CVD methods, the incorporation of fluorine into silica (F:SiO_2) is quite low, typically on the order of 1% (Schuster et al. 2014). While this is useful for conventional and selected double-clad fiber refractive index profiles, it does not afford much reduction in n_2. Secondly, while fluoride glasses are naturally more linear (lower nonlinearities) than oxides, they are not compatible with HEL demands.

It would therefore be reasonable to focus such "intrinsically low-n_2" glasses on refractory oxyfluoride compositions since, for practical reasons, HEL fibers must be silica based yet exhibit considerably higher fluorine levels than otherwise possible. As noted, the high temperatures ($\sim$2400 °C for final preform collapse) inherent in conventional CVD optical fiber processes facilitate evaporation of most fluorine doped into the glass. While the molten core process does necessarily imply that the core phase is molten, its temperature usually does not exceed about 2000 °C, and fiber often can be achieved at temperatures closer to 1900 °C such that more fluorine remains in the core glass. Further, high-melting point fluoride compounds with low vapor pressures that are combined with the intrinsically low Brillouin or Raman oxide compounds create a fascinating new class of oxyfluoride glasses with the potential for significant reductions in SBS, SRS, and n_2-related nonlinearities.

Materials Modeling

Parasitic nonlinear phenomena are light-matter interactions governed by the properties of the material in which they occur. Given the nearly boundless range of glass compositions that are possible, empirically isolating a particular material that is optimized for one or more applications is a tremendously daunting proposition. As such, modeling approaches are applied to experimental data in order to guide the selection of appropriate application-specific materials. Through a number of various oxide materials systems spanning a broad range of the periodic table (Ballato and Dragic 2013), models based on the Winkelmann–Schott (W–S) (Winkelmann and Schott 1894; Dragic and Ballato 2014; Ballato et al. 2016) method of addition (law of mixtures) have clearly been shown to be powerful tools that can be utilized compositionally to design glass fibers for a number of applications. This specifically includes those with suppressed parasitic phenomena, such as SBS and SRS, respectively, and TMI, and fibers with physical properties that are optimized for certain sensing applications. Specific achievements include prediction of a zero-SBS material and fibers whose Brillouin frequency shift is immune to changes in temperature, applicable to distributed sensing systems. This model has thus far only been applied to oxide glasses possessing at least some silica in their composition.

Numerous additive models for the design of glass with specific desirable properties have been developed, and a very nice review has been provided by Volf (1988). However, despite a rich history of glass modeling efforts by those studying the nature and design of glass, very few materials models have been applied to the design of optical fibers that utilize multicomponent glasses. One very well-known example of such a model is a form of the Sellmeier equation for mixed glasses, such as GeO_2–SiO_2 core glass, that can predict a refractive index as a function of composition (Fleming 1984). A key feature of this model type is the assumption that GeO_2 and SiO_2 are treated as being independent and that they together form a simple mixture. This is a reasonable expectation considering that these two compounds are network formers.

Turning to a visual representation, Fig. 5 shows how a unit length of the core glass may be "unmixed," giving rise to a region of pure SiO_2 and one of pure GeO_2. Since the refractive index is a measure of the speed of light (phase velocity), Fig. 5 suggests a means to calculate the refractive index of the mixed glass: calculate the total time of flight (TOF) of a photon through the glass then divide by the vacuum speed of light. The total TOF is $t = \frac{c}{n_1} L_1 + \frac{c}{n_2} L_2$ where n is the refractive index, c is the vacuum speed of light, and L is the length of the segment. For a normalized unit length ($L_1 + L_2 = 1$) of material, the Ls represent the volume fraction of the material represented by a particular segment. This summation could be extended with additional terms for each new component added to the glass. As such, this can be mathematically represented by a finite sum of the product of the refractive index multiplied by volume fraction. In other words, the material refractive index is modeled as a volume average of the refractive indices of the constituents.

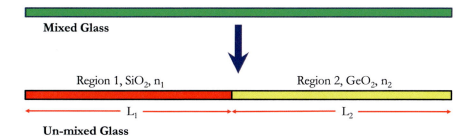

Fig. 5 A mixture representation of multicomponent glass. *Top*: a mixture of components gives rise to the aggregate glass. *Bottom*: aggregate glass separated into its pure constituents. If the segment is a normalized unit length ($L_1 + L_2 = 1$), the Ls represent fractional volume of a constituent

This approach may be generalized to material properties other than the refractive index. Furthermore, it may also include network modifier constituents provided there is caution exercised in the interpretation of data. In short, starting with the fused SiO_2 network, and its known bulk (and quenched) physical characteristics, material is added to the starting silica system via the following governing equation in its most basic form

$$G = \sum_{i=1}^{N} g_i x_i \qquad (2)$$

Here, x is the additivity parameter of constituent i, g is the physical property of that constituent, and G is the aggregate value. In the beginning example above, the refractive index was used as the physical property g and volume fraction the additivity parameter x. Taking the volume fraction as a more global x, g_B (the Brillouin gain coefficient) may be calculated for an arbitrary composition if each of its variables are known, namely, density ρ, refractive index n, photoelastic constant p_{12}, the acoustic velocity V_a, and Brillouin spectral width Δv_B (at some acoustic wave frequency v_B). V_a normalized to some reference velocity is sometimes referred to as the acoustic index, and as with the refractive index, the acoustic index is the material property g which is to be added.

In order to model the Pockels coefficients (p_{11} and p_{12}), the g's are derived from strain–stress relationships as (Ballato and Dragic 2013; Dragic and Ballato 2014; Ryan et al. 2015)

$$C_{1,i} = \frac{1}{2}n_i^3 \left[p_{12,i} - v_i \left(p_{11,i} - p_{12,i} \right) \right] \text{ and } C_{2,i} = \frac{1}{2}n_i^3 \left[p_{11,i} - 2v_i \left(p_{12,i} \right) \right] \qquad (3)$$

where v_i is the Poisson ratio for constituent i. With the assumption that the Poisson ratio adds (as g) according to Eq. 2, the p's for the aggregate glass may be determined. More specifically, Eq. 3 becomes that of the aggregate glass if the subscript i is dropped. In that case Eq. 3 becomes a system of two equations and two unknowns (p_{11} and p_{12} for the aggregate glass).

Equation 2 may be extended to include environmental conditions in a straight-forward way. Each property that can be evaluated (e.g., V_a or n) may be set to be functions of temperature or strain. For example, by making the replacement $n \rightarrow n_o + (dn/dT)\Delta T$, while recognizing that a change in temperature ΔT is a constant for all constituents, the thermo-optic coefficient (dn/dT) may be calculated. Other glass characteristics that can be calculated include the coefficient of thermal expansion, change of index with strain or stress, and change of acoustic velocity with temperature or strain. However, in doing so additional precautions must be taken to account for the presence of a cladding. More details are provided in the next section.

Such models clearly represent very powerful tools in the design of glasses that possess low nonlinearity, especially if one already knows how the addition of a constituent will change the aggregate physical property (see Table 1). If not, such models may be used "in the reverse" in order to determine empirical material values for the pure constituents. Working backward from experimental data, the g's can be used as fit parameters, which would then subsequently be determined and then employed as design variables. Once again, in the case of "mixed" glasses that form networks (such as germanosilicates), the various g's can be construed to be the bulk values of the individual constituents. For network modifiers (such as Group I or II oxides), they are likely more accurately interpreted to be their influence on the starting glass network, rather than some bulk value.

In principle, this additivity model seems to apply well to isotropic glass systems but over limited experimentally characterized compositional ranges. Moreover, it is not even necessary for an ion (e.g., Al in Al_2O_3) to have the same coordination, short-range order, etc. at every site. The additive model essentially provides an ensemble ("observed") average (or net) effect for that glass component, and that can depend strongly on how and by whom the glass was fabricated. Clearly, this removes the use of the model to gain physical insight into the short-range order of glass systems, thus to be able to predict accurately how the glasses form from their precursors. In other words, neither the x's nor g's are predicted by the additive model but once determined still serve as a powerful engineering tool in the design of tailored optical fibers of added value. To understand and even predict glass formation (the x's), an approach taken has been computational, based on a version of the Voigt–Reuss–Hill procedure (Ballato et al. 2016) to calculate "glassy" properties from those of the crystalline precursors. Excellent agreement was achieved between theory and experiment (i.e., those values determined using W–S methodologies) for alumina.

Modeling the Fiber Structure

Optical fibers come in many flavors and many levels of complexity. The simplest fiber arguably is the two-layer core-cladding structure. Usually, the cladding is much larger, more voluminous than the core. Given that the core and cladding glasses can be very different both compositionally and physically, the composite-like nature of

Fig. 6 The effect of CTE mismatch. *Left*: an air-bound core glass will expand to a new radius with an increase in temperature. *Right*: a core clad in a low-CTE material will experience pressure with an increase in temperature due to its expansion being restricted. The core is essentially compressed from its unclad position in a thermally expanded state

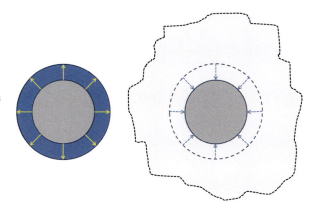

the fiber system must be taken into consideration. Of notable importance is that the geometric configuration of the optical fiber can have a strong influence on the material properties of the glass comprising it, especially in the cases where the fiber environmental conditions might be changing.

Along these lines, work has been done to reconcile a core-cladding mismatch in the coefficient of thermal expansion (CTE) (Dragic et al. 2016b). Take, for instance, a core glass (such as an aluminosilicate) possessing a CTE that is several times larger than that of the cladding (e.g., pure SiO_2, CTE $\sim 0.6 \times 10^{-6}/$ °C). A core that is bound entirely by air will expand at its rate of CTE with any increase in environmental temperature. This is illustrated in Fig. 6 in an end-view for a cylindrical glass rod (left image). However, the presence of a solid cladding (right image) will limit the extent of thermal expansion resulting in a net pressure on the core. This pressure can be construed to be a negative strain and therefore will influence any material properties that are dependent on it, such as refractive index and acoustic velocity.

It is possible that an optical fiber has a composition that is changing in the radial direction (such as a graded-index fiber). In this case, any parameters extracted from the fiber via measurements will be those of the waveguide modes and not necessarily those of the materials. In a limit, should an acoustic or optical wave be confined tightly to the center of the core, the mode values will approximately be the material values in that region. If this is not the case, modeling the system then becomes complicated by the need to adjust material parameters throughout the core region while solving for mode values that best match the measured values. If the composition of the fiber is well-known, these requisite parameters can easily be calculated anywhere in the fiber utilizing Eq. 2. In this case the problem again reduces to one where the physical characteristics of the constituents can become the fitting parameters.

Often, the determination of a fiber's composition will be one providing only a few points across the core. In this case, the core may be partitioned into sections, corresponding to the available measurements. Then, each layer would possess a unique composition and therefore unique physical attributes. The example of a

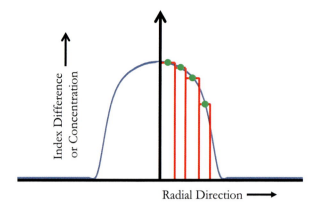

Fig. 7 Should a compositional analysis yield only a few data points (shown representatively in *green*), the core may be partitioned into segments. Each layer possesses a unique composition and therefore unique physical properties. A fictitious RIP is provided as an illustrative example. The RIP may be approximated by four layers (in *red*) in the core, corresponding to the four measured compositions. Solving for the mode is then a simple boundary condition problem

four-layer stepwise approximation to the refractive index profile (RIP) of a fiber is shown in Fig. 7. Outside of the outermost core layer is the cladding, which is typically a uniform material (such as pure silica). There are four boundaries in this example, and by enforcing the boundary conditions the modes can be calculated. A similar approach can be taken in characterizing the acoustic velocity relevant for Brillouin scattering. Importantly, the various layers will have material properties that are functions of temperature and/or strain, and therefore, the calculated modes will also have dependencies on these environmental conditions. This allows for the determination of material properties even from measurements on the modes of an optical fiber.

Applications of Intrinsically Low Nonlinearity Materials

The applications of intrinsically low nonlinearity materials can be classified into two main categories: passive and active. The primary goal of passive optical fibers is to deliver or transmit information or power. Active fibers are those typically utilized in fiber lasers and amplifiers. Considering Raman and parametric amplifiers, active fibers need not be doped with a rare-earth element, but this discussion will be limited largely to the rare-earth-doped variety.

Passive Fibers

The purpose of passive fibers in a light-wave system might be to deliver power, energy, or a signal to some other position or component in a system. Alternatively, it may serve as the medium from which a thermomechanical environment is characterized (described in a later section). For example, it may serve the purpose to deliver high continuous-wave (CW) power to the bottom of a well that has been laser-drilled (Hecht 2012), deliver pulses for a lidar system remotely from the laser source, or even act as a simple interconnect between components in a high-power system. It is important to clearly identify the fiber length needed for a particular application, as this will ultimately limit the power handling capability (Smith 1972). To a good approximation, a so-called critical or threshold power can be identified for both the Brillouin and Raman scattering processes, defining a turn-on point for the self-stimulated process. For the former $P_{crit}^{SBS} \approx 21\frac{A_{eff}}{g_B L}$ and for the latter $P_{crit}^{SRS} \approx 16\frac{A_{eff}}{g_R L}$ where g_B and g_R are the Brillouin and Raman gain coefficients (units of m/W), respectively, and A_{eff} is the effective mode area in the fiber. The effective area is defined to be

$$
A_{\text{eff}} = \frac{\left(\int\limits_{0}^{2\pi} \int\limits_{0}^{\infty} r E\left(r, \theta\right) E^*\left(r, \theta\right) dr d\theta \right)^2}{\int\limits_{0}^{2\pi} \int\limits_{0}^{\infty} r \left(E\left(r, \theta\right) E^*\left(r, \theta\right)\right)^2 dr d\theta}
\tag{4}
$$

where E is the electric field distribution. In the case where fiber attenuation is significant over the relevant lengths, L becomes an effective length defined as $L_{eff} = (1 - \exp(-\alpha L))/\alpha$ where α is the attenuation coefficient in the fiber (m^{-1}). In the case of pulses, the effective length to a reasonable approximation becomes (for $L_{eff} < L$) $L_{eff} = c\Delta t/2n_G$ where Δt is the pulse width and n_G is the group index. For Brillouin scattering, this approximation breaks down for pulses that are comparable to, or shorter than, the acoustic damping or build-up times (usually 10's of ns or less), and a more careful analysis is required. Finally, it should also be noted that these critical powers are more or less back-of-the-envelope estimates, which in practice serve as starting points in a design process. Ultimately, system requirements will drive power constraints.

By way of example, consider a single-mode optical fiber at an operating wavelength that gives rise to a mode diameter of 10 μm ($A_{eff} = 78$ μm^2). Using g_B typical of Brillouin scattering in communications fibers ($\sim 2.5 \times 10^{-11}$ m/W (Ballato and Dragic 2013)), one calculates a critical SBS threshold of 66 W-m. This suggests that in a 1-m length of fiber, 66 W of optical power may be transmitted prior to the onset of SBS. Presently, the most common method to enhance this threshold is through the broadening of the laser signal line width (where system constraints allow it). Also to good approximation, the Brillouin gain coefficient decreases by a factor $\frac{\Delta v_B + \Delta v_L}{\Delta v_B}$ for a laser line width (Δv_L) that is broad relative to the Brillouin line width (Δv_B). Consider, for example, the downhole application. Should the laser spectrum span 30 nm (9 THz) and assuming a Δv_B of 70 MHz (a Brillouin gain spectrum for SMF-28 is provided in Fig. 8; the gain coefficient at the peak is found from Eq. 1), the SBS threshold may be increased to ~ 8.5 MW-m. Therefore, one

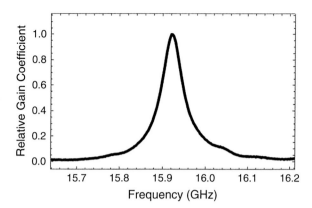

Fig. 8 Brillouin gain spectrum for SMF-28 fiber normalized to g_B and measured at 1064 nm. The small peaks on the higher-frequency side of the spectrum can be identified as higher-order acoustic modes (Koyamada et al. 2004)

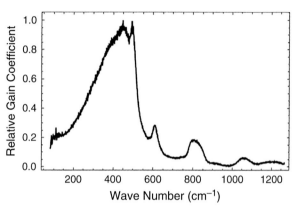

Fig. 9 Raman gain spectrum of SiO_2 normalized to g_R at the peak. The relative strength and widths of the various peaks give insight into the structure of the glass

could expect to be able to transmit on the order of 8.5 kW in a 1-km effective length. In order to increase the length and power, the fiber core could be made to be larger (i.e., multimode), and lengthwise stress or thermal gradients may also be introduced. However, the effectiveness of the latter methods diminishes when the laser spectrum is broad. This sheds light on the value and relevance of an intrinsically low Brillouin gain coefficient fiber, assuming the losses can be made to be low enough for this application.

The Raman gain coefficient is on the order of 1×10^{-13} m/W for typical mostly silica optical fibers. A Raman gain spectrum typical of such fibers is provided in Fig. 9. This puts the SRS threshold at roughly 200× higher than that of SBS (assuming no suppression methods have been utilized). In conventional silica fiber, the Raman gain bandwidth is broad and therefore influences both narrow- and broadband light sources. Due to the large wavelength shift, state-of-the-art techniques to suppress SRS include introducing a distributed loss to the Raman-scattered signal (Taru et al. 2007). This methodology does enable several dB of suppression of SRS relative to conventional fibers, likely comparable to what could be achieved with materials engineering alone. The key advantage of the latter is the potential for a much simpler fiber structure.

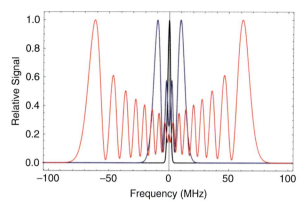

Fig. 10 Normalized spectrum of a 100-ns 1-kW-peak-power Gaussian pulse through a single-mode fiber at a wavelength of 1 μm of lengths 0.1 m (*black line*), 5 m (*blue line*), and 25 m (*red line*) illustrating the broadening due to SPM

The main n_2-related parasitic phenomenon manifests as self-phase modulation (SPM) in pulsed systems. Since the refractive index is a function of the intensity through n_2, the time-varying intensity (i.e., pulse shape) gives rise to a time-varying refractive index that modulates the phase of the optical signal. This leads to a broadening of the laser spectrum, which, depending on the application, may be fully tolerable, partly tolerable (e.g., where it might be used to suppress other parasitics such as SBS), or not at all tolerable (such as where a spectrum must be strictly maintained). To illustrate this, Fig. 10 shows the spectrum of a 100 ns, 1-kW-peak-power Gaussian pulse propagating through different lengths of a conventional single-mode fiber. The spectral distortion is very apparent but only tolerable if the system application allows for it. The nonlinear refractive index for conventional single-mode fibers is on the order of 2×10^{-20} m²/W $- 3 \times 10^{-20}$ m²/W (Boskovic et al. 1996). The nonlinear phase change can be determined from $\gamma = 2\pi n_2/\lambda_o A_{\mathrm{eff}}$ (Agrawal 1995), which has units of rad/W-m. For the single-mode fiber example above, this gives rise to a value of about 2×10^{-3} rad/W-m. As a general guideline, the nonlinear phase change should be much less than about 2π to avoid large spectral broadening. For the example provided here, this translates into a practical threshold on the order of 3 kW-m. While A_{eff} can certainly be increased in order to decrease the effects of SPM, n_2 may also be reduced through the judicious use of materials. As an example, the calculation of Fig. 10 is repeated in Fig. 11 for n_2 that is ½ of the original value. Spectral broadening is essentially cut to ½ the extent.

Active Fibers

The dominant application for rare-earth-doped fibers is fiber lasers. These fibers are of considerable interest for a myriad of applications because they are capable of producing multiple kilowatts of continuous-wave (CW) power in a single well-defined beam. However, due to the parasitic phenomena described above, namely, SBS, narrow line-width CW power has been limited to below about 1 kW (Zervas and Codemard 2014). SRS does not yet currently represent a fundamental limit to

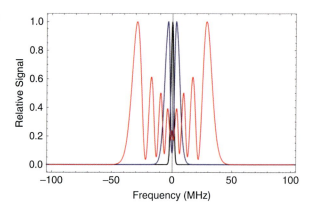

Fig. 11 Same calculation as Fig. 10 but with n_2 taking ½ the value

narrow line-width CW systems but may as laser power scales to beyond 10 kW (Dawson et al. 2008). In pulsed mode, especially as the pulse length is very long compared with the fiber length, these processes can limit *peak* power to values comparable to the CW limit.

Methods to suppress these parasitics generally have involved microstructuring of the fiber in order to achieve effectively single-mode performance while expanding the mode field diameter. The other methodologies for the suppression of nonlinear phenomena described in the previous section (such as thermal or strain distributions) can also be applied. Each of these methods ultimately is limited by the complexities surrounding their practical implementation (e.g., fiber fabrication complexity, unstable mechanics, cost, etc.) (Ballato and Dragic 2013). In this section, two interesting applications for fiber lasers, particularly narrow line-width ones, are briefly described in the context of the relevant power limiters. These applications are lidar and high-energy lasers.

Lidar

The field of lidar (light detection and ranging) is a rich and constantly evolving one with a history that predates the invention of the laser. There are a wide variety of lidar types (Fujii and Fukuchi 2005), including those based on both elastic and inelastic scattering. Applications span from simple range finders to wind speed sensors to systems that probe the composition of the ionosphere. Depending on the data that one wishes to retrieve, laser requirements can vary widely. Indeed, there currently is no single laser that can be used for all lidar applications. For instance, a range finder utilizing a laser operating at 1550 nm will not necessarily be able to detect the presence and quantity of CO_2 in that same wavelength range, despite the presence of absorption features there. As such, much of the area of lidar involves the development of a singular laser type, designed and optimized for one, very focused application, and often is one that pushes the technological limitations often associated with that laser type.

Fiber-based lasers are a particularly attractive light source for lidar due to several key features they possess: (1) they can be compact and lightweight and have (2) high

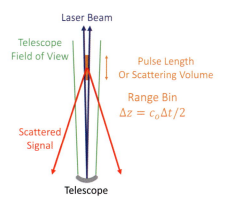

Fig. 12 Simple block diagram of the key elements of a pulsed monostatic lidar

efficiency, (3) tunability, and (4) access to wide wavelength ranges (including post-processes such as frequency conversion or Raman shifting). Features 1 and 2 enable portable and space-based systems, while features 3 and 4 suggest a wide application space. However, despite these apparent advantages that fiber laser technologies may possess over other types, they have not penetrated into areas that would seem to be natural for these laser types. It is the aforementioned parasitic nonlinearities that have been described above that have limited the reach of fiber laser sources for a broader range of lidar applications.

Figure 12 shows a very basic block diagram of a pulsed monostatic lidar. The laser beam is shown exiting through the center of the receiver aperture ("telescope"), although it could emerge from another position from within or outside that area. The laser beam has an associated divergence, which is usually further enhanced via propagation through a turbulent medium such as the atmosphere. Backscattered signal can be provided by any number of processes including Rayleigh and Mie (including aerosol) scattering, Raman scattering, and resonance fluorescence (RF). In RF lidars, a photon is absorbed and subsequently is reradiated into 4π steradians as spontaneous emission. In a pulsed system, the temporal resolution can be limited either by (1) the pulse length of the laser signal or (2) the integration time at the receiver. Since a photon must propagate a round trip through a scattering volume in one integration time (Δt, or pulse width whichever is larger), the range bin is given by $\Delta z = c\Delta t/2$.

The total received backscattered signal (typically measured in photon counts) as a function of range z is given by the lidar equation, which is described most generally as (Wandinger 2005)

$$P(z) = P_0 \frac{c\tau}{2} A\eta G(z)\beta(z) \exp\left[-2\int_0^z \alpha(z)\mathrm{d}z\right] \qquad (5)$$

where P_0 is the transmitted power, τ is the pulse width, A is the receive area, and η represents all basic system-level contributions to receiver efficiency (e.g., receiver quantum efficiency, transmittance of receiver optics, etc.). Furthermore, $G(z)$ represents a geometric factor between the scattered signal and the receiver field of view. It can be expressed generally as $G(z) = \Gamma(z)/z^2$ where $\Gamma(z)$ is an overlap factor between the laser beam and receiver field of view. The z^{-2} comes from the decreasing intensity the further from the scatterer. Finally $\beta(z)$ is the backscatter coefficient, which depends on the type or combination of scattering contributions, and $\alpha(z)$ the attenuation coefficient (m^{-1}) of the medium, with the factor of 2 once again accounting for a round trip.

Equation 5 suggests multiple means to increase signal in a lidar system. Since those quantities related to the atmosphere are usually fixed, an increase in laser power, receiver area, or range bin seem like attractive possibilities. Ultimately, the measurement sensitivity of a system is a function of the system signal-to-noise ratio (SNR), and for laser remote sensing systems, this is often at the shot noise limit. In other words, the SNR is proportional to the square root of the received signal. Hence, a twofold increase in SNR requires a fourfold increase in laser power or receiver area. The latter may not be practical, and the former may be limited by the capabilities of the laser itself. For a fiber laser, these can be clearly identified given the constraints of a lidar system. These are identified below:

Wavelength

Fiber-based light sources have a wide wavelength reach. Fiber lasers in the master-oscillator power amplifier (MOPA) configuration can be seeded by semiconductor lasers that are wavelength stabilized and appropriate for many spectroscopic lidars. These include RF and differential absorption lidar (DIAL).

Tunability

In the latter scheme above, differential absorption lidar (DIAL), the laser wavelength is tuned from on to off of an absorption peak of some species (e.g., H_2O or CO_2 in the atmosphere). This requirement is typically from the range of a few to 10's of GHz. Fiber-based lasers, therefore, are finely suited for this given their wide-gain bandwidth and the fact that the seed laser diodes may be internally or externally wavelength modulated.

Spectral Width and Pulsed Mode

In many applications, a very narrow spectral width is required of the laser. Of course, the moniker "very narrow" must somehow be related to the application. For spectroscopic systems (e.g., RF or DIAL), a requisite spectral width might be <100 MHz, and for coherent systems it could be set by the acceptable laser phase noise encountered in a particular application. Hence, these applications operate in a regime where SBS can become a significant issue. It is also important to remember the transform limit in pulsed operation. Should one need a very narrow spectral width, this could set a *minimum* allowable pulse width required to maintain it. It is not uncommon to encounter pulses that are 10–100's of ns in these systems.

Pulse Repetition Frequency

Aliasing in a lidar system measurement is the limiting factor to the pulse repetition frequency (PRF). When a pulse is launched to a target, one must wait one round trip to the target before launching another signal pulse lest that should obscure the measurement. As an example, an atmospheric target such as the ionosphere, say at a position of 150 km, would need to operate at a PRF at <1 kHz to prevent aliasing. Given upper-state lifetimes for the trivalent rare earths in glass, this alone represents a challenge possibly requiring the use of pulsed pumping.

Average Power

Average power required will depend on the speed with which an acquisition must occur for the measurement to be useful. This is driven largely by the number of samples or averages required to attain some sensitivity (e.g., temperature to within 1 K or concentration to 10 ppb).

Peak Power and Pulse Energy

Should a system require 1 W of average signal power at a PRF of 1 kHz, the pulse energy must be 1 mJ to satisfy this requirement. In a pulse width of 100 ns, this requires a peak power of 10 kW. For a Gaussian pulse, where the time-bandwidth product in the transform limit is $\Delta\tau\Delta\nu = 2(\ln 2)/\pi$, this represents a spectral width of ~4.4 MHz. Since the state-of-the-art fiber lasers with these temporal and spectral characteristics have peak powers still hovering around 1 kW (Nicholson et al. 2016), this places such systems squarely in the crosshairs of the SBS parasitic (and possibly even SPM), necessitating methods to suppress it. Should a temporally shorter pulse be needed for reasons of range resolution, this enhances peak power where SRS can become a concern.

Lidar, Prospects, and Future Directions

While conventional methods to suppress the relevant nonlinear effects in these systems (fibers with large mode area or applying thermal or strain gradients) are usually straightforward methods to enhance system power, scaling average powers to 10 W or more in these low-PRF systems represent a significant challenge. Optical fibers with low intrinsic nonlinearities, such as a 20 dB reduction in the Brillouin gain coefficient, appear to have the potential to serve this need and have a real and transformative impact. In many cases, the judicious use of materials can be done collaboratively with the more conventional methods to reduce the effects of parasitic power limiters.

Fig. 13 Block diagram of the double-clad fiber. The inner-cladding guides pump light (*blue lines*) of lower brightness. Pump light is slowly absorbed by the core (*light blue region*), and the signal (*red line*) is amplified as they propagate along the fiber. Performance-optimized fiber lengths often have some pump light leaking from the end which must be managed

Fiber-Based High-Energy Lasers (HEL)

Many reviews of high-power fiber lasers can be found in the literature (Dawson et al. 2008; Richardson et al. 2010; Jauregui et al. 2013; Zervas and Codemard 2014). The frequency with which they appear is clearly a testament to the rapidly evolving nature of this technological area. Therefore, this discussion will be limited to some brief background information regarding high-power fiber lasers followed by a short discussion of the relevance of a material-based approach to power scaling as they relate to future prospects. The state of the art in narrow line width relies on broadening, or even chirping, the laser spectrum to suppress SBS (White et al. 2017). While it is unlikely that these methods will be fully displaced by materials approaches, it is important to note where they can be used collaboratively.

Figure 13 provides a basic block diagram of the high-power laser scheme. Signal power is launched into the core of the fiber. The inner cladding is typically wrapped in a material typically some sort of polymer, of reduced refractive index (not shown in the figure) relative to the core. The inner cladding is also usually shaped (e.g., hexagon, octagon, D-shaped) in the transverse direction to break the circular symmetry and to enhance pump absorption efficiency. The inner cladding has much lower brightness than the core and serves to guide pump light, which is usually launched through a combiner (Dong and Samson 2017). As the pump light propagates, and rays pass through the core, it is slowly absorbed by the rare earth and reradiated into the signal via stimulated emission. For this reason, fiber lasers are sometimes referred to as "brightness converters." The single-pass amplifier configuration is the most common for high-power fiber lasers. These are known as the MOPA lasers described in the previous section. While pump absorption can be increased by increasing the core-to-cladding ratio (thereby shortening the fiber L_{eff}), that usually comes at the expense of a larger core and therefore potentially reduced beam quality.

Two key configurations can be utilized for the additive power scaling of fiber lasers. These are coherent beam combining (CBC) and spectral beam combining (SBC). The former is a phased array of single-fiber emitters that produce a high-quality beam in the combined far field. The latter stacks multiple laser beams of different wavelength through the use of a dispersive element. For CBC, since each element of the array must be coherently phased as a group, laser coherence length is a practical concern. For efficient phasing, the lengths of the fibers in the array should be controlled to well within a laser coherence length. This becomes much easier as the coherence length l_c increases, which is inversely proportional to the spectral width $l_c = Cc/\Delta\nu$ where C is a scaling factor depending on the spectral shape. Therefore, practical spectral widths in this case will likely be on the order of $\sim$ GHz or less. Since this is much broader than the Brillouin gain bandwidth, suppression of SBS is afforded by utilizing a laser emission spectrum that is as broad as is practical for phasing.

At this point, the value added by intrinsically low SBS fibers can be evaluated. Kilowatt-class fiber lasers have already been demonstrated in fiber systems where the spectral width is on the order of GHz or more. From a practical standpoint, it would be meaningful to enhance the power produced by one element of the array. Referring back to Eq. 1, the Brillouin gain coefficient is inversely proportional to the Brillouin spectral width Δv_B. It is important to note that the Brillouin spectral width itself is inversely proportional to the acoustic damping coefficient (viscous material damping). Since the acoustic attenuation coefficient is proportional to the square of the acoustic frequency, the spectral width tends to broaden with decreasing wavelength as λ_o^{-2}. While this has no influence on g_B (due to the presence of λ_o^2 in the denominator of Eq. 1), the effectiveness of SBS suppression is the laser spectral width relative to that of the Brillouin spectrum. Hence these approaches are more impactful for long-wavelength sources such as Tm-doped fiber lasers. Furthermore, *all* of the compounds identified in Table 1 have acoustic attenuation values larger than that of silica (clearly some more than others). Hence each of these glass components will broaden the Brillouin spectral width when added to silica. While this does not preclude or otherwise deleteriously affect the use of phase modulation or similar techniques to suppress SBS, the material sets a starting point. For example, a laser using a fiber with a Brillouin spectral width of 250 MHz would still benefit from a 1 GHz phase modulation.

More importantly, though, is that a large part of the suppression of SBS in the intrinsically low Brillouin gain optical fibers comes from a reduction in p_{12}, which does not manifest in the Brillouin spectral width. Therefore, phase-modulated systems can gain full advantage of Brillouin gain reduction or "zeroing" of the Brillouin gain coefficient altogether should that be necessary. The same is true of chirped wavelength systems that reduce the effective interaction length ($\sim$0.1 m was identified in White et al. (2017)). As an aside, from a practical perspective, the SBS threshold likely need only be enhanced to the next-highest parasitic limitation, namely, SRS. Of course, this neglects any limitations introduced by the availability of pump lasers.

Spectrally beam-combined systems seem to offer a means to combine several lasers of broad spectral width thereby circumventing the SBS limitation, although demonstrations of SBC systems have utilized relatively narrow laser line width (3 GHz, Honea et al. 2013) where SBS may represent a power limitation. Other system-level limitations might include thermal effects at the dispersive combiner element and limited gain bandwidth of the trivalent rare-earth ions. For example, A number of about 20 elements in an SBC system has been predicted to represent a maximum number of individual lasers in such a system (Bourdon et al. 2017). Regardless, both SBC and CBC systems potentially have the capability to be comprised of individual lasers that are limited by SRS (via suppression of SBS). In that case, material methods to suppress SRS may offer practical advantages to more complicated fiber structures that impart optical losses to the Stokes signal. However, assuming that the refractive index can be carefully controlled, such fibers may eventually be able to use intrinsically low-SRS materials.

Turning now briefly to transverse mode instability (TMI), depending on the fiber core diameter (Leidner and Marciante 2016), TMI can have a turn-on threshold that is similar to that of SBS. Therefore, power scaling in these laser systems, from a materials perspective, may require the development of glasses that are both low g_B and low dn/dT. The advantage in reducing the strength of Brillouin scattering, therefore, is the ability to move to smaller-core fibers where TMI itself can be somewhat less significant. Clearly, this is in a direction opposite to the further scaling of the core size and hence holds many practical advantages from the perspective of both fabrication and implementation. While very promising, the state of the art in these intrinsically low nonlinearity fibers, those engineered with focus on the material properties of the glass, still has relatively high losses for the fiber laser application. Additionally, they often give rise to a refractive index difference that creates a small fundamental mode diameter (<10 μm). Methods to reduce the index difference to enhance mode size are currently being investigated. More details are provided in a section below.

Prospects and Future Directions

While each nonlinearity described above requires different material considerations to be brought to bear, there are two general future issues that are reasonably consistent across all effects: (1) loss and (2) single-mode (or nearly so) operation. The lowest-loss molten core optical fibers reported to date exhibit losses on the order of 100 dB/km. While this is acceptable for base property determination (e.g., Brillouin gain coefficient, fiber dn/dT, etc.), lower loss is always preferred. It should be noted, though, that the near-intrinsic attenuation levels enjoyed by commercial telecommunication fiber are not required for the specialty applications where the materials approach to mitigating nonlinearities described herein are most useful. Indeed, HEL fibers are generally on the order of 10's of meters long. That said, a reasonable future goal is to reduce losses to a level of about 20 dB/km at the operational wavelength. Powder-derived fibers with losses below 20 dB/km have been realized previously (see references in Ballato and Dragic 2013), so this does not seem overly challenging with sufficient attention paid to reducing impurities in the core precursors.

As noted above, the focus on large mode area (LMA) fiber designs as relates to HEL applications has been driven by mitigating these nonlinearities geometrically, i.e., spreading the optical mode out over a larger cross-sectional area such that the threshold for a given (stimulated) nonlinearity is not reached. However, also as noted, such fibers are intrinsically multimoded and other parasitic effects, such as TMI, come into play. The ideal high-energy fiber in most cases is a single-mode design with a double-clad configuration for more efficient pumping. However, the materials largely studied to date that yield intrinsically (materially) low nonlinearities also have higher refractive indices than silica such that said fibers at conventional core sizes and operating wavelengths are multimoded (Ballato and Dragic 2013; Dragic et al. 2012, 2013a), particularly at higher modifier (i.e., non-silica) concentrations. The best approach to achieving single-mode operation at practical core sizes is to use greater concentrations of compounds possessing lower refractive indices in comparison to silica, such as B_2O_3, fluorine (F), and $AlPO_4$ (added into the core as Al_2O_3 and P_2O_5 individually, which react during the preform processing to form the $AlPO_4$ phase). As a result of the high temperatures associated with CVD processes (preform collapse typically >2300 °C), coupled with the volatility of B_2O_3, fluorine, and P_2O_5, conventionally prepared preforms and fibers have relatively low concentrations of these species. The molten core method largely employed to fabricate the intrinsically low nonlinearity fibers is more amenable to higher dopant concentrations as is described in more detail in the following section.

Applications in the Other Direction: Larger Nonlinearities

Perceptively, the process of tailoring the material properties of glass to minimize deleterious nonlinear effects casually hints that the opposite must also be true: that the glass and fiber system instead could be engineered to exploit scattering-related processes for improvements in system-level technologies.

Distributed Fiber Sensing

Distributed fiber optic sensors are analogous to lidar in that both are remote sensing schemes that probe some characteristic along some point in the beam path. Both are transmitted from a source of light, both rely on scattering or feedback from the sensing point, and signal-to-noise ratio is a leading performance benchmark. However, unlike lidar systems that rely mainly on line-of-sight propagation through a bulk material such as air or water, fiber-based distributed sensing systems have a propagation medium that can arbitrarily be defined. Perhaps more importantly, however, and unlike with the atmosphere, the optical fiber itself can be tailored or designed to enhance the performance of the system. In a broad sense, this could involve enhancing the scattering intensity or even the sensitivity of the fiber to changes in its thermomechanical environment. A number of various distributed fiber sensors have been demonstrated, including many that are commercially relevant. These include Brillouin-, Raman-, and Rayleigh-scattering-based sensors. For each of them, the driving feature is that some local environmental change will also alter key scattering characteristics at that position. By sampling the scattered light and by knowing how the fiber responds to thermomechanical changes from some rest state (e.g., $T = 23$ °C and zero strain), one can, for example, determine the local temperature.

A nice review of the state of these sensors can be found in Bao and Chen (2012), and most sensor systems rely on commercial off-the-shelf fibers. This requires engineering the system around the fiber and not the other way around. Raman scattering-based sensors rely on the fact that the inelastic Stokes scattering frequency is a function of both temperature (T) and strain (ε), and therefore both can be quantified along a fiber. The Brillouin scattering frequency (BSF) is found from $v_B = 2n_{eff}(T, \varepsilon)V_a(T, \varepsilon)/\lambda_o$ where n_{eff} is the mode effective area. The response of the BSF to either T or ε can be found by differentiating this equation by parts. The $dV_a/d(T, \varepsilon)$ term typically dominates that possessing $dn_{eff}/d(T, \varepsilon)$. Raman scattering-based sensors rely on the fact that the relative strength of Stokes and anti-Stokes scattering is a function of temperature. More specifically, $\frac{I_{a-S}}{I_S} \propto \exp\left(-\frac{h\Delta v_R}{kT}\right)$ where the I_S and I_{a-S} are the Stokes and anti-Stokes scattering intensities, respectively, and Δv_R is the Raman frequency shift. Rayleigh scattering-based sensors work on the principle that the distribution of Rayleigh scattering intensity along a segment of fiber is more or less a fingerprint of that fiber. Changes to the thermomechanical environment at some point deterministically influence that fingerprint. From this change the thermomechanical environment can be deduced.

In the case of Brillouin- and Rayleigh-based sensors, both strain and temperature can simultaneously influence a measurement often requiring schemes to distinguish them. For Brillouin scattering, this may involve the use of multiple fibers or single fibers that possess multiple BFSs. However, Brillouin scattering affords another approach, in principle similar to the minimization of dn/dT. Since the $dV_a/d(T, \varepsilon)$ term typically dominates the thermomechanical response, a mixture of materials possessing said coefficients of opposite sign could potentially zero this term. This would enable strain sensors that utilize fibers that are immune to either ambient temperature ("athermal") or to changes in strain ("atensic"). Considering silica, $dV_a/d(T, \varepsilon) > 0$ which would then require adding a glass constituent wherein $dV_a/d(T) < 0$ for an athermal fiber or $dV_a/d(\varepsilon) < 0$ for an atensic fiber. Fortunately, several materials are known to satisfy these conditions (Dragic et al. 2012, 2013a, 2016a). Figure 14 provides data on some examples. It should be noted that Fig. 9a in Dragic et al. (2013a) incorrectly gives those thermal coefficients in units of MHz/K, when, for those values in that work, the units should be kHz/K.

In most of these cases, a low-silica content fiber is required to achieve an athermal or atensic fiber. This may eventually prove to be inconsistent with the requirement of a low-NA fiber (for single- or few-moded propagation). It has been shown that the CTE mismatch in a two-layer composite-like fiber structure described above can prove to be very useful in this case. The design approach is simple. First, as was already described, is to recognize that a core with a CTE larger than that of the cladding will experience a positive pressure P, indeed one that increases with increasing temperature. Second is to consider the fact that SiO_2 is considered to be anomalous in that its acoustic velocity decreases with increasing pressure. Therefore, since $dV_a/dT > 0$ for silica, an increase in T will tend to increase V_a for bulk unclad silica. However, should a material be added to the core and that core be clad in a material with much lower CTE, the response of core glass will differ from the bulk case. More specifically, with increasing T, there will be an increasing pressure on the silica component which has $dV_a/dP < 0$. Hence, there are two competing effects: (1) increasing V_a with increasing T but (2) decreasing V_a due to the increasing P.

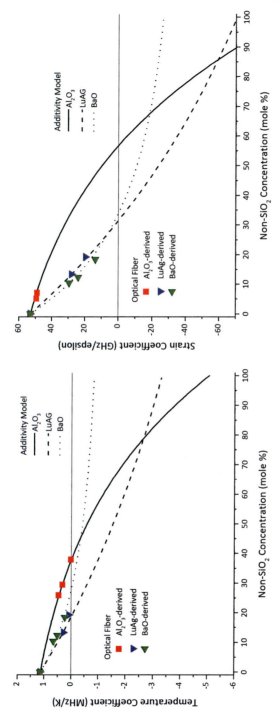

Fig. 14 Thermal (*left*) and strain (*right*) acoustic coefficients of Al_2O_3–SiO_2 ("Al_2O_3-derived"), Lu_2O_3–Al_2O_3–SiO_2 ("LuAG-derived"), and BaO–SiO_2 ("BaO-derived") fibers as a function of metal oxide content, which exhibit athermal and atensic Brillouin frequency behavior. *Symbols* represent compositions that have been fabricated into optical fibers (Dragic et al. 2012, 2013a, 2016a)

Within the constraints of these effects, a correction to the additivity of the acoustic velocity was developed (Dragic et al. 2016b). This is given as

$$V_a(T, \varepsilon) = V_{a,o} + \frac{dV_a}{dT} - \left(\alpha_L^{\text{core}} - \alpha_L^{\text{clad}}\right) \frac{dV_a}{d\varepsilon} \Delta T + 2\left(\alpha_L^{\text{core}} - \alpha_L^{\text{clad}}\right) G \frac{dV_a}{dP} \Delta T \tag{6}$$

where $V_{a,o}$ is the acoustic velocity at rest state (e.g., room temperature), α_L is the linear coefficient of thermal expansion, G is the bulk modulus (of the mixed glass), and strain ε is a fractional elongation. For a multicomponent glass, dV_a/dT and dV_a/dP can be balanced to produce no dependence of V_a on T. Should the CTE mismatch be large enough, it then becomes possible that this be realized in a single-mode fiber appropriate for the distributed sensing application. Lithium aluminosilicate glasses have been shown to satisfy these requirements (Dragic et al. 2015) with only a few mole% of Li_2O in silica leading to a fiber with BFS that is athermal and potentially single mode. These attributes likely also extend to other alkali metal oxide silicates with or without Al_2O_3. As such, these fibers represent ones that can be labeled as "CTE-assisted." It is worth pointing out that for a true zeroing of dv_B/dT, $dn_{\text{eff}}/d(T, \varepsilon)$ should also be considered, but mainly as a perturbation that can be corrected through further adjustment of $dV_a/d(T, \varepsilon)$.

Several fibers in the lithium aluminosilicate compositional family have been fabricated and characterized. Utilizing the modeling presented in a previous section, these experimental results can be extrapolated for arbitrary compositions. The results are provided in Fig. 15. The purple data points represent values determined for fabricated fiber, and the blue curve represents the fit and extrapolation. Tantalizingly, the theory predicts, at some composition, a value of dv_B/dT that is several times stronger than that encountered in conventional, mostly silica optical fibers. Conventional fiber has a thermal response on the order of $+1.1$ MHz/°C, whereas at its most sensitive the lithium aluminosilicate fiber has a thermal response on the order of -6.5 MHz/°C or about six times larger but negative valued. At $\sim$35 mole% Li_2O, this may again not satisfy single-mode behavior without the addition of a material that will lower the refractive index. By way of comparison, the orange curve in Fig. 15 provides the theoretical results for an unclad lithium aluminosilicate glass, demonstrating the very significant impact that the presence of a mechanical cladding layer has on the physical properties of the glass and fiber and the modes residing in them.

Prospects and Future Directions

The use of material science and engineering for the optimization of optical fiber for distributed sensing is a relatively new area. From the perspective of Brillouin-based sensors, truly single-mode athermal fibers have yet not been demonstrated. However, these appear to be well within sight. A disadvantage to the use of the lithium system is that it raises the acoustic velocity when added to silica. This creates an acoustic anti-wave-guiding fiber system that can significantly lower the Brillouin gain coefficient. As such the use of anions further down Group I, such

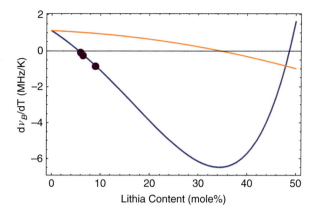

Fig. 15 Dependence of the BFS on lithia (Li_2O) content in lithium aluminosilicate glass at a vacuum optical wavelength of 1550 nm. The analysis assumes a 1:1 molar ratio of Li_2O and Al_2O_3 contents. *Purple circles* represent compositions that have been fabricated into optical fibers. Interesting points are a greatly enhanced sensitivity to T where the lithia content is roughly 35 mole% and a second athermal composition near 48 mole% lithia (although this is a highly improbable composition to realize in practice). The *orange curve* is the simulated response for an unclad lithium aluminosilicate glass showing the influence of the differential thermal expansion between the core and clad

as Na or K, rather than Li is expected to improve the performance of these fibers from a system-level perspective. Additionally, a truly atensic optical fiber has not yet been demonstrated. Much like with athermal fibers (without the assistance of CTE), these fibers also seem to require the use of low-silica content glasses. This may require the development of a novel method, perhaps somehow analogous to "CTE-assisting," to bring the silica content up. Finally, without doubt, the prospect of an athermal or atensic Rayleigh-based sensor is also quite appealing and would likely have significant commercial relevance. Finding such a niche for tailored Raman-based sensor appears to be a much greater challenge.

Notes on Fiber Fabrication

The marked reductions in SBS and SRS gain values achieved to date are the direct result of glass compositions that cannot be made using the conventional chemical vapor deposition (CVD) processes employed for making optical fiber (Ballato and Dragic 2013). While this might sound restrictive from a scalability and manufacturing perspective, the molten core method employed in these cases is indeed industrially relevant. The molten core fiber fabrication approach relies on a precursor core phase, which might be single- or polycrystalline, amorphous, or powdered, that is sleeved inside a (typically pure silica) glass cladding tube selected such that the glass draws into fiber at a temperature above the melting point of the core (Ballato and Snitzer 1995; Ballato et al. 2009; Ballato and Dragic 2013).

The core phase melts and solidifies as the fiber cools upon transit down the draw tower. This rapid quenching of the melt permits realization of core compositions that normally phase separate or devitrify (crystallize) as a result of the time/temperature profile of CVD (MCVD, OVD, VAD) processes. In all fibers discussed above and herein, telecommunication-grade silica is the cladding glass employed thus ensuring robustness, industrial confidence, and compatibility with existing HEL pump fibers and fiber components. A conventional fiber draw tower is employed further ameliorating concerns over a nonconventional fabrication process. The only substantive difference from a manufacturing perspective is that a CVD lathe is not required, which permits novel fiber material concepts and processes to be utilized without the need for a costly lathe.

One particularly useful advantage of the molten core method is that melt-phase chemical reactions occur between the molten core and the softened glass cladding, which can be tailored through fiber size and draw parameters (Ballato et al. 2009). Put another way, the resulting glass core is the interaction product of the molten precursor core material with the partially dissolved cladding material. With this concept of the core being an "interaction product," one can employ the cladding glass in much more subtle – and useful – ways.

Another practical benefit of the molten core approach is correlated with the note above that it enables the direct fabrication of the fiber without a need for a CVD lathe. As such, the high temperatures associated with preform consolidation and collapse are not experienced permitting higher concentrations in the resultant fiber of compounds exhibiting high volatilities. This includes the aforementioned lithium-containing glasses (see "Distributed Fiber Sensing" section) as well as fiber possessing high fluorine and boria concentrations. In addition to reducing the refractive index thus permitting single- or few-moded fibers, higher concentrations of these compounds also favorably impact nonlinearities by reducing Brillouin gain

Table 2 Best results to date for materially mitigating nonlinear optical phenomena

Phenomenon	Governing material property	Best result to date[a]	Materials system	References
SBS	Photoelastic coefficient, p_{12}	~20 dB reduction	Al_2O_3–SiO_2	Dragic et al. (2012)
SRS	Polarizability ($Im[n_2] \propto Im[\chi^{(3)}]$)	~3 dB reduction	Y_2O_3–Al_2O_3–SiO_2	Dragic and Ballato (2013)
TMI	Thermo-optic coefficient, dn/dT	~3 dB reduction	BaF_2–Al_2O_3–SiO_2	Unpublished
n_2	Nonlinear susceptibility, $\chi^{(3)}$		Presently under study	
Brillouin athermal	Thermal expansion, dV/dT	Athermal	Li_2O–Al_2O_3–SiO_2	Dragic et al. (2015)
Brillouin atensic	Strain optic and strain acoustic coefficient, dn/d_e and $dv_a/d\varepsilon$	~6 dB reduction	BaO–SiO_2	Dragic et al. (2013a)

[a]Relative to conventional fiber

(B_2O_3, though increasing the Brillouin line width, $\Delta \nu_B$), Raman gain (fluorine), n_2 (fluorine), and dn/dT (B_2O_3 and P_2O_5).

Thus, one now arrives at a scientifically fascinating and technologically meaningful benefit of the molten core approach: the ability to realize optical fibers, comprised of previously unattainable compositions, that mitigate performance-limiting parasitic phenomena through core/clad interaction products. As a baseline, Table 2 summarizes the best results to date based on this materials approach to mitigating these phenomena.

Conclusion

Presented here was a solution to and review of the associated state of the art for mitigating the optical nonlinearities that plague continued scaling to higher powers in both commodity and specialty/niche optical fiber applications. The approach is fundamental – attacking the nonlinearities at their origin – the materials through which the light is propagating and interacting. Specific materials and considerations for significant reductions in stimulated Brillouin scattering (SBS), stimulated Raman scattering (SRS), thermo-optic (dn/dT)-induced transverse mode instabilities (TMI), and nonlinear refractive index (n_2)-related wave-mixing phenomena. In the case of SBS and TMI, the effects can be completely eradicated through the use of glass compositions that exhibit zero p_{12} photoelastic or thermo-optic coefficients, respectively. While SRS and n_2 do not enjoy similar cancelation possibilities, significant reductions are possible through judicious materials approaches. If there is one memory of this chapter, let it be that a materials approach to mitigating optical nonlinearities is simpler, more manufacturable, and more powerful than the complex and costly approach of fiber geometry tailoring using microstructured or photonic crystal fiber designs.

References

G.P. Agrawal, Nonlinear fibre optics (1995)

J. Ballato, P. Dragic, Rethinking optical fibre: new demands, old glasses. J. Am. Ceram. Soc. **96**, 2675 (2013)

J. Ballato, P. Dragic, Materials development for next generation optical fibre. Materials **7**, 4411 (2014)

J. Ballato, P. Dragic, Glass: the carrier of light – a history of optical fibre. Int. J. Appl. Glas. Sci. **7**, 413 (2016)

J. Ballato, E. Snitzer, Appl. Opt. **34**, 6848 (1995)

J. Ballato, T. Hawkins, P. Foy, B. Kokuoz, R. Stolen, C. McMillen, M. Daw, Z. Su, T. Tritt, M. Dubinskii, J. Zhang, T. Sanamyan, M.J. Matthewson, On the fabrication of all-glass optical fibres from crystals. J. Appl. Phys. **105**, 053110 (2009)

A. Ballato, P. Dragic, S. Martin, J. Ballato, On the anomalously strong dependence of the acoustic velocity of alumina on temperature in aluminosilicate glass optical fibres, part II: acoustic properties of alumina and silica polymorphs, and approximations of the glassy state. Int. J. Appl. Glas. Sci. **7**, 11 (2016)

X. Bao, L. Chen, Recent progress in distributed fibre optic sensors. Sensors **12**, 8601 (2012)

N. Boling, A. Glass, A. Owyoung, Empirical relationships for predicting nonlinear refractive index changes in optical solids. IEEE. J. Quant. Electron. **14**, 601 (1978)

A. Boskovic, S.V. Chernikov, J.R. Taylor, L. Gruner-Nielsen, O.A. Levring, Direct continuous-wave measurement of n_2 in various types of telecommunication fibre at 1.55 μm. Opt. Lett. **21**, 1966 (1996)

P. Bourdon, L. Lombard, A. Durécu, J. Le Goüet, D. Goular, C. Planchat, Coherent combining of fibre lasers. Proc. SPIE **10254**, 1025402–1025401 (2017)

M. Cavillon, P. Dragic, J. Furtick, C. Kucera, C. Ryan, M. Tuggle, M. Jones, T. Hawkins, P. Dragic, J. Ballato, Properties of alkaline earth aluminosilicate glass optical fibres. J. Lightwave Technol. **34**, 1435 (2016)

J.W. Dawson, M.J. Messerly, R.J. Beach, M.Y. Shverdin, E.A. Stappaerts, A.K. Sridharan, P.H. Pax, J.E. Heebner, C.W. Siders, C.P.J. Barty, Analysis of the scalability of diffraction-limited fibre lasers and amplifiers to high average power. Opt. Express **16**, 13240 (2008)

L. Dong, Stimulated thermal Rayleigh scattering in optical fibres. Opt. Express **21**, 2642 (2013)

L. Dong, B. Samson, *Fibre Lasers. Basics, Technology, and Application* (CRC Press, Taylor and Francis, Boca Raton, 2017)

P. Dragic, Brillouin gain reduction via B_2O_3 doping. J. Lightwave. Technol. **29**, 967 (2011)

P. Dragic, J. Ballato, Characterization of the Raman gain spectra in Yb: YAG-derived optical fibres. Electron. Lett. **49**, 895 (2013)

P. Dragic, J. Ballato, 120 years of optical glass property calculations: from the law of mixtures and the birth of glass science to mixing the unmixable. Opt. Photon. News. **25**, 44 (2014)

P. Dragic, B. Ward, Accurate modeling of the intrinsic Brillouin linewidth via finite element analysis. IEEE Photon. Technol. Lett. **22**, 1698 (2010)

P. Dragic, P.-C. Law, J. Ballato, T. Hawkins, P. Foy, Brillouin spectroscopy of YAG-derived optical fibres. Opt. Express **18**, 10055 (2010)

P. Dragic, T. Hawkins, S. Morris, J. Ballato, Sapphire-derived all-glass optical fibres. Nat. Photon. **6**, 629 (2012)

P. Dragic, J. Ballato, S. Morris, T. Hawkins, Pockels' coefficients of alumina in aluminosilicate optical fibre. J. Opt. Soc. Am. B **30**, 244 (2013a)

P. Dragic, C. Kucera, J. Furtick, J. Guerrier, T. Hawkins, J. Ballato, Brillouin spectroscopy of a novel baria-doped silica glass optical fibre. Opt. Express **21**, 10924 (2013b)

P. Dragic, J. Ballato, S. Morris, T. Hawkins, The Brillouin gain coefficient of Yb-doped aluminosilicate glass optical fibres. Opt. Mater. **35**, 1627 (2013c)

P. Dragic, D. Litzkendorf, C. Kucera, J. Ballato, K. Schuster, Brillouin scattering properties of lanthano-aluminosilicate-core optical fibre. Appl. Opt. **53**, 5660 (2014)

P. Dragic, C. Ryan, C. Kucera, M. Cavillon, M. Tuggle, R. Stolen, J. Ballato, Single- and few-moded lithium aluminosilicate optical fibre for athermal Brillouin strain sensing. Opt. Lett. **40**, 5030 (2015)

P. Dragic, M. Pamato, V. Iordache, J. Ballato, C. Kucera, M. Jones, T. Hawkins, J. Ballato, Athermal distributed Brillouin sensors utilizing all-glass optical fibres fabricated from rare earth garnets: LuAG. New J. Phys. **18**, 015004 (2016a)

P. Dragic, S. Martin, A. Ballato, J. Ballato, On the anomalously strong dependence of the acoustic velocity of alumina on temperature in aluminosilicate optical fibres – Part I: Material modeling and experimental validation. Int. J. Appl. Glas. Sci. **7**, 3 (2016b)

J.W. Fleming, Dispersion in GeO_2-SiO_2 glasses. Appl. Opt. **23**, 4486 (1984)

T. Fujii, T. Fukuchi (eds.), *Laser Remote Sensing* (CRC Press, Taylor and Francis, Boca Raton, 2005)

J. Hecht, High-power lasers: fibre lasers drill for oil. Laser Focus. World **12**, 27 (2012)

E. Honea, R.S. Afzal, M. Savage-Leuchs, N. Gitkind, R. Humphreys, J. Henrie, K. Brar, D. Jander, Spectrally beam combined fibre lasers for high power, efficiency and brightness. Proc. SPIE **8601**, 860115–860111 (2013)

C. Jauregui, J. Limpert, A. Tünnermann, High-power fibre lasers. Nat. Photonics **7**, 861 (2013)

Y. Koyamada, S. Sato, S. Nakamura, H. Sotobayashi, W. Chujo, Simulating and designing Brillouin gain spectrum in single-mode fibres. J. Lightwave Technol. **22**, 631 (2004)

P.-C. Law, A. Croteau, P. Dragic, Acoustic coefficients of P_2O_5-doped silica fibre: the strain-optic and strain-acoustic coefficients. Opt. Mater. Express **2**, 391 (2012)

J.P. Leidner, J.R. Marciante, The impact of thermal mode instability on core diameter scaling in high-power fibre amplifiers, CLEO 2016, paper SM4Q.2

A. Mangognia, C. Kucera, J. Guerrier, J. Furtick, P. Dragic, J. Ballato, Spinel-derived single mode optical fibre. Opt. Mater. Express **3**, 511 (2013)

K. Nakajima, M. Ohashi, Dopant dependence of effective nonlinear refractive index in GeO2- and F-doped core single-mode fibres. IEEE Photon. Technol. Lett. **14**, 492 (2002)

J.W. Nicholson, A. DeSantolo, M.F. Yan, P. Wisk, B. Mangan, G. Puc, A.W. Yu, M.A. Stephen, High energy, 1572.3 nm pulses for CO_2 LIDAR from a polarization-maintaining, very-large-mode-area, Er-doped fibre amplifier. Opt. Express **24**, 19961 (2016)

D.J. Richardson, J. Nilsson, W.A. Clarkson, High power fibre lasers: current status and future perspectives. J. Opt. Soc. Am. B **27**, B63 (2010)

C. Ryan, P. Dragic, J. Furtick, C.J. Kucera, R. Stolen, J. Ballato, Pockels coefficients in multicomponent oxide glasses. Int. J. Appl. Glas. Sci. **6**, 387 (2015)

K. Schuster, S. Unger, C. Aichele, F. Lindner, S. Grimm, D. Litzkendorf, J. Kobelke, J. Bierlich, K. Wondraczek, H. Bartelt, Adv. Opt. Technol. **3**, 447 (2014)

R.G. Smith, Optical power handling capacity of low loss optical fibres as determined by stimulated Raman and Brillouin scattering. Appl. Opt. **11**, 2489 (1972)

A.V. Smith, J.J. Smith, A comparison of mode instability in Yb- and Tm-doped fibre amplifiers. Proc. SPIE **9728**, 97280C–972801 (2016)

T. Taru, J. Hou, J.C. Knight, Raman gain suppression in all-solid photonic bandgap fibre, in *European Conference and Exhibition on Optical Communication 2007* (Berlin 2007), paper 7.1.1

M.B. Volf, *Mathematical Approach to Glass* (Elsevier Science Publishers, Amsterdam, 1988)

U. Wandinger, in *Lidar. Range-Resolved Optical Remote Sensing of the Atmosphere*, ed. by C. Weitkamp (Ed), (Springer, Berlin, 2005), p. 1

J.O. White, M. Harfouche, J. Edgecumbe, N. Satyan, G. Rakuljic, V. Jayaraman, C. Burgner, A. Yariv, 1.6 kW Yb fibre amplifier using chirped seed amplification for stimulated Brillouin scattering suppression. Appl. Opt. **56**, B116 (2017)

A. Winkelmann, O. Schott, Uber die Elastizität und Uber die Zugund Druckfestigkeit Verschiedener Neuer Gläser in Ihrer Abhängigkeit von der Chemischen Zusammensetzung. Ann. Phys. (Berlin) **287**(697) (1894)

M. Zervas, C. Codemard, High power fiber lasers: a review. IEEE J. Sel. Top. Quant. Electron. **20**, 219 (2014)

Optoelectronic Fibers

34

Lei Wei

Contents

Abstract

Fibers are one of the most fundamental material forms, made by nature or by humans. In particular, optical fibers now are widely used in a multitude of applications, ranging from telecommunications to monitoring structural integrity of bridges. Integration of materials with disparate electrical, optical, thermal, or mechanical properties into a single fiber with complex architecture and diverse functionalities presents new opportunities for extending fiber applications in numerous fields, especially as optoelectronic devices. This chapter presents the development of optoelectronic fibers, from the fundamentals to in-fiber device demonstration. Especially, the integration of semiconductor materials into fiber geometries provides a unique route to introduce new optoelectronic functionality

L. Wei (✉)
School of Electrical and Electronic Engineering, Nanyang Technological University, Singapore, Singapore
e-mail: wei.lei@ntu.edu.sg

© Springer Nature Singapore Pte Ltd. 2019
G.-D. Peng (ed.), *Handbook of Optical Fibers*,
https://doi.org/10.1007/978-981-10-7087-7_40

1335

into existing glass fiber technologies. Firstly, as the core material, multi-material fibers made of semiconductor materials such as silicon, germanium, and compound semiconductors are developed, which offer different advantages in terms of the material, geometry, and waveguiding properties. Then, three main fabrication approaches to produce these fibers are summarized, in which the first approach is based on traditional drawing tower technique, the second approach involves chemical deposition inside glass capillary templates, and the third approach takes advantage of in-fiber fluid instability phenomenon. Finally, future prospects and applications of this new class of fibers are discussed.

Introduction

The integration of semiconductors, conductors, and insulators into precisely defined geometries with low-scattering interfaces and microscopic feature dimensions is essential to the realization of functional electronic, photonic, and optoelectronic devices (Pierret 1996). The platforms carrying these functional components have been extensively applied in wafer-based carriers using well-developed micro- or nanofabrication processes, which are the cornerstones of the electronics revolution, but fundamentally restricted to the wafer sizes, planar geometries, and the complexity associated with high-precision processing steps.

Fiber drawing from a preform has traditionally been exploited to produce with high precision and reliability the extended lengths of optical fibers that today span the globe, deliver telecommunications, and enable the Internet (Agrawal 2010; Russell 2003). One limitation of the current state of the art, however, is that the current fiber drawing has been limited to drawing of a single material and to features no smaller than several microns. Recently, fiber-drawing approaches have been extended to a new class of multi-material fibers containing functional structures that enable the development of optoelectronic fibers (Tao et al. 2012; Peacock et al. 2014; Schmidt et al. 2016). This new class of fibers centers on a method so-called preform-to-fiber fabrication by thermally drawing a macroscopic solid-state preform into extended lengths of uniform fibers, as shown in Fig. 1. Unlike the traditional optical fiber fabrication, this method features with three key elements in resulting fibers: (1) combining a multiplicity of solid materials with disparate electrical, optical, and mechanical properties into a single fiber, (2) realizing arbitrary nanometer-scale geometries in fibers with low-scattering interfaces, and (3) producing long lengths of fibers through the simple and scalable process of thermal drawing.

Fiber-like carriers have been demonstrated to achieve sophisticated functionalities at fiber optic length scales and uniformity (Bayindir et al. 2004; Abouraddy et al. 2007; Stolyarov et al. 2012; Canales et al. 2015; Zhang et al. 2016a, b, c, d, 2017a, b; Li et al. 2016; Shabahang et al. 2016; Wang et al. 2017). By introducing semiconductor functionalities into the fiber platform, new possibilities arise for the development of next-generation in-fiber optoelectronic devices both in terms of waveguide design and integration. The first challenge of the transition from traditional fibers made of single material to multi-material fibers is the selection of functional materials. Since multi-material preforms may contain materials that

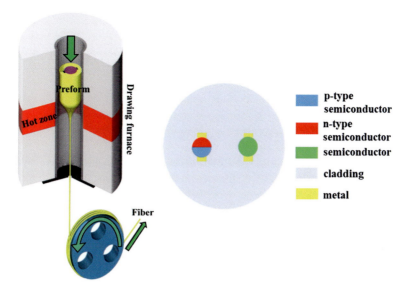

Fig. 1 The direct thermal fiber-drawing technique. The *left image* schematically illustrates the operation principle of direct fiber drawing (Adapted with permission Schmidt et al. (2016)). The *right image* shows the cross section of optoelectronic fiber with the integration of semiconductors, conductors, and insulators

are incompatible with thermal drawing when taken separately, such as crystalline semiconductors or metals, thermal drawing imposes some constraints on the materials combinations that are compatible with this fabrication strategy. Therefore, materials with disparate electronic and optical properties such as semiconductors and metals must be compatible in such a fabrication method that enables the production of kilometers of fibers to act as optoelectronic devices.

Material Selection

The main requirements in the materials used in these fibers are summarized as follows:

1. In general, semiconductor materials should be in the glassy or amorphous state to be thermally drawn at reasonable speeds in the fiber-drawing process with self-maintaining structural regularity. Thus, the selected materials (or at the very least, the majority component) must remain amorphous and not crystallize when cycled through softening and drawing temperatures.
2. As the core materials, crystalline semiconductors are commonly considered to offer superior optoelectronic performance, for example, higher carrier mobility compared to amorphous semiconductors. However, this will lead to a fabrication challenge to confine these materials during the thermal-drawing process, since the viscosities of such materials are low in the liquid phase.

3. The metals should also be in the liquid phase at the drawing temperature during the fiber-drawing process. The same challenge of confining the molten metals still exists, as molten metals have very low viscosities, especially between high-viscosity semiconducting and insulating interfaces.
4. To maintain clear interfaces between different materials, these materials should exhibit compatible viscosities at the softening and drawing temperatures of interest and should exhibit good adhesion/wetting in the viscous and solid states without cracking even when subjected to thermal quenching.

Taking silica glass, silicon, and gold as examples, these materials represent three distinct classes from the perspective of electronic properties: an amorphous insulator, a crystalline semiconductor, and a metal, respectively. Additionally, these three materials have very different optical, mechanical, and thermal properties. Silica is amenable to thermal drawing over a broad range of conditions since its softening temperature ranges from 1400 to 2350 °C. However, crystalline materials such as silicon and gold are characterized by an abrupt drop in viscosity above the melting temperature where a phase transition takes place. While this physical feature excludes the use of thermal drawing to produce a fiber from a single-material preform made of silicon or gold; nevertheless, a multi-material preform approach enables the use of such materials in fiber drawing. By making use of a crystalline material such as silicon or gold as a core embedded in an amorphous cladding such as silica, this multi-material preform has been successfully drawn into fibers. In this scenario, the cladding material is silica, which acts as a supporting scaffold that contains and restricts the flow of the low-viscosity core materials such as silicon or gold.

The wide range of possibilities for constructing multi-material preforms within the above-prescribed constraints may be appreciated from Fig. 2, which provides a guideline to select functional materials for low-, middle-, or high-temperature regions with the corresponding cladding materials to be polymer, soft glasses, or fused quartz (Ofte 1967; Glazov et al. 1969; Urbain et al. 1982; Kakimoto et al.

Fig. 2 Dynamic viscosity (logarithm of viscosity g in poise) of selected materials versus temperature, showing the viscosity for silicon, germanium, indium antimonide, tellurium, indium arsenide, indium, tin, and gold (Adapted with permission Tao et al. (2012))

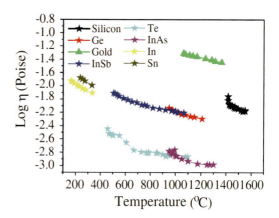

1989; Tverjanovich 2003; Koštál and Málek 2010). It is possible to choose potential pairs of amorphous-crystalline materials that may be combined in a preform and co-drawn into a multi-material fiber by consulting this figure. For example, it is possible to draw a fiber containing a core of silicon, germanium, or gold by using a cladding made of silica glass; a core of Bi_2Te_3 using a cladding made of borosilicate glass; or a core of selenium using a cladding made of polymers such as polycarbonate. Furthermore, it is possible for more than two materials to be chosen according to the above criteria and thus be incorporated in the same multi-material fiber, which enables the construction of optoelectronic fibers. As the core material, current research directions have focused on the following semiconductor materials.

Silicon

Silicon is by far the most pervasive electronic and photonic device material, and the ability to produce such material within a flexible fiber over a wide range of sizes is of significant technological and scientific importance (Reed and Knights 2004; Lipson 2005). As the fabrication process shown in Fig. 3, the high thermal conductivity, the high optical damage threshold, and the low loss transmission between 1 and 7 μm of crystalline silicon are particularly beneficial features.

Silicon has successfully acted as the core material to achieve meter lengths of fibers with the core diameters in the micrometer scale (Ballato et al. 2008,

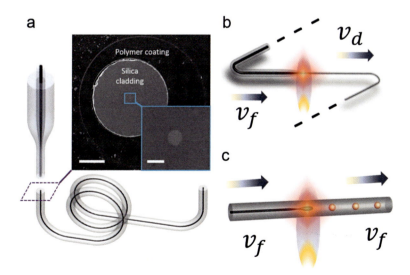

Fig. 3 Silicon-core fiber fabrication process (Adapted with permission Gumennik et al. (2013)). (**a**) Schematic representation and experimental results of thermally drawing a silicon-core fiber. (*Left scale bar*, 100 μm and inset, 5 μm) (**b**) Schematics of a high-tension redraw process that allows, upon repeated application, continuous reduction of the silicon core toward the submicron regime. (**c**) Generation of in-fiber particles using microfluidic instability

2009b; Scott et al. 2009; Gumennik et al. 2013; Healy et al. 2014; Peacock et al. 2016). Transmission losses for these fibers have been measured to be around 2.7 dB/cm at 1.5 μm and 4.3 dB/m at 2.9 μm, respectively. These losses are some of the lowest reported for a polysilicon material and point to the high quality of the single crystal grains. The core size of silicon fiber can be further reduced into submicron regime using a high-tension redraw process, as shown in Fig. 3b. However, as the dominant loss mechanism in these fibers is absorption due to impurities that cluster at the grain boundaries, the losses in the smaller core fibers are considerably larger owing to the increased defect density. Rapid photothermal annealing of these polysilicon fibers has resulted in increased local crystallinity of the material, which should improve the core uniformity and ultimately yield lower losses.

In-fiber junction structures are the fundamental roadblocks to the realization of a high-sensitivity optoelectronic fiber device by forming high-quality junctions between two or more domains. To prevent and/or control inter-domain mixing, one method is to introduce a viscous barrier in a two-material dual-core structure at the preform level to keep the constituent materials separated during the fiber draw process. As the preform necks down into a fiber, this barrier is controllably reduced in thickness to the desired dimensions by finely adjusting the drawing parameters and would ultimately be made to vanish at precisely the same time that the materials exit the fiber draw furnace hot zone. Following the above-described approach and dual-core structure, two initially separated regions of p-type and n-type crystalline-semiconductor rods within a glass fiber would be brought into contact during the thermal drawing. If mixing is prevented, the fiber should enable electrically rectifying behavior, along with high-sensitive photodetecting capabilities. Moreover, a three-domain junction structure would be realized by further managing the diffusion between p-type and n-type regions to bring more efficient and advantageous features in photodetecting systems. The intrinsic region can be resized during fiber fabrication and made much wider than the depletion region while sustaining a high electric field, making it possible to increase the quantum efficiency of the junction at the selected wavelengths.

An alternative way of forming in-fiber crystalline-semiconductor junction is to harness recent demonstration of in-fiber particles using microfluidic instability, as shown in Fig. 3c. To form in-fiber junction particles, heating the fiber post-draw could bring two adjacent p-type and n-type semiconductor cores to merge into a bi-spherical particle as the two cores simultaneously undergo breakup process. This direct physical self-assembly process would be exploited to form in-fiber particles with junction structures. A careful choice of the doping concentrations will allow for the tailoring of the junction width, which could range from a few tens to several hundreds of nanometers, thus going from negligible to comparable to the particle dimensions. This can result in unique electric field distributions and device behavior and would enable more efficient light absorption for silicon-based particles inside the fiber for optoelectronic applications.

Germanium

Germanium has been an important contributor for semiconductor technologies, especially in the areas of semiconductor photonics and optoelectronics. The optical transparency of germanium extends even further into the infrared, with a low loss window from 2 to 14 μm, covering the spectroscopic window for most characteristic vibrations of molecules. Compared to silicon, germanium has a higher refractive index, allowing for better mode confinement, and a larger third-order nonlinear susceptibility to enhance the nonlinear coefficient. Furthermore, the smaller energy difference between its indirect and direct bandgap allows the occurrence of indirect-to-direct bandgap transition under strain, leading to more efficient light emission. Its high absorption coefficients in the near-infrared combined with high carrier mobility also make germanium a suitable candidate for photodetectors.

Large core crystalline germanium fibers with silica cladding have been thermally drawn, and these fibers have a loss of 0.7 dB/cm at 3.39 μm (Ballato et al. 2009a). Extensive studies to determine the crystalline size and orientation of these fibers have revealed that this method can produce very large grain sizes, with single crystalline regions. On the other hand, small-core single-crystal germanium fibers have been annealed using a precisely controlled visible laser, which can produce fibers with low optical losses 1.33 dB/cm and good photoresponsivity in the near-infrared. These fibers can be further integrated with in-fiber silicon photonic devices for photodetection, operating especially at the optical telecommunication wavelength of 1550 nm.

Compound Semiconductors

Compound semiconductors offer a key advantage over the unary materials due to the fact that their properties can be tuned simply through the composition. Furthermore, many of these materials have a direct bandgap which is required for efficient light emission. Compared to the unary materials, incorporation of compound semiconductors into the optical fiber geometry is much more challenging as it is important to maintain the appropriate stoichiometry of the material during the fabrication process.

The first example of a crystalline III–V compound semiconductor optical fiber was fabricated via the thermal draw method to have an indium antimonide (InSb) core for electro-optic devices (Ballato et al. 2009b). Oxygen and phosphorous contamination of the core by diffusion from the cladding at the draw temperature was observed, an ongoing challenge in thermal draw method, but the material was phase pure. Further, with the help of high-pressure chemical vapor deposition, high-purity crystalline zinc selenide (ZnSe) compound semiconductor waveguides are fabricated inside optical fibers, as shown in Fig. 4a (Sparks et al. 2011). These fiber

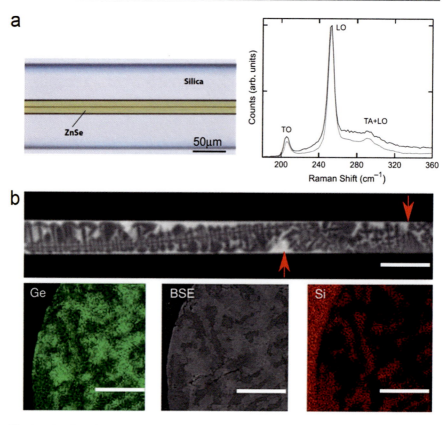

Fig. 4 (**a**) *Left*: optical microscope image showing the transparent, uniform ZnSe fiber core with a small nanoscale hole in the center. *Right*: Raman spectra of a ZnSe fiber core material (*black*) compared to an optical grade ZnSe reference wafer (*gray*) (Adapted with permission Sparks et al. (2011)). (**b**) XCT image of SiGe fiber, showing a large grain, and *red arrows* indicating grain boundaries (scale bar, 200 μm). (**b**). Cross-sectional BSE image (*gray*), and EDX compositional maps for Ge (*green*) and Si (*red*) (scale bar, 20 μm) (Coucheron et al. 2016)

waveguides exhibit very low loss (e.g., <1 dB/cm at 1550 nm wavelength). The superior optical and electronic properties of crystalline compound semiconductors can now be exploited in a fiber geometry.

Using silicon-germanium (SiGe) in the core extends the accessible wavelength range and potential optical functionality because the bandgap and optical properties can be tuned by changing the composition (Coucheron et al. 2016). Recently, the fabrication of SiGe-core optical fibers and the use of laser irradiation to heat the glass cladding and recrystallize the core have improved optical transmission, as shown in Fig. 4b. Tailoring the recrystallization conditions allows formation of long single crystals with uniform composition, as well as fabrication of compositional microstructures, such as gratings, within the fiber core.

Fabrication Approaches

Developments in fiber-drawing technology and the invention of new post-processing techniques have triggered strong advancements in the field of optoelectronic fibers. Here are the three major fabrication approaches.

Thermal Drawing

The first natural approach to integrating innovative functionalities inside optical fibers is to add novel materials at prescribed positions at the preform level and to directly thermally draw the full assembly into a long fiber. The principle of this approach mimics the way that conventional silica telecommunication fibers are fabricated. This approach exhibits the main advantage of retaining the simplicity and scalability of the drawing process.

This first generation of optoelectronic fibers was demonstrated in 2004, where a chalcogenide glass core contacted by four metallic electrodes and surrounded by a transparent polymer cladding showed photodetecting capabilities, as shown in Fig. 5. Further optical functionality was integrated when a Fabry-Perot cavity was added around the photodetecting structure to select the operating wavelength. Solid-core optoelectronic fibers were also shown to be sensitive to thermal radiation and

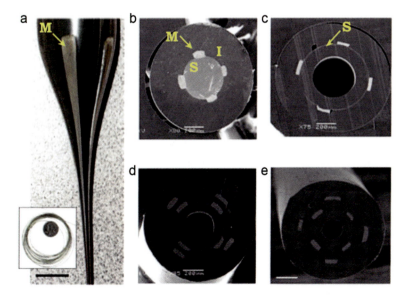

Fig. 5 (a) Thermal drawing of a metal-insulator-semiconductor (M-I-S) fibers (scale bar is 1 cm). Inset shows several meters of coiled fiber. (b–e) SEM micrographs of various M-I-S fiber architectures. (Scale bar for **b, c, d, e**: 200, 200, 200, 100 μm) (Adapted with permission Abouraddy et al. (2007), Tao et al. (2012))

to exhibit a great potential for large area interacting screens, meshes, and fabrics. In particular, a lensless imaging system was demonstrated using two grids made out of such solid-core optoelectronic fibers.

While these geometrical improvements have enabled the increase of the performance and flexibility of complex fiber devices, the materials and their phases need to be revisited to improve light-sensing fibers further. Indeed, thermally drawn fibers with amorphous materials reveal the relatively poor electronic and optoelectronic properties due to the glassy state of the semiconductors. To address this limitation, crystalline semiconductors have been successfully drawn inside fibers. However, rather than exploring their superior optoelectronic properties, the current main focus is their optical properties. The first attempt to integrate a silicon core in a silica cladding was done by introducing a Si rod inside a silica tube. The melting point of silicon is at around 1416 °C, well below the typical drawing temperature of silica around 2000 °C. Note that the approach of introducing silicon powder into a tube under vacuum has also been proposed. Other materials have been drawn with a similar approach such as sapphire-derived all-glass optical fibers (Dragic et al. 2012). At such high temperature, effects such as diffusion, chemical reactions, and the formation of grain boundaries upon cooling and solidification can be expected.

Using thermal drawing, an interesting phenomenon recently found in constructing optoelectronic fibers is the compound formation induced by the drawing process (Morris et al. 2011; Hou et al. 2013, 2015). During thermal drawing, materials are brought together at high temperature and confined in a small cross-sectional architecture. Depending on the materials in contact, the inter-diffusion coefficients, and the enthalpy of formation of various compounds, it can be anticipated that synthesis may occur and different materials end up in the fiber compared to those found in the preform. The first work resulting an optoelectronic fiber based on draw synthesis approach was the formation of ZnSe compound inside a polymer fiber after diffusion of the Zn from the electrode and reaction with Se, leading to the demonstration of in-fiber junction. Furthermore, a preform consisting of an aluminum rod in a silica tube was drawn into a silicon-core fiber via such a reactive process. To realize optoelectronic fibers with these materials, however, the difficulty remains of finding compatible materials to act as electrodes that do not mix or react with the molten semiconductor during drawing.

High-Pressure Chemical Vapor Deposition

High-pressure chemical vapor deposition (HPCVD) involves the use of high-pressure semiconductor precursor gases to deposit high-purity amorphous semiconductor materials inside a prefabricated hollow channel fiber, as shown in Fig. 6 (Sazio et al. 2006; Jackson et al. 2008; He et al. 2012). This technique was shown to be capable of producing semiconductor wires with diameters of the order of several micrometers down to hundreds of nanometers of various materials of extremely good optical and electrical quality. By controlling the fabrication process and applying particular post-processing techniques, the morphologies of the incorpo-

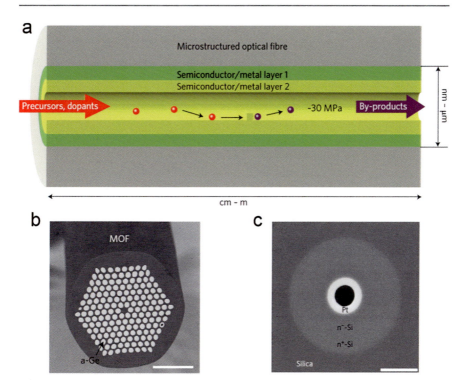

Fig. 6 (**a**) Illustration of HPCVD in the microstructured optical fiber (*MOF*) pores. (**b**) SEM image of an MOF filled with an array of amorphous germanium (a-Ge) wires (scale bar is 50 μm). (**c**) A Pt/n-Si Schottky junction formed inside an MOF pore (scale bar is 1 μm) (Adapted with permission He et al. (2012))

rated materials can be engineered, allowing the production of entirely amorphous, polycrystalline, or even single-crystal semiconductor wires inside microstructured fibers. For the formation of in-fiber amorphous and polycrystalline semiconductors using this method, a post-process annealing is commonly required to crystallize the core material for improving optoelectronic performance. The first experiments in the field of HPCVD were focused on doped and undoped single-component semiconductor wires such as germanium and silicon. One particularly interesting example is a concentric semiconductor multilayer of differently doped semiconductors in a PCF, which represented an entirely fiber-integrated semiconductor homojunction. More recent works extend toward integrating other semiconductor materials into microstructured fibers and capillaries such as II–VI semiconductors (e.g., ZnSe), which have promising applications in light-generation schemes or optoelectronics.

Post-processing Selective Breakup

Direct thermal drawing is commonly based on the use of thermally stable glassy semiconductors, which results in high densities of electronic defects and poor electrical properties of the resulting multi-material fiber. Can the scalability of the thermal-drawing process be harnessed to fabricate similar in-fiber optoelectronic microdevices using high-performance semiconductors such as silicon or germanium? This possibility has thus far seemed unlikely, since the thermal drawing of crystalline materials requires their liquefaction, which in turn precludes the making of functional structures with adjacent crystalline domains, lest they mix during the draw process and subsequently lose their desired functionalities.

A two-step fabrication method has been demonstrated to achieve high density of optoelectronic devices in a single silica fiber with the integration of crystalline semiconducting material and metal electrodes in a precisely defined metal-semiconductor-metal architecture. The first step is the co-drawing crystalline semiconducting and conducting materials into a fiber that defines the initial structure of one semiconductor core surrounded by two metal wires with silica barriers in between to prevent inter-domain mixing. The second step harnesses a fluid instability phenomenon in fiber drawing, which is related to the Plateau-Rayleigh capillary instability, the natural tendency of fluid cylinders to break up into spheres (Kaufman et al. 2012; Wei et al. 2017). Two materials that have distinct melting points and a specific breakup temperature are chosen so that only one of them is influenced by the fluid instability, while the other is not impacted. This step results in direct contacts between semiconductor spheres and two continuous metal wires by selectively amplifying crystalline semiconductor core into a series of spheres with larger sizes. The ladderlike structure forms an integrated device in a single silica fiber and exhibits good electrical and optoelectronic properties, as well as intensely increasing the device density over the entire fiber length. In particular, since the formation of optoelectronic structures inside a fiber relies on a physical breakup mechanism as opposed to chemical synthesis, this fabrication approach significantly broadens the repertoire of accessible materials and geometries toward advanced multifunctional fibers.

As shown in Fig. 7, platinum with the melting point of 1768 °C as the electrodes and doped germanium with the melting point of 938 °C as the semiconductor core are chosen to construct a triple-core preform. Both platinum and germanium are crystalline materials, and they show low viscosities and high diffusivities when temperature is above their melting points during thermal drawing. These lead to an uncontrollable inter-domain mixing and diffusion if direct contact between platinum and germanium is defined in preform and engaged during the drawing process. To prevent and control inter-domain mixing and diffusion, a viscous silica barrier is introduced intentionally in this two-material triple-core structure at the preform level to keep the constituent materials separated during the draw. Thus, all the rods are confined separately in silica tubes, and they are arranged in a row with the separation of around 1 mm, as illustrated in Fig. 7a. Then, the preform is loaded in a fiber draw tower and thermally drawn into extended length of fibers at a relatively low temperature of 2000 °C to confine low viscous materials with high viscous silica cladding. The optical micrograph in Fig. 7b depicts the cross section of the resulting fiber structure. It is then followed by fluidizing the resulting fiber to selectively induce the PRI at the semiconductor region along the entire fiber length. 1750 °C is

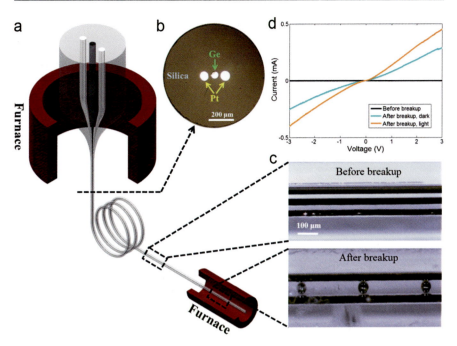

Fig. 7 (**a**) Schematic representation of the thermal drawing of a triple-core fiber and post-draw selective breakup process in a heating furnace. (**b**) Optical microscope image of the obtained fiber cross section. The resulting silica-cladding fiber contains one central semiconductor (germanium) core flanked by two metal (platinum) cores. (**c**) Side views of the triple-core fiber before and after the selective breakup process. (**d**) Photoconductance of a triple-core fiber before and after the selective breakup process in the dark and on the illumination of a 1550 nm laser (Adapted with permission Wei et al. (2017))

chosen to conduct selective breakup process. At this temperature, both germanium core and silica cladding are viscous, and only germanium core is influenced by the fluid instability to break up into a chain of germanium spheres, while platinum electrodes are not impacted. Due to the fact that the sphere diameter D_s is always larger than the diameter of the core from which it breaks D_c ($D_s = 3/2PD_c^2)^{1/3}$, where P is the particle spacing, the PRI-induced germanium spheres break the silica barriers and then directly contact to two solid platinum electrode wires, forming a ladderlike structure. The side view images of fibers before and after the selective breakup process are shown in Fig. 7c. This directed physical self-assembly process breaks the topological structure inside a fiber, allowing the in-fiber electrical contact between crystalline semiconductor and metal materials and therefore leading to the realization of integrated optoelectronic devices in silica fiber.

To investigate the optoelectronic performance in the resulting fiber configuration, a 1550 nm laser diode with the maximum output power of 130 mW is used to illuminate the germanium spheres. The germanium spheres undergo a change in electrical conductivity when externally illuminated, and a change in slope of the current-voltage (I-V) curve is obtained with respect to dark conditions, as plotted in Fig. 7d. The optical bandwidth of the resulting fiber is measured to be 470 kHz under a driving voltage of 3 V by modulating the input 1550 nm laser, which is three orders of magnitude higher than previously reported fiber-integrated optoelectronic devices made by glassy semiconductors. High-performance optoelectronic devices require high-quality semiconductors as the functional materials and junctions as the fundamental components. Therefore, further development in in-fiber material processing and the realization of in-fiber junction will lead to the improvements in device performance. For instance, the realization of silicon-germanium hetero-junction can significantly improve the efficiency and speed of in-fiber integrated optoelectronic devices.

Conclusions and Outlook

The concept of optoelectronic fibers enables the unique combination of metals, semiconductors, and insulators in the cross section of a uniform, potentially kilo-meter long flexible fiber. Optoelectronic fibers integrating several semiconductor devices can indeed detect light but also sense local heating, ultrasound waves, or chemicals. Moreover, the combination of photonic crystals with optoelectronic devices is an important thrust of research for photodetectors and photovoltaics. Future research directions in the field of optoelectronic fibers will focus on the integration of complex architectures to trap and absorb light efficiently. Another direction is to effectively solve the interface between metal and semiconductor, especially for crystalline semiconductors. The development of optoelectronic fibers also boosts the next generation of advanced fabrics and textiles. Fabrics that are woven either completely or partially from such integrated fibers can deliver a wide range of real-time, nontraditional functionalities over the full surface area of clothing, powered by electrical energy harvested from the ambient environment. The interplay between material properties and structure integration in these fibers, alongside fabric-array construction, is just beginning and promises to be an exciting field for fundamental and applied research.

References

A.F. Abouraddy, M. Bayindir, G. Benoit, S.D. Hart, K. Kuriki, N. Orf, O. Shapira, F. Sorin, B. Temelkuran, Y. Fink, Towards multimaterial multifunctional fibres that see, hear, sense and communicate. Nat. Mater. **6**, 336–347 (2007)
G.P. Agrawal, *Fiber-Optic Communication Systems*, 4th edn. (Wiley, Hoboken, 2010)
J. Ballato, T. Hawkins, P. Foy, R. Stolen, B. Kokuoz, M. Ellison, C. McMillen, J. Reppert, A.M. Rao, M. Daw, S.R. Sharma, R. Shori, O. Stafsudd, R.R. Rice, D.R. Powers, Silicon optical fiber. Opt. Express **16**, 18675–18683 (2008)
J. Ballato, T. Hawkins, P. Foy, B. Yazgan-Kokuoz, R. Stolen, C. McMillen, N.K. Hon, B. Jalali, R. Rice, Glass-clad single-crystal germanium optical fiber. Opt. Express **17**, 8029–8035 (2009a)
J. Ballato, T. Hawkins, P. Foy, C. McMillen, L. Burka, J. Reppert, R. Podila, A. Rao, R. Rice, Binary III–V core semiconductor optical fiber. Opt. Express **18**, 4972–4979 (2009b)

M. Bayindir, F. Sorin, S. Hart, O. Shapira, J.D. Joannopoulos, Y. Fink, Metal-insulator-semiconductor optoelectronic fibres. Nature **431**, 826–829 (2004)

A. Canales, X. Jia, U. Froriep, R. Koppes, C. Tringides, J. Selvidge, C. Lu, C. Hou, L. Wei, Y. Fink, P. Anikeeva, Multimodal fibres for simultaneous optical, electrical and chemical interrogation of neural circuits in vivo. Nat. Biotechnol. **33**, 277–284 (2015)

D. Coucheron, M. Fokine, N. Patil, D. Werner Breiby, O. Tore Buset, N. Healy, A.C. Peacock, T. Hawkins, M. Jones, J. Ballato, U. Gibson, Laser recrystallization and inscription of compositional microstructures in crystalline SiGe-core fibres. Nat. Commun. **7**, 13265 (2016)

P. Dragic, T. Hawkins, P. Foy, S. Morris, J. Ballato, Sapphire-derived all-glass optical fibres. Nat. Photonics **6**, 627–633 (2012)

V.M. Glazov, S.N. Chizhevskaya, N.N. Glagoleva, *Liquid Semiconductors* (Plenum Press, New York, 1969)

A. Gumennik, L. Wei, G. Lestoquoy, A.M. Stolyarov, X. Jia, P.H. Rekemeyer, M.J. Smith, X. Liang, S.G. Johnson, S. Gradeak, A.F. Abouraddy, J.D. Joannopoulos, Y. Fink, Silicon-in-silica spheres via axial thermal gradient in-fibre capillary instabilities. Nat. Commun. **4**, 2216 (2013)

R. He, P.J.A. Sazio, A.C. Peacock, N. Healy, J.R. Sparks, M. Krishnamurthi, V. Gopalan, J.V. Badding, Integration of gigahertz-bandwidth semiconductor devices inside microstructured optical fibres. Nat. Photonics **6**, 174–179 (2012)

N. Healy, S. Mailis, N.M. Bulgakova, P.J.A. Sazio, T.D. Day, J.R. Sparks, H.Y. Cheng, J.V. Badding, A.C. Peacock, Extreme electronic bandgap modification in laser-crystallized silicon optical fibres. Nat. Mater. **13**, 1122–1127 (2014)

C. Hou, X. Jia, L. Wei, A.M. Stolyarov, O. Shapira, J.D. Joannopoulos, Y. Fink, Direct atomic-level observation and chemical analysis of ZnSe synthesized by in situ high throughput reactive fiber drawing. Nano Lett. **13**, 975–979 (2013)

C. Hou, X. Jia, L. Wei, S. Tan, X. Zhao, J. Joannopoulos, Y. Fink, Crystalline silicon core fibres from aluminium core preforms. Nat. Commun. **6**, 6248 (2015)

B.R. Jackson, P.J.A. Sazio, J.V. Badding, Single-crystal semiconductor wires integrated into microstructured optical fibers. Adv. Mat. **20**, 1135–1140 (2008)

K. Kakimoto, M. Eguchi, H. Watanabe, T. Hibiya, Natural and forced convection of molten silicon during czochralski single crystal growth. J. Cryst. Growth **94**, 412–420 (1989)

J.J. Kaufman, G. Tao, S. Shabahang, E.-H. Banaei, D.S. Deng, X. Liang, S.G. Johnson, Y. Fink, A.F. Abouraddy, Structured spheres generated by an in-fibre fluid instability. Nature **487**, 463–467 (2012)

P. Koštál, J. Málek, Viscosity of selenium melt. J. Non-Crystal. Solid **356**, 2803–2806 (2010)

K. Li, T. Zhang, G. Liu, N. Zhang, M. Zhang, L. Wei, Ultrasensitive optical microfiber coupler based sensors operating near the turning point of effective group index difference. Appl. Phys. Lett. **109**, 101101 (2016)

M. Lipson, Guiding, modulating, and emitting light on silicon challenges and opportunities. J. Lightwave Technol. **23**, 4222–4238 (2005)

S. Morris, T. Hawkins, P. Foy, C. McMillen, J. Fan, L. Zhu, R. Stolen, R. Rice, J. Ballato, Reactive molten core fabrication of silicon optical fiber. Opt. Mater. Express **1**, 1141–1149 (2011)

D. Ofte, The viscosities of liquid uranium, gold and lead. J. Nucl. Mater. **22**, 28–32 (1967)

A.C. Peacock, J.R. Sparks, N. Healy, Semiconductor optical fibres: progress and opportunities. Laser Photonics Rev. **8**, 53–72 (2014)

A.C. Peacock, U. Gibson, J. Ballato, Silicon optical fibres – past, present, and future. Adv. Phys. X, 1–22 (2016)

R.F. Pierret, *Semiconductor Device Fundamentals*, 2nd edn. (Addison-Wesley, Boston, 1996)

G.T. Reed, A.P. Knights, *Silicon Photonics: An Introduction* (Wiley, Chichester, 2004)

P. Russell, Photonic crystal fibers. Science **299**, 358–362 (2003)

P.J.A. Sazio, A. Amezcua-Correa, C.E. Finlayson, J.R. Hayes, T.J. Scheidemantel, N.F. Baril, B.R. Jackson, D.-J. Won, F. Zhang, E.R. Margine, V. Gopalan, V.H. Crespi, J.V. Badding, Microstructured optical fibers as high-pressure microfluidic reactors. Science **311**, 1583–1586 (2006)

M. Schmidt, A. Argyros, F. Sorin, Hybrid optical fibers – an innovative platform for in-fiber photonic devices. Adv. Opt. Mater. **4**, 13 (2016)

B.L. Scott, K. Wang, G. Pickrell, Fabrication of n-type silicon optical fiber. IEEE Photon. Technol. Lett. **21**, 1798–1800 (2009)

S. Shabahang, G. Tao, J.J. Kaufman, Y. Qiao, L. Wei, T. Bouchenot, A. Gordon, Y. Fink, Y. Bai, R.S. Hoy, A.F. Abouraddy, Controlled fragmentation of multimaterial fibres and films via polymer cold-drawing. Nature **534**, 529–533 (2016)

J.R. Sparks, R. He, N. Healy, M. Krishnamurthi, A.C. Peacock, P.J.A. Sazio, V. Gopalan, J.V. Badding, Zinc selenide optical fibers. Adv. Mat. **23**, 1647–1651 (2011)

A.M. Stolyarov, L. Wei, O. Shapira, F. Sorin, S.L. Chua, J.D. Joannopoulos, Y. Fink, Microfluidic directional emission control of an azimuthally polarized radial fibre laser. Nat. Photon. **4**, 229–233 (2012)

G. Tao, A.M. Stolyarov, A.F. Abouraddy, Multimaterial fibers. I. J. Appl. Glass Sci. **3**, 349–368 (2012)

A.S. Tverjanovich, Temperature dependence of the viscosity of chalcogenide glass-forming melts. Glas. Phys. Chem. **29**, 532–536 (2003)

G. Urbain, Y. Bottinga, P. Richet, Viscosity of liquid silica, silicates and alumino-silicates. Geochim. Cosmochim. Acta **46**, 1061–1072 (1982)

S. Wang, T. Zhang, K. Li, S. Ma, M. Chen, P. Lu, L. Wei, Flexible piezoelectric fibers for acoustic sensing and positioning. Adv. Electron. Mater. **3**, 1600449 (2017)

L. Wei, C. Hou, E. Levy, G. Lestoquoy, A. Gumennik, A.F. Abouraddy, J.D. Joannopoulos, Y. Fink, Optoelectronic fibers via selective amplification of in-fiber capillary instabilities. Adv. Mater. **29**, 1603033 (2017)

N. Zhang, H. Liu, A.M. Stolyarov, T. Zhang, K. Li, P. Shum, Y. Fink, X. Sun, L. Wei, Azimuthally polarized radial emission from a quantum dot fiber laser. ACS Photon. **3**, 2275–2279 (2016a)

N. Zhang, G. Humbert, Z. Wu, K. Li, P. Shum, M. Zhang, Y. Cui, J. Auguste, X. Dinh, L. Wei, In-line optofluidic refractive index sensing in a side-channel photonic crystal fiber. Opt. Express **24**, 27674–27682 (2016b)

M. Zhang, D. Hu, P. Shum, Z. Wu, K. Li, T. Huang, L. Wei, Design and analysis of surface plasmon resonance sensor based on high-birefringent microstructured optical fiber. J. Opt. **18**, 65005–65011 (2016c)

N. Zhang, G. Humbert, T. Gong, P. Shum, K. Li, J. Auguste, Z. Wu, J. Hu, F. Luan, Q.X. Dinh, M. Olivo, L. Wei, Side-channel photonic crystal fiber for surface enhanced Raman scattering sensing. Sensors Actuators B Chem. **223**, 195–201 (2016d)

T. Zhang, K. Li, C. Li, S. Ma, H.H. Hng, L. Wei, Mechanically durable and flexible thermoelectric films from PEDOT:PSS/PVA/$Bi_{0.5}Sb_{1.5}Te_3$ nanocomposites. Adv. Electron. Mater. **3**, 1600554 (2017a)

M. Zhang, K. Li, P. Shum, X. Yu, S. Zeng, Z. Wu, Q. Wang, K. Yong, L. Wei, Hybrid graphene/gold plasmonic fiber-optic biosensor. Adv. Mater. Technol **2**, 1600185 (2017b)

Fiber Grating Devices

35

Christophe Caucheteur and Tuan Guo

Contents

Abstract

Biosensors made of an optical fiber section coated with a thin noble metal layer constitute a miniaturized counterpart to the Kretschmann-Raether prism configuration. They allow easy light injection and offer remote operation in very small volumes of analyte. They are perfectly suited to yield in situ (or even possibly in vivo) molecular detection. Usually, such biosensors are obtained

C. Caucheteur (✉)
Electromagnetism and Telecommunication Department, University of Mons, Mons, Belgium
e-mail: christophe.caucheteur@umons.ac.be

T. Guo
Institute of Photonics Technology, Jinan University, Guangzhou, China
e-mail: tuanguo@jnu.edu.cn

© Springer Nature Singapore Pte Ltd. 2019
G.-D. Peng (ed.), *Handbook of Optical Fibers*,
https://doi.org/10.1007/978-981-10-7087-7_42

from a gold-coated fiber segment for which the core-guided light is outcoupled and brought into contact with the surrounding medium, either by reducing the cladding diameter (through etching or side-polishing) or by using grating coupling. In the latter case, a refractive index modulation is photo-imprinted in the fiber core. Roughly 10 years ago, surface plasmon resonance (SPR) excitation was reported with gold-coated tilted fiber Bragg gratings (TFBGs). TFBGs are short-period gratings whose refractive index modulation is slightly angled with respect to the perpendicular to the optical fiber propagation axis. These devices probe the surrounding medium with narrowband (~200 pm) cladding mode resonances, which is compatible with the use of high-resolution interrogators as a read-out technique. These gratings remain the single configuration able to probe all the fiber cladding modes individually, with high Q-factors. These unique spectral features are used to sense various analytes, such as proteins and cells. Impressive limit of detection (LOD) and sensitivity are reported, which paves the way to the practical use of such immunosensors, in very small volumes of analytes or even possibly in vivo.

Keywords

Surface plasmon resonance · Optical fiber sensors · Bragg gratings · Immunosensing · Cells · Proteins

Introduction

Biosensors usually combine a biological receptor compound and a physical or physicochemical transducer. Among the different possible configurations, optical methods of transduction offer important advantages like:

- Non-invasive, safe and multi-dimensional detection based on wavelength, intensity, phase or polarization metrology
- Well-established technologies (lasers, optical sources, detectors, etc. . . .) available from the telecommunication field and the micro-nano technologies industries
- Used optical frequencies in the visible and near infrared regions coincide with a wide range of physical properties of bio-related materials.

Biosensors aim to provide a solution to the demand for direct, accurate and in situ monitoring techniques in various applications such as genomics, proteomics, environmental monitoring, food analysis, security and medical diagnosis. Label-free optical biosensors yield direct and real-time observation of molecular interactions, without requiring the use of labels since they sense binding-induced refractive index changes. Among the different optical biosensors configurations (measurements of absorbance, reflectance, fluorescence, refractive index changes, etc. . . .), those based on surface plasmon polaritons (SPPs) have been the subject of intense research and development during the past three decades.

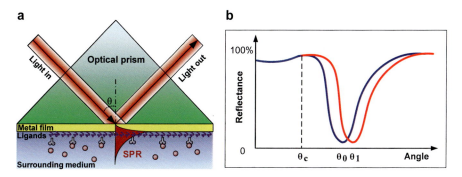

Fig. 1 Sketch of the Kretschmann-Raether prism used for SPPs excitation (**a**) and operating principle of its demodulation (**b**) to measure biomolecules binding

SPPs arise from collective excitations of electrons at the interface between a thin noble metal layer (either in the form of a continuous sheath or nanoparticles) and a dielectric. The strong sensitivity of the plasmon-propagation constant to the permittivity of bioreceptors grafted on the metallic surface has led to the monitoring of surface plasmon resonance (SPR) to detect biochemical changes resulting from molecules binding. Therefore, biochemical reactions are measured with SPR by monitoring their effective refractive index. To this aim, the so-called Kretschmann-Raether is used to launch light beams from a glass prism to a thin metallic interface at an angle such that light is totally reflected, as sketched in Fig. 1. An evanescent wave extends in both the metal layer and the surrounding dielectric medium, along a depth corresponding to a fraction of the light wavelength (λ) (Kretschmann and Raether 1968). When the component of the propagation constant of the light along the interface matches that of a plasmon excitation of the other side of the metal layer, part of the light couples to the plasmon, which decreases the reflection (at θ_0 in Fig. 1 where θ_c denotes the critical angle of incidence). The interrogation of this device can be made either by varying the wavelength and keeping the incidence angle constant or by using monochromatic light and modifying the angle. In both cases the polarization state of the light is set parallel to the incidence plane so that the plasmon wave is orthogonally polarized with respect to the interface. Biochemical reactions occurring on top of the metal surface slightly change the effective refractive index of the plasmon wave, which is detected through a shift of the SPR (from θ_0 to θ_1). The sensitivity to the surrounding refractive index often ranges in the order of 10^{-6}–10^{-7} RIU (refractive index unit) (Hecht et al. 1996; Barnes et al. 2003; Homola et al. 1999; Homola and Piliarik 2006).

Numerous transduction mechanisms have been developed, bringing additional assets with respect to the Kretschmann-Raether prism approach, which remains used in most commercial systems. Optical fiber based sensors are particularly attractive. With their lightweight, compactness and ease of connection, they provide remote operation in very small volumes of analyte of the order of microliter. And with their continuous development and optimization, they appear perfectly suited for in situ or even possibly in vivo diagnosis. Depending on the configuration, they also have the potential to assay different parameters simultaneously, either through the use of different fibers or by cascading functionalized regions along a single fiber.

To excite SPR from an optical fiber, the core-guided light has to be locally outcoupled and directed towards the surrounding medium. In practice, this can be achieved either from a geometrical alteration (polishing or etching of the cladding, totally or in part) so as to expose the evanescent wave to the surrounding medium or from a periodic refractive index modulation of the fiber core along the propagation axis. The latter configuration refers to the so-called fiber gratings that are permanently photo-inscribed in the fiber core. The main architectures used for SPP excitation are etched multimode optical fibers, side-polished, D-shaped, tapered or U-bent optical fibers, long period fiber gratings (LPFGs) and tilted fiber Bragg gratings (TFBGs) (Caucheteur et al. 2015). Configurations based on cladding removal/decrease can be quite easily obtained. In this case, the transmitted amplitude spectrum contains a broadband SPR resonance, usually characterized by a full width at half maximum (FWHM) of $\sim$50 nm or higher. When gold is used, the latter appears most often in the wavelength range [600–800 nm], depending on the used configuration. Such architectures tend to weaken optical fibers at the sensor head and may prevent their use in practical applications, out of laboratory settings. It is the reason why large core fibers (unclad 200–400 μm core fibers) are the most spread in practice (Pollet et al. 2009). These configurations operate at short wavelengths, which limits the extension of the evanescent wave in the surrounding medium. Indeed, its penetration depth usually ranges between λ/5 and λ/2, depending on the mode order (Baldini et al. 2012). Hence, operation at near-infrared telecommunication wavelengths enhance the penetration depth, which in turn improves the overall sensor sensitivity to large-scale targets such as proteins or cells. This can be achieved with in-fiber gratings.

Gratings preserve the fiber integrity while providing a strong coupling between the core-guided light and the cladding. LPFGs correspond to a periodic refractive index modulation of the fiber core with a uniform period of a few hundreds of μm. In single-mode optical fibers, they couple the forward-going core mode into forward-going cladding modes (Vengsarkar et al. 1996). Their transmitted amplitude spectrum is composed of a few wide resonances (FWHM $\sim$20 nm) spread over a wavelength range of several hundreds of nm. Conversely, TFBGs are short period ($\sim$500 nm) gratings with a refractive index modulation slightly angled with respect to the perpendicular to the optical fiber axis. In addition to the self-backward coupling of the core mode, they couple light into backward-going cladding modes. Their transmitted amplitude spectrum features several tens of narrow-band cladding mode resonances (FWHM $\sim$200 pm or even below) located at the left-hand side of the Bragg resonance (or core mode resonance) corresponding to the core mode self-coupling. According to phase matching conditions, each cladding mode resonance has its own effective refractive index and the maximum refractometric sensitivity is obtained when this effective index tends to the surrounding refractive index value. TFBGs act as spectral combs and constitute the only optical fiber configuration able to probe simultaneously but distinctively all the cladding modes supported by an optical fiber (Albert et al. 2013a). As for unclad optical fiber configurations, operation in reflection mode is possible by using a mirror deposited on the cleaved fiber end face beyond the sensing region.

SPR optical fiber sensors can be derived from the aforementioned structures when a thin film of gold or silver surrounds them. Sheaths of thickness ranging between 30 and 70 nm are most generally used. SPR generation is achieved when the electric field of the light modes is polarized mostly radially at the surrounding medium interface. The orthogonal polarization state is not able to excite the SPR, as the electric field of the light modes is polarized mostly azimuthally (i.e. tangentially to the metal) at the surrounding medium interface and thus cannot couple energy to the surface Plasmon waves. Depending on the configuration, impressive refractometric sensitivities in the range $[10^2–10^5$ nm/RIU] have been reported (Caucheteur et al. 2015).

When comparing the sensor performances between different configurations, it is not sufficient to compare only sensitivities (i.e. wavelength shifts), without considering the wavelength measurement accuracy. It is more convenient to refer to the figure of merit (FOM) of the device. The FOM corresponds to the ratio between the sensitivity and the linewidth of the resonance. In practice, a narrow resonance can be measured with a high resolution so that its exact location can be computed, which is not true for a broad one (Sharma et al. 2007; Offermans et al. 2011). As a result, in terms of experimentally demonstrated FOM, TFBGs outperform all other configurations by more than one order of magnitude (Caucheteur et al. 2015).

SPP excitation in gold-coated TFBGs has been pioneered in 2006 by the team of Prof. Jacques Albert at the *Carleton University of Ottawa*, Canada (Shevchenko and Albert 2007). Since then, as described in the remaining of this chapter, numerous experimental achievements have been obtained, demonstrating that gold-coated TFBGs are very well suited for immunosensing. In the following, the operating principle of plasmonic sensing with gold-coated TFBGs will be first described. Theses sensors are obtained from a single mode optical fiber in which a photoimprinted refractive index grating is formed over a short section of the core, and that is further coated with a bilayer coating: a very thin metal coating and a suitable biochemical recognition binding layer. Then, the main performances obtained in the case of cells and proteins sensing will be summarized.

Near-Infrared SPP Excitation with Gold-Coated TFBGs

Operating Principle of TFBG Refractometers

Gratings are usually manufactured in single-mode optical fibers, made of a 8 μm thick core surrounded by a 125 μm cladding. Single-mode operation happens at wavelengths beyond 1300 nm. Such fibers are widely available at low cost (less than 100 US$ per km) and are telecommunication-grade. In the same way, a large quantity of equipment is available to characterize, manipulate and use such fibers and devices made from them.

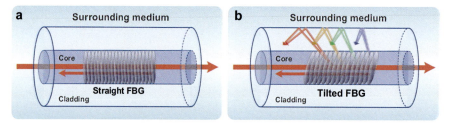

Fig. 2 Light coupling mechanisms in short-period uniform FBGs (**a**) and tilted FBGs (**b**)

As sketched in Fig. 2a, a uniform FBG is a periodic and permanent refractive index modulation of the fiber core that is imprinted perpendicularly to the propagation axis (Othonos and Kalli 1999; Erdogan 1997). It behaves as a selective mirror in wavelength for the light propagating in the core, reflecting a narrow spectral band centered on the so-called Bragg wavelength. According to phase matching condition, the latter is given by $\lambda_{Bragg} = 2n_{eff,core}\Lambda$ where $n_{eff,core}$ is the effective refractive index of the core mode (close to the refractive index of silica, 1.45 at 1550 nm) and Λ is the grating period. In practice, Λ is $\sim$530 nm to ensure that the Bragg wavelength falls in the band of minimum attenuation of the optical fiber centered on 1550 nm. The Bragg wavelength is inherently sensitive to axial strain and temperature, through a change of both n_{eff} and Λ (Othonos and Kalli 1999). In practice, a change of temperature of $+1$ °C yields a Bragg wavelength shift of $\sim$10 pm at 1550 nm. The sensitivity to axial strain is of the order of 1.2 pm/$\mu\varepsilon$, also at 1550 nm. Such changes are easily measured with standard telecommunication instruments since the full spectral width of the reflected light from a typical 1 cm-long grating is of the order of 100 pm. When the purpose is to measure events occurring on or near the fiber cladding surface, such gratings are useless because the penetration depth of the core-guided light into the cladding does not exceed a few micrometers, unless they are totally etched (Iadicicco et al. 2004). With a small modification consisting in tilting the refractive index modulation by a few degrees with respect to the perpendicular to the optical fiber axis (Fig. 2b), short-period gratings can be used to couple light from the core to the cladding and still benefit from narrowband spectral resonances that will reveal small changes at the cladding boundary.

In addition to the self-backward coupling of the core mode at the Bragg wavelength, TFBGs redirect some light to the cladding whose diameter is so large that several possible cladding modes can propagate, each with its own phase velocity (and subsequently effective refractive index $n_{eff,clad}$) (Erdogan and Sipe 1996). These possible modes of propagation correspond to different ray angles, as sketched in Fig. 2b. There is a one-to-one relationship between the wavelength at which coupling occurs for a given cladding mode and its effective index. This relationship is expressed by a similar phase matching condition as for uniform FBGs: $\lambda^i_{clad} = (n_{eff,core} + n^i_{eff,clad})\Lambda$ where the index "i" corresponds to the mode number. Figure 3 depicts the transmitted amplitude spectrum of a 1 cm-long 10°

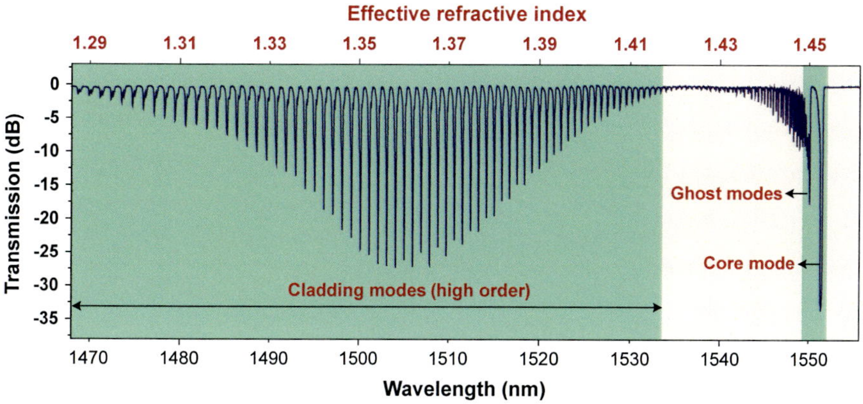

Effective refractive index

Fig. 3 Transmitted amplitude spectrum of a 1-cm long 10° TFBG. This graph corresponds to a bare grating measured in air using linearly polarized input light, as described in the third section

TFBG. Each resonance of the spectral comb corresponds to the coupling from the core mode to a group of backward propagating cladding modes. As a result of phase matching, the spectral position of a resonance now depends on the effective index of its associated cladding mode, which in turn depends on the optical properties of the medium over or near the cladding surface.

Therefore, spectral shifts of individual resonances can be used to measure changes in fiber coatings and surroundings. Laffont and Ferdinand were the first to demonstrate surrounding refractive index (SRI) sensing with TFBGs in 2001 (Laffont and Ferdinand 2001). For an SRI increase between 1.30 and 1.45, they observed a progressive smoothing of the transmitted amplitude spectrum starting from the shortest wavelengths. Several demodulation techniques can be used to quantitatively correlate the spectral content with the SRI value, either based on a global spectral evolution or a local spectral feature change. The first method involves monitoring the area delimited by the cladding mode resonance spectrum, through a computation of the upper and lower envelopes as resonances gradually disappear when the SRI reaches the cut-off points of each cladding mode (Laffont and Ferdinand 2001; Caucheteur and Mégret 2005). Another technique tracks the wavelength shift and amplitude variation of individual cladding mode resonances as they approach the cut-off, i.e. the wavelength at which their effective refractive index matches the one of the surrounding medium (Chan et al. 2007). Both techniques present minimum detectable SRI changes of $\sim 10^{-4}$ RIU when used with bare gratings. In terms of wavelength shift, this result corresponds to a sensitivity that peaks between 10 and 25 nm/RIU for the modes near cut-off. In all cases, the Bragg wavelength provides an absolute power and wavelength reference, which can therefore be used to remove uncertainties related to systematic fluctuations (such as unwanted power level changes from the light source) and even temperature changes. Indeed, all cladding mode resonances shift by the same amount as the Bragg resonance when the temperature changes, as demonstrated in

Chen and Albert (2006). Thus, TFBGs inherently provide temperature-insensitive SRI measurements and large signal-to-noise ratio. For biochemical experiments however, it is usually necessary to improve the LOD levels to at least 10^{-5} RIU, by increasing the wavelength shift sensitivity while keeping noise level down and spectral features narrow (White and Fan 2008). Fortunately, it has been recently demonstrated that the addition of a nanometric-scale gold coating overlay on the TFBG outer surface considerably enhances the refractometric sensitivity via SPR excitation (Shevchenko and Albert 2007; Caucheteur et al. 2011a, 2013). The main achievements obtained in this field will be summarized hereafter.

Influence of the Tilt Angle on the Transmitted Amplitude Spectral Content

As demonstrated in Erdogan and Sipe (1996), the tilt angle influences the cladding mode resonances distribution in the transmitted amplitude spectrum. Although the evolution is not monotonic, the global trend results in an increase of the coupling to high order modes (therefore at shorter wavelengths) for increasing tilt angle values. For tilt angle values less than 10°, gratings couple to cladding mode resonances with effective refractive indices ranging between ~1.30 and ~1.45, as shown in Fig. 4. Hence, such angles are preferred to operate when the SRI lies near the refractive index of water and aqueous solutions, which is often the case in biochemical research, because then the strongest cladding mode resonances are located near 1550 nm (when the Bragg wavelength is near 1610 nm), where they are easier to measure. For refractometric sensing purposes in gaseous media (Liedberg et al. 1984; Caucheteur et al. 2016), higher tilt angles can be envisaged

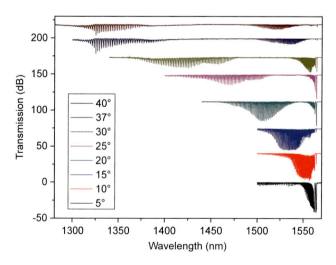

Fig. 4 Effect of the tilt angle on the cladding mode distribution in TFBGs

to couple light to cladding mode resonances with an effective refractive index close to 1.00. As depicted in Fig. 4, this is possible for tilt angles higher than 30°. In this case, the cladding mode resonances on the short wavelength side of the Bragg resonance can be divided into two main subsets:

1. In the wavelength range [1480–1550 nm], cladding mode resonances have effective refractive indices ranging between 1.30 and 1.44 and are therefore suited for measurement in aqueous solutions, similar to the case of weakly tilted FBGs with tilt angles limited to 10°.
2. In the wavelength range [1270–1410 nm], cladding mode resonances present effective refractive indices ranging between 0.92 and 1.18, according to the aforementioned phase matching condition.

Also, it was recently demonstrated that cascading TFBGs sections with different tilt angles can easily provide a continuous comb of cladding mode resonances with effective refractive indices ranging between 1.45 and less than 1.00 (Chen et al. 2017).

TFBGs Fabrication

TFBGs are fabricated using the same tools and techniques as standard FBGs, i.e. from a permanent refractive index change induced in doped glasses from the interference between two ultraviolet laser beams. Hence, all the technological improvements and reliability studies that have accompanied the development of a worldwide FBG industry for telecommunications and sensing applications over the last 25 years are directly applicable to TFBGs, including low cost mass production. We only describe here the modifications required to make tilted FBGs, as shown in Fig. 5.

With an interferometric set-up (Fig. 5a), it is sufficient to tilt the fiber relative to the fringe pattern. This was the approach used by Meltz and co-workers in their pioneering work of the early 1990s and it is still used (Zhou et al. 2003, 2006). The main advantage of this approach is its flexibility to change periods and tilt angles. However, when large numbers of identical gratings are needed, another approach is usually preferred, with the interference pattern generated by a phase mask located close to the fiber, as sketched in Fig. 4b. The period of the grating is fixed by the phase mask and because of the proximity of the fiber, low coherence ultraviolet sources can be used, such as high energy pulsed excimer lasers. In this case, tilting can be done in two ways: rotating the phase mask and fiber as in the previous case (Fig. 5b) or keeping the fiber and phase mask perpendicular to the incident writing beam but rotating the phase mask around the axis of the writing beam (Fig. 5c). Note that in the first case this requires also to tilt the cylindrical lens that is used to focus the writing light intensity along the fiber axis.

In all cases if a moderate power CW laser beam is used to write the grating (typically a continuous wave frequency-doubled Ar ion laser emitting at 244 nm), the laser beam must be scanned to obtain typical grating lengths of the order of

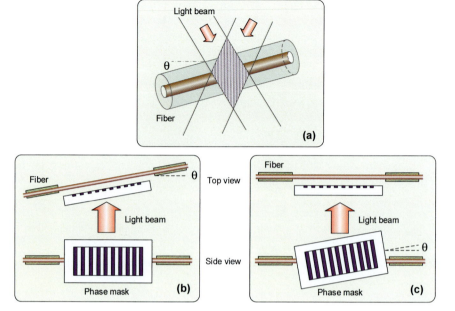

Fig. 5 Sketch of the main techniques used to photo-inscribe TFBGs in photosensitive optical fibers

1–2 cm. Scanning is not necessary with high energy pulsed excimer lasers (at 248 or 193 nm) that have energies per pulse ranging from 100 to 400 mJ and beam size of 10×40 mm^2 because the beam can be expanded in one dimension (along the fiber axis) and thus used to write long gratings (up to 10 cm and more) in a single exposure. We have found by experience that the fiber-phase mask assembly rotation technique provides the best spectral responses for strong TFBGs with tilt angles between 4 and 10°.

Gold Deposition on the Optical Fiber Surface

Thin noble metal sheaths can be successfully deposited on optical fibers using well-established technologies, such as electroless deposition, electroplating or sputtering (Albert et al. 2013b). The latter is used more routinely and provides very high quality metal surfaces. Although not really necessary as demonstrated in Feng et al. (2016), two consecutive depositions are usually made in the same conditions, with the optical fibers rotated by 180° between both processes, to ensure that the whole outer surface is covered by gold. The fiber could also be rotated during the gold covering process. To promote adhesion, a 2–3 nm buffer layer of chromium or titanium is often sandwiched between the optical fiber surface and the gold coating. Another option consists in thermally annealing the gold coating, which modifies its morphology and ensures robustness (Svorcik et al. 2011). What is important to realize is that it is quite difficult to obtain very uniform metal layers at the thicknesses required for optimum SPR excitation, and as a result that there may be some rugosity, or particles forming instead of smooth layers. This will have an effect on the SPR properties because the effective complex permittivity of the metal layers will be different from that of the bulk values (which are often used in the design of the sensors) (Sennett and Scott GD 1950; Cohen et al. 1973; Tu et al. 2003). The optimum gold thickness has been found to be approximately 50 nm, for the narrowest, deepest SPR attenuation.

A major recent development on the materials side relates to the use of non-metal layers for plasmonic applications, such as certain types of semiconductors and oxides, as reviewed recently (Naik et al. 2011). These materials have yet to be explored in fiber-based systems but represent interesting possibilities for the future.

Surface Functionalization

Finally, the gold-coated TFBGs are functionalized for biosensing purposes. The chemistry involved in this process depends on the target application. Most often, it is based on the antigen/antibody affinity. Whatever the analyte to be detected, a self-assembled monolayer (SAM) is manufactured. For this, gold-coated TFBGs are first cleaned with ethanol. They are then immersed in a solution of thiols dispersed in ethanol. Thiols incubations are usually done during 12 h at room temperature. Practically, this can be done in a 1 mm thick capillary tube sealed at both ends to prevent solvent evaporation. At the end of the incubation, the functionalized gold-coated TFBGs are again rinsed with ethanol, prior to the grafting of biomolecules on the activated surface.

These four consecutive steps (grating manufacturing, gold deposition, surface functionalization and bioreceptors grafting) yield the plasmonic probe sketched in Fig. 6.

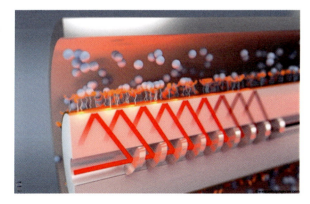

Fig. 6 Sketch of a gold-coated TFBG located at the fiber tip and prepared for use as biosensor

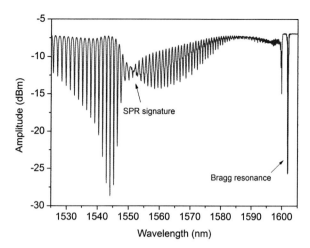

Fig. 7 Transmitted amplitude spectrum of an SPR-TFBG immersed in salted water (radial polarization – SRI = 1.358)

Interrogation of Gold-Coated TFBG Immunosensors

Most often, the interrogation of gold-coated TFBGs is based on transmitted amplitude spectra measurements. A polarization controller is usually placed behind the optical source to control and orient the state of polarization (SOP) of the light launched into the TFBG. In such a configuration, care is taken to avoid polarization instabilities (use of short fiber lengths, strong curvatures avoided and ambient temperature kept constant to within 1 °C). It is worth mentioning that these gratings can operate in reflection mode, provided that a gold or silver mirror is deposited on the cleaved fiber end face, after the TFBG section.

Figure 7 displays the typical transmitted amplitude spectrum of a gold-coated TFBG immersed in salted water (refractive index measured close to 1.356 at 589 nm). The latter was recorded with a linear input SOP optimized to maximize coupling to the SPR, corresponding to the radial polarization, as further explained in the following. This spectrum exhibits the typical SPR signature around 1551 nm, which is due to the maximum phase matching of the cladding mode to the surface plasmon mode of the gold water interface, according to Shevchenko and Albert (2007). The core mode resonance appears at the right end side, corresponding to a Bragg wavelength of 1602 nm. In practice, the Bragg wavelength is used to remove any effect from surrounding temperature changes by monitoring its shift. This intrinsic feature is very interesting in practice as a change of 0.1 °C induces an SRI change of 10^{-5}, thus prone to generate erroneous spectral modifications.

Figure 8 displays a zoom around the SPR signature for radially- (EH and TM modes) and azimuthally-polarized (HE and TE modes) amplitude spectra that yield antagonist behaviors in liquids, as explained in Baldini et al. (2012). It is obvious from this figure that EH and HE modes come in pairs and in order to facilitate the following explanations, the mode resonances are labeled as a function of their

Fig. 8 Transmitted amplitude spectra for two orthogonal SOPs (radial and azimuthal polarizations) of a TFBG immersed in salted water ($SRI = 1.338$)

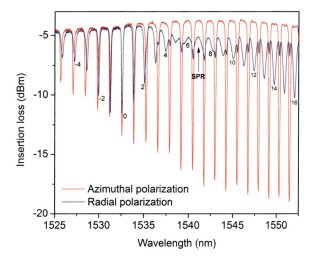

position with respect to the most important one (strongest peak-to-peak amplitude) in the EH spectrum (mode 0 in our numbering).

Figure 8 reveals that, for short wavelengths (mode-2), the behavior is similar to that of bare TFBGs in air, with EH modes at a longer wavelength than HE ones. Then, getting closer to the SPR mode and for each cladding mode resonances pair, the EH mode wavelength increases less than the HE one. The crossing point occurs for the 0 mode. Past this mode, EH resonances appear on the short wavelength side of HE ones. This peculiar behavior results from the fact that EH modes begin to localize in the gold sheath as they approach the SPR, which lowers their effective refractive index (due to the small value of the real part of the gold refractive index). HE modes are tangentially polarized at the gold boundary and hardly penetrate it, therefore they do not feel this effective index decrease.

Beyond the crossing, the +2 EH mode now lags significantly behind the HE mode, which points out to a strong localization of the EH mode field in the gold layer and the strong influence of the SPR (and hence of the SRI) on the mode resonance. The next modes show an even stronger lag of the EH mode but the associated resonance also becomes somewhat wider, because of the loss of energy to the metal. In the case of the +4 mode, the wavelength separation becomes so great that the EH and HE modes are completely dissociated. And further analysis of the modes in the vicinity of the SPR becomes meaningless.

The refractometric sensitivity of gold-coated TFBGs can be computed by immersing them in were immersed in different calibrated refractive index liquids. Figure 9 shows the transmitted amplitude spectra of a 50 nm gold-coated TFBG measured for three different refractive index values. For such large SRI changes, the SPR location can be unambiguously located by following the strongest attenuation in each spectrum. Therefore, the interrogation relies on the tracking of the wavelength shift of the center of the envelope of the most attenuated resonances.

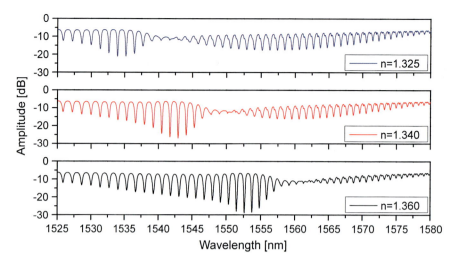

Fig. 9 SPR signature in the transmitted spectrum of gold-coated TFBG for coarse changes in SRI

Fig. 10 SPR wavelength
shift as a function of the SRI
value

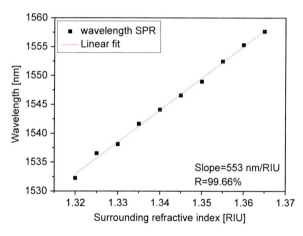

Using this technique, a linear response is obtained, as depicted in Fig. 10 and the
SRI sensitivity is ~550 nm/RIU in the range between 1.32 and 1.42.

For high-resolution refractometric sensing over an SRI range limited to 10^{-3}
typically – which corresponds to variations obtained in the case of biochemical
sensing –, accurate measurements of the SPR mode are not possible from the
radially polarized spectrum taken alone. Indeed, the SPP is only revealed by its
absence from the spectrum and it cannot be reliably measured for wavelength
shifts limited to a few picometers, due to the presence of noise in this part of the
spectrum. Hence, methods have been developed to track the SPR shift, mainly based
on a comparison between both orthogonally-polarized amplitude spectra (Voisin
et al. 2011; Caucheteur et al. 2011b). In practice, modes slightly off the SPR are
used, because they combine relatively high sensitivity with a narrow spectral width

and they can be "followed" by a combination of their changes in amplitude and wavelength. Indeed, as they stand on the shoulder of the SPR envelope, a slight change of the SPP location yields a modification of the peak-to-peak amplitude of these modes, as shown below.

Such high resolution sensing is demonstrated in Fig. 10 with spectra measured during a biochemical binding experiment, i.e. for a SPR shift close to the detection limit (a full report on these experiments can be found in Voisin et al. (2014)). By zooming in on resonances near the SPR and a few nm away, it becomes clear that the presence of a comb of resonances with widely different sensitivities allows for very small wavelength shifts to be detected unambiguously with high precision. Most often, the focus is made on the +2 mode among the spectral comb, as it appears to be the most sensitive in terms of both amplitude variation and wavelength shift (Figs. 8 and 11).

Optical fiber devices have two important advantages over bulky prism SPR configurations: light propagates essentially without loss in short lengths of fibers, resulting in very high signal-to-noise ratios, and interfacing devices to light sources and detectors consists essentially of plugging connectors into widely available fiber optic instrumentation, instead of having to carefully align optical beams through imaging systems. In terms of experimentally demonstrated FOM, TFBGs exhibit a value reaching 5000, which surpasses all configurations by more than one order of magnitude (Hecht et al. 1996). As shown in Fig. 12, this results from the fact that they exhibit narrow resonance bands (FWHM $\sim$0.1–0.2 nm) compared to even the best possible theoretical value ($\sim$5 nm obtained by calculating the reflection from the base of a prism in the Kretschmann-Raether configuration), also keeping in mind that the experimental SPR FWHM from other fiber configurations all exceed 20 nm and more.

Proteins and Cells Quantification with Gold-Coated TFBG Immunosensors

There are several experimental demonstrations confirming that gold-coated TFBGs can be succesfully used for biochemical sensing. In Shevchenko et al. (2011), the probe was associated with aptamers and a measurement of the dissociation constant was demonstrated. In Shevchenko et al. (2014), the probe was used to measure the intracellular density of non-physiological cells, namely human acute leukemia cells. In the following, we summarize our main achievements towards the demonstration of lab-on-fiber devices with TFBGs (both bare and gold-coated) suited for cancer diagnosis.

Sensing Density Alteration in Cells

These experiments allow us to demonstrate the powerful of polarization controllability of TFBG (simply using a bare TFBG) for highly sensitive biosample

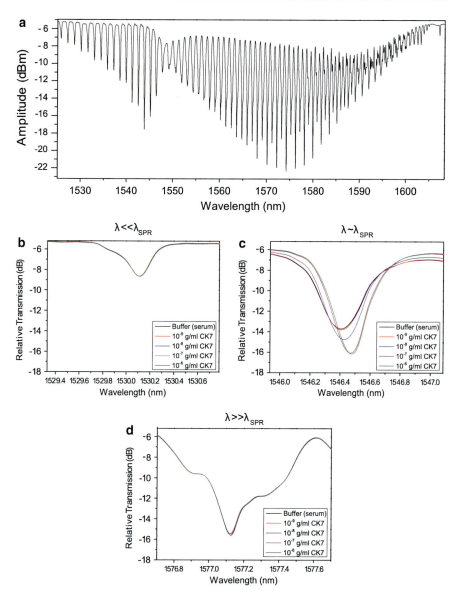

Fig. 11 Spectra measured during a biosensing experiment (**a**) with three spectral regions shown in detail: on the short wavelength side of the SPR (**b**); near the SPR (+2 mode) (**c**); and on the long wavelength side (**d**)

measurement. Human acute leukemia cells with different intracellular densities and refractive index ranging from 1.3342 to 1.3344 were clearly discriminated in-situ. The key point of this work lies in using the differential transmission spectrum between two orthogonal polarizations for the last guided mode resonance before

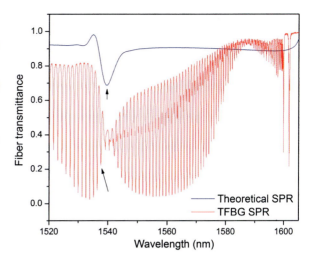

Fig. 12 Comparison between the best theoretical SPR response for 50 nm gold on silica in the Kretschmann-Raether configuration (*thick blue line*) and a measured TFBG-SPR spectrum with the same thickness of gold (*thin red line*). The *arrows* indicate the resonance to be followed in each case

the "cut-off" and leaky mode regime (Fig. 13a), which clearly identifies the "most" sensitive modal resonance just using a bare fiber inscribed with TFBG. Furthermore, the effect of temperature and power level fluctuations can be automatically cancelled by using the differential spectrum approach. In-situ discrimination of Leukemia cells at various stages of their lives has been achieved with a refractive index sensitivity measured to be 1.8×10^4 dB/RIU, corresponding to a LOD of 2×10^{-5} RIU. This study confirms the relationship between the intracellular density of cells and their refractive index (Fig. 13b). This provides a potential way to verify the hypothesis for "density alteration in non-physiological cells (DANCE)" in response to drugs and pathological changes in cells, and leads to helpful understanding of cell drug resistance of cells, and discovery of new physiological or pathological properties of cells (Guo et al. 2014).

Sensing Cytokeratins

A diverse range of tumor markers is associated to lung cancer. In this study, we focused on cytokeratins 7 (CK7) that are usually used for diagnosis in oncology. Indeed, profiling CK7 has proved to be a useful aid in the differential diagnosis of carcinomas, since primary and metastatic tumors present different profiles. In fact, primary lung tumors express cytokeratin 7 (CK7+) while secondary tumors are deficient in CK7 (CK7-). Moreover, it has been demonstrated that cytokeratin fragments can be released from malignant cells and consequently CK fragments can be located in blood circulation and are therefore easily accessible with an optical fiber properly modified.

The cytokeratin 7 antigens detection is based on the specific chemical reaction with their corresponding antibodies (AbCK7) previously immobilized on the optical fiber surface. The operating principle is based on the recognition of the cytokeratin 7

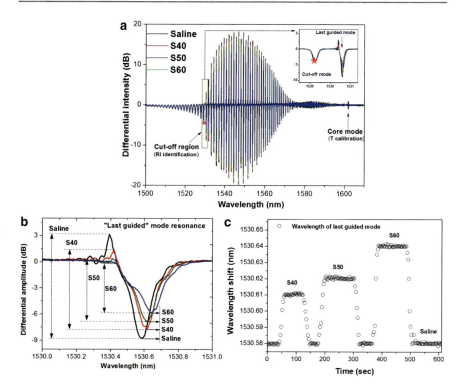

Fig. 13 In-situ detection of density alteration in cells (bare TFBG based differential transmission spectrum between two orthogonal polarizations for the last guided mode resonance before the "cut-off" and leaky mode regime)

antigen epitope by the fragment antigen-binding, a region present on the cytokeratin 7 antibody. In addition to the full protein (CK7FP) made of 469 amino acids, corresponding to a large biomolecule, we have also monitored a protein fragment called cytokeratin 7 peptide (CK7pep) made of only 23 amino acids. This allows to make a comparative study depending on the size of the target. This is driven by the fact that the detection of small molecule antigens remains a challenge in SPR immunosensing. Most of the SPR biosensors are based on large molecule detection since the SPR response in the presence of small mass usually suffers from low signal-to-noise ratio.

Gold-coated TFBGs, preliminary functionalized with cytokeratin 7 antibodies, have been first immersed in CK7 full protein solutions (Ribaut et al. 2016). To evaluate the sensitivity of the biosensor, initial measurements were conducted in a phosphate buffer (PBS) with increasing CK7FP concentrations from 1E-12 to 1E-6 g/mL. Recording of the transmission spectrum during experiments put in evidence an evolution of the SPR-TFBG starting from 3E-12 g/mL, as illustrated in Fig. 14a. Differences between CK7pep and CK7FP detection have been noticed

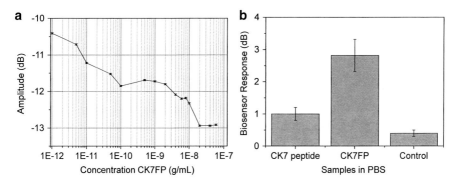

Fig. 14 Amplitude monitoring of the sensitive mode in the presence of CK7FP at different concentrations diluted in PBS (**a**). Biosensor response obtained for optical fiber functionalized with AbCK7, in the presence of CK7 peptide or CK7FP in comparison with OF functionalized without AbCK7 (**b**)

during experiments, as illustrated in Fig. 14b. Triplicates have confirmed the good reproducibility of the technique.

Since our aim is to develop these biosensors for medical diagnosis, further experiments have consisted in the peptide detection in complex media mimicking physiological conditions. In this context, experiments were performed in PBS mixed with 10% fetal bovine serum (FBS), resulting in a complex buffer full of proteins including albumin, enzymes, nutrients and others. The objective was to get close to physiological conditions, on the one hand, and to guarantee the selectivity of the biosensor, on the other hand. In the presence of FBS in the media, the optical fiber ended up with numerous compounds able to attach on the surface.

The amplitude monitoring, as presented in Fig. 15a, clearly confirms the evolution of SPR-TFBG signal as soon as the CK7pep was injected. Figure 15b presents the average output and deviations for a threefold repeated CK7pep detection on distinct optical fibers prepared in the same conditions. All three fibers have presented the identical amplitude evolution as a function of the CK7 concentration. The lowest concentration detected is estimated at 0.4 nM.

Sensing Transmembrane Receptors

In this section, we report on the use of gold-coated TFBGs for selective cellular detection through membrane protein targeting. The focus is made on the epithelial growth factor receptor (EGFR), which is a transmembrane receptor from the 4-tyrosine kinase receptors family (Malachovska et al. 2015). It is an important biomarker and therapeutic target that it is over-expressed by numerous cancer cells.

The sensor surface selectivity was ensured by bio-functionalization through a two-step approach. First, the clean waveguide surface was activated with the Carboxylic acid-SAM formation reagent (cat. n: C488) obtained from *Dojindo*

Fig. 15 Shift of the sensitive mode during incubation of the optical fiber functionalized with AbCK7 on CK7 peptide at different concentrations diluted in PBS + 10% FBS (**a**). Biosensor response obtained for optical fiber functionalized with AbCK7 immersed in PBS + 10% FBS, in presence or in absence of CK7 peptide (**b**).

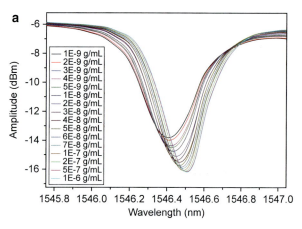

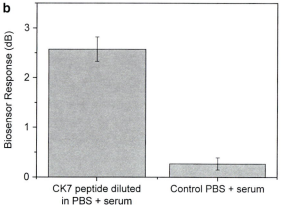

(Japan). Then, monoclonal mouse immunoglobulin G (IgG) raised against the human epidermoid carcinoma cell line (*Santa Cruz Biotechnology Inc.* and *American Type Culture Collection* (USA)) were immobilized on the surface through carbodiimide covalent biochemistry using *Dojindo*'s amine coupling kit (cat. n: A515–10). This antibody was diluted in the *Dojindo*'s reaction buffer to 0.01 μg/mL for efficient covalent immobilization during 30 min. The manufacturer's instructions were followed for each step for both processes.

Two cell lines were used: the first one with overexpressed human epidermal growth factor receptors (EGFRs) – A431 cell line, (EGFR positive, *EGFR (+)*). The second cell line was EGFR negative – OCM1 cell line, (*EGFR (−)*). The culture process will not be described here. It was done at 37 °C in a humidified incubator with 5% CO_2 atmosphere. No antibiotic was used and cell cultures were free of mycoplasma and pathogenic viruses. Cells suspensions were then prepared for SPR experiments. For this, A431 and OCM1 cells from a confluent monolayer were detached mechanically by a gentle scraping of cells from the growth surface into ice-cold phosphate-buffered saline (PBS). Cells were then washed three times in cold

Fig. 16 Different assays presenting specific cell interactions. *Dark grey bars* represent A431 cells, *light grey bars* represent OCM1 cells. *Blank bars* are washing and rinsing steps for RPMI-1640 cell culture media supplemented with FBS

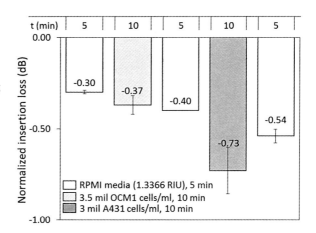

PBS by repeated centrifugation (at 2500 RPM and 4 °C for 3 min). Pellets were then again suspended to a concentration of 2–5×10^6 cells/mL. 0.5 mL volumes were use for experiments. A handheld automated cell counter from Millipore (Scepter 2.0, PHCC00000) was used to count the cells.

For immunosensing experiments, gold-coated TFBGs were hold straight between two clamps and were then approached towards the cells suspensions, using a vertical translation stage. To demonstrate that a differential response can be obtained between both cell lines, TFBGs were first immersed in RPMI-1640 cell culture media (*Roswell Park Memorial Institute* (RPMI) 1640, supplemented with 10% FBS – refractive index = 1.3366), until a stable baseline was reached. In the first step, sensors were incubated in a suspension of OCM1 during $\sim$10 min. After this assay, a secondary baseline of the media (RPMI-1640) was recorded. In the second step, sensors were incubated in a suspension of A431 during the same period. Finally, the third media baseline was recorded before, and after an extra rinsing step with the media. The conditioning in media was 5 min and assays lasted 10 min. Figure 16 shows the processed data (evolution of the peak-to-peak amplitude of the +2 mode) obtained after such binding experiments. It clearly figures out that an important response change is obtained with the A431 cell line (here at a concentration of 3×10^6 cells/mL), for which EGFRs are overexpressed. This is not the case for the other cell line for which the response remains comparable to the background noise level obtained in RPMI-1640 media. Error bars are the standard deviation obtained for three experiments made in the same conditions.

Different cells suspensions were used with various concentrations in order to estimate the LOD of the sensors. The latter was computed equal to $\sim$2 $\times$ 10^6 cells/mL, which is relevant with respect to the target application. Obviously, the sensor response is not linear with the cells concentration. The saturation has not been properly measured but it is beyond $\sim$5 $\times$ 10^6 cells/mL.

Finally, it is important to mention that a microscopy analysis could be used to precisely calibrate the sensor response. In practice, the number of cells attached to the sensor surface could be counted (or at least estimated) from images taken with a confocal fluorescence microscope, through the use of a fluorescent tag. This information could be used in the future to further refine the sensor response, allowing to better estimate the number of cells that have reacted.

Sensing Electroactive Biofilms

Electroactive biofilms (EABs) have received considerable attention due to their electrochemical connections with the substrate, which can be employed for harvesting energy from organic waste and converting waste into value-added chemicals. The electrochemical properties of biofilms primarily contribute to the presence of some specific strains (i.e. Shewanella and Geobacter) that are able to exchange electrons with solid substrate. The EABs commonly vary in thickness from few micrometers to several millimeters. Two layer bacterial aggregations have been identified in the thick EABs, including a live outer-layer (close to the solution interface) and a dead inner-layer of biofilm (close to the electrode interface). The live outer-layer of biofilm is used for current generation and the dead inner-layer provides an electrically conductive matrix. Nevertheless, specific interactions between electrochemically active microbes and solid electron acceptors are not fully understood. This results mainly from the lack of insight into the structure of the biofilm relative to its electrochemical properties. To date, electrochemical techniques have been primarily applied to study the extracellular electron transfer process in EABs, providing important insights into their redox process (Wang et al. 2010; Huang et al. 2010; Lu et al. 2012; Jiang et al. 2008). However, the traditional electrochemical measurement performs a "bulk" detection in which the current or change information originates from the whole electrode surface, which is difficult to provide structure and spatial information of the biofilm.

Spectroelectrochemical measurements with optical fiber sensors can be made in various hard-to-reach environments for in-situ detection, either as a hand-held probe or as a set of remotely operated devices along a fiber-optic cable, especially for environmental monitoring over the urban and suburban areas. Studies on spectroelectrochemical optical fiber sensors have been reported recently (Guan et al. 2014; Imai et al. 2015; Beam et al. 2008), also involving electrochemical SPR techniques (Nakamoto et al. 2012). In such cases, optical fibers are mainly used to transmit light.

The spectroelectrochemical SPR concept developed in our work provides a very "localized" and "surface" detection just between the up-side of metal film and the inner layer of biofilm (as the diagrammatic sketch shown in Fig.1a, NIR-SPR with about 1000 nm penetration depth and 1 mm propagation length above the gold film surface) (Yuan et al. 2016). Specifically, the Plasmon waves can mainly arise from the inner-layer of multilayered biofilm. As depicted in Fig. 17, the electrochemical plasmonic fiber-optic sensing system is composed of a compact plasmonic fiber-optic sensing system together with a traditional electrochemistry measurement apparatus for in-situ biofilm monitoring. The plasmonic optical fiber sensing probe

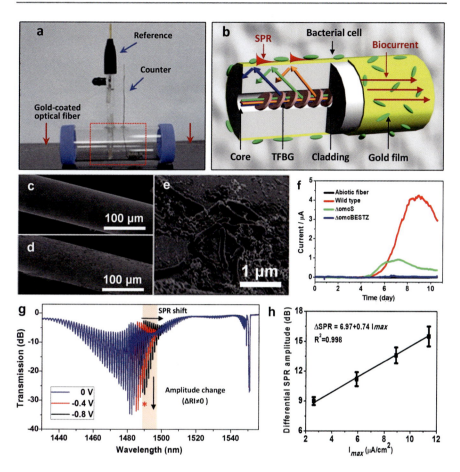

Fig. 17 Schematic of plasmonic fiber-optic sensing system (together with the electrochemistry measurement apparatus) for in-situ biofilm monitoring: (**a**) shows photographic image of the bio-electrochemical cell with the gold-coated optic fiber sensor; (**b**) presents the zoomed configuration of a gold-coated TFBG sensor probe. SEM images of the gold-coated optical fibers: (**c**) abiotic fiber; (**d**) optical fiber with G. sulfurreducens biofilm; (**e**) high resolution image of the biofilm. Electrochemical measurements of the EABs: (**f**) I-t curves of the bioelectrochemical systems with various inoculations. Plasmonic optical measurements of the EABs: (**g**) spectral response of gold-coated 20o TFBG with SPR excitations under different electrode polarizations, (**h**) linear relationship between the maximum current density (Imax) and the corresponding differential SPR amplitudes (ΔSPR) of various EABs (Yuan et al. 2016).

works as an electrode in the electrochemical system (Fig. 17a) while performing the optical detection function (Fig. 17b). EC-SPR measurements revealed the inner-layer of the EAB immediately adjacent to the metal surface. Any perturbation in that layer, such as bonding of analytes on receptor molecules or redox activity induced changes, modifies the local refractive index of the dielectric and the Plasmon phase velocity.

Prior to EC-SPR measurements, initially well cultured G. sulfurreducens cells were inoculated into the bioelectrochemical cell by polarizing the optical fiber working electrode at 0 V vs. saturated calomel electrode (SCE) to form a biofilm on the gold-coated optical fiber electrode using a multi-channel potentiostat (CHI1000C, China). Chronoamperometry (CA) and cyclic voltammetric (CV) measurements were performed with the gold-coated optical fiber as the working electrode (Surface area: 0.4 cm^2). Scanning electron microscope (SEM) images confirmed the formation of the biofilm on the optical fiber. As shown in Fig. 17c, a smooth optical fiber surface was observed before inoculation. After 10 days of culturing, microbial cells forming a biofilm were clearly observed on the optical fiber (Fig. 17d and zoomed Fig. 17e). During the biofilm formation, the electrochemical current was also recorded (Fig. 17f). After a lag phase, an electrochemical current was observed from the biofilm formed from wild-type G. sulfurreducens. In contrast, almost no current was obtained from the abiotic control.

Accompanying the electrochemical measurements, the optical spectral responses of the plasmonic fiber-optic sensor are recorded. With the variation of the redox state of the EAB (0, -0.4 V and -0.8 V vs. SCE), strong SPR wavelength shifts have been observed and their response varies in accordance with the applied potentials (Fig. 17g). Here, the amplitude of the "SPR" resonance increases about 15 dB when the applied potential (at the gold-coated fiber working electrode) changes from 0 to -0.8 V. Most importantly, a linear relationship between the variations of redox state of the EABs (maximum current density, Imax) and the changes of the SPR under potentiostatic conditions (differential SPR amplitude, ΔSPR) has been clearly demonstrated (Fig. 17h), pointing a very promising way to study the extracellular electron transfer mechanism of the EABs.

Conclusion

Optical fiber-based biochemical sensors fill an increasingly well-defined niche as they do not require elaborate light management schemes to probe molecules and materials. Light remains guided in the fiber from the source to the detector, apart from localized probe regions where it interacts with the immediate surroundings of the fiber. This feature is the dominant factor in making such sensors less expensive to fabricate and to use than conventional biosensors (especially those based on SPR and resonant waveguide gratings). However this comes with two main inconveniences: a higher (worse) limit of detection and the impossibility to scale up towards massively parallel testing. Therefore, researches and developments are made to overcome these limitations.

Our work features a robust biosensing platform with a simple four-step protocol enabling label-free cells sensing. It addresses the first issue by combining SPR effects with a grating-based approach in a single sensor that keeps all the advantages of the fiber solution. TFBGs provide a dense spectral comb of high Q-factor resonances that probe the SPR envelope and allow an order of magnitude improvement in the determination of the SPR shifts under biochemical interactions at the fiber surface. We have shown that we can separately excite resonances that enable SPR generation and others that do not (which can thus serve as reference channels). The sensor also provides absolute temperature information, from the wavelength position of Bragg core resonance that does not sense changes of the surrounding refractive index medium. This additional information can be used to average out noise and to eliminate many cross-sensitivity factors. We have also demonstrated that our biosensing platform can be used to investigate the binding of living human epithelial cells through specific interaction of transmembrane proteins and target extracellular membrane receptors in biological matrices.

The strength of our platform arises from its near-infrared operating wavelength, thus allowing improved penetration depth and full compatibility with telecommunication-grade equipment, including switch matrix devices that allow automatic interrogation of several tens of optical fibers in relatively rapid succession. Hence, different fibers coated with different bioreceptors could be easily interrogated to assay several parameters simultaneously, which would solve the second aforementioned limitation. Additionally for practical use, it could considerably enhance the reliability of a diagnosis.

Much of the development required for the exact quantification and deployment of practical sensors based on gold-coated TFBGs is still in progress. Therefore, it is hoped that the achievements presented herein will stimulate further research in this area.

References

J. Albert, L.-Y. Shao, C. Caucheteur, Tilted fiber Bragg grating sensors. Laser Photonics Rev. **7**, 83–108 (2013a)

J. Albert, S. Lepinay, C. Caucheteur, M.C. Derosa, High resolution grating-assisted surface plasmon resonance fiber optic aptasensor. Methods **63**, 239–254 (2013b)

F. Baldini, M. Brenci, F. Chiavaioli, A. Gianetti, C. Trono, Optical fiber gratings as tools for chemical and biochemical sensing. Anal. Bioanal. Chem. **402**, 109–116 (2012)

W.L. Barnes, A. Dereux, T.W. Ebbesen, Surface plasmon subwavelength optics. Nature **424**, 824–830 (2003)

B.M. Beam, N.R. Armstrong, S.B. Mendes, An electroactive fiber optic chip for spectroelectrochemical characterization of ultra-thin redox-active films. Analyst **134**, 454–459 (2008)

C. Caucheteur, P. Mégret, Demodulation technique for weakly tilted fiber Bragg grating refractometer. Photon. Technol. Lett. **17**, 2703–2705 (2005)

C. Caucheteur, Y.Y. Shevchenko, L.-Y. Shao, M. Wuilpart, J. Albert, High resolution interrogation of tilted fiber grating SPR sensors from polarization properties measurement. Opt. Express **19**, 1656–1664 (2011a)

C. Caucheteur, C. Chen, V. Voisin, P. Berini, J. Albert, A thin metal sheath lifts the EH to HE degeneracy in the cladding mode refractometric sensitivity of optical fiber sensors. Appl. Phys. Lett. **99**, 041118 (2011b)

C. Caucheteur, V. Voisin, J. Albert, Polarized spectral combs probe optical fiber surface plasmons. Opt. Express **21**, 3055–3066 (2013)

C. Caucheteur, T. Guo, J. Albert, Review of recent plasmonic fiber optic biochemical sensors: improving the limit of detection. Anal. Bioanal. Chem. **407**, 3883–3897 (2015)

C. Caucheteur, T. Guo, F. Liu, B.O. Guan, J. Albert, Ultrasensitive plasmonic sensing in air using optical fibre spectral combs. Nat. Commun. **7**, 13371 (2016)

C.F. Chan, C. Chen, A. Jafari, A. Laronche, D.J. Thomson, J. Albert, Optical fiber refractometer using narrowband cladding-mode resonance shifts. Appl. Opt. **46**, 1142–1149 (2007)

C. Chen, J. Albert, Stain-optic coefficients of individual cladding modes of single mode fibre: theory and experiment. Electron. Lett. **42**, 1027–1028 (2006)

X. Chen, J. Xu, X. Zhang, T. Guo, B.O. Guan, Wide range refractive index measurement using a multi-angle tilted fiber Bragg grating. IEEE Photon. Technol. Lett. **29**, 719–722 (2017)

R.W. Cohen, G.D. Cody, M.D. Coutts, B. Abeles, Optical properties of granular silver and gold films. Phys. Rev. B **8**, 3689 (1973)

T. Erdogan, Fiber grating spectra. J. Lightwave Technol. **15**, 1277–1294 (1997)

T. Erdogan, J.E. Sipe, Tilted fiber phase gratings. J. Opt. Soc. Am. A **13**, 296–313 (1996)

D. Feng, W. Zhou, X. Qiao, J. Albert, High resolution fiber optic surface plasmon resonance sensors with single-sided gold coatings. Opt. Express **24**, 16456–16464 (2016)

Y. Guan, X.N. Shan, S.P. Wang, P.M. Zhang, N.J. Tao, Detection of molecular binding via charge-induced mechanical response of optical fibers. Chem. Sci. **5**, 4375–4381 (2014)

T. Guo, F. Liu, Y. Liu, N.K. Chen, B.O. Guan, J. Albert, In situ detection of density alteration in non-physiological cells with polarimetric tilted fiber grating sensors. Biosens. Bioelectron. **55**, 452–458 (2014)

B. Hecht, H. Bielefeldt, L. Novotny, Y. Inouye, D.W. Pohl, Local excitation, scattering, and interference of surface plasmons. Phys. Rev. Lett. **77**, 1889–1892 (1996)

J. Homola, M. Piliarik, Surface plasmon resonance (SPR) sensors. Springer Ser. Chem. Sensors Biosens. **4**, 45–67 (2006)

J. Homola, S.S. Yee, G. Gauglitz, Surface plasmon resonance sensors: review. Sensors Actuators B **54**, 3–15 (1999)

X.P. Huang, S.P. Wang, X.N. Shan, X.J. Chang, N.J. Tao, Flow-through electrochemical surface plasmon resonance: detection of intermediate reaction products. J. Electroanal. Chem. **649**, 37–41 (2010)

A. Iadicicco, A. Cusano, A. Cutolo, R. Bernini, M. Giordano, Thinned fiber Bragg gratings as high sensitivity refractive index sensor. IEEE Photon. Technol. Lett. **16**, 1149–1151 (2004)

K. Imai, T. Okazaki, N. Hata, S. Taguchi, K. Sugawara, H. Kuramitz, Simultaneous multiselective spectroelectrochemical fiber-optic sensor: demonstration of the concept using methylene blue and ferrocyanide. Anal. Chem. **87**, 2375–2382 (2015)

X.Q. Jiang, Z.J. Cao, H. Tang, L. Tan, Q.J. Xie, S.Z. Yao, Electrochemical surface plasmon resonance studies on the deposition of the charge-transfer complex from electrooxidation of o-tolidine and effects of dermatan sulfate. Electrochem. Commun. **10**, 1235–1237 (2008)

E. Kretschmann, H. Raether, Radiative decay of non radiative surface plasmon excited by light. Z. Naturforsch. **23**, 2135 (1968)

G. Laffont, P. Ferdinand, Tilted short-period fiber Bragg grating induced coupling to cladding modes for accurate refractometry. Meas. Sci. Technol. **12**, 765–772 (2001)

B. Liedberg, C. Nylander, I. Lungstrom, Surface plasmon resonance for gas detection and biosensing. Sensors Actuators **4**, 299–304 (1984)

J. Lu, W. Wang, S.P. Wang, X.N. Shan, J.H. Li, N.J. Tao, Plasmonic-based electrochemical impedance spectroscopy: application to molecular binding. Anal. Chem. **84**, 327–333 (2012)

V. Malachovska, C. Ribaut, V. Voisin, M. Surin, P. Leclère, R. Wattiez, C. Caucheteur, Fiber-optic SPR immunosensors tailored to target epithelial cells through membrane receptors. Anal. Chem. **87**, 5957–5965 (2015)

G.V. Naik, J. Kim, A. Boltasseva, Oxides and nitrides as alternative plasmonic materials in the optical range. Opt. Express **1**, 1090–1099 (2011)

K. Nakamoto, R. Kurita, O. Niwa, Electrochemical surface plasmon resonance measurement based on gold nanohole array fabricated by nanoimprinting technique. Anal. Chem. **84**, 3187–3191 (2012)

P. Offermans, M.C. Shaafsma, S.R.K. Rodriguez, Y. Zhang, M. Crego-Calama, S.H. Brongersma, J.G. Rivas, Universal scaling of the figure of merit of plasmonic sensors. ACS Nano **5**, 5151–5157 (2011)

A. Othonos, K. Kalli, *Fiber Bragg Gratings: Fundamentals and Applications in Telecommunications and Sensing* (Artech House, Boston, 1999)

J. Pollet, F. Delport, K.P.F. Janssen, K. Jans, G. Maes, H. Pfeiffer, M. Wevers, J. Lammertyn, Fiber optic SPR biosensing of DNA hybridization and DNA-protein interactions. Biosens. Bioelectron. **25**, 864–869 (2009)

C. Ribaut, V. Voisin, V. Malachovska, V. Dubois, P. Mégret, R. Wattiez, C. Caucheteur, Small biomolecule immunosensing with plasmonic optical fiber grating sensor. Biosens. Bioelectron. **77**, 315–322 (2016). https://doi.org/10.1016/j.bios.2015.09.019

R.S. Sennett, G.D. Scott GD, The structure of evaporated metal films and their optical properties. J.Opt. Soc. Am. **40**, 203–211 (1950)

A.K. Sharma, J. Rajan, B.D. Gupta, Fiber-optic sensors based on surface plasmon resonance: a comprehensive review. IEEE Sensors J. **7**, 1118–1129 (2007)

Y.Y. Shevchenko, J. Albert, Plasmon resonances in gold-coated tilted fiber Bragg gratings. Opt. Lett. **32**, 211–213 (2007)

Y. Shevchenko, T.J. Francis, D.A.D. Blair, R. Walsh, M.C. DeRosa, J. Albert, In situ biosensing with a surface plasmon resonance fiber grating aptasensor. Anal. Chem. **83**, 7027–7034 (2011)

Y. Shevchenko, G. Camci-Unal, D.F. Cuttica, M.R. Dokmeci, J. Albert, A. Khademhosseini, Surface plasmon resonance fiber sensor for real-time and label-free monitoring of cellular behavior. Biosens. Bioelectron. **56**, 359–367 (2014)

V. Svorcik, J. Siegel, P. Sutta, J. Mistrik, P. Janicek, P. Worsch, Z. Kolská, Annealing of gold nanostructures sputtered on glass substrate. Appl. Phys. A **102**, 605–610 (2011)

J.J. Tu, C.C. Homes, M. Strongin, Optical properties of ultrathin films: evidence for a dielectric anomaly at the insulator-to-metal transition. Phys. Rev. Lett. **90**, 017402 (2003)

A.M. Vengsarkar, P.J. Lemaire, J.B. Judkins, V. Bhatia, T. Erdogan, J.E. Sipe, Long-period fiber gratings as band-rejection filters. J. Lightwave Technol. **14**, 58–65 (1996)

V. Voisin, C. Caucheteur, P. Mégret, J. Albert, Interrogation technique for TFBG-SPR refractometers based on differential orthogonal light states. Appl. Opt. **50**, 4257–4261 (2011)

V. Voisin, J. Pilate, P. Damman, P. Mégret, C. Caucheteur, Highly sensitive detection of molecular interactions with plasmonic optical fiber grating sensors. Biosens. Bioelectron. **51**, 249–254 (2014)

S.P. Wang, X.P. Huang, X.N. Shan, K.J. Foley, N.J. Tao, Electrochemical surface plasmon resonance: basic formalism and experimental validation. Anal. Chem. **82**, 935–941 (2010)

I.M. White, X.D. Fan, On the performance quantification of resonant refractive index sensors. Opt. Express **16**, 1020–1028 (2008)

Y. Yuan, T. Guo, X.H. Qiu, J.H. Tang, Y.Y. Huang, L. Zhuang, S.G. Zhou, Z.H. Li, B.O. Guan, X.M. Zhang, J. Albert, Electrochemical surface Plasmon resonance fiber-optic sensor: in-situ detection of electroactive biofilms. Anal. Chem. **88**, 7609–7761 (2016)

K. Zhou, A.G. Simpson, L. Zhang, I. Bennion, Side detection of strong radiation-mode outcoupling from blazed FBGs in single-mode and multimode fibers. IEEE Photon. Technol. Lett. **15**, 936–938 (2003)

K. Zhou, L. Zhang, X. Chen, I. Bennion, Optic sensors of high refractive-index responsivity and low thermal cross sensitivity that use fiber Bragg gratings of >80° tilted structures. Opt. Lett. **31**, 1193–1195 (2006)

CO$_2$-Laser-Inscribed Long Period Fiber Gratings: From Fabrication to Applications

36

Yiping Wang and Jun He

Contents

Y. Wang (✉) · J. He
Key Laboratory of Optoelectronic Devices and Systems of Ministry of Education and Guangdong Province, College of Physics and Optoelectronic Engineering, Shenzhen University, Shenzhen, China

Guangdong and Hong Kong Joint Research Centre for Optical Fibre Sensors, Shenzhen University, Shenzhen, China
e-mail: ypwang@szu.edu.cn; hejun07@szu.edu.cn

© Springer Nature Singapore Pte Ltd. 2019
G.-D. Peng (ed.), *Handbook of Optical Fibers*,
https://doi.org/10.1007/978-981-10-7087-7_78

Abstract

This chapter presents a systematic review of long period fiber gratings (LPFGs) inscribed by CO_2 laser irradiation. Firstly, various fabrication techniques based on CO_2 laser irradiations are introduced that inscribe LPFGs in different types of optical fibers such as conventional glass fibers, solid-core photonic crystal fibers, and air-core photonic bandgap fibers. Secondly, possible mechanisms, including residual stress relaxation, glass structure changes, and physical deformation, of refractive index modulations in the CO_2-laser-induced LPFGs are discussed. Asymmetrical mode coupling, resulting from single-side laser irradiation, is analyzed to understand the unique optical properties of the CO_2-laser-induced LPFGs. Thirdly, several pre- and posttreatment techniques for enhancing the grating efficiency are described. Fourthly, sensing applications of CO_2-laser-induced LPFGs for temperature, strain, bend, torsion, pressure, and biochemical sensors are presented. Finally, communication applications of CO_2-laser-induced LPFGs, such as band-rejection filters, gain equalizers, polarizers, couplers, and mode converters, are presented and discussed.

Keywords

Gratings · Long period fiber gratings (LPFGs) · Optical fiber sensors · CO2 laser

Introduction

Optical fiber gratings play a vital role in the field of optical communications and sensors. There are two types of in-fiber gratings: fiber Bragg gratings (FBGs) with period or pitch in the order of optical wavelength (Hill et al. 1978; Rao 1997, 1999; Kersey et al. 1997), and long period fiber gratings (LPFGs) with period or pitch in hundreds of wavelengths (Vengsarkar et al. 1996a, b; Bhatia and Vengsarkar 1996; Erdogan 1997; Eggleton et al. 1997, 1999, 2000; James and Tatam 2003; Bhatia 1999; Davis et al. 1998a, b). Since Hill et al. (1978) and Vengsarkar et al. (1996a) wrote the first FBG and LPFG in conventional glass fibers in 1978 and 1996, respectively, the fabrication and application of in-fiber gratings have developed rapidly. Various fabrication methods, such as ultraviolet (UV) laser exposure (Vengsarkar et al. 1996a, b; Bhatia and Vengsarkar 1996; Erdogan 1997; Eggleton et al. 1997, 1999, 2000; James and Tatam 2003; Bhatia 1999), CO_2 laser irradiation (Davis et al. 1998a, b; 1999; Kakarantzas et al. 2001, 2002; Rao et al. 2003a, 2004; Liu and Chiang 2008a; Wang et al. 2006a, b, 2007a, 2008a; Zhong et al. 2014a, b, 2015; Fu et al. 2015; Tan et al. 2013; Geng et al. 2012; Yang et al. 2011, 2017; Tian et al. 2013; Zhu et al. 2003a, 2004; Wu et al. 2011; Tang et al. 2015, 2017), electric arc discharge (Kosinski and Vengsarkar 1998; Karpov et al. 1998; Hwang et al. 1999; Humbert et al. 2003; Dobb et al. 2004; Rego et al. 2006), femtosecond laser exposure (Kondo et al. 1999; Fertein et al. 2001), mechanical microbends (Savin et al. 2000; S et al. 2001; Steinvurzel et al. 2006a; Su et al. 2005), etched corrugations (Lin and Wang 1999, 2001; Lin et al. 2001a, b, 2002; Yan et al. 2002), and ion beam implantation (Fujimaki et al. 2000; Bibra et al. 2001; Bibra and Roberts 2000), have been demonstrated to write LPFGs in different types of optical fibers. Numerous LPFG-based devices have also been developed for sensing and communication applications (Kersey et al. 1997; Davis et al. 1998a; Rao et al. 2004).

Davis et al. reported gratings written by CO$_2$ laser irradiation in a conventional glass fiber in 1998 (Davis et al. 1998a, b). Compared with the UV laser exposure, CO$_2$ laser irradiation is more flexible and is of lower cost because no photosensitivity and any other pretreated processes are required to write a grating in the glass fibers (Davis et al. 1998a, b, 1999; Kakarantzas et al. 2001, 2002; Rao et al. 2003a, 2004; Liu and Chiang 2008a; Wang et al. 2006a, b, 2007a, 2008a; Zhong et al. 2014a, b, 2015; Fu et al. 2015; Tan et al. 2013; Geng et al. 2012; Yang et al. 2011, 2017; Tian et al. 2013; Zhu et al. 2003a, 2004; Wu et al. 2011; Tang et al. 2015, 2017). Moreover, the CO$_2$ laser irradiation process can be controlled to generate complicated grating profiles via point-to-point exposure without any expensive masks. Hence, this technique could be used to write LPFGs in almost all types of fibers including pure-silica photonic crystal fibers (PCFs). Liu et al. presented development on the CO$_2$-laser-induced LPFGs (Liu and Chiang 2008a). The most exciting development of writing with CO$_2$ laser irradiation is to successfully inscribe the LPFG in an air-core photonic bandgap fiber (PBF) by using focused CO$_2$ laser light to periodically perturb or deform the air holes along the fiber axis (Wang et al. 2008a).

Here we present a systematic review of LPFGs written with CO$_2$ laser irradiation in different types of optical fibers. The methods for writing LPFGs in conventional glass fibers, solid-core PCFs, and air-core PBFs with a CO$_2$ laser are presented in section "CO$_2$ Laser Inscription Techniques." Then possible grating formation mechanisms, for example, residual stress relaxation, glass structural changes, and physical deformation that give rise to the refractive index modulations of the CO$_2$-laser-induced LPFGs, are analyzed in section "Grating Formation Mechanisms." Asymmetrical mode coupling in LPFGs is also discussed in this section. Various methods for fabricating better LPFGs, including pretreatment methods such as hydrogen loading and applying prestrain and posttreatment methods such as applying tensile strain and changing temperature, are described in section "Improvements of Grating Fabrications." Subsequently, sensing applications such as temperature, strain, bend, torsion, pressure, and biochemical sensors and communication applications such as band-rejection filters, gain equalizers, polarizers, couplers, and mode converters using the CO$_2$-laser-induced LPFGs are presented in sections "Sensing Applications" and "Communication Applications," respectively.

CO₂ Laser Inscription Techniques

Since Davis et al. reported the CO_2-laser-induced LPFG in a conventional glass fiber in 1998 (Davis et al. 1998a, b), various CO_2 laser irradiation techniques have been demonstrated and improved to write high-quality LPFGs in different types of optical fibers, including conventional glass fibers, PCFs, and PBFs, and to achieve unique grating properties. This section reviews the use of CO_2 laser irradiation for writing LPFGs in conventional glass fibers, solid-core PCFs, and air-core PBFs.

LPFG Inscription in Conventional Glass Fibers

Typically, in most of LPFG fabrication setups employing a CO_2 laser (Davis et al. 1998a, b; Zhu et al. 2003a; Braiwish et al. 2004), as shown in Fig. 1, the fiber is periodically moved along its axis direction via a computer-controlled translation stage, and the CO_2 laser beam irradiates periodically the fiber through a shutter controlled by the same computer. A light source and an optical spectrum analyzer are employed to monitor the evolution of the grating spectrum during the laser irradiation. This is a typical point-to-point technique for writing a grating in an optical fiber. Such an LPFG fabrication system usually requires an exact controlling of both the shutter and the translation stage to achieve a good simultaneousness of the laser irradiation and the fiber movement. Moreover, vibration from the periodic movement of the fiber during the irradiation with the CO_2 laser beam could affect the stability and repeatability of grating fabrication.

Rao et al. demonstrated a novel grating fabrication system based on two-dimensional scanning of the CO_2 laser beam (Rao et al. 2003a, 2004), as shown in Fig. 2. One end of the fiber is fixed and the other is attached to a small weight to provide a constant strain in the fiber. The strain could make it easier to fabricate LPFGs and improve the efficiency of the grating fabrication, as discussed in section "Sensing Applications" The focused high-frequency CO_2 laser pulses scanned periodically across the employed fiber along "x" direction and then shifted a grating pitch along "y" direction (the fiber axis) to create next grating period by means of two-dimensional optical scanners under the computer's control. Compared with typical point-to-point fabrication systems (Davis et al. 1998a, b; Zhu et al. 2003a; Braiwish et al. 2004), no synchronization is required in such a system because the

Fig. 1 Schematic diagram of an LPFG fabrication based on the point-to-point technique employing a CO_2 laser

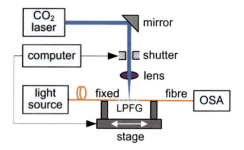

Fig. 2 Schematic diagram of LPFG fabrication based on two-dimensional scanning of focused high-frequency CO_2 laser pulses (Rao et al. 2003a)

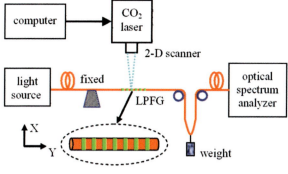

Fig. 3 Schematic diagram of an improved LPFG fabrication based on the point-to-point technique employing a CO_2 laser

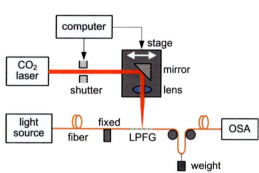

employed fiber is not periodically moved along the fiber axis. Such a system can write high-quality LPFGs with nearly zero insertion loss. A similar fabrication setup was also demonstrated to write an LPFG with a CO_2 laser, in which the laser beam, however, scanned one-dimensionally across the fiber that was periodically moved along the fiber axis (Chan et al. 2007).

The authors developed an improved LPFG fabrication system based on the point-to-point technique employing a CO_2 laser, as shown in Fig. 3, combining the advantages of the two fabrication systems illustrated in Figs. 1 and 2. The CO_2 laser beam is, through a shutter and a mirror, focused on the fiber by a cylindrical lens with a focus length of 254 mm. Both the mirror and the lens are mounted on a linear air-bearing motor stage (ABL 1500 from Aerotech). A LabVIEW program has been developed to control simultaneously the operations of both the linear motor stage and the shutter so that the fiber is exposed once as soon as the focused laser beam is shifted by a grating pitch via the mirror. In other words, the fiber is not moved in this system, which overcomes the disadvantage of the fiber vibration, resulting from periodic movement of the fiber, in normal point-to-point grating fabrication setups shown in Fig. 1.

Zhong et al. also developed a promising LPFG inscribing system based on an improved two-dimensional scanning technique with a focused CO_2 laser beam (Zhong et al. 2014a), as shown in Fig. 4a. This system consists of an industrial CO_2 laser with a maximum power of 10 W (SYNRAD 48-1) and a power stability

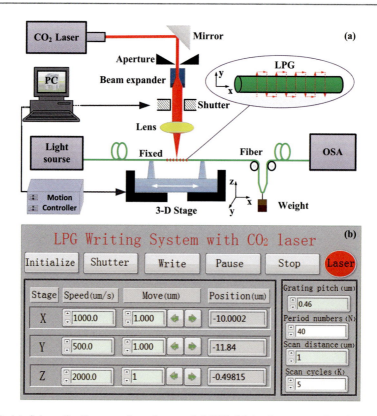

Fig. 4 (**a**) Schematic diagram of an improved LPFG fabrication system based on a two-dimensional scanning technique using a CO_2 laser. (**b**) Operation interface of the LPFG fabrication system (Zhong et al. 2014a)

of $\pm 10\%$, an electric shutter for turning on/off the laser beam, an infrared ZNSE PO/CX lens with a focused length of 63.5 mm, a four-times beam expander for decreasing the diameter of the focused laser spot, and a three-dimensional ultraprecision motorized stage (Newport XMS50,VP-25X, and GTS30V) with a minimum incremental motion of 10 nm and a bi-directional repeatability of 80 nm. A closed loop control system is employed to improve the power stability of the CO_2 laser to $\pm 2\%$, which is a huge advantage of this LPFG inscribing system. The power stability ($\pm 2\%$) of the CO_2 laser improves effectively the stability and reproducibility of grating inscription. For example, the success rate of grating inscription is almost 100% in this system. In contrast, the success rate was about 30% in the system with a CO_2 laser with power stability of $\pm 10\%$ (Rao et al. 2003a, 2004). A supercontinuum light source (NKT Photonics SuperK™ Compact) and an optical spectrum analyzer (YOKOGAWA AQ6370C) are employed to monitor the transmission spectrum of the CO_2-laser-inscribed LPFG during grating inscription. A control program with an easy-to-use operation interface is developed by use of LabVIEW software in order to control every device in the system and to inscribe

Fig. 5 Photograph of an asymmetric LPFG with periodic grooves (Wang et al. 2006a)

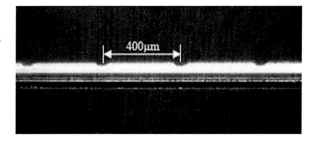

high-quality LPFGs. As soon as the grating parameters, such as grating pitch, number of grating periods, number of scanning cycles, are entered via the operation interface illustrated in Fig. 4b and the "Write" button is clicked, a high-quality LPFG could be obtained.

CO_2 laser irradiation may cause unexpected physical deformation, resulting from laser heating, of fiber structures during LPFG fabrication. Such physical deformations are usually avoided to decrease the insertion loss of the written LPFGs during early grating fabrications with a CO_2 laser (Davis et al. 1998a, b; Rao et al. 2003a). Wang et al. reported a novel technique for writing an asymmetric LPFG by means of carving periodic grooves on the surface of an optical fiber with a focused CO_2 laser beam (Wang et al. 2006a), as shown in Fig. 5. Physical deformations, i.e., periodic grooves, in such an asymmetric LPFG, do not cause a large insertion loss because these grooves are totally confined within the outer cladding and have no influence on the light transmission in the fiber core. Moreover, such grooves enhance the efficiency of grating fabrication and introduce unique optical properties, e.g., extremely high strain sensitivity, into the gratings (Wang et al. 2006a; 2007b). Further investigations discover that the insertion loss of LPFGs is mainly due to the nonperiodicity and the disorder of refractive index modulations in the gratings. Asymmetric LPFGs were also fabricated in thin core fibers using the same method (Fu et al. 2015). The proposed thin-core LPFG exhibits a high extinction ratio of over 25 dB at the resonant wavelength and a narrowed 3 dB-bandwidth of ~8.7 nm, which is nearly one order of magnitude smaller than that of LPFGs in conventional single mode fibers. It also exhibits high polarization-dependent loss of over 20 dB at resonant wavelength.

Additionally, a few other CO_2 laser irradiation methods have been demonstrated to write specialty LPFGs. For example, (1) edge-writing was reported to inscribe LPFGs with high-frequency CO_2 laser pulses (Zhu et al. 2007a, b, 2009a). Refractive index disturbance in such an edge-written LPFG mainly occurs at the edge region of the fiber cladding rather than in the fiber core. (2) A helical LPFG was fabricated in an optical fiber that continuously rotates and moves along the fiber axis during CO_2 laser irradiation (Oh et al. 2004a). Compared with a conventional LPFG, a helical LPFG has a very low polarization dependent loss (PDL) due to a screw-type index modulation in this grating, which thus could be a potential technique for achieving an LPFG with low PDL. (3) Microtaper-based LPFGs were also fabricated by means of tapering periodically a conventional glass fiber with a focused

CO_2 laser beam (Kakarantzas et al. 2001; Zhong et al. 2014b; Xuan et al. 2009; Zhu et al. 2005a; Lei et al. 2006). Negative temperature coefficient of the resonant wavelength was observed in the microtaper-based LPFGs (Xuan et al. 2009).

CO_2 laser irradiation has also been used to fabricate microtaper-based LPFGs on soft glass fibers. Soft glass fibers are developed to achieve flattened and near-zero dispersion profile and have potential bend sensing applications due to their unique light transmission ability. An all-solid soft glass fiber consisting of two types of commercial lead silicate glasses (core: Schott SF57 and cladding: Schott SF6) was designed and drawn. As shown in Fig. 6a, the core and cladding diameters of this fiber are 2.4 and 175 μm, respectively. Refractive index in the core and cladding are 1.80 and 1.76, respectively. The measured transmission loss of the soft glass fiber is about 6.1 dB per meter at the wavelength of 1550 nm. Compared with conventional silica glass fibers, the soft glass fiber has a lower drawing temperature of about 600–700 °C. The soft glass fiber was tapered periodically the by use of a CO_2 laser grating fabrication system, as shown in Fig. 6b. Such a soft glass fiber with periodic microtapers could be used to develop promising bend sensors with a sensitivity of -27.75 W/m^{-1} by means of measuring the bend-induced change of light intensity. The proposed bend sensor exhibits a very low measurement error of down to $\pm 1\%$ (Wang et al. 2012).

Apart from conventional single mode fibers (SMFs), there have been a number of studies on CO_2-laser writing of LPFGs in boron-doped fibers (Grubsky and Feinberg 2005; Liu et al. 2009a; Kim et al. 2001, 2002). The grating writing efficiency can be enhanced with a fiber annealed at a sufficiently high temperature (Liu and Chiang 2008a; Grubsky and Feinberg 2005; Liu et al. 2009a). Compared with normal UV laser exposure technique, the CO_2 laser irradiation technique is easily used to write special gratings such as phase-shifted LPFGs (Zhu et al. 2004, 2005a), chirped LPFGs (Yan et al. 2008), complicated apodized LPFGs (Gu et al. 2009), grating pairs (Chan et al. 2007), and ultra-long period fiber gratings with a period of up to several millimeters (Zhu et al. 2007c, 2005b; Rao et al. 2006). LPFGs have also been successfully achieved with a CO_2 laser during the fiber-drawing process (Hirose et al. 2007). So, it is possible for the CO_2 laser irradiation technique to continuously write numerous LPFGs with high quality and low cost in an optical fiber to develop potential distributed sensing systems and communication applications. By the way,

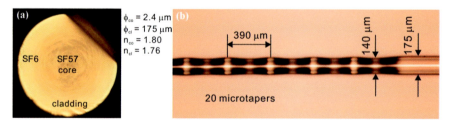

Fig. 6 (a) Cross-section microscope image of the all-solid soft glass fiber. (b) Microscope image of the soft glass fiber with periodic microtapers created by CO_2 laser irradiation technique (Wang et al. 2012)

CO$_2$ laser irradiation can be used to enhance the UV photosensitivity of GeO$_2$:SiO$_2$ optical fibers before grating writing (Brambilla et al. 1999).

LPFG Inscription in Solid-Core PCFs

Over the past decade, PCFs have attracted a great deal of interest due to their unique microstructures and optical properties (Knight et al. 1996; Smith et al. 2003). Since Eggleton et al. reported the grating in a photosensitive PCF with a Ge-doped core in 1999 (James and Tatam 2003), a large number of gratings have been written in different types of PCFs with or without photosensitivity by the use of various fabrication techniques such as UV laser exposure (Bhatia 1999), CO$_2$ laser irradiation (Kakarantzas et al. 2002; Wang et al. 2006b, 2007a; Tian et al. 2013; Zhong et al. 2014b, 2015; Zhu et al. 2003a, 2004; Yang et al. 2017), electric-arc discharge (Zhu et al. 2004), femtosecond laser exposure (Karpov et al. 1998; Humbert et al. 2003), and two-photon absorption (Groothoff et al. 2003). UV laser exposure is a common technique for writing a FBG/LPFG in a Ge-doped PCF with a photosensitivity (James and Tatam 2003; Wang et al. 2009a, b). In contrast, CO$_2$ laser irradiation is a highly efficient, low cost technique for writing an LPFG in a pure-silica PCF without photosensitivity (Kakarantzas et al. 2002; Wang et al. 2006b, 2007a; Tian et al. 2013; Zhong et al. 2014b, 2015; Zhu et al. 2003a, 2004; Yang et al. 2017).

Kakarantzas et al. reported, as shown in Fig. 7, the example of structural LPFGs written in pure-silica solid-core PCFs (Kakarantzas et al. 2002, 2003). The gratings are realized by periodic collapse of air holes in the PCF via heat treatment with a CO$_2$ laser. The resulting periodic hole-size perturbation produces core-to-cladding-mode conversion, thus creating a novel LPFG in the PCF (Kakarantzas et al. 2002). This technique, combining with periodic mechanical twisting, can be used to fabricate a rocking filter in a polarization-maintaining PCF (Kakarantzas et al. 2003).

As shown in Fig. 8, an asymmetrical LPFG with periodic grooves was written in a pure-silica large-mode-area PCF by the use of a focused CO$_2$ laser beam (Wang et al. 2006b, 2007a). The repeated scanning of the focused CO$_2$ laser beam creates a local high temperature in the fiber, which leads to the collapse of air holes

Fig. 7 LPFG written in a pure-silica solid-core PCF with a CO$_2$ laser (Kakarantzas et al. 2002)

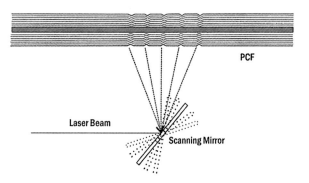

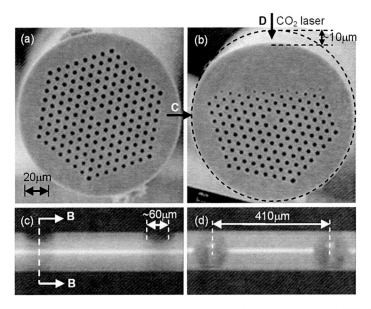

Fig. 8 Asymmetrical LPFG with periodic grooves in a pure-silica PCF (Wang et al. 2007a)

and the gasification of SiO_2 on the fiber surface. Consequently, periodic grooves with a depth of about 15 μm and a width of about 50 μm are created on the fiber, as shown in Fig. 8b. Such grooves, especially collapse of air holes, induce periodic refractive index modulations along the fiber axis due to the well-known photoelastic effect, thus creating an LPFG in the PCF. This asymmetrical LPFG has unique optical properties, e.g., high strain sensitivity, low temperature sensitivity, and high polarization dependence, as discussed in sections "Sensing Applications" and "Communication Applications."

As shown in Fig. 9, an inflated LPFG (I-LPFG) was also inscribed in a pure-silica PCF and could be used as a high-sensitivity strain sensor and a high-sensitivity gas pressure sensor (Zhong et al. 2014b, 2015). The I-LPFG was inscribed by use of the pressure-assisted CO_2 laser beam scanning technique to inflate periodically air holes of a PCF. Such periodic inflations enhanced the sensitivity of the LPFG-based strain sensor to −5.62 pm/μɛ. After high temperature annealing of the I-LPFG, moreover, a good repeatability and stability of temperature response with a sensitivity of 11.92 pm/°C was achieved (Zhong et al. 2014b). In addition, the I-LPFG with periodic inflations exhibits a very high gas pressure sensitivity of 1.68 nm/MPa, which is one order of magnitude higher than that, i.e., 0.12 nm/MPa, of the LPFG without periodic inflations. Moreover, the I-LPFG has a very low temperature sensitivity of 2.83.1 pm/°C due to the pure silica material in the PCF so that the pressure measurement error, resulting from the cross sensitivity between temperature and gas pressure, is less than 1.71.8 kPa/°C without temperature compensation. So, the I-LPFG could be used to develop a promising gas pressure sensor, and the achieved pressure measurement range is up to 10 MPa (Zhong et al. 2015).

Fig. 9 Microscope image of the cross section of the PCF (**a**) before and (**b**) after CO_2 laser irradiation, (**c**) side view of the CO_2-laser-inscribed inflated LPFG with periodic inflations (Zhong et al. 2015)

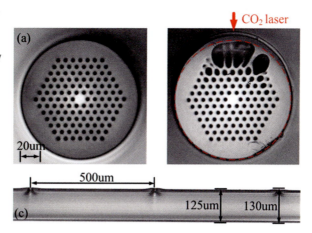

In addition to the asymmetrical LPFGs inscribed in PCFs, the symmetrical LPFGs were also inscribed in PCFs using the same method (Tian et al. 2013). Results show that symmetric index perturbation induced by laser irradiation with the aid of a 120° gold-coated reflecting mirror results in LP_{0n} symmetric mode coupling, while asymmetric irradiation without using the mirror leads to LP_{1n} asymmetric mode coupling. Because of the azimuthally anisotropic hexagonal cladding structure, symmetric irradiation yields far more reproducible LPFGs in PCFs than asymmetric irradiation. On the other hand, the irradiation symmetry has little effect on the reproducibility of LPFGs inscribed in SMFs due to the isotropy of its all-solid cladding structure.

Furthermore, a highly compact LPFG with only 8 periods and a short total length of 2.8 mm was written in a pure-silica large-mode-area PCF by the common point-by-point technique employing a CO_2 laser (Zhu et al. 2003a), in which clear physical deformation was also observed. In contrast, another LPFG without geometrical deformation and fiber elongation was written in an endlessly single-mode PCF by periodic stress relaxation resulting from CO_2 laser irradiation (Zhu et al. 2005c). Moreover, an LPFG pair has been successfully created in a pure-silica PCF with a CO_2 laser to develop a stain sensor (Shin et al. 2009). A novel coupled local-mode theory could be used to model and analyze this type of PCF-based LPFGs with periodic collapses of air holes (Jin et al. 2010). Such a theory is based on calculating the variations of local-mode profiles and propagation constants over the perturbed regions and on solving the coupled local-mode equations to obtain a quantitative description of the intermodal energy exchange.

The CO_2 laser irradiation technique could also be used to write an LPFG in a conventional or photonic crystal polarization-maintaining (PM) fibers (Lee et al. 2008). Such an LPFG has two clean polarization-splitting rejection bands. The writing efficiencies of gratings in the two types of PM fibers depend strongly on the fiber orientation, with the highest efficiency obtained when the irradiation direction is along the slow axis of the fiber (Lee et al. 2008). Such orientation dependence is much stronger for a conventional PM fiber than for a photonic crystal PM fiber and is attributed to the relaxation of mechanical stress in the stress-applying parts of the fiber.

LPFG Inscription in Air-Core PBFs

As discussed above, a large number of gratings have been demonstrated in different types of PCFs by the use of various fabrication techniques (Zhu et al. 2003a, 2004; Karpov et al. 1998; Humbert et al. 2003; Groothoff et al. 2003). All of these gratings, however, were written in index-guiding PCFs, instead of bandgap-guiding fibers. Recently, PBF-based gratings were also written in a new kind of bandgap-guiding fibers such as fluid-filled PBFs (Steinvurzel et al. 2006a, b; Wang et al. 2009c; Iredale et al. 2006; Kuhlmey et al. 2009) and all-solid PBFs (Jin et al. 2007a, b). However, PBF-based gratings have not been reported in air-core PBFs until the success in writing a high-quality LPFG in an air-core PBF (Wang et al. 2008a). After that, these air-core PBF-based LPFGs were further developed as gas-pressure sensors by the use of gas-pressure-induced deformation on the LPFG (Tang et al. 2015) or refractive index change in the air core of the PBF (Tang et al. 2017).

Since almost 100% of the light propagates in the air holes of an air-core PBF and not in the glass (Smith et al. 2003), PBF-based gratings offer a number of unique features including: high dispersion, low nonlinearity, reduced environmental sensitivity, unusual mode coupling, and new possibilities for long-distance light-matter interactions (by incorporating additional materials into the air-holes). Bandgap-based grating in air-core PBFs, therefore, represent an important platform technology with manifest applications in areas such as communications, fiber lasers and sensing. Periodic index modulations are usually required to realize mode coupling in in-fiber gratings. Although this presents no difficulties in conventional glass fibers (Davis et al. 1998a; Rao et al. 2003a), solid-core PCFs (James and Tatam 2003; Kakarantzas et al. 2002; Wang et al. 2007a), and solid-core PBFs (Steinvurzel et al. 2006a, b; Iredale et al. 2006; Kuhlmey et al. 2009; Jin et al. 2007a, b), it is very difficult, even impossible, to directly induce index modulations in an air-core PBF due to the air core structure, thereby seriously obstructing the development of PBF-based gratings over the past decade.

Wang et al. reported gratings written in an air-core PBF by the use of a focused CO_2 laser beam to periodically deform/perturb air holes along the fiber axis in 2008 (Wang et al. 2008a), as shown in Fig. 10. This reveals that it is experimentally possible to write a grating in an air-core PBF. Both the excellent stability of CO_2 laser power and the good repeatability of optical scanning are very critical to writing a high-quality grating in an air-core PBF. An experimental setup being similar to that in Fig. 2 was used to write an LPFG in an air-core PBF (Crystal-Fiber's HC-1500-02). Compared with the fabrication parameters for writing a grating in a solid-core PCF (Wang et al. 2006b, 2007a), a lower average laser power of about 0.2 W and shorter total time of laser irradiation are used to write an LPFG in an air-core PBF (Wang et al. 2008a). The focused CO_2 beam scans periodically the PBF with a line speed of scanning of 2.9 mm/s, causing the ablation of glass on the fiber surface and the partial or complete collapse of air holes in the cladding due

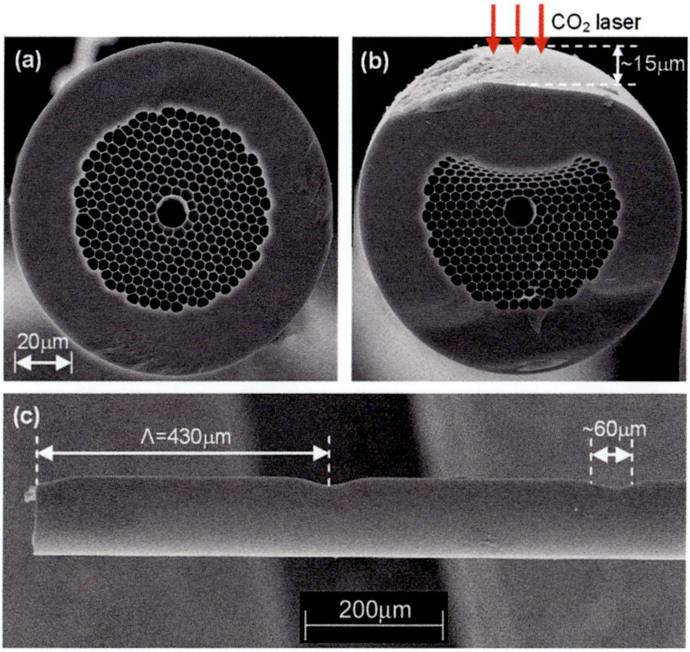

Fig. 10 Cross-section image of an air-core PBF (**a**) before and (**b**) after CO$_2$ laser irradiating, (**c**) side image of an LPFG written in the air-core PBF, where about two periods of the LPFG are illustrated (Wang et al. 2008a)

to the CO$_2$-laser-induced local high temperature, as shown in Fig. 10. The outer rings of air holes in the cladding, facing to the CO$_2$ laser irradiation, were largely deformed; however, little or no deformation were observed in the innermost ring of air-holes and in the air core. As a result, periodic index modulations are achieved along the fiber axis due to the photoelastic effect, thus creating a novel LPFG in the air-core PBF. For the LPFG written in air-core PBF, periodic perturbations of the waveguide (geometric) structure could be the dominant factor that causes resonant mode coupling, although the stress-relaxation-induced index variation may also contribute a little.

Wu et al. reported the inscription of high-quality LPFGs in simplified hollow-core PBFs using CO$_2$-laser-irradiation method (Wu et al. 2011). The PBFs are composed of a hollow hexagonal core and six crown-like air holes. These LPFGs are originated from the strong mode-coupling between the LP$_{01}$ and LP$_{11}$ core modes. A dominant physical mechanism for the mode coupling is confirmed to be the periodic microbends rather than the deformations of the cross section or other common factors. In addition, the LPFGs are highly sensitive to strain and nearly insensitive to temperature and are promising candidates for gas sensors and nonlinear optical devices.

Normal LPFGs written in the index-guiding fibers have a positive relationship between resonant wavelength and grating pitch (Vengsarkar et al. 1996a, b; Erdogan 1997; Eggleton et al. 1997). In contrast, the LPFGs written in the bandgap-guiding

Fig. 11 (**a**) Transmission spectra of six LPFGs, with different grating pitches, written in an air-core PBF. (**b**) The relationship between the pitch of each LPFG and the corresponding research wavelength, where two attenuation dips for each LPFG are observed from 1500 to 1680 nm, indicating that the fundamental mode is coupled to two different higher order modes (Wang et al. 2008a)

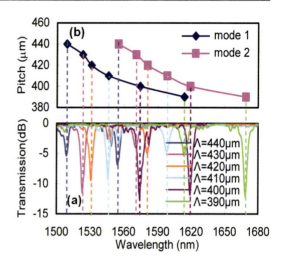

air-core PBF have distinct phase matching condition as function of wavelength. As shown in Fig. 11, the resonant wavelengths of the LPFGs written in an air-core PBF decrease with the increase in grating pitch, which is opposite to the LPFGs written in the index-guiding fibers (Vengsarkar et al. 1996a, b). Moreover, this PBF-based LPFG has unique optical properties such as very large PDL, large strain sensitivity, and very small sensitivity or insensitivity to temperature, bend and external refractive index, as shown in Fig. 4 in Wang et al. (2008a). Further investigations are being done to well understand resonant mode coupling and unique optical properties in the gratings written in air-core PBFs.

Grating Formation Mechanisms

Refractive Index Modulations

Possible mechanisms for refractive index modulation in the CO_2-laser-induced LPFGs could be attributed to residual stress relaxation (Kim et al. 2001, 2002; Li et al. 2008), glass densification (Grubsky and Feinberg 2005; Liu et al. 2009a; Morishita and Kaino 2005; Morishita et al. 2002), and/or physical deformation, depending on the types of the employed fibers and on the practical fabrication techniques. A few methods have been demonstrated to measure the refractive index modulation in the CO_2-laser-induced LPFGs (Li et al. 2008; Hirose et al. 2008).

Residual Stress Relaxation

Residual stress relaxation is found to be the main mechanism for the refractive-index change in the CO_2-laser-induced LPFGs written in optical fibers drawn at high drawing forces (Liu et al. 2009a; Kim et al. 2001, 2002; Li et al. 2008). Residual stress is formed in optical fibers during the drawing process, resulting mainly from a superposition of thermal stress caused by a difference in thermal expansion coefficients between core and cladding and mechanical stress caused by a difference in the viscoelastic properties of the two regions (Kim et al. 2001, 2002; Paek and Kurkjian 1975). Such residual stress can change refractive index in the fibers through the stress-optic effect and thus affect the optical properties of the fibers. Mechanical stress in a CO_2-laser-induced LPFG written in a Ge-B-codoped fiber can be fully relaxed by CO_2 laser irradiation. Consequently, only thermal stress remains in the fiber core due to a mismatch of the thermal expansion coefficients of the fiber core and cladding. Residual stress relaxation usually results in a decrease of refractive index in the fibers. And the efficiency of refractive-index decrease depends strongly on the types of fiber and can be enhanced linearly with the drawing force during the drawing process of the fiber (Kim et al. 2001, 2002). For example, the refractive-index change, resulting from residual stress relaxation, in a Ge-B-codoped fiber drawn at 0.53, 1.38, 2.50, and 3.43 N was measured to be -3.6×10^{-5}, -8.0×10^{-5}, -1.7×10^{-4}, and -2.1×10^{-4}, respectively (Kim et al. 2002). Another experiment shows that residual stress relaxation results in a larger refractive index change of -7.2×10^{-4} in a Corning SMF-28e fiber (Li et al. 2008).

Glass Structure Change

The changes of glass structure (glass volume increase and glass densification) play the dominant mechanisms in the CO_2-laser-induced LPFGs written in a commercial boron-doped single-mode fiber (Grubsky and Feinberg 2005; Liu et al. 2009a; Morishita and Kaino 2005; Morishita et al. 2002). For an unannealed fiber or a fiber annealed at a temperature lower than about 380 °C, glass densification, resulting in an increase in the refractive index, plays the dominant role (Liu et al. 2009a). For a fiber annealed at a higher temperature than about 400 °C, however, glass volume increase, resulting in a decrease in the refractive index, becomes more important (Liu et al. 2009a). On the other hand, residual stress relaxation in the fiber core, which is the dominant mechanism in a conventional Ge-doped or Ge-B-codoped fiber, plays only a minor role in the boron-doped fiber that has a core with a small residual stress and a low fictive temperature.

For the CO_2-laser-induced LPFGs in the Ge-doped or Ge-B-codoped fibers, e.g., Corning SMF-28 fiber, with large residual stress, the resonance wavelength shifts toward the shorter wavelength with the increase of the laser exposure dose due to the negative index modulation resulting from residual stress relaxation (Kim et al. 2001, 2002; Li et al. 2008). For the CO_2-laser-induced LPFGs in the boron-doped fibers with small residual stress, on the contrary, the resonance wavelength shifts toward the longer wavelength with the increase of the laser exposure dose due to the positive index modulation resulting from glass densification (Grubsky and Feinberg 2005; Liu et al. 2009a; Morishita and Kaino 2005; Morishita et al. 2002).

Physical Deformation

Physical deformation is considered as one of the main mechanisms of CO_2-laser-induced LPFGs. During CO_2 laser irradiation, the fiber usually elongates or tapers based on the so-called "self-regulating" mechanism (Grellier et al. 1998), resulting from constant axial tension and the CO_2-laser-induced local high temperature in the fiber. Thus, the fiber diameter decreases and eventually reaches a critical point at which the fiber elongation stops because no sufficient energy is absorbed to keep the softening temperature. Such physical deformation induces a change of effective refractive index in the fiber, and thus LPFGs are created in the periodically taped optical fibers (Kakarantzas et al. 2001; Xuan et al. 2009; Zhu et al. 2005a; Lei et al. 2006). Moreover, the CO_2-laser-induced high temperature in the fiber causes, not only the fiber elongation and the diameter decrease but also the ablation of glass on the fiber surface and the partial or complete collapses of air holes in the PCFs. As a result, different types of LPFGs are written in conventional glass fibers in which periodic grooves/microbends are created on the fiber surface (Kakarantzas et al. 2001; Wang et al. 2006a; Xuan et al. 2009; Zhu et al. 2005a; Lei et al. 2006), and in microstructured optical fibers, such as solid-core PCFs (Kakarantzas et al. 2002; Wang et al. 2006b, 2007a; Zhu et al. 2003a) and air-core PBF (Wang et al. 2008a), in which air holes are partially or completely collapsed.

Theoretical and experimental investigations show that the micro-deformation-induced LPFGs in pure-silica PCFs have discrete attenuation peaks whose spectral positions are correlated to the beat length between the fundamental mode and the first higher-order mode (Nielsen et al. 2003). Thus, simple description of the modal properties based on the perfectly uniform fiber structure may explain the mode-coupling properties of the micro-deformation-induced LPFGs in pure-silica PCFs (Nielsen et al. 2003).

Asymmetrical Mode Coupling

Asymmetrical mode coupling is one of the most distinctive features in the CO_2-laser-induced LPFGs, resulting in unique optical properties that have a large number of promising sensing and communication applications (Rao et al. 2003a; Wang et al. 2006a), as discussed in sections "Sensing Applications" and "Communication Applications." As shown in Fig. 12a, the laser energy is strongly absorbed on the incident side of the fiber, while the CO_2 laser light ($\lambda = 10.6$ μm) illuminates on the optical fiber (VanWiggeren et al. 2000). The nonuniform absorption results in an asymmetrical refractive index profile within the cross section of the CO_2-laser-induced LPFGs. In other words, a larger refractive index change is induced in the incident side of the fiber than in the opposite side. Consequently, as shown in Fig. 12b, asymmetrical mode coupling occurs in the CO_2-laser-induced LPFGs (Davis et al. 1998b; Wang et al. 2008b; Slavík 2007; Ryu et al. 2003a, b) so that a higher PDL usually are observed in the gratings (Wang et al. 2007a, b; Ryu et al. 2003a). Such asymmetrical mode coupling also results that the responses of the CO_2-laser-induced LPFGs to the applied bending (Rao et al. 2003a; VanWiggeren et al. 2000, 2001; Wang et al. 2004; Wang and Rao 2005), twisting (Wang and Rao 2004a, b; Wang et al. 2005), and transverse loading (Wang et al. 2007c, 2003a; Rao et al. 2003b) depend strongly on the fiber orientations, which is distinct from the gratings written by the UV exposure technique (Bhatia and Vengsarkar 1996; James and Tatam 2003; Bhatia 1999; Davis et al. 1998a). Moreover, the asymmetry in the

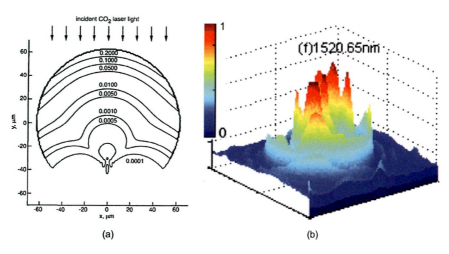

(a)   (b)

Fig. 12 (**a**) Calculated intensities (relative to unity incident intensity) within the cross section of optical fiber during the single-side irradiation of CO_2 laser light (VanWiggeren et al. 2000). (**b**) Asymmetrical mode coupling at the resonant wavelength in the CO_2-laser-induced LPFG (Wang et al. 2008b)

stress distribution of the cladding is found to be much larger than that in the core of the CO_2-laser-induced LPFG (Ryu et al. 2003a). So, the polarization-dependent transmission characteristics of the LPFG are affected mostly by the asymmetric stress distribution in the cladding region rather than the core region.

To solve the problem of asymmetrical mode coupling in LPFGs written by one-sided exposure of a CO_2 laser beam, various methods based on symmetric exposure of the laser beam were demonstrated to achieve axially symmetric LPFG by rotating the fiber during exposure (Oh et al. 2004b), by focusing a ring-shaped CO_2 laser beam axially on the fiber with a concave mirror (Oh et al. 2004b), and by placing a special reflector behind the fiber (Grubsky and Feinberg 2006). The achieved LPFGs with an axially symmetric refractive index profile exhibit clean spectra, very low insertion loss (<0.1 dB), and a low PDL of 0.21 dB (Oh et al. 2004b) or less than 0.12 dB (Grubsky and Feinberg 2006). These values are much lower than the reported measurements of standard CO_2-laser-induced gratings (whose PDL is typically 1 dB) (Davis et al. 1998b; Rao et al. 2003a; Ryu et al. 2003a) and are close to those of UV-written gratings (Vengsarkar et al. 1996a, b). Zhu et al. reported a multi-edge exposure method for decreasing the polarization dependence of the CO_2-laser-induced LPFG and demonstrated a triple-edge-written LPFG with a low PDL of 0.22 dB (Zhu et al. 2007b).

Improvements of Grating Fabrications

The grating fabrications employing a CO_2 laser can be significantly improved by pretreatment techniques, such as hydrogen loading and applying prestrain, and by posttreatment techniques such as applying tensile strain and changing temperature.

Pretreatment Techniques

Hydrogen Loading

As well-known, the use of hydrogen loading can enhance the photosensitivity of the Ge-doped fibers to UV exposure (Vengsarkar et al. 1996a; Lemaire et al. 1993). Hydrogen loading is also found to enhance the writing sensitivity of the CO_2-laser-induced LPFGs (Davis et al. 1998a; Drozin et al. 2000). As shown in Fig. 13, the spectrum characteristics of the CO_2-laser-induced LPFG in the H_2-loaded fiber are dramatically different from those obtained in the untreated fiber (Drozin et al. 2000). The spectrum of the grating written in an H_2-loaded fiber shows a clear coupling to the symmetric modes. In contrast, the coupling peaks for the grating written in the untreated fiber are much broader and the overall background losses are higher, especially in the 1550 nm wavelength range. The poor quality of the LPFG written in the non-H_2-loaded fiber results from the large laser exposure power required to create such a grating (Drozin et al. 2000).

Applying Prestrain

Viscoelastic strains can be frozen into the fiber during drawing an optical fiber, or by cooling down a drawn fiber from its fictive temperature under tension (Yablon et al. 2003, 2004; Yablon 2004). Such a frozen-in viscoelasticity has a significant impact on the refractive index profile of optical fibers. Hence it is possible to change frozen-in viscoelasticity to modulate refractive index in the fiber via heat treatment of a post-draw fiber under tension. Numerous experiments thus reveal that the efficiency of grating fabrication employing a CO_2 laser can be enhanced by means of applying a tensile strain to the employed fiber during CO_2 laser irradiation (Rao et al. 2003a; Liu and Chiang 2008a, b, Lee and Chiang 2009a, b; Wang et al. 2006a), which may be due to the frozen-in viscoelasticity in the fiber. For

Fig. 13 Transmission spectra of the CO_2-laser-induced LPFGs in H_2-loaded and untreated standard fiber (grating pitch: 450 mm), (i) H_2-loaded fiber with an exposure power of 0.5 W, (ii) untreated standard fiber with an exposure power of 2.3 W (Drozin et al. 2000)

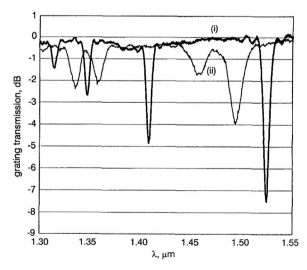

example, as shown in Figs. 2 and 3, one of the fiber ends was tensed by a small weight to provide an external tensile strain in the fiber in advance, enhancing the efficiency of the grating fabrication (Rao et al. 2003a; Wang et al. 2006a). Liu et al. investigated experimentally the effect of the applied tension on the writing efficiency (Liu and Chiang 2008a, b; Lee and Chiang 2009a, b). As shown in Fig. 14, the threshold energy density of CO$_2$ laser irradiation required to write a grating decreases dramatically with an increase in the applied tension (Liu and Chiang 2008a, b). The significant enhancement in the writing efficiency with an applied tension is attributed to the mechanism of frozen-in viscoelasticity in the fiber at the spots of the CO$_2$ laser irradiation (Liu and Chiang 2008b). To write a grating in an unstressed fiber, a sufficiently high temperature in the core is required to relax the residual stress in the core, which demands a high dose of CO$_2$ laser irradiation because the temperature inside the irradiated fiber decreases rapidly from the surface toward the core (Grellier et al. 1998; VanWiggeren et al. 2000). To write a grating in a strongly stressed fiber, however, only a sufficiently high temperature on the fiber surface is required to induce viscoelastic strains in the exposed cladding, which can be achieved with a relatively low energy density of CO$_2$ laser (Liu and Chiang 2008a, b).

However, Zhang et al. recently discovered that the applied tension during the writing process does not always increase the writing efficiency (Zhang and Chiang 2009). For example, during writing an LPFG in a commercial Ge-B-codoped fiber with CO$_2$ laser irradiation, without applying tension to the fiber, a low CO$_2$ laser dosage is sufficient to produce clean strong rejection bands for axially symmetric cladding modes (Zhang and Chiang 2009). On the contrary, with applying a sufficient tension, e.g., a 200 g applied tension, to the fiber, the rejection bands for axially symmetric cladding modes are suppressed due to the counteracting effects

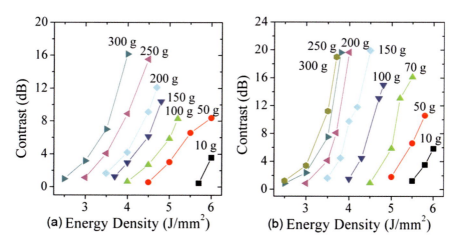

Fig. 14 Dependence of the grating contrast on the CO$_2$ laser energy density for LPFGs written in (**a**) SMF-130 V and (**b**) SMF-28e fibers subject to different applied weights using one scanning cycle of CO$_2$ laser irradiation (Liu and Chiang 2008a, b)

of glass densification and frozen-in strain in the core, while new rejection bands for nonaxially symmetric cladding modes are generated (Zhang and Chiang 2009).

Posttreatment Techniques

Applying Tensile Strain

Experimental investigations show that mode coupling in the CO_2-laser-indcued LPFGs (Zhu et al. 2005c; Slavik 2006), especially the CO_2-laser-carved LPFGs with periodic grooves (Wang et al. 2006a, 2007b), can be greatly enhanced by applying a suitable external stretching force to the LPFGs after writing gratings. For example, as shown in Fig. 15a, the peak transmission attenuation at the resonant wavelength of the CO_2-laser-carved LPFG with periodic grooves was increased from -28.422 to -54.335 dB when a tensile strain of about 355.0 $\mu\varepsilon$ was applied to the grating (Wang et al. 2007b). Such enhancement in mode coupling is attributed to the stretch-induced periodic microbends, as discussed in section "Temperature Sensors." It can also be seen from Fig. 15a that the peak transmission attenuation will reduce if a larger tensile strain, e.g., 525.5 $\mu\varepsilon$, is applied to the grating due to the over-coupling between the fundamental and cladding modes. Thus, this provides an alternative posttreatment technique for enhancing the efficiency of grating fabrication to achieve a desired attenuation spectrum. Moreover, as shown in Fig. 15b, the polarization dependence of the CO_2-laser-induced LPFG is also enhanced or reduced with an increase of the tensile strain. The maximum PDL was increased to 27.2 dB when a tensile strain about 355.0 $\mu\varepsilon$ was applied to the grating. Hence such a posttreatment technique employing a tensile strain can be used to develop an in-fiber polarizer, as discussed in section "Polarizers."

Changing Temperature

Another posttreatment technique has been demonstrated to enhance the efficiency of grating fabrication by means of changing temperature. As shown in Fig. 16a,

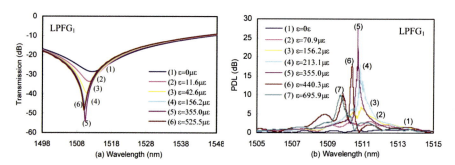

Fig. 15 Evolutions of (**a**) transmission spectrum and (**b**) polarization dependence in the CO_2-laser-carved LPFG with period grooves while increasing the applied tensile strain (Wang et al. 2007b)

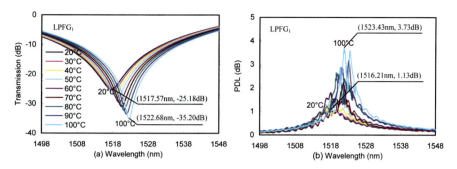

Fig. 16 Evolutions of (**a**) transmission spectra and (**b**) polarization dependence in the CO$_2$-laser-carved LPFG with periodic grooves while the temperature is increased from 20 °C to 100 °C in steps of 10 °C (Wang et al. 2009d)

the peak transmission attenuation at the resonant wavelength of the CO$_2$-laser-carved LPFG with periodic grooves was increased from −25.18 to −35.20 dB when the temperature of the grating rose to 100 °C, resulting from additional refractive index perturbation induced by the thermal strain in the asymmetric LPFG (Wang et al. 2009d). Thus, this provides another alternative posttreatment technique for enhancing the efficiency of grating fabrication to achieve a desired attenuation spectrum. This unique temperature characteristic in the CO$_2$-laser-carved LPFG can be used to develop a practical temperature sensor based on intensity modulation, as discussed in section "Temperature Sensors." Moreover, as shown in Fig. 16b, the polarization dependence of the CO$_2$-laser-carved LPFG was also enhanced with temperature rise, which could find potential sensing or communication applications based on polarization dependence modulation.

Sensing Applications

The CO$_2$-laser-induced LPFGs have many promising sensing applications including for temperature, strain, bend, torsion, pressure, and biochemical sensors. This is because of the sensitivities of resonance wavelength and attenuation amplitude to these measured parameters (Kersey et al. 1997; Bhatia and Vengsarkar 1996; Davis et al. 1998a, b; Rao et al. 2003a; Wang et al. 2003b).

Temperature Sensors

Transmission spectra of the CO$_2$-laser-induced LPFGs shift linearly with the change of temperature and are exceptionally stable even when subjected to very high temperatures of up to 1200 °C (Davis et al. 1999; Zhu et al. 2005c). Therefore, the CO$_2$-laser-induced LPFGs are excellent temperature sensing elements, especially high temperature sensors . In contrast, the sensors employing a UV-laser-induced grating usually have to work under a temperature of about 300 °C. Temperature sensitivities of resonance wavelength of LPFGs depend strongly on the types of the fibers employed and fabrication techniques (Bhatia and Vengsarkar 1996). For example, temperature sensitivities of 58, 10.9, and 2.9 pm/°C were demonstrated in the CO_2-laser-induced LPFGs written in a conventional glass fiber (Rao et al. 2003a), a solid-core PCF (Zhu et al. 2005c), and an air-core PBF (Wang et al. 2008a), respectively. Especially, the resonance wavelength of the LPFG written in a solid-core PCF has a very low strain sensitivity of -0.192 pm/$\mu\varepsilon$ (Zhu et al. 2005c), thus solving the cross-sensitivity problem between temperature and strain in sensing applications.

The coupling from the fundamental mode to the cladding modes in normal LPFGs is hardly affected by the change of temperature except for the shift of transmission spectrum. So the temperature sensors based on these types of LPFGs usually have to measure the shift of resonant wavelength (Bhatia and Vengsarkar 1996; Rao et al. 2003a; Khaliq et al. 2002; Lee and Nishii 1998), which needs expensive equipment, e.g., an optical spectrum analyzer, and thus is not suitable for practical applications. As shown in Fig. 16, the mode coupling in the CO_2-laser-induced LPFG with periodic grooves can be enhanced by increasing the ambient temperature, and the PDL in the LPFG is also changed (Wang et al. 2009d, c). Hence, such an LPFG with periodic grooves is a potential temperature sensor based on the intensity modulation and/or on the PDL modulation. For example, as shown in Fig. 17, a practical temperature sensing system based on intensity modulation was developed by the use of a CO_2-laser-induced LPFG with periodic grooves as a sensing element (Wang et al. 2006c), in which two FBGs were employed to select two different single wavelengths. Such a temperature sensing system exhibits a number of advantages such as convenient intensity measurement, double temperature sensitivity, high resolution of 0.1 °C, and simple configuration.

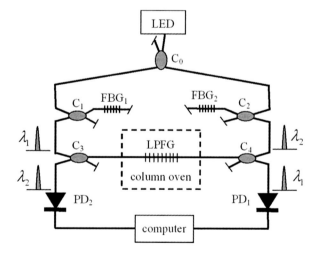

Fig. 17 Schematic diagram of the temperature sensor system based on intensity modulation (Wang et al. 2006c)

Strain Sensors

The CO_2-laser-induced LPFGs with physical deformations exhibit unique optical properties while a tensile strain is applied (Wang et al. 2006a, b, 2007b; Zhu et al. 2005a; Lei et al. 2006; Slavik 2006), thus being excellent strain sensing elements. As shown in Fig. 18, periodic microbends will be induced while a CO_2-laser-induced LPFG with asymmetric grooves is stretched (Wang et al. 2006a, b, 2007b). Such stretch-induced microbends effectively enhance refractive index modulation in the gratings. As a result, such an LPFG have an extremely high strain sensitivity of -102.89 nm/$\mu\varepsilon$ (Wang et al. 2006a, 2007b), which is two orders of magnitude higher than that of other CO_2-laser-induced LPFGs without physical deformations in the same type of fibers (Rao et al. 2003a).

An LPFG strain sensor with a high strain sensitivity of -7.6 pm/$\mu\varepsilon$ and a very low temperature sensitivity of 3.91 pm/°C has been developed by the use of focused CO_2 laser beam to carve periodic grooves on the large mode area PCF (Wang et al. 2006b). Such a strain sensor can effectively reduce the cross-sensitivity between strain and temperature, and the temperature-induced strain error obtained is only 0.5 $\mu\varepsilon$/°C without temperature compensation. Another strain sensor based on a CO_2-laser-induced LPFG pair in a PCF exhibits a high stain sensitivity of about -3 pm/$\mu\varepsilon$ and a low temperature sensitivity of about 4.6 pm/°C (Shin et al. 2009). Theoretical analysis reveals that a simple, low-cost LPG sensor with approximately zero temperature sensitivity but large strain sensitivity could be realized by selecting an appropriate grating period (Zhao et al. 2008). Moreover, temperature-independent strain sensors were demonstrated by measuring the separation change between the two spit peaks of LPFGs written in the twisted optical fibers with a CO_2 laser and exhibited a high strain sensitivity of about 106.7 pm/$\mu\varepsilon$ (Zhu et al. 2009b, 2007d).

Bend Sensors

As discussed in section "Grating Formation Mechanisms," the single-side irradiation of the CO_2 laser beam results in an asymmetrical index profile within

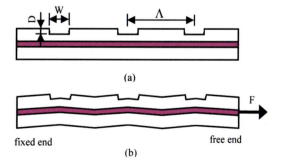

Fig. 18 Schematic diagram of the CO_2-laser-carved LPFG with asymmetric grooves (**a**) before and (**b**) after a stretching force is applied to the grating (Wang et al. 2006a)

(a)

fixed end   free end

(b)

Fig. 19 The shift of resonant wavelength while the CO_2-laser-induced LPFG is bent to different fiber orientations (Wang et al. 2004)

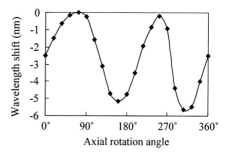

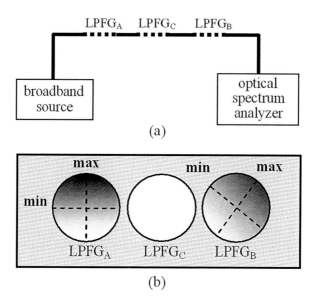

Fig. 20 Schematic diagram of LPFG sensor that can not only measure curvature but also determine bending direction, (**a**) sensor scheme, (**b**) cross section of three LPFGs (Wang and Rao 2005)

the cross section of the achieved gratings. As a result, distinct bend-sensitivity of resonant wavelength is observed while a CO_2-laser-induced LPFG is bent to different fiber orientations (VanWiggeren et al. 2000; Wang et al. 2004; Wang and Rao 2005), as shown in Fig. 19. In other words, the bending responses of the CO_2-laser-induced LPFGs depend strongly on the bending direction, which is distinct from that of the UV-laser-written LPFGs in which nearly symmetrical index modulations are limited in the fiber core. As shown in Fig. 20, a promising bend-sensing system based on the axial rotation dependence of the bending response in the CO_2-laser-induced LPFGs has been developed to measure directly the applied curvature of the engineering structures and to determine simultaneously the bending direction (Wang and Rao 2005). This bend-sensing system consists of one UV-laser-induced LPFG and two CO_2-laser-induced LPFGs. The curvature of the engineering structure is measured by the UV-laser-induced LPFG whose bend-sensitivity is independent on the bending directions, and the bending direction is determined by the two CO_2-laser-induced LPFGs whose bend-sensitivities depend strongly on the bending directions. Similar orientation dependence of the bending response also was observed in FBGs written in the Ge-doped microstructured optical fibers (Wang et al. 2009b).

It is interesting to see from Fig. 19 that the resonant wavelength is insensitive to the curvature change or has a very low bend-sensitivity, while the CO_2-laser-induced LPFG is bent toward special fiber orientations, e.g., about 90° and about 270° orientations. Thus such an LPFG can be used to develop a bend-insensitive sensor to solve the problem of cross-sensitivity between bend and other measurands, such as temperature, strain, or refractive index, which is an unsolved problem for normal grating-based sensors in practice (Wang et al. 2004). On the other hand, the resonant wavelength shifts linearly while the CO_2-laser-induced LPFG is bent toward other fiber orientations, as shown in Fig. 3 in Wang and Rao (2005) and Fig. 7 in Jin et al. (2009). Hence such a grating can also be used to measure directly the applied curvature of engineering structures.

Torsion Sensors

All-fiber torsion sensors have been demonstrated by the use of a corrugated LPFG (Lin et al. 2001a; Wang et al. 2001). These LPFG torsion sensors, however, cannot determine the twisting directions, i.e., whether clockwise or anticlockwise. It is interesting to see from Fig. 21a that the resonant wavelength shifts linearly toward the longer wavelength while a CO_2-laser-induced LPFG is twisted clockwise, whereas it shifts linearly toward the shorter wavelength while the grating is twisted anticlockwise (Wang and Rao 2004a; Wang et al. 2005). Hence such a CO_2-laser-induced LPFG is an excellent in-fiber torsion sensing element that can not only measure directly the applied twisting rate but also determine simultaneously the twisting direction (Wang and Rao 2004a). Furthermore, providing the twisted fiber including a CO_2-laser-induced LPFG is much longer than the grating, as shown in Fig. 21b, the resonant wavelength shifts wavelike toward the longer and shorter wavelength, while the LPFG is twisted clockwise and anticlockwise, respectively (Wang and Rao 2004b; Wang et al. 2005). Such unique torsion characteristics of the CO_2-laser-induced LPFGs owe to the twist-induced right- and left-rotatory elliptical birefringence, resulting from asymmetric refractive index distribution within the cross section of the gratings.

Moreover, Rao et al. reported that the high-order resonant wavelengths of the CO_2-laser-induced ultra-long period fiber gratings have higher torsion sensitivities, which are several times higher than that of the normal LPFG (Rao et al. 2006). An intensity-type demodulation approach used to realize real-time torsion measurement has been demonstrated based on the edge filtering effect of ultra-long period fiber gratings (Rao et al. 2006). The helical LPFG written by CO_2 laser irradiation can also be used as an optical torsion sensor because its attenuation peak shifts with

Fig. 21 (**a**) Resonant
wavelength change while a
CO_2-laser-induced LPFG is
twisted clockwise or
anticlockwise (Wang and Rao
2004a). (**b**) The response of
resonant wavelength to the
torsion providing the twisted
fiber including a
CO_2-laser-induced LPFG is
much longer than the grating
(Wang and Rao 2004b)

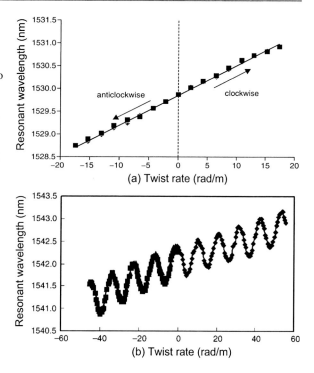

codirectional or contra-directional torsion to the helix (Oh et al. 2004a). In addition, the rejection band of an LPFG written in a heavily twisted single-mode fiber by a CO_2 laser can split into two bands when the twist applied to the fiber is removed after the writing of the grating (Zhu et al. 2009c). Such a grating could find applications as a temperature-insensitive torsion sensor.

Pressure Sensors

Resonant wavelength of optical fiber gratings usually shifts linearly while a transverse load (a pressure) is applied. Thus various grating-based pressure sensors have been developed to measure the applied transverse load (Wang et al. 2009b, e; Liu et al. 1999; Shu et al. 2003; Bjerkan et al. 1997; Lawrence et al. 1997; Bosia et al. 2002). Pressure sensors based on conversional UV-laser-induced gratings usually, however, fail to determine the direction of the applied force due to their symmetric transverse-loading characteristics. The CO_2-laser-induced LPFGs exhibit asymmetric transverse-loading characteristics due to the load-induced birefringence that leads to the rotation of optical principal axes in the gratings (Wang et al. 2007c; Zhu et al. 2006). As shown in Fig. 22, the responses of resonant wavelength of the CO_2-laser-induced LPFG to an applied transverse load depend strongly on the directions of the load, whereas those of the peak transmission attenuation are independent on the loading directions (Wang et al. 2007c). It can be seen from Fig. 23 that while a

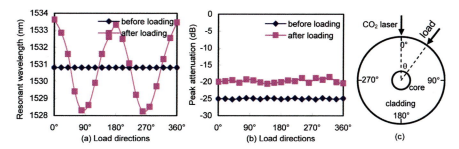

Fig. 22 (**a**) Resonant wavelength (**b**) peak transmission attenuation of the CO$_2$-laser-induced LPFG before and after a constant transverse load of 820.75 N·m^{-1} is applied to the grating along different fiber orientations illustrated by an angle of θ in (**c**), where the direction of CO$_2$ laser irradiation is defined as the 0° orientation at the circle of the LPFG (Wang et al. 2007c)

transverse load is applied to a CO$_2$-laser-induced LPFG along different orientations of the grating, the resonant wavelength may be shifted linearly toward the longer wavelength, the shorter wavelength, or hardly shifted, whereas the absolute value of peak transmission attenuation is linearly decreased with an increase of the applied transverse load, with almost no sensitivity to the loading direction (Wang et al. 2007c). Hence, such a CO$_2$-laser-induced LPFG can be used to develop a promising pressure sensor that can not only measure the pressure strain applied to engineering structures but also determine the direction of the applied pressure (Wang et al. 2007c). A similar pressure sensor was also demonstrated to realize simultaneous measurement of a transverse load (pressure) and its orientation by the use of a FBG written in a multicore fiber (Silva-Lopez et al. 2005). Similar orientation dependence of the transverse-loading characteristics also was observed in FBGs written in the Ge-doped microstructured optical fibers (Wang et al. 2009b, e). A twisted CO$_2$-laser-induced LPFG can weaken the strong orientation-dependence of the transverse load and enhance the transverse-load sensitivity considerably, when compared with a torsion-free LPFG, offering the potential for use as a dynamic load sensor by means of simple intensity detection (Zhu et al. 2006).

For health monitoring of engineering structures and civil infrastructures, measurement accuracy is often affected by the cross-sensitivity between temperature and transverse loading; hence, simultaneous measurement of temperature and transverse loading using a single sensor element is very attractive from the practical point of view. It is interesting to see from Fig. 23 that the resonant wavelength is insensitive to the loading applied to special orientations, e.g., 0°, while the peak transmission attenuation linearly decreases with an increase of the loading. On the other hand, as discussed in section "Band-Rejection Filters," the resonant wavelength linearly shifts with temperature whereas the peak transmission attenuation hardly changes during temperature varying. Based on these unique transverse-load and temperature characteristics, a novel sensor employing a CO$_2$-laser-induced LPFG was developed to measure simultaneously temperature and transverse loading by detecting the resonant wavelength and the peak attenuation of the grating, respectively (Rao et al. 2003a, b). In operation, the side, corresponding to the load-insensitive orientation, of

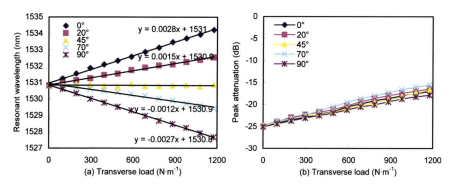

Fig. 23 (a) Resonant wavelength and (b) peak transmission attenuation of the CO_2-laser-induced LPFG as functions of the transverse load applied along orientations of $0°$, $20°$, $45°$, $70°$, and $90°$ on the grating (Wang et al. 2007c)

the LPFG is mounted on the surface of a base. The applied transverse loading alters linearly the peak attenuation and hardly affects the resonant wavelength, whereas the change of temperature results in a linear shift of the resonant wavelength and has no influence on the peak attenuation. Therefore, such a sensor employing only one LPFG avoids the cross-sensitivity problem between temperature and transverse loading.

Biochemical Sensors

Both resonant wavelength and attenuation peak of LPFGs are very sensitive to the change of external refractive index due to the dependence of the phase matching condition upon the effective refractive index of the cladding modes (Bhatia and Vengsarkar 1996; Davis et al. 1998a). $LPFG_S$ can thus be used for biochemical sensing based on the evanescent-wave detection principle (Rindorf et al. 2006; DeLisa et al. 2000; Fini 2004; Jensen et al. 2004, 2005). As shown in Fig. 24, a promising biochemical sensor based on an LPFG written in a PCF by a CO_2 laser has been demonstrated to detect the average thickness of a layer of biomolecules within a few nm (Rindorf et al. 2006). By measuring the thickness, the technique may thus be used for label-free detection of selective binding of biomolecules such as DNA and proteins. A detecting sensitivity of approximately 1.4 nm/1 nm was achieved in terms of the shift in resonance wavelength in nm per nm thickness of biomolecule layer. The presented experiments open up new possibilities for PCFs to be used in surface chemistry studies, such as in drug discovery and as sensors in point-of-care devices and other laboratory equipment.

Various techniques have been demonstrated to enhance the sensitivities of LPFG-based refractometers (Zhu et al. 2003b; 2007a, 2009a) and to solve the cross-sensitivity between temperature and refractive index (Zhu et al. 2005b). A CO_2-laser-induced LPFG in a microfiber drawn by taping technique has a high sensitivity of about 1900 nm/RI to external refractive index (Xuan et al. 2009).

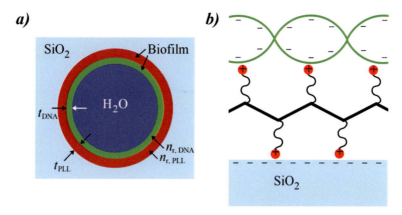

Fig. 24 Schematic diagram of biochemical sensing with a CO_2-laser-induced LPFG in a PCF. (**a**) A hole of a PCF. The side is coated with poly-L-lysine (PLL) and DNA in monolayers of various thickness (tDNA and tPLL) and refractive indices (nr,DNA and nr,PLL). (**b**) The molecular structure of poly-L-lysine (triangle line with '+' circles) with positive charges immobilized onto the negatively charged silica surface (SiO_2) (Rindorf et al. 2006)

This value is about 60 times bigger than an in-fiber Michelson interferometer made from abrupt taper of SMF (Tian et al. 2008) and 5 times bigger than an LPFG specially designed for sensitivity enhancement (Minkovich et al. 2005). Another highly sensitive refractometer based on a long-period grating in a large-mode-area PCF was reported, where the maximum sensitivity is 1500 nm/refractive index unit at a refractive index of 1.33, and the minimal detectable index change is 2×10^{-5} (Rindorf and Bang 2008).

Communication Applications

The CO_2-laser-induced LPFGs have found many promising applications in optical telecommunications such as band-rejection filters (Vengsarkar et al. 1996a; Braiwish et al. 2004; Shin et al. 2007; Zhu et al. 2009d; Rao and Zhu 2006), gain equalizers for erbium-doped fiber amplifiers (EDFAs) (Vengsarkar et al. 1996b; Rao et al. 2002, 2004, 2005; Zhu et al. 2007e), in-fiber polarizers (Wang et al. 2007a; Yang et al. 2009), couplers (Liu et al. 2007a; Sasaki 2001), and mode converters (Hill et al. 1990) because of their bandstop filtering capability and selective mode coupling between the fundamental mode and the cladding modes.

Band-Rejection Filters

Attenuation property of LPFGs at the desired wavelength makes them to be used as promising band-rejection filters , as shown in Fig. 25. Many all-fiber band-rejection filters based on an LPFG have been reported (Vengsarkar et al. 1996a;

Fig. 25 Typical transmission
spectrum of an LPFG in
which the fundamental
guided mode is coupled into
forward propagating cladding
modes (Vengsarkar et al.
1996a)

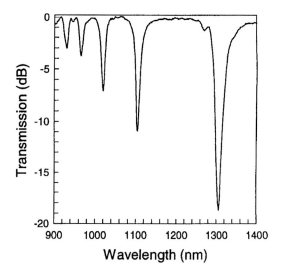

Fig. 25 Typical transmission
spectrum of an LPFG in
which the fundamental
guided mode is coupled into
forward propagating cladding
modes (Vengsarkar et al.
1996a)

Fig. 26 Schematic of the bandwidth-tunable all-fiber band-rejection filters based on a CO_2-laser-induced helicoidal LPFG pair of opposite helicities (Shin et al. 2007)

Eggleton et al. 1997; Rao et al. 2004; Wang et al. 2010a, b). Normal LPFG-based filters, however, allow tuning of only resonance wavelength or attenuation amplitude (Vengsarkar et al. 1996a; Rao et al. 2004; Lin and Wang 2001). Shin et al. developed a bandwidth-tunable band-rejection filter consisting of a CO_2-laser-induced helicoidal LPFG pair of opposite helicities (Shin et al. 2007), as shown in Fig. 26. Such a filter enables unique rejection bandwidth tuning over more than 14 nm at the rejection level of 15 dB, with low insertion loss and PDL achieved by adjusting torsion stress. Another bandwidth-tunable band-rejection filter was demonstrated by putting a normal LPFG in series with a rotary LPFG written in a twisted single mode fiber by CO_2 laser pulses (Zhu et al. 2009d). As shown in Fig. 27, a bandwidth tuning of about 16.3 nm at a rejection level of about 15 dB with a low PDL of less than 0.9 dB was realized by applying a suitable torsion to the rotary LPFG, resulting from the wavelength splitting effect of the grating. In addition, a CO_2-laser-induced LPFG with special apodization was demonstrated to achieve a filter with a top rejection bandwidth of up to 10 nm and a top flatness of less than 0.5 dB (Rao and Zhu 2006).

The LPFG-based filters with an extremely deep attenuation dip recently find interesting applications, e.g., subpicosecond pulse-shaping (Kulishov and Azana

Fig. 27 Bandwidth tuning of the LPFG-based filter by applying a suitable torsion to the rotary LPFG (Zhu et al. 2009d)

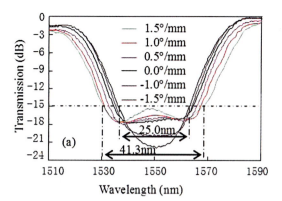

2005; Azana and Kulishov 2005). However, the experimentally attainable level of control, especially for the coupling coefficient, usually does not allow an LPFG to reach the resonant attenuation values over 35 dB (Dubov et al. 2005). Fortunately, as discussed in section "Improvements of Grating Fabrications," the peak attenuation of the CO$_2$-laser-induced LPFGs can be enhanced to an extremely large value by applying a suitable tensile strain to obtain the optimum coupling, $kL = 2\pi$, at the resonant wavelength, where k is the coupling coefficient and L is the grating length (Wang et al. 2007b; Slavik 2006). For example, an extremely deep LPFG with a peak attenuation of more than 60 dB have been achieved to realize an exact π-phase shift at the resonance wavelength by the use of the prestrain technique (Slavik 2006). Another interest application of CO$_2$ laser irradiation is to produce a phase-shifted bandpass filter on a FBG (Xia et al. 2005). Such a filter has one or multiple narrow passbands within the stopband of the FBG, depending on the position and number of the CO$_2$ laser pulses irradiation on the FBG.

Gain Equalizers

A typical application of LPFGs is for tunable gain equaliz ation of EDFAs (Vengsarkar et al. 1996b; Rao et al. 2002, 2004, 2005; Zhu et al. 2007e; Costantini et al. 1999). However, the resonant wavelength and the attenuation amplitude of normal LPFG-based gain equalizers are changed simultaneously during tuning operation. As a result, after one parameter is adjusted, another parameter has to be readjusted. And such a readjustment usually has to be repeated many times to achieve a desired gain spectrum. Resonant wavelength of a CO$_2$-laser-induced LPFG hardly shifts, while its attenuation amplitude linearly changes with the increase of a transverse load applied to special fiber orientation (Rao et al. 2004; Wang et al. 2003a), as shown in Fig. 28. On the other hand, the attenuation amplitude of the LPFG is hardly changed while the resonant wavelength linearly shifts during temperature varying. Based on these unique temperature and transverse-load characteristics of the CO$_2$-laser-induced LPFG, a promising tunable gain equalizer

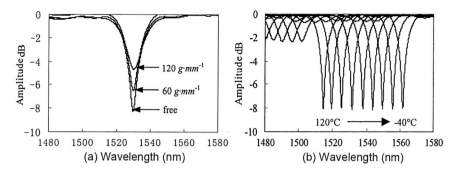

Fig. 28 Transmission spectrum evolution of a CO_2-laser-induced LPFG while (**a**) a transverse-load is applied to the load-insensitive fiber orientation and (**b**) the temperature is changed (Wang et al. 2003a)

Fig. 29 Typical gain profiles of an EDFA employing an LPFG-based gain equalizer: (**a**) original gain spectrum; (**b**) flattened gain spectrum (Wang et al. 2003a)

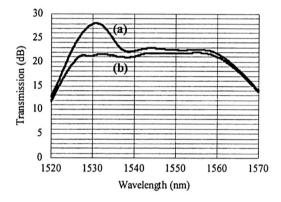

has been developed to realize dynamic gain equalization in EDFA systems (Wang et al. 2003a), in which the resonant wavelength and the attenuation amplitude of the LPFG can be independently tuned by means of changing the applied load and the temperature, respectively. As shown in Fig. 29, a gain flatness of 0.5 dB was achieved over a bandwidth of 33 nm in an EDFA system by the use of such an LPFG-based gain equalizer (Wang et al. 2003a).

The equalization efficiency of such a dynamic gain equalizer can be enhanced by employing a twisted LPFG (Zhu et al. 2007e). Another all-fiber dynamic gain equalizer based on an LPFG written in a bend-insensitive fiber was developed to flatten the gain of EDFAs by controlling the bending curvature of the LPFG directly (Rao et al. 2005). A similar tuning technique for an LPFG-based filter was proposed in Braiwish et al. (2004), in which prototype devices capable of variable attenuation at a fixed wavelength, wavelength tuning at a constant attenuation, and combinations of these spectral characteristics were demonstrated in the CO_2-laser-induced LPFGs.

Polarizers

Compared with bulk waveguide polarizers, in-fiber polarizers are desirable devices in all-fiber communication systems because of their low insertion loss and compatibility with optical fiber (Bergh et al. 1980; Dyott et al. 1987; Zhou et al. 2005). As discussed in section "Asymmetrical Mode Coupling," the CO$_2$-laser-induced LPFGs have clear polarization dependence due to their asymmetric refractive index profile, resulting from single-side laser irradiation, within the cross section of the gratings (Wang et al. 2006a; 2007b; Ryu et al. 2003a), thus being potential in-fiber polarizing device. Moreover, as discussed above in section "Posttreatment Techniques," the polarization dependence of the CO$_2$-laser-induced LPFG with periodic grooves can be greatly enhanced by applying a tensile strain (Wang et al. 2007b) or increasing temperature (Wang et al. 2009d). So a promising in-fiber polarizer based on an LPFG was developed by the use of a focused CO$_2$ laser beam to collapse or perturb periodically air holes in a pure-silica PCF (Wang et al. 2007a). In practical operation, a stretch strain is applied to the LPFG-based polarizer to enhance the polarization dependence of the grating. As a result, the maximum PDL and the maximum polarization extinction ratio of the LPFG are increased to 27.27 and 22.83 dB, respectively, as shown in Fig. 30. Such an LPFG-based polarizer thus exhibits a high polarization extinction ratio of more than 20 dB over a wide wavelength range of about 11 nm near the communication wavelength of 1550 nm (Wang et al. 2007a). Moreover, this polarizer has a very low temperature sensitivity of 3.9 pm/°C, which overcomes the disadvantages of the temperature sensitivity in other in-fiber polarizers created in conventional glass fibers.

An alternative fabrication method has been developed to enhance the polarization dependence of the LPFG by the use of defocused CO$_2$ laser pulse scanning, which is realized by moving the laser beam focus plane slightly away from the fiber cladding surface (Yang et al. 2009). For the cascaded LPFGs fabricated by this technique, the

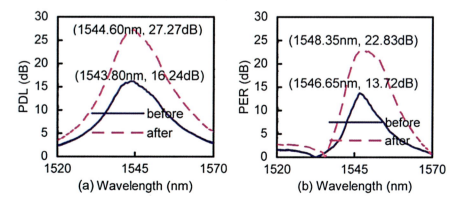

Fig. 30 (**a**) Polarization dependent loss (PDL) and (**b**) polarization extinction ratio (PER) of the LPFG-based polarizer before (dashed curve) and after (solid curve) a stretch strain of 500 $\mu\varepsilon$ is applied (Wang et al. 2007a)

maximum PDL obtained can be largely increased to 17.8 dB (Yang et al. 2009), which is larger than that reported in Lee et al. (2002). A rocking filter with an extinction ratio as high as -23.5 dB and an insertion loss of less than 0.02 dB was demonstrated in a polarization-maintaining PCF (Kakarantzas et al. 2003). Such a filter was fabricated by periodic mechanical twisting of the fiber and heating with a scanned CO_2 laser beam.

Couplers

To enhance the functionality of LPFGs, the structure of two or three parallel LPFGs has been proposed as a broadband optical coupler or optical add/drop multiplexer for applications in coarse wavelength division multiplexing systems (Chiang 2009; Chiang et al. 2000, 2004; Grubsky et al. 2000; Chan and Chiang 2006; Liu and Chiang 2006; Han et al. 2006; Liu et al. 2007b; b; Kim et al. 2007), where light launched into one fiber is coupled into another fiber through evanescent-field coupling between the cladding modes of the two parallel gratings. All of these couplers employ gratings written by the conventional UV laser exposure technique (Chiang 2009; Chiang et al. 2000, 2004; Grubsky et al. 2000; Chan and Chiang 2006; Liu and Chiang 2006; Han et al. 2006; Liu et al. 2007b, 2009b; Kim et al. 2007). Liu et al. investigated experimentally all-fiber couplers formed by two parallel LPFGs written by high-frequency CO_2-laser pulses and found asymmetrical coupling efficiency (Liu et al. 2007a), as shown in Fig. 31. For gratings written in the standard SMFs with a CO_2 laser, the coupling efficiency depends strongly on the fiber orientations with the strongest coupling obtained when the exposed sides of the fibers face each other. For gratings written in the boron-doped SMFs with a CO_2 laser, in contrast, the coupling efficiency is independent of the fiber orientation. A record-high peak coupling efficiency of about 86% was achieved in a coupler formed with two LPFGs written in boron-doped fibers by using a suitable surrounding refractive index and offset distance between such two gratings (Liu et al. 2007a). Hence, the CO_2-laser-induced LPFGs in the boron-doped fiber could be employed potentially for the realization of efficient broadband optical couplers. Additionally, the CO_2 laser irradiation technique can also be used to fabricate a fused fiber microcoupler with an extremely short interaction length of 200 μm (Kakarantzas et al. 2001) and a FBG-based coupler (Sasaki 2001).

Mode Converters

Mode-division multiplexing (MDM), which utilizes spatial linear polarization (LP) modes or orbital angular momentum (OAM) in the few-mode fiber (FMF) to carry independent data channels or enable efficient multiplexing, has the potential to increase the transmission capacity in optical communication systems. A key component in the MDM systems is a mode converter, which makes conversion between two spatial modes. To date, many mode converters have been demonstrated,

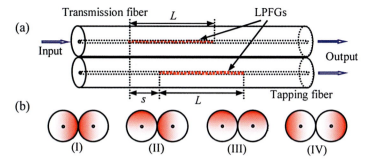

Fig. 31 (**a**) Schematic diagram of two parallel identical LPFGs. (**b**) Four different fiber orientations for the two LPFGs, in which the shadow areas represent the sides exposed to the CO$_2$ laser pulses (Liu et al. 2007a)

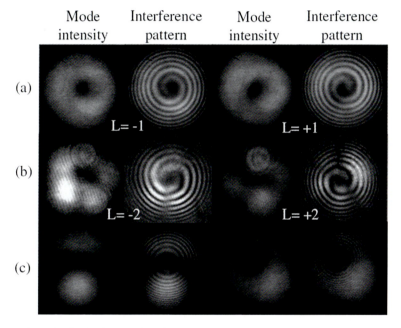

Fig. 32 Intensity distributions and interference patterns of the three single LPFG mode converters: (**a**) $LP_{01} - LP_{11}$; (**b**) $LP_{01} - LP_{21}$; and (**c**) $LP_{01} - LP_{02}$ (Zhao et al. 2017)

which can be divided into three categories: bulk-optic, waveguide, and all-fiber mode converters. All-fiber mode converters use fiber-based photonic lanterns, mode selective couplers, FBGs, and LPFGs. Recently, various all-fiber mode converters based on LPFGs written in different types of FMFs have been reported (Wang et al. 2015; Dong and Chiang 2015; Zhao et al. 2016, 2017; Zhang et al. 2018).

For example, Zhao et al. reported an all-fiber mode converter based on CO$_2$-laser-inscribed LPFGs in the FMF (four-mode step index fiber, OFS). As shown in Fig. 32, mode conversion between the fundamental core mode and different higher-order core modes (LP_{11}, LP_{21}, and LP_{02} modes) can be realized via a single LPFG with an efficiency of 99% at the resonance wavelength. Moreover, optimized mode conversion between the LP_{01} and LP_{21} modes can be realized by cascading two LPFGs with different grating pitches. The maximum conversion efficiency is estimated to be $\sim$99.5% at 1553 nm. The orbital angular momentum states with different topological charges (± 1; ± 2) are demonstrated experimentally (Zhao et al. 2017).

Conclusions

Compared with UV laser exposure, CO_2 laser irradiation is much more flexible and is of lower cost because no photosensitivity or any complicated pretreated process is required for fabricating long period gratings in the glass fibers. Moreover, the CO_2 laser irradiation process can be controlled to generate complicated grating profiles via the well-known point-to-point technique without using any expensive masks. CO_2 laser can write high-quality LPFGs in different types of optical fibers such as conventional glass fibers, solid-core photonic crystal fibers, and air-core photonic bandgap fibers. Possible mechanisms of refractive index modulations in CO_2-laser-induced LPFGs are due to residual stress relaxation, glass structure changes, and physical deformation. Single-side irradiation of CO_2 laser results in asymmetrical mode coupling in the gratings. The LPFG fabrications could be improved by various techniques: pretreatment such as hydrogen loading and applying prestrain, and by posttreatment techniques such as applying tensile strain and changing temperature. CO_2-laser-induced LPFGs have found promising sensing applications, such as temperature, strain, bend, torsion, pressure, and biochemical sensors, and communication applications such as band-rejection filters, gain equalizers, polarizers, couplers, and mode converters.

Acknowledgments This work was supported by National Natural Science Foundation of China (NSFC) (grant nos. 61425007, 61635007, 61875128, 61377090); Guangdong Natural Science Foundation (grant nos. 2015B010105007, 2014A030308007); Science and Technology Innovation Commission of Shenzhen (grant nos. JCYJ20170412105604705, JCYJ20160427104925452); and Development and Reform Commission of Shenzhen Municipality Foundation.

References

J. Azana, M. Kulishov, All-fibre ultrafast optical differentiator based on π-phase-shifted long-period grating. Electron. Lett. **41**, 1368–1369 (2005)

R.A. Bergh, H.C. Lefevre, H.J. Shaw, Single-mode fiber optic polarizer. Opt. Lett. **5**, 479–481 (1980)

V. Bhatia, Applications of long-period gratings to single and multi-parameter sensing. Opt. Express **4**, 457–466 (1999)

V. Bhatia, A. Vengsarkar, Optical fiber long-period grating sensors. Opt. Lett. **21**, 692–694 (1996)

M. Bibra, A. Roberts, Long period fibre gratings made with focused ion beam irradiation, in *The 14th International Conference on Optical Fiber Sensors*, (2000), pp. 604–607

M. Bibra, A. Roberts, J. Canning, Fabrication of long-period fiber gratings by use of focused ion-beam irradiation. Opt. Lett. **26**, 765–767 (2001)

L. Bjerkan, K. Johannessen, X. Guo, Measurements of Bragg grating birefringence due to transverse compressive forces, in *The 12th International Conference on Optical Fiber Sensors, OTuC7*, (1997), pp. 60–63

F. Bosia, P. Giaccari, M. Facchini, J. Botsis, H. Limberger, R. Salathe, Characterization of embedded fibre Bragg grating sensors written in high-birefringent optical fibres subjected to transverse loading, in *Smart Structures and Materials: Smart Sensor Technology and Measurement Systems*, Proc. of SPIE, vol. 4694, (2002), pp. 175–186

M. Braiwish, B. Bachim, T. Gaylord, Prototype CO$_2$ laser-induced long-period Fiber grating variable optical attenuators and optical tunable filters. Appl. Opt. **43**, 1789–1793 (2004)

G. Brambilla, V. Pruneri, L. Reekie, D. Payne, Enhanced photosensitivity in germanosilicate fibers exposed to CO2 laser radiation. Opt. Lett. **24**, 1023–1025 (1999)

F.Y.M. Chan, K.S. Chiang, Transfer-matrix method for the analysis of two parallel dissimilar nonuniform long-period fiber gratings. J. Lightwave Technol. **24**, 1008–1018 (2006)

H. Chan, E. Perez, F. Alhassen, I. Tomov, H. Lee, Ultra-compact long-period fiber grating and grating pair fabrication using a modulation-scanned CO$_2$ laser, in *Optical Fiber Communication and the National Fiber Optic Engineers Conference (OFC/NFOEC 2007)*, (2007), pp. 1–3

K.S. Chiang, Development of long-period fiber grating coupling devices. Appl. Opt. **48**, F61–F67 (2009)

K.S. Chiang, Y. Liu, M.N. Ng, S. Li, Coupling between two parallel long-period fibre gratings. Electron. Lett. **36**, 1408–1409 (2000)

K.S. Chiang, F.Y.M. Chan, M.N. Ng, Analysis of two parallel long-period fiber gratings. J. Lightwave Technol. **22**, 1358–1366 (2004)

D.M. Costantini, C.A.P. Muller, S.A. Vasiliev, H.G. Limberger, R.P. Salathe, Tunable loss filter based on metal-coated long-period fiber grating. IEEE Photonic Tech. L. **11**, 1458–1460 (1999)

D. Davis, T. Gaylord, E. Glytsis, S. Kosinski, S. Mettler, A. Vengsarkar, Long-period fibre grating fabrication with focused CO$_2$ laser pulses. Electron. Lett. **34**, 302–303 (1998a)

D. Davis, T. Gaylord, E. Glytsis, S. Mettler, CO$_2$ laser-induced long-period fibre gratings: spectral characteristics, cladding modes and polarisation independence. Electron. Lett. **34**, 1416–1417 (1998b)

D. Davis, T. Gaylord, E. Glytsis, S. Mettler, Very-high-temperature stable CO$_2$-laser-induced long-period fibre gratings. Electron. Lett. **35**, 740–742 (1999)

M.P. DeLisa, Z. Zhang, M. Shiloach, S. Pilevar, C.C. Davis, J.S. Sirkis, W.E. Bentley, Evanescent wave long-period fiber Bragg grating as an immobilized antibody biosensor. Anal. Chem. **72**, 2895–2900 (2000)

H. Dobb, K. Kalli, D. Webb, Temperature-insensitive long period grating sensors in photonic crystal fibre. Electron. Lett. **40**, 657–658 (2004)

J. Dong, K. Chiang, Temperature-insensitive mode converters with CO$_2$-laser written long-period fiber gratings. IEEE Photon. Technol. Lett. **27**, 1006–1009 (2015)

L. Drozin, P.Y. Fonjallaz, L. Stensland, Long-period fibre gratings written by CO$_2$ exposure of H$_2$-loaded, standard fibres. Electron. Lett. **36**, 742–744 (2000)

M. Dubov, I. Bennion, S.A. Slattery, D.N. Nikogosyan, Strong long-period fiber gratings recorded at 352 nm. Opt. Lett. **30**, 2533–2535 (2005)

R.B. Dyott, J. Bello, V.A. Handerek, Indium-coated D-shaped-fiber polarizer. Opt. Lett. **12**, 287–289 (1987)

B. Eggleton, R. Slusher, J. Judkins, J. Stark, A. Vengsarkar, All-optical switching in long-period fiber gratings. Opt. Lett. **22**, 883–885 (1997)

B. Eggleton, P. Westbrook, R. Windeler, S. Spalter, T. Strasser, Grating resonances in air-silica microstructured optical fibers. Opt. Lett. **24**, 1460–1462 (1999)

B. Eggleton, P. Westbrook, C. White, C. Kerbage, R. Windeler, G. Burdge, Cladding-mode-resonances in air-silica microstructure optical fibers. J. Lightwave Technol. **18**, 1084–1100 (2000)

T. Erdogan, Fiber grating spectra. J. Lightwave Technol. **15**, 1277–1294 (1997)

E. Fertein, C. Przygodzki, H. Delbarre, A. Hidayat, M. Douay, P. Niay, Refractive-index changes of standard telecommunication fiber through exposure to femtosecond laser pulses at 810 cm. Appl. Opt. **40**, 3506–3508 (2001)

J.M. Fini, Microstructure fibres for optical sensing in gases and liquids. Meas. Sci. Technol. **15**, 1120–1128 (2004)

C. Fu, X. Zhong, C. Liao, Y.P. Wang, Y. Wang, J. Tang, S. Liu, Q. Wang, Thin-core-fiber based long period fiber grating for high sensitivity refractive index measurement. IEEE Photonics J. **7**, 7103208 (2015)

M. Fujimaki, Y. Ohki, J. Brebner, S. Roorda, Fabrication of long-period optical fiber gratings by use of ion implantation. Opt. Lett. **25**, 88–89 (2000)

P. Geng, W. Zhang, S. Gao, H. Zhang, J. Li, S. Zhang, Z. Bai, L. Wang, Two-dimensional bending vector sensing based on spatial cascaded orthogonal long period fiber. Opt. Express **20**, 28557–28562 (2012)

A.J.C. Grellier, N.K. Zayer, C.N. Pannell, Heat transfer modelling in CO_2 laser processing of optical fibres. Opt. Commun. **152**, 324–328 (1998)

N. Groothoff, J. Canning, E. Buckley, K. Lyttikainen, J. Zagari, Bragg gratings in air-silica structured fibers. Opt. Lett. **28**, 233–235 (2003)

V. Grubsky, J. Feinberg, Rewritable densification gratings in boron-doped fibers. Opt. Lett. **30**, 1279–1281 (2005)

V. Grubsky, J. Feinberg, Fabrication of axially symmetric long-period gratings with a carbon dioxide laser. IEEE Photon. Technol. Lett. **18**, 2296–2298 (2006)

V. Grubsky, D.S. Starodubov, J. Feinberg, Wavelength-selective coupler and add-drop multiplexer using long-period fiber gratings, in *Optical Fiber Communication Conference,***24**, 28–30 (2000)

Y. Gu, K. Chiang, Y. Rao, Writing of apodized phase-shifted long-period fiber gratings with a computer-controlled CO_2 laser. IEEE Photon. Technol. Lett. **21**, 657–659 (2009)

Y.-G. Han, S.B. Lee, C.-S. Kim, M.Y. Jeong, Tunable optical add-drop multiplexer based on long-period fiber gratings for coarse wavelength division multiplexing systems. Opt. Lett. **31**, 703–705 (2006)

K. Hill, Y. Fujii, D. Johnson, B. Kawasaki, Photosensitivity in optical fiber waveguides: application to reflection filter fabrication. Appl. Phys. Lett. **32**, 647–649 (1978)

K.O. Hill, B. Malo, K.A. Vineberg, F. Bilodeau, D.C. Johnson, I. Skinner, Efficient mode conversion in telecommunication fibre using externally written gratings. Electron. Lett. **26**, 1270–1272 (1990)

T. Hirose, K. Saito, S. Kojima, B. Yao, K. Ohsono, S. Sato, K. Takada, A. Ikushima, Fabrication of long-period fibre grating by CO_2 laser-annealing in fibre-drawing process. Electron. Lett. **43**, 443–445 (2007)

T. Hirose, K. Saito, K. Takada, Mid-infrared spectroscopic detection of refractive index in CO_2 laser-written long-period fibre grating. Electron. Lett. **44**, 1187–1188 (2008)

G. Humbert, A. Malki, S. Fevrier, P. Roy, D. Pagnoux, Electric arc-induced long-period gratings in Ge-free air-silica microstructure fibres. Electron. Lett. **39**, 349–350 (2003)

I. Hwang, S. Yun, B. Kim, Long-period fiber gratings based on periodic microbends. Opt. Lett. **24**, 1263–1265 (1999)

T.B. Iredale, P. Steinvurzel, B.J. Eggleton, Electric-arc-induced long-period gratings in fluid-filled photonic bandgap fibre. Electron. Lett. **42**, 739–740 (2006)

S. James, R. Tatam, Optical fibre long-period grating sensors: characteristics and application. Meas Sci Technol **14**, R49–R61 (2003)

J.B. Jensen, L.H. Pedersen, P.E. Hoiby, L.B. Nielsen, T.P. Hansen, J.R. Folkenberg, J. Riishede, D. Noordegraaf, K. Nielsen, A. Carlsen, A. Bjarklev, Photonic crystal fiber based evanescent-wave sensor for detection of biomolecules in aqueous solutions. Opt. Lett. **29**, 1974–1976 (2004)

J. Jensen, P. Hoiby, G. Emiliyanov, O. Bang, L. Pedersen, A. Bjarklev, Selective detection of antibodies in microstructured polymer optical fibers. Opt. Express **13**, 5883–5889 (2005)

L. Jin, Z. Wang, Q. Fang, B. Liu, Y. Liu, G. Kai, X. Dong, B.-O. Guan, Bragg grating resonances in all-solid bandgap fibers. Opt. Lett. **32**, 2717–2719 (2007a)

L. Jin, Z. Wang, Q. Fang, Y. Liu, B. Liu, G. Kai, X. Dong, Spectral characteristics and bend response of Bragg gratings inscribed in all-solid bandgap fibers. Opt. Express **15**, 15555–15565 (2007b)

L. Jin, W. Jin, J. Ju, Directional bend sensing with a CO_2-laser-inscribed long period grating in a photonic crystal Fiber. J. Lightwave Technol. **27**, 4884–4891 (2009)

L. Jin, W. Jin, J. Ju, Y. Wang, Coupled local-mode theory for strongly modulated long period gratings. J. Lightwave Technol. **28**, 1745–1751 (2010)

G. Kakarantzas, T. Dimmick, T. Birks, R. Roux, P. Russell, Miniature all-fiber devices based on CO_2 laser microstructuring of tapered fibers. Opt. Lett. **26**, 1137–1139 (2001)

G. Kakarantzas, T. Birks, P. Russell, Structural long-period gratings in photonic crystal fibers. Opt. Lett. **27**, 1013–1015 (2002)

G. Kakarantzas, A. Ortigosa-Blanch, T.A. Birks, P.S. Russell, L. Farr, F. Couny, B.J. Mangan, Structural rocking filters in highly birefringent photonic crystal fiber. Opt. Lett. **28**, 158–160 (2003)

V. Karpov, M. Grekov, E. Dianov, K. Golant, S. Vasiliev, O. Medvedkov, R. Khrapko, Mode-field converters and long-period gratings fabricated by thermo-diffusion in nitrogen-doped silica-core fibers, in *Optical Fiber Communication Conference and Exhibit (OFC '98)*, (1998), pp. 279–280

A. Kersey, M. Davis, H. Patrick, M. LeBlanc, K. Koo, C. Askins, M. Putnam, E. Friebele, Fiber grating sensors. J. Lightwave Technol. **15**, 1442–1463 (1997)

S. Khaliq, S.W. James, R.P. Tatam, Enhanced sensitivity fibre optic long period grating temperature sensor. Meas Sci Technol **13**, 792–795 (2002)

B. Kim, Y. Park, T.J. Ahn, D. Kim, B. Lee, Y. Chung, U. Paek, W. Han, Residual stress relaxation in the core of optical fiber by CO2 laser irradiation. Opt. Lett. **26**, 1657–1659 (2001)

B. Kim, T. Ahn, D. Kim, B. Lee, Y. Chung, U. Paek, W. Han, Effect of CO_2 laser irradiation on the refractive-index change in optical fibers. Appl. Opt. **41**, 3809–3815 (2002)

M.J. Kim, Y.M. Jung, B.H. Kim, W.-T. Han, B.H. Lee, Ultra-wide bandpass filter based on long-periodfiber gratings and the evanescent field couplingbetween two fibers. Opt. Express **15**, 10855–10862 (2007)

J. Knight, T. Birks, P. Russell, D. Atkin, All-silica single-mode optical fiber with photonic crystal cladding. Opt. Lett. **21**, 1547–1549 (1996)

Y. Kondo, K. Nouchi, T. Mitsuyu, M. Watanabe, P. Kazansky, K. Hirao, Fabrication of long-period fiber gratings by focused irradiation of infrared femtosecond laser pulses. Opt. Lett. **24**, 646–648 (1999)

S. Kosinski, A. Vengsarkar, Splicer-based long-period fiber gratings, in *Optical Fiber Communication Conference and Exhibit (OFC '98)*, (1998), pp. 278–279

B.T. Kuhlmey, B.J. Eggleton, D.K.C. Wu, Fluid-filled solid-core photonic bandgap fibers. J. Lightwave Technol. **27**, 1617–1630 (2009)

M. Kulishov, J. Azana, Long-period fiber gratings as ultrafast optical differentiators. Opt. Lett. **30**, 2700–2702 (2005)

C.M. Lawrence, D.V. Nelson, E. Udd, Measurement of transverse strains with fiber Bragg gratings. Proc. of SPIE **3042**, 218–228 (1997)

H.W. Lee, K.S. Chiang, CO_2 laser writing of long-period fiber grating in photonic crystal fiber under tension. Opt. Express **17**, 4533–4539 (2009a)

H.W. Lee, K.S. Chiang, CO_2-laser writing of long period fiber grating in photonic crystal fiber by frozen-in viscoelasticity, in *European Conference on Lasers and Electro-Optics 2009 and the European Quantum Electronics Conference (CLEO Europe – EQEC 2009)*, 1–1 (2009b)

B.H. Lee, J. Nishii, Self-interference of long-period fibre grating and its application as temperature sensor. Electron. Lett. **34**, 2059–2060 (1998)

B.H. Lee, J. Cheong, U.-C. Paek, Spectral polarization-dependent loss of cascaded long-period fiber gratings. Opt. Lett. **27**, 1096–1098 (2002)

H.W. Lee, Y. Liu, K.S. Chiang, Writing of long-period gratings in conventional and photonic-crystal polarization-maintaining fibers by CO_2-laser pulses. IEEE Photon. Technol. Lett. **20**, 132–134 (2008)

S. Lei, C. Seng, L. Chao, CO_2-laser-induced long-period gratings in graded-index multimode fibers for sensor applications. IEEE Photon. Technol. Lett. **18**, 190–192 (2006)

P.J. Lemaire, R.M. Atkins, V. Mizrahi, W.A. Reed, High pressure H_2 loading as a technique for achieving ultrahigh UV photosensitivity and thermal sensitivity in GeO_2 doped optical fibres. Electron. Lett. **29**, 1191–1193 (1993)

Y. Li, T. Wei, J.A. Montoya, S.V. Saini, X. Lan, X. Tang, J. Dong, H. Xiao, Measurement of CO_2-laser-irradiation-induced refractive index modulation in single-mode fiber toward long-period fiber grating design and fabrication. Appl. Opt. **47**, 5296–5304 (2008)

C. Lin, L. Wang, Loss-tunable long period fibre grating made from etched corrugation structure. Electron. Lett. **35**, 1872–1873 (1999)

C. Lin, L. Wang, A wavelength- and loss-tunable band-rejection filter based on corrugated long-period fiber grating. IEEE Photon. Technol. Lett. **13**, 332–334 (2001)

C. Lin, L. Wang, G. Chern, Corrugated long-period fiber gratings as strain, torsion, and bending sensors. J. Lightwave Technol. **19**, 1159–1168 (2001a)

C. Lin, G. Chern, L. Wang, Periodical corrugated structure for forming sampled fiber Bragg grating and long-period fiber grating with tunable coupling strength. J. Lightwave Technol. **19**, 1212–1220 (2001b)

C. Lin, Q. Li, A. Au, Y. Jiang, E. Lyons, H. Lee, A loss tunable long-period fiber gratings on corrugated silicon with on-chip microheater and temperature sensor, in *Optical Fiber Communication Conference and Exhibit (OFC 2002)*, (2002), pp. 193–194

Y. Liu, K.S. Chiang, Broad-band optical coupler based on evanescent-field coupling between three parallel long-period fiber gratings. IEEE Photon. Technol. Lett. **18**, 229–231 (2006)

Y. Liu, K. Chiang, Recent development on CO_2-laser written long-period fiber gratings, in *Asia-Pacific Optical Communications (APOC 2008)*. Proc. SPIE, vol 7134, Paper no. 713437 (2008a)

Y. Liu, K.S. Chiang, CO_2 laser writing of long-period fiber gratings in optical fibers under tension. Opt. Lett. **33**, 1933–1935 (2008b)

Y. Liu, L. Zhang, I. Bennion, Fibre optic load sensors with high transverse strain sensitivity based on long-period gratings in B/Ge co-doped fibre. Electron. Lett. **35**, 661–663 (1999)

Y. Liu, K.S. Chiang, Y.J. Rao, Z.L. Ran, T. Zhu, Light coupling between two parallel CO_2-laser written long-period fiber gratings. Opt. Express **15**, 17645–17651 (2007a)

Y. Liu, K.S. Chiang, Q. Liu, Symmetric 3 × 3 optical coupler using three parallel long-period fiber gratings. Opt. Express **15**, 6494–6499 (2007b)

Y. Liu, H.W. Lee, K.S. Chiang, T. Zhu, Y.J. Rao, Glass structure changes in CO_2-laser writing of long-period fiber gratings in boron-doped single-mode fibers. J. Lightwave Technol. **27**, 857–863 (2009a)

Y. Liu, Q. Liu, K.S. Chiang, Optical coupling between a long-period fiber grating and a parallel tilted fiber Bragg grating. Opt. Lett. **34**, 1726–1728 (2009b)

V. Minkovich, J. Villatoro, D. Monzon-Hernandez, S. Calixto, A. Sotsky, L. Sotskaya, Holey fiber tapers with resonance transmission for high-resolution refractive index sensing. Opt. Express **13**, 7609–7614 (2005)

K. Morishita, A. Kaino, Adjusting resonance wavelengths of long-period fiber gratings by the glass-structure change. Appl. Opt. **44**, 5018–5023 (2005)

K. Morishita, Y. Shi Feng, Y. Miyake, Refractive-index changes and long-period fiber gratings made by rapid solidification, in *Proceedings of 2002, IEEE/LEOS Workshop on Fibre and Optical Passive Components*, (2002), pp. 98–103

M.D. Nielsen, G. Vienne, J.R. Folkenberg, A. Bjarklev, Investigation of microdeformation-induced attenuation spectra in a photonic crystal fiber. Opt. Lett. **28**, 236–238 (2003)

S. Oh, K. Lee, U. Paek, Y. Chung, Fabrication of helical long-period fiber gratings by use of a CO_2 laser. Opt. Lett. **29**, 1464–1466 (2004a)

S.T. Oh, W.T. Han, U.C. Paek, Y. Chung, Azimuthally symmetric long-period fiber gratings fabricated with CO_2 laser. Microw. Opt. Technol. Lett. **41**, 188–190 (2004b)

U.C. Paek, C.R. Kurkjian, Calculation of cooling rate and induced stresses in drawing of optical fibers. J. Am. Ceram. Soc. **58**, 330–335 (1975)

Y. Rao, In-fibre Bragg grating sensors. Meas. Sci. Technol. **8**, 355–375 (1997)

Y. Rao, Recent progress in applications of in-fibre Bragg grating sensors. Opt. Lasers Eng. **31**, 297–324 (1999)

Y.J. Rao, T. Zhu, A novel flat-band long period grating with special apodization induced by high frequency CO_2 laser pulses, in *Optical Fiber Communication Conference, 2006 and the 2006 National Fiber Optic Engineers Conference. OFC 2006*, (2006), p. 3

Y.J. Rao, T. Zhu, Z.L. Ran, J.A. Jiang, An all-fibre dynamic gain equalizer based on a novel long-period fibre grating written by high-frequency CO_2 laser pulses. Chin. Phys. Lett. **19**, 1822–1824 (2002)

Y. Rao, Y. Wang, Z. Ran, T. Zhu, Novel fiber-optic sensors based on long-period fiber gratings written by high-frequency CO_2 laser pulses. J. Lightwave Technol. **21**, 1320–1327 (2003a)

Y.J. Rao, Y.P. Wang, T. Zhu, Z.L. Ran, Simultaneous measurement of transverse load and temperature using a single long-period fibre grating element. Chin. Phys. Lett. **20**, 72–75 (2003b)

Y. Rao, T. Zhu, Z. Ran, Y. Wang, J. Jiang, A. Hu, Novel long-period fiber gratings written by high-frequency CO_2 laser pulses and applications in optical fiber communication. Opt. Commun. **229**, 209–221 (2004)

Y.J. Rao, A.Z. Hu, Y.C. Niu, A novel dynamic LPFG gain equalizer written in a bend-insensitive fiber. Opt. Commun. **244**, 137–140 (2005)

Y. Rao, T. Zhu, Q. Mo, Highly sensitive fiber-optic torsion sensor based on an ultra-long-period fiber grating. Opt. Commun. **266**, 187–190 (2006)

G. Rego, J. Santos, H. Salgado, Polarization dependent loss of arc-induced long-period fibre gratings. Opt. Commun. **262**, 152–156 (2006)

L. Rindorf, O. Bang, Highly sensitive refractometer with a photonic-crystal-fiber long-period grating. Opt. Lett. **33**, 563–565 (2008)

L. Rindorf, J.B. Jensen, M. Dufva, L.H. Pedersen, P.E. Hoiby, O. Bang, Photonic crystal fiber long-period gratings for biochemical sensing. Opt. Express **14**, 8224–8231 (2006)

H.S. Ryu, Y. Park, S.T. Oh, Y. Chung, D.Y. Kim, Effect of asymmetric stress relaxation on the polarization-dependent transmission characteristics of a CO_2 laser-written long-period fiber grating. Opt. Lett. **28**, 155–157 (2003a)

H. Ryu, Y. Park, D. Kim, Asymmetric stress distribution analysis on the polarization dependent loss in a CO_2-laser-written long period fiber grating, in *Optical Fiber Communications Conference (OFC 2003)*, vol. 562, (2003b), pp. 569–570

T. S, N. George, P. Sureshkumar, P. Radhakrishnan, C. Vallabhan, V. Nampoori, Chemical sensing with microbent optical fiber. Opt. Lett. **26**, 1541–1543 (2001)

Y. Sasaki, Optical fiber grating couplers and their applications, in *Proceedings of 2001, 3rd International Conference on Transparent Optical Networks*, (2001), pp. 123–126

S. Savin, M. Digonnet, G. Kino, H. Shaw, Tunable mechanically induced long-period fiber gratings. Opt. Lett. **25**, 710–712 (2000)

W. Shin, B.-A. Yu, Y.-C. Noh, J. Lee, D.-K. Ko, K. Oh, Bandwidth-tunable band-rejection filter based on helicoidal fiber grating pair of opposite helicities. Opt. Lett. **32**, 1214–1216 (2007)

W. Shin, Y.L. Lee, T.J. Eom, B.A. Yu, Y.C. Noh, Temperature insensitive strain sensor based on long period fiber grating pair in photonic crystal fibers, in *The 14th OptoElectronics and Communications Conference (OECC 2009)*, (2009), pp. 1–2

X. Shu, K. Chisholm, I. Felmeri, K. Sugden, A. Gillooly, L. Zhang, I. Bennion, Highly sensitive transverse load sensing with reversible sampled fiber Bragg gratings. Appl. Phys. Lett. **83**, 3003–3005 (2003)

M. Silva-Lopez, W.N. MacPherson, C. Li, A.J. Moore, J.S. Barton, J.D.C. Jones, D. Zhao, L. Zhang, I. Bennion, Transverse load and orientation measurement with multicore fiber Bragg gratings. Appl. Opt. **44**, 6890–6897 (2005)

R. Slavik, Extremely deep long-period fiber grating made with CO_2 laser. IEEE Photon. Technol. Lett. **18**, 1705–1707 (2006)

R. Slavík, Coupling to circularly asymmetric modes via long-period gratings made in a standard straight fiber. Opt. Commun. **275**, 90–93 (2007)

C. Smith, N. Venkataraman, M. Gallagher, D. Muller, J. West, N. Borrelli, D. Allan, K. Koch, Low-loss hollow-core silica/air photonic bandgap fibre. Nature **424**, 657–659 (2003)

P. Steinvurzel, E. Moore, E. Magi, B. Kuhlmey, B. Eggleton, Long period grating resonances in photonic bandgap fiber. Opt. Express **14**, 3007–3014 (2006a)

P. Steinvurzel, E.D. Moore, E.C. Magi, B.J. Eggleton, Tuning properties of long period gratings in photonic bandgap fibers. Opt. Lett. **31**, 2103–2105 (2006b)

L. Su, K. Chiang, C. Lu, Microbend-induced mode coupling in a graded-index multimode fiber. Appl. Opt. **44**, 7394–7402 (2005)

Y. Tan, L. Sun, L. Jin, J. Li, B. Guan, Microfiber Mach-Zehnder interferometer based on long period grating for sensing applications. Opt. Express **21**, 154–164 (2013)

J. Tang, G. Yin, S. Liu, X. Zhong, C. Liao, Z. Li, Q. Wang, J. Zhao, K. Yang, Y. Wang, Gas pressure sensor based on CO_2-laser-induced long-period fiber grating in air-core photonic bandgap fiber. IEEE Photonics J. **7**, 6803107 (2015)

J. Tang, Z. Zhang, G. Yin, S. Liu, Z. Bai, Z. Li, M. Deng, Y. Wang, C. Liao, J. He, W. Jin, G. Peng, Y. Wang, Long period fiber grating inscribed in hollow-core photonic bandgap fiber for gas pressure sensing. IEEE Photonics J. **9**, 7105307 (2017)

Z. Tian, S.S.H. Yam, H.-P. Loock, Refractive index sensor based on an abrupt taper Michelson interferometer in a single-mode fiber. Opt. Lett. **33**, 1105–1107 (2008)

F. Tian, J. Kanka, B. Zou, K. Chiang, H. Du, Long-period gratings inscribed in photonic crystal fiber by symmetric CO_2 laser irradiation. Opt. Express **21**, 13208–13218 (2013)

G.D. VanWiggeren, T.K. Gaylord, D.D. Davis, E. Anemogiannis, B.D. Garrett, M.I. Braiwish, E.N. Glytsis, Axial rotation dependence of resonances in curved CO_2-laser-induced long-period fibre gratings. Electron. Lett. **36**, 1354–1355 (2000)

G.D. VanWiggeren, T.K. Gaylord, D.D. Davis, M.I. Braiwish, E.N. Glytsis, E. Anemogiannis, Tuning, attenuating, and switching by controlled flexure of long-period fiber gratings. Opt. Lett. **26**, 61–63 (2001)

A. Vengsarkar, P. Lemaire, J. Judkins, V. Bhatia, T. Erdogan, J. Sipe, Long-period fiber gratings as band-rejection filters. J. Lightwave Technol. **14**, 58–65 (1996a)

A. Vengsarkar, J. Pedrazzani, J. Judkins, P. Lemaire, N. Bergano, C. Davidson, Long-period fiber-grating-based gain equalizers. Opt. Lett. **21**, 336–338 (1996b)

Y.P. Wang, Y.J. Rao, Long period fibre grating torsion sensor measuring twist rate and determining twist direction simultaneously. Electron. Lett. **40**, 164–166 (2004a)

Y.P. Wang, Y. Rao, CO_2-laser induced LPFG torsion characteristics depending on length of twisted fibre. Electron. Lett. **40**, 1101–1103 (2004b)

Y.-P. Wang, Y.-J. Rao, A novel long period fiber grating sensor measuring curvature and determining bend-direction simultaneously. IEEE Sensors J. **5**, 839–843 (2005)

L.A. Wang, C.Y. Lin, G.W. Chern, A torsion sensor made of a corrugated long period fibre grating. Meas. Sci. Technol. **12**, 793–799 (2001)

Y.P. Wang, Y.J. Rao, Z.L. Ran, T. Zhu, A.Z. Hu, A novel tunable gain equalizer based on a long-period fiber grating written by high-frequency CO_2 laser pulses. IEEE Photon. Technol. Lett. **15**, 251–253 (2003a)

Y.P. Wang, Y.J. Rao, Z.L. Ran, T. Zhu, Unique characteristics of long-period fibre gratings fabricated by high-frequency CO2 laser pulses. Acta Phys. Sin. **52**, 1432–1437 (2003b)

Y.P. Wang, Y.J. Rao, Z.L. Ran, T. Zhu, X.K. Zeng, Bend-insensitive long-period fiber grating sensors. Opt. Lasers Eng. **41**, 233–239 (2004)

Y.-P. Wang, J.-P. Chen, Y.-J. Rao, Torsion characteristics of long-period fiber gratings induced by high-frequency CO_2 laser pulses. J. Opt. Soc. Am. B **22**, 1167–1172 (2005)

Y. Wang, D. Wang, W. Jin, Y. Rao, G. Peng, Asymmetric long period fiber gratings fabricated by use of CO_2 laser to carve periodic grooves on the optical fiber. Appl. Phys. Lett. **89**, 151105 (2006a)

Y. Wang, L. Xiao, D. Wang, W. Jin, Highly sensitive long-period fiber-grating strain sensor with low temperature sensitivity. Opt. Lett. **31**, 3414–3416 (2006b)

Y.P. Wang, D.N. Wang, W. Jin, CO_2 laser-grooved long period fiber grating temperature sensor system based on intensity modulation. Appl. Opt. **45**, 7966–7970 (2006c)

Y. Wang, L. Xiao, D.N. Wang, W. Jin, In-fiber polarizer based on a long-period fiber grating written on photonic crystal fiber. Opt. Lett. **32**, 1035–1037 (2007a)

Y. Wang, W. Jin, D. Wang, Strain characteristics of CO$_2$-laser-carved long period fiber gratings. IEEE J. Quantum Electron. **43**, 101–108 (2007b)

Y. Wang, D.N. Wang, W. Jin, Y. Rao, Asymmetric transverse-load characteristics and polarization dependence of long-period fiber gratings written by a focused CO$_2$ laser. Appl. Opt. **46**, 3079–3086 (2007c)

Y. Wang, W. Jin, J. Ju, H. Xuan, H.L. Ho, L. Xiao, D. Wang, Long period gratings in air-core photonic bandgap fibers. Opt. Express **16**, 2784–2790 (2008a)

Y. Wang, D.N. Wang, W. Jin, J. Ju, H.L. Ho, Mode field profile and polarization dependence of long period fiber gratings written by CO$_2$ laser. Opt. Commun. **281**, 2522–2525 (2008b)

Y. Wang, H. Bartelt, M. Becker, S. Brueckner, J. Bergmann, J. Kobelke, M. Rothhardt, Fiber Bragg grating inscription in pure-silica and Ge-doped photonic crystal fibers. Appl. Opt. **48**, 1963–1968 (2009a)

Y. Wang, H. Bartelt, W. Ecke, R. Willsch, J. Kobelke, M. Kautz, S. Brueckner, M. Rothhardt, Sensing properties of fiber Bragg gratings in small-core Ge-doped photonic crystal fibers. Opt. Commun. **282**, 1129–1134 (2009b)

Y. Wang, W. Jin, L. Jin, X. Tan, H. Bartelt, W. Ecke, K. Moerl, K. Schroeder, R. Spittel, R. Willsch, J. Kobelke, M. Rothhardt, L. Shan, S. Brueckner, Optical switch based on a fluid-filled photonic crystal fiber Bragg grating. Opt. Lett. **34**, 3683–3685 (2009c)

Y. Wang, W. Jin, D.N. Wang, Unique temperature sensing characteristics of CO$_2$-laser-notched long-period fiber gratings. Opt. Lasers Eng. **47**, 1044–1048 (2009d)

Y. Wang, H. Bartelt, W. Ecke, K. Schroeder, R. Willsch, J. Kobelke, M. Rothhardt, I. Latka, S. Brueckner, Investigating transverse loading characteristics of microstructured fiber Bragg gratings with an active fiber depolarizer. IEEE Photon. Technol. Lett. **21**, 1450–1452 (2009e)

Z. Wang, K.S. Chiang, Q. Liu, All-fiber tunable microwave photonic filter based on a cladding-mode coupler. IEEE Photon. Technol. Lett. **22**, 1241–1243 (2010a)

Z. Wang, K.S. Chiang, Q. Liu, Microwave photonic filter based on circulating a cladding mode in a fiber ring resonator. Opt. Lett. **35**, 769–771 (2010b)

Y. Wang, D. Richardson, G. Brambilla, X. Feng, M. Petrovich, M. Ding, Z. Song, Intensity-measurement bend sensors based on periodically-tapered soft glass fibers. Opt. Lett. **36**, 558–560 (2012)

B. Wang, W. Zhang, Z. Bai, L. Wang, L. Zhang, Q. Zhou, L. Chen, T. Yan, CO$_2$-laser-induced long period fiber gratings in few mode fibers. IEEE Photon. Technol. Lett. **27**, 145–148 (2015)

Z. Wu, Z. Wang, Y. Liu, T. Han, S. Li, H. Wei, Mechanism and characteristics of long period fiber gratings in simplified hollow-core photonic crystal fibers. Opt. Express **19**, 17344–17349 (2011)

L. Xia, P. Shum, C. Lu, Phase-shifted bandpass filter fabrication through CO$_2$ laser irradiation. Opt. Express **13**, 5878–5882 (2005)

H. Xuan, W. Jin, M. Zhang, CO$_2$ laser induced long period gratings in optical microfibers. Opt. Express **17**, 21882–21890 (2009)

A.D. Yablon, Optical and mechanical effects of frozen-in stresses and strains in optical fibers. IEEE J. Sel. Top. Quantum Electron. **10**, 300–311 (2004)

A.D. Yablon, M.F. Yan, P. Wisk, F.V. DiMarcello, J.W. Fleming, W.A. Reed, E.M. Monberg, D.J. DiGiovanni, J.R. Jasapara, M.E. Lines, Anomalous refractive index changes in optical fibers resulting from frozen-in viscoelastic strain, in *Optical Fiber Communications Conference (OFC 2003)*, 3, PD6-1-3 (2003)

A.D. Yablon, M.F. Yan, P. Wisk, F.V. DiMarcello, J.W. Fleming, W.A. Reed, E.M. Monberg, D.J. DiGiovanni, J. Jasapara, M.E. Lines, Refractive index perturbations in optical fibers resulting from frozen-in viscoelasticity. Appl. Phys. Lett. **84**, 19–21 (2004)

J. Yan, L. Qun, L. Chien-Hung, E. Lyons, I. Tomov, H. Lee, A novel strain-induced thermally tuned long-period fiber grating fabricated on a periodic corrugated silicon fixture. IEEE Photon. Technol. Lett. **14**, 941–943 (2002)

M. Yan, S. Luo, L. Zhan, Y. Wang, Y. Xia, Z. Zhang, Step-changed period chirped long-period fiber gratings fabricated by CO2 laser. Opt. Commun. **281**, 2784–2788 (2008)

M. Yang, Y. Li, D.N. Wang, Long-period fiber gratings fabricated by use of defocused CO_2 laser beam for polarization-dependent loss enhancement. J. Opt. Soc. Am. B **26**, 1203–1208 (2009)

J. Yang, C. Tao, X. Li, G. Zhu, W. Chen, Long-period fiber grating sensor with a styrene-acrylonitrile nano-film incorporating cryptophane a for methane detection. Opt. Express **19**, 14696–14706 (2011)

J. Yang, X. Che, R. Shen, C. Wang, X. Li, W. Chen, High-sensitivity photonic crystal fiber long-period grating methane sensor with cryptophane-A-6Me absorbed on a PAA-CNTs/PAH nanofilm. Opt. Express **25**, 20258–20267 (2017)

C. Zhang, K.S. Chiang, CO_2 laser-written long-period fiber gratings in a germanium-boron codoped fiber: effects of applying tension during the writing process. IEEE Photon. Technol. Lett. **21**, 1456–1458 (2009)

X. Zhang, Y. Liu, Z. Wang, J. Yu, H. Zhang, LP_{01}-LP_{11a} mode converters based on long-period fiber gratings in a two-mode polarization-maintaining photonic crystal fiber. Opt. Express **26**, 7013–7021 (2018)

C.-L. Zhao, L. Xiao, J. Ju, M.S. Demokan, W. Jin, Strain and temperature characteristics of a long-period grating written in a photonic crystal fiber and its application as a temperature-insensitive strain sensor. J. Lightwave Technol. **26**, 220–227 (2008)

Y. Zhao, Y. Liu, L. Zhang, C. Zhang, J. Wen, T. Wang, Mode converter based on the long-period fiber gratings written in the two-mode fiber. Opt. Express **24**, 6186–6195 (2016)

Y. Zhao, Y. Liu, C. Zhang, L. Zhang, G. Zheng, C. Mou, J. Wen, T. Wang, All-fiber mode converter based on long-period fiber gratings written in few-mode fiber. Opt. Lett. **42**, 4708–4711 (2017)

X. Zhong, Y. Wang, C. Liao, G. Yin, J. Zhou, G. Wang, B. Sun, J. Tang, Long period fiber gratings inscribed with an improved two-dimensional scanning technique. IEEE Photonics J. **6**, 2201508 (2014a)

X. Zhong, Y. Wang, J. Qu, C. Liao, S. Liu, J. Tang, Q. Wang, J. Zhao, K. Yang, Z. Li, High-sensitivity strain sensor based on inflated long period fiber grating. Opt. Lett. **39**, 5463–5466 (2014b)

X. Zhong, Y. Wang, C. Liao, S. Liu, J. Tang, Q. Wang, Temperature-insensitivity gas pressure sensor based on inflated long period fiber grating inscribed in photonic crystal fiber. Opt. Lett. **40**, 1791–1794 (2015)

K.M. Zhou, G. Simpson, X.F. Chen, L. Zhang, I. Bennion, High extinction ratio in-fiber polarizers based on 45 degrees tilted fiber Bragg gratings. Opt. Lett. **30**, 1285–1287 (2005)

Y. Zhu, P. Shum, J.-H. Chong, M.K. Rao, C. Lu, Deep-notch, ultracompact long-period grating in a large-mode-area photonic crystal fiber. Opt. Lett. **28**, 2467–2469 (2003a)

Y.Zhu, J.H. Chong, M.K. Rao, H. Haryono, A. Yohana, P. Shum, C. Lu, A long-period grating refractometer: measurements of refractive index sensitivity, in *Microwave and Optoelectronics Conference, 2003. IMOC 2003. Proceedings of the 2003 SBMO/IEEE MTT-S International*, 902, 901–904 (2003b)

Y. Zhu, P. Shum, H. Bay, X. Chen, C. Tan, C. Lu, Wide-passband, temperature-insensitive, and compact pi-phase-shifted long-period gratings in endlessly single-mode photonic crystal fiber. Opt. Lett. **29**, 2608–2610 (2004)

Y. Zhu, P. Shum, X. Chen, C.-H. Tan, C. Lu, Resonance-temperature-insensitive phase-shifted long-period fiber gratings induced by surface deformation with anomalous strain characteristics. Opt. Lett. **30**, 1788–1790 (2005a)

T. Zhu, Y. Rao, Q. Mo, Simultaneous measurement of refractive index and temperature using a single ultralong-period fiber grating. IEEE Photon. Technol. Lett. **17**, 2700–2702 (2005b)

Y. Zhu, P. Shum, H.-W. Bay, M. Yan, X. Yu, J. Hu, J. Hao, C. Lu, Strain-insensitive and high-temperature long-period gratings inscribed in photonic crystal fiber. Opt. Lett. **30**, 367–369 (2005c)

T. Zhu, Y.J. Rao, J.L. Wang, M. Liu, Transverse-load characteristics of twisted long-period fibre gratings written by high-frequency CO_2 laser pulses. Electron. Lett. **42**, 451–452 (2006)

T. Zhu, Y. Rao, J. Wang, Y. Song, A highly sensitive fiber-optic refractive index sensor based on an edge-written long-period fiber grating. IEEE Photon. Technol. Lett. **19**, 1946–1948 (2007a)

T. Zhu et al., Multi-edge-written long-period fibre gratings with low PDL by using high-frequency CO$_2$ laser pulses. Chin. Phys. Lett. **24**, 1971 (2007b)

T. Zhu, Y. Rao, J. Wang, Characteristics of novel ultra-long-period fiber gratings fabricated by high-frequency CO$_2$ laser pulses. Opt. Commun. **277**, 84–88 (2007c)

T. Zhu, Y.J. Rao, J.L. Wang, Y. Song, Strain sensor without temperature compensation based on LPFG with strongly rotary refractive index modulation. Electron. Lett. **43**, 1132–1133 (2007d)

T. Zhu, Y.J. Rao, J.L. Wang, All-fiber dynamic gain equalizer based on a twisted long-period grating written by high-frequency CO$_2$ laser pulses. Appl. Opt. **46**, 375–378 (2007e)

T. Zhu, Y. Song, Y. Rao, Y. Zhu, Highly sensitive optical refractometer based on edge-written ultra-long-period fiber grating formed by periodic grooves. IEEE Sensors J. **9**, 678–681 (2009a)

T. Zhu, Y.-J. Rao, Y. Song, K.S. Chiang, M. Liu, Highly sensitive temperature-independent strain sensor based on a long-period fiber grating with a CO$_2$-laser engraved rotary structure. IEEE Photon. Technol. Lett. **21**, 543–545 (2009b)

T. Zhu, K.S. Chiang, Y.J. Rao, C.H. Shi, Y. Song, M. Liu, Characterization of long-period fiber gratings written by CO$_2$ laser in twisted single-mode fibers. J. Lightwave Technol. **27**, 4863–4869 (2009c)

T. Zhu, C.H. Shi, Y.J. Rao, L.L. Shi, K.S. Chiang, All-fiber bandwidth-tunable band-rejection filter based on a composite grating induced by CO$_2$ laser pulses. Opt. Express **17**, 16750–16755 (2009d)

Micro-/Nano-optical Fiber Devices

37

Fei Xu

Contents

Abstract

Recently, there has been an increasing interest in the study of micro-/nano-optical fibers (MNOFs) with submicron transverse dimensions. The MNOFs are usually fabricated from standard optical fibers and have interesting optical properties like large evanescent fields, strong confinement of electromagnetic fields, as well as low interconnection loss when coupled to other optical fiber components and systems. In addition to the excellent optical properties, they are also extremely flexible and lightweight and offer a high degree of configurability. Several applications have been developed to take advantage of MNOF's unique optical and mechanical properties, including the miniaturization of conventional fiber devices (gratings, couplers, interferometers, etc.) and special stereo devices

F. Xu (✉)
National Laboratory of Solid State Microstructures and College of Engineering and Applied Sciences, Nanjing University, Nanjing, Jinagsu, P. R. China
e-mail: feixu@nju.edu.cn

© Springer Nature Singapore Pte Ltd. 2019
G.-D. Peng (ed.), *Handbook of Optical Fibers*,
https://doi.org/10.1007/978-981-10-7087-7_41

based on the wrap-on-a-rod technique. With the incorporation of new materials like graphene, MNOFs are ideally placed for the fabrication of several linear and nonlinear optical devices with immense potential for applications in optical telecommunications, lasers, and sensors. In this chapter, we present an overview of the latest results from the theoretical and experimental studies of MNOFs. The waveguide model, fabrication techniques, device fabrication, and applications are discussed.

Keywords
Microfiber · Nanofiber · Sensor · Integration · Taper

Introduction

Since Kao and Hockham proposed the use of glass waveguides as a practical medium for optical communication in 1965, optical fibers and related devices have been widely employed in telecommunication, sensing, optics, biology, medicine, etc. The optical fibers usually have diameters that are significantly larger than the wavelength of transmitted light. The most common optical fiber is the standard telecom single-mode fiber (SMF-28), with a cladding diameter of 125 μm and a core of 10 μm. In the nineteenth century, Boys was the first to report the application of micrometer-size glass fibers for mechanical issues (Boys 1887). In 1959, Kapany published the first report on the utilization of image-guiding fiber bunches for optical applications (Kapany 1959). Before 2003, there was little interest in the study of thin optical fibers with subwavelength diameters. Only a handful number of researchers have explored attempts to fabricate subwavelength microwires using a top-down process (Bilodeau et al. 1987; Bures and Ghosh 1999). However, in the last couple of decades, there has been an extraordinary interest in the synthesis, device fabrication, and applications of non-fiber photonic structures, namely, nanowires of several materials, for example, made from metals, insulators, semiconductors, and polymers.

In 2003, Tong and Mazur presented a two-step method for the fabrication of low-loss subwavelength silica-based micro-/nano-optical fibers (MNOFs) by wrapping and drawing a re-tapered fiber on a sapphire probe. The low loss of the fiber allowed for the realization of several MNOF-based devices. Subsequently, Birks, Gilberto, Leon-Saval, and Sumetsky have reported the fabrication of MNOFs with a waist diameter of hundreds of nanometers and a lower loss (lesser by an order of magnitude), by using the conventional flame-brushing tapering technique (Brambilla et al. 2004; Leon-Saval et al. 2004; Sumetsky et al. 2004). After these reports, the interest in this research topic has continuously increased with several publications on the fabrication, properties, and applications of MNOFs. Several novel MNOF-based devices have been theoretically modeled and experimentally demonstrated, including interferometers, gratings, couplers, and resonators. In the future, MNOFs can be considered as an important building block for all fiber-based circuits.

Fig. 1 Schematic diagram of a complete MNOF

An MNOF, as illustrated in Fig. 1, has a taper in its structure comprising of a narrow-stretched filament, which is called as the taper waist. Each end of the taper waist is linked to upstretched fiber by a conical tapered section (the taper transition). The waist diameter typically ranges between several hundred nanometers and several micrometers. The waist length is normally several millimeters to centimeters, and the transition region is often longer than the waist. However, it is possible to optimize the fabrication process to obtain ultrashort transition. In the literature, MNOFs are also called as microfiber, nanofiber, nanotaper, subwavelength-diameter fiber, fiber nanowire, etc.

When compared to standard SMFs, MNOFs have attracted a range of interesting and emerging fiber-optic applications as they provide unique optical and mechanical properties. In particular, they allow the following:

(a) Large evanescent fields, with a considerable fraction of the transmitted power propagating in the evanescent field outside of the MNOF.
(b) Nonlinear effects – the enhancement of nonlinear interactions is possible by the strong confinement of light in a very small area over relatively long lengths.
(c) Extreme flexibility and configurability – MNOFs can be easily bent and manipulated, and yet remain relatively strong mechanical strength. A bend radius of the order of a few microns can be easily achieved with relatively low bend loss, thereby allowing for the construction of highly compact devices with complex geometry.
(d) Low-loss interconnection with other optical fibers and other fiber-based optical components – MNOFs are fabricated by adiabatically stretching optical fibers; hence, the original dimensions of the optical fiber at the input and output are preserved, therefore allowing for splicing to standard fibers.
(e) Functionalization – silica-based MNOFs can be easily functionalized for coupling to new materials (Brambilla et al. 2009).

Due to the extraordinary properties, MNOFs have attracted a wide variety of applications. Several reports on the theoretical modeling, experimental fabrication, optical properties (e.g., linear and nonlinear effects), and applications (components and devices) of MNOFs have been published in the last 20 years. These recent advances in MNOFs have been categorized into five areas by Tong: waveguide and near-field optics, nonlinear optics, quantum and atomic optics, plasmonics, and opto-mechanics (Tong et al. 2012). Brambilla has classified the different MNOF devices and applications into three groups according to the property being exploited: transition region, where the mode guidance changes from core bound to cladding bound; confinement properties, which enhance the nonlinear effects and optomechanical effects; and evanescent field, which leads to the interaction between the light in the MNOF and ambient matter (Brambilla 2010).

In this chapter, we review the recent research on MNOFs with the goal of summarizing the wealth of information on the properties and applications of MNOFs reported in the last few years. In the following sections, the fabrication methodologies, the linear and nonlinear properties of MNOFs, and the different MNOF-based devices are reviewed and discussed in detail.

Manufacture of MNOFs

Fabrication of MNOFs by Top-Down Techniques

The ability to manufacture smooth, low-loss, and arbitrary-profile MNOFs is an essential precondition for any practical application. In the last 10 years, optical MNOFs have been fabricated with the help of different methods, often based on either heat-and-pull or etching.

The "two-step process" is the first method proposed by Tong et al. (2003). A pre-tapered optical fiber with a diameter of several micrometers is broken and drawn by a conventional flame technique. The optical fiber is first wound on a tip of a sapphire taper, which is heated by a flame at a distance from the fiber. The sapphire taper is used to transport the heat from the flame to the MNOF in a controlled manner. The sapphire tip also confines the heat to a small volume, and therefore the temperature distribution in the drawing region remains stable. MNOFs with extremely small diameters of tens of nanometers have been fabricated with this approach. However, the fabrication procedure is complex and the MNOFs have a relatively high loss ($\sim$0.1 dB/mm for a radius $\sim$430 nm and wavelength $\sim$1.55 μm).

Based on the first method, the second approach involves the direct drawing technique from bulk glass (Tong et al. 2006). In this method, a sapphire fiber is heated and put in contact with a bulk glass. A small amount of the bulk glass melts and is removed on top of the sapphire fiber. The nanowire is then drawn by placing another sapphire fiber in contact with the molten glass, and then the two sapphire fibers are pulled apart. Although the fabrication methodology is simple, the reproducibility and uniformity remain a challenge. A typical loss of 0.1 dB/mm has been observed for MNOFs fabricated with this approach.

The third method called as the "flame brushing technique" was developed for the manufacture of conventional fiber tapers and couplers. In this method, a small flame moves under an optical fiber which is being stretched (Bilodeau et al. 1988; Birks and Li 1992; Brambilla et al. 2004). The control of the flame movement and the fiber tension is used to define the taper shape with a high degree of accuracy. MNOFs with a radii of $\sim$30 nm have been fabricated with this approach (Brambilla et al. 2006). This method by far provides the longest and most uniform MNOFs (Brambilla et al. 2004), with the lowest measured loss to date (0.001 dB/mm) (Clohessy et al. 2005; Leon-Saval et al. 2004). Moreover, this approach enables the access of the MNOFs from both the pigtailed ends, unlike other nanowires. We have adopted this method for our research and the studies using this approach will be discussed in detail.

In the modified flame brushing technique, the flame is replaced by a different heat source such as an electrical microheater or a sapphire capillary tube (Brambilla et al. 2005; Sumetsky et al. 2004). The temperature of the microheater can be controlled by tuning the current flowing through the resistive element. A CO_2 laser beam is used to heat the sapphire capillary tube, and the temperature can be controlled by changing the focus of the CO_2 laser beam (Brambilla et al. 2005; Sumetsky et al. 2004). It is safer to use an electrical heater instead of a flammable gas, but at the cost of a longer heating zone. However, this technique has been used to manufacture MNOFs made from a wide range of low softening temperature glasses. Moreover, this technique is helpful for the fabrication of silica-based MNOFs with extremely low -OH content (more than three orders of magnitude smaller than the flame brushing technique, where water vapor is a combustion by-product) (Brambilla 2010).

Finally, there are a few reports on the fabrication of MNOFs with wet chemical etching using hydrofluoric acid (HF). However, the main challenges associated with acid etch techniques include the high optical losses and the difficulty in producing thin MNOFs. Generally, chemically etched MNOFs have exhibited insertion losses more than 10 dB even for micron-scale waist diameters due to the formation of surface corrugations. Zhang et al. (2010a) proposed a modified etch method based on surface tension-driven flows of hydrofluoric acid microdroplets. They were able to fabricate a 1 μm diameter MNOF with $\sim$0.1 dB/mm loss, corresponding to an order of magnitude increase in the optical transmission over previously reported acid etch techniques. However, the waist length is seriously limited to only few millimeters.

Of all the methods discussed here, the flame brushing technique is the most widespread approach for the fabrication of low-loss MNOFs. A simplified diagram of the "flame brushing" fabrication setup is illustrated in Fig. 2a. A millimeter-sized flame heats a small fraction of fiber, which is being pulled by two translation stages. The flame is scanned along the fiber over a length of several tens of millimeters to produce tapers with an extremely uniform waist diameter and taper transitions of well-defined length and shape.

A constant gas flow is ensured to guarantee uniform taper heating during all stages of the process. Two sets of gas regulators and flow meters are used for the hydrogen (or isopropane) and oxygen. The whole fabrication rig is typically housed in a Plexiglas box to avoid air turbulence. By controlling the flame movement and the fiber stretch, the desired taper shape with an extremely high degree of accuracy can be produced. This method provides the longest ($\sim$110 mm) and most uniform MNOFs (Brambilla et al. 2004) with the lowest measured loss ($\sim 10^{-3}$ dB/mm) (Clohessy et al. 2005; Leon-Saval et al. 2004). This method can also be used to draw fibers made from diverse materials, including chalcogenide and fluoride glasses (Brambilla et al. 2005). However, most of the non-silica fibers produce toxic gases during heating and pulling.

The quality of MNOFs is indicated by the loss, which can be monitored in situ during the drawing process. The losses in transmission arise from the surface roughness and nonuniformity of the MNOF diameter. The typical loss in a 1 μm

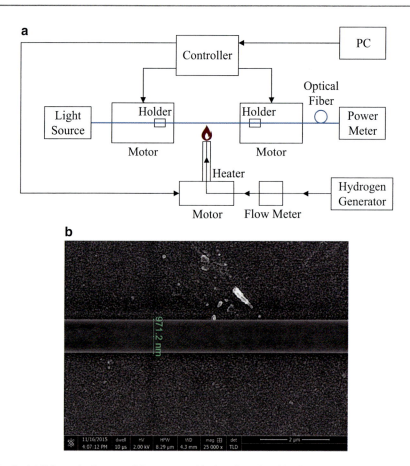

Fig. 2 (**a**) Schematic diagram of the setup used in the "flame brushing" technique. (**b**) SEM image of a silica MNOF with diameter of ~971 nm

diameter MNOF is ~0.001 dB–0.01 dB/mm. The loss is extremely sensitive to changes in the cleanliness of the environment and decays rapidly in air due to the accumulation of dirt on the surface of MNOFs. Cleaning MNOFs using common solvents such as acetone, isopropanol, methanol, and water leads only to a partial recovery in the induced loss. However, the induced loss can be recovered by post-fabrication flame brushing treatment (Xu et al. 2011).

The MNOFs are mechanically strong enough to be handled in the laboratory environment. The axial strength of a 750 nm diameter MNOF has been measured. The fracture tensile stress is roughly 10–11 GPa, which is significantly higher than those measured in bare optical fibers (~5 GPa), commercially available high-strength materials like Kevlar (3.88 GPa), and the high-strength steel ASTM A514 (0.76 GPa) (Brambilla et al. 2009). It is interesting to note that MNOFs fabricated by flame brushing technique have better mechanical strength than those manufactured by the two-step technique. This difference can be explained by the better surface quality of the MNOFs produced by the flame brushing technique (Brambilla et al. 2009).

Figure 2b shows an electron microscope image of the waist of a MNOF created from a 125 μm diameter standard SMF. The waist is uniform and smooth and has a diameter of ∼971 nm.

Manufacture of MNOF Tips by a Pipette Puller

A tapered MNOF tip is a type of tip with a continuously decreasing diameter at one end, which can be considered as a half of a MNOF. There are several methods for the fabrication of MNOF tips:

The first approach to produce an MNOF tip is to cut an MNOF; however, it is difficult to obtain a sharp tip. The second method involves chemical etching, which is time- and energy-consuming. The quickest and most convenient way to manufacture MNOF tip relies on the use of a commercially available pipette puller. However, the reproducibility is poor for the pipette puller method. A commercially available pipette puller (P-2000) has been often used for the manufacture of very short and sharp MNOF tips. P-2000 is a microprocessor-controlled CO_2 laser-based micropipette puller, with the bare fiber held on two puller stages. The primary advantage of using the CO_2 laser as a heat source is the ability to work with fused silica glass, a much stronger and purer glass formulation than standard glass capillary tubing. P-2000 can also be used to pull tubing and optical fibers to exceedingly small diameters for research applications such as scanning near-field optical microscopy.

Embedding

MNOFs can undergo optical and mechanical degradation, which results in the optical spectra exhibiting an increasing and unrecoverable loss after fabrication. The optical loss in MNOF is proportional to the decreases in the mechanical strength. The degradation can be attributed to the formation of cracks on the MNOF surface upon exposure to air (Brambilla et al. 2009). Therefore, it is necessary to protect the MNOF with additional coating for long-term reliability.

Standard optical fibers are coated with a polymeric multilayer coating during their fabrication. The coating has a higher refractive index than silica to strip the modes propagating in the cladding. In MNOFs, the use of higher refractive index polymers can induce huge loss, as the propagating mode has a substantial fraction of the total power at the interface between silica and air. The use of low refractive index materials is a better approach to avoid losses. The low refractive index materials investigated for this purpose include silicone rubber (n∼1.4 at λ = 1.55 μm) (Caspar and Bachus 1989a), Efiron UV373 (n ∼1.373 at λ = 1.55 μm), and Teflon (n∼1.3 at λ = 1.55 μm) (Brambilla et al. 2009).

Silicone rubber is an elastomer (rubberlike material) consisting of silicone and a thermocurable polymer. Silicone rubber has been widely adopted partly due to its stability, resistance to degradation in extreme environments, operating temperatures ranging from −55 °C to 300 °C, and a refractive index close to that of silica. It has also been employed to embed MNOFs with a diameter of 8.5 µm and construct MNOF resonators (Caspar and Bachus 1989a).

Efiron UV373 is a UV-curable polymer manufactured by Luvantix (South Korea, also known as SSCP USA) and has been widely used in cladding-pumped fiber lasers. They have a series of low-index polymer names such as UV 340 (1.340 at 852 nm, 0.55 NA) and SH 380 (1.380 at 852 nm, high temperature).

Teflon is a fluoropolymer with excellent thermal and chemical properties. However, due to its poor solubility in common solvents, pure Teflon is difficult to handle/deposit. Nevertheless, modified amorphous Teflon in a highly volatile solvent is commercially available with the concentration of 6% or 18% (Teflon AF, DuPont, United States). Modified amorphous Teflon represents the best option to the efficient embedding of optical fiber nanowires. Teflon AF is a polymer with lowest-known refractive index. Hence, the large refractive index difference between the optical fiber nanowire and the Teflon coating results in a stronger mode confinement when compared to other polymers (Brambilla et al. 2009; Fei and Gilberto 2008).

The Shape of Fiber Taper

Birks has reported a detailed study on the shape of MNOF, with his results being efficiently used to predesign the waist diameter, length, and transition profile (Birks and Li 1992). Figure 3 illustrates the quantities used to describe the shape of a complete MNOF. In this case, a symmetric MNOF with identical transitions is considered. The radius of the untapered fiber is r_0, the uniform taper waist length is l_w, and a radius is r_w. Each identical taper transition has a length, z_0, and shape described by a decreasing local radius function $r(z)$, where z is the longitudinal coordinate. The origin of z is at the beginning of each taper transition (point P for the representative left-hand transition in Fig. 3); hence, $r(0) = r_0$, and $r(z_0) = r_w$. The taper extension X is the net distance through which the taper has been stretched.

At any instant time t during taper elongation, a symmetrically placed length L of the taper waist is uniformly heated. Outside the hot zone, the glass is cold and solid. The ends of the taper are steadily pulled apart such that at time $t + dt$ the hot glass cylinder stretches to form a narrower cylinder of length $L+dx$, where dx is the increase in extension during the interval dt. The hot zone length changes to $L+dL$ in the same time, where dL may be negative. As the taper is elongated, the extremities of the stretched heated cylinder may leave the hot zone and solidify, forming the new elements of the taper transitions.

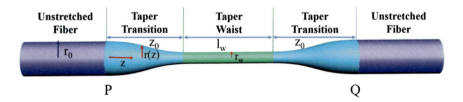

Fig. 3 Schematic diagram displaying the different parts of a MNOF structure

Mathematically, the relationship between the shape of the tapered fiber [expressed by r_w, l_w, z_0, and r (z)] and the elongation conditions [expressed by X_0 and $L(X)$)] is entirely determined by three equations:

First, the instantaneous length l_w of the taper waist at time t is equal to the hot zone length at that time:

$$l_w(t) = L(t) \tag{1}$$

and we define $L_0 = L(0)$, the hot zone length at the beginning of the tapering process.

Consideration of the mass implies that the volume of the fiber does not change during the tapering process, so we can obtain the second equation, termed as the "volume law":

$$\frac{dr_w}{dX} = -\frac{r_w}{2L} \tag{2}$$

The third equation is the "distance law":

$$2z + L = X + L_0 \tag{3}$$

With the help of these three equations, Birks et al. (Birks and Li 1992) demonstrated that $L(r)$ (the hot zone length as point z was pulled out of the hot zone) and $z(r)$ can be related by the following equations:

$$L(z) = \frac{r_w^2}{r^2(z)} L_w + \frac{2}{r^2(z)} \int_z^{z_0} r^2\left(z'\right) dz' \tag{4}$$

or

$$L(r) = \frac{1}{r^2} \int_{r_0}^{r} 4z\left(r'\right) dr' + L_0 \frac{r_0^2}{r^2} - 2z(r) \tag{5}$$

$L(z)$ or $L(r)$ is now completely known; hence, the distance law (3) gives x as a known function of either z or r. With Eqs. 3, 4, and 5, we have a framework with which MNOF of any taper shape can be fabricated.

Linear and Nonlinear Characteristic Properties of MNOFs

Basic Waveguide Theory

To design MNOF devices, it is necessary to understand the theory of electromagnetic wave propagation in MNOFs. By solving the Maxwell's equations for the optical structure, one can obtain the local field profiles. In this chapter, the optical properties of common single-mode fibers and tapers are discussed. The waveguide structure model of the fiber without coating is illustrated in Fig. 4. The three-

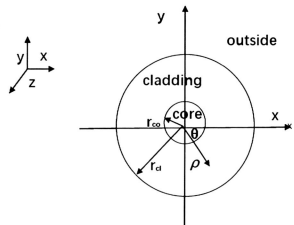

Fig. 4 Schematic diagram of the cross-sectional profile of a step fiber or MNOF

layer waveguide has three regions, namely, the core, cladding, and air. Optical nanowires drawn from standard single-mode fibers have the same structure but different diameters. To describe a position along the fiber, cylindrical coordinates (z, ρ, θ) are used; r_{cl} and r_{co} are the cladding and core radii, respectively.

When the diameter is large, the mode is confined in the core and the cladding can be assumed to be infinite. In this scenario, the fiber can be simplified as a two-layer model, and the weak guidance theory can be applied as commonly treated in most books. When the diameter decreases, the fiber should be taken as a three-layer waveguide with a finite cladding.

When we consider light having angular frequency ω and propagating in the z-direction, the electromagnetic mode field **e** and **h** can be expressed in cylindrical coordinates as:

$$\mathbf{e} = \overline{E}(\rho, \theta)\, e^{j(\omega t - \beta z)} \tag{6}$$

$$\mathbf{h} = \overline{H}(\rho, \theta)\, e^{j(\omega t - \beta z)} \tag{7}$$

where β is the propagation constant and ω is the frequency of light.

Substituting into Maxwell equations, two sets of wave equations for the longitudinal electromagnetic field can be obtained:

$$\frac{\partial^2 E_z}{\partial \rho^2} + \frac{1}{\rho}\frac{\partial E_z}{\partial \rho} + \frac{1}{\rho^2}\frac{\partial^2 E_z}{\partial \theta^2} + \left[k^2 n^2(\rho, \theta) - \beta^2\right] E_z = 0 \tag{8a}$$

$$\frac{\partial^2 H_z}{\partial \rho^2} + \frac{1}{\rho}\frac{\partial H_z}{\partial \rho} + \frac{1}{\rho^2}\frac{\partial^2 H_z}{\partial \theta^2} + \left[k^2 n^2(\rho, \theta) - \beta^2\right] H_z = 0 \tag{8b}$$

where $k = \frac{2\pi}{\lambda}$ and $n(\rho, \theta)$ is the refractive index. Hence, the transverse electromagnetic field can be written by the following equations:

$$\begin{cases} E_\rho = \frac{-j}{\left[k^2 n^2(\rho) - \beta^2\right]} \left(\beta \frac{\partial E_z}{\partial \rho} + \frac{\omega \mu_0}{\rho} \frac{\partial H_z}{\partial \theta} \right) \\ E_\theta = \frac{-j}{\left[k^2 n^2(\rho) - \beta^2\right]} \left(\frac{\beta}{r} \frac{\partial E_z}{\partial \theta} - \omega \mu_0 \frac{\partial H_z}{\partial \rho} \right) \\ H_\rho = \frac{-j}{\left[k^2 n^2(\rho) - \beta^2\right]} \left(\beta \frac{\partial H_z}{\partial \rho} - \frac{\omega \varepsilon_0 n^2(\rho)}{\rho} \frac{\partial E_z}{\partial \theta} \right) \\ H_\theta = \frac{-j}{\left[k^2 n^2(\rho) - \beta^2\right]} \left(\frac{\beta}{\rho} \frac{\partial H_z}{\partial \theta} + \omega \varepsilon_0 n^2(\rho) \frac{\partial E_z}{\partial \rho} \right) \end{cases} \qquad (9)$$

In axially symmetric fibers, the refractive index distribution is not dependent on θ and can be expressed by:

$$n = \begin{cases} n_{co} & 0 \leq \rho \leq r_{co} \\ n_{cl} & r_{co} \leq \rho \leq r_{cl} \\ n_{ou} & \rho \geq r_{cl} \end{cases} \qquad (10)$$

where $n_{eff} = \frac{\beta}{k}$ is the effective refractive index. The modes in MNOFs are classified as a cladding mode if $n_{cl} > n_{eff} > n_{ou}$ and as a core mode for $n_{eff} > n_{cl}$.

Modes corresponding to the solutions in Eq. 9 and the appropriate boundary conditions [the continuity of $E_z(H_z)$ and $\frac{dE_z}{d\rho}\left(\frac{dH_z}{d\rho}\right)$] can be classified as either TE modes ($E_z = 0$), TM modes ($H_z = 0$), or hybrid modes ($E_z \neq 0$, $H_z \neq 0$), respectively.

The solutions of Eq. 9 in each region of the fiber depend on the magnitude of n_{eff} with respect to the refractive index n. If $n_{eff} < n$, the two independent solutions are the vth-order Bessel functions of the first and second kind J_v and Y_v (where v is integer), while if $n_{eff} > n$, the two independent solutions are the vth-order-modified Bessel functions of the first and second kinds I_v, and K_v, respectively. Love has presented some solutions in ref. (Love et al. 1991).

When the diameter of the MNOF is very small or the cladding is very large ($r_{co} \ll \lambda$, $r_{cl} \sim \lambda$), the core can be ignored and a two-layer model applies:

$$n = \begin{cases} n_{MNOF} & \text{in MNOF} \\ n_{ou} & \text{out MNOF} \end{cases} \qquad (11)$$

In most cases the MNOF is in air, called as air-cladding MNOF, and therefore $n_{ou} = 1$. In this case, the eigenvalue equations for various modes are as follows:
for HE_{vm} and EH_{vm} modes (Tong et al. 2004):

$$\left[\frac{J'_v(U)}{U J_v(U)} + \frac{K'_v(U)}{W K_v(U)} \right] \left[\frac{J'_v(U)}{U J_v(U)} + \left(\frac{n_{ou}}{n_{MNOF}} \right)^2 \frac{K'_v(U)}{W K_v(U)} \right] = \left(\frac{v \beta}{k n_{MNOF}} \right)^2 \left(\frac{V}{U W} \right)^4$$

$$(12)$$

where $U = d/2\sqrt{k_0^2 n_{MNOF}^2 - \beta^2}$, $W = d/2\sqrt{\beta^2 - k_0^2 n_{ou}^2}$, and $V = k_0 \cdot d/2$ $\sqrt{n_{MNOF}^2 - n_{ou}^2}$ and d is the MNOF diameter. The propagation constants (β) of the air-cladding MNOFs can be obtained numerically using these equations. However, only finite-element or the finite-difference time-domain methods can be used to numerically model MNOFs such as slot MNOFs. To obtain the index of fused silica, the Sellmeier-type dispersion formula (at room temperature) can be applied (Okamoto 2006):

$$n_{silica}(\lambda) = \sqrt{1 + c_1\lambda^2/(\lambda^2 - c_4) + c_2\lambda^2/(\lambda^2 - c_5) + c_3\lambda^2/(\lambda^2 - c_6)} \qquad (13)$$

The wavelength λ is in μm, and $c_1 = 0.6965325$, $c_2 = 0.4083099$, $c_3 = 0.8968766$, $c_4 = 4.368309 \times 10^{-3}$, $c_5 = 1.394999 \times 10^{-2}$, and $c_6 = 9.793399 \times 10^1$.

In practical applications, it is important to know the profile of the power distribution across the waveguide. For the fiber mode considered here, the average energy flow in the radial (r) and azimuthal (θ) directions is zero; therefore, we only consider the energy flow in z-direction. The z-components of the Poynting vectors can be obtained with the following equation:

$$S_z = \frac{1}{2}\left(\vec{E} \times \vec{H}^*\right) \cdot \vec{u_z} = \frac{1}{2}\left(E_\rho H_\theta^* - E_\theta H_\rho^*\right) \qquad (14)$$

where u_z is a unit vector in the z-direction.

The power fraction carried in the optical core for the two-layer model can be defined as:

$$\tau = \frac{\int_0^{2\pi}\int_0^{d/2} S_z dA}{\int_0^{2\pi}\int_0^{d/2} S_z dA + \int_0^{2\pi}\int_{d/2}^{\infty} S_z dA} \qquad (15)$$

where $dA = \rho d\rho d\theta$.

For example, the effective index in MNOFs with different diameters has been calculated by applying Eq. 12. The diameter-dependent effective index of silica MNOFs at the wavelength of 800 nm is displayed in Fig. 5, where the MNOF diameter (d) is directly related to the V-number: $V = k_0 d \sqrt{\left(n_{MNOF}^2 - n_{ou}^2\right)/2}$. Clearly, when the MNOF diameter is reduced to a certain value (corresponding to $V = 2.405$), only the HE_{11} mode exists, corresponding to single-mode operation.

The fractional power as a function of the MNOF diameter for 800 nm and 1550 nm wavelengths is shown in Fig. 6. At longer wavelengths, more power is located outside the MNOF. At 1550 nm wavelength, more than 20% of the power is outside the MNOF when the diameter is smaller than 1 μm.

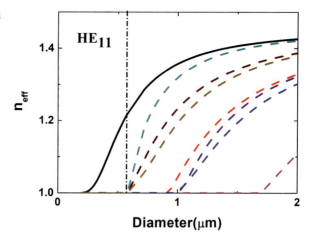

Fig. 5 Effective index values for the fundamental and high-order modes propagating in the air-cladding MNOF as a function of the diameter. Results were obtained by calculating the exact solution of the Maxwell's equations

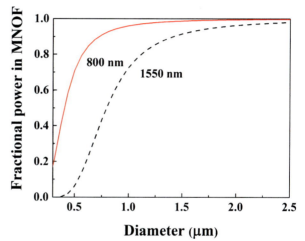

Fig. 6 Fractional power in the MNOF as a function of the diameter

Dispersion and Nonlinearity

Group velocity dispersion (GVD) or chromatic dispersion is a primary cause for concern in high bit-rate single-mode wavelength-division multiplexing (WDM) systems as the light from a typical optical source will contain different wavelength components with different speeds. Therefore, the pulse will be disordered due to the dispersion. There are two types of material dispersion, waveguide dispersion or profile dispersion. In nonlinear applications of fiber, GVD is an important parameter to be considered, especially in supercontinuum generation, which has several applications in pulse compression, parametric amplifiers, supercontinuum-based WDM telecom sources, etc.

GVD is defined as:

$$D = -\frac{2\pi c}{\lambda^2} \frac{d^2\beta}{d\omega^2} \tag{16}$$

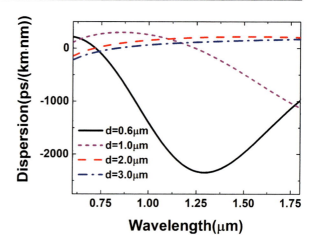

Fig. 7 GVD of MNOF as function of light wavelength for four different diameters (0.6, 1.0, 2.0, and 3.0 μm)

The confinement of light to a very small nonlinear core region in MNOFs facilitates in the enhancement of nonlinear effects at moderate power levels. When 1 W of optical power propagates in a 10 μm core SMF, the optical power density is higher than 1 MW/cm^2 and increases by a factor of more than 100 in a 1 μm diameter MNOF. At such high electromagnetic intensity, the optical Kerr effect arises resulting in the nonlinear behavior of the refractive index of silica, which increases with the intensity.

The GVD of MNOFs as function of wavelength for four different diameters is displayed in Fig. 7. The GVD values are quite large in the anomalous-dispersion regime, ranging in the order of 1000 ps/km/nm.

The nonlinear refractive index is expressed as n_2E^2, where n_2 is the Kerr coefficient. Typically, n_2 is 2.6×10^{-20} m^2/W for silica (Afshar V and Monro 2009) and 5.0×10^{-23} m^2/W for air (Boyd 1992). The Kerr effect leads to self- and cross-phase modulations (SPM, XPM) of the propagating signal. Other nonlinear phenomena can also be produced by utilizing the Kerr effect, for example, optical solitons, optical pulse compression, and modulation instabilities.

The nonlinear coefficient is defined as (Okamoto 2006):

$$\gamma = \frac{n_2\omega}{cA_{\text{eff}}} \tag{17}$$

where A_{eff} is the effective area, which is given by the following equation:

$$A_{\text{eff}} = \frac{\left\{ \int\limits_0^{2\pi}\int\limits_0^\infty |\overline{E}(\rho,\theta)|^2 \rho\, d\rho\, d\theta \right\}^2}{\int\limits_0^{2\pi}\int\limits_0^\infty |\overline{E}(\rho,\theta)|^4 \rho\, d\rho\, d\theta} \tag{18}$$

Generally, a larger nonlinear coefficient can be obtained with a smaller diameter, as the field intensity of the fiber increases as the diameter decreases.

Post-processing Techniques

The "heat-and-pull" fabrication processes allow for the manufacturing of MNOF with different waist diameters and different transition shapes. However, additional post-processing techniques are needed for the realization of MNOFs with complex geometry. Much of the semiconductor device fabrication processes can be modified and adapted for the production of complex MNOFs. These processes include material deposition, lithography, electron beam/laser writing system, focused ion beam (FIB) milling, laser etching, annealing, slicing, and polishing. Additional challenges are introduced due to the non-planar nature of the MNOF surface. Therefore, some unique techniques such as "wrap-on-a-rod" have been developed to deal with MNOFs. In this section, we introduce several popular micromachining and wrap-on-a-rod techniques for the construction of complex MNOFs.

Micromachining Techniques

Wet and Dry Etching
Etching is used to chemically remove layers from the surface of a MNOF. For many etching processes, a part of the MNOF is protected from the etchant by a "masking" material, which is resistant to etching. In some cases, the masking material is a photoresist patterned using photolithography. In wet etching, hydrofluoric acid (HF) solution generally serves as the etchant for the removal of silica. The etching rate is determined by the HF concentration, and the etching depth is controlled by the etching time. In dry etching, the MNOFs are placed in a sputtering apparatus and the energetic ion or RF discharge plasma is bombarded on the sample. A reproducible etching rate can be achieved by controlling the RF power, ion voltage, current, and other instrument parameters. Etching can be used to pattern or reshape the endface or the side wall of MNOFs.

FIB
By far, FIB is the most flexible and powerful tool for mask-less patterning, cross sectioning, or functionalizing subwavelength circular MNOF due to its small and controllable spot size and high beam current density. FIB milling provides great flexibility in the creation of innumerable nanostructures with high precision, which has been used for the fabrication of Fabry-Perot cavities and fiber Bragg gratings (FBGs). The FIB method makes use of accelerated gallium ions to mill several or periodic multiple nanoscale grooves on MNOFs, thereby resulting in the formation of surface-corrugated MNOF devices.

MNOF is made from electrically insulating material; therefore, pretreatments are necessary to enhance its conductivity before FIB milling. For example, the MNOF to be machined can be coated with a thin film of metal (aluminum or gold) or can be placed on a conductive substrate (Liu et al. 2011; Luo et al. 2012) to prevent the accumulation of surface charges during the milling process. During the FIB milling process, the MNOF should be fixed firmly in the vacuum chamber to avoid displacements. The entire fabrication process can last between a few minutes to hours depending on the milling area and the beam current used. After milling, the metal film on the MNOFs can be easily removed by a suitable etchant (Kou et al. 2011a). Typical parameters for FIB machining (for Strata FIB 201, FEI Company, Ga ions) include a 30.0 kV gallium ion beam with current 60–300 pA. We have been able to fabricate structures with high accuracy and sharp endfaces using these settings.

Femtosecond (fs) Laser Irradiation

High-power fs laser pulses can also be employed for the fabrication of complex MNOF devices. When high peak intensity femtosecond laser pulses are focused on a MNOF, the light intensity at the focal volume is sufficient to induce multiphoton absorption, thereby resulting in the permanent structural or refractive index change in the material. This enables the direct integration of three-dimensional nanostructures in the MNOF (Rihakova and Chmelickova 2015). The main parameter is the laser pulse duration, which significantly affects the quality of the micro-features and the material removal rate. A schematic diagram of a typical fs laser fabrication setup was demonstrated by Nayak and Hakuta (2013). With the help of this setup, a grating consisting of thousands of periodic nanocraters has been inscribed onto a MNOF by the irradiation of just a single fs laser pulse. As the effective RI changes by the introduction of the periodic nanostructures, this technique also allows for the inscription of Bragg gratings in MNOFs without photosensitivity.

In addition, other types of lasers have also been used for the micromachining of materials. These include microsecond carbon dioxide lasers with wavelengths ranging between 9.3 and 11 μm, nanosecond laser with wavelengths between 1030 and 1064 nm, and excimer lasers emitting in the UV region of the electromagnetic spectrum (157–353 nm) (Karnakis 2008).

Wrap-on-a-Rod

Very long MNOFs (typically a few microns) can be bent tightly and coiled into complex 3D stereo structures, which cannot be achieved by traditional planar fabrication techniques (Sumetsky 2004; Xu and Brambilla 2007a, 2008a). However, it is difficult to achieve a freestanding coiled geometry in air. Generally, a micro-rod is employed to assist and wrap the MNOF on a low-index dielectric rod, which is the so-called wrap-on-a-rod or lab-on-a-rod technology.

The 3D warp-on-a-rod devices are manufactured with the setup illustrated in Fig. 8. In the first step, the MNOF is connected to an optical source via its pigtails and so that the optical spectra can be monitored in real time with a spectral analyzer. Next, with the help of a microscope, the MNOF is wrapped onto a low refractive index rod with surface functionalization, while one of its ends of the MNOF is fixed on a 3D stage. In the final step, the other MNOF end is fixed to another 3D stage, and both the MNOF ends are tuned to find the optimum spectra. The rod

Fig. 8 Schematic diagrams depicting the steps involved in the fabrication of 3D MNOF devices

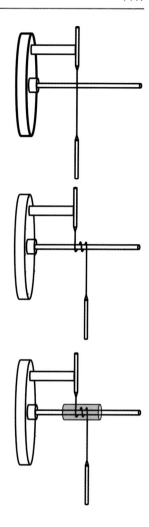

size typically ranges between tens of micrometers and several millimeters. The low-index polymers are often Teflon, UV glue, poly(dimethylsiloxane) (PDMS), and any other polymer with a refractive index smaller than the MNOF. The rod surface can be modified with any possible structure and materials. The whole device can be embedded in the low refractive index polymer to protect it from environmental damages (see Fig. 9).

As the MNOFs are manufactured from a SMF with low stiffness and micrometric bending radii, they can be wrapped multiple times on a thin micro-rod (hundreds of micrometers) and do not experience any input/output coupling problems. By tuning the pitch between the turns and modifying the micro-rod surface, a variety of functions (gratings, resonators, etc.) can be realized. It is possible to integrate multiple functions on a single device, which is the futuristic concept of the "lab-on-a-microrod."

Fig. 9 Schematic of an
embedded 3D MNOF device

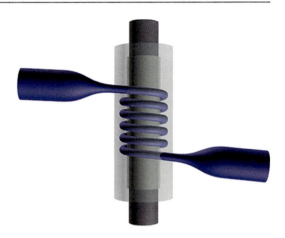

Surface Functionalization with External Materials

The functionalization and applications of MNOFs are limited by silica. Although,
it is possible to use several non-silica fibers to fabricate MNOFs; however, it is
still preferable for the MNOF to be perfectly compatible with the standard single-
mode fiber system. An alternative method is the integration of the external material
on the MNOF surface. The external materials can interact with the light in the
MNOF through which the evanescent field on the side or even directly interacts
on the endface. Depending on the application requirements, almost all types of
materials can be used to functionalize MNOF. The external materials include metals,
semiconductors, liquids, and low-dimensional materials. Especially, polymer and
two-dimensional materials are popular as they can be easily integrated with MNOFs.

Typical polymers used in MNOFs include poly(methyl methacrylate) (PMMA),
Efiron PC-373, PDMS, photo-responsive liquid crystals, high-substituted hydrox-
ypropyl cellulose (H-HPC), poly(ethylene oxide) (PEO), etc. It is important to
note that the viscosity of the HPC solution is very high. A HPC solution with a
mass fraction of only 2% at room temperature has a viscosity of 40 Pa·s. The high
viscosity can overcome the effects of gravity, yielding a uniform coating layer along
the fiber. Several studies have demonstrated the water vapor-induced modification
of the refractive index of a thin PEO layer. The high adhesion of PEO on silica
surfaces allows for the easy coating onto optical fibers.

Two-Dimensional Material Integration

Two-dimensional materials including graphene, transition metal dichalcogenides,
silicene, and phosphorene are low-dimensional materials with a thickness of a few
nanometers or less. There has been a widespread interest in the study of two-
dimensional materials due to their excellent optical, electrical, and mechanical
properties. Graphene, a monolayer of carbon atoms, exhibits a constant absorption
over a wide spectral range from the visible to the infrared due to its unique electronic
structure. The other unique merits of graphene include the exceptionally high

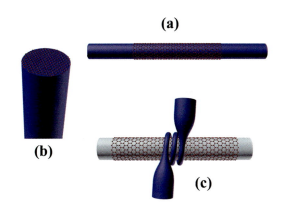

nonlinearity over a broad spectral range, tunability of chemical potential by doping
or electrical gating, and the high carrier mobility. Several graphene-based devices
have been developed including on-chip broadband modulators, photodetectors,
mode-lock lasers, optical limiters, saturable absorbers, gas and biosensors, etc.
Moreover, graphene acts as an attractive platform for plasmon waveguides, thereby
leading to stronger confinement of electromagnetic fields and lower loss compared
to noble metals such as gold and silver.

Graphene can be easily transferred and integrated onto a MNOF, therefore
allowing for the utilization of novel optoelectronic properties of graphene in fiber
devices. There are three different types of integration approaches based on the
light-graphene interaction, side integration, endface integration, and 3D integration,
as illustrated in Fig. 10. 3D stereo integration has the maximum light-graphene
interaction length but is complicated to fabricate. Side integration has a moderate
light-graphene interaction length. The endface integration is the easiest one in terms
of fabrication, but has a poor light-graphene interaction length. The integration
methods described here are also suitable for other two-dimensional materials.

Figure 11 illustrates the fabrication process involved in the 3D stereo integration
of graphene in MNOFs. First, a graphene Cu foil (Six Carbon Technology,
Shenzhen, China) was spin-coated with 4 wt.% PMMA-anisole solution and later
dried in an oven with 60 °C for 30 min. In the second step, the Cu film was dissolved
by using 1M $FeCl_3$ and the PMMA-graphene film was washed in deionized (DI)
water for several times. In the third step, the PMMA-graphene film was laminated
on a pretreated rod. A thin low-index Teflon layer (Teflon AF 601S1-100-6, DuPont,
tens of micrometers in thickness and a refractive index of $\sim$1.31) was dipcoated on
a rod's surface and dried in air for several hours. In the fourth step, the rod with
PMMA-graphene film was heated in an oven for 30 min with 80 °C. Subsequently,
the PMMA layer was removed by acetone, leaving the rod coated only with the
graphene layer. In the final step, a tapered MNOF was wrapped around the graphene-
functionalized rod. The self-coupling coil resonator was fabricated by carefully
controlling the adjacent coil's distance.

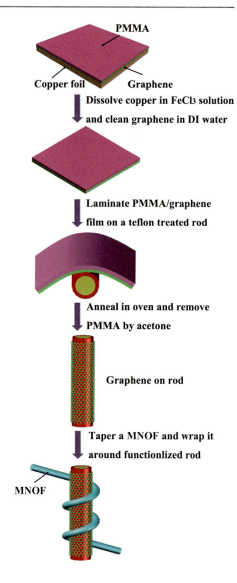

Fig. 11 Schematic diagram illustrating the fabrication process of 3D stereo integration of graphene in MNOFs

The fabrication process for the side integration of graphene to MNOFs is illustrated in Fig. 12. In the first step of this process, the graphene on copper foil was etched in a 1 M FeCl₃ solution and cleaned in deionized water for several times. In the second step, a MNOF was tapered into diameter of 5–10 μm directly from standard SMF-28. The MNOF's pigtails were fixed by scaffolds and the waist part of MNOF was freely hung in air. In the third step, the graphene layer floating on water surface in a container was placed right under the MNOF. By carefully tuning the distance and displacement between the graphene layer and MNOF, the graphene geometry was transferred on MNOF. When the MNOF touched graphene, we slowly

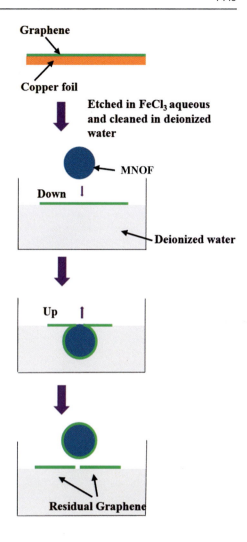

Fig. 12 Schematic diagram illustrating the different steps in the fabrication of side-integrated graphene MNOF

and carefully pressed down the MNOF until the graphene wrapped around MNOF completely as illustrated in Fig. 12. The final step involved the slow removal of the MNOF until the MNOF left the water surface. Due to the ultrahigh strength of graphene layer, when the MNOF was pressed against the graphene layer, the graphene membrane deformed rather than break apart.

The fabrication process of the endface integration is similar to the side integration process; the tip end can be moved to attach the graphene layer floating on water surface in a container. Similarly, graphene powder can be dispersed in a solution and drop cast on the endface. Gravity and surface tension of the drop would induce a semilunar-like structure, which encapsulated the facet of patch cord. Finally, the sample was heated in an oven at 100 °C for an hour to evaporate the solvent. A thin film of graphene nanosheet was naturally coated on the endface of the MNOF.

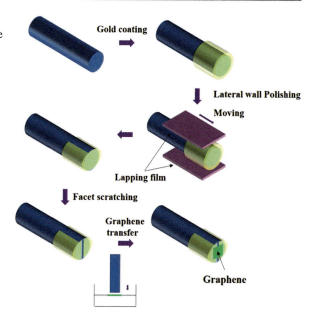

Fig. 13 Sequential fabrication process of endface bonding with electrodes

For certain optoelectronic application such as photodetectors, there is a necessity to fabricate electrodes on the MNOF. Figure 13 illustrates the typical process of graphene bonding with electrodes to a fiber facet. In the first step in this process, the SMF protective coating was peeled off and the optical fiber was ultrasonically treated in ethanol for a few minutes. Then, the fiber's end surface was cleaved and a flat edge on the facet was created. In the second step, the fiber was placed in a gold deposition chamber under vacuum (5×10^{-2} mbar). A thin gold film was deposited on the surface of the optical fiber. Subsequently the gold film was polished. In the next step, the gold film on the fiber facet was scratched into a narrow channel using a tapered tungsten probe under an optical microscope, while the lateral electrodes were directly obtained by using a lapping film. In our scratching process, we first input a red-light source into an optical fiber for guidance. Since the gold film's thickness is only ~30 nm, the red light can easily transmit over the thin film. The transmitted red-light signal can be used for orienting the optical fiber core's position, with which we can quite precisely scratch the gold film into a small channel right on the fiber core position. In the fourth step, the graphene sheet was transferred to a water surface by etching copper substrate with 1M $FeCl_3$. The graphene sheet was transferred to deionized water with a glass slide to ensure cleanliness. In the fifth step, the graphene was directly bonded to the fiber end surface using the dipcoating method. The as-fabricated device was annealed in a hot stage at 120 °C in air for 30 min to enhance the contact between the graphene and electrodes.

Passive and Active Devices

MNOFs provide a significant platform for the miniaturization of fiber-based devices and the integration of complex functionality. Currently, almost all conventional passive fiber components and devices can be realized with MNOFs including gratings, couplers, interferometers, and resonators. With the help of the wrap-on-a-rod technique, 3D devices can be obtained, which are impossible to be fabricated with conversational optical fibers. By adding extrinsic materials like graphene, new active devices can be realized, for example, lasers, modulators, photodetectors, and micro-electromechanical systems (MEMS).

Grating

Fiber Bragg gratings (FBGs) are periodic modulations of the refractive index along the length of the fiber. FBGs have been widely used as a sensing element, a filter, and a delay line. Standard FBGs are primarily manufactured by modifying the core refractive index using a phase mask and an ultraviolet (UV) laser. For the fabrication of MNOF-based Bragg gratings (MFBGs), several techniques have been reported. The following is a summary of these methods.

Wet Chemical Etching

For a conventional standard size FBG, the periodic refractive index modulations are inscribed in the photosensitive germanium-doped (Ge-doped) core with a diameter of several microns. Thus, a MFBG can be obtained by removing its cladding (Iadicicco et al. 2004; Liang et al. 2005). Generally, HF solution is used as the etchant for removing the cladding. The etching rate is determined by the concentration of the HF solution, and the diameter of the grating is controlled by the etching time. The whole fabrication process can be monitored in situ, by recording the transmission loss and the reflected spectrum of the grating. The advantages of the wet chemical etching technique for the fabrication of MFBG are the simplicity of the method and the low cost. However, the HF solution used in this method is poisonous and highly corrosive; therefore, the etching procedure should be handled with extreme care.

Ultraviolet (UV) Laser Irradiation

UV laser irradiation with a phase mask is considered as a standard writing method for the fabrication of conventional FBGs (Hill et al. 1993) and can be adopted in the manufacture of MFBGs. For example, Zhang et al. have demonstrated the fabrication of MFBGs using a 248 nm KrF excimer laser with the help of a uniform phase mask (Zhang et al. 2010b). However, the MNOFs in their experiments were highly Ge-doped and needed presoaking in hydrogen to guarantee adequate photosensitivity, therefore significantly complicating the fabrication process. To avoid additional sensitization treatments, 248 nm KrF excimer laser can be replaced by 193 nm ArF excimer laser in the MFBG inscription procedure, as the two-photon process at 193 nm can induce refractive index variation in materials without photosensitivity (Bilodeau et al. 1993). The inscription of Bragg gratings by the use of ArF excimer laser irradiation has been reported in MNOFs drawn from standard single-mode fibers (Ran et al. 2011) and standard multimode fibers (Ran et al. 2012), respectively.

Femtosecond Laser Irradiation
High-power femtosecond laser pulses may also be employed for the fabrication of MFBGs. A typical fs laser fabrication setup for MFBGs includes a femtosecond laser having a wavelength of 400 nm with 120 fs pulse width at a repetition rate of 1 kHz and maximum pulse energy of 1.3 mJ. Nayak et al. have demonstrated a Talbot interferometer, consisting of a phase mask as a beam splitter and two folding mirrors to create the two-beam interference pattern on the MNOF (Nayak and Hakuta 2013). Using this setup, a grating consisting of thousands of periodic nanocraters has been inscribed onto a MNOF by the irradiation of just a single fs laser pulse. Since an effective refractive index change is introduced by the structural fluctuation, this technique also allows the inscription of Bragg gratings in MNOFs made with materials without photosensitivity.

FIB Milling
FIB milling is a powerful mask-less patterning method that provides great flexibility to create various nanostructures with high degree of precision and has been employed for the fabrication of MFBGs. This method utilizes accelerated gallium ions to mill periodic nanoscale grooves on MNOFs, therefore resulting in the formation of surface-corrugated gratings. During the FIB process, the MNOF is tightly fixed in the vacuum chamber to avoid displacements. Depending on the milling area and the beam current used, the whole FIB fabrication process may take minutes to hours. After milling, the metal film on the MFBG can be easily removed by an appropriate etchant (Kou et al. 2011a).

Figure 14a displays an SEM image of a typical MFBG manufactured by the FIB milling technique. The grating was manufactured using the Helios 600i system. The MNOF diameter is 1.745 μm, the grating period is 580 nm, the milling depth is 150 nm, the number of grooves is 50, and the grating length is only ~29 μm. This grating has a strong refractive index modulation ($\sim 10^{-2}$ RIU). Figure 14b displays the reflectance spectra of the MFBG. A reflectance of 50% reflection can be achieved with a ~29 μm length MFBG. (RIU is the refractive index unit.)

Other Techniques
In addition to the methods mentioned above, other techniques have been explored for the fabrication of MFBG. Using the interaction between the evanescent field and external periodic structure, MFBGs can be manufactured by wrapping an MNOF on a microstructured low refractive index rod (see Fig. 15a) (Xu et al. 2009, 2010) or laying a MNOF on a low refractive index substrate with a surface-corrugated

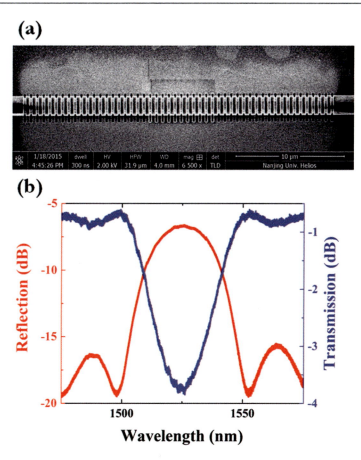

Fig. 14 (**a**) SEM image of a typical MFBG manufactured by the FIB milling technique. (**b**) Optical spectra of the grating (*black*, reflection; *purple*, transmission)

planar grating (see Fig. 15b) (Sadgrove et al. 2013). This method prevents the direct handling of fragile MNOFs and provides greater flexibility.

Ding et al. have demonstrated the fabrication of MFBGs by an interference-based lithography technique (Ding et al. 2006, 2007). The MFBGs can also be metallic gratings without plasma etch post-processing (Ding et al. 2006) or surface-corrugated gratings with plasma etch post-processing (Ding et al. 2007).

In the grating, the momentum mismatch between the forward and backward propagating modes is compensated by the reciprocal vector provided by the periodic index modulation. For the first-order diffraction commonly observed in MFBGs, the Bragg resonance condition can be obtained, namely, $\lambda_B = 2n_{eff}\Lambda$.

Before designing the grating, n_{eff} should be calculated first. For a uniform MFBG, the index modulation is relatively weak (usually on the order of 10^{-4}), and the cross section of the MNOF is a symmetrical circle. n_{eff} can be easily obtained by

Fig. 15 Schematic diagram of MFBGs manufactured by (**a**) wrapping a MNOF on a microstructured low refractive index rod (Xu et al. 2010) and (**b**) laying the MNOF on a low refractive index substrate with a surface-corrugated planar grating (Sadgrove et al. 2013)

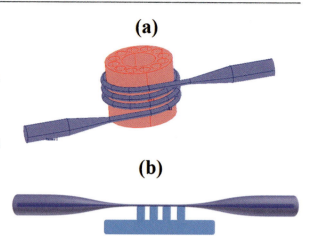

solving the dispersion equations numerically. However, for a structured-modulated MFBG, the effective index difference between the MNOF milled and un-milled cross section can be as large as $\sim10^{-3}$–10^{-1}, orders of magnitude larger than that in conventional FBGs. In such a case, a weighted average effective index of the grating region must be calculated.

Coupler

Coupler is one of the simplest passive optical fiber devices. The fabrication of MNOF coupler was first reported by Jung et al. (2009). In contrast to conventional couplers, MNOF coupler has a micrometer-size-diameter waist, typically 1–3 μm in diameter. To fabricate the MNOF coupler, two conventional optical fibers are placed side by side, tapered, and fused together with the help of a hydrogen-oxygen flame using the flame brushing method.

The coupler has one uniform waist region, two transition regions, two input ports, and two output ports. Due to the coupling of the evanescent field between the two clingy MNOFs, the power from one fiber can be coupled to the other fiber. The schematic diagram of the fused MNOF coupler is illustrated in Fig. 16. When light is injected into port 1, the normalized power transmitted outside of port 3 and 4 could be described with the following formula (Chen et al. 2014; Jung et al. 2009; Yan et al. 2015):

$$
\begin{aligned}
P_3 &= \tfrac{1}{2}\left\{1 + \cos\left[\left(\tilde{C}_x + \tilde{C}_y\right) L_C\right] \cdot \cos\left[\left(\tilde{C}_x - \tilde{C}_y\right) L_C\right]\right\}, \\
P_4 &= \tfrac{1}{2}\left\{1 - \cos\left[\left(\tilde{C}_x + \tilde{C}_y\right) L_C\right] \cdot \cos\left[\left(\tilde{C}_x - \tilde{C}_y\right) L_C\right]\right\}.
\end{aligned}
\tag{19}
$$

where L_C is the coupling length and C_x, C_Y are the averaged coupling coefficients for x and y polarizations. As the coupling coefficient is different for different wavelengths, the power distribution between the two fibers is dependent on the wavelength from 1:0 to 0:1.

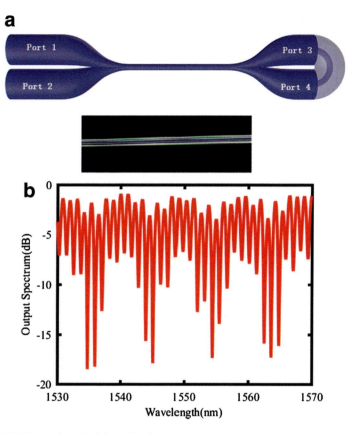

Fig. 16 MNOF coupler. (**a**) Schematic diagram of the MNOF coupler. Port 3 and 4 can be connected to form a reflective loop mirror. The insert figure is a microscope image of the sample with diameter of ∼3 μm. (**b**) Spectrum of the fabricated MNOF coupler

The coupling spectra are strongly dependent on the coupler diameter and coupling coefficient. A high extinction ratio and large period are observed when the diameter has a reasonable thickness. Both extinction ratio and period decrease when the diameter is reduced to several micrometers. The development of the envelope is obviously modulated by the polarization-dependent coupling, as shown in Fig. 16b. When the diameter of the coupler is less than 1–2 μm, the spectra become flat.

Resonator

MNOFs can be easily coiled with negligible bending loss, and they are benefit to form coil microresonators. There are three different types of MNOF resonators including the loop, knot, and coil resonators, as illustrated in Fig. 17a–c. The loop and knot resonators are example of 2D structures, while the microcoil resonator is a 3D resonator.

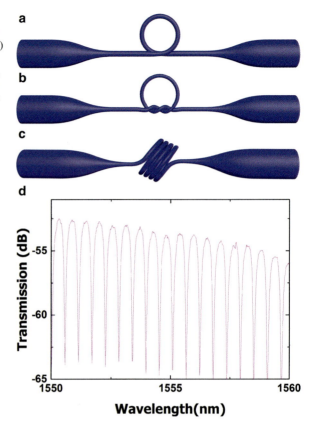

Fig. 17 Schematic diagram of the (**a**) loop, (**b**) knot, and (**c**) MNOF coil resonators. (d) Transmission spectrum of a typical single-loop resonator, which consisted of a 750 nm diameter MNOF, Q ∼15,500

The MNOF loop resonator is the simplest fiber coil resonator often constructed with just a single turn. It is a miniaturized version of the fiber loop resonator and was created in 1982 using a conventional SMF and a directional coupler (Stokes et al. 1982). Due to the bending losses of the weakly guiding single-mode optical fiber and the dimensions of the fiber coupler, the maximum value of the free spectral range of the fiber resonator was limited to the order of a gigahertz. In 1982, Caspar et al. fabricated a 2 mm diameter self-coupling fiber loop resonator by using a 8.5 μm diameter optical fiber taper (Caspar and Bachus 1989b). The fiber diameter was too large to ensure sufficient inter-fiber coupling; therefore, the MNOF was imbedded in a silicone rubber having a refractive index close to the index of the fiber in order to enhance the coupling efficiency. After the development of low-loss, thin MNOF fabrication techniques, Sumetsky demonstrated a small loop resonator using ∼1 μm diameter MNOF (Sumetsky et al. 2004, 2005, 2006). The Q-factors for the MNOF-based loop resonator were in the range 1500–120,000. A typical spectrum of the loop resonator is displayed in Fig. 17d. The fabrication of the loop resonator involves the drawing of the optical fiber nanowire and then bending it into a self-coupling loop with the input and output ends touching each other. The touching probably arises due to surface attraction forces (Van der Waals and electrostatic forces), which keeps the ends together. The surface attraction forces help the structure to overcome the elastic forces that would straighten the nanowire out. The characteristic diameters of the nanowire used for fabrication of a loop resonator are in the range of 600–1000 nm. The diameters are usually uniform along the 2 mm length of the optical nanowire. A major drawback of the self-touching loop resonator is its geometrical stability in air. The coupling is strongly affected by the microcoil geometry, and a small change in its shape can result in a large change in the transmission properties.

Knot resonators are challenging to fabricate, as the MNOFs need to be split to make a knot. In 2011, Xiao and Birks from Bath University demonstrated a knot resonator constructed using a whole 1 μm diameter MNOF (Xiao and Birks 2011). The knot resonators benefit from the long and stable coupling region. However they present severe drawbacks including the complexity in fabrication, necessity for an additional coupler at the output or input of the resonator, high loss, and the presence of only one standard telecom fiber pigtail for most cases.

MNOF coil resonator (MCR) is fabricated by wrapping the MNOF around a rod to realize the self-coupling of the evanescent field with several circles (see Fig. 17c) (Chen et al. 2011; Xu and Brambilla 2007a, 2008b; Xu et al. 2007). The coupling coefficient can be tuned by changing the spacing between circles and the number of circles. The interaction length can be tuned by changing the size of the circle. Due to the coupling between the circles, the light is repeatedly transmitted in the circle, and the optical power is confined in the resonators.

MCR can overcome stability issues and have the potential to act as a basic functional element for the MNOF-based photonics. MCR-based optical devices have two significant advantages over planar devices: smaller loss and compact nature. However, it is still difficult to manufacture MCR, as the MNOF is prone to breakage during the wrapping process. Prior to 2007, the experimental demonstrations of MCRs were reported in liquid (Sumetsky 2007) and in air (Xu and Brambilla 2007b). Similar to bare silica-based MNOFs, uncoated MCRs also suffer from degradation. To obtain a practical device, MCR needs to be coated with a low refractive index polymer such as Teflon (Xu and Brambilla 2007a). Embedding MCR and loop resonators in low refractive index materials has allowed for the construction of high-sensitivity microfluidic sensors (Xu and Brambilla 2008a; Xu et al. 2007). In 2008, two such sensors were experimentally demonstrated (Xu and Brambilla 2008b). Sensitivities as high as 700 nm/RIU have been predicted.

The resonance condition for the loop and knot resonators is similar to that of a conventional ring resonator and can be expressed with the following equation (Chen et al. 2011):

$$2\pi n_{\mathrm{eff}} L_R = u\lambda_R \qquad (20)$$

where u is an integer, L_R is the cavity length, n_{eff} is the effective refractive index, and λ_R is the wavelength. The spectra of a resonator depend on the loss, coupling coefficient, and cavity length. When there is no loss and gain, an ideal cavity would confine light indefinitely and the group delay is unlimited. The deviation from this ideal condition is described by the cavity Q factor (which is proportional to the confinement time in units of the optical period). The cavity Q factor can be expressed as (Schwelb 2004):

$$Q = \frac{\lambda}{\text{FWHM}} \tag{21}$$

where FWHM is the full width at half maximum of the transmission spectrum.

For 3D MCRs, the spectra are complicated due to the multiple coupling between the different coils. Therefore the coupled wave equations need to be solved numerically. However, the two-turn MCR is similar to the loop resonator; therefore, the resonator condition for the two-turn MCR can be expressed as:

$$\begin{aligned} K &= (2u - 1)\frac{\pi}{2} \\ \beta 2\pi R_0 &= 2v\pi + \frac{\pi}{2} \end{aligned} \tag{22}$$

where R_0 is the radius of coil and K is coupling parameter. $K = 2\pi R_0 \kappa$, κ is the coupling coefficient between coils and u, v are integers.

For three-turn coils, the resonance condition can be expressed with the following equation:

$$\begin{aligned} K &= (2u - 1)\frac{\pi}{\sqrt{2}} \\ \beta 2\pi R_0 &= v\pi \end{aligned} \tag{23}$$

Interferometers

Interferometers are one of the most popular fiber devices for sensing applications. There are different types of MNOF-based interferometers, including MNOF mode interferometer, MNOF Fabry-Perot interferometer, and MNOF Sagnac interferometer (MSI). There are two methods for the fabrication of mode interferometers. In the first method, the MNOF is tapered under non-adiabatic condition (Ji et al. 2012; Luo et al. 2014; Salcedadelgado et al. 2012; Zheng et al. 2015b). The higher-order mode is generated due to the steep angle of the transition region. Higher-order and fundamental modes are both transmitted in the waist region, with the interference occurring at the other transition region. In the second method, the waist region is separated as several parts with different diameters. The mismatch generates the higher-order mode (Jaddoa et al. 2016; Jasim et al. 2013a, b). The Sagnac interferometer can be fabricated by splitting the incident light beam into two sub-beams transmitting in opposite directions (Lim et al. 2010; Sun et al. 2012). Interference is observed when the two sub-beams meet each other. MNOF-based Fabry-Perot interferometer can be constructed using two reflective mirrors which are generally fabricated by milled a micrometer-size hole cavity in the MNOF by FIB or lasers, splicing a short piece of capillary or PCF between two SMFs.

The two-beam interference is the basic working principle of an interferometer. $I = I_1 + I_2 + 2\sqrt{I_1 I_2} \cos \frac{2\pi}{\lambda} \Delta$, where I_1 I_2 are the intensities of two beams and Δ is the optical path difference between the two beams. The interference wavelength is $\lambda_i = \Delta/u$, where u is an integer. Any changes in the optical path which the two beams experience induce a phase difference, which can be detected by monitoring the shift of interference wavelength.

Highly Birefringent MNOF

The asymmetry of the fiber in two orthogonal directions produces birefringence naturally with a magnitude of $\sim 10^{-4}$ or smaller. Although, the elimination of birefringence is important, few researchers have explored new approaches for high birefringence and associated phenomena.

Based on our literature survey, we found a variety of high-birefringence fiber studies. Photonic crystal fiber has become a hot research topic because of its special structural properties. With photonic crystals, it is easy to break the circular symmetry of optical fibers. A high birefringence can be achieved by designing a special air hole matrix structure on the photonic crystal fiber, therefore introducing two large air holes around the core layer and so on. On the other hand, high birefringence can be introduced by induced stress in Panda fiber, a tie-end fiber and other polarization-maintaining fiber. In addition, the researchers have also designed and fabricated high-birefringence MNOF. In 2010, Jung et al. obtained a high-birefringence fiber by drawing a polarization-maintaining fiber to a diameter of about 1 μm (Jung et al. 2010). In the same year, Hong Kong Polytechnic University researchers obtained a high-birefringence MNOF with an oval-shaped cross section by femtosecond processing to remove and pre-drawn ordinary SMF on the opposite side of the partial cladding (Xuan et al. 2010). Subsequently, Kou et al. from Nanjing University have designed a new type of microfiber structure by introducing an air slot in the microfiber to achieve a birefringence of 4×10^{-2} (Kou et al. 2011b).

High-birefringence MNOFs have several important applications. Due to the separation of the fundamental modes in the ordinary fiber, it is possible to obtain single-mode single-polarization transmission by forming high-loss geometry for only one polarization state, therefore reducing the polarization-dependent loss in the fiber transmission process. On the other hand, due to the difference in the refractive index of the two modes, it is possible to achieve the interference between the modes and therefore the ability to sense external variables.

A popular configuration of the use of highly birefringent MNOF is based on the Sagnac polarimetric interferometer, where the MNOF is inserted into a loop. In order to minimize the loop size, a Sagnac loop can be fabricated directly by twisting the highly birefringent and non-birefringent MNOFs (Sun et al. 2010), where the birefringence of coupling region is used. There is a considerable amount of power in the evanescent field of a MNOF; for a Sagnac loop, the braided MNOF takes a part of the coupler, and light propagating in the MNOF can couple from one section to another. A miniature Sagnac interferometer can be widely used in the construction of photonic devices such as linear edge filter, optical variable attenuator, gain flattening filter, and beat-wave interferometer (Chen et al. 2013; Sun et al. 2012).

Graphene-Integrated Devices

Graphene-fiber integration was first to be investigated for applications in laser and nonlinear optics. Conventionally, saturable absorbers are expensive semiconductors and have limited absorption bandwidth. Half-metallic graphene is a zero-bandgap material and can absorb considerable amount of radiation in the visible, infrared, and even THz waves without any bandwidth limitations. Moreover, graphene has high nonlinearity with $n_2 \sim 10^{-11}$ m²/W at 1550 nm, which is nine orders of magnitude larger than silica (Zhang et al. 2012). These advantages have resulted in the fast development of ultrafast pulse fiber laser based on graphene integrated on the fiber endface. However, the light interaction length is poor in this direct integration method. Most practical applications require a sufficient length and strength of interaction between graphene and the optical fields. Such integration allows for the novel optoelectronic properties of graphene to be effectively utilized in photonic devices. The integration can be achieved by transferring and laminating graphene on top of the MNOFs with an accessible evanescent field or the wrapping a MNOF on a graphene-coated rod. Currently, there are several reports that combine MNOF with graphene or graphene-like 2D materials [transition metal dichalcogenides (TMDs), topological insulators, etc.], to achieve functional applications such as fiber lasers (Bao et al. 2009; Du et al. 2014; Lee et al. 2015; Zhao et al. 2012), optical modulators (Gan et al. 2015; Li et al. 2014; Liu et al. 2013), photodetectors (Sun et al. 2015), polarizers (Bao et al. 2011), and sensors (Ma et al. 2012, 2013; Wu et al. 2014a; Zhang et al. 2014).

The light-graphene interaction length is the main difference among the different integration methods. As mentioned earlier, 3D integration can maximize the interaction length. Kou and Chen have demonstrated the manipulation of broadband polarization by wrapping a MNOF ($\sim$3 µm in diameter) on a rod-supported graphene sheet (Kou et al. 2014). The extinction ratio (ER) is $\sim$8 dB/turn, and by employing a two-coil structure, an ER as high as $\sim$16 dB was obtained over a 450 nm bandwidth in the telecommunication wavelength range. Subsequently, we reported a major step toward the realization of a high-Q graphene-based single-polarization resonator with excellent suppression of polarization noise. Utilizing the Pauli blocking effect in MNOF platform, Chen et al. have demonstrated an in-line, all-optical fiber modulator with a modulation depth of 7.5 dB (2.5 dB) and a modulation efficiency of 0.2 dB/mW (0.07 dB/mW) for the two polarization states (Chen et al. 2015). The modulation depth and modulation efficiency are more than one order of magnitude larger than those of other graphene-straight-MNOF hybrid all-optical modulators, although at the cost of a higher insertion loss. Complex functionality, such as electrical/optical modulation, high-speed photon detection, pulse lasing, and mechanical or biochemical sensing, can be performed by these rods with further design and processing of the coil geometry and rod surface. Additionally, in multi-coil resonators, new phenomena may be highly enhanced and observed due to the unique geometry and enhanced light-graphene interaction length. For example, the second-order nonlinearity stimulated Brillouin scattering and the magneto-optic effect, which have only been investigated in theory or demonstrated in the THz range. All of these phenomena are worth investigating.

Side integration has moderate light-graphene integration length and can be applied for strain engineering, sensing of gas molecules, refractive index and pressure, etc. The single-mode optical fiber has a poor theoretical sensing performance, as most of the light field is confined in the inner core structure, thereby inhibiting the light-environment interactions. Usually, researchers introduce microstructures into the core part of fiber, such as Bragg gratings, tilted gratings, and long period gratings. The core mode can be transferred into the cladding mode or the forward mode into the backward mode with mismatched momentum compensated by the reciprocal vector of the spatial microstructures. When the surrounding environment changes, the wave vector matching condition will change accordingly. The sensing function is based on the shift in the coupling wavelength. It is well-known that the MNOF has a strong evanescent field and good mechanical strength. From theoretical calculations and experiments, the sensitivity of traditional fiber gratings can be greatly improved when structures are inscribed into the MNOFs (Kou et al. 2012). However, these techniques cannot be applied to trace gas sensing, as the gas molecule is difficult to adsorb on the fiber/MNOF surface; therefore, only a small change in the refractive index occurs. Fortunately, graphene has a very large surface area per unit volume and high adsorption capability for gas molecules. In addition, graphene's conductivity can be easily manipulated by the adsorption of gas molecules. Hence, graphene has immense potentials in gas molecule sensing. Yunjiang Rao's group was the first to report on the integration of graphene and MNOF waveguide for high sensitivity, fast speed, and selective detection of gas molecules (Wu et al. 2014b). Recently, few researchers have deposited reduced graphene oxide (rGO) onto the surface of D-shaped fiber to measure the humidity and detect toluene (Xiao et al. 2014). These methods take advantage of the interaction between D-shaped fiber's evanescent field and rGO. Electrochemical methods for biomolecular sensing in solutions are an important area of research (Shao et al. 2010). In fact, if one could modify the surface properties of graphene and combine with optical fiber, theoretically, it might be possible to detect biomolecules with high sensitivity.

Endface integration is the simplest way to achieve weak light-graphene interaction. However, it is really challenging to transfer a graphene sheet on to a MNOF tip with a diameter less than 1 μm. Most of the reports are based on the fiber tip with diameters of tens of micrometers or even hundreds of micrometers. Chen et al. have demonstrated an all-fiber photodetector by directly bonding a piece of molybdenum disulfide (MoS_2) and electrodes to the fiber facet (Chen et al. 2017). The photodetector exhibited a high photoresponsivity of $\sim$0.6 A/W at 4 V and $\sim$0.01 A/W at 0 V bias for 400 nm light. Another interesting device making use of endface integration is the graphene NEMS, which was fabricated by the transfer of a suspended graphene membrane on a hollow endface. The hollow endface was prepared by etching or splicing a piece of capillary. A Fabry-Perot cavity is formed between the membrane and the base of the hole.

Atomically thick membranes are ultra-sensitive to ambient pressure. Changes in the geometry of the atomically thin membrane can be detected from the intensity variation of reflective interference signal. Ma et al. have presented an air pressure sensor (39.4 nm/kPa) based on a graphene-membrane Fabry-Perot cavity (Ma et al. 2012). Zheng et al. added additional electrodes and realized a current and magnetic sensor. The conventional optical fiber-based current sensors are built on thermal effects and therefore have several shortcomings such as relatively low sensitivity, long response time, and huge device size. The MNOF-based current sensor has simultaneously a high sensitivity of 2.2×10^5 nm/A^2, a short response time of $\sim$0.25 s, and a compact device size of $\sim$15 μm (Zheng et al. 2015a).

Sensing Applications

Using MNOF-based devices, a variety of physical, chemical, or biological optical sensors have been proposed and demonstrated. Different sensing principles have been employed in MNOF devices, for example, fluorescence effects, nonlinear effects, and linear optical path effects. The main sensing mechanisms of MNOF-based sensors include shifts in the characteristic wavelength, change of the output intensity as a function of change in the optical path. The change in the optical path can be due to a change in the refractive index, temperature, or strain/force. In such mechanisms, the sensitivity is primarily dependent on the material, size, and length of the MNOF. The different configurations of the optical device (e.g., resonator, grating, coupler, etc.) result only in different resolutions and detection ranges. Hence, we have only investigated the sensing characteristics of gratings.

As the effective index and period of grating is a function of d, n_{ou}, T, and ε, the Bragg condition can be rewritten as:

$$\lambda_B = 2n_{\text{eff}}\left(d, n_{\text{MNOF}}, n_{\text{ou}}, T, \varepsilon_e\right) \Lambda\left(T, \varepsilon_e\right) \tag{24}$$

The refractive index of fused silica and the ambient medium surrounding the MFBG are denoted by n_f and n_a, respectively. T is the operating temperature and ε is the strain applied to the MFBG (Kou et al. 2012). The sensitivity to refractive index from reference (Kou et al. 2012) can be expressed by the following equation:

$$S_{RI} = \frac{d\lambda_B}{dn_{ou}} = \frac{\partial \lambda_B}{\partial n_{\text{eff}}(n_{ou}, d)} \frac{\partial n_{\text{eff}}(n_{ou}, d)}{\partial n_{ou}} = 2\Lambda \frac{\partial n_{\text{eff}}}{\partial n_{ou}} \tag{25}$$

Typically, the MFBG sensor is immersed in an ambient liquid with RI in the 1.32–1.46 range. Depending on the radius of the MF and ambient liquid used, S_{RI} varies from 10^1 nm/RIU (refractive index unit) to 10^3 nm/RIU. Usually, a smaller radius and a larger RI of the ambient medium result in a higher sensitivity regardless of the fabrication method. In the literature, a sensitivity of 16 nm/RIU at a RI $\sim$1.35 with a MF 6 μm in diameter has been observed (Liang et al. 2005), which agrees well with the prediction from Eq. 25.

The temperature-induced thermo-optical and thermal expansion effect have obvious influence on the grating spectra in three ways: the temperature-induced index variation, the temperature-induced variation of the MF radius, and the grating period; each of the quantities is represented in the right hand of Eq. 7 where sensitivity of temperature (S_T) is defined by the following equation:

$$S_T = \frac{d\lambda_B}{dT} = 2\Lambda \left(\sigma_T \frac{\partial n_{\text{eff}}}{\partial n_f} + d\alpha_T \frac{\partial n_{\text{eff}}}{\partial d} + n_{\text{eff}}\alpha_T \right) + 2n_{\text{eff}}\Lambda\alpha_T \qquad (26)$$

where σ_T (1.2×10^{-5}/°C) is the thermo-optics coefficient and α_T (5.5×10^{-7}/°C) is the thermal expansion coefficient of fused silica. Nevertheless, the thermal expansion effect contributes less than 2 pm/°C to the total sensitivity mainly due to the low thermal expansion coefficient of silica and therefore can be neglected. S_T resulting from the thermo-optical effect is ~10–20 pm/°C and plays a dominant role in temperature sensing. The UV laser written grating is not suitable for high-temperature sensing applications due to the photosensitized index modulation, which is unstable at high temperature. Pure silica gratings with a surface congregation can survive up to ~1000 °C (Feng et al. 2011; Kou et al. 2011a). The sensitivity of these components is around 20 pm/°C and can be predicted with Eq. 26.

Based on the mechanics of materials, the strain sensitivity (S_S) is reduced to:

$$S_s = \frac{\Delta\lambda_B}{\varepsilon_e} = \lambda_B (1 - p_{\text{eff}}) \qquad (27)$$

where ε is the applied strain and p_{eff} the effective photoelastic coefficient, which is ~0.21 for an MFBG strain sensor. Therefore, S_S at a Bragg wavelength of 1550 nm is ~1.2 pm/$\mu\varepsilon$. This value is in excellent agreement with the experimental results (Sang-Mae et al. 2010; Wieduwilt et al. 2011). Another approach to characterize the capability of MFBG sensors is to use force sensitivity (S_F):

$$S_F = \frac{S_s}{E_e \pi r_{\text{MF}}^2} \qquad (28)$$

where E represents the Young' modulus. From Eq. 28, it can be concluded that S_F scales inversely with the fiber diameter. Luo et al. have milled a MFBG with a diameter of 2.5 μm using FIB. Their sensor achieved a force sensitivity of ~1900 nm/N, which is more than three orders of magnitude when compared to that of a conventional fiber (Wieduwilt et al. 2011). If the Bragg wavelength can be detected to an accuracy of 0.05 nm, it would be possible to measure forces in the order of 10 μN.

Besides the basic parameters like refractive index, temperature, and force, there are diverse ambient parameters that have real-world applications. For example, magnetic field, flow rate, electrical current, velocity, accelerator, and liquid pressure can be converted to these basic parameters. Generally, we need to convert changes in these parameters into the changes in the basic parameters. For example, electrical current can be measured by measuring the temperature via electric heating effects.

For resonators, couplers, and interferometers, the sensing performance is similar to that of gratings. However, different configurations bring unique advantages. For a resonator, the Q factor is relatively high and detection limit is low. Multiple channels in couplers allow for the signal processing of multiple parameters. For an interferometer, the size can be as small as a micrometer. Overall, there is a plethora of literature reports on the use of MNOF devices for sensing applications.

Fluidic Applications

Fiber-based optofluidics is an active research area involving the convergence of fiber optics and microfluidics, with great potential in medical and biochemical applications. One of the key issues in fiber-based optofluidics platforms is the conflict between the optical path and liquid path. However, with the help of the wrap-on-a-rod method, it is possible to construct a fiber-fluidic platform wherein a MNOF is wrapped around an ultrathin capillary. The capillary serves as the microchannel with tunable circles to enlarge the contact length and, therefore, enhance the interaction between the evanescent field associated with the light propagating in the MNOF and the liquid in the hollow channel.

Due to the unique geometry, the optical and liquid channels can be combined and separated smoothly, without conflict in input and output coupling, as presented in Fig. 18. Different optofluidic devices can be constructed with the integration of surface-functionalized capillaries and by tuning the stereo-geometry of the MNOF coils.

Fig. 18 MNOF fluidic platform prepared by wrapping a MNOF around an ultrathin capillary, which acts as the microchannel

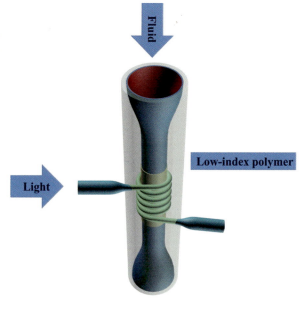

Based on this platform, Xu et al. have demonstrated a fluidic biosensor, which was realized by wrapping a MNOF to form a MCR on a PMMA rod. Subsequently, the PMMA rod was removed after embedding the whole device in Teflon (Xu and Brambilla 2008a; Xu et al. 2009). Lorenzi et al. have used a similar device to measure the absorption of gases (Lorenzi et al. 2011). By wrapping a MNOF coupler around a gold-functionalized capillary, a high-sensitivity "hot-wire" microfluidic flowmeter has been reported by Yan et al. (Yan et al. 2016). This sensor exhibited an ultrahigh sensitivity of 2.183 nm (μl/s) at a flow rate of 1 μl/s. Other physical and optical characteristics such as temperature, particle count, pressure, stimulated Raman scattering, and fluorescence signal can be measured with the help of fiber-based optofluidic devices. By integrating more coils in a single-hole capillary or even multi-hole capillary, it is possible to construct multipoint, multiparameter, and multichannel measurements for applications in online chemical and bio-detectors as well as tunable devices.

Conclusion

In conclusion, MNOFs with subwavelength diameters have significant optical and mechanical advantages over conventional optical fibers. MNOFs provide a unique and flexible platform for the integration of miniaturized and multifunctional optical fiber devices. With the help of different micromachinings, post-processing techniques, and addition of external materials like graphene, "lab-in-a-MNOF" can be fabricated with a diverse range of configurations, functions, and applications. Although this chapter mainly focuses on the recent developments in silica-based MNOFs, there is a vast and growing area of research focusing on the study of non-silica MNOFs fabricated from polymer fibers, chalcogenide glass fibers, and microstructured fibers. The non-silica MNOFs focus on new functionality and applications, especially in the mid-infrared region of the electromagnetic spectrum.

With all these developments, MNOF-based techniques have become an interdisciplinary research topic involving materials, optics, electronics, and biochemistry, thereby opening a range of new opportunities and challenges. We believe that further research on MNOFs will be focused on practical applications of MNOF and novel and interesting physical effects in MNOF devices.

Acknowledgments The author acknowledges financial support from the National Natural Science Foundation of China (61535005 and 61475069). The author gratefully acknowledges Zhen-xin Wu, Meng-tao Mao, Wei Luo, Jin-hui Chen, Jin-hong Li, and Shao-cheng Yan for their help in preparing the manuscripts.

References

Q. Bao, H. Zhang, Y. Wang, Z. Ni, Y. Yan, Z.X. Shen, K.P. Loh, D.Y. Tang, Adv. Funct. Mater. **19**(19), 3077–3083 (2009)
Q. Bao, H. Zhang, B. Wang, Z. Ni, C.H.Y.X. Lim, Y. Wang, D.Y. Tang, K.P. Loh, Nat. Photonics **5**(7), 411–415 (2011)
F. Bilodeau, K.O. Hill, D.C. Johnson, S. Faucher, Opt. Lett. **12**(8), 634–636 (1987)

F. Bilodeau, K.O. Hill, S. Faucher, D.C. Johnson, J. Lightwave Technol. 6(10), 1476–1482 (1988)
F. Bilodeau, Y. Hibino, M. Abe, B. Malo, J. Albert, M. Kawachi, D. Johnson, K. Hill, Opt. Lett. 18(12), 953–955 (1993)
T.A. Birks, Y.W. Li, J. Lightwave Technol. 10(4), 432–438 (1992)
R.W. Boyd, *Nonlinear Optics* (Academic, San Diego, 1992)
C.V. Boys, Philos. Mag. Ser.. 5 23(145), 489–499 (1887)
G. Brambilla, J Opt.-Uk 12(4), 043001 (2010)
G. Brambilla, V. Finazzi, D.J. Richardson, Opt. Express 12(10), 2258–2263 (2004)
G. Brambilla, E. Koizumi, X. Feng, D.J. Richardson, Electron. Lett. 41(7), 400–402 (2005)
G. Brambilla, F. Xu, X. Feng, Electron. Lett. 42(9), 517–519 (2006)
G. Brambilla, F. Xu, P. Horak, Y. Jung, F. Koizumi, N.P. Sessions, E. Koukharenko, X. Feng, G.S. Murugan, J.S. Wilkinson, et al., Adv. Opt. Photon. 1(1), 107–161 (2009)
J. Bures, R. Ghosh, J. Opt. Soc. Am. A 16(8), 1992–1996 (1999)
C. Caspar, E.J. Bachus, Electron. Lett. 25(22), 1506–1508 (1989a)
C. Caspar, E.J. Bachus, Electronnics Lett. 25(11), 1506–1508 (1989b)
Y. Chen, F. Xu, Y. Lu, Opt. Express 19(23), 22923–22928 (2011)
N.K. Chen, G.L. Cheng, Z.Y. Chen, IEEE Photon. Technol. Lett. 25(13), 1211–1213 (2013)
Y. Chen, S. Yan, X. Zheng, F. Xu, Y. Lu, Opt. Express 22(3), 2443–2450 (2014)
J.-H. Chen, B.-C. Zheng, G.-H. Shao, S.-J. Ge, F. Xu, Y.-Q. Lu, Light-Sci. Appl. 4, e360 (2015)
J.-h. Chen, Z.-h. Liang, L.-r. Yuan, C. Li, M.-r. Chen, Y.-d. Xia, X.-j. Zhang, F. Xu, Y.-q. Lu, Nanoscale 9(10), 3424–3428 (2017)
A.M. Clohessy, N. Healy, D.F. Murphy, C.D. Hussey, Electron. Lett. 41(17), 954–955 (2005)
W. Ding, S. Andrews, T. Birks, S. Maier, Opt. Lett. 31(17), 2556–2558 (2006)
W. Ding, S. Andrews, S. Maier, Opt. Lett. 32(17), 2499–2501 (2007)
J. Du, Q. Wang, G. Jiang, C. Xu, C. Zhao, Y. Xiang, Y. Chen, S. Wen, H. Zhang, Sci. Rep. 4(4), 6346 (2014)
X. Fei, B. Gilberto, Jpn. J. Appl. Phys. 47(8S1), 6675 (2008)
J. Feng, M. Ding, J.-l. Kou, F. Xu, Y.-q. Lu, IEEE Photon. J. 3(5), 810–814 (2011)
X. Gan, C. Zhao, Y. Wang, D. Mao, L. Fang, L. Han, J. Zhao. Opt. 2(5), 468 (2015)
K.O. Hill, B. Malo, F. Bilodeau, D. Johnson, J. Albert, Appl. Phys. Lett. 62(10), 1035–1037 (1993)
A. Iadicicco, A. Cusano, A. Cutolo, R. Bernini, M. Giordano, IEEE Photon. Technol. Lett. 16(4), 1149–1151 (2004)
M.F. Jaddoa, A.A. Jasim, M.Z.A. Razak, S.W. Harun, H. Ahmad, Sens. Actuator A-Phys. 237, 56–61 (2016)
A.A. Jasim, S.W. Harun, H. Arof, H. Ahmad, IEEE Sensors J. 13(2), 626–628 (2013a)
A.A. Jasim, S.W. Harun, M.Z. Muhammad, H. Arof, H. Ahmad, Sens. Actuator A-Phys. 192, 9–12 (2013b)
W.B. Ji, H.H. Liu, S.C. Tjin, K.K. Chow, A. Lim, IEEE Photon. Technol. Lett. 24(20), 1872–1874 (2012)
Y. Jung, G. Brambilla, D.J. Richardson, Opt. Express 17(7), 5273–5278 (2009)
Y.M. Jung, G. Brambilla, D.J. Richardson, Opt. Lett. 35(12), 2034–2036 (2010)
N.S. Kapany, Nature 184(4690), 881–883 (1959)
D. Karnakis, *Ultrafast Laser Nanomachining: Doing More with Less, Commercial MicroManufacturing.* (Oxford, 2008). https://www.researchgate.net/publication/266217841_Ultrafast_Laser_Nanomachining_Doing_More_With_Less
J.-l. Kou, S.-j. Qiu, F. Xu, Y.-q. Lu, Opt. Express 19(19), 18452–18457 (2011a)
J.-l. Kou, F. Xu, Y.-q. Lu, IEEE Photon. Technol. Lett. 23(15), 1034–1036 (2011b)
J.-L. Kou, M. Ding, J. Feng, Y.-Q. Lu, F. Xu, G. Brambilla, Sensors 12(7), 8861–8876 (2012)
J.-l. Kou, J.-h. Chen, Y. Chen, F. Xu, Y.-q. Lu, Optica 1(5), 307–310 (2014)
E.J. Lee, S.Y. Choi, H. Jeong, N.H. Park, W. Yim, M.H. Kim, J.-K. Park, S. Son, S. Bae, S.J. Kim, Nat. Commun. 6, 6068 (2015)
S.G. Leon-Saval, T.A. Birks, W.J. Wadsworth, P.S.J. Russell, M.W. Mason, Opt. Express 12(13), 2864–2869 (2004)

W. Li, B. Chen, C. Meng, W. Fang, Y. Xiao, X. Li, Z. Hu, Y. Xu, L. Tong, H. Wang, Nano Lett. **14**(2), 955–959 (2014)

W. Liang, Y. Huang, Y. Xu, R.K. Lee, A. Yariv, Appl. Phys. Lett. **86**(15), 151122 (2005)

S.D. Lim, K.J. Park, B.Y. Kim, K. Lee, S.B. Lee, *IEEE. 2010 Conference on Optical Fiber Communication Ofc Collocated National Fiber Optic Engineers Conference Ofc-Nfoec* (2010)

S.D. Lim, K.J. Park, B.Y. Kim, K. Lee, S.B. Lee, An optical microfiber sagnac interferometer with adjustable transmission, in *Optical Fiber Communication Conference, OSA Technical Digest (CD) (Optical Society of America, 2010), Paper JWA7. Optical Fiber Communication Conference 2010*, San Diego, 21–25 March 2010, pp. 1–3

Y. Liu, C. Meng, A.P. Zhang, Y. Xiao, H. Yu, L. Tong, Opt. Lett. **36**(16), 3115–3117 (2011)

Z.B. Liu, M. Feng, W.S. Jiang, W. Xin, P. Wang, Q.W. Sheng, Y.G. Liu, D.N. Wang, W.Y. Zhou, J.G. Tian, Laser Phys. Lett. **10**(6), 065901 (2013)

R. Lorenzi, Y. Jung, G. Brambilla, Appl. Phys. Lett. **98**(17), 173504 (2011)

J.D. Love, W.M. Henry, W.J. Stewart, R.J. Black, S. Lacroix, F. Gonthier, Iee Proc.-J. Optoelectron. **138**(5), 343–354 (1991)

W. Luo, J.-l. Kou, Y. Chen, F. Xu, Y.-Q. Lu, Appl. Phys. Lett. **101**(13), 133502 (2012)

H. Luo, Q. Sun, Z. Xu, D. Liu, L. Zhang, Opt. Lett. **39**(13), 4049–4052 (2014)

J. Ma, W. Jin, H.L. Ho, J.Y. Dai, Opt. Lett. **37**(13), 2493–2495 (2012)

J. Ma, H. Xuan, H.L. Ho, W. Jin, Photon. Technol. Lett. IEEE **25**(10), 932–935 (2013)

K. Nayak, K. Hakuta, Opt. Express **21**(2), 2480–2490 (2013)

K. Okamoto, *Fundamentals of Optical Waveguides* (Elsevier, Amsterdam, 2006)

Y. Ran, Y.-N. Tan, L.-P. Sun, S. Gao, J. Li, L. Jin, B.-O. Guan, Opt. Express **19**(19), 18577–18583 (2011)

Y. Ran, L. Jin, Y.-N. Tan, L.-P. Sun, J. Li, B.-O. Guan, IEEE Photon. J. **4**(1), 181–186 (2012)

L. Rihakova, H. Chmelickova, Adv. Mater. Sci. Eng. **2015**, 6 (2015)

V. S Afshar, T.M. Monro, Opt. Express **17**(4), 2298–2318 (2009)

M. Sadgrove, R. Yalla, K.P. Nayak, K. Hakuta, Opt. Lett. **38**(14), 2542–2545 (2013)

G. Salcedadelgado, D. Monzonhernandez, A. Martinezrios, G.A. Cardenassevilla, J. Villatoro, Opt. Lett. **37**(11), 1974–1976 (2012)

L. Sang-Mae, S.S. Saini, J. Myung-Yung, IEEE Photon. Technol. Lett. **22**(19), 1431–1433 (2010)

O. Schwelb, J. Lightwave Technol. **22**(5), 1380–1394 (2004)

Y. Shao, J. Wang, H. Wu, J. Liu, I.A. Aksay, Y. Lin, Electroanalysis **22**(10), 1027–1036 (2010)

L.F. Stokes, M. Chodorow, H.J. Shaw, Opt. Lett. **7**(6), 288–290 (1982)

M. Sumetsky, Opt. Express **12**(10), 2303–2316 (2004)

M. Sumetsky, *Demonstration of a Multi-turn Microfiber Coil Resonator*. Optical Fiber Communication Conference (San Diego, 2007)

M. Sumetsky, Y. Dulashko, A. Hale, Opt. Express **12**(15), 3521–3531 (2004)

M. Sumetsky, Y. Dulashko, J.M. Fini, A. Hale, Appl. Phys. Lett. **86**(16), 161108 (2005)

M. Sumetsky, Y. Dulashko, J.M. Fini, A. Hale, D.J. DiGiovanni, J. Lightwave Technol. **24**(1), 242–250 (2006)

L. Sun, J. Li, Y. Tan, X. Shen, X. Xie, S. Gao, B. Guan, Opt. Express **20**(9), 10180–10185 (2012)

X. Sun, C. Qiu, J. Wu, H. Zhou, T. Pan, J. Mao, X. Yin, R. Liu, W. Gao, Z. Fang, Opt. Express **23**(19), 25209–25216 (2015)

L.M. Tong, R.R. Gattass, J.B. Ashcom, S.L. He, J.Y. Lou, M.Y. Shen, I. Maxwell, E. Mazur, Nature **426**(6968), 816–819 (2003)

L.M. Tong, J.Y. Lou, E. Mazur, Opt. Express **12**(6), 1025–1035 (2004)

L. Tong, L. Hu, J. Zhang, J. Qiu, Q. Yang, J. Lou, Y. Shen, J. He, Z. Ye, Opt. Express **14**(1), 82–87 (2006)

L.M. Tong, F. Zi, X. Guo, J.Y. Lou, Opt. Commun. **285**(23), 4641–4647 (2012)

T. Wieduwilt, S. Bruckner, H. Bartelt, Meas. Sci. Technol. **22**(7), 075201 (2011)

Y. Wu, B. Yao, A. Zhang, Y. Rao, Z. Wang, Y. Cheng, Y. Gong, W. Zhang, Y. Chen, K. Chiang, Opt. Lett. **39**(5), 1235–1237 (2014a)

Y. Wu, B.C. Yao, Y. Cheng, Y.J. Rao, Y. Gong, W. Zhang, Z. Wang, Y. Chen, IEEE J. Sel. Top. Quantum Electron. **20**(1), 49–54 (2014b)

L. Xiao, T.A. Birks, Opt. Lett. **36**(7), 1098–1100 (2011)

Y. Xiao, J. Zhang, X. Cai, S. Tan, J. Yu, H. Lu, Y. Luo, G. Liao, S. Li, J. Tang, et al., Opt. Express **22**(25), 31555–31567 (2014)

F. Xu, G. Brambilla, Opt. Lett. **32**(15), 2164–2166 (2007a)

F. Xu, G. Brambilla, IEEE Photon. Technol. Lett. **19**(19), 1481–1483 (2007b)

F. Xu, G. Brambilla, Appl. Phys. Lett. **92**(10), 093112 (2008a)

F. Xu, G. Brambilla, Appl. Phys. Lett. **92**(10), 101126 (2008b)

F. Xu, P. Horak, G. Brambilla, Opt. Express **15**(12), 7888–7893 (2007)

F. Xu, G. Brambilla, Y. Lu, Opt. Express **17**(23), 20866–20871 (2009)

F. Xu, G. Brambilla, J. Feng, Y.-Q. Lu, IEEE Photon. Technol. Lett. **22**(4), 218–220 (2010)

F. Xu, Q. Wang, J.F. Zhou, W. Hu, Y.Q. Lu, IEEE J. Sel. Top. Quantum Electron. **17**(4), 1102–1106 (2011)

H.F. Xuan, J. Ju, W. Jin, Opt. Express **18**(4), 3828–3839 (2010)

S. Yan, Y. Chen, C.H. Li, F. Xu, Y. Lu, Opt. Express **23**(7), 9407–9414 (2015)

S.C. Yan, Z.Y. Liu, C. Li, S.J. Ge, F. Xu, Y.Q. Lu, Opt. Lett. **41**(24), 5680–5683 (2016)

E.J. Zhang, W.D. Sacher, J.K.S. Poon, Opt. Express **18**(21), 22593–22598 (2010a)

Y. Zhang, B. Lin, S.C. Tjin, H. Zhang, G. Wang, P. Shum, X. Zhang, Opt. Express **18**(25), 26345–26350 (2010b)

H. Zhang, S. Virally, Q. Bao, L.K. Ping, S. Massar, N. Godbout, P. Kockaert, Opt. Lett. **37**(11), 1856–1858 (2012)

J. Zhang, G. Liao, S. Jin, D. Cao, Q. Wei, H. Lu, J. Yu, X. Cai, S. Tan, Y. Xiao, Laser Phys. Lett. **11**(3), 035901 (2014)

C. Zhao, Y. Zou, Y. Chen, Z. Wang, S. Lu, H. Zhang, S. Wen, D. Tang, Opt. Express **20**(25), 27888–27895 (2012)

B.-C. Zheng, S.-C. Yan, J.-H. Chen, G.-X. Cui, F. Xu, Y.-Q. Lu, Laser Photonics Rev. **9**(5), 517–522 (2015a)

Y. Zheng, X. Dong, C.C. Chan, P.P. Shum, H. Su, Opt. Commun. **336**, 5–8 (2015b)

Part VIII
Optical Fiber Device Measurement

Measurement of Optical Fiber Grating

38

Zhiqiang Song, Jian Guo, Haifeng Qi, and Weitao Wang

Contents

Abstract

As one of the key photonic devices, optical fiber grating has been playing an important role in the fiber communications and remote sensing. In research, development, and application of fiber gratings, it is necessary to apply a range of measurement techniques for characterization and evaluation. This chapter introduces the major types of optical fiber gratings and describes related characterization and measurement techniques. Firstly, the history of fiber grating is briefly reviewed and different types of fiber gratings are introduced. Then the theoretical definitions and experimental measurements of typical parameters for fiber gratings have been described in details.

Z. Song (✉) · J. Guo · H. Qi · W. Wang
Shandong Key Laboratory of Optical Fiber Sensing Technologies, Qilu Industry University (Laser Institute of Shandong Academy of Sciences), Jinan, China
e-mail: szq821214@163.com; guojianphy@163.com; qihf@sdlaser.cn; wangweitao@sdlaser.cn

© Springer Nature Singapore Pte Ltd. 2019
G.-D. Peng (ed.), *Handbook of Optical Fibers*,
https://doi.org/10.1007/978-981-10-7087-7_44

Introduction

Fiber Grating Development

In 1978, Hill K.O. et al. first observed optical fiber photosensitivity in Ge-doped fiber, where he injected 488 nm argon ion laser into a smooth end optical fiber (Hill et al. 1978) and the laser generated standing wave effect then inscribed the grating structure in the fiber core. Then in 1989, Melts G. et al. reported the fiber grating fabrication based on lateral exposure of high intensity ultra-violet laser generated interference fringes on the optical fiber (Meltz et al. 1989), since then the fiber grating technology has been developed rapidly. In 1993, advances in technologies such as phase mask application and photosensitivity enhancement significantly simulated fiber grating fabrication technology (Hill et al. 1993), thereby enabled the mass production of fiber gratings.

Fiber grating has many advantages such as compact size, good wavelength selectivity, nonlinear effects immunity, polarization insensitivity, fiber system inherent compatibility, ease to use and maintenance, wide bandwidth range, and low additional loss, combined with highly developed fiber grating fabrication process, which made low cost mass production of fiber gratings possible. Thus, it has good practicality and irreplaceable superiority over many other devices. All these above features make fiber grating and fiber grating-based devices ideal components for all-optical network (Baumann et al. 1996; Dong et al. 1996; Ball and Glenn 1992).

Photosensitive Fiber

After the first demonstration of photosensitivity of optical fibers, Russel et al. found that the ultraviolet radiation-induced germanium color-centers re-organization is the origin of fiber index change (Russell et al. 1991). Then in 1993, Lemaire P.J. et al. proposed hydrogen loading as a technique which can greatly increase the photosensitivity in optical fibers (Lemaire 1993). However, the loaded hydrogen leads to degeneration of the fiber structure strength. In the meantime, Williams D.L. et al. discovered that the Boron ions co-doped germanosilicate fiber can enhance the photosensitivity through the relaxation of photon-induced elastic force in the fiber core (Williams et al. 1993). Other types of rare earth doped fibers also proved effective at photosensitivity enhancement (Dong et al. 1995a, b; Myslinski et al. 1999). Recently, Peng G.D. et al. showed dye-doped polymer fiber can acquire photosensitivity with relatively small pump power (Peng et al. 1999). With the advancement of photosensitive specialty fiber technology, fiber grating inscription for multiple applications including fiber laser, fiber sensor, and fiber amplifier received great development.

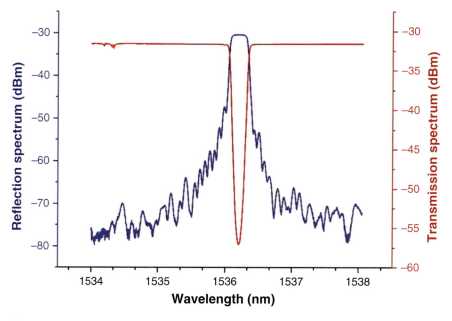

Fig. 1 Typical reflection and transmission spectrum of a fiber Bragg grating

Fiber Grating Types

Classification Based on Grating Structure

Fiber Bragg grating (FBG): including apodized fiber Bragg grating, which is the first developed fiber grating with the widest applications nowadays. Its refractive index modulation depth and the grid period are constant, and its grating wave vector direction is consistent with the fiber axis. Such gratings have important applications in fiber lasers, optical fiber sensors, optical fiber wavelength division multiplexing/demultiplexing, and other fields. Its typical spectrum is shown in Fig. 1.

   Phase shift grating: created by interrupting the spatial distribution at some point in the Bragg grating to generate a discontinuous index of refraction variation. It can be seen as a discontinuous connection of two (or more) gratings. It can open a transmission window in the periodic grating spectral band, so that the grating has higher degree of selectivity for a certain wavelength, which can be used to construct multichannel filter devices and narrow linewidth fiber lasers. Its typical spectrum is shown in Fig. 2.

   Chirped grating: fiber gratings with unequal grid spacing, including linear chirp and nonlinear chirped grating, mainly used for dispersion compensation and fiber amplifier gain flat. Its typical spectrum is shown in Fig. 3.

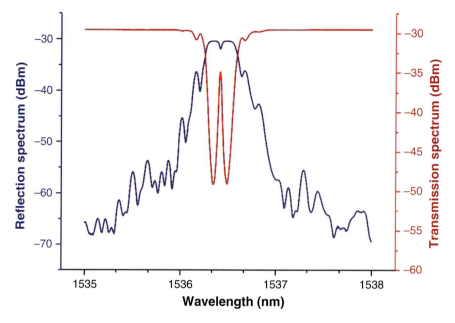

Fig. 2 Typical reflection and transmission spectrum of a phase shift grating

Tilted grating: during grating fabrication process, the ultra violet beam is inclined to the fiber axis, resulting in a small angle between the variation of refractive index and fiber axis. It is characterized by the fact that the grating has not only the reflection wavelength of the core mode, but also a series of loss wavelengths due to the existence of the cladding mode, which made it useful as an optical filter. Its typical spectrum is shown in Fig. 4.

Besides the above four types of gratings, there are other types of gratings such as long period grating, sampled fiber grating, superstructure grating, the Tapered grating, and Tophat grating.

Classification Based on Grating Fabrication Mechanism

Type I grating: the most common fiber grating, can be made in any type of photosensitive fiber. Main feature is that the reflection spectrum and transmission spectrum of its waveguide modes compensates with each other, with negligible absorption or cladding coupling loss, another feature is its susceptibility to "bleaching," i.e., the grating will start to deteriorate at a relatively low temperature (about 200 °C).

Type II grating: fabricated by laser exposure employing a single high energy (typically greater than 0.5 J/cm^2) pulse. Its transmission spectrum only allows light with wavelength greater than the Bragg wavelength to transmit, while the other portion is coupled into the cladding and damped. The grating inscription mechanism can be understood as the result of physical damage from the laser pulse nonuniform

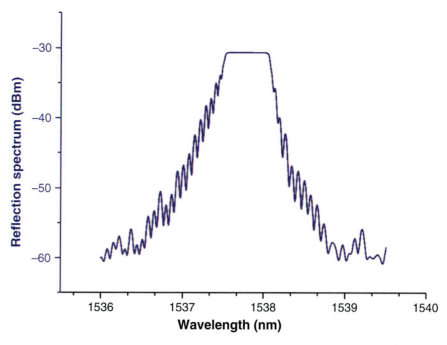

Fig. 3 Typical reflection spectrum of a chirped grating

energy which is strongly amplified by the fiber core quartz. It has superior high temperature stability, receives no significant change at 800 °C in 24 h, and can withstand 4 h in 1000 °C environment before grating gradually disappears.

Type IIA grating: also known as regenerative grating, inscribed on high germanium-doped (15% mol) photosensitive fiber or boracium-germanium co-doped photosensitive fiber. It is fabricated using similar mechanism as Type I grating but with a much longer exposure time, which will partially or completely erase the preformed Type I grating with the increase of the exposure time and regenerate a second grating on its ruin, that is, the Type IIA grating. It has better temperature stability than Type I grating and can sustain until 500 °C before the grating is "bleached," which is more suitable for usage at high temperature field, such as high temperature sensing.

Classification Based on Fiber Materials

Silica fiber grating: Fiber gratings written on silica fibers are the most conventional grating type for mainstream research, and industry usage. The common fiber materials include pure silica, germanium, aluminum, and fluorine doped silica as well as rare earth ions doped silica. Silica fiber gratings are relatively easy to produce, vastly adjustable, and inherently compliant with general silica fiber-based systems.

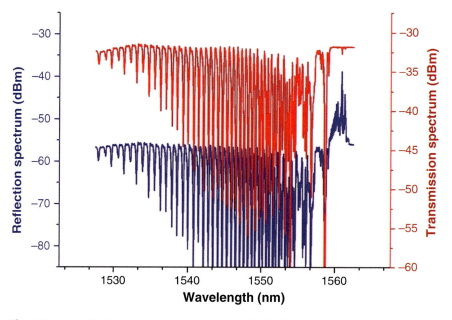

Fig. 4 Typical reflection and transmission spectrum of a tilted grating

Polymer fiber grating : Recent advances has made grating inscription on polymer fibers possible (Peng et al. 1999). Different from silica fiber grating, the fiber materials are made of plastic, such as acrylic and polystyrene. Polymer fiber gratings benefit from the superior elastic and high refractive index properties of polymer optical fiber over common silica fiber gratings, which offers them an ultra-high strain sensitivity. Such features are crucial to various medical and biological applications. On the other hand, polymer fiber gratings have great mechanical flexibility and robustness; however, they are also limited by the high attenuation loss and low reflectivity.

Fiber Grating Fabrication

Since the emergence of first fiber grating, many kinds of fiber grating inscription methods have been proposed according to the characteristics of different gratings, including standing wave interferometry, lateral holographic exposure method, point-by-point exposure method, and phase mask direct exposure method (Loh et al. 1995; Hill and Meltz 1997; Abdullina et al. 2012). With the continuous development of automated control technology, the inflexible traditional methods that require high stability become obsolete; currently the most widely used inscription method is dynamic scan phase mask process (Poladian et al. 2003). It employs a conventional phase mask installed on a piezoelectric displacement platform that can precisely control the position of the action, during the scanning exposure inscription, the phase mask moves in accordance with the piezoelectric platform in a preset motion pattern, in this way it can flexibly change the modulation distribution function of fiber grating refraction index, in order to achieve different types of fiber grating production.

Fiber grating production methods can also be classified into UV exposure and femtosecond laser exposure method (Jiangand and Wang 2008; Kawamura et al. 2000). UV exposure method utilizes ultraviolet photosensitivity of optical fiber. Germanium-doped silica fiber material has two significantly strong absorption band at 195 nm and 240 nm band; thus, the refractive index of the fiber material under UV exposure will change to form a grating structure. Over the years UV excimer laser has become the most commonly used tool for writing FBGs, but it has some drawbacks, first of all, fiber core of FBGs is highly dependent on the nature of the doped material; secondly, at high temperature, the refractive index grating modulation can be easily erased. Writing fiber gratings using UV exposure method is a single-photon absorption process; on the other hand, the femtosecond laser exposure inscription is a multiphoton absorption process, such as 800 nm femtosecond laser irradiation on fiber which goes through five photon absorption processes to excite the material. Thus, femtosecond laser can inscribe gratings in pure silica fiber with no photosensitivity, or writing type II grating with higher temperature stability.

Fiber Grating Measurement

Measurement of Fiber Grating Structural Parameters

The structural parameters of the fiber grating mainly depend on design, which includes size, refractive index modulation, and modulation distribution function. The size of the grating and the refractive index modulation distribution function can be precisely controlled in the grating fabrication process, but the existing optical instruments cannot measure the grating directly. The refractive index modulation can be calculated by measuring the spectral changes in the grating inscription process.

One method is to measure the Bragg wavelength variation of the fiber grating. The Bragg wavelength of the fiber grating λ_B is expressed as:

$$\lambda_B = 2 \left(n_0 + \Delta n_{eff} \right) \Lambda \qquad (1)$$

where n_0 is the original refraction index of the fiber, Δn_{eff} is the refraction index modulation of fiber grating, and Λ is the grating period. Since Δn_{eff} gradually increases during the grating inscription process, therefore after the grating inscription, the Bragg wavelength will experience red-shift. According to Eq. 1, the Δn_{eff} can be calculated as:

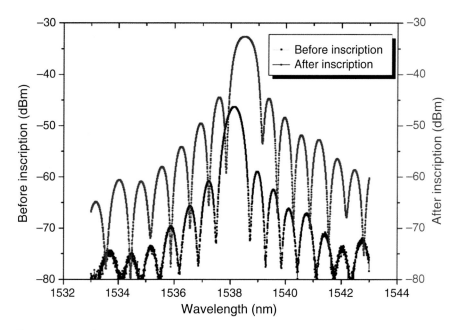

Fig. 5 Example fiber Bragg grating reflection spectrum before and after the inscription process

$$\Delta n_{eff} = \frac{\lambda_B}{2\Lambda} - n_0 \tag{2}$$

This method is suitable for direct exposure fabrication method. Examples are shown in Fig. 5: a 2 mm long grating is prepared with a grating period of 529.41 nm and refraction index of 1.44704555, its initial Bragg wavelength (black line) and final Bragg wavelength (blue line) is 1532.16 nm and 1532.525 nm, respectively. Then the Δn_{eff} is calculated to be 3.44×10^{-4}.

Another method is to retrieve the refraction index modulation through measuring the peak reflectivity of a fiber grating. The maximum reflectivity of the fiber grating is:

$$r_{max} = \tanh^2(kL) \tag{3}$$

where $k = \frac{\pi}{\lambda}\nu\Delta n$, L is the length of the grating, ν is the contrast ratio of the interference fringe, which can be set to 1 in case of good ultra violet laser coherence and system stability, Δn is the grating refraction index modulation, which can be calculated through measuring the grating transmission spectrum to obtain its reflectivity r_{max}, and then substitute it into:

$$\Delta n = \frac{arc \tanh\left(\sqrt{r_{max}}\right)\lambda}{\pi \nu L} \tag{4}$$

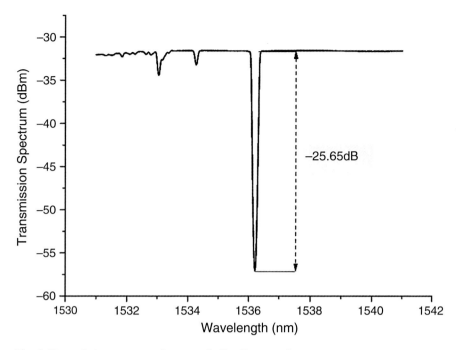

Fig. 6 Transmission spectrum of an example fiber Bragg grating

This method is suitable for scanning exposure inscription method. Example is shown in the following Fig. 6: a 10 mm long grating is fabricated with a grating wavelength of 1530.764 nm, its grating reflectance is measured to be −25.65 dB, using Eq. 4, Δn is calculated as 1.95×10^{-4}.

Measurement of Fiber Grating Frequency Specifications

The frequency specifications of the fiber grating, also known as spectral characteristics, mainly including center wavelength, bandwidth, side mode suppression ratio, and reflectivity, etc., associated measurement is usually carried out employing spectrometer, fiber grating demodulator, and swept laser methods.

Spectrometer Measurement
This method employs a broadband light source, e.g., an Amplified Spontaneous Emission (ASE) module, with two Optical Spectrum Analyzer (OSA) to measure the transmission and reflection spectrum of the grating under test, respectively. The setup schematic is shown in Fig. 7.

ASE generated flat broadband light goes through the Optical Circulator into the fiber grating, the light with wavelength other than fiber grating resonant wavelength will transmit through the grating, while light at the resonant wavelength will

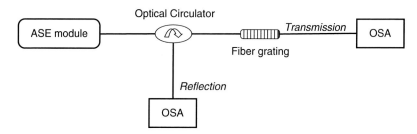

Fig. 7 Setup schematic of spectrometer measurement scheme

be reflected back or become cladding mode loss. Its transmission spectrum and reflection spectrum can be measured by the OSA. For the wide band gratings such as long-period fiber grating, tilted grating, and sampling grating, in order to accurately measure the grating spectrum, it is usually required to do spectral processing to eliminate the broadband light source induced uneven effect. For example, in Fig. 8 the **Bragg Wavelength** is defined as the center wavelength of the main mode, which is measured by taking 3 dB down from the highest reflection spectrum peak, the wavelength at the middle of the top 3 dB band is the Bragg wavelength, which is 1536.20 nm. And the width of this 3 dB band is defined as the **Grating Bandwidth**, which is 0.25 nm. On the other hand, the **Side Mode Suppression Ratio** is defined as the intensity difference between the main mode and the first side mode peak, which is marked as 17.74 dB in Fig. 8 with red line.

Fiber Grating Demodulator Measurement

Fiber grating demodulator mainly comprises broadband scanning light source, wavelength calibration, data acquisition, and scanning control modules plus some optical auxiliary devices. The ASE output goes through an optical tapping element into each sensing channel, respectively, while the reflected signals from fiber grating sensors are received by a set of photodiodes which transfer them into electrical signals, then after magnifying and filtering process, the demodulated electrical signal becomes wavelength data with associated sensor information.

Figure 9 shows a typical fiber grating demodulator. Its demodulation wavelength range is generally 1525–1565 nm, with an accuracy of 10 pm and a resolution up to 2 pm, and the number of sensing channels as well as scan frequency can be set, so large-capacity sensor demodulation can be achieved with calibration.

Fiber demodulator based on fiber grating sensors is widely used in various structural health monitoring systems, such as bridges, dams, aerospace, petrochemical, and other industries that involves fiber grating temperature sensor, fiber grating pressure sensor, and fiber grating strain sensor. However, the fiber grating demodulator can only measure the wavelength variation of the

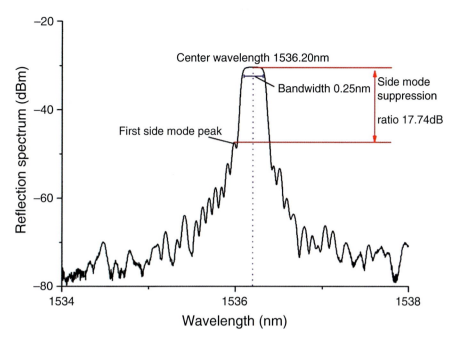

Fig. 8 Examples of OSA measured grating spectra

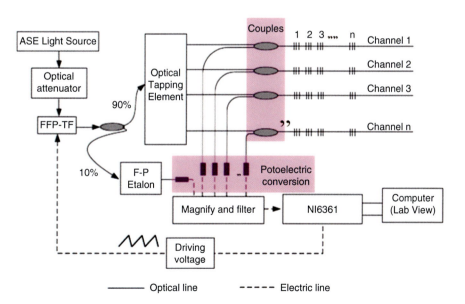

Fig. 9 Schematic Setup of a typical fiber grating demodulator

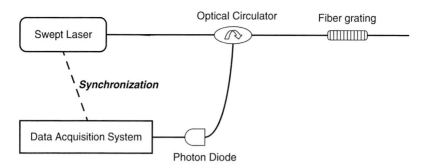

Fig. 10 Schematic setup of swept laser measurement

grating, which also requires a higher side-mode rejection ratio of the fiber grating. The nonchirped fiber gratings are often inaccurate in wavelength measurement.

Swept Laser Measurement

For ordinary fiber gratings, the spectrum measurement generally does not need very high spectral resolution; however, for grating devices which have very narrow filter characteristics, such as phase-shifted fiber gratings, strongly coupled fiber gratings formed FP cavity and tilted gratings, the resolution of the instruments like spectrometer or fiber grating demodulator cannot guarantee accurate measurement of the spectrum, which requires narrow linewidth laser sweep measurement method.

As shown in Fig. 10, the test system uses a tunable laser and a power meter to complete the wavelength scanning process. Before starting the wavelength scan, you need to set the parameters of the tunable laser. First set the wavelength range of tunable laser, which should be able to cover the center wavelength of the measured fiber grating. And then set the step size of the wavelength scan, the step size choice is a trade-off choice between the measurement accuracy and scanning speed. Finally, the output power of the tunable laser can be set, and sufficiently high power can ensure that the measured reflection spectrum has a good signal to noise ratio, although it cannot exceed the maximum receivable optical power accepted by the optical power meter. After the setting is completed, the test software sends instructions to the tunable laser and starts the wavelength spectrum scan; finally the scanned data are collected by data acquisition system to obtain the reflection spectrum of the fiber grating under test.

The scan spectra examples are shown in Fig. 11. It is the measurement of π-phase shifted fiber gratings. Due to the high grating strength, the transmission window is very narrow and much smaller than the resolution of the spectrometer, which must be measured by swept laser measurement. The tunable laser in the system has an output wavelength range of 1450–1590 nm with a wavelength resolution of 0.1 pm and an absolute wavelength accuracy of ± 3.6 pm. The power meter has a working

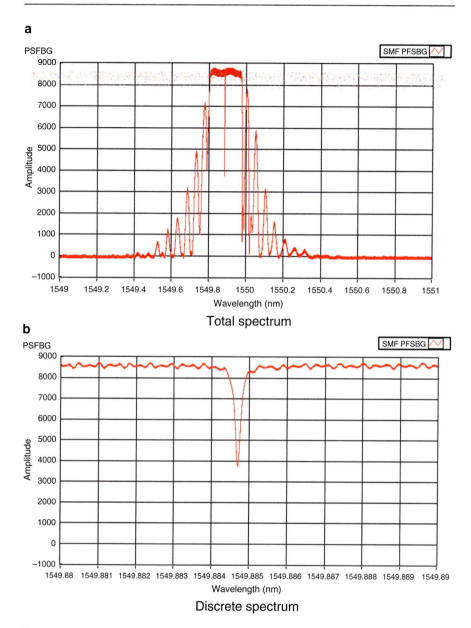

Fig. 11 Scan spectrum examples for swept laser measurement

wavelength of 800–1700 nm, the measurement accuracy is ±0.1 dB and the power range of −90–10 dBm. These parameters increased the measurement accuracy at low input power level, which ensures that the system has a large measurement dynamic range.

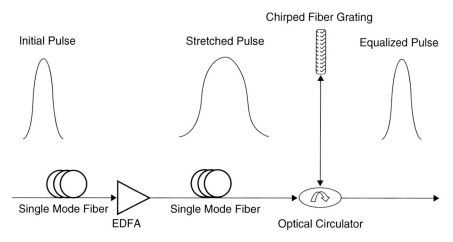

Fig. 12 Principle of fiber grating dispersion compensation

Measurement of Fiber Grating Time Domain Specifications

Generally, light source output is not monochromatic, and their modulation signal has a certain bandwidth, which leads to different phase velocity of optical signals with different frequency in the same medium. This phenomenon induces light spatial separation, which is called dispersion.

The time delay τ_g of fiber grating reflection spectrum can be written as (Pei et al. 2005):

$$\tau_g = \frac{d\varphi_\rho}{d\omega} = -\frac{\lambda^2}{2\pi c}\frac{d\varphi_\rho}{d\lambda} \tag{5}$$

where φ_ρ is the light phase, ω is the radian frequency of light. The dispersion of the reflection spectrum is defined as the rate of change of the delay with wavelength, and its expression is:

$$D_g = \frac{d\tau_g}{d\lambda} = -\frac{2\pi c}{\lambda^2}\frac{d^2\varphi_\rho}{d\omega^2} \tag{6}$$

In the chirped fiber grating, the resonant wave is a function of position, and the incident light of different wavelengths are reflected at different positions of the chirped fiber grating with different time delays. The short wavelength component suffers larger extension than long wavelength component; the delay introduced by the grating is opposite to the delay caused by the transmission in the fiber. Hence, the delay difference between the two is canceled and the pulse width is restored. That means, chirped fiber grating has dispersion compensation ability, which is essential for long distance communication using fiber cables, and this is shown in Fig. 12.

The mainstream methods of fiber dispersion measurements are simulated Raman scattering method, phase shift measurement method, interference measurement

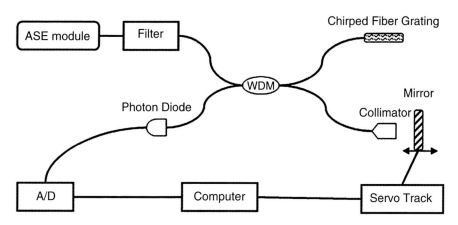

Fig. 13 System schematic of low correlation interference fiber grating dispersion measurement

method, various pulse delay measurement method, and Hill Porter transformation method. Figure 13 shows the schematic diagram of low coherence interferometric measurement based on Fourier transform spectroscopy (Jiang 2006). The test fiber grating is connected to one end of the reference arms of the interferometer; there is a mirror at the other end of the reference arm that moves along the optical axis to achieve optical path variation. The light emitted by the ASE source goes through the filter and Wavelength Division Multiplexer (WDM) into the reference arm, and then the Chirped FBG reflected light combines with the moveable mirror reflected light and transmits to the photodetector (PD), which then transfers the interference signal into electrical signal and sends it to the computer for further analysis.

According to the basic formula of classical Fourier transform spectroscopy of the optical fiber interferometer, there will be:

$$\varnothing(v) = \arg F\left[I_{as}(\tau)\right] \tag{7}$$

where distortion phase $\varnothing(v)$ equals to the radian angle of the Fourier transform of the interference pattern, which is the basis for the recovery of the phase dispersion in discrete Fourier transform spectroscopy. According to $\tau = \Delta/c$, the above equation can be rewrite to:

$$\varnothing(\sigma) = \arg F\left[I_{as}(\Delta)\right] \tag{8}$$

where $\sigma = 1/\lambda = v/c$ is the space frequency. The dispersion of the fiber grating is generally expressed by the relative time delay of the different frequency components:

$$\tau_g = \frac{d\phi(\omega)}{d\omega} = \frac{1}{2\pi}\frac{d\phi(v)}{dv} = \frac{1}{2\pi c}\frac{d\phi(\sigma)}{d\sigma} = -\frac{\lambda^2}{2\pi c}\frac{d\phi(\lambda)}{d\lambda} \tag{9}$$

By measuring the intensity of the interference light which varies with the optical path difference, the dispersion of the fiber grating under test can be calculated using Eqs. 6 and 9.

Acknowledgments Authors are thankful for the financial support of National Natural Science Foundation of China (61605103), Shandong Provincial Natural Science Foundation of China(ZR2016FM33) and Key Research and Development Program of Shandong Province (2018GGX101030).

References

S.R. Abdullina, I.N. Nemov, S.A. Babin, Suppression of side lobes in a spectrum of fiber Bragg gratings due to the transverse displacement of phase mask with respect to the optical fiber. Quant. Elect. **42**(9), 794–798 (2012)

G.A. Ball, W.H. Glenn, Design of a single mode linear cavity Erbium fiber laser utilizing bragg reflectors. J. Lightwave Technol. **10**, 1338–1344 (1992)

I. Baumann, J. Seifert, W. Nowak, M. Sauer, Compact all-fiber add-drop-multiplexer using Fiber Bragg gratings. IEEE Photon. Technol. Lett. **8**, 1331–1333 (1996)

L. Dong, J.L. Archambault, E. Taylor, M.P. Roc, Photosensitivity in tantalum-doped silica optical fibers. J Opt. Soc. Am. B **12**(9), 1746–1750 (1995a)

L. Dong, J.L. Cruz, L. Reckie, Enhanced photosensitivity in tin-docodped Germanosilicate optical Fibers. IEEE Photon Tech. Lett. **7**(9), 1048–1052 (1995b)

L. Dong, P. Hua, T.A. Birks, J.P. Russell, Novel add/drop filters for wavelength-division-multiplexing optical fiber systems using a Bragg grating assisted mismatched coupler. IEEE Photon. Technol. Lett. **8**, 1656–1658 (1996)

K.O. Hill, G. Meltz, Fiber Bragg grating technology fundamentals and overview. J. Lightwave Tecnol. **15**(8), 1263–1276 (1997)

K.O. Hill, Y. Fujii, D.C. Johnson, B.S. Kawasaki, Photo-sensitivity in optical fiber wave-guild: Application to reflection filter fabrication. Appl. Phys. Lett. **32**, 647–649 (1978)

K.O. Hill, B. Malo, F. Bilodeau, D.C. Johnson, J. Albert, Bragg gratings fabricated in Monomode photosensitive optical Fiber by UV exposure through a phase mask. Appl. Phys. Lett. **62**, 1035–1037 (1993)

F. Jiang, Research on measurement of fiber grating dispersion by low coherence interferometry. PhD Thesis. Beijing Insititute of Machinery Industry, Beijing, 2006

C. Jiangand, D.N. Wang, Research progress of femtosecond laser pulse inscription of fiber bragg gratings. Laser Optoelect Progr. **6**, 59–66 (2008)

K. Kawamura, T. Ogawa, N. Sarukura, M. Hirano, H. Hosono, Fabrication of surface relief gratings on transparent dielectric materials by two-beam holographic method using infrared femtosecond laser pulses. Appl. Phys. B Lasers **71**(1), 119–121 (2000)

P.J. Lemaire, High-pressure H2 loading as a technique for achieving ultrahigh UV photosensitivity and thermal sensitivity in GeO2 doped fibres. Electron. Lett. **29**, 1191–1193 (1993)

W.H. Loh, M.J. Cole, M.N. Zervas, Complex grating structures with uniform phase masks based on the moving fiber-scanning beam technique. Opt. Lett. **20**(20), 2051–2053 (1995)

G. Meltz, W. Moreyw, N.J. Doran, Formation of Bragg gratings in optical Fibers by a transverse holographic method. Opt. Lett. **14**, 823–825 (1989)

P. Myslinski, C. Szubert, A.J. Bruce, Performance of high-concentration erbium-doped fiber amplifiers. IEEE Photon Tech. Lett. **11**(8), 973–975 (1999)

L. Pei, T.G. Ning, T.J. Li, X.W. Dong, S.S. Jian, Studies on the dispersion compensation of fiber Bragg grating in high speed optical communication system. Acta Phys. Sin. **54**(4), 1630–1635 (2005)

G.D. Peng, Z. Xiong, P.L. Chu, Photosensitivity and gratings in dye-doped polymer optical Fibers. Opt. Fiber Technol. **5**(2), 242–251 (1999)

L. Poladian, B. Ashton, W.E. Padden, C. Marra, Characterization of phase-shifts in gratings fabricated by over-dithering and simple displacement. Opt. Fiber Technol. **9**, 173–188 (2003)

P.S. Russell, D.P. Hand, Y.T. Chow, L.J. Poyntzwright, Optically induced creation, transformation, and organization of defects and color centers in optical fibers. Proc. SPIE **1516**(2), 47–54 (1991)

D.L. Williams, B.J. Ainslie, J.R. Armitage, R. Kashyap, Enhanced UV photosensitivity in boron codoped germanosilicate fibers. Electron. Lett. **29**, 45–47 (1993)

Measurement of Optical Fiber Amplifier

39

Yanhua Luo, Binbin Yan, Jianxiang Wen, Jianzhong Zhang, and Gang-Ding Peng

Contents

Y. Luo (✉)
Photonics and Optical Communications, School of Electrical Engineering and
Telecommunications, University of New South Wales, Sydney, NSW, Australia

Key Laboratory of Optoelectronic Devices and Systems of Ministry of Education and Guangdong
Province, Shenzhen University, Shenzhen, China
e-mail: yanhua.luo1@unsw.edu.au

G.-D. Peng
Photonics and Optical Communications, School of Electrical Engineering and
Telecommunications, University of New South Wales, Sydney, NSW, Australia
e-mail: g.peng@unsw.edu.au

B. Yan
State Key Laboratory of Information Photonics and Optical Communications, Beijing University
of Posts and Telecommunications, Beijing, China
e-mail: yanbinbin@bupt.edu.cn

J. Wen
Key Laboratory of Specialty Fiber Optics and Optical Access Networks, Shanghai University,
Shanghai, China
e-mail: wenjx@shu.edu.cn

J. Zhang
Key Lab of In-fiber Integrated Optics, Ministry of Education, Harbin Engineering University,
Harbin, China
e-mail: zhangjianzhong@hrbeu.edu.cn

© Springer Nature Singapore Pte Ltd. 2019
G.-D. Peng (ed.), *Handbook of Optical Fibers*,
https://doi.org/10.1007/978-981-10-7087-7_45

Abstract

As one of the key photonic devices, optical amplifier, especially optical fiber amplifier, has been playing an important role in the optical communications and laser physics. In research, development, and application of optical amplifiers, it is necessary to apply a range of measurement techniques for characterization and evaluation. This chapter introduces the main characteristics of optical amplifier and describes related characterization and measurement techniques.

Introduction

When the optical signal travels in a waveguide, it will be attenuated due to the absorption, scattering, etc. For a very long distance, the optical signal will be attenuated to be too weak to be detected by the photodetector. To reach destinations that are hundreds of kilometers away, optical signal must be amplified by a key optical device-optical amplifier. It is a device that amplifies an optical signal directly as shown in Fig. 1, without the need to convert it to an electrical signal. It may be thought of as a laser without an optical cavity or one in which feedback from the

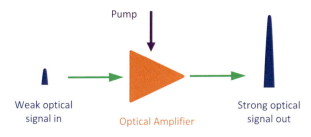

Fig. 1 Function scheme of an optical amplifier

cavity is suppressed (Wikipedia 2017a). Optical amplifiers play an important role in optical communication and laser physics.

Especially in modern optical telecommunications, to transmit the signal/information over hundreds of kilometers, the optical power of the signal must be periodically conditioned. Optical fiber amplifiers are key devices that reconstitute the attenuated optical signal, thus expanding the effective fiber span between the data source and the destination (Stamatios 2000).

Optical Fiber Amplifier

Optical amplifiers directly amplify optical signals through a stimulated emission process in an optical gain medium. Optical fiber amplifiers and semiconductor amplifiers are two kinds of optical amplifiers used in optical transmission systems. The former may be classified into *doped fiber amplifiers*, which are based on doped active centers, e.g., rare earth ions, introduced in doped optical fibers and *nonlinear fiber amplifiers,* which are based on nonlinear optical effects, e.g., stimulated Raman scattering, which intrinsically existed in conventional optical fibers.

Doped Fiber Amplifier

Doped fiber amplifiers are the most popular optical amplifiers. They are using optical fibers doped with rare earth ions or other elements to amplify optical signals.

According to the doping materials or relevant active centers, there are erbium-doped fiber amplifier (EDFA), ytterbium-doped fiber amplifier (YDFA), thulium-doped fiber amplifier (TDFA), and praseodymium-doped fiber amplifier (PDFA) as well as recently developed bismuth-doped fiber amplifier (BDFA).

So far EDFA is the most widely used optical fiber amplifier which consists typically of erbium-doped fiber (EDF), pump light source, isolator, and optical coupler. EDF is the most important part of EDFA. In an EDF, a certain amount of Er^{3+} is doped into the optical fiber core. The typical energy levels of erbium ions and the relevant energy transition processes are shown in Fig. 2. When an Er^{3+} in its ground state ($^4I_{15/2}$) is illuminated (pumped) with light at a suitable wavelength (either 980 nm or 1480 nm), it could be excited to its excited state ($^4I_{11/2}$ or $^4I_{13/2}$ with 980 nm or 1480 nm, respectively). While pumped at 1480 nm wavelength, the ion is directly excited to the quasi-stable state ($^4I_{13/2}$). Pumped at 980 nm wavelength, the ion stays the unstable, short lifetime state ($^4I_{11/2}$) briefly before decaying to a quasi-stable, longer lifetime state ($^4I_{13/2}$). When optical signal transmits through an EDF with sufficient high number of excited ions (population inversion), it is amplified through a stimulated emission process.

Different optical fiber amplifiers operate at different bands or wavelength ranges. For example, EDFAs operate in C- and L -bands (1520–1620 nm), TDFAs in S-band (1450–1490 nm), YDFA in wavelength range of around 1055–1070 nm, and PDFA in around 1300 nm region (Wikipedia 2017a). BDFA under development has the

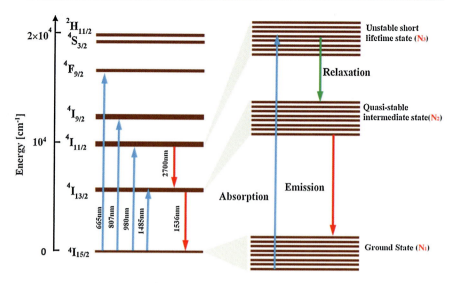

Fig. 2 The energy levels of Er^{3+} ion and usual energy transition processes

potentially to operate ultra-broadband, from 1100 nm to 1700 nm, depending upon doping materials and pumping conditions as well as fabrication conditions (Dvoyrin et al. 2006, 2008; Bufetov et al. 2008; Firstov et al. 2016).

Nonlinear Optical Fiber Amplifier

Nonlinear optical fiber amplifier includes Raman and Brillouin fiber amplifiers. The coefficients related to nonlinear optical effects, especially Raman effect, in silica optical fibers are typically very weak. Raman amplification requires very long fibers (in the order of several kilometers) and pump lasers with high optical power (>1 W). These amplifiers take advantage of low loss of un-doped conventional optical fibers and high-power pump lasers that make effective use weak nonlinear effects, i.e., stimulated Raman scattering (SRS) and stimulated Brillouin scattering (SBS), over long fiber lengths (Stamatios 2000).

Raman Fiber Amplifier

The operation of Raman fiber amplifier is based on the SRS effect, while the gain medium is un-doped conventional optical fiber. Optical signal is amplified involving a stimulated Raman process as shown in Fig. 3 (Finisar Corporation 2012). An incident pump photon excites an electron to a virtual state, and, at the presence of a signal photon, the stimulated emission occurs when the electron de-excites down to the vibrational state and produces a duplicated signal photon. For Raman amplification, the signal gain is spectrally shifted by a Stokes frequency with regard to that of the pump. The typical Raman gain versus Stokes shift curve, with two gain peaks at 440 cm^{-1} and 490 cm^{-1}, of silica optical fibers is shown in Fig. 4

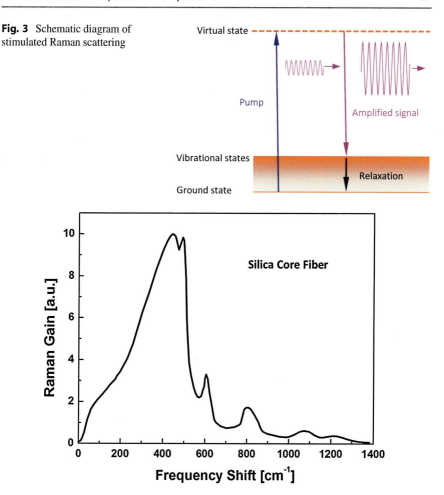

Fig. 3 Schematic diagram of stimulated Raman scattering

Fig. 4 Raman gain of silica optical fiber (Stolen et al. 1984)

(Stolen et al. 1984). The main Stokes shift peak at 440 cm^{-1} for silica optical fibers is approximately 13.2 THz. Raman amplifiers use normally backward pumping that their pump near the receiver and the pump light travels in opposite direction toward the source. Thus, the pump is strongest at the receiver and weakest at the source. This arrangement has an important advantage: the pump power is the highest where it is the most needed at the remotest distance from the source and less needed in the vicinity of the source. Consequently, the signal is amplified most where it is weakest and least where it is strongest. Of course other pumping configurations such as forward pumping and dual pumping – one forward pumping at the transmitting side and the other backward pumping at the receiving side – can be used.

The most important feature of Raman amplifiers is the flexibility of their operation bandwidth that is relative to the pump wavelength and, by selecting pumps at different wavelengths, can be extended over the whole transmission spectrum

Fig. 5 Schematics of Brillouin amplification

of silica fiber from 1100 nm to 1700 nm – sufficient to accommodate 500 optical channels at 100 GHz spacing.

Brillouin Fiber Amplifier

Brillouin amplification is also a distributed process and relies on the interaction of intense light with stimulated acoustic vibrations (phonons). In Brillouin amplification, the signal, different from the pump with a Brillouin frequency shift that involves the vibration states in a process similar to that shown in Fig. 3, is amplified through a SBS process. However, backward pumping that enables the interference of the counter-propagating pump and signal gives rise to a moving density grating, as schemed in Fig. 5, is required by the phase-match condition to produce Brillouin gain. For silica fiber, the Brillouin gain bandwidth is very narrow and around the Brillouin frequency shift, typically about 11 GHz (Physikalisch-Technische Bundesanstalt 2017).

Fiber Brillouin amplification is characterized with narrow bandwidth and, compared with Raman amplification, higher small-signal gain. In contrast to broadband amplifiers, like EDFAs, this allows taking advantage of a single-stage gain of up to 50 dB or even more (Physikalisch-Technische Bundesanstalt 2017). With a gain bandwidth of around 10 MHz, Brillouin amplification may act as a "built-in" narrow band optical filter.

Semiconductor Optical Amplifiers

Optical amplifiers also include semiconductor optical amplifiers (SOAs) which use the semiconductor material as the gain medium. SOAs are typically made from groups III to V compound semiconductors such as GaAs/AlGaAs, InP/InGaAs, InP/InGaAsP, or InP/InAlGaAs which cover different wavelength ranges. Typical SOAs have a similar structure to Fabry-Pérot laser diodes (LDs) but with anti-reflection elements at the end faces, as shown in Fig. 6. New designs of anti-reflective coatings can reduce end-face reflection to <0.001% and, by minimizing the optical feedback to the gain region, prevent the amplifier from acting as a laser while achieving the highest possible gain.

SOAs are readily used in telecommunication systems in the form of fiber-pigtailed components, operating at signal wavelengths between 850 nm and 1600 nm and generating gains of up to 30 dB (Wikipedia 2017a). SOA is of small size and compact, able to be easily integrated with other semiconductor and optical components. With electrical pumping, it is normally less expensive and more

Fig. 6 Structure of the typical SOA

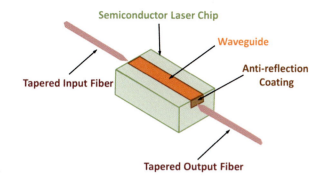

efficient than EDFA. In addition, "linear optical amplifiers" using gain-clamping techniques have been developed (Wikipedia 2017a).

However, the performance of SOAs for optical telecommunication systems is still not comparable to EDFA in general. Typically SOAs have higher noise, lower gain, more polarization dependence, narrower bandwidth, and higher nonlinearity. The high nonlinearity originates from the upper state with short lifetime that the gain reacts rapidly to changes of pump or signal power. The gain change could significantly distort the signals. This nonlinearity leads the most severe problem for optical communication applications. However, SOA provides gain in alternative wavelength regions from fiber amplifiers. One advantage of SOA worth noting is that it could be easily linked to useful nonlinear operations such as cross-gain modulation, cross-phase modulation, wavelength conversion, and four-wave mixing (FWM). These nonlinear optical operations enable SOAs for all optical signal processing processes such as all-optical switching, wavelength conversion, clock recovery, signal de-multiplexing, and pattern recognition (Wikipedia 2017a).

According to their respective role or position in the optical transmission lines, optical amplifiers can also be classified into the following categories:

Booster amplifiers: Booster optical amplifiers are used immediately after the optical transmitters or signal sources to boost the optical power level of signals.
In-line amplifiers: In-line optical amplifiers amplify signal in the middle of optical transmission lines. They are often cascaded for a very long optical link.
Preamplifiers: Optical preamplifiers are often used to amplify weak transmitted signals before they are detected at optical receivers.

Properties

Spectral Properties

The key physical phenomenon behind signal amplification in EDFA is stimulated emission of radiation by atoms in the presence of an electromagnetic field as

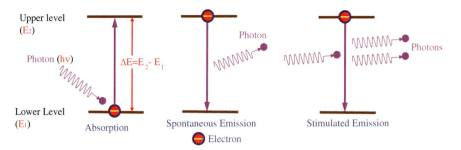

Fig. 7 Absorption, spontaneous emission, and stimulated emission under the mechanism of the interaction between an atom and a photon ($h\nu = E_2 - E_1$)

schemed in Fig. 7. Consider an atom and two of its energy levels, E_2 (upper level) and E_1 (lower level). A photon, whose frequency ν satisfies $h\nu = E_2 - E_1$ (h is Planck constant), induces electronic transitions of atoms between the upper level (E_2) and the lower level (E_1). Electronic transitions of $E_1 \rightarrow E_2$ or $E_2 \rightarrow E_1$ can occur. $E_1 \rightarrow E_2$ transitions are accompanied by *absorption* of photons from the incident photon. An atom in the upper energy level (E_2) can spontaneously decay to the lower level (E_1) and emit a photon of frequency ν if the transition between E_2 and E_1 is radiative. This photon has a random direction, polarization, and phase. Such kind of emitting photon is called *spontaneous emission*. In addition, if an incident photon with energy of $h\nu$ is present, the decay transitions of $E_2 \rightarrow E_1$ are accompanied by emitting a "stimulated" photon whose properties are identical to those of the incident photon. Thus if *stimulated emission* dominates over absorption, a net increase in the number of photons of energy leads to an amplification of the signal (otherwise, the signal will be attenuated). At thermal equilibrium, there exists only absorption of the input signal. In order for amplification to occur, the relationship between the populations of levels must be inverted, which prevails under thermal equilibrium. Population inversion can be achieved by supplying additional energy from pump (OPI online course Laser: Fundamentals 2017; Rosenkranz 2016). The stimulated emission is unlike the spontaneous emission process, where the emitted photons not only have the same energy as the incident photons but also the same direction of propagation, phase, and polarization (*stimulated emission* process is coherent, whereas the *spontaneous emission* process is incoherent). When the amplifier treats spontaneous emission as another electromagnetic field at the frequency, the spontaneous emission will be amplified, in addition to the incident optical signal. This *amplified spontaneous emission* (ASE) generally appears as noise at the output of the amplifier (Rosenkranz 2016). Generally, interpolation source subtraction is one of the methods to estimate the ASE power at signal wavelength. In the interpolation source subtraction method, the ASE power at signal wavelength will be obtained through the interpolation of the power close the signal channel as well as the subtraction the amplified source noise (Amonics Ltd 2004).

Figure 8 shows a typical ASE spectrum of EDF with and without the input signal. Seen from Fig. 8, when the input signal is turned off, then a big ASE will be measured. However, when the input signal is turned on, a big signal can be measure,

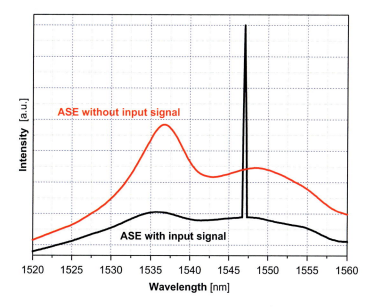

Fig. 8 ASE output spectra with and without input signal

while ASE will be restrained. The photons generated from ASE travelling along the fiber will trigger the stimulated emission that of course will have their wavelength, phase, etc. At the end almost all energy pumped into an amplifier without any input signal reappears as ASE. However, if an input signal consumes electrons in the metastable state through stimulated emission process, then fewer are left for spontaneous emission, therefore reducing the ASE as shown in Fig. 8.

Population Inversion in Fiber Amplifier

The main task of an optical amplifier is the optical gain, realized when the amplifier is pumped to achieve population inversion. Generally, the inversion coefficient is defined to quantify the fraction of ions inverted (in $^4I_{13/2}$ state) averaged along the fiber length L as (Wysocki 2001):

$$Inv = \frac{1}{L} \int_0^L \frac{N_2(z)}{N_0(z)} dz \qquad (1)$$

where $N_0(z) = N_1(z) + N_2(z)$ is the total radially averaged Er^{3+} concentration and $N_1(z)$ and $N_2(z)$ are the fractional populations of the ground state and quasi-stable intermediate state. In order to achieve certain population inversion, additional energy will be provided from pump (OPI online course Laser: Fundamentals 2017; Rosenkranz 2016).

Further, the shape of the gain spectrum produced by an EDFA is relatively simple if homogeneity, and occupation of only the $^4I_{13/2}$ and $^4I_{15/2}$ states are

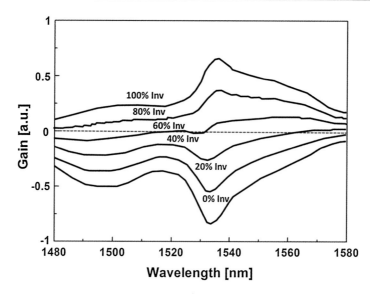

Fig. 9 Gain and absorption (negative gain) of Er^{3+} for excitation levels from 0 to 100% (Paschotta 2017b)

assumed. All possible spectra can then be predicted by fractional combinations of the gain spectrum measured when all ions are pumped to the $^4I_{13/2}$ state and the absorption spectrum measured when all ions occupy the $^4I_{15/2}$ state (Wysocki 2001). Neglecting background loss, the gain spectrum produced by a length of EDF with a given ion inversion (fraction in upper state) can be written as (Wysocki 2001):

$$G\,(\lambda, Inv) = \{[g^*\,(\lambda) + \alpha\,(\lambda)]\,Inv - \alpha\,(\lambda)\}\,L \qquad (2)$$

where $g^*(\lambda)$ is the measured gain per unit length produced by the EDF when all ions are in the $^4I_{13/2}$ state, $\alpha(\lambda)$ is the measured loss per unit length produced by the EDF when all ions are in the $^4I_{15/2}$ state, and L is the EDF length (Wysocki 2001). Figure 9 shows typical gain spectra with different inversion coefficients (Paschotta 2017b). It will have some variant of silica with additional dopants, e.g., to avoid clustering of Er^{3+}. Other glass compositions can lead to substantially different gain spectra (Paschotta 2017b).

Parameters

Amplifier Gain

The most basic property of an optical amplifier is its amplifier gain (gain, g, or G), which represents the amplification amount of the input signal. It is typically expressed by the ratio between the input signal power (P_{in}) and the output signal power (P_{out}) as followed:

$$G = \frac{P_{out}}{P_{in}} \tag{3}$$

Different meanings of gain can occur in the literature (Paschotta 2017a):

1. The gain can simply be an amplification factor or the ratio of output power and input power, like Eq. 3.
2. For small gains, the gain is often specified as a percentage. For example, 3% correspond to an amplification factor of 1.03.
3. Particularly large gains typically measured in decibels (dB), which is the most common definition used for optical amplifier. It is ten times the logarithm (to base ten) of the amplification factor, i.e., $10 \times \log_{10}G$. For example, a fiber amplifier may have a small-signal gain of 30 dB, corresponding to an amplification factor of $10^3 = 1000$.
4. One also often specifies a gain per unit length, or more precisely the natural logarithm of the amplification factor per unit length, or alternatively the decibels per unit length.

For EDFA, under 980 nm pumping, Er^{3+} will be excited from ground state ($^4I_{15/2}$) to unstable short lifetime state ($^4I_{11/2}$, several μs) by absorption as shown in Fig. 2. And then through the rapid non-radiative decay, Er^{3+} will come to a quasi-stable state ($^4I_{13/2}$). $^4I_{13/2}$ is an inversion state with high population density, where ions have a long lifetime of several ms. Therefore, when signal light at $\sim$1530 nm passed through it, it will be amplified by stimulated emission ($^4I_{13/2} \rightarrow {}^4I_{15/2}$). Due to the Stark splitting of each energy level, EDFA will have a broad gain bandwidth in the range of 1520–1570 nm with wavelength dependent as shown in Fig. 10.

Due to the existence of ASE noise, the gain of EDFA is defined as the ratio of output signal power corrected by ASE noise (Amonics Ltd 2004):

$$G = \frac{P_{out} - P_{ASE}}{P_{in}} \tag{4}$$

where P_{ASE} is the ASE noise power at the signal wavelength. However, in most cases where the input signal power is sufficiently large, the ASE noise adjustment is negligible and can be set to zero, then Eq. 4 is often simplified into the common Eq. 3.

In addition, when the gain is measured through the on-off gain technique, the concept of the on-off gain and net gain will be introduced. The on-off gain ($G_{on/off}$) is often defined as the ratio between the output power with pumps on and output power with pumps off, while the net gain is defined as usual as the relationship between the signal at the output and at the input of the amplifier in Eq. 3. The relation between the net gain (G) and the on-off gain can be written as follows (Iannone 2017):

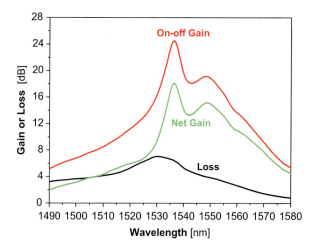

Fig. 10 Schematic on-off gain, net gain, and loss spectra of a typical EDFA

$$G = G_{on/off} - \alpha \tag{5}$$

where α is the loss of the fiber and all the gains and losses are expressed in dB. In Fig. 10, the schematic on-off gain, net gain, and loss spectra of a typical EDFA are shown, and the difference between net gain and on-off gain is very evident due to the loss.

Gain Efficiency

For an amplifier, the *gain efficiency* is often defined for the steady state and in a situation with low signal intensities (negligible stimulated emission). It may be negative for a quasi-three-level system due to the reabsorption and then become positive for sufficiently high pump powers as schemed in Fig. 9. Normally, the gain rises in proportion to the pump power, until ASE becomes strong and reduces the further rise of gain. The differential gain efficiency in a simple case (without ASE) is expressed as (Paschotta 2010):

$$\frac{\partial g_{av}}{\partial P_p} = \frac{\eta_p \left(\sigma_{em} + \sigma_{abs}\right) \tau_2}{h\nu A_{eff} L} \tag{6}$$

where g_{av} is the average small-signal gain coefficient in units of m^{-1}, P_p is the incident pump power, L is the length of the gain medium, η_p is the pumping efficiency (including the pump absorption efficiency, the quantum defect, and the pump quantum efficiency, so an increased length L also increases η_p via the pump absorption efficiency), σ_{em} and σ_{abs} are the emission and absorption cross sections of the signal, τ_2 is the upper-state lifetime, $h\nu$ is signal photon energy, and A_{eff} is the effective mode area.

Especially, the small-signal gain efficiency can further be expressed as (Paschotta 2010):

$$\frac{\partial g_{av}}{\partial P_p} = \frac{\eta_p}{P_{sat} L} \tag{7}$$

where $P_{sat} = \frac{h\nu A_{eff}}{(\sigma_{em} + \sigma_{abs})\tau_2}$ is the signal saturation power. Equation 7 shows that a high gain efficiency inevitably implies a low saturation power so that low signal powers can cause significant saturation effects.

Conversion Efficiency

When high output power is required (e.g., power amplifier), it is essential that large percentage of pump power is converted into signal power. For a measure of system efficiency in terms of power, power conversion efficiency (PCE) is defined as a measure of pump power converted to signal power for amplification, which is given by (Becker et al. 1999):

$$PCE = \frac{P_{signal-out} - P_{signal-in}}{P_{pump}} \tag{8}$$

where $P_{signal-out}$, $P_{signal-in}$, and P_{pump} are the power of the output signal, the input signal, and the pump in linear unit.

In addition, the quantum conversion efficiency (QCE) provides a more direct measure of efficiency in terms of photon transfer from pump to signal, which is given by (Becker et al. 1999):

$$QCE = PCE \times \frac{\lambda_{signal}}{\lambda_{pump}} \tag{9}$$

where λ_{signal} and λ_{pump} are the wavelength of the signal and the pump.

Design Considerations of Fiber Amplifier

As mentioned above, the fiber amplifier is designed based on the phenomenon of stimulated emission. So a certain kind of conditions is required to achieve population inversion, including the pump (pump configuration, wavelength, and power), fiber property (fiber length, dopant concentration, and fiber composition), signal (signal wavelength and power), etc (Giles and Desurvire 1991; Finisar Corporation 2010).By using commonly three-level rate equation model, the total amplifier gain G with a L long EDF can be given by (Rosenkranz 2016):

$$G = \Gamma_S \exp\left[\int_0^L \left(\sigma_s^e N_2 - \sigma_s^a N_1\right) dz\right] \tag{10}$$

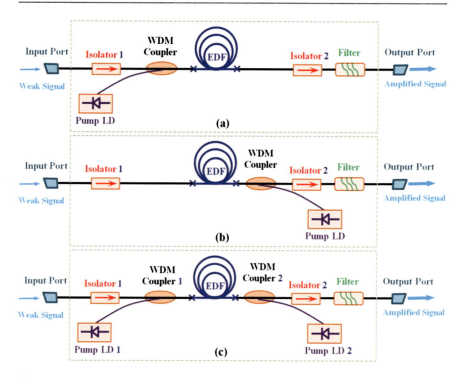

Fig. 11 Block diagram of three typical EDFA configurations, (**a**) co-propagating pump and signal, (**b**) counter-propagating pump and signal, and (**c**) bi-directional pump

where Γ_S is the overlap factor that the doped region within the core provides the gain for the entire fiber mode, N_1 and N_2 are the populations densities of the ground state and first excited state (quasi-stable intermediate state), σ_s^e and σ_s^a are the emission and absorption cross section at the frequency of signal.

Pump

Pump configuration: So far, there are mainly three pump configurations for EDFAs, co-propagating pump and signal (Fig. 11a), counter-propagating pump and signal (Fig. 11b), and bidirectional pump (Fig. 11c). The principal difference between these configurations is in the ASE pattern and inversion profile. For the small-signal amplification, co- and counter-propagating configurations yield the same gain since pump patterns are the mirror images of each other and the average population inversion is the same. However, ASE power and spectra will be different at the output. When the signal is increased, so that it can affect seriously the inversion, pump direction becomes important for signal gain. Generally, the most advantageous situation is when the signal is strong where the inversion is high so that the signal, not the ASE, will deplete the gain. Compared with counter-propagating pump schemes, co-propagating pump schemes for moderate and small signals will result in lower noise figure (NF). It is attributed by two reasons. First, forward ASE is smaller in co-propagating schemes. Second, when signals enter the amplifier with counter-propagating pump, the inversion at the beginning is low so that signal may experience some loss, which contributes to the signal-to-noise ratio (SNR) degradation. What is more, bidirectional pump gives better gain and noise performance (Lu 2005).

Pump wavelength: For modern EDFAs, 980 nm and 1480 nm LDs are the typical pump sources as shown in Fig. 2, and other pumping wavelength at 532, 664 and 827 nm will have lower pumping efficiency compared with them (Desurvire 1994). Pumping at 1480 nm yields a better pump conversion efficiency, given a larger number of photons at 1480 nm for a specific power, compared with pumping at 980 nm. However, the NF is smaller with 980 nm pump since it can offer higher inversion population. So the multistage amplifiers have practically been used to achieve the high power conversion efficiency and low NF. It is often advantageous to use 980 nm pump at the first stage and 1480 nm pump at the second stage. 980 nm pumping has the benefit of low NF at the input portion, while 1480 nm pumping offers much higher pump conversion efficiency for the power portion. The simplest two-stage EDFA contains two sections of EDF, separated by an isolator or a filter. The isolator eliminates the backward ASE from the second portion that would otherwise deplete inversion in the first section. The first stage needs to be well-inverted so that a moderate amount of gain is obtained with minimum NF. The second stage acts as a power amplifier (Lu 2005).

Pump power: For ideal EDFA, we have a simple analytical expression of the pump power (P_p) as a function of gain, given by (Schiopu and Vasile 2004):

$$P_p = \alpha_p L \frac{Q_s P_{sat}\left(\lambda_p\right)}{1 - \exp\left[(Q_s - 1)\,\alpha_p L\right]} \tag{11}$$

where L is the fiber length, P_{sat} is the pump saturation power, λ_p is the pump wavelength, α_p and α_s are the pump and signal absorption coefficient, and

$$Q_s = \frac{1 + \eta_p}{1 + \eta_s}\left[1 + \frac{\lg G}{\alpha_s L}\right],$$

where η_p and η_s are the ratio of emission and absorption cross section at pump and signal wavelength, respectively. Based on Eq. 11, when there is no pump and $Q_s = 0$, the gain represents the loss transmission of the EDF with signal absorption coefficient α_s, which means that all Er^{3+} are in the ground level. When G = 1, P_p is called transparency pump power, which is proportional to the pump saturation power and increases with EDFA length. When $P_p \to \infty$, the gain will reach the maximum saturation. Figure 12 shows a typical gain response to the pump power when the input signal, fiber samples, and other conditions are the same. As shown in Fig. 12, the gain will be negative at the beginning due to the absorption. When pump power is sufficient, the gain will become positive, which will keep on increasing and achieve the asymptotic value.

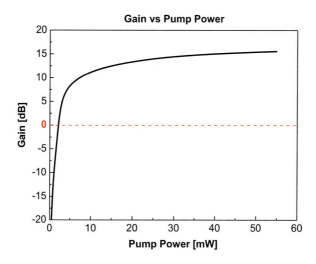

Fig. 12 Typical gain response to the pump power with fixed length and input signal power

Pump saturation: Pump saturation (P_{sat}) occurs when the pump power is large enough to significantly deplete the ground energy levels, which reduces the pumping rate of Er^{3+}. Gain will increase with pump power but will flatten out as pump saturation takes over (Amonics Ltd 2004), as shown in Fig. 12.

Active Fiber

Length: As expressed in Eqs. 6 and 7, the gain is much related with fiber length. Figure 13 shows a typical gain versus fiber length with other conditions fixed. Seen from Fig. 13, the maximum gain is obtained for a wider range of fiber length. However, as length further increases, gain value decreases due to the loss of the fiber. The optimum length of the EDF will depend upon the input signal power, pump configuration and power, Er^{3+} density, and the signal and pump wavelength (Shukla and Kaur 2013).

Composition and concentration: Generally, high gain per unit length of EDF requires highly doped fibers. The conversion efficiency of EDFA is strongly dependent on Er^{3+} concentration, due to the concentration-dependent effects. These effects include interactions between Er^{3+}, which result in a reduction in the lifetime of $^{4}I_{13/2}$ metastable state and up-conversion emission at visible wavelength under IR pumping. In order to reduce these deleterious high concentration effects, it is necessary to choose suitable host materials and fabrication techniques (de Barros et al. 1996). Although it provides higher gain per unit length for amplifier with higher Er^{3+} concentration, it will have a higher pump power threshold, because the threshold is inversely proportional to the fluorescence lifetime. For lower Er^{3+} concentration, the lifetime will be longer, resulting in the lower threshold of the pump (de Barros et al. 1996). In addition, according to Eq. 10, the gain is related to the absorption and emission cross section of the fiber. Due to the existed difference

Fig. 13 Typical gain response to the fiber length with fixed pump and input signal power

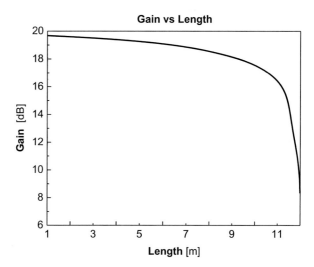

of the cross section for different fiber composition (Barnes et al. 1991), the gain will also be much dependent upon the fiber composition.

Signal

Signal wavelength: The gain dependence upon the signal wavelength is one of the most important EDFA characteristics. Such feature is resulted from two factors: the particular absorption and emission cross-section line shapes of the EDF and the variation of the gain coefficient along the EDF length due to the nonuniform medium inversion (Desurvire 1994). The gain spectrum in Fig. 10 shows a typical gain dependence upon the signal wavelength for EDFA, where the maximum gain is achieved at ∼1535 nm.

Signal power: For a homogeneously broadened two-level system, the optical gain depends not only on the optical frequency of the incident signal but also on the local signal intensity at the point of amplifier. The local gain coefficient can be expressed as:

$$g\left(\omega\right) = \frac{g_0}{1 + \left(\omega - \omega_0\right)^2 \tau_2^2 + P/P_s} \tag{12}$$

where g_0 is the peaked gain, ω is the frequency of the incident signal, ω_0 is the atomic transition frequency, and P is the optical power of the signal being amplified. The saturation power P_s depends on the gain medium. The parameter τ_2, known as the dipole relaxation time, is typically quite small (<1 ps) (Rosenkranz 2016). The saturated output power is also a key specification of the amplifier, which is the maximum output power and often critical in determining the amplifiers' cost (Finisar Corporation 2010).

When gain is not saturated (i.e., $P < <P_s$), the gain coefficient $g(\omega)$ becomes (Kostuk 2008):

$$g(\omega) = \frac{g_0}{1 + (\omega - \omega_0)^2 \tau_2^2} \tag{13}$$

According to Eq. 13, when $\omega = \omega_0$, gain will reach the maximum. Eq. 13 also indicates that gain should follow the homogeneously broadened characteristic of a two-level atom (i.e., Lorentzian profile) at non-resonant frequencies (Kostuk 2008).

Gain Bandwidth

The concept of amplifier bandwidth is commonly used in place of the gain bandwidth. The difference becomes clear when one considers the amplifier gain G defined in Eq. 3. For an amplifier based on a section of EDF with the length of L, by noting that $P(L) = P_{out}$ and $P(0) = P_{in}$ and using $P(z) = P_{in}e^{gz}$, the amplification factor is given by (Agrawal 1997): $G(\omega) = e^{g(\omega)L}$. Both the amplifier gain $G(\omega)$ and $g(\omega)$ is shown explicitly. When $\omega = \omega_0$, both the amplifier gain $G(\omega)$ and the gain coefficient $g(\omega)$ are maximal and decrease with the signal detuning. However, $G(\omega)$ decreases much faster than $g(\omega)$ (Rosenkranz 2016). According to Eq. 13, the gain bandwidth of gain spectrum is typically expressed as the full width at half maximum (FWHM) and given by (Kostuk 2008):

$$\Delta\omega_g = 2/\tau_2 \text{ and } \Delta v_g = \Delta\omega_g \Big/ 2\pi , \tag{14}$$

while the amplifier bandwidth Δv_A is defined as the FWHM of $G(\omega)$ (Kostuk 2008):

$$\Delta v_A = \Delta v_g \left(\frac{\ln 2}{\ln (G_0/2)} \right)^{1/2} \tag{15}$$

where Δv_g is the gain bandwidth and $G_0 = \exp(g_0 L)$. If the amplifier bandwidth is smaller than the gain bandwidth, the difference depends on the amplifier gain characteristics.

Gain Saturation

When the input signal power is sufficiently high, gain saturation will be experienced where the population of metastable states is severely depleted by a high rate of stimulated emission. Since $g(\omega)$ is reduced when P_{in} becomes comparable to P_s, the amplification factor G decrease with the input signal increasing. Consider the case in which input signal frequency is exactly tuned to the gain peak ($\omega = \omega_0$). The detuning effects can be incorporated in a straightforward manner. The following implicit relation for the large-signal amplifier gain can be achieved by using the relation $\frac{dP}{dz} = \frac{g_0 P}{1 + P/P_s}$ (Rosenkranz 2016):

$$G = G_0 \exp\left[-\frac{(G-1)}{G} \frac{P_{out}}{P_s} \right] \qquad (16)$$

which shows that the amplification factor G decreases from its unsaturated value G_0 when P_{out} becomes comparable to P_s. For a fixed pump power, the rate of Er^{3+} excited for population inversion will be a constant. As the input signal power is increased past the small-signal region, more photons will enter the EDF stimulating emission and depleting the metastable energy level faster than it can be filled. Therefore, with increasing input signal power, the amplification will reach a limit and then the gain will decrease (Amonics Ltd 2004).

Saturation Output Power

The saturation output power is defined as the output signal power at which the gain has been reduced by 3 dB of its small-signal gain (G = G_0/2) (Stamatios 2000). It can be given by the following equation (Kostuk 2008):

$$P_{sat}^{out} = \frac{G_0 \ln 2}{G_0 - 2} P_s \qquad (17)$$

It shows the maximum output power capabilities of the EDFA for a fixed pump power and is an outcome of gain saturation. So as the input signal power is increased past the small-signal region, the gain will decrease even as the output signal power increases (Amonics Ltd 2004).

Noise Figure

All amplifiers, including optical amplifiers, introduce noise during the amplification process so that the output signal is always noisier than the input signal. The noise performance of an optical amplifier is characterized by its NF (F_n), which is defined as the ratio of the SNR at the amplifier output to an ideal SNR at the input, given by:

$$F_n = \frac{SNR_{in}}{SNR_{out}} \qquad (18)$$

A high NF indicates that the SNR has been impaired by the amplification process, so it is essential that the NF should be kept as low as possible to achieve higher performance of amplifier in optical link. The NF depends on the technology used for the amplifier, as well as the gain, with higher gain amplifier usually having lower NF (Finisar Corporation 2010). As defined in Eq. 18, SNR degradation is quantified through the amplifier NF F_n (Kostuk 2008). As SNR is based on the electrical power after converting the optical signal to an electrical current, so F_n will be referenced to the detection measurement and depend on parameters such as detector bandwidth, thermal, and shot noise.

The amplifier noise is the ultimate limiting factor for optical communication links and systems. EDFA noise mainly originates from ASE noise, which will degrade the SNR by adding to the noise during the amplification process. The total ASE noise power (P_{ASE}) is summed over all the spatial modes that the optical fiber supports in an optical bandwidth (Rosenkranz 2016):

$$P_{ASE} = 2n_{sp}hv_c (G - 1) B_o \qquad (19)$$

where G stands for optical gain of the EDFA, B_o is the optical bandwidth, v_c is the carrier frequency, h is the Planck constant, and n_{sp} is the spontaneous emission population inversion factor and depends on the relative populations N_1 and N_2 of the ground and excited states of the amplifying medium, given by (Rosenkranz 2016):

$$n_{sp} = \frac{N_2}{N_2 - \frac{\sigma_a(\lambda)}{\sigma_e(\lambda)} N_1} \qquad (20)$$

Furthermore, the NF F_n can be given by (Rosenkranz 2016):

$$F_n = \frac{P_{ASE}}{Ghv_c B_o} + \frac{1}{G} = 2n_{sp}\frac{G - 1}{G} + \frac{1}{G} \approx 2n_{sp} \qquad (21)$$

For EDFA operating with the three-level pumping scheme, $N_1 \neq 0$ and $n_{sp} > 1$. Like the amplifier gain, F_n depends both on the amplifier length L and pump power P_p. Especially, F_n decreases dramatically as the pump power increases until the pump saturation occurs (Rosenkranz 2016). In theory the best F_n achievable by an EDFA is 3 dB (Amonics Ltd 2004).

So far, there are a number of the optical methods and optoelectronic/electrical methods for accurate NF measurement as listed in Table 1 (Baney and Gallion 2000). Their merits and disadvantages are also listed in Table 1. In all these measurements with the optical methods, the amplifier gain is measured at the signal wavelengths, and the amplifier generated ASE is measured preferably at the same wavelength or estimated through wavelength interpolation. In addition, all of them require accurate information regarding the effective noise bandwidth of the OSA. The nominal measurement resolution bandwidth is not necessarily equal to the effective optical noise bandwidth. Moreover, the effective noise bandwidth typically varies with optical wavelength. Fortunately, manufacturers typically supply the wavelength-dependent effective noise bandwidths of their OSAs (Baney and Gallion 2000). Unlike optical methods, all amplifier-generated intensity noise that would occur in an actual receiver such as multipath interference noise, double Rayleigh scattering, and pump-induced noise are measured by optoelectronic/electrical technique. Optoelectronic methods therefore provide the only means to completely characterize the intensity noise generated by an optical amplifier (Baney and Gallion 2000).

Table 1 Summary of the merits and disadvantages of the various measurements methods for the noise figure (Baney and Gallion 2000)

Methods	Advantages	Disadvantages
Optical methods		
Time domain extinction	Provides rapid measurement. Can be used in conjunction with small-signal probe to measure dynamic gain. Works well with the reduced-source technique. Allows possibility for noise measurement at the signal wavelength	Insertion loss of gating switches, more complicated measurement setup
Source subtraction	Simple to implement	Sensitive to additive SSE from dense wavelength division multiplexing (DWDM) channels, requires stable SSE
Polarization extinction	Allows possibility for noise measurement at the signal wavelength	Longer measurement time, sensitive to polarization mode dispersion (PMD), polarization stability and polarization hole burning
Signal substitution	Reduces wavelength resolution requirement of the optical spectrum analyzer (OSA), maintains constant power to the EDFA, simple to implement	Sensitive to spectral hole burning in certain applications
Reduced source	Reduces size of DWDM test laser array	Iterative method ($\sim$2 iterations), largest benefit obtained where the SHB widths are wider than DWDM channel separations
Optoelectronic/electrical methods		
Relative intensity noise (RIN) transfer technique	Provides complete noise figure with baseband frequency dependence of noise factor. Measures all intensity noise generated by the EDFA	If shot noise limited source is not available, source output noise must be stable to permit subtraction
Constant power	Provides complete noise figure with baseband frequency dependence of noise factor. Measures all intensity noise generated by the EDFA	If shot noise limited source is not available, source output noise must be stable to permit subtraction. Also requires ASE to be small compared to amplified signal

Gain with WDM

Gain Peak Wavelength

The gain peak wavelength is the wavelength within a certain wavelength range of the interest that achieves the maximum gain as illustrated in Fig. 14a (Wysocki 2001). It depends on the ion inversion, which is determined by the operation conditions

Fig. 14 The definition illustrating of the gain peak wavelength (**a**), gain flatness(**b**) and gain slope(**c**) of an EDFA

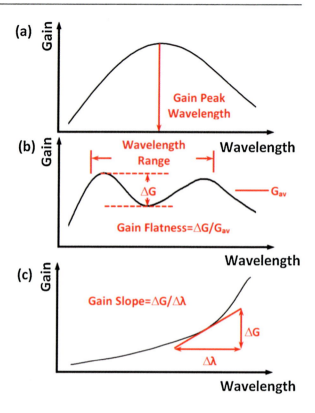

of the EDFA. These operating conditions include the pump wavelength, pump power, signal wavelength, signal power, fiber length, and temperature as well as the spectral characteristics of the components used in the EDFA. The wavelength range must be specified as some gain spectra have multiple gain peak wavelengths and operation may occur near a secondary gain peak wavelength. In single-channel long-haul communications, EDFAs operate with a signal at a particular wavelength to produce the desired level of dispersion. To produce gain compression, high output power, and automatic gain control with substantial signal power, EDFA operating conditions must be precisely set to determine the gain peak wavelength of the EDFA. Especially, the gain peak wavelength of the EDFA and the spectral characteristics of the system components can be used to determine the gain peak wavelength of the entire system (Wysocki 2001).

Gain Flatness

Broadly speaking, optical amplifiers can be classified as either single channel or multichannel (wavelength-division multiplexing (WDM)). As their name implies, single-channel amplifiers are designed to amplify only a single optical channel, which can be located anywhere within a specified band, such as 1530–1560 nm.

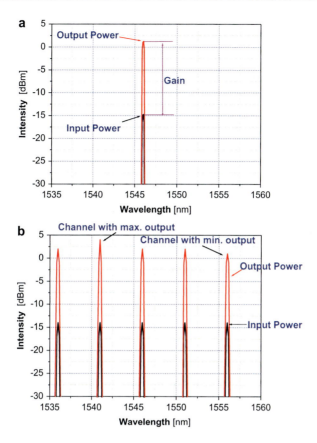

Fig. 15 Schematic input (black) and output (red) spectra of a single-channel amplifier (**a**) and a WDM multichannel amplifier (**b**)

Single-channel amplifiers can usually operate over a wide range of operating gains and require relatively low levels of output power (Finisar Corporation 2010), as shown in Fig. 15a. In contrast, WDM amplifiers are designed to operate when any number of channels (within a specified band) is the input to the amplifier. In WDM systems, the most important spectral characteristic is gain flatness as defined in Fig. 14b (Wysocki 2001). Although several different definitions exist for gain flatness, it is conceptually achieved when the gain of an EDFA is constant across a specific wavelength range under a set of fixed operating conditions like gain peak wavelength. Any measure of gain flatness should quantify the gain variation across a wavelength range under fixed conditions (Finisar Corporation 2010). If the gain is not flat, different WDM channels will have different gain as illustrated by Fig. 15b (Finisar Corporation 2010), which can accumulate along a chain of amplifiers leading to a large mismatch between channels at the end of the link (Finisar Corporation 2010).

It is critical as the channel with the most gain eventually dominates over the channels with less gain. Especially, the SNR of the weakest channel determines the maximum system length before significant transmission errors appear. In general, the more gain each EDFA produces, the fewer EDFAs are needed to cover a given span. Hence, gain flatness must be normalized to the gain level achieved as shown in Fig. 14b (Wysocki 2001).

In order to maintain flat gain, most low-end WDM amplifiers only support a single operating gain or a relatively narrow gain range. WDM amplifiers providing both flat gain and a large operating gain range require a more complex design (Finisar Corporation 2010). In addition to gain flatness, WDM amplifiers are required to provide a large dynamic input power range, to support different input conditions where any number of channels from 1 to 80 may be present. Additionally, in order to support the maximum amount of channels, WDM amplifiers require a relatively high saturated output power, typically in the range 17–23 dBm (Finisar Corporation 2010).

Related to gain flatness, the bandwidth is another important parameter used to describe EDFAs for WDM systems. It indicates the spread of wavelengths over which the EDFA gain is constant within some specified tolerance (often 3 dB). It is normally used to quantify the possible useful wavelength range of a gain medium or characterize the number of channels that may be supported by a given fiber. An EDFA with an excellent flatness figure of merit over a specified wavelength range is often characterized as having a bandwidth far greater than the flat-gain wavelength range (Wysocki 2001).

Gain Slope

The gain slope is the first derivative of the gain versus wavelength at a given wavelength under the fixed operating conditions as depicted in Fig. 14c (Wysocki 2001). Such concept is mostly used in analog applications, particularly when a signal has been modulated and broadened to a significant bandwidth. So any difference in gain across the bandwidth will lead to the distortion of the analog signal. If the slope is computed or measured without disturbing the EDFA operating point, it is truly just a derivative of a particular gain curve for fixed operating conditions. However, in real analog systems, the important value is the gain slope at the signal wavelength as the signal wavelength is varied across the band. This does not produce the same result except in the small signal region because tuning a large signal alters the operating point of the EDFA (Wysocki 2001).

Dynamic Response

Another important property is the response to dynamic changes in input power of optical amplifiers. The gain of an amplifier should not change at all when the input power changes; however, this is impossible when the amplifier operates at or near the maximum output power. In this case, it is essential that the amplifier responds slowly enough so that its gain is determined only by the average input power and is not affected by fast changes (Finisar Corporation 2010).

Amplifiers that respond too fast may be noisy and do not handle multiple channels well. This is because when there are multiple channels, the gain of one channel may change according to whether the other channels have a 0 or 1, an affect known as cross-gain modulation. Even if there is a single high-power channel near saturation, then distortion could occur since the 0s will experience different gain than the 1s (Finisar Corporation 2010).

On the other hand, even if the amplifier has a slow response, it should also be able to handle sudden long-term changes in the average input power. Such sudden changes can occur, for example, due to channel add/drop (especially in dynamically reconfigurable networks) or protection and restoration switching. In such cases the amplifier may experience large temporary gain variations (known as "transients") that need to be suppressed as much as possible by the amplifier control mechanism. In the absence of suitable transient suppression, the gain transients could accumulate over a chain of amplifier, leading to large power and/or SNR surges at the receiver (Finisar Corporation 2010).

Polarization Mode Dispersion
In the ideal optical fiber, two different polarizations of light normally travel at the same speed as shown in Fig. 16. However, in a realistic fiber, they will travel at different speeds due to random imperfections and asymmetries. In this case, the two polarization components of a signal will slowly separate, e.g., causing pulses to spread and overlap. Such kind of phenomenon is called polarization mode dispersion (PMD, shown in Fig. 16) (Wikipedia 2017b). Because the imperfections are random, the pulse spreading effects correspond to a random walk and thus have a mean polarization-dependent time-differential $\Delta\tau$ (also called the differential group delay, or DGD, $\Delta\tau = \tau_x - \tau_y$) proportional to the square root of propagation distance L (Wikipedia 2017b):

$$\Delta\tau = D_{PMD}\sqrt{L}, \tag{22}$$

where D_{PMD} is the PMD parameter of the fiber, typically measured in ps/$\sqrt{}$km, a measure of the strength and frequency of the imperfections (Wikipedia 2017b).

So far, there are various methods for the measurement of PMD of optical fibers and their systems. They can be divided into two broad categories. For the first category, PMD is determined from the measured variation of output polarization with wavelength. It can further be divided into narrowband and broadband methods. Narrowband methods determine PMD from the variation of output polarization over narrow wavelength intervals, typically narrower than the principal state bandwidth of the fiber under test. Methods of this type include the Jones matrix eigenanalysis (JME), Poincaré sphere analysis (PSA), and Mueller matrix method (MMM). Broadband methods determine PMD from the variation of output polarization

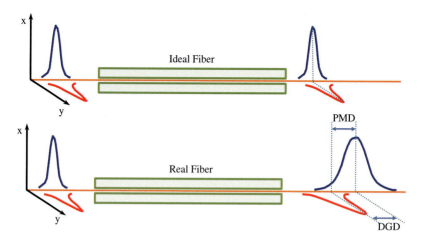

Fig. 16 Schematic diagram of polarization mode dispersion

over a broad wavelength range. Methods of this type include the fixed analyzer (or wavelength scanning) method and its variations (Photonics Media 2017). For the second category, PMD is determined from the measured difference in the propagation time associated with the two input polarization modes. This category can be further divided into low-coherence interferometric and modulation phase-shift methods (Photonics Media 2017).

In high bit rate, ultra-long-haul optical amplifier systems, especially when the chromatic dispersion effects of the signal close to the zero dispersion wavelength become very small, the optical noise due to PMD will be amplified together with the optical signals, as the signals are transmitted over ultra-long distances. Therefore, it is very important to measure the PMD of long-distance EDFA transmission systems (Namihira and Nakajima 1994). Generally, the PMD of the EDFA system ($\Delta\tau_{EDFA}$) can be expressed as (Namihira and Nakajima 1994):

$$\Delta\tau_{EDFA} = \sqrt{\sum_{i=1}^{N} \left(\Delta\tau_{ISO_i}^2 + \Delta\tau_{WDM_i}^2 + \Delta\tau_{EDF_i}^2\right)} \qquad (23)$$

where $\Delta\tau_{ISO}$, $\Delta\tau_{WDM}$, and $\Delta\tau_{EDF}$ are the PMD values of optical isolator, WDM fiber coupler, and EDF and N is the number of EDFAs in the system. As the PMD of isolator and WDM are often negligibly smaller than that of EDF, so for the same EDFA modules, Eq. 23 will further be simplified into:

$$\Delta\tau_{EDFA} \approx \Delta\tau_{EDFA}\sqrt{N} \qquad (24)$$

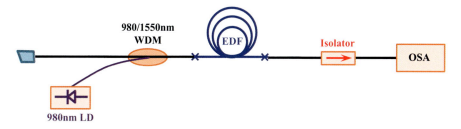

Fig. 17 ASE broadband light source

Measurements

Amplified Spontaneous Emission

Before setting up the EDFA, an ASE broadband light source is built using a 980 nm LD pumping an EDF as shown in Fig. 17. Connect the output of EDF to OSA after the isolator, gradually increasing pump LD power. The ASE spectra at different pump powers are shown in Fig. 18. Figure 18 shows that ASE power spectra cover a broadband wavelength range and total ASE power increases with the pump power. Furthermore, with the pump power increasing, the maximum ASE power will increase fast at beginning and then increase slowly due to the saturation effect as shown in Fig. 12.

Optical Gain

On/Off Gain

Figure 19 shows a typical experimental setup for the on-off gain measurement based on the lock-in amplifier. Chopped white light from a halogen lamp was launched into one end of EDF. The transmitted signal passed through a 980 nm/1550 nm WDM coupler and then the monochromators (MONO) and finally was detected with an InGaAs photo detector (PD). The lock-in amplifier was used to amplify the detected signal with the same frequency of the chopper. A fiber-pigtailed 980 nm laser diode was served as the pump source.

On-off gain measured by the change in transmission with and without pumping of the ground state is determined by (Lu 2005):

$$G_{on-off} = \frac{10}{L} \lg\left(\frac{T_{on}}{T_{off}}\right) \tag{25}$$

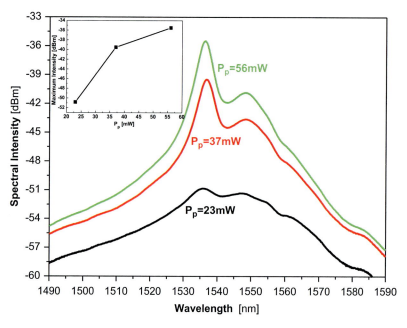

Fig. 18 ASE spectrum with different pump powers and the inset is the maximum intensity of the peak versus the pump power

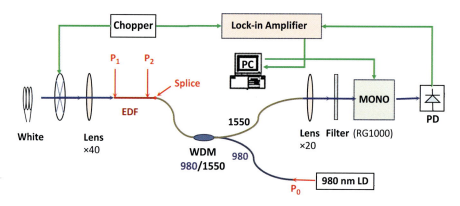

Fig. 19 Experimental setup of on-off gain measurement under 980 nm pumping

where L is a fiber length and T_{on} and T_{off} are signal transmission with pump on and off, respectively. In Fig. 10, the red curve shows a typical on-off gain spectrum measured.

Small-Signal Gain

According to Fig. 11a, co-propagating schemed EDFA is assembled. Tunable laser source or distributed feedback (DFB) laser source is used as the signal source, which is connected with the input port of the first isolator as the input of the EDFA. OSA or power meter is used to measure the output signal, which is connected with the output port of the optical filter. Common parameters of EDFA like gain, gain saturation, saturation output power, pump saturation, etc., will be measured and displayed in the following sections (Rosenkranz 2016; Amonics Ltd 2004).

1. Turn on the power of the pump and signal source, and set the pump LD to certain fixed value, $P_p = 145$ mW.
2. Adjust the signal display power to be -40 dBm, which is the input signal power of EDFA, P_{in}.
3. The output optical power will be measured by the power meter and displayed in dBm. This is the EDFA output signal power, P_{out}.
4. The gain will be the EDFA output signal power, P_{out}, minus the input signal power, P_{in}: $G = P_{out} - P_{in}$. The units will be in dB.
5. Repeat steps 2–4 for input signal powers from -35 dBm to -10 dBm, in 5 dBm steps.
6. Plot gain in dB versus input signal power in dBm as shown in Fig. 20 (Amonics Ltd 2004).

The plot shows that the gain is relatively flat until the input signal power is increased to a certain level. The flat region from -40 to -20 dB can be assigned as the small-signal gain region (Amonics Ltd 2004).

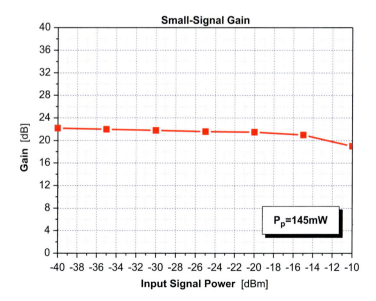

Fig. 20 Small-signal gain of EDFA with pump power of 145 mW at 980 nm (Amonics Ltd 2004)

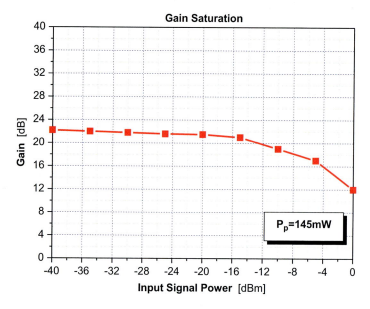

Fig. 21 Gain saturation of EDFA with pump power of 145 mW at 980 nm (Amonics Ltd 2004)

Gain Saturation

Similarly with the small-signal gain measurement, to reach the gain saturation, repeat steps 2–4 for input signal powers from −35 dBm to 0 dBm, in 5 dBm steps. As it is shown in Fig. 21 (Amonics Ltd 2004), the gain would begin to decrease when the input signal power is increased beyond −20 dBm. This gain decrease indicates the effect of gain saturation.

Saturation Output Power

1. Turn on the power of the pump and signal source, and set the pump laser to a certain fixed value, P_p (mW).
2. Adjust the signal display power to be −40 dBm, which is the input signal power of EDFA, P_{in}.
3. The output optical power will be measured by the power meter and displayed in dBm. This is the EDFA output signal power, P_{out}.
4. The gain will be the EDFA output signal power, P_{out}, minus the input signal power, P_{in}. $G = P_{out} − P_{in}$. The units will be in dB.
5. Repeat steps 2–4 for input signal powers from −35 dBm to −10 dBm, in 5 dBm steps.
6. Plot gain in dB versus output signal power in dBm as shown in Fig. 22 (Amonics Ltd 2004).

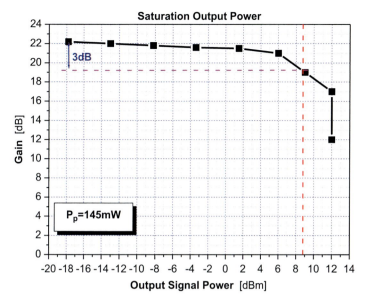

Fig. 22 Saturation output power of EDFA with pump power of 145 mW at 980 nm. The gain drop 3 dB when $P_{out} = 9$ dBm (Amonics Ltd 2004)

Figure 22 shows that as the input signal power is increased beyond a certain level, the gain would start to decrease. When the gain decreased 3 dB, the corresponding output signal power is the saturation output power. The case in Fig. 22 shows that the saturation output power is about 9 dBm (Amonics Ltd 2004).

Pump Saturation

1. Turn on the power of the pump and signal source, and set the pump laser to the maximum value, $P_{p,max}$ (mW).
2. Adjust the signal display power to be -20 dBm, which is the input signal power of EDFA, P_{in}.
3. The output optical power will be measured by the power meter and displayed in dBm. This is the EDFA output signal power, P_{out}.
4. The gain will be the EDFA output signal power, P_{out}, minus the input signal power, P_{in}: $G = P_{out} - P_{in}$. The units will be in dB.
5. Repeat steps 2–4 for pump powers from 10, 15, 20, 40, 60, 80, 100, 120, 140, to 145 mW.
6. Plot gain in dB versus pump power in mW as shown in Fig. 23 (Amonics Ltd 2004).

Figure 23 shows that the gain increases linearly when the pump power is lower than 20 mW. But with the pump power further increasing, the gain slope begins to get smaller and flatten out. This decrease in the gain slope is the effect of

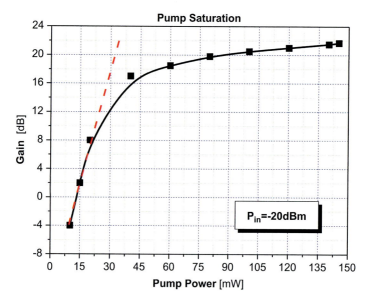

Fig. 23 Pump saturation at 980 nm of EDFA with input signal power of −20 dBm (Amonics Ltd 2004)

pump saturation (Amonics Ltd 2004), given out the pump saturation power of $P_{sat} = 20$ mW.

Noise Figure

When measuring the NF of EDFAs, noise from the laser source will be a problem encountered. Like EDFAs generate ASE, laser source also generates noise, taken as source spontaneous emission (SSE). So when the laser source is connected to the EDFA, the SSE will be amplified and contribute to the ASE noise of the EDFA, causing an error in the NF measurement. Here, the source subtraction approach is used to deal with the SSE influence as an example, where the SSE is estimated by interpolation and subtracted from the total noise emitted by the EDFA to obtain the true ASE (Amonics Ltd 2004). Therefore, after the SSE correction, Eq. 21 will be changed into:

$$F_n = \frac{P_{ASE}}{G h v_c B_o} + \frac{1}{G} - \frac{P_{SSE}}{h v B_o}$$

or

$$F_n(dB) = 10 \lg \left(\frac{P_{ASE}}{G h v_c B_o} + \frac{1}{G} - \frac{P_{SSE}}{h v B_o} \right) \tag{26}$$

where P_{ASE} is the total noise spectral density, including SSE at the signal wavelength, and P_{SSE} is the SSE spectral density at the signal wavelength. Please note that all the values should be converted into the linear form before substituting into Eq. 26 as well as all the units, which should also be unified. For the NF measurement, it is best to interpolate ASE and SSE noise. The ASE power is measured at wavelengths below and above the signal wavelength and then interpolated. In addition, the measurement of the ASE power always takes place between the main mode and the first secondary mode (Rosenkranz 2016).

Although some OSAs have functions to measure and calculate the gain and NFs of the optical amplifier, most do not apply the SSE correction.

1. Set OSA resolution to 0.1 nm and wavelength span to 8 nm.
2. Turn on the power of the pump and signal source, and set the pump laser to the maximum value.
3. Adjust the signal display power to be -40 dBm, which is the input signal power of EDFA, P_{in}.
4. Connect the laser output port to the OSA using a fiber patch cord.
5. Measure the optical power on the OSA at the signal wavelength in dBm with the help of the OSA marker, which is the EDFA input signal power, P_{in}, as shown in Fig. 24 (Amonics Ltd 2004).
6. Measure the optical powers on the OSA at 1 nm on both sides of the signal wavelength in dB/(resolution) with the help of the OSA marker. Use these powers to interpolate a value at the signal wavelength, which is the SSE spectral density, P_{SSE}, as shown in Fig. 24 (Amonics Ltd 2004).

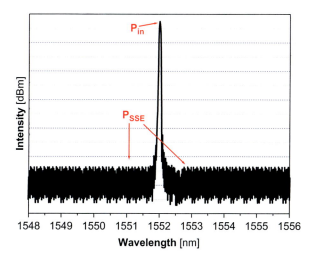

Fig. 24 Measure input signal power at signal wavelength and SSE power levels 1 nm on both sides of signal wavelength

7. Swap the signal laser source output from OSA input port to the input port of the EDFA (the input port of the first isolator).
8. Connect the output port of the second isolator to OSA, which is also the output signal port of the EDFA to the OSA.
9. Measure the optical powers at 1 nm on both sides of the signal wavelength in dB/(resolution). Use these powers to interpolate a value at the signal wavelength, which is the ASE spectral density, P_{ASE}, as shown in Fig. 25 (Amonics Ltd 2004).
10. Measure the optical power on the OSA at the signal wavelength in dBm with the help of the OSA marker, which is the EDFA output signal power, P_{out}, as shown in Fig. 25 (Amonics Ltd 2004).
11. According to the Eq. 26, calculate the NF with the measured values: $P_{in} = -40$ dBm, $\lambda = 1552$ nm, $G = 30.7$ dB, $P_{SSE} = -65.4$ dBm/(resolution), and $P_{ASE} = -21.31$ dBm/(resolution). Convert G, P_{SSE} and P_{ASE} into the linear unit before substituting into the Eq. 26.

$$G = 10^{\frac{30.7}{10}} \approx 1174.90$$

$$P_{SSE} = 10^{\frac{-65.4}{10}} \approx 2.88 \times 10^{-10} W/resolution$$

$$P_{ASE} = 10^{\frac{-21.31}{10}} \approx 7.40 \times 10^{-10} W/resolution$$

The effective noise bandwidth B_0 is considered to be 0.1 nm for the used resolution of OSA, so:

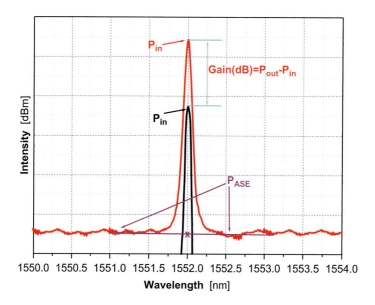

Fig. 25 Measure ASE power levels 1 nm on both sides of the signal wavelength

$$B_0 = \frac{c \times \Delta\lambda}{\lambda^2} = \frac{(3 \times 10^8) \times (0.1 \times 10^{-9})}{(1552 \times 10^{-9})^2} \approx 1.25 \times 10^{10}\,Hz$$

From the signal wavelength at 1552 nm, the optical frequency is calculated:

$$v = \frac{c}{\lambda} = \frac{3 \times 10^8}{1552 \times 10^{-9}} \approx 1.93 \times 10^{14}\,Hz$$

Substitute all these values into Eq. 26 to calculate the NF when $P_{in} = -40$ dBm:

$$F_n(dB) = 10\lg\left(\frac{P_{ASE}}{Ghv_c B_o} + \frac{1}{G} - \frac{P_{SSE}}{hv B_o}\right)$$

$$= 10\lg\left[\frac{7.40 \times 10^{-10}}{1174.90 \times (6.63 \times 10^{-34}) \times (1.93 \times 10^{14}) \times (1.25 \times 10^{10})}\right.$$

$$\left. + \frac{1}{1174.9} - \frac{2.88 \times 10^{-10}}{(6.63 \times 10^{-34}) \times (1.93 \times 10^{14}) \times (1.25 \times 10^{10})}\right]$$

$$\approx 5.8 dB$$

12. Repeat steps 3–11 for input signal powers from −35 dBm to 0 dBm, in 5 dBm steps.
13. Plot the noise figure in dB versus input signal power in dBm as shown in Fig. 26 (Amonics Ltd 2004), given out an averaged NF of ∼6.0 dB.

Gain in WDM System

Gain Peaking in Multiple-EDFA Systems

The accurate measurement of the gain peak wavelength for long, multi-amplifier chains is difficult in a single EDFA. However, measurement in a long amplifier chain requires the construction and connection of many identical amplifiers. For the simplification of the measurement, the EDF can be inserted in a fiber ring laser with a known loss as shown in Fig. 27. The amplification loop consists of a single EDFA with a counter-propagating pump at 1480 nm, 0.8 km of dispersion-shifted fiber (DSF) with a small effective area, an optical isolator (ISO) to force signal flow in one direction, a polarization controller (PC), and a variable attenuator (HP attenuator) to vary the loop loss. When the pump power is over the threshold, laser emission will occur at the wavelength with the maximum gain for the operating level of saturation. Similar with any other laser, the gain of the EDF equals the loss in the rest of the loop. To eliminate polarization-induced instability and accurately measure the gain peak wavelength, polarization must be incorporated in the fiber loop as in Fig. 27 (Wysocki 2001). A total of six scramblers were used, because each component is

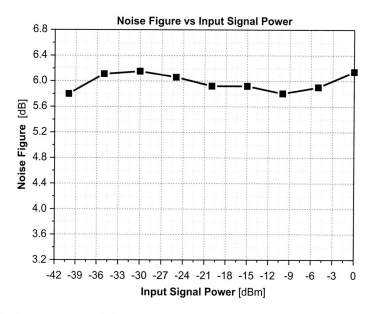

Fig. 26 Noise figure versus different input signal powers at 1552 nm (Amonics Ltd 2004)

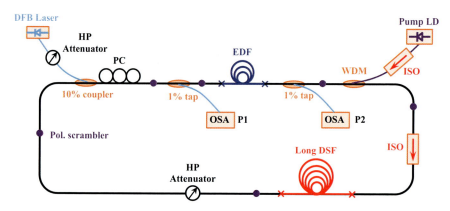

Fig. 27 Experimental setup to measure the gain peak wavelength using a depolarization fiber loop (Wysocki 2001)

slightly polarization-dependent and slightly repolarizes the signal. Taps have been introduced on both sides of the EDFA to monitor the signal power and compute the gain. All components must be carefully selected to minimize the wavelength dependence of the loop loss (Wysocki 2001). Under the assumption of homogeneous broadening, the gain peak wavelength of an EDFA is always the same for a certain average inversion as the average inversion and the gain per unit length are almost linear as shown in Fig. 28 (Wysocki 2001).

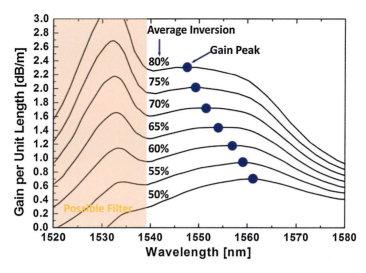

Fig. 28 Demonstration of the shift in gain peak wavelength with increasing inversion level for EDFA. A postulated ASE filter near the 1530 nm is shown. The blue dots indicate the peak locations (Wysocki 2001)

In EDFA chains, the gain peak wavelength is determined not only by the EDFA performance but also by the wavelength-dependent losses of the components in each span. Owing to the flatness of the EDFA gain, even slight wavelength-dependent losses can shift the gain peak wavelength. This dependence can be used to the optimization of the system design. If the ring laser of Fig. 27 is configured to incorporate representative components of a single span in a real amplifier chain, the measured gain peak wavelength is the peak of the actual system. The only complication is that the result cannot be plotted in terms of gain per unit length as the losses do not all scale with amplifier length. Such a measurement is most useful for fine-tuning the EDFA design once the span loss, fiber length, signal power, and pump power have been finalized (Wysocki 2001).

Gain Flatness in WDM Systems

WDM is an efficient way to increase the capacity of fiber communication systems. Consequently, approaches to flattening the gain of EDFAs are important (Wysocki 2001). Generally, a gain flattening filter is usually placed following the output isolator in order to flatten the gain spectrum in a multichannel WDM amplifier as shown in Fig. 29. The attenuation spectrum of the filter is designed to match the gain spectrum of the output EDF (operating at a given fixed gain), such that the combination of the two produces a flat gain. Following the filter, the signal passes through an output tap used to divert a small percentage of the output power (typically 1–2%) to the output detector. The output and input detectors are used to monitor the input and output power, respectively, and thus provide feedback to the control unit, which controls the amplifier by setting the pump laser current and thus the amount

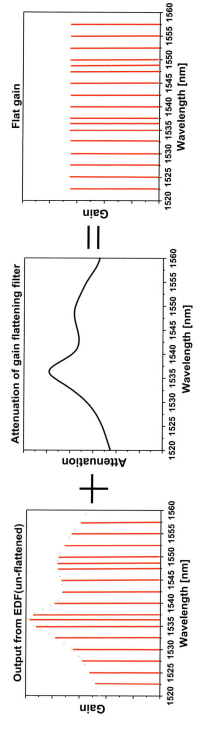

Fig. 29 Use of a gain-flattened filter to achieve a flat gain spectrum. The attenuation spectrum of the filter matches well with the gain spectrum at the output of the EDF, providing a flat gain spectrum (Finisar Corporation 2009)

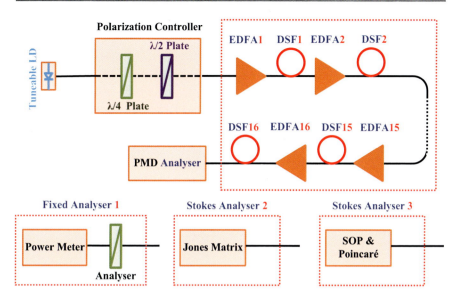

Fig. 30 Polarization mode dispersion (PMD) measurement setup for long-haul EDFA system

of pump power injected into the EDF (Finisar Corporation 2009). According to the definition shown in Fig. 15b, the gain flatness before and after the flattening can be calculated and compared to achieve the wide and flattened optical gain.

PMD for Long-Distance EDFA System

A schematic diagram of the experimental setup for the wavelength scanning PMD measurements in a 1600 km EDFA system is shown in Fig. 30 (Namihira and Nakajima 1994). Four types of PMD measurement methods have been used, including: the fixed analyzer method and three polarimetric methods using a Stokes analyzer (Jones matrix method, states of polarization (SOP) method, and Poincad sphere method) (Namihira and Nakajima 1994). Figure 31 shows a typical wavelength dependence of PMD measurement result in 1600 km system using the Jones matrix method (Namihira and Nakajima 1994).

For the schematic EDFA system in Fig. 30, the estimated PMD $\Delta\tau_{sys}$ can be expressed as (Namihira and Nakajima 1994):

$$
\Delta\tau_{sys} = \sqrt{\sum_{i=1}^{N} \Delta\tau_{EDFA_i}{}^2 + \sum_{1}^{j} \left(\Delta\tau_{DSF_j}\sqrt{L_j}\right)^2}
$$

$$
\approx \sqrt{\left(\Delta\tau_{EDFA}\sqrt{N}\right)^2 + \left(\Delta\tau_{DSF}\sqrt{L}\right)^2}
$$

(27)

where $\Delta\tau_{DSFj}$ is the PMD of jth optical DSF with the length of L_j, PMD of all the EDFA and PMD of all the DSF are the same with each other. Substituting all the

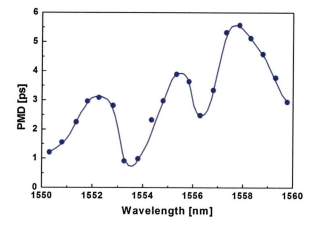

Fig. 31 Example of wavelength dependence of PMD measurement result in 1600 km system using the Jones matrix method (Namihira and Nakajima 1994)

parameters: L = 1600 km, $\Delta\tau_{EDFA}$ = 0.2 ps, $\Delta\tau_{DSF}$ = 0.1 ps/$\sqrt{km}$, and N = 16 into Eq. 27, $\Delta\tau_{sys}$ = 4.3ps, which is in good agreement with the experimental results-averaged value ranged from 3.6 to 5.8 ps.

Measurement Uncertainty

It is most probable that the optical power values stated and those measured will be different. Some of this uncertainty can be attributed to using different equipment and calibration sources. However, there are other sources of uncertainty that will affect the accuracy of these measurements (Amonics Ltd 2004). An error contribution is determined for each of the uncertainty terms described below. The total uncertainties are then calculated using the following equation (Agilent Product Note 2017):

$$uncerta\text{ int } y = 2\sqrt{\sum \frac{U^2}{3}} \qquad (28)$$

where "U" is the uncertainty of each individual term. All uncertainties are expressed as peak values. Source stability, source repeatability, and source spontaneous emission repeatability will directly affect the measurement accuracy (Agilent Product Note 2017). The OSA and power meter used in measurements will have an amount of uncertainty that is dependent on the scale used for measurement. For example, the absolute amplitude accuracy, polarization sensitivity, scale fidelity, flatness, resolution bandwidth accuracy, internal etalons, and dynamic range of OSA all related to the measurement uncertainties (Agilent Product Note 2017). The connectors used to connect the fiber pigtails will also introduce an uncertainty of ±0.25 dB for each connector. Using fusion splices in real EDFAs reduces this connector uncertainty significantly, especially for the mismatched fusion. The polarization dependency of the equipment will also introduce an uncertainty, causing the optical power to fluctuate up and down. Using polarization scramblers and measurement averaging will reduce this kind of uncertainty (Amonics Ltd 2004).

Typical Optical Amplifiers

As described above, each type of amplifier has advantages and disadvantages depending on their application. Their properties have typically been summarized

Table 2 Summary of the property for EDFA, Raman amplifier, and SOA (Tutorials of Fiber Optic Products 2017)

Property	EDFA	Raman	SOA
Gain [dB]	>40	>25	>30
Wavelength [nm]	1530–1560	1280–1650	1280–1650
Bandwidth [3 dB]	30–60	Pump dependent	60
Max. saturation [dBm]	22	$0.75 \times$ pump	18
Polarization sensitivity	No	No	Yes
Noise figure [dB]	5	5	8
Pump power	25	>30 dBm	<400 mA
Time constant [s]	10^{-2}	10^{-15}	2×10^{-9}
Size	Rack mounted	Bulk module	Compact
Switchable	No	No	Yes
Cost factor	Medium	High	Low

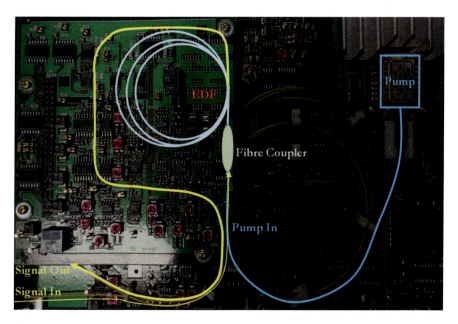

Fig. 32 The schematic diagram of EDFA

in Table 2. So far, the most common and widely used optical amplifier is EDFA, where a typical EDFA is shown in Fig. 32.

Conclusion

Since the commercialization in the early 1990s, optical amplifier, especially EDFA, has become a key enabling technology for the optical communications system. The characterization of their properties of EDFA is very important in terms of the photonic network application. A series of parameters like gain, gain saturation, pump saturation, noise figure, etc., have been used and measured to assess their performance. Furthermore, proper measurement has been designed and described for high accuracy and good repeatability as well as low uncertainty. Although there is still much work performed in the measurement of optical amplifier, the content in this chapter mainly concentrates on the most common and simple methods for good instruction and standardization. These wide and uniformed measurement methods are important for easier and better understanding and development of optical amplifier for future network communications.

Acknowledgements Authors are thankful for the support of National Natural Science Foundation of China (61520106014, 61405014 and 61377096), Key Lab of In-fiber Integrated Optics, Ministry Education of the People's Republic of China, State Key Laboratory of Information Photonics and Optical Communications (Beijing University of Posts and Telecommunications) (IPOC2016ZT07), Key Laboratory of Optical Fiber Sensing and Communications (Education Ministry of China), Key Laboratory of Optoelectronic Devices and Systems of Ministry of Education and Guangdong Province (GD201702), and Science and Technology Commission of Shanghai Municipality, China (SKLSFO2015-01 and 15220721500).

References

Agilent Product Note, 71452-1: EDFA Testing with the Interpolation Technique, http://notes-application.abcelectronique.com/018/18-27242.pdf. Accessed 13 July 2017

G.P. Agrawal, *Fiber-Optic Communication Systems* (Wiley, New York, 1997)

Amonics Ltd, Erbium-doped fiber amplifier education kit manual (2004), https://www.tau.ac.il/~lab3/OPTICOM/EDFA_Kit_upgrade_manual.pdf. Accessed 13 July 2017

D.M. Baney, P. Gallion, Theory and measurement techniques for the noise figure of optical amplifiers. Opt. Fiber Technol. **6**, 122–154 (2000)

W.L. Barnes, R.I. Laming, E.J. Tarbox, P.R. Morkel, Absorption and emission cross section of Er^{3+} doped silica fibers. IEEE J. Quantum Electron. **27**, 1004–1010 (1991)

M.R.X. de Barros, G. Nykolak, R. Ghosh, C.F. Kane, J. Shmulovich, D.J. DiGiovanni, A. Bruce, W.H. Grodkiewicz, P.C. Becker, High concentration Er^{3+}-doped fiber and planar waveguide amplifiers (1996), http://www.eletrica.ufpr.br/anais/sbrt/SBrT14/SBrT_1996v1_020.pdf. Accessed 30 July 2017

P.M. Becker, A.A. Olsson, J.R. Simpson, *Erbium-Doped Fiber Amplifiers: Fundamentals and Technology* (Elsevier, 1999), p. 460

I.A. Bufetov, S.V. Firstov, V.F. Khopin, O.I. Medvedkov, A.N. Guryanov, E.M. Dianov, Bi-doped fiber lasers and amplifiers for a spectral region of 1300–1470 nm. Opt. Lett. **33**, 2227–2229 (2008)

E. Desurvire, *Erbium-Doped Fiber Amplifiers: Principles and Applications* (Wiley, 1994), pp. 306–440

V.V. Dvoyrin, V.M. Mashinsky, E.M. Dianov, A.A. Umnikov, M.V. Yashkov, A.N. Guryanov, Bi-doped silica fibers- a new active medium for tunable fiber lasers and broadband fiber amplifiers, in *OFC 2006*, 5–10 Mar 2006 (IEEE, Anaheim, 2006)

V.V. Dvoyrin, O.I. Medvedkov, V.M. Mashinsky, A.A. Umnikov, A.N. Guryanov, E.M. Dianov, Optical amplification in 1430–1495 nm range and laser action in bi-doped fibers. Opt. Express **16**, 16971–16976 (2008)

Finisar Corporation, Introduction to EDFA technology (2009), https://www.finisar.com/sites/default/files/resources/Introduction%20to%20EDFA%20technology.pdf. Accessed 13 July 2017

Finisar Corporation, Introduction to optical amplifiers (2010), https://www.finisar.com/sites/default/files/resources/Introduction%20to%20Optical%20Amplifiers.pdf. Accessed 13 July 2017

Finisar Corporation, Applications for distributed Raman amplification (2012), https://www.finisar.com/sites/default/files/resources/Applications%20for%20Distributed%20Raman%20Amplification.pdf. Accessed 13 July 2017

S.V. Firstov, S.V. Alyshev, K.E. Riumkin, V.F. Khopin, A.N. Guryanov, M.A. Melkumov, E.M. Dianov, A 23-dB bismuth-doped optical fiber amplifier for a 1700 nm band. Sci. Rep. **6**, 28939 (2016)

C.R. Giles, E. Desurvire, Modeling erbium-doped fiber amplifiers. J. Lightwave Technol. **9**, 271–283 (1991)

E. Iannone, *Telecommunication Networks* (CRC Press, 2017), p. 918

R.K. Kostuk, Section 5: Optical amplifiers (2008), http://www2.engr.arizona.edu/~ece487/opamp1.pdf. Accessed 13 July 2017

Y.-H. Lu, Design of optically gain-clamped erbium-doped fiber amplifiers with ring laser structure, Dissertation, National Chiao-Tung University, 2005

Y. Namihira, K. Nakajima, Comparison of various polarisation mode dispersion measurement methods in 1600 km EDFA system. Electron. Lett. **30**, 1157–1159 (1994)

OPI online course Laser: Fundamentals, http://www.optique-ingenieur.org/en/courses/OPI_ang_M01_C01/co/Contenu_05.html. Accessed 30 July 2017

R. Paschotta, Optical fiber technology: Active fiber devices, in *Field Guide to Optical Fiber Technology*, ed. by R. Paschotta (SPIE Press, Bellingham, 2010), pp. 86–87

R. Paschotta, Gain (2017a), https://www.rp-photonics.com/gain.html. Accessed 13 July 2017

R. Paschotta, Erbium-doped fiber amplifiers (2017b), https://www.rp-photonics.com/erbium_doped_fiber_amplifiers.html. Accessed 31 July 2017

Photonics Media, Polarization mode dispersion: concepts and measurement, https://www.photonics.com/EDU/Handbook.aspx?AID=25153. Accessed 13 July 2017

Physikalisch-Technische Bundesanstalt, Fibre-Brillouin-Amplifiers, https://www.ptb.de/cms/en/ptb/fachabteilungen/abt4/fb-43/ag-434/fibre-brillouin-amplifiers.html. Accessed 13 July 2017

W. Rosenkranz, *P5: EDFA and Measurement Techniques* (LNT, 2016), pp. 1–38

P. Schiopu, F. Vasile , The EDFA performance with gain versus pump power, in 2004 International Semiconductor Conference. CAS 2004 Proceedings (IEEE Cat. No.04TH8748), Sinaia, 4–6 Oct 2004

P. Shukla, K.P. Kaur, Performance analysis of EDFA for different pumping configurations at high data rate. Int. J. Eng. Adv. Technol. **2**, 487–490 (2013)

V.K. Stamatios, Light amplifiers, in *Introduction to DWDM Technology: Data in a Rainbow*, (Wiley-IEEE Press, 2000), pp. 119–130

R.H. Stolen, C. Lee, R.K. Jain, Development of the stimulated Raman spectrum in single-mode silica fibers. J. Opt. Soc. Am. B. **1**, 652–657 (1984)

Tutorials of Fiber Optic Products, Comparison of different optical amplifiers (2017), http://www.fiber-optic-tutorial.com/comparison-of-different-optical-amplifiers.html. Accessed 13 July 2017

Wikipedia, Optical amplifier (2017a), https://en.wikipedia.org/wiki/Optical_amplifier. Accessed 13 July 2017

Wikipedia, Polarization mode dispersion (2017b), https://en.wikipedia.org/wiki/Polarization_mode_dispersion. Accessed 13 July 2017

P.F. Wysocki, Erbium-doped fiber amplifiers: Advanced topics, in *Rare-Earth-Doped Fiber Lasers and Amplifiers, Revised and Expanded*, ed. by M.J.F. Digonnet (CRC Press, Baton Rouge. ch. 11, 2001), p. 98

Measurement of Optical Fiber Laser

40

Haifeng Qi, Weitao Wang, Jian Guo, and Zhiqiang Song

Contents

Abstract

As one of the key photonic devices, optical fiber laser has been playing an important role in the fiber communications and laser physics. In research, development, and application of fiber lasers, it is necessary to apply a range of measurement techniques for characterization and evaluation. This chapter introduces the main characteristics of optical fiber laser and describes related characterization and measurement techniques. Firstly, the history of fiber laser is briefly reviewed, and different types of fiber lasers are introduced. Then the theoretical definitions and experimental measurements of typical parameters for fiber lasers have been described in details.

H. Qi (✉) · W. Wang · J. Guo · Z. Song
Shandong Key Laboratory of Optical Fiber Sensing Technologies, Qilu Industry University
(Laser Institute of Shandong Academy of Sciences), Jinan, China
e-mail: qihf@sdlaser.cn; wangweitao@sdlaser.cn; guojianphy@163.com; szq821214@163.com

© Springer Nature Singapore Pte Ltd. 2019   1529
G.-D. Peng (ed.), *Handbook of Optical Fibers*,
https://doi.org/10.1007/978-981-10-7087-7_46

Introduction

Fiber laser is a kind of laser whose active gain medium is an optical fiber doped with rare-earth or any other elements, like erbium, ytterbium, neodymium, praseodymium, thulium, holmium, or bismuth (Fig. 1). As shown in Fig. 1, all-fiber laser is usually pumped by a laser diode with fiber output or other fiber lasers, while the resonant cavity is constructed with active gain optical fiber and other optical fiber components, e.g., optical fiber Bragg gratings (FBGs) as reflectors. Fiber laser with a section of fiber containing the laser medium has a long interaction length, resulting in high photon conversion efficiency. When fiber components are spliced together, there is no discrete optics to adjust or to get out of alignment, so that fiber laser is usually rugged and compact.

Fiber laser has been researched and developed for many years since the 1960s. In 1961, the first fiber laser with Nd-doped glass waveguide as gain medium and xenon flash lamp as pump was invented by E. Snitzer (1961). In the following 10 years, fiber laser was developed slowly for the high loss in optical fiber and having no high-performance semiconductor laser as pump source. Since the great development of low-loss optical fiber and ion doping technology in the 1970s, the fiber amplifier and fiber laser have made great progress. Especially in the 1980s, Er-doped single-mode fiber with low loss was produced in the University of Southampton, UK (Pools et al. 1985), and was promptly applied in optical fiber laser. Since then, optical fiber laser starts to become practical. In 1987, optical fiber grating was invented (Hill et al. 1978) and then widely used in fiber laser and optical fiber sensing. And the application of fiber Bragg grating boosted the development of single-frequency fiber laser. Before the double-cladding fiber was presented in the late 1980s, it is hard to obtain the high-power output from fiber laser due to the small core diameter in normal single-cladding fiber and the low coupling efficiency of the pump into fiber core. Since the invention of double-cladding fiber by E. Snitzer in 1988 (Snitzer et al. 1988), high-power fiber laser has made great progress. Employing the double-cladding fiber, sufficient pump power could be coupled into fiber core through the reduplicative reflection in inner cladding of the fiber, so that the doped fiber in core region could absorb enough energy from pump to enable high laser output. With the development of semiconductor laser as pump source and pump coupling technology, the attainable laser power in double-cladding fiber laser becomes higher and higher. Currently, fiber laser with several kilowatt power outputs is common in many laser products.

In fiber laser, there are several widely used ion dopants, such as Nd^{3+}, Yb^{3+}, Er^{3+} Tm, and Ho. Nd- or Yb-doped fiber laser emits laser in 1.06 μm wavelength

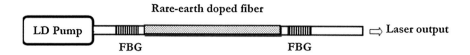

Fig. 1 The basic configuration of the laser diode-pumped fiber laser

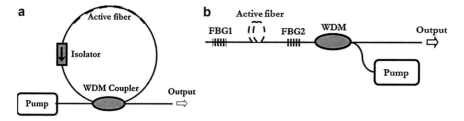

Fig. 2 Ring cavity and linear cavity fiber laser. (**a**) Ring laser, (**b**) Linear cavity laser

region which is widely used in laser process and laser weapon. Er-doped fiber laser emits laser in 1.55 μm wavelength region which is suitable for optical communication and optical sensing. Tm- or Ho-doped fiber laser emits laser in 2 μm wavelength region which is eye-safe and widely used in special laser communication, atmosphere sensing, and medical application.

Fiber laser has many advantages over other conventional laser types.

1. High beam quality: The fiber's waveguide properties can reduce or eliminate thermal distortion of the optical path, typically producing a diffraction-limited, high-quality optical beam.
2. High power: The active region in fiber laser can be several kilometers long, so that it can provide very high laser gain. The fiber laser with a very long gain section can generate very high output power over several kilowatts due to efficient cooling based on its high surface-to-volume ratio.
3. Compact size: Fiber lasers are very compact compared with rod or gas lasers of comparable power because of its structure principle.
4. Convenience of application: Fiber is a flexible component. Fiber laser contains the laser light in fiber core, and it can be easily delivered to a movable output focusing element, thus made it very suitable in laser cutting, welding, and folding of metals and polymers.
5. Reliability: Fiber lasers exhibit high temperature and vibrational stability, electromagnetic interference immunity, and extended lifetime.

Based on above features and advantages, fiber lasers have great potential in many application fields as shown in Fig. 2, such as material processing (marking, engraving, and cutting), telecommunications, spectroscopy, clinical medicine, and remote sensing.

Types of Fiber Laser

Since its versatility in fiber material, dopant, fiber structure, and cavity configuration, there are many different types of fiber lasers. In addition, fiber lasers can run in different modes and wavelengths. Due to these differences, the fiber lasers have been classified in this section.

Classification by Operating Conditions

Cavity Configuration

According to the cavity configuration used, fiber lasers can be classified into **ring cavity fiber laser** and **linear cavity fiber laser**. **Ring cavity fiber laser** (Fig. 2a) usually utilizes an optical coupler and connects its two ends together with a gain optical fiber so that the optical feedback is formed. It is a traveling wave cavity laser and has a relatively long cavity resulting in a small longitudinal mode gap. Usually the separate component is needed in the ring cavity to achieve single-frequency operation. **Linear cavity fiber laser** (Fig. 2b) usually utilizes the fiber gratings as reflectors replacing the traditional optic reflectors. Since fiber Bragg grating (FBG) can be written in the passive fiber or active fiber directly, the linear cavity fiber laser can be very composite. By adjusting the bandwidth and reflectivity of the FBGs and the cavity length, single-frequency operation is easily obtained in a short-cavity fiber laser. Both distributed Bragg reflector fiber laser (DBR-FL) (Ball et al. 1993) and distributed feedback fiber laser (DFB-FL) (Kringlebotn et al. 1994) belong to this linear cavity fiber laser.

Fiber Structure

According to the structure of optical fiber used, fiber lasers can be classified into **single-cladding fiber laser, double-cladding fiber laser** (Fig. 3), and **photonic crystal fiber laser**. **Single-cladding fiber** has one cladding around the fiber core. In the single-cladding fiber laser, the pump light and signal light are both propagating in the core. For **double-cladding fiber** (Fig. 3a), it has two claddings around the core. Fiber core is doped with active dopant material, and the signal light is propagating and amplified in the core. The inner cladding and core together guide the pump light, which provides the energy needed to allow amplification in the core. In these fibers, the core has the highest refractive index and the outer cladding has the lowest. In most cases, the outer cladding is made of a polymer material rather than glass. The pump light can easily be coupled into the large inner cladding and propagates through the inner cladding, while the signal propagates in the smaller core. The doped core gradually absorbs the cladding light as it propagates, driving the amplification process. This pumping scheme is often called cladding pumping, which is an alternative to the conventional core pumping, in which the pump light is coupled with the small core. Using this method, modern fiber lasers can produce continuous power up to several kilowatts, while the signal light in the core maintains near diffraction-limited beam quality. **Photonic crystal fiber** (PCF) is a new class of optical fiber based on the properties of photonic crystals. Because of its ability to confine light in hollow cores or with confinement characteristics not possible in conventional optical fiber (Fig. 4), PCF is now finding applications in fiber-optic communications, fiber lasers, nonlinear devices, high-power transmission, highly sensitive gas sensors, and other areas.

Fig. 3 Double-cladding fiber and its fiber laser. (**a**) Double-cladding fiber, (**b**) Double-cladding fiber laser

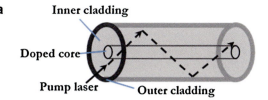

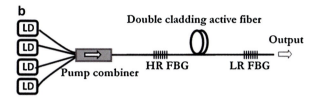

Fig. 4 Photonic crystal fiber structure and the laser propagating mode

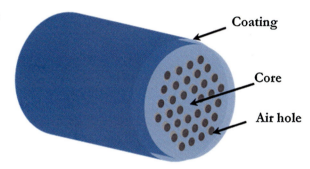

Operation Material

According to the operation material in fiber laser, fiber lasers can be classified into **crystal fiber laser, rare-earth-doped fiber laser, and nonlinear effect fiber laser**. **Crystal fiber laser** uses laser crystal fiber as operation material, such as ruby crystal fiber laser and Nd:YAG crystal fiber laser. **Rare-earth ion-doped fiber laser** uses rare-earth ion-doped fiber as operation material whose host is usually silica-based glass. **Nonlinear effect fiber laser** is based on some nonlinear effect to provide gain for laser instead of the upper and lower-population inversion. The two main nonlinear effect fiber lasers are **Raman fiber laser** (Headley et al. 2004) based on stimulated Raman scattering effect and **Brillouin fiber laser** (Smith et al. 1991) based on stimulated Brillouin scattering effect. A **Raman fiber laser** is a specific type of laser in which the fundamental light-amplification mechanism is stimulated Raman scattering. Raman lasers are optically pumped. However, this pumping does not produce a population inversion as in conventional lasers. Rather, pump photons are absorbed and "immediately" reemitted as lower-frequency laser light photons ("Stokes" photons) by stimulated Raman scattering. The difference between the two photon energies is fixed and corresponds to a vibrational frequency of the gain medium. This makes it possible, in principle, to produce arbitrary laser

Fig. 5 A typical first-order
Raman fiber laser

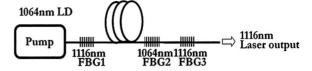

Fig. 6 A Brillouin fiber laser

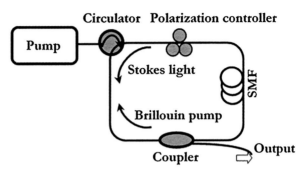

output wavelengths by choosing the pump laser wavelength appropriately. This is in contrast to conventional lasers, in which the possible laser output wavelengths are determined by the emission lines of the gain material. In optical fibers made of silica, for example, the frequency shift corresponding to the largest Raman gain is about 13.2 THz. In the near-infrared region, this corresponds to a wavelength separation between pump light and laser output light of about 100 nm. For Ge-doped optical fiber, the first-order Stokes wavelength shift is 52 nm. A typical first-order Raman fiber laser pumped by 1064 nm laser is demonstrated in Fig. 5. **Brillouin fiber laser** utilizes stimulated Brillouin scattering (SBS) effect of the fiber as gain. SBS process can be described as a nonlinear interaction between pump light and Stokes light through the elastic wave. An opposite ring cavity for SBS gain is shown in Fig. 6. The Brillouin gain bandwidth is only 20 GHz, and a single-frequency operation in Brillouin fiber laser can be obtained conveniently.

Classification by Output Type

Output Wavelength Number

According to the wavelength numbers emitted from the fiber laser, fiber lasers can be classified into **single-wavelength fiber laser** and **multiwavelength fiber laser**. Single-wavelength fiber laser can be generated by using some frequency selection components, such as bandpass filter in a ring cavity and narrow bandwidth reflectors in a short linear cavity. Without the special frequency-selective components, fiber laser is always working in the multi-longitudinal mode, i.e., multiwavelength mode, but not stable. A stable multiwavelength fiber laser can be achieved with some special configuration. For example, a comb filter is applied in the ring or linear cavity to generate multiwavelength operation; cascade SBS in Brillouin fiber laser can emit stable multiple wavelengths; multiphase-shifted distributed feedback fiber

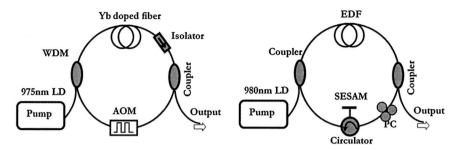

Fig. 7 Active AOM Q-switched fiber laser and passive SESAM Q-switched fiber laser

laser (DFB-FL) or superstructure FBG-based fiber laser can also provide stable multiwavelength operation.

Operation Mode

According to the operation mode of fiber laser, fiber lasers can be classified into **continuous wave fiber laser** and **pulsed fiber laser**. In the common ring and linear cavities, fiber laser emits continuous wave laser. Pulsed fiber laser can be achieved with Q-switch mode or master oscillator power-amplifier (MOPA) mode. The active Q-switch mode is performed with acousto-optic Q-switch (AOM) or electro-optic Q-switch (EOM) in the cavity, while the passive Q-switch mode is performed with saturable absorber, e.g., semiconductor saturable mirror (SESAM) or graphene film in the laser cavity, as shown in Fig. 7.

Special Fiber Lasers

There are some special fiber lasers with some special performances: **narrow linewidth fiber laser**, **random fiber laser**, and **ultrafast fiber laser**. The common narrow linewidth fiber lasers are ring cavity fiber laser with frequency selection, **distributed feedback fiber laser (DFB-FL)**, **distributed Bragg reflector fiber laser (DBR-FL)**, and composite cavity fiber laser (CCFL). Especially, a DFB-FL, as shown in Fig. 8, is usually made of a single continuous FBG with a π-phase shift inside the grating written in an active fiber. The cavity length of a DBF-FL ranges from a few millimeters to centimeters. The phase shift in the grating opens an ultra-narrow transmission band in the Bragg reflection region that makes the DFB-FL lasing with only an extremely narrow linewidth. The effective cavity length is inversely proportional to the grating strength. Hence, in good DFB-FLs, the laser mode is tightly confined around the phase shift, free from mode hops, and exhibiting extremely narrow linewidth, long coherent length, and excellent environmental stability. The linewidth of a DFB-FL usually is about several kilohertz or even below 1 kHz.

Random fiber laser (de Matos et al. 2007) is a new type fiber laser without traditional cavity reflectors. It utilizes the random distributed feedback mechanism

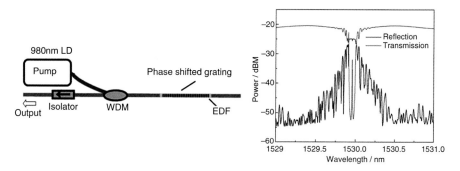

Fig. 8 Distributed feedback fiber laser structure and the optical spectrum of a π-phase-shifted grating

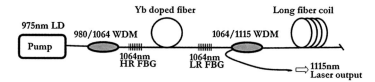

Fig. 9 A random fiber laser

generated by continuous Raleigh scattering in fiber. Although the Raleigh scattering coefficient is small about 4.5×10^{-5} km^{-1}, the laser can be realized by continuously accumulation and amplification on stimulated scattering to overcome the fiber loss in a long fiber, such as several kilometers. As shown in Fig. 9, the random fiber laser has an output wavelength at 1115 nm which is 13 THz Raman shift from its pump wavelength at 1064 nm. Since the small backward Raleigh scattering coefficient, a long fiber is needed to overcome the loss, and the pump threshold of the random fiber laser is usually high over 1 W. **Ultrafast fiber laser** mainly contains picoseconds or femtoseconds pulse laser produced with some mode-locking technique (Fermann and Hartl 2009).

Parameters and Measurement of Fiber Laser

Parameters for Typical Fiber Laser

Spectral Parameters

Peak wavelength: The peak wavelength λ_p of the laser is the wavelength at which the laser output has the maximum intensity in the laser spectrum. For the multiwavelength laser, it may have several peak wavelengths. The laser spectrum usually is measured by an optical spectrum analyzer (OSA).

Spectral width: The spectral width of the laser is the full width at half maximum (FWHM) of the laser spectrum. Generally, the spectrum of the laser is measured

Fig. 10 Diagram of laser spectrum measurement

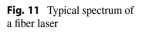

Fig. 11 Typical spectrum of a fiber laser

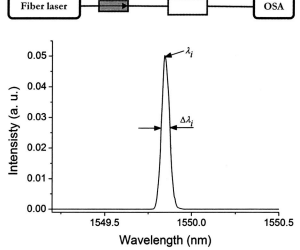

by an OSA. For the measurement of the spectral parameters, it should have high resolution to ensure the accuracy of their measurement.

The scheme of the spectrum measurement is illustrated in Fig. 10. The isolator is utilized to prevent the damage of the laser under test by the backscattering and reflected light. An optical attenuator is required to avoid the damage to the OSA due to the high laser power, especially the pulsed laser. The measured typical spectrum is shown in Fig. 11. Generally, the laser spectrum will be measured by N times. The peak wavelength and the spectral width of the spectrum in the ith measurement are noted as λ_i and $\Delta\lambda_i$, respectively. The peak wavelength λ_p and the linewidth $\Delta\lambda$ of the fiber laser are $\lambda_p = \frac{1}{N}\sum_i^N \lambda_i$ and $\Delta\lambda = \frac{1}{N}\sum_i^N \Delta\lambda_i$, respectively.

Laser Beam Parameters

Laser beam profile: The spatial intensity profile (laser beam profile) of a laser beam is often recorded by a laser beam profiler at a particular transverse to the beam propagation path. Instruments and techniques to obtain the beam characteristics include knife-edge technique (Suzaki and Tachibana 1975), camera technique (Ruff and Siegman 1992; Siegman 1998), and phase-front technique (Bélanger 1991). The knife-edge technique is to detect the power distribution in the beam transverse section which is cut by a spinning blade or slit. By taking the integrated intensity profiles in number of cuts, the original beam profile could be reconstructed using algorithms developed for tomography. The knife-edge technique is simple, but it does not provide a true 2D beam profile and is not applied for the pulsed lasers. The camera technique is to use a camera sensor to image the light directly. The maximum spot size that will fit onto a charge-coupled device (CCD) sensor is on the order of 10 mm. For larger laser spot size, a lens is usually introduced to focus the laser beam to fit the CCD sensor. But it requires that the lens has uniform reflectivity over the laser beam. The phase-front technique is that the laser beam passed through a 2D array of tiny lenses in a Shack-Hartmann wavefront sensor. Each lens will redirect its portion of the beam, and from the position of the defected beamlet, the phase of the original beam can be reconstructed.

The spot size of laser beam is the most important characteristic of a laser beam profile. Several definitions of beamwidth are in common use, including FWHM, $1/e^2$ width, D4σ width, and 10/90 or 20/80 knife-edge width (Siegman 1998).

The simplest way to define the width of a beam is to choose two diametrically opposite points at which the irradiance is a specified fraction of the beam's peak irradiance and take the distance between them as a measure of the beam's width. An obvious choice for this fraction is 1/2 (-3 dB), in which the diameter obtained is the full width of the beam at half its maximum intensity.

The $1/e^2$ width is equal to the distance between the two points on the marginal distribution that are $1/e^2 = 0.135$ times the maximum value. If there are more than two points that are $1/e^2$ times the maximum value, then the two points closest to the maximum are chosen. The $1/e^2$ width is important in the Gaussian beam, in which the intensity profile is described by

$$I(r) = \exp\left(-2\frac{r^2}{\omega^2}\right) \tag{1}$$

where r is spot radius and ω is $1/e^2$ width.

For a Gaussian beam, the relationship between the $1/e^2$ width and the FWHM is $2\omega = \frac{\sqrt{2}\text{FWHM}}{\sqrt{\ln 2}}$, where 2ω is the full width of the beam at $1/e^2$.

The D4σ width or second-moment width of a beam in the horizontal or vertical direction is 4 times σ, where σ is the standard deviation of the horizontal or vertical marginal distribution, respectively. Mathematically, the D4σ beamwidth in the x-dimension for the beam profile $I(x, y)$ is expressed as

$$D4\sigma = 4\sigma = 4\sqrt{\frac{\iint I(x.y)(x - \overline{x})^2 \mathrm{d}x\mathrm{d}y}{\iint I(x.y)\mathrm{d}x\mathrm{d}y}} \tag{2}$$

where $\overline{x} = \frac{\iint I(x.y)x\mathrm{d}x\mathrm{d}y}{\iint I(x.y)\mathrm{d}x\mathrm{d}y}$ is the centroid of the beam profile in the x direction.

When a beam is measured with a laser beam profiler, the wings of the beam profile influence the D4σ value more than the center of the profile since the wings are weighted by the square of its distance, x^2, from the center of the beam. If the beam does not fill more than a third of the beam profiler's sensor area, then there will be a significant number of pixels at the edges of the sensor that register a small baseline value (the background value). If the baseline value is large or it is not subtracted out of the image, then the computed D4σ value will be larger than the actual value because the baseline value near the edges of the sensor is weighted in the D4σ integral by x^2. Therefore, baseline subtraction is necessary for accurate

D4σ measurements. The baseline is easily measured by recording the average value for each pixel when the sensor is not illuminated. The D4σ width, unlike the FWHM and $1/e^2$ widths, is meaningful for multimodal marginal distributions – that is, beam profiles with multiple peaks – but requires careful subtraction of the baseline for accurate results. The D4σ is the ISO international standard definition for beamwidth.

The measured beam curve using the knife-edge technique is the integral of the marginal distribution and starts at the total beam power and decreases monotonically to zero power. The width of the beam is defined as the distance between the points of the measured curve that are 10% and 90% (or 20% and 80%) of the maximum value. If the baseline value is small or subtracted out, the knife-edge beamwidth always corresponds to 60%, in the case of 20/80, or 80%, in the case of 10/90, of the total beam power no matter what the beam profile. On the other hand, the D4σ width, $1/e^2$ width, and FWHM encompass fractions of power that are beam shape dependent. Therefore, the 10/90 or 20/80 knife-edge width is a useful metric when the user wishes to be sure that the width encompasses a fixed fraction of total beam power.

Beam quality factor: The beam quality factor M^2 represents the degree of variation of a beam from an ideal Gaussian beam. It is calculated from the ratio of the beam parameter product (BPP) of the beam to that of a Gaussian beam with the same wavelength. It relates the beam divergence of a laser beam to the minimum focused spot size that can be achieved. M^2 cannot be determined from a single beam profile measurement. The M^2 parameter is determined experimentally as follows (Siegman 1998):

1. Measure the D4σ widths at five axial positions near the beam waist (the location where the beam is narrowest).
2. Measure the D4σ widths at five axial positions at least one Rayleigh length away from the waist.
3. Fit the ten measured data points to

$$\sigma^2(z) = \sigma_0^2 + M^4 \left(\frac{\lambda}{\pi \sigma_0} \right)^2 (z - z_0)^2 \qquad (3)$$

where $\sigma^2(z)$ is the second moment of the distribution in the x or y direction and z_0 is the location of the beam waist with second moment width of $2\sigma_0$. Fitting the ten data points yields M^2, z_0, and σ_0.

Polarization: Light, as electromagnetic wave traveling in free space or another homogeneous isotropic non-attenuating medium, are properly described as transverse wave. It means that a plane wave's electric field vector E and magnetic field H are in directions perpendicular to the direction of wave propagation; E and H are also perpendicular to each other. Generally, polarization is represented by the oscillation of the electric field E emitted by a single-mode laser. So a propagating light wave can be represented by electromagnetic field vectors in the plane that is perpendicular to the wave propagation direction. The complex electrical field envelope could be expressed as (Born and Wolf 1999; Collett 1992)

$$\vec{E} = \left[E_{x0}e^{-jkz} \quad E_{y0}e^{-j(kz+\varphi)} \right] \begin{bmatrix} \vec{a}_x \\ \vec{a}_y \end{bmatrix} \tag{4}$$

where E_{x0} and E_{y0} are the amplitudes in the x and y directions, a_x and a_y are the unit vectors, and φ is the initial phase delay between in the x and y directions. $E_x = E_{x0}e^{-j(kz-\omega t)}$ and $E_y = E_{y0}e^{-j(kz-\omega t + \varphi)}$ are the scalars of the optical field over time in the x and y directions, respectively. On the xy-plane, they could form a fixed pattern described by a polarization ellipse:

$$\frac{E_x^2}{E_{x0}^2} + \frac{E_y^2}{E_{y0}^2} - 2\frac{E_x E_y}{E_{x0} E_{y0}} \cos\varphi = \sin^2\varphi \tag{5}$$

According to the amplitudes E_{x0} and E_{y0}, and the delay phase φ, the polarization states could be classified as linear polarization, circular polarization, and elliptical polarization, as shown in Fig. 12. The first two graphs on the left are fully polarized, whereas an optical signal may not be fully polarized. It means that E_x and E_y are not completely correlated with each other. Their phase difference φ may be random and not constant. Many light sources in nature are unpolarized, such as sunlight, whereas optical signals from lasers are mostly polarized. Generally, the light from an optical source could be divided into a fully polarized part and a completely unpolarized part. The degree of polarization (DOP) is often introduced to describe the polarization

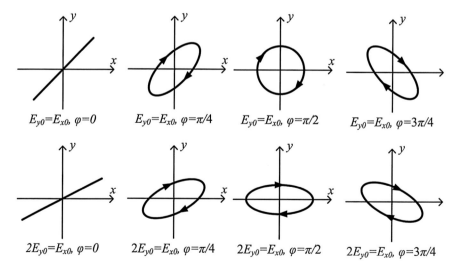

Fig. 12 Polarization ellipse view for different polarization states

characteristics of a partially polarized light, which is defined as

$$\text{DOP} = \frac{I_{\text{polarized}}}{I_{\text{polarized}} + I_{\text{unpolarized}}} \tag{6}$$

where $I_{\text{polarized}}$ and $I_{\text{unpolarized}}$ are the light intensities of the polarized part and unpolarized part, respectively. Therefore, DOP is equal to unity for a fully polarized light while DOP is zero for a completely unpolarized light.

For the fully polarized light, its polarization state could be well represented by the polarization ellipse shown by complex electric field envelope. If the light is not fully polarized, the polarization ellipse is not applicable. A Stokes vector $S = [S_0\ S_1\ S_2\ S_3]$ could give a general and convenient description mathematically, introduced by George Gabriel Stokes in 1852. The Stokes parameters could describe incoherent or partially polarized radiation in terms of its total intensity I, degree of polarization DOP, and the shape parameters of the polarization ellipse.

A fully polarized light could be described by the four Stokes parameters, which are given by

$$\begin{aligned}
S_0 &= I = E_{x0}^2 + E_{y0}^2 \\
S_1 &= E_{x0}^2 - E_{y0}^2 \\
S_2 &= 2\,|E_{x0}|\,|E_{y0}|\cos\varphi \\
S_3 &= 2\,|E_{x0}|\,|E_{y0}|\sin\varphi
\end{aligned} \tag{7}$$

where I is the total light intensity. For the fully polarized light,

$$S_1^2 + S_2^2 + S_3^2 = \left(E_{x0}^2 + E_{y0}^2\right)^2 = I^2 \tag{8}$$

which represents the total light intensity. However, it only represents the light intensity of the polarized part I_{polaried} for a partially polarized light, while $S_0 = I = I_{\text{polaried}} + I_{\text{unpolaried}}$ represents the total light intensity. According to the definition of DOP,

$$\text{DOP} = \frac{\sqrt{S_1^2 + S_2^2 + S_3^2}}{S_0} \tag{9}$$

The normalized Stokes parameters are, respectively,

$$s_0 = 1 \quad s_1 = \frac{S_1}{S_0} \quad s_2 = \frac{S_2}{S_0} \quad s_3 = \frac{S_3}{S_0} \tag{10}$$

Therefore, any polarization state could be represented as the normalized Stokes vector $s = [s_0\ s_1\ s_2\ s_3]$. The degree of polarization DOP is represented

Fig. 13 Three-dimensional
representation of a
normalized Stokes vector

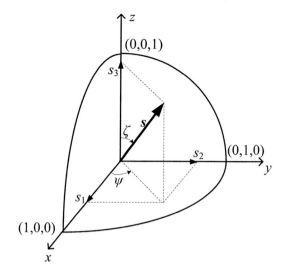

by $\sqrt{S_1^2 + S_2^2 + S_3^2} \leq 1$. The normalized Stokes parameters s_1, s_2, and s_3 could
be displayed in three-dimensional Cartesian coordinates, as shown in Fig. 13. The
normalized Stokes vector endpoint is not over the sphere with unit radius, which
is commonly referred to as Poincare sphere. For the fully polarized light, the
normalized Stokes vector endpoint is on Poincare sphere while inside the Poincare
sphere for the partially polarized light. The length of this normalized Stokes vector
is the DOP of the light. In a spherical polar coordinate, the normalized Stokes
parameters could be expressed by

$$s_1 = \text{DOP} \sin\zeta \cos\psi \quad s_2 = \text{DOP} \sin\zeta \sin\psi \quad s_3 = \text{DOP} \cos\zeta \tag{11}$$

where ψ $(0 \leq \psi < 2\pi)$ is the azimuthal angle and ζ $(0 \leq \zeta \leq \pi)$ is the polar angle.

An optical polarimeter is an instrument used to measure the Stokes parameters
of the optical signals (Chipman 1994; Madsen et al. 2004; Weiner and Shang 2006;
Heffner 1992). According to the definition of the Stokes parameters, they could
be calculated through measuring the optical powers after the light passes through
polarization sensors such as linear polarizer and wave plates, and they could be
given by

$$\begin{aligned}
S_0 &= P_{0°} + P_{90°} \\
S_1 &= P_{0°} - P_{90°} \\
S_2 &= P_{+45°} - P_{-45°} \\
S_3 &= P_{\lambda/4+45°} - P_{\lambda/4-45°}
\end{aligned} \tag{12}$$

where P is total optical power, $P_{0°}$ is optical power measured after a linear horizontal
polarizer, $P_{90°}$ is optical power measured after a linear vertical polarizer, $P_{+45°}$

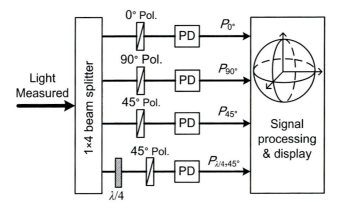

Fig. 14 Diagram of a typical polarimeter with parallel measurement. Pol., polarizer; PD, photodiode

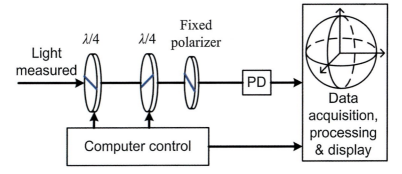

Fig. 15 Diagram of a typical polarimeter with sequential measurement. PD, photodiode

is optical power measured after a linear $+45°$ polarizer, $P_{-45°}$ is optical power measured after a linear $-45°$ polarizer, $P_{\lambda/4,+45°}$ is optical power measured after a $\lambda/4$ wave plate and a linear $+45°$ polarizer, and $P_{\lambda/4,-45°}$ is optical power measured after a $\lambda/4$ wave plate and a linear $-45°$ polarizer. Among these measurements, $P = P_{0°} + P_{90°} = P_{+45°} + P_{-45°} = P_{\lambda/4,+45°} + P_{\lambda/4,-45°}$. Based on their relations, most polarimeters only depend on four independent power measurements. A typical polarimeter diagram with parallel measurement is shown in Figs. 14 and 15. The measured light is divided into four equal parts by a 1×4 beam splitter. Using polarizers with $0°, 90°$, and $45°$ orientation angles, $P_{0°}, P_{90°}$, and $P_{+45°}$ are obtained in the first three beams. With a $\lambda/4$ wave plate and a polarizer oriented at $45°$, $P_{\lambda/4,+45°}$ is measured in the fourth beam. These measurements are analyzed by the signal processing system. Then the Stokes parameters could be obtained, and the Stokes vector could be displayed on a Poincare sphere.

Another type of polarimeter is to use sequential measurement, as shown in Fig. 15. Only one polarizer and two wave plates are required. The light field

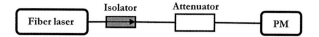

Fig. 16 Diagram of Laser power measurement. PM: Power Meter

measured could be expressed as $\begin{bmatrix} E_x \\ E_y \end{bmatrix}$. The $\lambda/4$ wave plate could be expressed by Jones matrix:

$$
\begin{bmatrix}
\cos\frac{\delta}{2} - i\sin\frac{\delta}{2}\cos 2\theta & -i\sin\frac{\delta}{2}\sin 2\theta \\
-i\sin\frac{\delta}{2}\sin 2\theta & \cos\frac{\delta}{2} + i\sin\frac{\delta}{2}\cos 2\theta
\end{bmatrix}
\tag{13}
$$

where δ is the delay phase between the fast and slow axes of wave plate and is equal to $\pi/2$ for $\lambda/4$ wave plate and θ is the angle between the fast axis of wave plate and x axis. The principle axis of the polarizer is fixed along the x axis. Through specific combinations of optic axis of the two $\lambda/4$ wave plates, $P_{0°}$, $P_{90°}$, $P_{+45°}$, and $P_{\lambda/4,+45°}$ could be selected by the fixed polarizer and detected by the photodiode.

Specified Parameters for Continuous Wave (CW) Fiber Laser

Output Power
The measurement scheme of the output power is illustrated in Fig. 16. The output power of the laser is measured by a power meter at a certain operation condition. The power meter includes the pyroelectric detector and photoelectric detector. The laser beam should be aimed to the probe of the power meter, to ensure all the laser power is detected. And the PM wavelength should be set at the nearest range to the fiber laser. The power over the time $P(t)$ is recorded in the measure time of T. The average power P_{out} is

$$
P_{\text{out}} = \frac{1}{T}\int_0^T P(t)\,dt
\tag{14}
$$

Output Power Stability
At a certain laser operation condition, the output power stability characterizes the output power influence during a specified interval of time. $P_{\max}$ and $P_{\min}$ are the maximum and minimum of the output power $P(t)$ and their difference. So the power stability S_p is defined by

$$
S_P = \pm\frac{\Delta P}{2P} \times 100\%
\tag{15}
$$

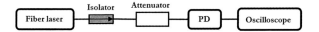

Fig. 17 Diagram of laser power measurement. PD, photon diode

while the power relative maximum stability $S_p{'}$ is expressed as

$$S_p{'} = \pm \frac{P_{\text{max}} - P_{\text{min}}}{P_{\text{max}} + P_{\text{min}}} \times 100\% \qquad (16)$$

Specified Parameters for Pulsed Fiber Laser

Average Output Power
The average output power of the pulsed laser is the average value of the measured output power by N times during a specified interval of time in a certain operation condition. The measurement scheme is similar to the output power measurement of the CW laser as shown in Fig. 16.

Repetition Rate and Pulse Width
The repetition rate f is the number of laser pulses during the time of 1 s. The pulse width τ is the FWHM of the single pulse under a certain operation condition. Their measurement scheme is illustrated in Fig. 17. The spectral response range of the photodiode should satisfy the wavelength of the laser. The response of the photodiode and the digital oscilloscope should be fast enough to ensure accuracy of the measured pulse width. The pulse width is the average value of the pulse widths measured by N times:

$$\tau = \frac{1}{N} \sum_{i}^{N} \tau_i \qquad (17)$$

where τ_i is the ith measured pulse width. T is the interval time between the adjacent pulses. The repetition rate is $f = 1/T$.

Peak Output Power
The peak power is the ratio of the single pulse energy E and the pulse width τ:

$$P_{\text{peak}} = \frac{E}{\tau} \qquad (18)$$

where $E = P/f$.

Specified Parameters for Single-Frequency Fiber Laser

High-performance single-frequency fiber lasers such as distributed feedback fiber laser, distributed Bragg reflector fiber laser, and ring cavity fiber laser have attracted significant research interests due to their unique distinguished characteristics and great application potential in fields like distributed sensing, laser spectroscopy, and fiber-based hydrophone (Yuliang et al. 2011). Thus precise and comprehensive measurements of specified parameters such as relative intensity noise, linewidth, and frequency noise become crucial to the characteristic investigation and performance improvement of single-frequency fiber lasers. Unlike normal optic parameters mentioned above, which can be measured using standard device like spectrometer or optical power meter, currently there are no established standards or devices to directly measure these specific parameters. The following parts of these parameter measurements are based on associated research papers and laboratory practice, which might be limited because of lack of experiences.

Relative Intensity Noise

Relative intensity noise (RIN) describes the fluctuation of laser intensity noise, which is the power noise normalized to the average power level. It is defined as

$$\text{RIN} = \frac{(\delta P)^2}{\tilde{p}^2} \tag{19}$$

where δP is laser output power noise and $\tilde{p}$ is laser average output power. The RIN has been a concern since the early beginning of laser optics (McCumber 1966). It originates from the random perturbations of laser cavity, gain medium, and pump source. It is often associated with transmission of optical data; total RIN within the system bandwidth determines the signal-to-noise ratio, which in turn will set the limit of the lowest detectable signal. In high-speed digital systems, RIN can affect the bit error rate and system performance under certain conditions. Knowledge of RIN can be used when designing new lasers to improve their performance for specific applications. Very-low-RIN lasers are used to determine the noise figure of optical fiber amplifiers, which are essential for building faster and more efficient optical communications systems. Demand for better techniques to calibrate the response and sensitivity of a RIN measurement system is also increasing with the appearance of commercial distributed feedback lasers and diode-pumped Nd:YAG lasers exhibiting very low RIN.

Generally, the RIN is measured by the fluctuation of the direct current at the output of a photodetector, which can be statistically described with the power spectral density (PSD) of the laser under test. Figure 18 shows the typical setup to measure the fiber laser RIN. Fiber laser and corresponding pump source are connected with suitable wavelength division multiplexer (WDM), and an optical isolator is added between the WDM and a photodetector (PD) to prevent back reflection interference. The RIN is measured by launching the laser emission onto

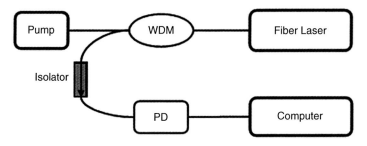

Fig. 18 Typical fiber laser RIN measurement setup

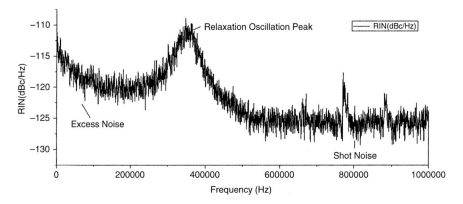

Fig. 19 RIN spectrum of a distributed feedback fiber laser with 200 μW output power

the PD and converting it into an electric voltage; thus the laser output can be acquired by an acquisition card. A fast Fourier transform process is completed through a computer program to transform the received time domain data set into frequency domain, and then the RIN spectrum can be displayed.

Figure 19 shows the measured RIN spectrum of a distributed feedback fiber laser pumped with 100 mW 980 nm diode laser. There is a pronounced relaxation oscillation peak at around 380 kHz. The appearance of this peak is due to pump power modulation-induced damping effects, and the relaxation oscillation peak will glide toward higher frequency and lower level as pump power increases (Pengpeng et al. 2013); thus a fixed stable pump source is crucial to the RIN measurement precision. In addition, based on the relation between pump power and RIN relaxation oscillation peak of the fiber laser under test, the measured relaxation oscillation peak level and frequency can be used to determine the actual pump power received by the fiber laser.

Besides, for a given fiber laser signal, its RIN is also influenced by the system Excess Noise and Shot Noise. As indicated in Fig. 20, the Excess Noise comprises the RIN low-frequency part before relaxation oscillation peak. It is related to the thermal noise generated through electron Brownian motion in passive electrical

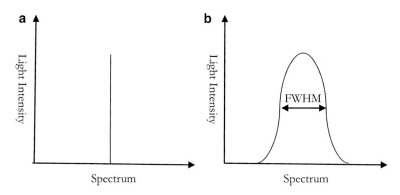

Fig. 20 Comparison between ideal laser and real laser on the demonstration of linewidth. (**a**) Ideal laser, (**b**) real laser

circuit components such as resistors and feeder; hence it can be reduced under proper cooling devices. For high-frequency part of RIN, it is mainly affected by photon detector Shot Noise. This is a quantum noise effect, related to the discreteness of photons and electrons, and interpreted as arising from the random occurrence of photon absorption events in a photon detector. Thus sub-Shot Noise measurement can be obtained by carefully selecting photon detectors with high quantum efficiency and appropriate bandwidth.

In a word, the key of RIN measurement is the mensuration of relaxation oscillation peak level and frequency, which can be considered as one of the signature characteristics of fiber lasers under different pump configurations. To maximally lower the system noise level, additional temperature reduction devices such as wind fan or hydrocooling installment to electrical circuit parts of the system are necessary; in the meantime, proper choice of photon detector can effectively suppress the Shot Noise from the total RIN spectrum.

Linewidth

One of the most distinctive characteristics of fiber laser is its spectral coherence, which is evaluated by measuring the laser linewidth of the radiation. For ideal single-frequency laser, the laser output from stimulated emission should be mono-lithic light with zero linewidth, as shown in Fig. 20a. However, in reality, the stimulated emission is influenced by thermal disequilibrium generated sublevels between the ground state and metastable state, thus leading to photon emission with various frequencies around the original single frequency with different intensities and the effects of quantum noise, cavity fluctuation, and pump noise. The broadened output spectrum is shown in Fig. 20b, where the FWHM of the laser spectrum is defined as laser linewidth.

Laser linewidth is directly related to laser coherence length; lower linewidth means longer coherent transmission distance; hence narrow linewidth of laser source became crucial for laser applications in long-distance communication and sensing,

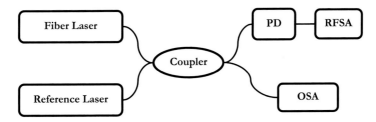

Fig. 21 Schematic of the setup for heterodyne method

which also brings up the exigent needing for precise linewidth measurement. For large linewidth (≥ 20 MHz), traditional techniques of optical spectrum analysis based on diffraction gratings are suitable, although a high-frequency resolution might be required for the spectrometer. For narrow linewidth measurement, many different schemes with unique advantages and disadvantages have been developed. Two typical methods will be described in details in the following.

Heterodyne method: Schematic setup of heterodyne method is shown in Fig. 21. This method employs two lasers, one is the fiber laser under test and the other is a tunable reference laser with stable narrow linewidth. The signal from both lasers was combined by a 2×2 coupler. In order to form a beat note under interference, the central wavelength of the reference laser must be tuned close to the test laser wavelength to allow the mixing product to fall within the detection bandwidth of PD, which is realized through monitoring the course wavelength using an OSA at one of the 2×2 coupler output arms. The other output arm connects with the PD, which converts the interference beat note into an electrical signal, and then directs it to a radio-frequency spectrum analyzer (RFSA) to retrieve the linewidth spectrum.

Heterodyne method is one of the most classical linewidth measurement schemes. It can provide not only laser linewidth data but also optical power spectrum. This method is the only technique that is capable of characterizing asymmetrical spectral line shape. This method also offers high sensitivity and high resolution. The main difficulty to use heterodyne method is that two lasers must be used. And the linewidth of the reference laser must be narrower than or at least comparable to that of the fiber laser under test in order to achieve reasonable measurement accuracy. For extremely narrow linewidth measurements, the characterization of the reference laser itself is very difficult. In addition, the restricted bandwidth of the PD limits the measurement frequency range to at most tens of gigahertz vicinity of the reference laser central frequency, which greatly diminishes the applicability of this method to lasers with a wide range of different frequencies.

Delayed self-heterodyne/homodyne method: To overcome the disadvantages of heterodyne method, in 1980, T. Okoshi et al. proposed the delayed self-heterodyne method to measure the linewidth of semiconductor lasers without additional reference laser (Okoshi et al. 1980). The basic idea of the technique is to convert the optical phase or frequency fluctuations of the test laser into variations of light intensity in a Mach-Zehnder interferometer. In the interferometer, the optical field

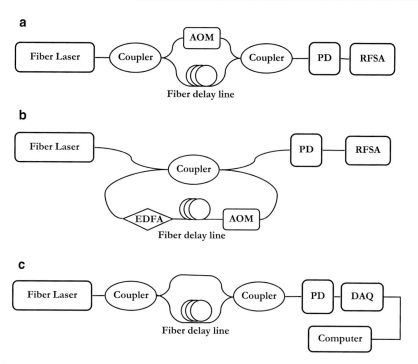

Fig. 22 Typical setup for delayed self-heterodyne/homodyne and loss-compensated delayed self-heterodyne linewidth measurement scheme. (**a**) Delayed self-heterodyne setup, (**b**) loss-compensated delayed self-heterodyne setup, (**c**) delayed self-homodyne setup

is mixed with a delayed replica of itself, and the interference signal is detected with a fast PD. The laser linewidth is then deduced from the recorded power spectrum of the fluctuations of the photocurrent. An acousto-optic modulator (AOM) is used to shift the spectrum up in frequency in order to reject the detected direct current signal in the PD and to allow the use of a standard RFSA to measure the spectrum of the photocurrent fluctuations. A typical setup is shown in Fig. 22a.

The key component of delayed self-heterodyne method is the optical fiber delay line in the Mach-Zehnder interferometer. To ensure the uncorrelated interference at the PD, the optical path delay must be longer than the coherence length of the test laser; hence the two combining beams can interfere as if they originated from two independent lasers offset in frequency determined by the AOM. This technique is simple, inherently self-calibrated, and capable of measuring a large range of laser frequencies where the fiber loss is tolerable. Compared with heterodyne detection, another advantage of the delayed self-heterodyne method is the auto wavelength tracing. Since the local oscillator signal in these measurements is provided by the laser under test, slow drift in wavelength is usually tolerable. However, it is only suited for measurement of laser linewidth on the order of or above 10 kHz because otherwise the required fiber delay line would become impractically long.

In order to increase the linewidth measurement capability to sub-kilo hertz region, in 1992, Jay W. Dawson et al. reported the implementation of a loss-compensated reticulating delayed self-heterodyne interferometer (LC-RDSHI) detection by including an AOM and an erbium-doped fiber amplifier (EDFA) in the loop that circulates the input and output arm of a 2×2 coupler (Dawson et al. 1992). The schematic LC-RDSHI is shown in Fig. 22b, in which a short loss-compensated fiber loop is used as a fiber delay line. This method allows the same fiber delay to be used multiple times in order to improve the resolution of conventional delayed self-heterodyne detection method, and then by measuring the multiple orders of beat notes on the RFSA, the linewidth with associated delay length can be derived. For example, for a delay line of 10 km, the 100th order of the beat note has gone through 1000 km delay path, which means a linewidth as narrow as 100 Hz can be measured.

On the other hand, although this method is very promising in terms of ultra-narrow linewidth measurement, there are many drawbacks induced by the recirculation fiber loop such as frequency instability, EDFA spontaneous emission noise, RIN overlap, and fiber birefringence effect. Also the complex system setup increased the operating and maintenance difficulty, for different fiber laser inputs, the system parameters have to be separately adjusted to obtain an acceptable high-order beat note spectrum.

As a simplified version of delayed self-heterodyne linewidth measurement, K. Liyama et al. proposed a delayed self-homodyne configuration based on Mach-Zehnder interferometer without additional frequency shifter (Liyama et al. 1991). Its principle is similar to delayed self-heterodyne interferometry, while the only difference is instead of getting a full linewidth spectrum at shifted frequency, this method will only get a half linewidth spectrum at zero frequency, and the FWHM of this spectrum is twice of the laser linewidth. To read and interpret the linewidth, nowadays the RFSA is usually replaced with a data acquisition (DAQ) card and associated computer software. Figure 22c shows a typical setup of such method.

Figure 23a shows a linewidth spectrum of a 200 mW 980 nm diode laser-pumped distributed feedback fiber laser, which is measured with delayed self-homodyne method. The displayed full linewidth spectrum is achieved through combining the measured half spectrum and its mirror copy in the opposite frequency region. To reduce the effect of 1/f noise and white noise, a Lorentzian fit is performed on the linewidth spectrum to increase the signal-to-noise ratio; the typical result is shown in Fig. 23b, and the linewidth is retrieved through measuring the FWHM of the fitted line.

Generally speaking, delayed self-homodyne method is simple and effective at narrow linewidth measurement. Although its resolution is still limited by the conflict between the finite fiber delay line length and ultra-long transmission distance required by uncorrelated interference, the system robustness and ease to use make it preferable over delayed self-heterodyne method. For ultra-narrow linewidth measurement, either LC-RDSHI or heterodyne method with reference laser which has significantly narrower linewidth is reliable. Besides, since fiber laser linewidth

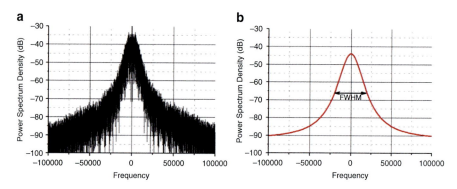

Fig. 23 Linewidth spectrum and its Lorentzian fit of a distributed feedback fiber laser measured with delayed self-homodyne method. (**a**) Measured linewidth spectrum, (**b**) Lorentzian fit of measured spectrum

is very sensitive to environmental perturbations such as acoustic noise, temperature variation, and mechanical vibration, a very stable and quiet test environment is strictly required for both the fiber laser under test and the linewidth measurement device, especially the Mach-Zehnder interferometer part. And the environment parameters should be recorded as a sidenote to the measured linewidth to ensure the repeatability of the experiment.

Phase Noise

Fiber laser phase noise describes how the phase of the output electrical field of a fiber laser deviates from the classical electric field wave form. Similar to linewidth, it also represents time or spectral coherence of the laser, however, unlike linewidth with relatively long measurement period; phase noise characterizes the short-term stability of the laser, and it may occur in the form of a continuous frequency drift or as sudden phase jumps or as a combination of both.

In many applications, such as long-distance coherent optical communication, optical fiber interferometric sensors, high-resolution spectroscopy, ultralow phase noise photonic microwave generation, and optical atomic clock, the laser phase noise can profoundly impact the limitation of a system. Thus, fiber lasers with ultralow phase noise are actively studied, while the precise characterization of such ultra-stable fiber lasers becomes more important.

One of the most established and simple way to measure the fiber laser phase noise is cross correlation method based on the unbalanced Michelson interferometer composed of a 3×3 optical fiber coupler (Xu et al. 2015). The basic idea is to transfer the fiber laser phase fluctuations into intensity fluctuations of the laser interference signal and retrieve the phase noise through measuring the power spectrum density of the interference signal. According to theoretical analysis proposed by R. Yang et al. (2010), laser phase noise measured by unbalance interferometer can be described as

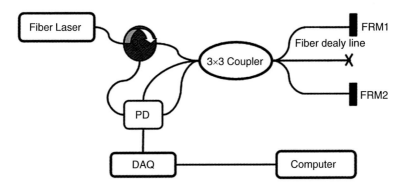

Fig. 24 Setup schematic for phase noise measurement based on Michelson interferometer composed of a 3 × 3 optical fiber coupler

$$\Delta\varphi = \frac{2\pi D}{c}\Delta v \tag{20}$$

where D is the optical path difference of the unbalanced interferometer, c is the light velocity in free space, Δv is the magnitude of laser frequency fluctuation, and $\Delta\varphi$ is the phase fluctuation. Hence, through performing fast Fourier transform on the received intensity power spectrum density, both phase and frequency fluctuation of the laser under test can be demodulated.

The typical setup of the 3 × 3 optical fiber coupler-based Michelson interferometer method is shown in Fig. 24. The fiber laser output goes through an optical circulator (OC); two thirds of the laser goes into a 3 × 3 optical fiber coupler. Two Faraday rotation mirrors (FRM) and a certain length of optical fiber delay line are installed on two arms at the other side of the 3 × 3 optical fiber coupler to form the unbalanced Michelson interferometer while canceling the polarization fading effect induced by external disturbances. The reflected interference signal and rest unreflected laser signal are received by the three PDs, respectively. PD transfers the optical signal into electrical signal, which is picked up by the DAQ and processed with associated computer software to retrieve the fiber laser phase noise. Figure 25 shows the phase noise spectrum of a distributed feedback fiber laser with 180 μW output power pumped by 200 mW 980 nm diode laser.

Compared to other phase noise measurement methods, the unbalanced Michelson interferometer composed of a 3 × 3 optical fiber coupler saves the use of additional phase demodulator, active/passive phase control devices, and oscilloscope. In the meantime, since the frequency power spectrum density contains full information of the laser frequency/phase/wavelength fluctuation, it is possible to theoretically calculate the laser linewidth from the phase noise spectrum, and this has been confirmed experimentally by N. Bucalovic et al. (2012). However, this method is based on comparison between laser self-interference signals from sub arms of a 3 × 3 optical fiber coupler; the injection losses are relatively high, which makes the detection of small laser signal (<20 μW) unreliable.

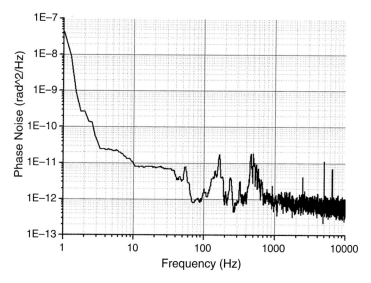

Fig. 25 Phase noise spectrum of a distributed feedback fiber laser with 180 μW output power

Conclusion

In this chapter, the history and development of optical fiber lasers have been concisely reviewed. Then fiber lasers are classified into several types according to their operating conditions and output type. Furthermore, the focus of this chapter is placed on the fiber laser characteristics and associated measurements, including general parameters such as wavelength, polarization and output power, as well as specified parameters like relative intensity noise, linewidth and phase noise. After all, the mensuration of optical fiber laser characteristics is a fast-developing technology as the fiber laser research progresses further into uncharted region in the future.

Acknowledgments Authors are thankful for the financial support of National Natural Science Foundation of China (61605103), Shandong Provincial Natural Science Foundation of China(ZR2016FM33) and Key Research and Development Program of Shandong Province (2018GGX101030).

References

G.A. Ball, W.H. Glenn, W.W. Morey, P.K. Cheo, Modeling of short, single-frequency, fiber lasers in high-gain fiber. IEEE Photon. Technol. Lett. **5**(6), 649–651 (1993)

P.A. Bélanger, Beam propagation and the ABCD ray matrices. Opt. Lett. **16**(4), 196–198 (1991)

M. Born, E. Wolf, *Principles of Optics*, 7th edn. (Cambridge University Press, Cambridge, MA, 1999)

N. Bucalovic, V. Dolgovskiy, C. Schori, P. Thomann, D.G. Di, Experimental validation of a simple approximation to determine the linewidth of a laser from its frequency noise spectrum. Appl. Opt. **51**(20), 4582–4588 (2012)

R.A. Chipman, Polarimetry, in *Handbook of Optics*, ed. by M. Bass (Ed), vol. 2, 2nd edn., (McGraw-Hill, New York, 1994)

E. Collett, *Polarized Light: Fundamentals and Applications, Optical Engineering* (Dekker, New York, 1992), p. c-1

J.W. Dawson, N. Park, K.J. Vahala, An improved delay self-heterodyne interferometer for linewidth measurement. IEEE Photon. Technol. Lett. **4**(9), 1063–1066 (1992)

C.J. de Matos, S.M.L. De, A.M. Brito-Silva, M.A.M. Gamez, A.S. Gomez, Random fiber laser. Phys. Rev. Lett. **99**(15), 153903 (2007)

M.E. Fermann, I. Hartl, Ultrafast fiber laser technology. IEEE J. Sel. Top. Quant. Electron. **15**(1), 191–206 (2009)

C. Headley, M. Mermelstein and J.C. Bouteiller, Chapter 11: Raman fiber lasers, in *Raman Amplifiers for Telecommunications 2*, ed. by N. I. Mohammed (Ed), (Springer, New York, 2004), pp. 353–382

B.L. Heffner, Automated measurement of polarization mode dispersion and using Jones matrix eigenanalysis. IEEE Photon. Technol. Lett. **4**(9), 1066–1069 (1992)

K.O. Hill, Y. Fujii, D.C. Johnson, B.S. Kawasaki, Photo-sensitivity in optical fiber wave-guild: Application To reflection filter fabrication. Appl. Phys. Lett. **32**, 647–649 (1978)

J.T. Kringlebotn, J.L. Archambault, L. Reekie, D.N. Payne, Er^{3+}:Yb^{3+}-codoped fiber distributed feedback laser. Opt. Lett. **19**(24), 2101–2103 (1994)

K. Liyama, K. Hayashi, Y. Ida, H. Ikeda, Reflection-type delayed self-homodyne/heterodyne method for optical linewidth measurements. J. Lightwave Technol. **9**(5), 635–640 (1991)

C. Madsen, P.O. Swald, M. Cappuzzo, E. Chen, L. Gomez, Integrated optical spectral polarimeter for signal monitoring and feedback to a PMD compensator. J. Opt. Netw. **3**(7), 490–500 (2004)

D.E. McCumber, Intensity fluctuations on the output of cw laser oscillator. Phys. Rev. **141**(1), 306–322 (1966)

T. Okoshi, K. Kikuchi, A. Nakayama, Novel method for high resolution measurement of laser output spectrum. Electron. Lett. **16**, 630–631 (1980)

W. Pengpeng, C. Jun, Z. Cungang, Z. Yanjie, Z. Xiaolei, The relative intensity noise and relaxation oscillation characteristics of a distributed-feedback fiber laser. Laser Phys. **23**(9), 095108 (2013)

S.B. Pools, D.N. Payne, M.E. Fermann, Fabrication of low-loss optical fibers containing rare-earth ions. Electron. Lett. **21**(17), 737–738 (1985)

J.A. Ruff, A.E. Siegman, Single-pulse laser beam quality measurements using a CCD camera system. Appl. Opt. **31**(24), 4907–4909 (1992)

A.E. Siegman, How to (maybe) measure laser beam quality, in *DPSS (Diode Pumped Solid State) Lasers: Applications and Issues*, OSA Trends in Optics and Photonics, ed. by M. Dowley (Ed), vol. 17, (Optical Society of America, 1998). paper MQ1

S.P. Smith, F. Zarinetchi, S. Ezekiel, Narrow-linewidth stimulated Brillouin fiber laser and applications. Opt. Lett. **16**(6), 393–395 (1991)

E. Snitzer, Optical master action of Nd^{3+} in a barium crown glass. Phys. Rev. Lett. **7**(12), 444–446 (1961)

E. Snitzer, H. Po, F. Hakami, R. Tumminelli, B.C. McCollum, Double-clad offset core Nd fiber laser, in *Optical Fiber Sensors*, OSA Technical Digest Series (Optical Society of America, 1988), Vol. 2, paper PD5

Y. Suzaki, A. Tachibana, Measurement of the μm sized radius of Gaussian laser beam using the scanning knife-edge. Appl. Opt. **14**(12), 2809–2810 (1975)

A.M. Weiner, X. Shang, A complete spectral polarimeter design for lightwave communication systems. J. Lightwave Technol. **24**(11), 3982–3991 (2006)

D. Xu, F. Yang, D. Chen, F. Wei, H. Cai, Laser phase and frequency noise measurement by Michelson interferometer composed of a 3 × 3 optical fiber coupler. Opt. Express **23**(17), 22386–22393 (2015)

R. Yang, P. Yang, L. Dong, M. Ao, B. Xu, A strip extracting algorithm for phase noise measurement and coherent beam combining of fiber amplifiers. Appl. Phys. B Lasers Opt. **99**(1–2), 19–22 (2010)

L. Yuliang, Z. Wentao, X. Tuanwei, H. Jun, Z. Faxiang, L. Fang, Fiber laser sensing system and its applications. Photon. Sens. **1**(1), 43–53 (2011)